NEUROBIOLOGY of
CHEMICAL COMMUNICATION

FRONTIERS IN NEUROSCIENCE

Series Editor
Sidney A. Simon, Ph.D.

Published Titles

Apoptosis in Neurobiology
Yusuf A. Hannun, M.D., Professor of Biomedical Research and Chairman, Department of Biochemistry and Molecular Biology, Medical University of South Carolina, Charleston, South Carolina
Rose-Mary Boustany, M.D., tenured Associate Professor of Pediatrics and Neurobiology, Duke University Medical Center, Durham, North Carolina

Neural Prostheses for Restoration of Sensory and Motor Function
John K. Chapin, Ph.D., Professor of Physiology and Pharmacology, State University of New York Health Science Center, Brooklyn, New York
Karen A. Moxon, Ph.D., Assistant Professor, School of Biomedical Engineering, Science, and Health Systems, Drexel University, Philadelphia, Pennsylvania

Computational Neuroscience: Realistic Modeling for Experimentalists
Eric DeSchutter, M.D., Ph.D., Professor, Department of Medicine, University of Antwerp, Antwerp, Belgium

Methods in Pain Research
Lawrence Kruger, Ph.D., Professor of Neurobiology (Emeritus), UCLA School of Medicine and Brain Research Institute, Los Angeles, California

Motor Neurobiology of the Spinal Cord
Timothy C. Cope, Ph.D., Professor of Physiology, Wright State University, Dayton, Ohio

Nicotinic Receptors in the Nervous System
Edward D. Levin, Ph.D., Associate Professor, Department of Psychiatry and Pharmacology and Molecular Cancer Biology and Department of Psychiatry and Behavioral Sciences, Duke University School of Medicine, Durham, North Carolina

Methods in Genomic Neuroscience
Helmin R. Chin, Ph.D., Genetics Research Branch, NIMH, NIH, Bethesda, Maryland
Steven O. Moldin, Ph.D., University of Southern California, Washington, D.C.

Methods in Chemosensory Research
Sidney A. Simon, Ph.D., Professor of Neurobiology, Biomedical Engineering, and Anesthesiology, Duke University, Durham, North Carolina
Miguel A.L. Nicolelis, M.D., Ph.D., Professor of Neurobiology and Biomedical Engineering, Duke University, Durham, North Carolina

The Somatosensory System: Deciphering the Brain's Own Body Image
Randall J. Nelson, Ph.D., Professor of Anatomy and Neurobiology, University of Tennessee Health Sciences Center, Memphis, Tennessee

The Superior Colliculus: New Approaches for Studying Sensorimotor Integration
William C. Hall, Ph.D., Department of Neuroscience, Duke University, Durham, North Carolina
Adonis Moschovakis, Ph.D., Department of Basic Sciences, University of Crete, Heraklion, Greece

New Concepts in Cerebral Ischemia
Rick C.S. Lin, Ph.D., Professor of Anatomy, University of Mississippi Medical Center, Jackson, Mississippi

DNA Arrays: Technologies and Experimental Strategies
Elena Grigorenko, Ph.D., Technology Development Group, Millennium Pharmaceuticals, Cambridge, Massachusetts

Methods for Alcohol-Related Neuroscience Research
Yuan Liu, Ph.D., National Institute of Neurological Disorders and Stroke, National Institutes of Health, Bethesda, Maryland
David M. Lovinger, Ph.D., Laboratory of Integrative Neuroscience, NIAAA, Nashville, Tennessee

Primate Audition: Behavior and Neurobiology
Asif A. Ghazanfar, Ph.D., Princeton University, Princeton, New Jersey

Methods in Drug Abuse Research: Cellular and Circuit Level Analyses
Barry D. Waterhouse, Ph.D., MCP-Hahnemann University, Philadelphia, Pennsylvania

Functional and Neural Mechanisms of Interval Timing
Warren H. Meck, Ph.D., Professor of Psychology, Duke University, Durham, North Carolina

Biomedical Imaging in Experimental Neuroscience
Nick Van Bruggen, Ph.D., Department of Neuroscience Genentech, Inc.
Timothy P.L. Roberts, Ph.D., Associate Professor, University of Toronto, Canada

The Primate Visual System
John H. Kaas, Department of Psychology, Vanderbilt University, Nashville, Tennessee
Christine Collins, Department of Psychology, Vanderbilt University, Nashville, Tennessee

Neurosteroid Effects in the Central Nervous System
Sheryl S. Smith, Ph.D., Department of Physiology, SUNY Health Science Center, Brooklyn, New York

Modern Neurosurgery: Clinical Translation of Neuroscience Advances
Dennis A. Turner, Department of Surgery, Division of Neurosurgery, Duke University Medical Center, Durham, North Carolina

Sleep: Circuits and Functions
Pierre-Hervé Luppi, Université Claude Bernard, Lyon, France

Methods in Insect Sensory Neuroscience
Thomas A. Christensen, Arizona Research Laboratories, Division of Neurobiology, University of Arizona, Tuscon, Arizona

Motor Cortex in Voluntary Movements
Alexa Riehle, INCM-CNRS, Marseille, France
Eilon Vaadia, The Hebrew University, Jerusalem, Israel

Neural Plasticity in Adult Somatic Sensory-Motor Systems
Ford F. Ebner, Vanderbilt University, Nashville, Tennessee

Advances in Vagal Afferent Neurobiology
Bradley J. Undem, Johns Hopkins Asthma Center, Baltimore, Maryland
Daniel Weinreich, University of Maryland, Baltimore, Maryland

The Dynamic Synapse: Molecular Methods in Ionotropic Receptor Biology
Josef T. Kittler, University College, London, England
Stephen J. Moss, University College, London, England

Animal Models of Cognitive Impairment
Edward D. Levin, Duke University Medical Center, Durham, North Carolina
Jerry J. Buccafusco, Medical College of Georgia, Augusta, Georgia

The Role of the Nucleus of the Solitary Tract in Gustatory Processing
Robert M. Bradley, University of Michigan, Ann Arbor, Michigan

Brain Aging: Models, Methods, and Mechanisms
David R. Riddle, Wake Forest University, Winston-Salem, North Carolina

Neural Plasticity and Memory: From Genes to Brain Imaging
Frederico Bermudez-Rattoni, National University of Mexico, Mexico City, Mexico

Serotonin Receptors in Neurobiology
Amitabha Chattopadhyay, Center for Cellular and Molecular Biology, Hyderabad, India

TRP Ion Channel Function in Sensory Transduction and Cellular Signaling Cascades
Wolfgang B. Liedtke, M.D., Ph.D., Duke University Medical Center, Durham, North Carolina
Stefan Heller, Ph.D., Stanford University School of Medicine, Stanford, California

Methods for Neural Ensemble Recordings, Second Edition
Miguel A.L. Nicolelis, M.D., Ph.D., Professor of Neurobiology and Biomedical Engineering,
    Duke University Medical Center, Durham, North Carolina

Biology of the NMDA Receptor
Antonius M. VanDongen, Duke University Medical Center, Durham, North Carolina

Methods of Behavioral Analysis in Neuroscience
Jerry J. Buccafusco, Ph.D., Alzheimer's Research Center, Professor of Pharmacology and Toxicology,
    Professor of Psychiatry and Health Behavior, Medical College of Georgia, Augusta, Georgia

In Vivo Optical Imaging of Brain Function, Second Edition
Ron Frostig, Ph.D., Professor, Department of Neurobiology, University of California,
Irvine, California

Fat Detection: Taste, Texture, and Post Ingestive Effects
Jean-Pierre Montmayeur, Ph.D., Centre National de la Recherche Scientifique, Dijon, France
Johannes le Coutre, Ph.D., Nestlé Research Center, Lausanne, Switzerland

The Neurobiology of Olfaction
Anna Menini, Ph.D., Neurobiology Sector International School for Advanced Studies, (S.I.S.S.A.),
    Trieste, Italy

Neuroproteomics
Oscar Alzate, Ph.D., Department of Cell and Developmental Biology, University of North
    Carolina, Chapel Hill, North Carolina

Translational Pain Research: From Mouse to Man
Lawrence Kruger, Ph.D., Department of Neurobiology, UCLA School of Medicine, Los Angeles,
    California
Alan R. Light, Ph.D., Department of Anesthesiology, University of Utah, Salt Lake City, Utah

Advances in the Neuroscience of Addiction
Cynthia M. Kuhn, Duke University Medical Center, Durham, North Carolina
George F. Koob, The Scripps Research Institute, La Jolla, California

Neurobiology of Huntington's Disease: Applications to Drug Discovery
Donald C. Lo, Duke University Medical Center, Durham, North Carolina
Robert E. Hughes, Buck Institute for Age Research, Novato, California

Neurobiology of Sensation and Reward
Jay A. Gottfried, Northwestern University, Chicago, Illinois

The Neural Bases of Multisensory Processes
Micah M. Murray, CIBM, Lausanne, Switzerland
Mark T. Wallace, Vanderbilt Brain Institute, Nashville, Tennessee

Neurobiology of Depression
Francisco López-Muñoz, University of Alcalá, Madrid, Spain
Cecilio Álamo, University of Alcalá, Madrid, Spain

Astrocytes: Wiring the Brain
Eliana Scemes, Albert Einstein College of Medicine, Bronx, New York
David C. Spray, Albert Einstein College of Medicine, Bronx, New York

Dopamine–Glutamate Interactions in the Basal Ganglia
Susan Jones, University of Cambridge, United Kingdom

Alzheimer's Disease: Targets for New Clinical Diagnostic and Therapeutic Strategies
Renee D. Wegrzyn, Booz Allen Hamilton, Arlington, Virginia
Alan S. Rudolph, Duke Center for Neuroengineering, Potomac, Maryland

The Neurobiological Basis of Suicide
Yogesh Dwivedi, University of Illinois at Chicago

Transcranial Brain Stimulation
Carlo Miniussi, University of Brescia, Italy
Walter Paulus, Georg-August University Medical Center, Göttingen, Germany
Paolo M. Rossini, Institute of Neurology, Catholic University of Rome, Italy

Spike Timing: Mechanisms and Function
Patricia M. Di Lorenzo, Binghamton University, Binghamton, New York
Jonathan D. Victor, Weill Cornell Medical College, New York City, New York

Neurobiology of Body Fluid Homeostasis: Transduction and Integration
Laurival Antonio De Luca Jr., São Paulo State University–UNESP, Araraquara, Brazil
Jose Vanderlei Menani, São Paulo State University–UNESP, Araraquara, Brazil
Alan Kim Johnson, The University of Iowa, Iowa City, Iowa

Neurobiology of Chemical Communication
Carla Mucignat-Caretta, University of Padova, Padova, Italy

NEUROBIOLOGY of CHEMICAL COMMUNICATION

Edited by

Carla Mucignat-Caretta

Department of Molecular Medicine
University of Padova
Padova, Italy

CRC Press
Taylor & Francis Group
Boca Raton London New York

CRC Press is an imprint of the
Taylor & Francis Group, an **informa** business

First published in paperback 2024

First published 2014
by CRC Press
2385 NW Executive Center Drive, Suite 320, Boca Raton FL 33431

and by CRC Press
4 Park Square, Milton Park, Abingdon, Oxon, OX14 4RN

CRC Press is an imprint of Taylor & Francis Group, LLC

© 2014, 2024 Taylor & Francis Group, LLC

Library of Congress Cataloging-in-Publication Data

Neurobiology of chemical communication / editor, Carla Mucignat-Caretta.
        pages cm. -- (Frontiers in neuroscience)
    Includes bibliographical references and index.
    ISBN 978-1-4665-5341-5 (alk. paper)
    1. Molecular neurobiology. 2. Neurocommunication. 3. Brain chemistry. I. Mucignat-Caretta, Carla, editor of compilation.

QP356.2.N478 2014
573.8'4--dc23   2013031195

ISBN: 978-1-4665-5341-5 (hbk)
ISBN: 978-1-03-291821-1 (pbk)
ISBN: 978-0-429-16202-2 (ebk)

DOI: 10.1201/b16511

Visit the Taylor & Francis Web site at
http://www.taylorandfrancis.com

and the CRC Press Web site at
http://www.crcpress.com

Contents

Series Preface xi
Preface xiii
Editor xv
Contributors xvii

Chapter 1 Introduction to Chemical Signaling in Vertebrates and
Invertebrates 1

Tristram D. Wyatt

Chapter 2 Pheromones and General Odor Perception in Insects 23

Michel Renou

Chapter 3 First Investigation of the Semiochemistry of South African
Dung Beetle Species 57

Barend (Ben) Victor Burger

Chapter 4 Pheromone Reception in Insects: The Example of Silk Moths 99

Karl-Ernst Kaissling

Chapter 5 Chemical Communication in the Honey Bee Society 147

Laura Bortolotti and Cecilia Costa

Chapter 6 *Drosophila* Pheromones: From Reception to Perception 211

Wynand van der Goes van Naters

Chapter 7 How *Drosophila* Detect Volatile Pheromones: Signaling,
Circuits, and Behavior 229

Samarpita Sengupta and Dean P. Smith

Chapter 8 Chemical Signaling in Amphibians 255

Sarah K. Woodley

Chapter 9 Vomeronasal Organ: A Short History of Discovery and an
Account of Development and Morphology in the Mouse 285

Carlo Zancanaro

Chapter 10 Vomeronasal Receptors and Signal Transduction in the
Vomeronasal Organ of Mammals 297

*Simona Francia, Simone Pifferi, Anna Menini, and
Roberto Tirindelli*

Chapter 11 Central Processing of Intraspecific Chemical Signals in Mice 325

Carla Mucignat-Caretta

Chapter 12 Molecular and Neural Mechanisms of Pheromone Reception in
the Rat Vomeronasal System and Changes in the Pheromonal
Reception by the Maturation and Sexual Experiences 347

Makoto Kashiwayanagi

Chapter 13 Social Cues, Adult Neurogenesis, and Reproductive Behavior 367

Paolo Peretto and Raúl G. Paredes

Chapter 14 Influence of Cat Odor on Reproductive Behavior and Physiology
in the House Mouse (*Mus Musculus*) 389

Vera V. Voznessenskaya

Chapter 15 Pheromone of Tiger and Other Big Cats 407

Mousumi Poddar-Sarkar and Ratan Lal Brahmachary

Chapter 16 Cattle Pheromones 461

*Govindaraju Archunan, Swamynathan Rajanarayanan, and
Kandasamy Karthikeyan*

Chapter 17 Pheromones for Newborns 489

Benoist Schaal

Chapter 18 Pheromone Processing in Relation to Sex and Sexual Orientation 523

Ivanka Savic

Chapter 19 Human Pheromones: Do They Exist? 535

Richard L. Doty

Index 561

Series Preface

The *Frontiers in Neuroscience Series* presents the insights of experts on emerging fields and theoretical concepts that are, or will be, at the vanguard of neuroscience.

The books cover new and exciting multidisciplinary areas of brain research and describe breakthroughs in fields like visual, gustatory, auditory, olfactory neuroscience, as well as aging and biomedical imaging. Recent books cover the rapidly evolving fields of multisensory processing, glia, depression, and different aspects of reward.

Each book is edited by experts and consists of chapters written by leaders in a particular field. The books are richly illustrated and contain comprehensive bibliographies. The chapters provide substantial background material relevant to the particular subject.

The goal is for these books to be the references every neuroscientist uses in order to acquaint themselves with new information and methodologies in brain research. I view my task as series editor to produce outstanding products that contribute to the broad field of neuroscience. Now that the chapters are available online, the effort put in by me, the publisher, and the book editors hopefully will contribute to the further development of brain research. To the extent that you learn from these books, we will have succeeded.

Sidney A. Simon
Duke University

Preface

The chemical senses evolved to gain information about the external world. This book presents the current view on a peculiar function of the chemical senses that is their role in mediating intraspecific communication.

Several species, both invertebrates and vertebrates, collect data about members of the same species to gain information about various traits of the releaser animal. These signals, whether they completely match the definition of "pheromones" or not, are received and decoded by animals of the same species, and ultimately induce a variety of effects, from modulation of behavior to alteration of hormone release.

The complex link between the signal and the response is mediated by specific brain circuits. This book is intended to give an overview of the peripheral and central processing of intraspecific chemical signals in some key laboratory species or model organisms, from invertebrate to humans. Some chapters deal with wild species, for which the most data is available on behavior while neurobiological knowledge is still lacking. Other chapters represent an open window on related fields; for example communication between species or neural plasticity. However, for comparisons to the main olfactory processing the readers are referred to a previous book in this series, *The Neurobiology of Olfaction*, edited by Anna Menini. Each chapter presents a specific topic in the context of the most recent literature, giving strong emphasis to the neural processing of chemical signals.

I thank Sid Simon for inviting me to join this project and pushing it from the very beginning, Kathryn Everett for her immense patience in driving me through the mysteries of copyediting, Lance Wobus for his support in all the steps of the project, and all the persons who participated: B. Norwitz, J. Lynch, J. Gibbons, M. Cregan and A. Nanas.

Last but not least, I am deeply indebted to all the contributors who were so kind as to dedicate their time to this book, surviving my countless messages. Some of them are long-lasting collaborators; others are colleagues from all over the world that believed in the value of sharing knowledge. My warmest thanks to all of them.

Editor

Carla Mucignat-Caretta, PhD, is associate professor of physiology at the Department of Molecular Medicine, Medical School, University of Padova, Padova, Italy. She earned a PhD in neuroscience at the University of Parma, Italy, working on the role of major urinary proteins as pheromones in mice. Dr. Mucignat-Caretta next moved to the University of Padova as a researcher and from 2007 as associate professor. Her research focus is on the intraspecific chemical communication in mice, from chemical identification of the stimuli to the behavioral and neurohormonal effects induced in the recipient mice, and to the modifications in the brain circuits implied in deciphering chemical signals throughout the lifespan.

Contributors

Govindaraju Archunan
Centre for Pheromone Technology
Department of Animal Science
Bharathidasan University
Tiruchirappalli, India

Laura Bortolotti
Consiglio per la Ricerca e la
    Sperimentazione in Agricoltura
Unità di ricerca di apicoltura e
    bachicoltura (CRA-API)
Bologna, Italy

Ratan Lal Brahmachary
Indian Statistical Institute
Motijheel
Calcutta, India

Barend (Ben) Victor Burger
Department of Chemistry and Polymer
    Science
Stellenbosch University
Stellenbosch, South Africa

Cecilia Costa
Consiglio per la Ricerca e la
    Sperimentazione in Agricoltura
Unità di ricerca di apicoltura e
    bachicoltura (CRA-API)
Bologna, Italy

Richard L. Doty
Smell & Taste Center
University of Pennsylvania Medical
    Center
Philadelphia, Pennsylvania

Simona Francia
Department of Neuroscience
University of Parma and Italian Institute
    of Technology
Parma, Italy

and

Institute of Veterinary Physiology,
    University of Zürich
Zürich, Switzerland

Karl-Ernst Kaissling
Max-Planck-Institute für
    Verhaltensphysiologie
Seewiesen
Starnberg, Germany

Kandasamy Karthikeyan
Centre for Pheromone Technology
Department of Animal Science
Bharathidasan University
Tiruchirappalli, India

Makoto Kashiwayanagi
Department of Sensory Physiology
Asahikawa Medical University
Asahikawa, Japan

Anna Menini
International School for Advanced
    Studies (SISSA)
Trieste, Italy

Carla Mucignat-Caretta
Department of Molecular Medicine
University of Padova
Padova, Italy

Raúl G. Paredes
Instituto de Neurobiología
Universidad Nacional Autónoma de
    México
Querétaro, México

Paolo Peretto
Department of Life Sciences and
    Systems Biology
University of Turin
and
Neuroscience Institute Cavalieri-
    Ottolenghi (NICO) Orbassano
Turin, Italy

Simone Pifferi
International School for Advanced
    Studies (SISSA)
Trieste, Italy

Mousumi Poddar-Sarkar
Chemical Signal Lab.
Department of Botany
(Centre of Advanced Study)
University of Calcutta
Kolkata, India

Swamynathan Rajanarayanan
Centre for Pheromone Technology
Department of Animal Science
Bharathidasan University
Tiruchirappalli, India

and

Department of Biotechnology
St. Michael College of Engineering and
    Technology
Kalayarkoil, India

Michel Renou
Research Unit "Physiology of Insect:
    Signalling and Communication"
    (PISC)
INRA
Versailles, France

Ivanka Savic
Stockholm Brain Institute
Department of Womens and Childrens
    Health
Karolinska Institute
Stockholm, Sweden

Benoist Schaal
Developmental Ethology and Cognitive
    Psychology Group
Center for Smell, Taste and Food
    Science
CNRS
Dijon, France

Samarpita Sengupta
Departments of Pharmacology and
    Neuroscience
University of Texas Southwestern
    Medical Center
Dallas, Texas

Dean P. Smith
Departments of Pharmacology and
    Neuroscience
University of Texas Southwestern
    Medical Center
Dallas, Texas

Roberto Tirindelli
Department of Neuroscience
University of Parma and Italian Institute
    of Technology
Parma, Italy

Wynand van der Goes van Naters
School of Biosciences
Cardiff University
Cardiff, United Kingdom

Vera V. Voznessenskaya
A.N. Severtzov Institute of Ecology &
    Evolution
Russian Academy of Sciences
Moscow, Russia

Sarah K. Woodley
Department of Biological Sciences
Duquesne University
Pittsburgh, Pennsylvania

Tristram D. Wyatt
Department of Zoology
University of Oxford
Oxford, United Kingdom

Carlo Zancanaro
Anatomy and Histology Section
Department of Neurological
   and Movement Sciences
University of Verona
Verona, Italy

1 Introduction to Chemical Signaling in Vertebrates and Invertebrates

Tristram D. Wyatt

CONTENTS

1.1 Introduction 1
1.2 Discovering Pheromones 5
1.3 Pheromones Evolve from Chemical Cues 6
1.4 Pheromone Specificity and Speciation 8
1.5 Reception and Processing of Chemical Signals 11
1.6 Pheromones and Olfaction 13
1.7 Releaser and Primer Effects 14
1.8 Are Pheromones Innate? 14
1.9 Human Pheromones? 15
1.10 Applications of Pheromones 15
1.11 Conclusions 16
Acknowledgments 16
References 16

1.1 INTRODUCTION

Chemical senses are probably among the first senses to have evolved and all cellular life forms from bacteria to animals are sensitive to chemical information, whether it comes from potential food, predators, the environment, or other members of the same species.

Chemicals from outside the organism that provide information are termed semiochemicals (Figure 1.1). With the chemical senses in place it was perhaps inevitable that chemical communication would evolve. **Pheromones** are evolved chemical signals between members of the same species. The molecules are emitted by an individual and received by a second individual of the same species, in which they cause a specific reaction, for example, a stereotyped behavior or a developmental process (Wyatt 2010, after Karlson & Lüscher 1959). The word **pheromone** was coined from the Greek *pherein*, to carry or transfer, and *hormōn*, to excite or stimulate. While Karlson and Lüscher (1959) proposed the word after the first pheromone had been identified in an insect, the silk moth *Bombyx mori*, they suggested the term would apply to chemical signals in all types of animal from crustaceans to

1

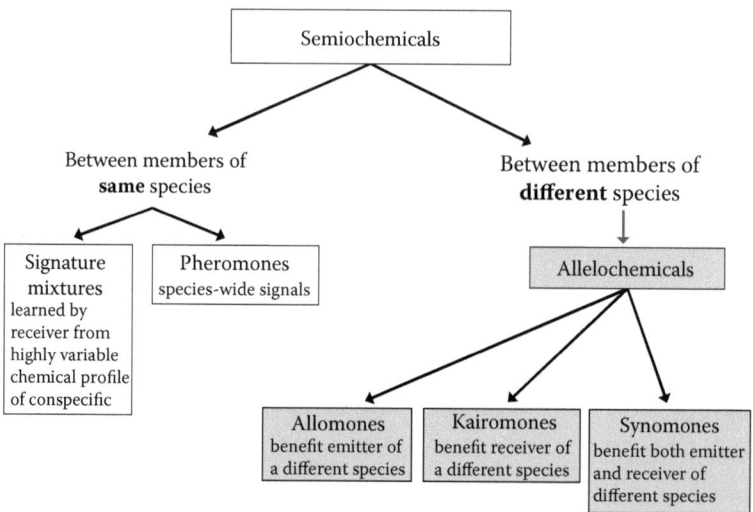

FIGURE 1.1 Diagram showing the relationships between different kinds of semiochemicals. Pragmatically, when considering semiochemical interactions within a species, it is useful to distinguish species-wide pheromone signals, such as male sex pheromones produced by any dominant male, from phenomena that rely on the learning of signature mixtures for individuals (for recognition of siblings, for example) based on differences in chemical profile between individuals, and thus there is no species-wide molecule(s) to find. "Signature mixture" comes from Wyatt (2010), and is derived from Johnston's "mosaic signal" *sensu* (2003, 2005), Hölldobler and Carlin's (1987) ideas, and Wyatt's (2005) "signature odor." (From Wyatt, TD, *Pheromones and Animal Behavior: Chemical Signals and Signature Mixes,* Second Edition. Cambridge, UK: Cambridge University Press, 2014.)

fish to terrestrial mammals. Equally, while the silk moth pheromone was a single molecule, pheromones consisting of many molecules together were not excluded. Karlson and Lüscher anticipated that different species might share some of the same molecules. No requirement was made for pheromone responses to be innate. I emphasize these points only because some authors writing about mammals raise these as objections to applying the term to mammal pheromones (e.g., Doty 2010; Petrulis 2013). As I argue below, pheromones are found across the animal kingdom and while mammal pheromones *are* hard to study, we do have many examples that seem robust (see for example Schaal et al. 2003; Schaal, Chapter 17). Even so, criticisms of some claims in the literature are justified. We should also be careful to distinguish phenomena mediated by pheromones from olfactory phenomena that have more to do with differences between individuals' chemical profiles, allowing individuals to be distinguished (see below).

As well as pheromones, animals also receive chemical information· about the identity of another individual. The receiving animal may learn a **signature mixture**: a variable chemical mixture (a subset of the molecules given off in an animal's chemical profile) learned as a template by other conspecifics and used to recognize an animal as an individual (e.g., lobsters, mice) or as a member of a particular social group such as a family, clan, or colony (e.g., ants, bees, badgers) (Figures 1.1 and 1.2) (Wyatt 2010, 2014).

The signature mixture is the mix of molecules (and likely, their relative ratios) that are learned. The template is the neural representation of the signature mixture stored in the memory of the learner (after van Zweden & d'Ettorre 2010). There are two distinguishing characteristics of signature mixtures: first, a requirement for learning, and second, the variability of the cues learned, allowing other individuals to be distinguished by their different chemical profiles (Wyatt 2014).

There are many biological systems where distinguishing between pheromones and learned highly variable signature mixtures can help our understanding. For example, each ant colony has different combinations of molecules, largely cuticular hydrocarbons, which can be learned as signature mixtures allowing ants to discriminate between colony members and noncolony members (Bos & d'Ettorre d'Ettorre 2012). Every queen of the species, however, has the same queen pheromone (Liebig 2010). In the ant *Lasius niger*, the queen pheromone is the cuticular hydrocarbon 3-methylhentriacontane, missing from worker profiles (Holman et al. 2010). Similarly, explaining a phenomenon such as the Bruce effect in mice (Brennan 2009; Mucignat-Caretta, Chapter 11) is easier if we distinguish between the individually distinct odors of different males and the pheromone(s) that all male mice produce. The female's memory of the signature mixture of the particular male she mated with has the effect of blocking any stimulus from later contact with his male pheromones whereas the stimulus from the same male pheromones of other, unfamiliar males passes through to the hypothalamus, triggering the Bruce effect.

Pheromones and the molecules learned as signature mixtures appear in the cloud of molecules that make up the chemical profile of an organism (Figure 1.2). Much of the chemical profile is highly variable from individual to individual. The sources of the molecules in the chemical profile include the animal's own secretions as well as its environment, food, bacteria, and other individuals. It is this complex background that makes identifying pheromones so challenging in many organisms.

Semiochemicals can also be significant to individuals from other species and these are termed **allelochemicals** (Figure 1.1) (Nordlund & Lewis 1976). Broadcast signals can be eavesdropped, as kairomones. Some of the most spectacular examples of using such kairomones come from predatory beetles homing in on the aggregation pheromones released by their prey, bark beetles (Raffa 2001). Eavesdropping can occur across taxa. Nematode-trapping fungi detect the pheromones of their prey nematodes and produce more traps in response (Hsueh et al. 2013). Aggressive chemical mimicry (allomones) can exploit the responses of organisms to their own pheromones (Vereecken & McNeil 2010). For example, most orchids offer no nectar reward but instead, by mimicking the female pheromones of the insect, they attract bee and wasp males to pollinate them. Bolas spiders lure male moths by producing the moths' female sex pheromone.

In this introductory chapter, I will touch on many classic systems such as moths, *Drosophila,* and mice, which are explored in more detail in the other chapters of this book, but I will also take the opportunity to illustrate points with the pheromones of other animal taxa such as nematodes, mollusks, and fish. For this

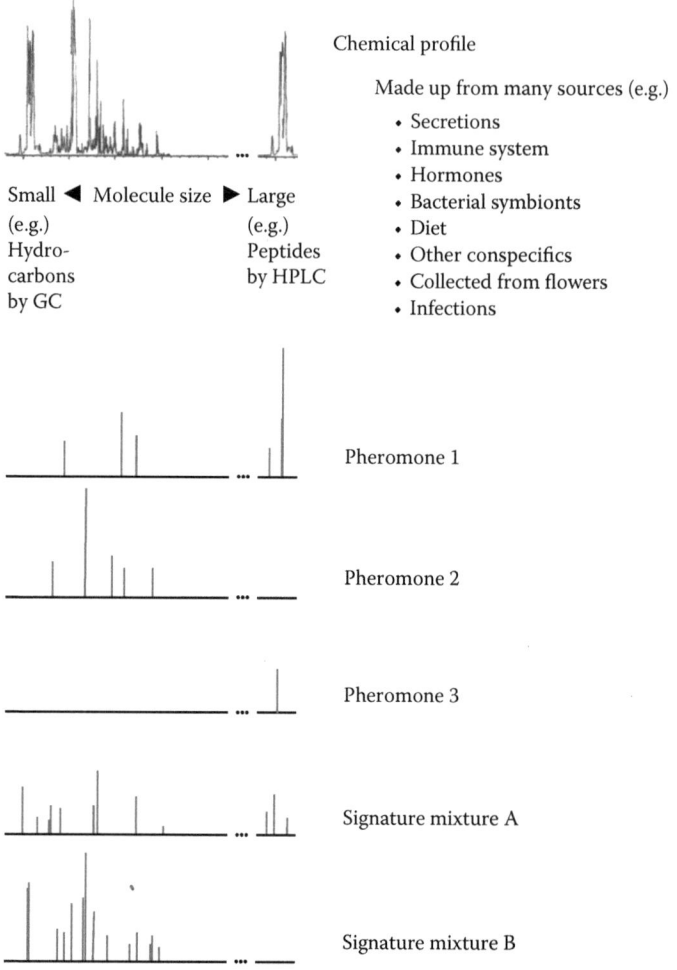

Chemical profile

Made up from many sources (e.g.)

- Secretions
- Immune system
- Hormones
- Bacterial symbionts
- Diet
- Other conspecifics
- Collected from flowers
- Infections

Small ◄ Molecule size ► Large
(e.g.)   (e.g.)
Hydro-   Peptides
carbons   by HPLC
by GC

Pheromone 1

Pheromone 2

Pheromone 3

Signature mixture A

Signature mixture B

FIGURE 1.2 Pheromones occur in a background of molecules that make up the chemical profile consisting of all the molecules extractable from an individual. The chemical profile (top) is an imaginary trace from an imaginary column capable of analyzing all the molecules (at one side is gas chromatography (GC) with small volatile molecules; at the other side is high-performance liquid chromatography (HPLC) with large proteins). Each peak represents at least one molecule. As examples, I have included some possible kinds of pheromones that are known from organisms (not necessarily in the same species): a specific combination of large and small molecules (Pheromone 1), a combination of small molecules (Pheromone 2), or a particular large molecule by itself such as a peptide (Pheromone 3). The signature mixtures (A and B) are subsets of variable molecules from the chemical profile that are learned as a template for distinguishing individuals or colonies. Different receivers might learn different signature mixtures of the same individual. For example, a male might learn a different signature mixture of their mate from what one of her offspring might learn. (Figure and caption from Wyatt, TD, *Pheromones and Animal Behavior: Chemical Signals and Signature Mixes,* Second Edition. Cambridge, UK: Cambridge University Press, 2014; adapted from Wyatt, TD, *J Comp Physiol A* 196: 685–700, 2010.)

introductory chapter, I have usually chosen recent references that will lead you to the relevant literature. More detail on the topics covered in this chapter can be found in Wyatt (2014).

1.2 DISCOVERING PHEROMONES

The history of the 1959 identification of the first pheromone, bombykol, the sex pheromone of female silk moths, is well known. Butenandt and his team spent more than 20 years on the project, taking the pheromone glands from more than 500,000 female moths (Butenandt et al. 1959; Hecker & Butenandt 1984). The advances in techniques since then, including mass spectroscopy and nuclear magnetic resonance, mean that the analysis could now be attempted with the secretions of just a few insects, at the level of picogram or femtogram quantities (Meinwald 2009; Wyatt 2014). Many techniques use the sensory system of the organism as the sensor to identify the likely components in the pheromone, using an electroantennogram (EAG) or single-cell recording from olfactory sensilla on the male moth antenna (Schneider 1999) or electroolfactograms (EOG) in fish (Chung-Davidson et al. 2011). Chapters in Touhara (2013) cover modern experimental techniques for studying pheromones in a variety of organisms.

While identification of pheromones has traditionally involved fractionating samples, techniques that can work on whole samples may offer a way forward for some biological systems, especially if the number of potential compounds to test in combination is high or if the molecules are labile or otherwise easy to lose in processing. One such technique is differential analysis by 2D-NMR spectroscopy (DANS) (Forseth & Schroeder 2011), which has been used to identify the many molecules of the multicomponent pheromones of the nematode *Caenorhabditis elegans* (Pungaliya et al. 2009; Srinivasan et al. 2012).

Whatever the route to identifying the pheromone, a crucial factor is still the bioassay, which might be a behavior such as the wing-fluttering of the male silk moth in response to the active molecule, or a physiological measure. Moreover, identifying a molecule(s) as a possible (or "putative") pheromone is just the beginning. The experimenters must establish that this really is the molecule (or combination of molecules) and that other similar molecules to which the animal might be exposed are not equally stimulating. In the elegant study of the male mouse pheromone (Z)-5-tetradecen-1-ol, Yoshikawa et al. (2013) showed that while the particular olfactory receptor that was responsible for the female's response was activated by related molecules, (Z)-5-tetradecen-1-ol was the only molecule found in male mouse exocrine secretions that had this activity at natural concentrations. The molecule is produced by the male preputial glands and is under androgen control.

Stimulus concentration is also crucially important in bioassays. In both vertebrate and invertebrate olfactory systems, at very high concentrations other molecules may start to activate olfactory receptors that would not normally be stimulated, thus giving a false positive to a molecule that is not a pheromone (Firestein 2001; Kreher et al. 2008; Leal 2013). Vertebrate and invertebrate pheromones have optimum concentrations for their specific response. This is shown for example by the behavioral response of rabbit pups to the mammary pheromone, 2-methylbut-2-enal, which is limited to a range of concentrations (from 2.5×10^{-9} to 2.5×10^{-5} g/ml in milk) (Coureaud et al. 2004; Schaal, Chapter 17). In other situations, changing concentration can change the meaning of a signal. For example, in *C. elegans,* overlapping sets of ascaroside molecules act as sex pheromones at picomolar concentrations and at concentrations about 10,000 greater they stimulate larvae to enter the dauer (resting stage) (Srinivasan et al. 2012).

Other key aspects of good experimental design have been slow to come to studies of chemical senses. It is essential that the allocation of animals to treatments is truly random and that, wherever possible, observers or scorers should be blind as to which treatment they are assessing, whether it a behavior or scoring characters such as *c-Fos* staining in brain sections (Burghardt et al. 2012; Evans et al. 2011).

1.3   PHEROMONES EVOLVE FROM CHEMICAL CUES

The enormous variety of molecules acting as pheromones across the animal kingdom suggests that almost any kind of molecule can evolve into a pheromone if it gives selective advantage (Wyatt 2014). Small volatile molecules are used as pheromones by terrestrial vertebrates and invertebrates, from the familiar female moth sex pheromones to the small molecule pheromones of mice. Underwater, solubility is the equivalent of volatility in air and both small and large molecules can be soluble. The small molecule amino acid L-kynurenine is the female sex pheromone of masu salmon, *Oncorhynchus masou* (Yambe et al. 2006). Soluble peptides or proteins have evolved as sex pheromones in the marine mollusk *Aplysia* (Cummins et al. 2007; Cummins & Degnan 2010) and as sex pheromones in the underwater courtship of Japanese *Cynops* newts (Woodley 2010; Woodley, Chapter 8). Less volatile molecules such as cuticular hydrocarbons are also used as contact pheromones in animals such as *Drosophila* (Ferveur & Cobb 2010; van der Goes van Naters, Chapter 6; Sengupta & Smith, Chapter 7) and long chain unsaturated methyl ketones in garter snakes (Mason & Parker 2010). Large peptides or proteins can be used as pheromones by terrestrial vertebrates if the molecules are transferred directly to the nose in courtship, as in plethodontid salamanders (Woodley 2010; Woodley, Chapter 8) and the male ESP1 tear secretion pheromone in mice (Haga et al. 2010).

It seems there are likely to be two main routes for chemical cues to evolve into pheromone signals: from sender precursors or from receiver sensory bias (Bradbury & Vehrencamp 2011; Wyatt 2014). The first, from sender precursors, is by the evolution of increasing sensitivity in the receiver to molecules that are reliable cues to a resource or behavior; for example, hormone molecules leaking from a female as her eggs reach maturity. This can then be followed by selection on females to produce and release larger quantities of these or related molecules, to attract males. This scenario seems to describe the evolution of the many fish pheromones that are steroid hormones or their derivatives (Stacey & Sorensen 2011). In the second, from receiver sensory bias, senders are selected to produce more molecules of a type to which the receiver is already sensitive. For example, in many species the male moth pheromones used in the final stages of courtship are the same as, or related to, molecules produced by the host plants of that insect species.

Females already have sensitive receptors and neural circuits for these host plant molecules as they must search out these plants to lay their eggs on (Hansson & Stensmyr 2011; Renou, Chapter 2). Male moth pheromones appear to have evolved to exploit this preexisting sensory bias (Birch et al. 1990; Phelan 1997). Similarly, the male sex pheromone of the European beewolf wasp *Philanthus triangulum*, (Z)-11-eicosen-1-ol, may exploit a preexisting female sensory bias: female beewolves were already sensitive to this molecule as it is given off by their honeybee prey (Kroiss et al. 2010; Steiger et al. 2011).

Sometimes the same molecules have evolved to be used as pheromones in widely different taxa (Kelly 1996; Wyatt 2014). This may be explained by a shared origin of life leading to a shared biochemistry combined with the characteristics that make some molecules good for chemical communication (such as nontoxicity, volatility, and stability). For example, the female sex pheromone of the Asian elephant, (Z)-7-dodecen-1-yl acetate, is used as one component of their multicomponent pheromones by some 140 species of moth and the Asian male elephant's pheromone, the terpene frontalin, is also used by some bark beetles as an aggregation pheromone (Rasmussen et al. 1997; Rasmussen et al. 2003). The male mouse pheromone (Z)-5-tetradecen-1-ol (Yoshikawa et al. 2013) differs only by the position and orientation of its double bond from one of the *Ostrinia* moth pheromone components (E)-11-tetradecen-1-ol (*E*11-14:OH) (Figure 1.3).

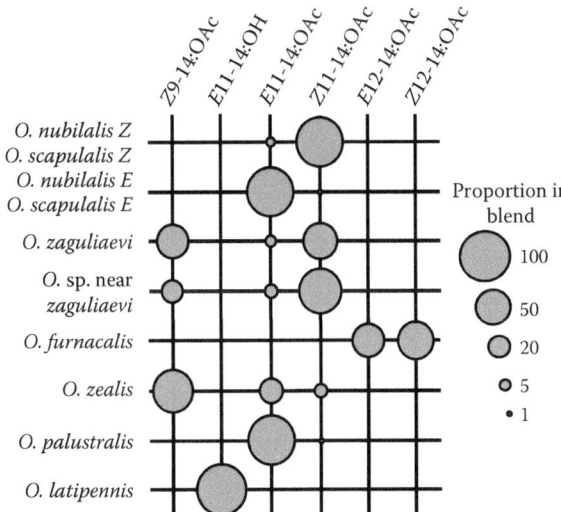

FIGURE 1.3 Related moth species gain specificity for their female sex pheromones by different combinations and quantities of a range of similar molecules. Here the pheromone components of different *Ostrinia* species are presented. Some species have *E* and *Z* races with different ratios of E/Z isomers of a component. The size of a dot is proportional to the ratio of an individual pheromone component (Lassance et al. 2013). See Figure 1.4 for the pathways for producing these molecules. (From Lassance, J-M et al., *Proc Natl Acad Sci USA* 110: 3967–3972, 2013.)

1.4   PHEROMONE SPECIFICITY AND SPECIATION

If there is no selective disadvantage to responding to the same pheromone, species may share one. For example, many different aphid species share the sesquiterpene (*E*)-β-farnesene as their alarm pheromone (Dewhirst et al. 2010). A coccinellid beetle eating an aphid of a different species nearby is likely to eat you too, so it is worth responding to any aphid's alarm pheromone. By contrast, aphid sex pheromones are species-specific multicomponent blends because of the high reproductive cost of responding to the sex pheromone of the wrong species (Dewhirst et al. 2010).

There are two main routes to species-specificity in pheromones: unique molecules, or more commonly, different combinations of molecules in a multicomponent pheromone. Unique molecules are illustrated by animals using a peptide as their pheromone: different species can use different amino acid sequences to give specificity. Two related *Cynops* newt species have peptide pheromones that differ by just two amino acids (Toyoda et al. 2004). Unlike other cockroaches (see below), the brown-banded cockroach *Supella longipalpa* appears to use a single unique molecule, the (2*R*,4*R*)-isomer of supellapyrone (Gemeno et al. 2003).

Most species-specific sex pheromones, if they are not peptides, are likely to be multicomponent. Such pheromones gain their species-specificity from a combination of molecules that only works as a whole (synergy, see below). The molecules themselves need not be unusual and some may even be shared with other species: it is the particular combination that makes the pheromone specific. Moth female sex pheromones tend to be particular combinations of a number of short chain hydrocarbons including unbranched fatty acids, acetates, or aldehydes (Figures 1.3 and 1.4) (Allison and Cardé 2014; de Bruyne & Baker 2008). Multicomponent nematode pheromones appear to gain specificity from overlapping sets of different acarosides (Choe et al. 2012). Vertebrates also have multicomponent pheromones. For example, the goldfish, *Carassius auratus,* and closely related carp, *Cyprinus carpio,* share many of their hormonal pheromone components but species-specificity comes from combinations of these plus other molecules as yet unidentified (Levesque et al. 2011; Lim & Sorensen 2012). Some mammal pheromones may be multicomponent (for example, in male mice urine, dehydro-*exo*-brevicomin and 2-*sec*-butyl-4,5-dihydrothiazole act together (Novotny et al. 1999; Novotny 2003)). Most other mammal pheromones found so far are single molecules such as the rabbit mammary pheromone (Schaal et al. 2003; Schaal, Chapter 17). However, single molecule pheromones are likely to be the first identified ones. We may expect that more multicomponent pheromones will be found in future years in mammals, just as they were in insects and other organisms.

Molecules are carried together from the source in parcels or patches of fluid, whether that is air or water. This means that the concentration and ratios of multicomponent pheromone blends remain largely unchanged in the parcel of air or water as it is carried downwind or downstream (Vickers 2000; Webster & Weissburg 2009; Wyatt 2014). Male oriental fruit moth, *Grapholita molesta,* 100 m downwind from the source showed little response to the main component released alone but responded strongly to the full three-component blend released from the source (demonstrating both synergism and the arrival of the components together) (Linn et al. 1987).

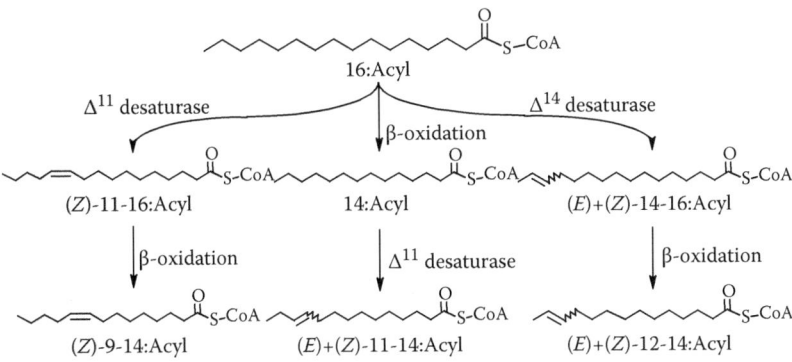

FIGURE 1.4 The synthesis of the multicomponent pheromone blends used by different *Ostrinia* moth species (seen in Figure 1.3) can be largely explained by which enzyme genes are expressed and the differences in substrate specificities of these enzymes between species (Lassance et al. 2013). De novo biosynthesis of all precursors starts from palmitoyl-CoA (16:Acyl). A Δ11desaturase catalyzes the production of (Z)-11-hexedecenoyl [(Z)-11–16:Acyl], which undergoes one cycle of β-oxidation to give (Z)-9-tetradecenoyl [(Z)-9–14:Acyl]. The same Δ11desaturase acts on myristoyl-CoA (14:Acyl) to produce (E)-and (Z)-11-tetradecenoyl [(E)+(Z)-11–14:Acyl]. In *O. latipennis*, a different Δ11 desaturase uniquely catalyzes the stereo-specific production of (E)-11-tet-radecenoyl [(E)-11–14:Acyl]. Finally, in *O. furnacalis*,(E)- and (Z)-12-tetra-decenoyl [(E)+(Z)-12–14:Acyl] moieties are produced through chain-shortening of (E)- and (Z)-14-hexedecenoyl [(E)+(Z)-14–16:Acyl], respectively, the production of which is catalyzed by a Δ14 desaturase (Lassance et al. 2013). The conversion of these fatty-acyl phero-mone precursors into fatty alcohol pheromone components is catalyzed by the pheromone gland-specific fatty-acyl reductase pgFAR that differs between species. (From Lassance, J-M et al., *Proc Natl Acad Sci USA* 110: 3967–3972, 2013.)

Pheromones may play an important role in speciation, the formation of new species that can no longer interbreed successfully (Smadja & Butlin 2009). Over evolutionary time, a population that is diverging from its parent population may gain specificity for its own pheromone by adding related molecules to the pheromone molecule(s) used by its parent species (or in the case of species with peptide phero-mones, drive change in amino acid sequence as shown in plethodontid salamanders (Woodley 2010; Woodley, Chapter 8).

Sibling species often have variations on a pheromone theme, as if exploring chemical space from a new starting point (Wyatt 2014). For example, each genus of cockroach tends to have long range female sex pheromones from a different class of compound (Eliyahu et al. 2012). Within the genus, related species of *Periplaneta* cockroach use different combinations of molecules based on the unusual molecule periplanone. Similarly, the cockroach *Parcoblatta lata* uses the unusual macrocyclic lactone (4Z,11Z)-oxacyclotrideca-4,11-dien-2-one as a main component of its female sex pheromone (Eliyahu et al. 2012). Other species in the genus have this as one of their pheromone components. The pheromones of *Ostrinia* moth species involve a number of related molecules (Figure 1.3).

An advantage that studies of chemical senses have over those of other senses is that the genetics of the production of chemical signals as well as their reception (next section) can be dissected in detail. For example, in mice some of the sex-dependent changes in expression of enzymes that lead to production of a male chemosignal, trimethylamine, have been identified along with its specific receptor (trace amine-associated receptor 5, TAAR5) (Li et al. 2013).

In insects the changes in pheromone blend can be due to changes in a small number or in many genes (polygenic systems) (Allison & Cardé 2014; Symonds & Elgar 2008; Wyatt 2014). I start with examples of systems that involve change in a few genes. In *Drosophila* species, cuticular hydrocarbons are important in species isolation at courtship (van der Goes van Naters, Chapter 6). In a classic study, Shirangi et al. (2009) looked at the rapid evolution of changes in the expression of long chain cuticular hydrocarbon pheromones across 24 species of the genus *Drosophila*. Diene cuticular hydrocarbons produced by a key desaturase DESAT-F are characteristic of female *D. melanogaster*. Across the genus, there have been a large number of evolutionary transitions in the gene *desatF* including six independent gene inactivations, three losses of expression without gene loss, and two transitions in sex specificity. In addition, activation/inactivation of binding sites for the sex-determination transcription factor *doublesex* were involved in transitions between species with the same dienes in males and females and those that were dimorphic.

The moth genus *Ostrinia* has become another model system with pheromone change due to a few genes (Lassance 2010; Lassance et al. 2013). The *Ostrinia* female sex pheromones are typically blends of monounsaturated tetradecenyl (C14) acetates that vary in double bond position (Δ^9, Δ^{11}, or Δ^{12}) and geometry [*cis* (Z) or *trans* (E)] (Figure 1.4) (Lassance 2010). Specificity comes from presence and absence of components and their ratios in the pheromone blend. The production of different pheromone blends in different species and races has been tracked down to changes in enzymes involved in biosynthesis of the pheromones (Lassance 2010). Starting with palmitic acid (C16), enzymatic steps include desaturation and chain shortening, leading to C14 alcohols that in some species are turned into acetates. Differences in the enzyme pheromone gland-specific fatty-acyl reductase gene *pgFAR*, which encodes the reductase catalyzing the specific reduction of the fatty-acyl pheromone precursors into fatty alcohols underlie the differences in the final blend in different species and races of *Ostrinia* species (Figure 1.4) (Lassance et al. 2013). The amino acid sequence differences in the reductase cause differences in substrate specificity that produce both the subtle and dramatic differences in the ratios of components of female pheromone blends in the different species and races (see legend, Figure 1.4). Changes in the olfactory receptors in sibling *Ostrinia* species help explain how the males of the different species respond to their respective females' blends (Section 1.4) (Leary et al. 2012).

However, in many other moth genera the changes in pheromone blend as species diverge appear to result from many genes, each making small changes. For example, the pheromone blend differences between the sympatric sibling species *Heliothis virescens* and *H. subflexa* result from polygenic changes involving genes spread out on at least nine different chromosomes (Groot et al. 2009). The *Heliothis* version of the *pgFAR* gene mentioned above is one of the genes involved (Hagström et al. 2012). As in *Drosophila* some of the changes have occurred in regulatory sequences as well as to the DNA sequences coding for enzymes themselves.

How do male responses follow the changes in female pheromone blend as species' signals diverge? A serendipitous observation of lab cultures of the cabbage looper moth *Trichoplusia ni* revealed mutant females producing a new blend with 50 times more of one of the minor components (through a change in a gene for a single biosynthetic enzyme in the pheromone gland) (Allison & Cardé 2014; Domingue et al. 2009). A few wild males nonetheless responded to the new blend. By selecting responding males in the laboratory population over 50 generations, the researchers found that most males' response spectrum was broadened to include the mutant female blend.

Tracing the evolution of male pheromone signals and speciation in *Nasonia* parasitoid wasps, Niehuis et al. (2013) suggest that in these wasps the mutations leading to the addition of a new molecule to the male blend in one species, *N. vitripennis*, may have come before selection on the females of this species to detect the new molecule.

Currently we know much less about the evolution of chemical signals in vertebrates. However, there are examples in lizards, fish, and rodents among others, in which chemical signals likely to be involved in speciation have been demonstrated, even if the molecules involved and the genetics have been elucidated in only a very few species (Smadja & Butlin 2009; Wyatt 2014).

1.5 RECEPTION AND PROCESSING OF CHEMICAL SIGNALS

In all animals, olfaction starts with the interaction of an odorant molecule with a chemoreceptor, an olfactory receptor (OR) protein in the membrane of an olfactory sensory neuron (OSN, sometimes called an olfactory receptor neuron (ORN)) (Wyatt 2014; Renou, Chapter 2; Kaissling, Chapter 4; Francia et al., Chapter 10). Taste also starts similarly, with activation of a gustatory receptor (GR) protein but taste works in importantly different ways. I will discuss olfaction first.

Richard Axel and Linda Buck received their Nobel Prize in 2004 for their earlier identification of the diverse mammalian family of ORs in the main olfactory epithelium (Buck & Axel 1991). Their prize was also for discovering how olfaction works, in a combinatorial way, and how this is enabled by the way OSNs are organized in the brain (Malnic et al. 1999). In vertebrates, the ORs are G-protein coupled receptor (GPCRs) proteins. In the mouse there are about 1000 ORs that differ in their odorant specificity (which odorant molecules will stimulate them) and most are broadly tuned (Malnic et al. 2010). The ORs have overlapping sensitivities so a given odorant will likely stimulate more than one OR and a given OR will be stimulated by a variety of molecules with similar characteristics, for example by most alcohols with 12 carbons and a double bond toward one end of the molecule. Which molecules interact with an OR is determined by the amino acid sequence of its binding site (Kato & Touhara 2009; Reisert & Restrepo 2009). Even a single amino acid change can result in a very different specificity, as shown in an *Ostrinia* moth OR in different species (Figure 1.5).

A given OSN expresses only one kind of OR, and crucially, all OSNs expressing the same kind of OR send their axon to the same spot (glomerulus) in the olfactory bulb in the brain. The glomerulus is where the OSNs synapse with interneurons and projection neurons with connections to other glomeruli and higher processing

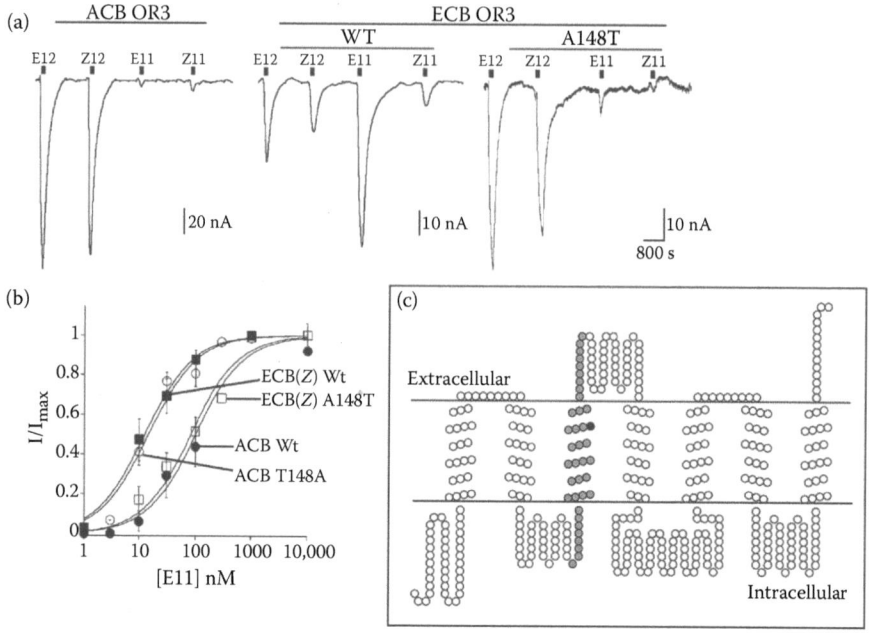

FIGURE 1.5 Some of the differences in response to pheromone components by male moths of closely related species can be tracked down to differences as small as a single amino acid in the sequence of the binding site of their respective olfactory receptors (ORs) (Leary et al. 2012). The European corn borer (ECB, *Ostrinia nubilalis*), responds to (*E*)-11-tetradecenyl acetate (E11) but males of its sibling species, the Asian corn borer (ACB, *O. furnacalis*), do not, instead responding to their key molecules E12 ((*E*)-12-tetradecenyl acetate) and Z12. This is explained by a difference in amino acid residue 148 of ACB and ECB(Z) OR3. (a) Representative traces show membrane currents in *Xenopus* oocytes coexpressing Orco with ACB OR3, wild type (WT)ECB(Z) OR3, or modified ECB(Z) OR3^{A148T}, in response to 10^{-n} M doses of E11, Z11, E12, and Z12. (b) Concentration-dependence of ECB(Z) OR3 (filled squares), ACB OR3 (filled circles), ECB(Z) OR3^{A148T} (open squares), and ACB OR3^{T148A} (open circles) receptors to E11. (c) A TOPCONS.net model of *Ostrinia* OR3 sequences showing the unique threonine (T) for alanine (A) substitution at position 148 (black circle) within the ACB OR3 binding site. (Adapted from Leary, GP et al., *Proc Natl Acad Sci USA* 109: 14081–14086, 2012.)

areas of the brain. Different odorants are distinguished by a combinatorial code of activation of glomeruli so the brain can distinguish different odorants: an odorant stimulating OR1, OR2, and OR3 can be distinguished from an odorant stimulating OR2 and OR7, and so forth.

Remarkably, in insects, although the insect olfactory receptors have evolved completely independently and are not GPCRs like the vertebrate ORs, the organization of OSNs into glomeruli and the combinatorial mechanisms seem very similar to those of vertebrates (Martin et al. 2011; Renou, Chapter 2). Insect ORs come from completely different families of membrane proteins from vertebrate ones, sit with the opposite orientation in the membrane, and appear to work by ionotropic mechanisms rather than G-protein mediated cascades (Silbering & Benton 2010). Insect ORs are dimers consisting of a variable ORx and an OR, called Orco, common to all dimers.

1.6 PHEROMONES AND OLFACTION

As cues evolve into pheromones, there is selection for increasing sensitivity and specificity of ORs for the molecules involved. The ORs lose their broad tuning and instead respond only to one molecule at physiological concentrations. The change in the specificity of the ORs comes from selection for changes in the sequence of amino acids in the binding site. A change of just a single amino acid changes the response of one of the ORs sensitive to pheromone components in male *Ostrinia* moths and largely explains the differences in response between species (Figure 1.5) (Leary et al. 2012). The change in sensitivity to a pheromone due to OR changes at the level of amino acid sequences has also been explored in the nematode *C. elegans* (McGrath et al. 2011). We do not know as much about mammal pheromone receptors but the vomeronasal receptor for the male peptide pheromone ESP1 in the mouse is likely to be highly specific (Haga et al. 2010), as is the mouse receptor for (Z)-5-tetradecen-1-ol in the main olfactory epithelium (Yoshikawa et al. 2013). Incidentally, these receptor examples are a reminder that in the mouse, pheromones can be detected by either or both the main olfactory or accessory olfactory systems, depending on the pheromone. The outputs of the two olfactory systems are integrated in higher brain regions including the hypothalamus (Mucignat-Caretta et al. 2012; Mucignat-Caretta, Chapter 11).

Most species-specific pheromones are likely to be multicomponent. Male moths have ORs sensitive to each of the components of their females' sex pheromone. The response to detecting the pheromone is combinatorial: a male will only fly upwind if all the components are present and in the right ratios, thus stimulating the right glomeruli and higher levels of the brain correctly (Haupt et al. 2010; Martin et al. 2011). The observed effect of a response to all the components together being greater than the sum of the responses to each of the components presented separately is often called "synergy." I would suggest it is the natural outcome of having a combinatorial response to a multicomponent pheromone, an inevitable consequence of the way blends are processed in the brain (Wyatt 2014). Similar synergistic effects from multicomponent pheromones are found in nematodes (even though they have simpler circuits, without glomeruli) (Srinivasan et al. 2012).

In vertebrates and invertebrates, most pheromones appear to be detected by the olfactory system acting via its glomerular organization (in mammals olfaction includes both the main olfactory system and vomeronasal organ-accessory olfactory system (Francia et al., Chapter 10; Mucignat-Caretta et al. 2012; Wyatt 2014; Zancanaro, Chapter 9). That pheromones are largely detected by the olfactory system may be due to the flexibility that the combinatorial olfactory organization gives for changing components gradually, particularly in the early stages of transition from cue to specific pheromone.

Fewer pheromones are detected by taste. Taste uses different families of chemosensory receptors from olfaction in both vertebrates and invertebrates, and more importantly the taste receptor neurons follow simpler circuits and do not end up in glomeruli in the brain (Wyatt 2014). Thus the opportunity for subtle combinatorial processing is not there (though there may be integration of information from different senses in a combinatorial fashion; for example, pheromones and sound cues in *Drosophila* courtship (Dickson 2008)). A number of important *Drosophila* pheromones are detected by taste receptors; for example, the female hydrocarbon pheromones detected by gustatory receptors on the male's front leg (e.g., Inoshita et al. 2011). The boundaries between taste and smell may be less distinct in Crustacea (Breithaupt & Thiel 2011).

Not all pheromones act via olfaction or taste. The exceptions are allohormone pheromones, which bypass the conventional sensory system and act on tissues of the recipient (Koene & ter Maat 2001; Wyatt 2014). For example, a glycoprotein, royalactin, in the royal jelly fed to honeybee larvae acts on fat body tissues, influencing the development of larvae into queens (Kamakura 2011). Accessory gland peptides and other molecules in the ejaculate of male *Drosophila melanogaster* switch the behavior and physiology of the female so she rejects male courtship and starts to lay eggs (Avila et al. 2011; Wolfner 2009). One molecule, the sex peptide, activates a specific receptor on chemosensory neurons in her uterus and oviduct (Rezával et al. 2012).

1.7 RELEASER AND PRIMER EFFECTS

Pheromones may elicit immediate behavioral responses (releaser effects) and longer-term responses (primer effects), such as physiological or developmental changes (Wilson & Bossert 1963). For example, the primer effect of pheromone(s) in royal jelly in determining the developmental path of honeybee larvae into workers or queens (above) has been contrasted with the prompt aggressive behavior of worker honeybees exposed to alarm pheromone. However, as anticipated by Wilson and Bossert (1963), many pheromones have both releaser and primer effects (Wyatt 2014). For example, the principal molecule of alarm pheromone, isopentyl acetate, causes honeybees to be more aggressive and quicker to respond to the pheromone for many hours after exposure and these changes in behavior are correlated with changes in gene expression in the antennal lobes in their brains (Alaux & Robinson 2007; Alaux et al. 2009; Bortolotti and Costa, Chapter 5). Similar multiple effects are found in pheromone responses of goldfish (Stacey & Sorensen 2011) and mice (Mucignat-Caretta, Chapter 11; Novotny 2003).

1.8 ARE PHEROMONES INNATE?

Most pheromones appear to be innate, not requiring learning. This seems to be a reasonable generalization but molecules could be pheromones if, in normal development, for example, all males of a species learned to respond to the same particular molecules as a female pheromone. Tested as adults, all males would respond to the pheromone. While unexpected, this would be consistent with studies showing species identity for adult mate choice can be imprinted at juvenile stages (Verzijden et al. 2012). Being innate was not part of the original pheromone definition (Karlson and Lüscher 1959) or its updated version (Wyatt 2010, 2014).

It may be counterintuitive but many seemingly innate behaviors are dependent on certain developmental and environmental conditions at critical stages (Bateson

& Mameli 2007; Mameli & Bateson 2011). Only experimental interventions to study the development (ontogeny) of a behavior or sense will reveal the necessary conditions. For example, for its visual cortex to develop correctly, a young mammal's eyes have to be exposed to visual stimuli at a critical period after birth (Hensch 2004). As this happens in normally raised animals we are not aware of the conditionality. In most animals the development (ontogeny) of responses to pheromones has not been studied. The development of the response is not known in full detail but all young rabbit pups respond to the mammary pheromone, 2-methylbut-2-enal (Schaal et al. 2003; Schaal, Chapter 17). By contrast, it seems that mice do not have an equivalent mammary pheromone as, instead, the mouse pups learn olfactory cues for suckling based on their mother's individual chemical profile that can be influenced by her diet (Logan et al. 2012).

1.9 HUMAN PHEROMONES?

I remain unconvinced that any human pheromones have been chemically identified so far (Doty 2010; Wyatt 2014; Doty, Chapter 19). Sadly the steroid molecules often cited have no biologically demonstrated validity despite their widespread use in studies (helpfully tabulated in Havlíček et al. 2010; e.g., Savic, Chapter 18). The kind of detailed bioassay-led research needed to establish that these or any other molecules are pheromones has simply not been done with humans.

However, I do think that human pheromones may yet be found in the future. Currently the most promising example is the areolar secretion of lactating women, secreted by glands around the nipple (Doucet et al. 2009; Schaal, Chapter 17). It appears to contain what may be a human suckling/mammary pheromone that causes human babies to turn their head and stick out their tongue towards it in response. The authors are cautious but I would count this as a possible pheromone as it is not a signature mixture recognition of their own mother's smell since the secretion taken from any lactating mother elicits the response. The molecules involved have not been identified though the bioassay is a good one.

1.10 APPLICATIONS OF PHEROMONES

A key reason for the long-standing interest in pheromones, going back to the 19th century and before, has been the recognition that pheromones might be used to control pests or influence the biology of domestic animals (Wyatt 2014). The greatest successes so far have been control of agricultural moth pests, largely using synthetic sex pheromones to disrupt mating (Allison & Cardé 2014; Witzgall et al. 2010). Lure-and-kill, combining a pheromone attractant and a trap or poison station, is another method (El-Sayed et al. 2009). It is important for controlling introduced pests or ones with a slower reproductive rate—for example, large weevils *Cosmopolites sordidus*, which are a pest of bananas (Alpizar et al. 2012). This approach might offer an effective control of a vertebrate fish pest, the invasive sea lamprey *Petromyzon marinus*, using either the sex or larval migration pheromones (Johnson et al. 2012; Meckley et al. 2012).

A greater understanding of nematode pheromones was initially explored in the widely cultured *C. elegans* (Srinivasan et al. 2012) but is now extended to a wide variety of nematode species including those with important agricultural and human health impacts (Choe et al. 2012). There is the prospect, distant at this stage, of using their pheromones to control nematode pests and diseases in similar ways to moth pests.

The use of pheromones with beneficial insects, companion animals (cats and dogs), and farm animal husbandry is surprisingly limited (Wyatt 2014). Pheromones orchestrate all aspects of the life of the honeybee hive (Bortolotti and Costa, Chapter 5; Grozinger 2013). Among the interventions are the use of synthetic queen mandibular pheromone to delay swarming behavior and synthetic Nasonov pheromone to attract swarms into trap hives. Other uses are likely to be found as we understand more about the complex ways pheromones affect honeybee physiology, behavior, and the organization of labor in the hive.

The male pig pheromone 5α-androstenone is available commercially as Boar Mate® spray. Female pigs (sows) in estrus respond with a mating posture to the pheromone and pressure on the back, helping pig farmers to identify sows ready for artificial insemination. However, while proprietary "pheromones" have been claimed to exist for cats, dogs, and chickens, no rigorous bioassay data has been published to establish that these are indeed the important molecules in each of these species (Wyatt 2014). In the case of proprietary cat and dog pheromones, a systematic review suggested that further work would be needed to demonstrate efficacy (Frank et al. 2010).

An understanding of chemical signaling in sheep and goats is showing promise as a way of allowing farmers to influence the timing of breeding by using olfactory cues and pheromones rather than treatment with synthetic hormones (Hawken & Martin 2012). Possible pheromones in cattle are discussed by Archunan et al. (Chapter 16).

1.11   CONCLUSIONS

Chemical signaling seems to be ubiquitous across the animal kingdom. We might expect it to evolve as a natural consequence of the universal importance of chemical senses even in animal taxa such as birds, which traditionally had not been thought to use semiochemical information (Campagna et al. 2012; Caro & Balthazart 2010; Zhang et al. 2010). Perhaps more than signals perceived by the other senses, semiochemicals offer opportunities to dissect animal communication at every level, from behavior and physiology down to the genetics of signal production and perception.

ACKNOWLEDGMENTS

I thank Darren Logan and Joan Wyatt for helpful comments on the manuscript. I also thank them and the many colleagues who have helped me by discussing ideas about pheromones over the years.

REFERENCES

Alaux, C & Robinson, G (2007) Alarm pheromone induces immediate–early gene expression and slow behavioral response in honey bees. *J Chem Ecol* 33: 1346–1350.

Alaux, C, Sinha, S, Hasadsri, L, Hunt, GJ, Guzman-Novoa, E, Degrandi-Hoffman, G, Uribe-Rubio, JL, Southey, BR, Rodriguez-Zas, S & Robinson, GE (2009) Honey bee aggression supports a link between gene regulation and behavioral evolution. *Proc Natl Acad Sci U S A* 106: 15400–15405.

Allison, JD & Cardé, RT (eds.) (2014) *Pheromone Communication in Moths: Evolution, Behavior and Application.* Berkeley, CA: University of California Press.

Alpizar, D, Fallas, M, Oehlschlager, A & Gonzalez, L (2012) Management of *Cosmopolites sordidus* and *Metamasius hemipterus* in banana by pheromone-based mass trapping. *J Chem Ecol* 38: 245–252.

Avila, FW, Sirot, LK, Laflamme, BA, Rubinstein, CD & Wolfner, MF (2011) Insect seminal fluid proteins: identification and function. *Annu Rev Entomol* 56: 21–40.

Bateson, P & Mameli, M (2007) The innate and the acquired: useful clusters or a residual distinction from folk biology? *Dev Psychobiol* 49: 818–831.

Birch, MC, Poppy, GM & Baker, TC (1990) Scents and eversible scent structures of male moths. *Annu Rev Entomol* 35: 25–58.

Bos, N & d'Ettorre, P (2012) Recognition of social identity in ants. *Front Psychol* 3: 83.

Bradbury, JW & Vehrencamp, SL (2011) *Principles of Animal Communication,* Second Edition. Sunderland, MA: Sinauer.

Breithaupt, T & Thiel, M (eds.) (2011) *Chemical Communication in Crustaceans.* New York: Springer.

Brennan, PA (2009) Outstanding issues surrounding vomeronasal mechanisms of pregnancy block and individual recognition in mice. *Behav Brain Res* 200: 287–294.

Buck, L & Axel, R (1991) A novel multigene family may encode odorant receptors—a molecular-basis for odor recognition. *Cell* 65: 175–187.

Burghardt, GM, Bartmess-LeVasseur, JN, Browning, SA, Morrison, KE, Stec, CL, Zachau, CE & Freeberg, TM (2012) Perspectives—Minimizing observer bias in behavioral studies: A review and recommendations. *Ethology* 118: 511–517.

Butenandt, A, Beckmann, R, Stamm, D & Hecker, E (1959) Über den Sexuallockstoff des Seidenspinners *Bombyx mori*—Reindarstellung und Konstitution. *Zeitschrift für Naturforschung Part B—Chemie Biochemie Biophysik Biologie und Verwandten Gebiete* 14: 283–284.

Campagna, S, Mardon, J, Celerier, A & Bonadonna, F (2012) Potential semiochemical molecules from birds: A practical and comprehensive compilation of the last 20 years studies. *Chem Senses* 37: 3–25.

Caro, SP & Balthazart, J (2010) Pheromones in birds: Myth or reality? *J Comp Physiol A* 196: 751–766.

Choe, A, von Reuss, SH, Kogan, D, Gasser, RB, Platzer, EG, Schroeder, FC & Sternberg, PW (2012) Ascaroside signaling is widely conserved among nematodes. *Curr Biol* 22: 772–780.

Chung-Davidson, Y-W, Huertas, M & Li, W (2011) A review of research in fish pheromones. In: Breithaupt, T & Thiel, M (eds.) *Chemical Communication in Crustaceans.* pp. 467–482. New York: Springer.

Coureaud, G, Langlois, D, Sicard, G & Schaal, B (2004) Newborn rabbit responsiveness to the mammary pheromone is concentration-dependent. *Chem Senses* 29: 341–350.

Cummins, SF & Degnan, BM (2010) Sensory sea slugs: Towards decoding the molecular toolkit required for a mollusc to smell. *Commun Integr Biol* 3: 423–426.

Cummins, SF, Xie, F, de Vries, MR, Annangudi, SP, Misra, M, Degnan, BM, Sweedler, JV, Nagle, GT & Schein, CH (2007) *Aplysia* temptin—The 'glue' in the water-borne attractin pheromone complex. *FEBS J* 274: 5425–5437.

de Bruyne, M & Baker, TC (2008) Odor detection in insects: Volatile codes. *J Chem Ecol* 34: 882–897.

Dewhirst, SY, Pickett, JA & Hardie, J (2010) Aphid pheromones. In: Gerald, L (ed.) *Pheromones.* pp. 551–574. London: Academic Press.

Dickson, BJ (2008) Wired for sex: The neurobiology of *Drosophila* mating decisions. *Science* 322: 904–909.

Domingue, MJ, Haynes, KF, Todd, JL & Baker, TC (2009) Altered olfactory receptor neuron responsiveness is correlated with a shift in behavioral response in an evolved colony of the cabbage looper moth, *Trichoplusia ni. J Chem Ecol* 35: 405–415.

Doty, RL (2010) *The Great Pheromone Myth.* Baltimore, MD: Johns Hopkins University Press.

Doucet, S, Soussignan, R, Sagot, P & Schaal, B (2009) The secretion of areolar (Montgomery's) glands from lactating women elicits selective, unconditional responses in neonates. *PLoS ONE* 4: e7579.

El-Sayed, AM, Suckling, DM, Byers, JA, Jang, EB & Wearing, CH (2009) Potential of "lure and kill" in long-term pest management and eradication of invasive species. *J Econ Entomol* 102: 815–835.

Eliyahu, D, Nojima, S, Santangelo, RG, Carpenter, S, Webster, FX, Kiemle, DJ, Gemeno, C, Leal, WS & Schal, C (2012) Unusual macrocyclic lactone sex pheromone of *Parcoblatta lata*, a primary food source of the endangered red-cockaded woodpecker. *Proc Natl Acad Sci U S A* 109: E490–E496.

Evans, I, Thornton, H, Chalmers, I & Glasziou, P (2011) *Testing Treatments: Better Research for Better Healthcare*, Second Edition. London: Pinter and Martin.

Ferveur, J-F & Cobb, M (2010) Behavioral and evolutionary roles of cuticular hydrocarbons in Diptera. In: Blomquist, GJ & Bagnères, A-G (eds.) *Insect Hydrocarbons: Biology, Biochemistry, and Chemical Ecology.* pp. 325–343. Cambridge, UK: Cambridge University Press.

Firestein, S (2001) How the olfactory system makes sense of scents. *Nature* 413: 211–218.

Forseth, RR & Schroeder, FC (2011) NMR-spectroscopic analysis of mixtures: From structure to function. *Curr Opin Chem Biol* 15: 38–47.

Frank, D, Beauchamp, G & Palestrini, C (2010) Systematic review of the use of pheromones for treatment of undesirable behavior in cats and dogs. *J Am Vet Med Assoc* 236: 1308–1316.

Gemeno, C, Leal, WS, Mori, K & Schal, C (2003) Behavioral and electrophysiological responses of the brownbanded cockroach, *Supella longipalpa*, to stereoisomers of its sex pheromone, supellapyrone. *J Chem Ecol* 29: 1797–1811.

Groot, AT, Estock, ML, Horovitz, JL, Hamilton, J, Santangelo, RG, Schal, C & Gould, F (2009) QTL analysis of sex pheromone blend differences between two closely related moths: Insights into divergence in biosynthetic pathways. *Insect Biochem Mol Biol* 39: 568–577.

Grozinger, CM (2013) Honey bee pheromones In: Graham, J (ed.) *The Hive and the Honey Bee.* Hamilton, IL: Dadant & Sons Inc.

Haga, S, Hattori, T, Sato, T, Sato, K, Matsuda, S, Kobayakawa, R, Sakano, H, Yoshihara, Y, Kikusui, T & Touhara, K (2010) The male mouse pheromone ESP1 enhances female sexual receptive behaviour through a specific vomeronasal receptor. *Nature* 466: 118–122.

Hagström, ÅK, Liénard, MA, Groot, AT, Hedenström, E & Löfstedt, C (2012) Semi-selective fatty acyl reductases from four heliothine moths influence the specific pheromone composition. *PLoS ONE* 7: e37230.

Hansson, BS & Stensmyr, MC (2011) Evolution of insect olfaction. *Neuron* 72: 698–711.

Haupt, SS, Sakurai, T, Namiki, S, Kazawa, T & Kanzaki, R (2010) Olfactory information processing in moths. In: Menini, A (ed.) *The Neurobiology of Olfaction.* Boca Raton, FL: CRC Press.

Havlíček, J, Murray, AK, Saxton, TK & Roberts, SC (2010) Current issues in the study of androstenes in human chemosignaling. In: Gerald, L (ed.) *Pheromones.* pp. 47–81. London: Academic Press.

Hawken, P & Martin, G (2012) Sociosexual stimuli and gonadotropin-releasing hormone/luteinizing hormone secretion in sheep and goats. *Domest Anim Endocrinol* 43: 85–94.

Hecker, E & Butenandt, A (1984) Bombykol revisited—Reflections on a pioneering period and on some of its consequences. In: Hummel, HE & Miller, TA (eds.) *Techniques in Pheromone Research.* pp. 1–44. New York: Springer.

Hensch, TK (2004) Critical period regulation. *Annu Rev Neurosci* 27: 549–579.

Hölldobler, B & Carlin, NF (1987) Anonymity and specificity in the chemical communication signals of social insects. *J Comp Physiol A* 161: 567–581.

Holman, L, Jørgensen, C, Nielsen, J & d'Ettorre, P (2010) Identification of an ant queen pheromone regulating worker sterility. *Proc R Soc B* 277: 3793–3800.

Hsueh, Y-P, Mahanti, P, Schroeder, FC & Sternberg, PW (2013) Nematode-trapping fungi eavesdrop on nematode pheromones. *Curr Biol* 23: 83–86.

Inoshita, T, Martin, JR, Marion-Poll, F & Ferveur, JF (2011) Peripheral, central and behavioral responses to the cuticular pheromone bouquet in *Drosophila melanogaster* males. *PLoS ONE* 6: e19770.

Johnson, NS, Yun, SS, Buchinger, TJ & Li, W (2012) Multiple functions of a multi-component mating pheromone in sea lamprey *Petromyzon marinus. J Fish Biol* 80: 538–554.

Johnston, RE (2003) Chemical communication in rodents: from pheromones to individual recognition. *J Mammal* 84: 1141–1162.

Johnston, RE (2005) Communication by mosaic signals: Individual recognition and underlying neural mechanisms. In: Mason, RT, LeMaster, MP & Müller-Schwarze, D (eds.) *Chemical Signals in Vertebrates 10.* pp. 269–282. New York: Springer.

Kamakura, M (2011) Royalactin induces queen differentiation in honeybees. *Nature* 473: 478–483.

Karlson, P & Lüscher, M (1959) "Pheromones": A new term for a class of biologically active substances. *Nature* 183: 55–56.

Kato, A & Touhara, K (2009) Mammalian olfactory receptors: pharmacology, G protein coupling and desensitization. *Cell Mol Life Sci* 66: 3743–3753.

Kelly, DR (1996) When is a butterfly like an elephant? *Chem Biol* 3: 595–602.

Koene, JM & ter Maat, A (2001) "Allohormones": A class of bioactive substances favoured by sexual selection. *J Comp Physiol A* 187: 323–326.

Kreher, SA, Mathew, D, Kim, J & Carlson, JR (2008) Translation of sensory input into behavioral output via an olfactory system. *Neuron* 59: 110–124.

Kroiss, J, Lechner, K & Strohm, E (2010) Male territoriality and mating system in the European beewolf *Philanthus triangulum* F. (Hymenoptera: Crabronidae): Evidence for a "hotspot" lek polygyny. *J Ethol* 28: 295–304.

Lassance, J-M (2010) Journey in the *Ostrinia* world: From pest to model in chemical ecology. *J Chem Ecol* 36: 1155–1169.

Lassance, J-M, Liénard, MA, Antony, B, Qian, S, Fujii, T, Tabata, J, Ishikawa, Y & Löfstedt, C (2013) Functional consequences of sequence variation in the pheromone biosynthetic gene *pgFAR* for *Ostrinia* moths. *Proc Natl Acad Sci U S A* 110: 3967–3972.

Leal, WS (2013) Odorant reception in insects: roles of receptors, binding proteins, and degrading enzymes. *Annu Rev Entomol* 58: 373–391.

Leary, GP, Allen, JE, Bunger, PL, Luginbill, JB, Linn, CE, Macallister, IE, Kavanaugh, MP & Wanner, KW (2012) Single mutation to a sex pheromone receptor provides adaptive specificity between closely related moth species. *Proc Natl Acad Sci U S A* 109: 14081–14086.

Levesque, HM, Scaffidi, D, Polkinghorne, CN & Sorensen, PW (2011) A multi-component species identifying pheromone in the goldfish. *J Chem Ecol* 37: 219–227.

Li, Q, Korzan, WJ, Ferrero, DM, Chang, RB, Roy, DS, Buchi, M, Lemon, JK, Kaur, AW, Stowers, L, Fendt, M & Liberles, SD (2013) Synchronous evolution of an odor biosynthesis pathway and behavioral response. *Curr Biol* 23: 11–20.

Liebig, J (2010) Hydrocarbon profiles indicate fertility and dominance status in ant, bee, and wasp colonies. In: Blomquist, GJ & Bagnères, A-G (eds.) *Insect Hydrocarbons: Biology, Biochemistry, and Chemical Ecology.* pp. 254–281. Cambridge, UK: Cambridge University Press.

Lim, H & Sorensen, PW (2012) Common carp implanted with prostaglandin $F_{2\alpha}$ release a sex pheromone complex that attracts conspecific males in both the laboratory and field. *J Chem Ecol* 38: 127–134.

Linn, CE, Campbell, MG & Roelofs, WL (1987) Pheromone components and active spaces: What do moths smell and where do they smell it? *Science* 237: 650–652.

Logan, DW, Brunet, JL, Webb, WR, Cutforth, T, Ngai, J & Stowers, L (2012) Learned recognition of maternal signature odors mediates the first suckling episode in mice. *Curr Biol* 22: 1998–2007.

Malnic, B, Hirono, J, Sato, T & Buck, LB (1999) Combinatorial receptor codes for odors. *Cell* 96: 713–723.

Malnic, B, Gonzalez-Kristeller, DC & Gutiyama, LM (2010) Odorant receptors. In: Menini, A (ed.) *The Neurobiology of Olfaction.* Boca Raton, FL: CRC Press.

Mameli, M & Bateson, P (2011) An evaluation of the concept of innateness. *Phil Trans R Soc B* 366: 436–443.

Martin, JP, Beyerlein, A, Dacks, AM, Reisenman, CE, Riffell, JA, Lei, H & Hildebrand, JG (2011) The neurobiology of insect olfaction: Sensory processing in a comparative context. *Prog Neurobiol* 95: 427–447.

Mason, RT & Parker, MR (2010) Social behavior and pheromonal communication in reptiles. *J Comp Physiol A* 196: 729–749.

McGrath, PT, Xu, YF, Ailion, M, Garrison, JL, Butcher, RA & Bargmann, CI (2011) Parallel evolution of domesticated *Caenorhabditis* species targets pheromone receptor genes. *Nature* 477: 321–325.

Meckley, TD, Wagner, CM & Luehring, MA (2012) Field evaluation of larval odor and mixtures of synthetic pheromone components for attracting migrating sea lampreys in rivers. *J Chem Ecol* 38: 1062–1069.

Meinwald, J (2009) The chemistry of biotic interactions in perspective: small molecules take center stage. *J Org Chem* 74: 1813–1825.

Mucignat-Caretta, C, Redaelli, M & Caretta, A (2012) One nose, one brain: contribution of the main and accessory olfactory system to chemosensation. *Front Neuroanat* 6: 46.

Niehuis, O, Buellesbach, J, Gibson, JD, Pothmann, D, Hanner, C, Mutti, NS, Judson, AK, Gadau, J, Ruther, J & Schmitt, T (2013) Behavioural and genetic analyses of *Nasonia* shed light on the evolution of sex pheromones. *Nature* 494: 345–348.

Nordlund, DA & Lewis, WJ (1976) Terminology of chemical releasing stimuli in intraspecific and interspecific interactions. *J Chem Ecol* 2: 211–220.

Novotny, MV (2003) Pheromones, binding proteins and receptor responses in rodents. *Biochem Soc Trans* 31: 117–122.

Novotny, MV, Ma, W, Zidek, L & Daev, E (1999) Recent biochemical insights into puberty acceleration, estrus induction and puberty delay in the house mouse. In: Johnston, RE, Müller-Schwarze, D & Sorensen, PW (eds.) *Advances in Chemical Signals in Vertebrates.* pp. 99–116. New York: Kluwer Academic/Plenum Press.

Petrulis, A (2013) Chemosignals, hormones and mammalian reproduction. *Horm Behav* 63:723–741.

Phelan, PL (1997) Evolution of mate-signalling in moths: Phylogenetic considerations and predictions from the asymmetric tracking hypothesis. In: Choe, JC & Crespi, BJ (eds.) *The Evolution of Mating Systems in Insects and Arachnids.* pp. 240–256. Cambridge, UK: Cambridge University Press.

Pungaliya, C, Srinivasan, J, Fox, B, Malik, R, Ludewig, A, Sternberg, P & Schroeder, F (2009) A shortcut to identifying small molecule signals that regulate behavior and development in *Caenorhabditis elegans. Proc Natl Acad Sci U S A* 106: 7708–7713.

Raffa, KF (2001) Mixed messages across multiple trophic levels: the ecology of bark beetle chemical communication systems. *Chemoecology* 11: 49–65.

Rasmussen, LEL, Lee, TD, Zhang, AJ, Roelofs, WL & Daves, GD (1997) Purification, identification, concentration and bioactivity of (*Z*)–7-dodecen-1-yl acetate: sex pheromone of the female Asian elephant, *Elephas maximus. Chem Senses* 22: 417–437.

Rasmussen, LEL, Lazar, J & Greenwood, DR (2003) Olfactory adventures of elephantine pheromones. *Biochem Soc Trans* 31: 137–141.

Reisert, J & Restrepo, D (2009) Molecular tuning of odorant receptors and its implication for odor signal processing. *Chem Senses* 34: 535–545.

Rezával, C, Pavlou, HJ, Dornan, AJ, Chan, Y-B, Kravitz, EA & Goodwin, SF (2012) Neural circuitry underlying *Drosophila* female postmating behavioral responses. *Curr Biol* 22: 1155–1165.

Schaal, B, Coureaud, G, Langlois, D, Ginies, C, Semon, E & Perrier, G (2003) Chemical and behavioural characterization of the rabbit mammary pheromone. *Nature* 424: 68–72.

Schneider, D (1999) Insect pheromone research: Some history and 45 years of personal recollections. *IOBC-WPRS Bull* 22. Available at: http://phero.net/iobc/dachau/bulletin99/schneider.pdf [accessed January 21, 2013].

Shirangi, TR, Dufour, HD, Williams, TM & Carroll, SB (2009) Rapid evolution of sex pheromone-producing enzyme expression in *Drosophila. PLoS Biol* 7: e1000168.

Silbering, AF & Benton, R (2010) Ionotropic and metabotropic mechanisms in chemoreception: "Chance or design"? *EMBO Rep* 11: 173–179.

Smadja, C & Butlin, RK (2009) On the scent of speciation: the chemosensory system and its role in premating isolation. *Heredity* 102: 77–97.

Srinivasan, J, von Reuss, SH, Bose, N, Zaslaver, A, Mahanti, P, Ho, MC, O'Doherty, OG, Edison, AS, Sternberg, PW & Schroeder, FC (2012) A modular library of small molecule signals regulates social behaviors in *Caenorhabditis elegans. PLoS Biol* 10: e1001237.

Stacey, NE & Sorensen, PW (2011) Hormonal pheromones In: Farrell, AP (ed.) *Encyclopedia of Fish Physiology: From Genome to Environment.* pp. 1553–1562. San Diego, CA: Academic Press.

Steiger, S, Schmitt, T & Schaefer, HM (2011) The origin and dynamic evolution of chemical information transfer. *Proc R Soc B* 278: 970–979.

Symonds, MRE & Elgar, MA (2008) The evolution of pheromone diversity. *Trends Ecol Evol* 23: 220–228.

Touhara, K (ed.) (2013) *Pheromone Signaling: Methods and Protocols.* New York, Humana Press (Springer).

Toyoda, F, Yamamoto, K, Iwata, T, Hasunuma, I, Cardinali, M, Mosconi, G, Polzonetti-Magni, AM & Kikuyama, S (2004) Peptide pheromones in newts. *Peptides* 25: 1531–1536.

van Zweden, JS & d'Ettorre, P (2010) Nestmate recognition in social insects and the role of hydrocarbons. In: Blomquist, GJ & Bagnères, A-G (eds.) *Insect Hydrocarbons: Biology, Biochemistry, and Chemical Ecology.* pp. 222–243. Cambridge, UK: Cambridge University Press.

Vereecken, NJ & McNeil, JN (2010) Cheaters and liars: chemical mimicry at its finest. *Can J Zool* 88: 725–752.

Verzijden, MN, ten Cate, C, Servedio, MR, Kozak, GM, Boughman, JW & Svensson, EI (2012) The impact of learning on sexual selection and speciation. *Trends Ecol Evol* 27: 511–519.

Vickers, NJ (2000) Mechanisms of animal navigation in odor plumes. *Biol Bull* 198: 203–212.

Webster, DR & Weissburg, MJ (2009) The hydrodynamics of chemical cues among aquatic organisms. *Annu Rev Fluid Mech* 41: 73–90.

Wilson, EO & Bossert, WH (1963) Chemical communication among animals. *Recent Prog Horm Res* 19: 673–716.

Witzgall, P, Kirsch, P & Cork, A (2010) Sex pheromones and their impact on pest management. *J Chem Ecol* 36: 80–100.

Wolfner, MF (2009) Battle and ballet: molecular interactions between the sexes in *Drosophila*. *J Hered* 100: 399–410.

Woodley, SK (2010) Pheromonal communication in amphibians. *J Comp Physiol A* 196: 713–727.

Wyatt, TD (2005) Pheromones: convergence and contrasts in insects and vertebrates. In: Mason, RT, LeMaster, MP & Müller-Schwarze, D (eds.) *Chemical Signals in Vertebrates 10.* pp. 7–20. New York: Springer.

Wyatt, TD (2010) Pheromones and signature mixtures: defining species-wide signals and variable cues for identity in both invertebrates and vertebrates. *J Comp Physiol A* 196: 685–700.

Wyatt, TD (2014) *Pheromones and Animal Behavior: Chemical Signals and Signature Mixes,* Second Edition. Cambridge, UK: Cambridge University Press.

Yambe, H, Kitamura, S, Kamio, M, Yamada, M, Matsunaga, S, Fusetani, N & Yamazaki, F (2006) L-Kynurenine, an amino acid identified as a sex pheromone in the urine of ovulated female masu salmon. *Proc Natl Acad Sci USA* 103: 15370–15374.

Yoshikawa, K, Nakagawa, H, Mori, N, Watanabe, H & Touhara, K (2013) An unsaturated aliphatic alcohol as a natural ligand for a mouse odorant receptor. *Nat Chem Biol* 9: 160–162.

Zhang, J-X, Wei, W, Zhang, J-H and Yang, W-H (2010) Uropygial gland-secreted alkanols contribute to olfactory sex signals in budgerigars. *Chem Senses* 35: 375–382.

2 Pheromones and General Odor Perception in Insects

Michel Renou

CONTENTS

2.1 Introduction 24
2.2 Peripheral and Central Architectures of the Insect Olfactory System 25
    2.2.1 Antennae and ORNs 25
    2.2.2 Antennal Lobes, Primary Olfactory Centers 26
    2.2.3 Second-Order Olfactory Areas 28
2.3 Neural Coding of Odor Signals 28
    2.3.1 Quality Coding of Pheromones and General Odors 28
    2.3.2 How Does Inhibition Contribute to Quality Coding? 30
    2.3.3 Intensity Coding 31
    2.3.4 Coding of Signal Temporality 32
    2.3.5 Temporal Codes 33
2.4 Odor Interactions and the Coding of Complex Blends 34
    2.4.1 Measuring Interactions 35
    2.4.2 Mixture Effects at the ORN Level 36
    2.4.3 Recognition of Complex Plant Bouquets 36
    2.4.4 Interactions between Odorant Signals Released from Different
         Sources 37
2.5 Variable Responses to Fixed Olfactory Signals 39
    2.5.1 Changes in Responsiveness to Olfactory Signals According to
         Physiological Status 39
    2.5.2 Plurimodality Interactions 40
    2.5.3 Plasticity in Response: Better Facing the Risks of
         Communication? 41
2.6 Facing Poorly Specific or Variable Signals 41
    2.6.1 Dealing with the Ubiquity of Some Chemical Signals: Context
         Dependence 42
    2.6.2 Dealing with Variable Odors: Generalization 42
    2.6.3 Sensitization and Learning 43
2.7 Concluding Remarks 45
Acknowledgments 46
References 46

2.1 INTRODUCTION

For insects, finding of a mate or a food source relies often chiefly on olfactory information. The identification of a specific mate or host involves the recognition of a specific odor blend and its discrimination from a complex and changing background. Their highly efficient olfactory systems have evolved to detect behaviorally relevant compounds with a high sensitivity and properly decode the olfactory message to finally lead to adapted behaviors (Martin et al. 2011). Moth sex pheromones, for instance, are precisely defined blends that trigger innate and stable behavioral responses in physiologically competent individuals. We begin this chapter with a description of the general organization of the insect olfactory system from the olfactory organs to the brain (Section 2.2). We will then consider the neuronal bases of the coding of pheromone and general odors (Section 2.3). The chemical specificity of the detection is at first based on the specificity of olfactory receptors (ORs) expressed in precise types of olfactory receptor neurons (ORNs) to detect individual compounds. But to insects as to other organisms, biologically relevant odors are often blends of volatile compounds whose perception involves progressive integration of the input in the central nervous system (CNS). In the wild, different sources release their volatiles simultaneously and their components intermingle to form a complex and fluctuating olfactory environment. Interactions between odor components, or their resulting neural codes, take place at the different levels of the sensory system and are intrinsic to olfaction. Section 2.4 will consider how insects recognize odor multicomponent blends in such chemical complexity. To be pertinent, the response of insects to odors must integrate the physiological and sensory contexts. Even strongly determined odor-guided behaviors are modulated by underlying changes in the internal physiological state of the animal and many behaviors are multimodal (Section 2.5). Very often chemical signals vary in an unpredictable way for the receiver and the chemical environment in which the signal is released is rather complex and changing. Previous experiences also modify the processing of odor signals and the associated behavioral responses. In the last section (Section 2.6), we consider the processing of olfactory signals under the influences of the ecological context and the individual history.

Outstanding progress in our understanding of insect olfaction has been accomplished through numerous experimental studies conducted in parallel on several model species, like bees, flies, or moths with their specificities. Although belonging to quite different groups, with diversity in their biology and phylogenetical origin, these model species share many of the general traits of organization of their olfactory system. However, making a synthesis would have been impossible due to the abundance of the literature and the risks of raising specific adaptations to general facts. I arbitrarily chose to focus this chapter on moths. Reproduction in that nocturnal Lepidoptera offers the advantage of providing cases for a highly determined sex pheromone communication system, and a more plastic use of olfaction for feeding and oviposition. Thus, most of the examples herein have been taken from moths and completed by data collected from other classical model insect species when relevant.

2.2 PERIPHERAL AND CENTRAL ARCHITECTURES OF THE INSECT OLFACTORY SYSTEM

2.2.1 ANTENNAE AND ORNs

As in vertebrates, odors are first transformed into a neural signal by the ORNs. In contrast to the vertebrate nasal epithelium, insect ORNs are not randomly distributed but are housed in morphofunctional units, the olfactory sensilla. The olfactory sensilla can take different shapes, appearing as long or short hairs, plates, pegs, or cavities. Odor molecules penetrate into the sensilla through numerous pores, a characteristic of the cuticular wall of the olfactory sensilla whatever their external morphology is. Insect ORNs are bipolar neurons whose distal processes (the dendrite) extend into the lumen of the hair or right below the surface plate, and whose axons project to the primary olfactory centers of the insect brain, the antennal lobes (ALs). According to their response spectra, insect ORNs can be sorted into functional classes that are housed in stereotyped combinations within functional types of olfactory sensilla.

Olfactory sensilla are concentrated on the main olfactory organs, the antennae, which may totalize several tens of thousands of ORNs. But antennae are not the unique site of olfaction on the insect body. Other head appendages, the labial and maxillary palps, can bear olfactory sensilla with various odor specificity depending on insect orders. The odor tuning of maxillary palp ORNs has been well described in mosquitoes (Syed and Leal 2007) and in *Drosophila melanogaster* (de Bruyne et al. 1999). Adult Lepidoptera possess reduced maxillary palps but their labial palps bear a labial pit organ housing neurons highly sensitive to CO_2 (Kent et al. 1986).

The challenges of detection of the huge chemical diversity of volatile molecules are met by a repertoire of ORs. Individual insect ORNs generally express a single functional type of OR (Couto et al. 2005) that is the determinant of their specificity. ORs vary in their degree of specificity, from narrowly to very broadly tuned, so even if ORNs express only one OR type they can show very different width of chemical tuning (Hallem and Carlson 2006; Hallem et al. 2004).

Insect olfactory receptors differ from those of vertebrates and belong to different families varying in their mode of coupling with transduction pathways. The first family of insect ORs that have been identified was thought to belong to the GPCR family. It is now admitted that insect ORs are composed of an odorant-sensitive protein belonging to the G-protein coupled receptor family (ORx), dimerized with a protein akin to GPCRs (Orco, formerly named Or83 in *Drosophila*) that have ion channel properties. Like ORx, Orco has inverted orientation in the cell membrane. This molecular complex serves both as a GPCR and an ion channel, allowing very rapid detection of high odor concentrations by means of the ionotropic pathway, and slower but prolonged and highly sensitive odor detection via the G-protein-mediated signal amplification (Wicher et al. 2008). A second family of receptors, first described in *D. melanogaster*, has been called ionotropic receptors (IRs). IRs are odor-gated ion channels that might have derived from ionotropic glutamate receptors (iGluRs) (Benton et al. 2009). IRs and ORs possess distinct ligand spectra in *D. melanogaster* (Silbering et al. 2011).

The sensillum lymph that bathes the ORN dendrites contains odorant binding proteins (OBPs), a class of extracellular proteins postulated to carry the lipophilic odorant molecules in a hydrophilic medium. A subclass of OBPs, the pheromone binding proteins (PBPs) bind specifically the pheromone components. Apart from acting as carriers, OBPs and PBPs might also contribute to the specificity of the detection by their selective binding of a fraction of the volatile compounds that enter the sensillum. The olfactory tissues also contain odorant degrading enzymes (ODEs) that catabolize the odor and other volatile molecules.

The transduction of the chemical signal into a neural code involves sequential steps. The binding of odorant molecules to their corresponding type of ORs leads to a change in the dendritic transmembrane potential called the receptor potential. The membrane depolarization opens voltage-gated ion channels involved in the generation of action potentials whose firing frequency constitutes the neural code of the chemical signal. Although several of the molecular actors and ionic mechanisms underlying this complex transduction process remain elusive or controversial, a reasonably consistent picture emerges from modeling (Gu et al. 2009; Kaissling 2001). The dynamic properties of this complex process are critical to the functioning of the system for its sensitivity as well as its capacity to follow rapidly changing odor stimuli. For more details, see Chapter 4.

2.2.2 ANTENNAL LOBES, PRIMARY OLFACTORY CENTERS

The axons of the ORNs form an antennal nerve that enters into a specific part of the insect deutocerebrum, the ALs (Homberg et al. 1989). The ALs are constituted of spherical condensed neuropil structures called glomeruli. The numbers of glomeruli vary between species. Each AL of the silk moth, *Bombyx mori* (Kazawa et al. 2009) and the sphingid *Manduca sexta* (Rospars and Hildebrand 1992, 2000) for instance, contains about 60 ordinary glomeruli, while circa 50 glomeruli have been identified in *D. melanogaster* (Laissue et al. 1999). Within a species, glomeruli have been shown to be invariant, with each glomerulus having the same shape, size, and location across individuals of same species, sex, and developmental stage (Rospars 1988). Investigations of the anatomical features and molecular receptive ranges of the central neurons associated with identified glomeruli in different species, and particularly in *M. sexta*, have established that glomeruli show odor tuning and are functionally significant structures (Hildebrand 1996). The functional specialization of glomeruli is striking in moths, whose ALs present a strong sexual dimorphism correlated with the high specialization of the male olfactory system in sex pheromone perception. As in other moth species, ALs of male *M. sexta* possess a cluster of identified glomeruli forming the macro glomerulus complex (MGC) in which project exclusively the ORNs tuned to the components of the sex pheromone blend (Hildebrand 1995, 1996; Rössler et al. 1998). A sexually dimorphic olfactory glomerulus has also been identified in female ALs (Rössler et al. 1998) and the neurons it contains are activated when the ipsilateral antenna is stimulated with a general odorant, linalool (King et al. 2000). Electrophysiological and imaging data indicate that the glomeruli present in males and females ALs (ordinary glomeruli (OG)) receive inputs from general odorant ORNs (Carlsson et al. 2002). OR expression and ORN targeting maps have furthermore established in *D. melanogaster* that the axons of the ORNs expressing the same OR type typically converge onto the same glomerulus (Couto et al. 2005; Laissue and Vosshall 2008).

Within glomeruli ORNs synapse with projection neurones (PNs) that transmit odor-evoked neural activity into higher regions of the brain, as well as with local neurones (LNs) whose arborizations do not leave the AL (Hansson and Anton 2000; Hansson et al. 1992; Homberg et al. 1989). The AL neurons are far less numerous than the ORNs. The total number of PNs in *B. mori*, for instance, is estimated to be about 430 for a total of 710 LNs (Namiki and Kanzaki 2011a).

In *M. sexta* dendritic arborization of the AL PNs are usually restricted either to OG or to MGCs showing maintained segregation between pheromone and general odor information. PNs innervating OG have arborization in a single glomerulus or are multiglomerular (Kanzaki et al. 1989). Most PNs are cholinergic (Homberg et al. 1995; Waldrop and Hildebrand 1989). In *B. mori*, most PNs innervating OG have extraglomerular ramifications (Namiki and Kanzaki 2011a), while those innervating sexually dimorphic glomeruli have little or no extraglomerular ramifications, suggesting different coding strategies for pheromone and general odors. Although a majority (75.6%) of them are uniglomerular in *B. mori*, PNs show diversity in the number of innervated glomeruli, their innervation pattern in the AL, their soma position, and the antennocerebral tract through which they project to higher olfactory centers (Namiki and Kanzaki 2011a). The PN and LN somata are clustered. There are two major cell clusters in *B. mori*, the medial and the lateral clusters, and one small cluster, the anterior cluster (Namiki and Kanzaki 2011a).

LNs branch exclusively within the ALs and interconnect glomeruli. In bees, homo-LNs uniformly innervate many glomeruli but hetero-LNs innervate one glomerulus densely and several sparsely (Meyer and Galizia 2012). LNs modulate the ORN input and the PN output activities within and between glomeruli. LNs are generally broadly tuned and morphologically diverse, innervating several glomeruli; 61% of the 360 LNs of *M. sexta* were identified as GABA-immunoreactive, (Reisenman et al. 2011). Most of the LNs in the AL of *B. mori* are also GABA-ergic (Seki and Kanzaki 2008). The dominant model of interglomerular interaction has long been "lateral inhibition," serving a putative narrowing of the PN tuning (Olsen and Wilson 2008) and presynaptic gain control (Root et al. 2008). However, the dense AL network might be even more complex. Non-GABA-ergic LNs have been found in the ALs (Iwano and Kanzaki 2005). Accordingly, subsequent electrophysiological studies have established the existence of lateral excitation (Tootoonian and Laurent 2010). Presynaptic peptidergic modulation of ORNs has also been found in the ALs of *Drosophila* (Ignell et al. 2009) and the ORNs express a *Drosophila* tachykinin receptor allowing presynaptic inhibitory feedback from peptidergic LNs.

The olfactory information processed in the ALs is sent to the protocerebral areas via several fiber tracts. Although insects share a common plan of the antenno-protocerebral tracts (APTs), there are notable variations among taxa (Martin et al. 2011). There are five APTs in Lepidoptera (Homberg et al. 1988; Kanzaki et al. 2003). In turn, some protocerebral neurons project into the ALs (Homberg et al. 1988).

2.2.3 SECOND-ORDER OLFACTORY AREAS

The second-order olfactory areas are situated in the insect protorocebrum and comprise the mushroom bodies (MBs), the lateral and superior protocerebrum (LP and SP), and the lateral horn (LH) (Galizia and Rössler 2010; Kanzaki et al. 1991).

The insect MBs have been particularly studied, notably in a comparative context (Strausfeld et al. 2009). In the MBs, PNs synapse with hundreds of thousands of intrinsic neurons, the Kenyon cells (KCs), in the calyces. The MB calyces are not homogeneous structures but contain multiple subsystems (Martin et al. 2011). The KCs, far more numerous compared to AL neurons, are targeted by AL PNs. KCs project to two regions called the alpha and beta lobes where they synapse onto small populations of "extrinsic" neurons. Behavioral and genetic experiments in *D. melanogaster* and the honeybee have revealed that the MBs are critical for associative learning of odor stimuli (Heisenberg 2003). Detailed analyses of the fine organization of the MB calyces and targeting of PNs in *D. melanogaster* (Jefferis et al. 2002, 2007) suggest that classes of KCs integrate information from a small group of PNs (Turner et al. 2008). Detailed data on the fine organization of the MB calyces as comparable to those in *D. melanogaster* are not yet available in other insect groups. In moths, functional domains are also visible in the MBs of *B. mori* (Namiki et al. 2013) with some segregation of pheromone information (Kanzaki et al. 2003). However, although the axons of the PNs that innervate pheromonal and nonpheromonal glomeruli are organized into concentric zones and show regional localization, the dendritic fields of KCs mixed pheromonal and nonpheromonal inputs (Namiki et al. 2013).

The LH of the PC appears in to be a diffuse aglomerular neuropil. However, PN inputs in the LH are stereotyped and segregated according to their originating glomerulus in *D. melanogaster*. Fruit odors are represented mostly in the posterior dorsal LH, whereas pheromone-responsive PNs project to the anterior ventral LH (Jefferis et al. 2007; Marin et al. 2002) showing persistent spatial organization in higher olfactory centers. Some regionalization can still be observed for the projections of pheromone-responsive PNs in moth LHs (Seki et al. 2005).

2.3 NEURAL CODING OF ODOR SIGNALS

2.3.1 QUALITY CODING OF PHEROMONES AND GENERAL ODORS

Female moths produce and release a sex pheromone that stimulates and attracts males from a distance. The sex pheromone is a relatively simple blend comprising a small number of components released in stable ratios. Each pheromone component is individually recognized by specialist ORNs (Pher-ORNs). Pher-ORNs are narrowly tuned to one component, which means that this component triggers excitatory responses at very low concentration although higher concentrations of related compounds also activate them. Pher-ORNs are housed in male-specific types of sensilla: long hair sensilla or trichoid sensilla. Pheromone sensilla generally house two Pher-ORNs, each tuned to one different component. The functional types of sensilla defined according to the chemical tuning of the ORNs they house are constant within a species. The distribution of the types of sensilla on the antennal segments is also constant within species. When stimulated with the proper pheromone component, Pher-ORNs show a phasic-tonic excitatory response, increasing their firing frequency in a concentration-dependent way. Pher-ORNs project into a specialized area of the male AL, the MGC (Christensen et al. 1995; Hildebrand 1996). The moth MGC comprises several subunits. Generally a larger cumulus is accompanied by several smaller glomeruli that vary in number and shape according to species and are called the toroid and the horseshoe in *B. mori* (Koontz and Schneider 1987). Pher-ORNs expressing different ORs project to different subparts of the MGC showing a spatial organization called odotopy (Hansson and Anton 2000; Hansson et al. 1992; Hildebrand 1996). In some moth species, a type of Pher-ORNs is specifically tuned to one component of the pheromone blend of a related species that, when added to the proper blend, inhibits behavioral responses. The detection of such interspecific antagonist reinforces sex pheromone specificity in sympatric species sharing some of their pheromone components. In the noctuids *Heliothis virescens* and *Helicoverpa zea*, these antagonists evoke excitatory firing activity in PNs restricted to a third glomerulus in the MGCs that is clearly distinct from the two glomeruli in which the attractive binary blend is represented in both species (Vickers et al. 1998). With its well-individualized lines, coding of sex pheromones is a near-perfect example of a labeled line type of coding.

How individual pheromone components and blends are represented in the MBs has not yet been fully elucidated. The spatial distribution of input and output neurons in the MB calyx was recently investigated in *B. mori* (Namiki et al. 2013). The authors compared the distribution of the presynaptic buttons of AL-PNs, which transfer odor information from the antennal lobe to the calyx. PNs for pheromone components and plant odors enter the calyx in a concentric fashion and they are read out by the dendritic fields of KCs. The axons of PNs that innervate the MGC are confined to a relatively small area within the calyx. In contrast, the axons of PNs innervating nonpheromonal glomeruli are more widely distributed. PN axons for the minor pheromone components cover a larger area than those for the major pheromone component and partially overlap with those innervating nonpheromonal glomeruli, suggesting the integration of the minor pheromone component with plant odors by KCs.

Host plant volatiles and other general odors comprise generally a higher number of components compared to pheromones. How are their components detected individually? Although less narrowly tuned than Pher-ORNs, GO-ORNs still show selectivity and are not as broad "generalists" as most vertebrate ORNs. Their molecular receptive ranges generally allow sorting of odorants according to their functional groups (Carlsson and Hansson 2006). Inside functional groups selective ORNs can be found that respond selectively to a more limited range of molecules (Shields and Hildebrand 2001). The general principles by which olfactory stimuli are represented in the ALs are nevertheless not so different in the pheromone and general odorant subsystems (Christensen and Hildebrand 2002). Functional types of GO-ORNs project into specific glomeruli in the AL. Calcium imaging has shown in several different species, including moths (Carlsson et al. 2002; Galizia et al. 2000), that more than one glomerulus is activated by a general odorant and the same glomerulus is activated by several odorants. All odorants, however, evoke unique patterns of glomerular activity that are reproducible at repeated stimulation within an individual. Intracellular recordings from the PNs specifically associated to one glomerulus are difficult to obtain and scarce. The PNs arborizing in one identified glomerulus of the AL of *M. sexta* were found to have a narrow receptive range, responding to cis-3-hexenyl acetate, while the propionate and butyrate homologues needed higher doses to elicit responses in the same glomerulus (Reisenman et al. 2005). In turn, these PNs were hyperpolarized by linalool, a compound that excited PNs in an adjacent glomerulus, providing evidence for lateral inhibitory interactions between glomeruli. Across fiber mode of coding and network functioning in the CNS are determinant in recognition of general odors. Disruption of GABA$_A$ receptors in the AL increases discrimination thresholds in *M. sexta* (Mwilaria et al. 2008). By analogy with other sensory modalities, it has been suggested that lateral inhibition narrows the tuning of PNs in the ALs. However, PNs are more broadly tuned compared to their ORN input in *Drosophila* showing nonlinear transformation of the olfactory code (Bhandawat et al. 2007; Wilson et al. 2004). In various insect species odors are represented by very small assemblies of highly specific KCs realizing a sparse coding, while odor codes are compact into the ALs (Perez-Orive et al. 2002; Szyszka et al. 2005) indicating that integration of the olfactory information is far from being linear from the periphery to the higher centers. Sparse coding seems to be the rule in moth MBs too. KCs in *M. sexta*, show very low prestimulus activity; a defined olfactory stimulus triggers the firing of a few spikes mainly at stimulus onset, sometimes at its offset, from a few cells (Ito et al. 2008). Most of the KCs respond to only a few of the 21 compounds that were tested, although a small subset of them show broader tuning. Responses to general odor in the MBs are spatially distributed and vary over the stimulus course.

2.3.2 How Does Inhibition Contribute to Quality Coding?

In various insect species, certain odorants elicit an increase in the firing activity in some ORNs but a decrease in the firing frequency in other ORNs. Reciprocally, the firing activity of some ORNs is strongly increased by certain odorants but decreased by others. For instance, among a panel of 102 general odorants presented to the antennae of *M. sexta* females, some compounds elicited excitatory responses (measured as an increase of firing frequency) in some sensilla, but inhibitory (decrease in firing frequency) responses in other (Shields and Hildebrand 2001).

To which extent the inhibition of ORNs involves a specific inhibition pathway or results from the modulation of the excitatory pathway is still not clear. In insects, Dubin and Harris (1997) found that odor-modulated conductances differed among cells in semi-intact *D. melanogaster* antennae. Pupal and adult ORNs responded with increased as well as decreased action potential frequency and hyperpolarization with a concomitant decrease in conductance was observed (Dubin and Harris 1997). Odorant-induced suppression of inward transduction or voltage-dependent currents have been frequently observed in vitro with ORNs isolated from different vertebrate species. However, whether such effects play a role in natural odor coding may be questionable due to their generally poor specificity.

Inhibitions by odorants were observed in *D. melanogaster* after prolonged stimulation of ORNs with low concentration of their best ligand to increase their firing (de Bruyne et al. 2001). Linalool, but not citronellol, inhibited responses to sustained (25 sec) stimulations by ethyl acetate, for example. Due to the conditioning of cells with low concentrations of their excitatory odorant, the hypothesis that this type of inhibition results from molecular masking rather than an inhibitory pathway cannot be ruled out. An intracellular mechanism for inhibition of the response in mixtures has been described in rodents but not yet in insects. Certain odorants can activate the phosphatidylinositol 3-kinase (PI3K) pathway and induce the generation of phosphatidyl-inositol 3,4,5-triphosphate (PIP3). PIP3 is then able to inhibit primary signal transduction and calcium signaling, leading to inhibition of response to excitatory olfactory stimuli when two odorants are presented together (Wetzel et al. 2005). Alternatively, cases of inhibition by general odorants of the response to pheromone could be attributed to molecular interactions in several moth species. First, linalool and other terpenoids inhibit the firing response of Pher-ORNs in *Spodoptera littoralis* without hyperpolarizing the ORN when presented alone (Party et al. 2009). Second, plant odorants significantly inhibited the binding of pheromone to HR13, a pheromone receptor of *H. virescens*, expressed in a cell line (Pregitzer et al. 2012).

2.3.3 INTENSITY CODING

Odor intensity (i.e., the concentration of odorant molecules in the air) is important in many respects for insect olfactory communication. Intensity may not only provide information on the proximity and value of the source but also alter the specificity of detection, or it may even modify the valence of the stimulus. The dose behavioral-response curves to aggregation pheromones of mites are convex, no aggregation is observed at low doses, aggregation is observed at the optimal midconcentration, and pheromone compounds like neral elicit alarm and escape at the highest doses (Kuwahara 2004).

The intensity of the concentration of an odor decreases rapidly with the distance from the source and under stable conditions and short distance concentration gradients could provide orientation cues to a moving insect. Navigation strategies to olfactory cues may be different among species, and vary within a species between circumstances (Gaudry et al. 2012). However, due to the turbulent regime of dissemination of airborne signals, uniform gradients of concentration rapidly break into discontinuous filaments of odorized air at a relatively short distance from the source (Cardé and Willis 2008). Thus, the orientation strategy of flying male moths is based on the dynamic perception of events of finding or loss of the target odor filaments and seems well adapted to the intermittent nature of the pheromone plume (Cardé and Willis 2008).

The capacity of insect ORNs to code the concentration of stimulating molecules into a spike firing frequency over a surprisingly large range of concentrations is well documented. Dose response curves in ORNs present a typical sigmoid curve. Above threshold, the firing frequency increases linearly as the log of the concentration, and it reaches a plateau. Central neurons in the MGC of *Agrotis ipsilon* male moths respond to the pheromone blend with a biphasic excitatory-inhibitory response (Jarriault et al. 2009b). Pheromone stimulus intensity is encoded by the firing frequency, the number of spikes, and the latency of the excitatory phase. In turn intensity has no effect on the duration of the response. MGC neurons present a lower threshold compared to ORNs. This lowering of the threshold is generally considered as the direct consequence of the convergence of many ORNs onto a lower number of PNs.

Encoding odor intensity, however, interferes with that of stimulus identity. Specificity and intensity are indeed interdependent as chemical specificity decreases when concentrations of odorants increase. High concentrations of a poorly efficient compound may trigger the same level of firing activity in ORNs as a low concentration of a very efficient compound. Globally, the total afferent input to the AL increases with odor concentration. Broadening of glomerular activation with increasing doses has been shown in several species. In *A. ipsilon*, for example, more and more AL glomeruli show calcium activity with increasing doses of a specific plant volatile, heptanal, causing a loss of specificity in the representation of this odor (Deisig et al. 2012).

2.3.4 Coding of Signal Temporality

Because of the nature of the stimulus, olfaction is not a fast communication channel compared to either audition or vision. The timing of the stimulus largely depends on the transport of molecules by air movement and its turbulence (Murlis and Jones 1981), although sometimes emission might be pulsed at the source. Still, time is a more important parameter for olfaction than it could appear at a first glance. First, as underlined in the previous section, fast detection of the plume finding and loss events is crucial to moth orientation to pheromone sources. Second, temporal patterns of firing activity within neuronal networks are a component of the coding of odor quality.

The intermittent character of airborne odors due to the turbulence in carrying air and its importance for male moths flying in pheromone plumes have been acknowledged for a long time (see, for instance, Mafra-Neto and Cardé 1994; Willis and Baker 1984). Numerous studies in the wind tunnel with different species resulted in building a generally admitted strategy for orientation of male moths to sex pheromone in flight that involves a succession of upwind surges when encountering pheromone filaments and zigzag flight when losing odor plume (Cardé and Willis 2008). During flight, moth ORNs must be able to follow these fast changes in pheromone stimulus. The characterization of the dynamic of ORNs has primarily concentrated on their capacity to discriminate pulsed stimuli (Barrozo and Kaissling 2002; Bau et al. 2002). The upper experimental limits to repetition rate resolution are probably limited by the strong hysteresis in the olfactometers used to deliver olfactory pulses at high rates and the methods used to analyze the resulting ORN activity. Dedicated statistical methods showed that the neuronal activity in male antennae was able to follow repetition rates up to 33 pulses/sec (Bau et al. 2002). The intermittency of the stimulus is reproduced by the bursting activity of PNs in the ALs (Chaffiol et al. 2012; Lei et al. 2009). Central neurons in the AL of *M. sexta* can track pulsed odor up to 30 Hz (Tripathy et al. 2010). Disruption of the GABA inhibition in the AL suppressed the ability of PNs to produce bursting responses reproducing the intermittency of pulsed stimulation (Lei et al. 2009). Accordingly, GABA inhibition blocked male orientation to the pheromone in the wind tunnel.

The temporal performances of insect olfactory systems are also important for a fast discrimination of odors. Concurrent airborne odorants intermingle and fluctuate at fast time scales and the olfactory system needs to segregate odors from independent sources as different odor objects. Honeybees can segregate two odors presented with a 6-ms temporal difference showing that the temporal resolution of the olfactory system of insects is faster than previously thought (Szyszka et al. 2012).

To maintain these dynamic performances the odor molecules must be cleared off the antennae by odorant ODEs The enzymatic activity directly impacts the dynamic of the detection. For instance, ORN firing responses to cis-vaccenyl acetate were both stronger and longer in mutant *D. melanogaster* flies lacking the gene for an extracellular carboxylesterase, esterase-6 (Chertemps et al. 2012). In turn, responses to heptanone were not altered in mutants. ODEs should not only quickly inactivate the molecules of the specific signal (the half-life of a pheromone molecule has been estimated to be 15 ms), but also reduce the concentration of general odorants or xenobiotics that otherwise would alter the detection or be toxic to neurons with broad-spectrum biotransformation enzymes. Four categories of ODEs have been recognized (Vogt 2005): (1) soluble extracellular ODEs are present in the sensillum lumen (esterases, aldehyde oxydases, alcohol dehydrogenases, etc.) and show a diversity corresponding to the diverse functional groups of the pheromone components, (2) membrane-bound ODEs (epoxyde hydrolase), (3) cytosolic ODEs are multifunctional enzymes involved in the biotransformation of potentially toxic chemicals that can also attack odor molecules (for instance cytochrome P450 oxygenases, glutathionine-S-transferase), and (4) cuticular ODEs can degrade airborne pheromone and other odors while adsorbed on the waxy insect cuticle. Since these adsorbed molecules constitute second sites of odor release, their surface degradation might significantly reduce the chemical noise.

Insect ORNs show quickly decreasing response to prolonged or repetitive stimulation (Zack-Strausfeld and Kaissling 1986). The concentration of odorants probably stay at a high level for a long time in the antennae so that sensory adaptation might constitute a protection mechanism against overstimulation and improve response dynamics of ORNs. Adaptation is a form of plasticity at the neurone level that results from different cellular mechanisms affecting the transduction pathway, each involved in different types of adaptation (Zufall and Leinders-Zufall 1997, 2000). Although adaptation has been proven in laboratory conditions, how it contributes to the temporal shaping of the response to the low concentration levels normally faced in natural conditions is still unknown.

2.3.5 TEMPORAL CODES

Temporality in the olfactory system is not only important for detection of the fluctuations of olfactory stimuli, but convergent data from different species show that temporal patterns of neuronal activities contribute to quality coding. Temporal coding means that the spike firing fluctuation carries specific information on odor identity beyond merely reproducing the temporal pattern of the stimulus (Wilson and Mainen 2006).

Compared to ORNs, the responses of PNs in the ALs present far more complex temporal patterns in the different insect species studied so far. The PNs show change in firing behavior with successive increases and decreases in the instantaneous firing rates, which often greatly outlast the stimulus duration. Many PN responses are inhibitory, with an inhibitory period followed by a period of increased firing. Furthermore, the time pattern of this offset activity is different among responses to different odors so it could contribute to odor coding in *B. mori* (Namiki and Kanzaki 2011b).

Odor-evoked synchronized oscillations of PN ensembles have been observed in several insect species as local field potential (LFP) oscillations (Laurent and Davidowitz 1994; MacLeod et al. 1998; Wehr and Laurent 1996). Synchronous firing among neurons is believed to facilitate the integration of signals in the olfactory neural network by creating a consistent representation of a given odor in the insect brain. Such oscillatory synchrony is an emergent property of the neural network and must be distinguished from a synchrony that simply reflects the coactivation of neurons by the same stimulus (Wilson and Mainen 2006). There is evidence that emergent synchronatory oscillations arise from dendro-dendritic connections between PNs and LNs in the locust ALs (MacLeod and Laurent 1996). In the absence of odors, moth PNs typically fire action potentials sporadically and their firing is asynchronous. In *M. sexta*, simultaneous recordings of the responses evoked by pheromone components from pairs of PNs innervating the same or different glomeruli within the MGC showed that PNs that branched in the same glomerulus and were activated by the same pheromone component also showed the strongest synchronization of their firing activity. Stimulation with a two-component blend evoked increased synchrony between intraglomerular pairs of PNs, supporting the idea that lateral inhibition between glomeruli shape the representation of pheromone blend (Lei et al. 2002).

The synchronized input from PNs is further processed in the protocerebron. There, contrasting to the high spontaneous activity and high probability of response of the AL-PNs, KC responses to odors in the locust protocerebron are rare and made of a small number of spikes lacking the temporal patterning of PNs (Perez-Orive et al. 2002). KCs integrate incoming spikes over brief, 50-ms time windows that are periodically reset by the oscillation cycle and are particularly sensitive to phase-locked spikes. Perez-Orive and colleagues concluded that KCs act as selective coincidence detectors from the fraction of AL-PNs that show synchronized activity over the staggered activity of PNs from which they receive input.

2.4   ODOR INTERACTIONS AND THE CODING OF COMPLEX BLENDS

Although single compounds may trigger behavior, most of the insect real odor world is made up of complex mixtures that may include many odorants, and insects generally respond better to blends (Riffell et al. 2009a). How are these blends coded to be recognized? Various behavioral, psychological, and neurophysiological experiments have shown that the response to an odorant mixture is not a simple function of the responses to its individual components. Olfaction is thus considered a "synthetic" sense because the ability to segment the perception of odor mixtures into distinct components is limited. The perception of an odor mixture is either dominated by a salient component or acquires a new quality. Thus, the perception of a mixture results from the interactions between its elements, giving rise to a neural representation that loses information about individual components but acquires new characteristics relevant to the mixture. Odor mixture perception can be configural (when the mixture is qualitatively different from the components), or elemental (if the individual components are still recognizable).

In nature, insects rarely encounter olfactory stimuli from single sources in isolation but they must extract meaningful information from complex streams of overlapping signals. Thus, interactions also arise between odorants emanating concurrently from different sources releasing blends, eventually sharing common components. Knowing that interactions between odor components may occur at different levels of odor processing, it makes identifying their mechanisms a difficult task. Masking occurs when the perception of one stimulus can be diminished by the close temporal proximity of another. A decrease in behavior for instance may result, either from the activation of ORNs and PNs signaling for a repellent or from the antagonistic interaction of an aversive signal with the agonist path.

2.4.1 Measuring Interactions

Because of the interactions between its components, the amplitude of the response (response being either the firing of a single ORN or a very integrated behavior) to an odor mixture is not entirely predictable by the responses to the components presented individually. It can be less than (suppression) or greater than (enhancement) the value inferred from the responses to the individual components (Laing et al. 1984). The first methodological problem is to properly define how much is less or greater. Interactions may be synergistic or antagonistic. Synergy means broadly "working together" and antagonism means "working against each other." However, these generally understood meanings imply quantitative criteria that do not receive general agreement (Berenbaum 1989). A review of the methods aiming to quantify synergy or antagonism is out the scope of this chapter. Since it is a general problem in pharmacology, biochemistry, or even ecotoxicology, useful information for olfaction may be found in papers from these research fields (Berenbaum 1989; Jonker et al. 2005). This difficulty in quantification is reflected by the great diversity of terms appearing in the literature. Many definitions have been proposed, such as hypoadditivity, complete additivity, hyperadditivity; synergy and inhibition; agonism, partial agonism, complete agonism, synergistic agonism. Because the relation between individual stimulus intensity and response is not linear and its parameters depend on the compound, calculating the response to a blend requires computing the equation from a dose-response function after measuring its main parameters (threshold, maximum response, slope), a very strict condition that is not often reached in most experimental studies (Duchamp-Viret et al. 2003).

It is worth a reminder that the purely physical mixture interactions arising at the source well upstream should not be neglected. Hygrometry, for instance, changes the vapor pressure of organic volatile compounds. Preevaporative effects have been reported for interactions with DEET (Syed and Leal 2008). Such effects must be taken into account when a precise quantification of the response to mixture is needed.

2.4.2 Mixture Effects at the ORN Level

Mixture effects are commonly observed at the level of single ORNs (for inhibitory interactions, see also Section 2.3.2). ORs are often differentially sensitive to a variety of odorants. Interaction is said to be competitive when two or more agonist odorants bind to the main receptor site and trigger receptor activation, although only one can be bound at a time. Noncompetitive effects may result from different mechanisms including one of the odorants binding to another site, which modifies the receptor properties at the main binding site. A high frequency of noncompetitive interactions were found in rat ORNs (Rospars et al. 2008), suggesting that such interactions play a major role in the perception of natural odorant mixtures. Competitive interactions may be discriminated from noncompetitive interactions because they usually can be counteracted by increasing the amount of agonist relative to that of the competitor. Recording the responses of ORNs to single compounds and their blend revealed a majority of mixture suppressions in various insect species (Carlsson and Hansson 2002; De Jong and Visser 1988; Hillier and Vickers 2011). Prevalence of inhibition was also found in response to binary mixtures compared to pure odorants in the cockroach *Periplaneta americana* during testing a series of aliphatic alcohols on the ORNs found in one type of sensilla housing generalist neurones (Getz and Akers 1997).

A second mode of interactions at the peripheral level results from the specific organization of the insect antenna compared to the olfactory epithelium of vertebrates. In insects, two or more ORNs with different odor tunings are cohoused into single functional units, the sensilla. This colocalization of neurones with different rather than similar sensitivities was postulated to be highly adaptive, serving a more precise ratio discrimination of pheromone blends (Baker 2009; Vermeulen and Rospars 2004). This seducing hypothesis was lacking experimental support for interactions between neighbor ORNs, however. Such interactions by edaphic coupling have been demonstrated only recently in *D. melanogaster*. A conjugation of electrophysiological and molecular biology approaches showed that the sustained response of one ORN is inhibited by transient activation of its neighboring ORN within the same sensillum (Su et al. 2012).

2.4.3 Recognition of Complex Plant Bouquets

Phytophagous insects localize their host plants through emitted odors (Bruce et al. 2005). Volatile blends differ between plant species both quantitatively and qualitatively. The specific combinations of compounds in these blends, many of which are ubiquitous, as well as their ratios, are assumed to drive host plant finding. Natural plant odors are complex blends of tens of components, most of them common to several plant species. The composition of these blends may greatly vary between individuals within a plant species according to its metabolism, physiological state, and developmental stage. Thus, to recognize and efficiently locate their host plants, herbivorous insects are faced with three problems: the complexity of the aroma, the ubiquity of components, and the variability of the composition, these last two compromising the specificity of the odor cues. About 120 different compounds are present in the volatile emissions of one major host plant of *M. sexta*, and female antennae possess ORNs able to detect 60% of them (Spathe et al. 2012). The majority of these ORNs are broadly tuned to a number of the host volatiles, but some ORNs respond to compounds specific to only one of the host plants. This indicates that their sensory equipment enable females to select plant species and plant quality for oviposition sites on the basis of olfactory cues (Spathe et al. 2012).

Adults of many moth species visit flowers for feeding on nectar, and flower volatiles may trigger innate behavioral responses. The floral bouquet from *Datura wrightii* flowers evokes foraging behavior in naïve *M. sexta* (Raguso and Willis 2002). To find out how behaviorally relevant floral mixtures are encoded in the olfactory system, Riffel et al. used a combination of chemical analysis of flower aroma, wind tunnel experiments, and multielectrode extracellular recordings in the ALs (Riffell et al. 2009a,b). Mixtures were more efficient than individual compounds in eliciting behavioral response, but although the whole floral bouquet comprised more than 60 compounds that varied in identity and concentration, moths responded with an equal frequency to natural flowers and a blend of only nine odorants. The behavioral responses to mixtures were consistent over a 1000-fold range in concentration. The ensemble analysis of the multiunit responses revealed a strong preference for only a small subset of the compounds emitted by *D. wrightii* flowers. Many single units were activated by one or more of the nine behaviorally active odorants. Single-unit responses to the blend were found to be suppression (11%), hypoadditivity (42%), or synergy (7%). The proportion of units exhibiting synergy or responding to mixtures was higher at high concentrations compared to responses to single odorants, but not at low concentrations. Spatial distribution, correlation coefficient, and Euclidian distance were calculated to compare the ensemble representations between different odor mixtures and single compounds. The representations for a mixture at lower concentrations were not statistically dissimilar to those of the single odorants, suggesting that the spatial distribution pattern of ensemble responses alone does not fully explain the behavioral activity of the mixtures relative to the single odorant or their consistency across concentrations. Neither the percentage of neurons that exhibited synergistic responses to the mixture, nor percentage of responsive units alone could explain the behavioral consistency of the attractive mixture across the concentration range. In turn, stimulation with mixtures greatly enhanced the synchronous firing between pairs of neurons to form a distinct temporal activity that did not change over concentration range, suggesting that neural synchrony between certain cells encodes the mixtures in a spatiotemporal representation.

2.4.4 Interactions between Odorant Signals Released from Different Sources

The detection of a specific signal and its perception may be altered by odorants from the odorant background. Pheromones and plant odors provide a good model example for such interactions between signal and environment. The responses of male moths to sex pheromone are innate. However, it has been repeatedly observed that male moths are attracted in larger numbers to pheromone traps baited with a blend of synthetic pheromone plus some plant-related volatiles compared to pheromone alone (Light et al. 1993; Meagher 2001). They also fly more readily in the wind tunnel in response to such blends (Deng et al. 2004; Schmidt-Büsser et al. 2009; Yang et al. 2004). Their increased attraction to blends has been interpreted as evolutionary adaptive because of the high probability of finding sexually receptive females on host plants so that host odor provides supplementary cues to find a mate. Correspondingly, in *Eupoecilia ambiguella*, the synergistic effects were observed mainly at low or high dosages of the pheromone when orientation to males becomes more difficult (Schmidt-Büsser et al. 2009). The mechanisms of these interactions remain not well understood but several levels of the olfactory system are involved. Several authors have recorded and analyzed the firing responses of Pher-ORNs to mixtures of pheromone and plant odors with contradictory results. The Pher-ORNs of *A. segetum* were found to respond to high doses of plant compounds (Hansson et al. 1989). Synergistic effects, the firing response being increased in response to the main pheromone component plus linalool or cis-3-hexenyl acetate, have been observed only in the noctuid moth *Helicoverpa zea* (Ochieng et al. 2002). Most studies, however, conducted on other moth species using linalool or other plant compounds revealed a prevalence of antagonistic interactions (Kaissling et al. 1989; Party et al. 2009; Van der Pers et al. 1980). Stimulation of Pher-ORNs of the noctuid moth, *Heliothis virescens*, with mixtures comprised of the cognate pheromone component and either another pheromone component or a host plant volatile resulted most frequently in attenuation of the firing response (mixture suppression) (Hillier and Vickers 2011). The antagonism between linalool and pheromone was recently convincingly proven to arise at the OR levels in *H. virescens* (Pregitzer et al. 2012). In *Spodoptera littoralis*, antagonism between linalool and the main pheromone compound resulted in an increase in the capacity of Pher-ORNs to follow pulsed stimuli and the background was postulated to increase the capacity of the olfactory system to follow the fast changes in pheromone concentration during the moth flight (Rouyar et al. 2011).

Because pheromone and general odors interact mostly antagonistically in the antennae, the mechanisms to explain the increased behavioral responses to mixtures must be searched for in their integration in the brain. While the marked spatial separation between both types of inputs in the moth ALs has long been considered a sign that intraspecific (pheromone) and interspecific (plant volatiles) odor signals were processed separately, there is now evidence for some convergence in the ALs of different species (Kanzaki et al. 1989; Namiki et al. 2008; Trona et al. 2010). Among the multiglomerular PNs that were found in *M. sexta*, one of them arborized within the MGC (the pheromone-specific structure) and the OGs and showed responses to pheromone components and a plant volatile (Kanzaki et al. 1989). Calcium imaging (Deisig et al. 2012) and electrophysiological recording of pheromone-sensitive projection neurones in the MGC of male *A. ipsilon* (Chaffiol et al. 2012) showed that the response to the mixture pheromone + heptanal was generally weaker than to the pheromone alone, indicating a suppressive effect of heptanal. However, these neurones responded with a better resolution to pulsed stimuli. Conversely in *B. mori*, bombykol responses in PNs of the MGC were enhanced in the presence of Z3-hexenol, a host plant odor (Namiki et al. 2008). In contrast, the responses of PNs innervating ordinary glomeruli to Z3-hexenol were unaffected when bombykol was applied along with the plant odor.

2.5 VARIABLE RESPONSES TO FIXED OLFACTORY SIGNALS

Highly determined responses to a specific signal are a very efficient way of communicating when the signal nature is precisely controlled and stable, as with a sex pheromone. However, it might be adaptive for an insect to modulate its responses according to its physiological state. Correspondingly, fixed responses might be a disadvantage for less constant signals. Evidence is accumulating that insects have developed a variety of mechanisms that allow them to modulate their odor-driven behaviors through neuronal plasticity in their olfactory system. We will examine three examples showing that responsiveness to sex pheromone or floral odors is influenced by physiological or sensory contexts and how this behavioral flexibility is adaptive.

2.5.1 CHANGES IN RESPONSIVENESS TO OLFACTORY SIGNALS ACCORDING TO PHYSIOLOGICAL STATUS

It is advantageous to insects to reach maximal responsiveness to olfactory signals at the right time. For instance, the maximal responsiveness of a male moth to the sex pheromone should coincide to the development of its reproductive organs and the presence of receptive females (Gadenne et al. 2001). Similarly, female moths should be more sensitive to larval host-plant odors after mating, when they are physiologically ready to oviposit. Male *A. ipsilon* copulate only once in a single scotophase and mating induces a transient inhibition of the attraction to the sex pheromone in newly mated males (Gadenne et al. 2001), but does not affect their responses to plant odor (Barrozo and Gadenne 2010; Barrozo et al. 2010). In turn, mating turns on attraction to larval host plant-odor in female *Lobesia botrana* (Masante-Roca et al. 2007) or *S. littoralis* (Saveer et al. 2012). These changes in behavioral responses are correlated to changes in the responsiveness of the AL neurons to sex pheromones or plant odors (Anton et al. 2007). There is a reduction in the sensitivity to pheromone of AL neurons in male *A. ipsilon*; however, their sensitivity is not completely shut off. Newly mated males not only do not respond any more to pheromone and still fly toward a linden flower extract, but are also not attracted to mixtures of pheromone and flower extract at a high ratio of the pheromone (Barrozo et al. 2010), indicating a change in hedonic valence of the pheromone.

Besides these mating effects, an age-dependent plasticity has been described in moths. In spite of the conservative organization of the ALs, a certain degree of variability has been observed in the shape, volume, or location of their glomeruli in *M. sexta* (Couton et al. 2009; Huetteroth and Schachtner 2005) or *S. littoralis* (Couton et al. 2009), whose functional signification remains unclear (Guerrieri et al. 2012) but could result from the morphological changes that occur epigenetically in the olfactory centers of adult holometabolous insects. Persistent neurogenesis has been evidenced in the MBs of adult males and females, but not in their ALs (Dufour and Gadenne 2006). Hormones and neuromodulators are also involved in the modulation of the responsiveness during adult maturation. The sensitivity of AL neurons to pheromone increases with age and is dependent on the levels of juvenile hormone (JH) and octopamine in *A. ipsilon* (Jarriault et al. 2009a).

While in moths the regulation of olfactory sensitivity seems to operate mainly at a central level whereas the sensitivity of the periphery remains stable, female mosquitoes show a reduced sensitivity of their lactic acid ORNs after a blood meal (Davis 1984, 1986; Qiu et al. 2006). Correspondingly, a downregulation of a putative OR gene (Fox et al. 2001) or other olfactory genes (OBPs) has been observed (Biessmann et al. 2005). The sensitivity of functional classes of ORNs involved in the detection of cues for oviposition sites was not downregulated after a blood meal (Siju et al. 2010).

2.5.2 PLURIMODALITY INTERACTIONS

While communication between sexes in moths and several other insect groups is largely moderated by olfaction as a dominant sensory modality, communication between sexes in other groups involves multimodality. Pentatomid bugs integrate both pheromone and vibratory signals in their mating communication system. In the green stink bug *Nezara viridula* long-range attraction has been generally attributed to the male emitted pheromone (Brézot et al. 1994), while short-range mate localization relies on substrate-borne vibration (Cokl et al. 1999). However, the emission of pheromones by the male is modulated by the female song (Miklas et al. 2003b). Bugs respond to a large range of component ratios of the pheromone blend so that integrating vibratory songs and pheromones contribute to increase the species-specificity of the communication between sexes (Brézot et al. 1994; Miklas et al. 2003a).

Pollinators show innate responses to colors and learn more rapidly to associate a reward with a color they innately prefer. Orientation to flowers in *M. sexta* combines visual and olfactory stimuli (Balkenius and Dacke 2010). Calcium imaging of responses to bimodal stimuli consisting of odor and color in *M. sexta* showed that color could either enhance or suppress odor-induced responses in the MBs. A blue stimulus suppressed the response to phenylacetaldehyde, a general flower scent (Balkenius et al. 2009). By contrast, it enhanced the response to a green leaf volatile (1-octanol). Hawk moths can learn combinations of stimuli, which suggests that they are capable of configurational learning (Balkenius and Kelber 2006).

Using multiple modalities offers several advantages to the insect. It can increase the detection of relevant objects against background noise and reduce the risks of confusion in a changing environment. It can also decrease the reaction time and response threshold. More generally, integrating information from different sensory modalities might indeed improve the reliability of olfactory signaling, a global context being often more specific than individual cues.

2.5.3 PLASTICITY IN RESPONSE: BETTER FACING THE RISKS OF COMMUNICATION?

The context in which olfactory communication takes place also involves risks from predation or deception by concurrent organisms so that a certain degree in plasticity in the response may be advantageous. Innate and stereotyped communication systems may be subject to exploitation of perceptual biases by other organisms that will take advantage of the strong attractivity of an odor to the receiver for attracting it to their own and only advantage (deceptive communication). Plants in the Araceae and Orchidaceae families are known to use deceptive pollination by which they attract specific pollinators (Dafni 1984) by mimicking sex pheromones or food and prey signals (Brodmann et al. 2009; Stokl et al. 2010, 2011). Predators can also use sensory traps. Bolas spiders, for instance, produce volatiles mimicking a sex pheromone to lure and catch male moths (Stowe et al. 1987).

Communication makes animals conspicuous to their predators. Flying to pheromone sources exposes male moths to bat predation. Male moths react to bat echolocation sounds by escape maneuvers such as diving to the soil (Svensson et al. 2007). Males of *A. segetum* and *Plodia interpunctella* orienting towards a sex pheromone source in a flight tunnel were exposed to ultrasound mimicking the echolocation calls of a bat (i.e., high predation risk). Males of both species accepted the predation risk when attracted to a pheromone source of high quality (a female gland extract or the complete synthetic blend at high dose). In contrast, a lower proportion of ultrasound-exposed males than unexposed ones located the pheromone source when it was of low quality (an incomplete synthetic blend or the complete blend at low doses) (Svensson et al. 2004). Thus, when confronted with various contexts male moths make a trade-off between reproduction and predator avoidance depending on the relative strength of the perceived conflicting stimuli showing behavioral plasticity (Svensson et al. 2007).

2.6 FACING POORLY SPECIFIC OR VARIABLE SIGNALS

The plant volatile emissions, which have not evolved specifically as communication signals, may be very variable in their composition compared to pheromones. Nectar-feeding moths attracted to the odors of their floral hosts, for instance, are faced with complex blends of tens to hundreds of components with different biosynthetic pathways up to their release. Production rates of molecules issuing from different metabolisms are not as precisely controlled by the emitter as is the pheromone. The composition of the blend can vary from flower to flower within and between plants because of a variety of genetic or environmental causes. A pollinator that narrowly focuses only on blend qualities would risk perceiving each flower as unique, so that generalization from one flower to the next one containing the same resources might be difficult (Hosler and Smith 2000). In addition, a source of nectar can get exhausted so that it is counteradaptive for an insect to continue to respond to an odorant that is no longer positively correlated to a resource. Because olfactory cues may be inconsistent or context-dependent and the animal's needs may change, there is an advantage to different forms of flexibility. In this last section we will review some cases of plasticity and context dependence in the responses.

2.6.1 Dealing with the Ubiquity of Some Chemical Signals: Context Dependence

Many chemicals are released by a large variety of organisms so that their presence alone would not allow an insect to reliably identify or even locate a resource. Yet, several of these ubiquitous compounds have proven to play key roles as resource cues, even by specialist insects. Carbon dioxide is naturally present in the atmosphere as a background, but nevertheless it plays multiple roles in insect foraging behavior (Guerenstein and Hildebrand 2008). Carbon dioxide is released into the atmosphere from various sources, for instance, respiration of living organisms or decomposition of plant matter. Variations in the rate of plant respiration lead to large intraday fluctuations of its atmospheric concentrations from 350 ppm during the day to up to 1000 ppm in dense vegetation at night. Atmospheric turbulences also modify the pattern of CO_2 concentrations. Nevertheless, 10 to 30 ppm bursts of CO_2 could be detected in natural habitats by a gas analyzer several tens of meters downwind from natural and artificial sources (Zöllner et al. 2004). Insects possess receptor cells highly sensitive to such changes of CO_2 concentrations in their antennae or their labial or maxillary palps. The CO_2 receptor cells of moths for instance can encode fast increases or decreases in the CO_2 concentration (Guerenstein et al. 2004). Upwind orientation in airstreams enriched with CO_2 have been demonstrated in different hematophagous insects, like mosquitoes, tsetse flies, and triatomine bugs, as well as in moths (Guerenstein and Hildebrand 2008). Differences in concentrations as small as 50 ppm above the ambient level are sufficient to trigger tsetse fly oriented flight (Evans and Gooding 2002).

Because CO_2 is everywhere, high sensitivity to CO_2 is not enough to make it a reliable resource indicator. The capacity to orient toward CO_2 enriched airflows is synergistically enhanced by organic volatile compounds released by the hosts in triatomine bugs (Barrozo 2004) and mosquitoes (Takken and Knols 1999). Sensitivity of female yellow fever mosquitoes to human skin odors increases significantly after a brief exposure to a plume of 4% CO_2 (Dekker et al. 2005). Dependence of the response to CO_2 on the odorant context has been established also in phytophagous moths and flies. Female *M. sexta* showed significant bias in approach of artificial flowers with above-ambient CO_2 only in the presence of volatiles of the leaves of their host plant, the tomato (Goyret et al. 2008). Experienced *M. sexta* when offered surrogate flowers without nectar do not maintain their spontaneous preference for flowers associated with high CO_2 level (Thom et al. 2004). Interestingly, the context might even modify the valence of one odorant compound for the insect. As a potential toxic, CO_2 has an inherent negative hedonic value to *D. melanogaster* flies that spontaneously avoid it. In turn, cider vinegar has an inherent positive value (attraction) (Faucher et al. 2006). But in a four-field olfactometer in which the flies could smell CO_2 in one field and some other odors or pure air in the other fields, their responses to CO_2 were found to depend on the odors presented in the other fields (Faucher et al. 2006).

2.6.2 Dealing with Variable Odors: Generalization

Showing a narrow specificity of responses to odors may be disadvantageous when the composition of the signal being not strictly controlled by the emitter varies unpredictably. Tolerance for ratio changes in host plant effluvia has been observed in moths. Mated females of the oriental fruit moth, *Cydia molesta*, are innately attracted to synthetic mixtures mimicking the natural blend of their host plants. Female attraction is maintained while the ratio of one of the constituents, benzonitrile, is increased up to 100 times (Najar-Rodriguez et al. 2010). Calcium imaging showed that odor evoked responses in one glomerulus of the female ALs mirror the behavioral effects of the manipulation of the benzonitrile ratio. A second glomerulus responded to changes in benzonitrile ratios, but small levels of benzonitrile in the mixture were inhibitory. These differences between glomeruli show that although the blend representation starts in the AL, final processing of the ratio must take place in higher-order olfactory centers. This relatively broad tolerance to the constituent ratios could offer the moth the advantage of recognizing its host plants across phenological stages in spite of quantitative and qualitative significant fluctuations in their volatile emissions.

The ability for an organism to learn that perceptually distinct olfactory stimuli lead to common outcomes is called generalization. Honeybees conditioned to individual compounds or to mixtures generalize their responses to stimuli belonging to the same chemical classes or with the same functional value (Sandoz et al. 2001). Components of alarm pheromones induce more generalization than floral compounds (Sandoz et al. 2001). Bees trained to respond to a binary odor mixture by being rewarded with sugar respond to novel proportions of the same compounds even when the interblend differences are substantial (Wright et al. 2008). The resulting generalization depends on the rewarding paradigm and the variability of the conditioned stimulus. This suggests that this outcome-dependent generalization is cognitive rather than perceptual. The bee brain can construct perceptual qualities of odors that depend, at least in part, on previous experiences. This capacity is necessary for maintaining sensitivity to interodor differences in complex olfactory scenes and it is not restricted to social hymenoptera. *M. sexta* conditioned to 2-hexanone or 1-decanol also responded to other alcohols and ketones and the generalization of the conditioned response decreased as a function of the chain length and functional group (Daly et al. 2001).

2.6.3 SENSITIZATION AND LEARNING

The basic knowledge on learning in insect and its neural basis comes first from the honeybee. The past decades have generated a wealth of novel research on the cognitive capabilities of bees and other insect species (see, for instance, reviews in Giurfa 2013; Menzel and Giurfa 2006; Menzel et al. 2006). Here I will focus on examples showing that experience can also modulate olfactory behavior generally considered as highly determined in nonsocial insects and in which structures in the olfactory systems are involved. Lepidoptera may be innately attracted to floral odors from flowers they feed on, but in addition they can learn new odorant-reward associations. *Helicoverpa armigera* moths trained to feed on flowers providing sugar sources that were odor-enhanced using phenylacetaldehyde or alpha-pinene showed a significant preference for the flower odor type on which they were trained (Cunningham et al. 2004). In turn, moths conditioned on flowers that were not odor-enhanced showed no preference for either of the odor-enhanced types. These results imply that moths not only show learned responses to floral volatiles, but they can also discriminate among odor profiles of individual flowers from the same species.

Like bees, moths respond to the presentation of sucrose by extending their mouth parts, which is called the proboscis extension reflex (PER). Effective associative conditioning of the PER occurs when the unconditioned stimulus (US), a sucrose reward, is proposed a few seconds after the onset of a sustained odor pulse, the conditioned stimulus (CS). Interestingly, associative learning has revealed experience-dependent plasticity in the ALs of *M. sexta* (Daly et al. 2004). PER was monitored by recording the feeding muscle activity by electromyography while recording neuronal activity in the ALs. More and more responsive neural units were found in the AL during conditioning and this neuronal recruitment persisted after conditioning. Recruitment occurred when odor reliably predicted food (CS+, not CS−, so that sensitization can be excluded). Conversely when odor did not predict food, a loss of responsive units was observed. This demonstrates that odor representations in the first olfactory centers show experience-dependent plasticity and are involved in olfactory memory.

Nonassociative learning, or nonassociative plasticity, is also largely present in insect olfactory-driven behaviors although its neuronal bases have received less attention than those of classical conditioning. The experience of a type of larval food at emergence influences flight orientation and oviposition choices of females in two species, *Plodia interpunctella* and *Ephestia cautella*, that do not feed as adults, (Olsson et al. 2005). Plasticity in the development of the olfactory system in relation to olfactory experience has been reported. For instance, deprivation from antennal sensory input causes changes of AL volume (Sanes and Hildebrand 1976). Males of *S. littoralis* briefly exposed to the sex pheromone are more responsive to it than naïve ones 27 hours after the preexposure. This increase of behavioral responsiveness is correlated with an increased sensitivity of AL neurons to pheromones (Anderson et al. 2003, 2007). Preexposure to linalool and geraniol, two volatile plant compounds, and to attractive and repulsive gustatory stimuli also modified the response level of males to pheromones (Minoli et al. 2012). The increased behavioral response to pheromone following brief exposure to a stimulus mimicking echo-locating sounds from predatory bats is accompanied by an increase in the sensitivity of AL neurons (Anton et al. 2011). These cross-modality effects have been attributed to a general sensitization (Minoli et al. 2012).

Before more recent studies showed that ALs are also involved in olfactory learning, their areas of projection, the MBs, have long been linked to memory and associative learning. Many gene products with roles in learning are expressed at high levels in fly MB neurons (Turner et al. 2008). MBs are required for acquisition, storage, and recall of olfactory memories in *Drosophila* and different lobes of the MBs probably have different roles in memory (Krashes et al. 2007). However, we are still far from a comprehensive view of the neural circuit involved in olfactory learning. In *M. sexta* stimulus time-dependent plasticity in the KCs does not constitute the odor representation that coincides with reinforcement (Ito et al. 2008). Odor presentations that supported associative conditioning elicited only one or two spikes on the odor's onset (and sometimes offset) in a small fraction of KCs. Moths learned to associate individual odors with the reward even when unconditioned stimulus were temporally separated from the time window of KC firing by several seconds. These spikings ended well before the reinforcement was delivered.

2.7 CONCLUDING REMARKS

Insect olfaction is remarkably efficient in terms of sensitivity and specificity. The insects' capacities to localize their vital resources by detecting minute amounts of volatile compounds fascinate us, so a large research effort has been devoted to understanding the mechanisms underlying the performances of their olfactory systems. Molecular biology together with functional investigations of perireceptor events and transduction pathways now reveal a complex world at the subcellular level. Current active debates on the nature of ORs and molecular transductory cascades show that we are still far from having exhausted the topic and there is far more to unravel. We nevertheless have built a robust picture: odors are coded within a structured neuronal network resulting in overlapping but different maps of activities.

At the same time, we have gained a more realistic picture of the physical structure of odor plumes as they evolve in the air in spite of the difficulties to trace in real time odor molecules at concentrations below the detection thresholds of our measurement systems. Due to their dispersion by aerial turbulent currents, natural odors are highly intermittent and timing of their detection matters for orientation. Time also matters in the coding of the signal. Spatiotemporal patterns built in the deuto- and proto-cerebrum during odor coding are the best images we can reconstruct to correlate with discrimination capacities of insects. Adding the temporal dimension results in a more refined vision of how neuronal networks process complex sensory input, which can now be implemented in some fascinating practical issue in terms of artificial noses and robotics.

Most interestingly, evidence has grown in the last few years that a chemical signal may also take its full meaning from the general context. A largely ubiquitous compound has some value as a cue if its concentration significantly increases above a background level around a specific resource. This context-dependent communication opens new perspectives in chemical ecology. We will still identify specific compounds, or unique blend ratios, serving specific communication like sex pheromones, or in the case of the strong adaptation to a monophagous insect, the volatile bouquet of its host plants, but we will also need to look carefully at unspecific compounds. Olfaction appears to be an integrative sensory modality with a diversity of interactions between odorant molecules and between neural activities that issue from their detection and taking place from the periphery. This makes it a more complex sensory modality to study, but also a very rich research field. It is also important to keep in mind that small insect brains integrate sensory inputs from multiple sources or even multiple modalities and pleads for the necessity to include cognitive approaches in chemical ecology.

The capacity of insects to learn odor stimuli has been acknowledged for years now. Experience-related plasticity appears to be an essential component of olfaction and has some major issues for cognitive ecology. The final behavioral response is dependent on the physiological state and previous experience of the insect. Associative learning is essential to social and nonsocial pollinators. Nonassociative learning has been comparatively less investigated. A combination of innate preferences and cognitive abilities gives insects the ability to engage in specialized interactions with their host plants while keeping the flexibility to adapt to their natural variability or to exploit new resources. There is no need to emphasize how the understanding of the cognitive aspects of olfaction may be essential to behavioral ecology and practical uses of semiochemicals for crop plant protection.

Finally, recent progress in olfaction research has been made possible only by multidisciplinary approaches. In turn, they open fascinating perspectives not only in neurosciences, but more largely to cognitive and agroecology.

ACKNOWLEDGMENTS

I am highly indebted to Dr. Sylvia Anton and Dr. Philippe Lucas for their helpful comments on an earlier version of the manuscript.

REFERENCES

Anderson, P., B. S. Hansson, U. Nilsson, Q. Han, M. Sjoholm, N. Skals and S. Anton. 2007. Increased behavioral and neuronal sensitivity to sex pheromone after brief odor experience in a moth. *Chem Senses* 32: 483–91.

Anderson, P., M. M. Sadek and B. S. Hansson. 2003. Pre-exposure modulates attraction to sex pheromone in a moth. *Chem Senses* 28: 285–91.

Anton, S., M. C. Dufour and C. Gadenne. 2007. Plasticity of olfactory-guided behaviour and its neurobiological basis: Lessons from moths and locusts. *Entomol Exp Appl* 123: 1–11.

Anton, S., K. Evengaard, R. B. Barrozo, P. Anderson and N. Skals. 2011. Brief predator sound exposure elicits behavioral and neuronal long-term sensitization in the olfactory system of an insect. *Proc Natl Acad Sci U S A* 108: 3401–5.

Baker, T. C. 2009. Nearest neural neighbors: Moth sex pheromone receptors HR11 and HR13. *Chem Senses* 34: 465–68.

Balkenius, A., S. Bisch-Knaden and B. Hansson. 2009. Interaction of visual and odour cues in the mushroom body of the hawkmoth, *Manduca sexta*. *J Exp Biol* 212: 535–41.

Balkenius, A. and M. Dacke. 2010. Flight behaviour of the hawkmoth *Manduca sexta* towards unimodal and multimodal targets. *J Exp Biol* 213: 3741–7.

Balkenius, A. and A. Kelber. 2006. Colour preferences influences odour learning in the hawkmoth, *Macroglossum stellatarum*. *Naturwiss* 93: 255–8.

Barrozo, R. 2004. The response of the blood-sucking bug *Triatoma infestans* to carbon dioxide and other host odours. *Chem Senses* 29: 319–29.

Barrozo, R. and C. Gadenne. 2010. Post-mating sexual abstinence in a male moth. *Commun Integr Biol* 3: 629–30.

Barrozo, R., C. Gadenne and S. Anton. 2010. Switching attraction to inhibition: Mating-induced reversed role of sex pheromone in an insect. *J Exp Biol* 213: 2933–9.

Barrozo, R. B. and K. E. Kaissling. 2002. Repetitive stimulation of olfactory receptor cells in female silkmoths *Bombyx mori* L. *J Insect Physiol* 48: 825–34.

Bau, J., K. A. Justus and R. T. Cardé. 2002. Antennal resolution of pulsed pheromone plumes in three moth species. *J Insect Physiol* 48: 433–42.

Benton, R., K. S. Vannice, C. Gomez-Diaz and L. B. Vosshall. 2009. Variant ionotropic glutamate receptors as chemosensory receptors in *Drosophila*. *Cell* 136: 149–62.

Berenbaum, M. C. 1989. What is synergy? *Pharmacol Rev* 41: 93–141.

Bhandawat, V., S. R. Olsen, N. W. Gouwens, M. L. Schlief and R. I. Wilson. 2007. Sensory processing in the *Drosophila* antennal lobe increases reliability and separability of ensemble odor representations. *Nat Neurosci* 10: 1474–82.

Biessmann, H., Q. K. Nguyen, D. Le and M. F. Walter. 2005. Microarray-based survey of a subset of putative olfactory genes in the mosquito *Anopheles gambiae*. *Insect Mol Biol* 14: 575–89.

Brézot, P., C. Malosse, K. Mori and M. Renou. 1994. Bisabolene epoxies in sex pheromone in *Nezara viridula* (L.) (Heteroptera: Pentatomidae): role of cis isomer and relation to specificity of pheromone. *J Chem Ecol* 20: 3133–47.

Brodmann, J., R. Twele, W. Francke, L. Yi-bo, S. Xi-qiang and M. Ayasse. 2009. Orchid mimics honey bee alarm pheromone in order to attract hornets for pollination. *Curr Biol* 19: 1368–72.

Bruce, T. J., L. J. Wadhams and C. M. Woodcock. 2005. Insect host location: A volatile situation. *Trends Plant Sci* 10: 269–74.

Cardé, R. T. and M. A. Willis. 2008. Navigational strategies used by insects to find distant, wind-borne sources of odor. *J Chem Ecol* 34: 854–66.

Carlsson, M. A., C. G. Galizia and B. S. Hansson. 2002. Spatial representation of odours in the antennal lobe of the moth *Spodoptera littoralis* (Lepidoptera: Noctuidae). *Chem Senses* 27: 231–44.

Carlsson, M. A. and B. S. Hansson. 2002. Responses in highly selective sensory neurons to blends of pheromone components in the moth *Agrotis segetum*. *J Insect Physiol* 48: 443–51.

Carlsson, M. A. and B. S. Hansson 2006. Detection and coding of flower volatiles in nectar-foraging insects. In *Biology of Floral Scent*, N. Duradeva and E. Pichersky (eds.). Boca Raton, FL, CRC Press: 243–61.

Chaffiol, A., J. Kropf, R. B. Barrozo, C. Gadenne, J. P. Rospars and S. Anton. 2012. Plant odour stimuli reshape pheromonal representation in neurons of the antennal lobe macroglomerular complex of a male moth. *J Exp Biol* 215: 1670–80.

Chertemps, T., A. Francois, N. Durand, G. Rosell, T. Dekker, P. Lucas and M. Maibeche-Coisne. 2012. A carboxylesterase, Esterase-6, modulates sensory physiological and behavioral response dynamics to pheromone in *Drosophila*. *BMC Biol* 10: 56.

Christensen, T. A., I. D. Harrow, C. Cuzzocrea, P. W. Randolph and J. G. Hildebrand. 1995. Distinct projections of two populations of olfactory receptor axons in the antennal lobe of the sphinx moth *Manduca sexta*. *Chem Senses* 20: 313–23.

Christensen, T. A. and J. G. Hildebrand. 2002. Pheromonal and host-odor processing in the insect antennal lobe: How different? *Curr Opin Neurobiol* 12: 393–9.

Cokl, A., M. Virant Doberlet and A. McDowell. 1999. Vibrational directionality in the southern green stink bug *Nezara viridula* (L.), is mediated by female song. *Anim Behav* 58: 1277–83.

Couto, A., M. Alenius and B. J. Dickson. 2005. Molecular, anatomical, and functional organization of the *Drosophila* olfactory system. *Curr Biol* 15: 1535–47.

Couton, L., S. Minoli, K. Kieu, S. Anton and J. P. Rospars. 2009. Constancy and variability of identified glomeruli in antennal lobes: Computational approach in *Spodoptera littoralis*. *Cell Tissue Res* 337: 491–511.

Cunningham, J. P., C. J. Moore, M. P. Zalucki and S. A. West. 2004. Learning, odour preference and flower foraging in moths. *J Exp Biol* 207: 87–94.

Dafni, A. 1984. Mimicry and deception in pollination. *Ann Rev Ecol Syst* 15: 259–78.

Daly, K. C., S. Chandra, M. L. Durtschi and B. H. Smith. 2001. The generalization of an olfactory-based conditioned response reveals unique but overlapping odour representations in the moth *Manduca sexta*. *J Exp Biol* 204: 3085–95.

Daly, K. C., T. A. Christensen, H. Lei, B. H. Smith and J. G. Hildebrand. 2004. Learning modulates the ensemble representations for odors in primary olfactory networks. *Proc Natl Acad Sci U S A* 101: 10476–81.

Davis, E. E. 1984. Regulation of sensitivity in the peripheral chemoreceptor systems for host-seeking behaviour by a haemolymph-borne factor in *Aedes aegypti*. *J Insect Physiol* 30: 179–83.

Davis, E. E. 1986. Peripheral chemoreceptors and regulation of insect behaviour. In *Mechanisms in Insect Olfaction*, T. L. Payne, M. C. Birch and C. E. J. Kennedy (eds.). Oxford, Clarendon Press: 243–51.

de Bruyne, M., P. J. Clyne and J. R. Carlson. 1999. Odor coding in a model olfactory organ: The *Drosophila* maxillary palp. *J Neurosci* 19: 4520–32.

de Bruyne, M., K. Foster and J. R. Carlson. 2001. Odor coding in the *Drosophila* antenna. *Neuron* 30: 537–52.

De Jong, R. and J. H. Visser. 1988. Specificity-related suppression of responses to binary mixtures in olfactory receptors of the Colorado potato beetle. *Brain Res* 447: 18–24.

Deisig, N., J. Kropf, S. Vitecek, D. Pevergne, A. Rouyar, J. C. Sandoz, P. Lucas, C. Gadenne, S. Anton and R. Barrozo. 2012. Differential interactions of sex pheromone and plant odour in the olfactory pathway of a male moth. *PLoS One* 7: e33159.

Dekker, T., M. Geier and R. T. Carde. 2005. Carbon dioxide instantly sensitizes female yellow fever mosquitoes to human skin odours. *J Exp Biol* 208: 2963–72.

Deng, J.-Y., H. Wei, Y.-P. Huang and J.-W. Du. 2004. Enhancement of attraction to sex phero-mones of *Spodoptera exigua* by volatile compounds produced by host plants. *J Chem Ecol* 30: 2037–45.

Dubin, A. E. and G. L. Harris. 1997. Voltage-activated and odor-modulated conductances in olfactory neurons of *Drosophila melanogaster*. *J Neurobiol* 32: 123–37.

Duchamp-Viret, P., A. Duchamp and M. A. Chaput. 2003. Single olfactory sensory neurons simultaneously integrate the components of an odour mixture. *Eur J Neurosci* 18: 2690–6.

Dufour, M. C. and C. Gadenne. 2006. Adult neurogenesis in a moth brain. *J Comp Neurol* 495: 635–43.

Evans, W. and R. Gooding. 2002. Turbulent plumes of heat, moist heat, and carbon diox-ide elicit upwind anemotaxis in tsetse flies *Glossina morsitans morsitans* Westwood (Diptera: Glossinidae). *Can J Zool* 80: 1149–55.

Faucher, C., M. Forstreuter, M. Hilker and M. de Bruyne. 2006. Behavioral responses of *Drosophila* to biogenic levels of carbon dioxide depend on life-stage, sex and olfactory context. *J Exp Biol* 209: 2739–48.

Fox, A. N., R. J. Pitts, H. M. Robertson, J. R. Carlson and L. J. Zwiebel. 2001. Candidate odorant receptors from the malaria vector mosquito *Anopheles gambiae* and evidence of down-regulation in response to blood feeding. *Proc Natl Acad Sci U S A* 98: 14693–7.

Gadenne, C., M. C. Dufour and S. Anton. 2001. Transient post-mating inhibition of behav-ioural and central nervous responses to sex pheromone in an insect. *Proc R Soc Lond B Bio* 268: 1631–35.

Galizia, C. and W. Rössler. 2010. Parallel olfactory systems in insects: Anatomy and function. *Annu Rev Entomol* 55: 399–420.

Galizia, C. G., S. Sachse and H. Mustaparta. 2000. Calcium responses to pheromones and plant odours in the antennal lobe of the male and female moth *Heliothis virescens*. *J Comp Physiol A* 186: 1049–63.

Gaudry, Q., K. I. Nagel and R. I. Wilson. 2012. Smelling on the fly: Sensory cues and strate-gies for olfactory navigation in *Drosophila*. *Curr Opin Neurobiol* 22: 216–22.

Getz, W. M. and R. P. Akers. 1997. Response of American cockroach (*Periplaneta americana*) olfactory receptors to selected alcohol odorants and their binary combinations. *J Comp Physiol* 180: 701–9.

Giurfa, M. 2013. Cognition with few neurons: Higher-order learning in insects. *Trends Neurosci* 36: 285–94.

Goyret, J., P. M. Markwell and R. A. Raguso. 2008. Context- and scale-dependent effects of floral CO_2 on nectar foraging by *Manduca sexta*. *Proc Natl Acad Sci U S A* 105: 4565–70.

Gu, Y., P. Lucas and J. P. Rospars. 2009. Computational model of the insect pheromone trans-duction cascade. *PLoS Comput Biol* 5: e1000321.

Guerenstein, P. G., T. A. Christensen and J. G. Hildebrand. 2004. Sensory processing of ambient CO_2 information in the brain of the moth *Manduca sexta*. *J Comp Physiol A Neuroethol Sens Neural Behav Physiol* 190: 707–25.

Guerenstein, P. G. and J. G. Hildebrand. 2008. Roles and effects of environmental carbon dioxide in insect life. *Annu Rev Entomol* 53: 161–78.

Guerrieri, F., C. Gemeno, C. Monsempes, S. Anton, E. Jacquin-Joly, P. Lucas and J. M. Devaud. 2012. Experience-dependent modulation of antennal sensitivity and input to antennal lobes in male moths (*Spodoptera littoralis*) pre-exposed to sex pheromone. *J Exp Biol* 215: 2334–41.

Hallem, E. A. and J. R. Carlson. 2006. Coding of odors by a receptor repertoire. *Cell* 125: 143–60.

Hallem, E. A., M. G. Ho and J. R. Carlson. 2004. The molecular basis of odor coding in the *Drosophila* antenna. *Cell* 117: 965–79.

Hansson, B. S. and S. Anton. 2000. Function and morphology of the antennal lobe: New developments. *Annu Rev Entomol* 45: 203–31.

Hansson, B. S., H. Ljunberg, E. Hallberg and C. Lofstedt. 1992. Functional specialization of olfactory glomeruli in a moth. *Science* 256: 1313–15.

Hansson, B. S., J. N. C. Van der Pers and J. Löfqvist. 1989. Comparison of male and female olfactory cell response to pheromone compounds and plant volatiles in the turnip moth, *Agrotis segetum*. *Physiol Entomol* 14: 147–55.

Heisenberg, M. 2003. Mushroom body memoir: From maps to models. *Nat Rev Neurosci* 4: 266–75.

Hildebrand, J. G. 1995. Analysis of chemical signals by nervous systems. *Proc Natl Acad Sci U S A* 92: 67–74.

Hildebrand, J. G. 1996. Olfactory control of behavior in moths: Central processing of odor information and the functional significance of olfactory glomeruli. *J Comp Physiol [A]* 178: 5–19.

Hillier, N. K. and N. J. Vickers. 2011. Mixture interactions in moth olfactory physiology: examining the effects of odorant mixture, concentration, distal stimulation, and antennal nerve transection on sensillar responses. *Chem Senses* 36: 93–108.

Homberg, U., T. A. Christensen and J. G. Hildebrand. 1989. Structure and function of the deutocerebrum in insects. *Annu Rev Entomol* 34: 477–501.

Homberg, U., S. G. Hoskins and J. G. Hildebrand. 1995. Distribution of acetylcholinesterase activity in the deutocerebrum of the sphinx moth *Manduca sexta*. *Cell Tissue Res* 279: 249–59.

Homberg, U., R. Montague and J. Hildebrand. 1988. Anatomy of antenno-cerebral pathways in the brain of the sphinx moth *Manduca sexta*. *Cell Tissue Res* 254: 255–81.

Hosler, J. S. and B. H. Smith. 2000. Blocking and the detection of odor components in blends. *J Exp Biol* 203: 2797–806.

Huetteroth, W. and J. Schachtner. 2005. Standard three-dimensional glomeruli of the *Manduca sexta* antennal lobe: A tool to study both developmental and adult neuronal plasticity. *Cell Tissue Res* 319: 513–24.

Ignell, R., C. M. Root, R. T. Birse, J. W. Wang, D. R. Nassel and A. M. Winther. 2009. Presynaptic peptidergic modulation of olfactory receptor neurons in *Drosophila*. *Proc Natl Acad Sci U S A* 106: 13070–5.

Ito, I., R. C. Ong, B. Raman and M. Stopfer. 2008. Sparse odor representation and olfactory learning. *Nat Neurosci* 11: 1177–84.

Iwano, M. and R. Kanzaki. 2005. Immunocytochemical identification of neuroactive substances in the antennal lobe of the male silkworm moth *Bombyx mori*. *Zoolog Sci* 22: 199–211.

Jarriault, D., R. B. Barrozo, C. J. de Carvalho Pinto, B. Greiner, M. C. Dufour, I. Masante-Roca, J. B. Gramsbergen, S. Anton and C. Gadenne. 2009a. Age-dependent plasticity of sex pheromone response in the moth, *Agrotis ipsilon*: Combined effects of octopamine and juvenile hormone. *Horm Behav* 56: 185–91.

Jarriault, D., C. Gadenne, J. P. Rospars and S. Anton. 2009b. Quantitative analysis of sex-pheromone coding in the antennal lobe of the moth *Agrotis ipsilon*: A tool to study network plasticity. *J Exp Biol* 212: 1191–201.

Jefferis, G. S., E. C. Marin, R. J. Watts and L. Luo. 2002. Development of neuronal connectivity in *Drosophila* antennal lobes and mushroom bodies. *Curr Opin Neurobiol* 12: 80–6.

Jefferis, G. S., C. J. Potter, A. M. Chan, E. C. Marin, T. Rohlfing, C. R. Maurer, Jr. and L. Luo. 2007. Comprehensive maps of *Drosophila* higher olfactory centers: Spatially segregated fruit and pheromone representation. *Cell* 128: 1187–203.

Jonker, M. J., C. Svendsen, J. J. Bedaux, M. Bongers and J. E. Kammenga. 2005. Significance testing of synergistic/antagonistic, dose level-dependent, or dose ratio-dependent effects in mixture dose-response analysis. *Environ Toxicol Chem* 24: 2701–13.

Kaissling, K. E. 2001. Olfactory perireceptor and receptor events in moths: A kinetic model. *Chem Senses* 26: 125–50.

Kaissling, K. E., L. Z. Meng and H.-J. Bestmann. 1989. Responses of bombykol receptor cells to *(Z,E)*-4,6-hexadecadiene and linalool. *J Comp Physiol A* 165: 147–54.

Kanzaki, R., E. Arbas and J. Hildebrand. 1991. Physiology and morphology of protocerebral olfactory neurons in the male moth *Manduca sexta*. *J Comp Physiol A* 168: 281–98.

Kanzaki, R., E. A. Arbas, N. J. Strausfeld and J. G. Hildebrand. 1989. Physiology and morphology of projection neurons in the antennal lobe of the male moth *Manduca sexta*. *J Comp Physiol A* 165: 427–53.

Kanzaki, R., K. Soo, Y. Seki and S. Wada. 2003. Projections to higher olfactory centers from subdivisions of the antennal lobe macroglomerular complex of the male silkmoth. *Chem Senses* 28: 113–30.

Kazawa, T., S. Namiki, R. Fukushima, M. Terada, K. Soo and R. Kanzaki. 2009. Constancy and variability of glomerular organization in the antennal lobe of the silkmoth. *Cell Tissue Res* 336: 119–36.

Kent, K., I. Harrow, P. Quartararo and J. Hildebrand. 1986. An accessory olfactory pathway in Lepidoptera: The labial pit organ and its central projections in *Manduca sexta* and certain other sphinx moths and silk moths. *Cell Tissue Res* 245: 237–45.

King, J. R., T. A. Christensen and J. G. Hildebrand. 2000. Response characteristics of an identified, sexually dimorphic olfactory glomerulus. *J Neurosci* 20: 2391–99.

Koontz, M. A. and D. Schneider. 1987. Sexual dimorphism in neuronal projections from the antennae of silk moths (*Bombyx mori, Antheraea polyphemus*) and the gypsy moth (*Lymantria dispar*). *Cell Tissue Res.* 249: 39–50.

Krashes, M. J., A. C. Keene, B. Leung, J. D. Armstrong and S. Waddell. 2007. Sequential use of mushroom body neuron subsets during drosophila odor memory processing. *Neuron* 53: 103–15.

Kuwahara, Y. 2004. Chemical ecology of astigmatid mites. In *Advances in Chemical Ecology*, R. T. Cardé and J. G. Millar (eds.). Cambridge, New York, Cambridge University Press: 76–109.

Laing, D. G., H. Panhuber, M. E. Willcox and E. A. Pittman. 1984. Quality and intensity of binary odor mixtures. *Physiol Behav* 33: 309–19.

Laissue, P. P., C. Reiter, P. R. Hiesinger, S. Halter, K. F. Fischbach and R. F. Stocker. 1999. Three-dimensional reconstruction of the antennal lobe in *Drosophila melanogaster*. *J Comp Neurol* 405: 543–52.

Laissue, P. P. and L. B. Vosshall. 2008. The olfactory sensory map in *Drosophila*. *Adv Exp Med Biol* 628: 102–14.

Laurent, G. and H. Davidowitz. 1994. Encoding of olfactory information with oscillating neural assemblies. *Science* 265: 1872–5.

Lei, H., T. A. Christensen and J. G. Hildebrand. 2002. Local inhibition modulates odor-evoked synchronization of glomerulus-specific output neurons. *Nat Neurosci* 5: 557–65.

Lei, H., J. A. Riffell, S. L. Gage and J. G. Hildebrand. 2009. Contrast enhancement of stimulus intermittency in a primary olfactory network and its behavioral significance. *J Biol* 8: 21.

Light, D. M., R. A. Flath, R. G. Buttery, F. G. Zalom, R. E. Rice, J. C. Dickens and E. B. Jang. 1993. Host-plant green-leaf volatiles synergize the synthetic sex pheromones of the corn earworm and codling moth (Lepidoptera). *Chemoecology* 4: 145–52.

MacLeod, K., A. Backer and G. Laurent. 1998. Who reads temporal information contained across synchronized and oscillatory spike trains? *Nature* 395: 693–8.

MacLeod, K. and G. Laurent. 1996. Distinct mechanisms for synchronization and temporal patterning of odor-encoding neural assemblies. *Science* 274: 976–9.

Mafra-Neto, A. and R. T. Cardé. 1994. Fine-scale structure of pheromone plumes modulates upwind orientation of flying moths. *Nature* 369: 142–4.

Marin, E. C., G. S. Jefferis, T. Komiyama, H. Zhu and L. Luo. 2002. Representation of the glomerular olfactory map in the *Drosophila* brain. *Cell* 109: 243–55.

Martin, J. P., A. Beyerlein, A. M. Dacks, C. E. Reisenman, J. A. Riffell, H. Lei and J. G. Hildebrand. 2011. The neurobiology of insect olfaction: Sensory processing in a comparative context. *Prog Neurobiol* 95: 427–47.

Masante-Roca, I., S. Anton, L. Delbac, M. C. Dufour and C. Gadenne. 2007. Attraction of the grapevine moth to host and non-host plant parts in the wind tunnel: Effects of plant phenology, sex, and mating status. *Entomol Exp Appl* 122: 239–94.

Meagher, J. R. L. 2001. Trapping fall armyworm (Lepidoptera: Noctuidae) adults in traps baited with pheromone and a synthetic floral compound. *Fla Entomol* 84: 288–92.

Menzel, R. and M. Giurfa. 2006. Dimensions of cognition in an insect, the honeybee. *Behav Cogn Neurosci Rev* 5: 24–40.

Menzel, R., G. Leboulle and D. Eisenhardt. 2006. Small brains, bright minds. *Cell* 124: 237–9.

Meyer, A. and C. G. Galizia. 2012. Elemental and configural olfactory coding by antennal lobe neurons of the honeybee (*Apis mellifera*). *J Comp Physiol A Neuroethol Sens Neural Behav Physiol* 198: 159–71.

Miklas, N., A. Cokl, M. Renou and M. Virant-Doberlet. 2003a. Variability of vibratory signals and mate choice selectivity in the southern green stink bug. *Behav Proc.* 61: 131–42.

Miklas, N., T. Lasnier and M. Renou. 2003b. Male bugs modulate pheromone emission in response to vibratory signals from conspecifics. *J Chem Ecol* 29: 561–74.

Minoli, S., I. Kauer, V. Colson, V. Party, M. Renou, P. Anderson, C. Gadenne, F. Marion-Poll and S. Anton. 2012. Brief exposure to sensory cues elicits stimulus-nonspecific general sensitization in an insect. *PLoS One* 7: e34141.

Murlis, J. and C. D. Jones. 1981. Fine-scale structure of odour plumes in relation to distant pheromone and other attractant sources. *Physiol Entomol* 6: 71–86.

Mwilaria, E. K., C. Ghatak and K. C. Daly. 2008. Disruption of GABAA in the insect antennal lobe generally increases odor detection and discrimination thresholds. *Chem Senses* 33: 267–81.

Najar-Rodriguez, A. J., C. G. Galizia, J. Stierle and S. Dorn. 2010. Behavioral and neurophysiological responses of an insect to changing ratios of constituents in host plant-derived volatile mixtures. *J Exp Biol* 213: 3388–97.

Namiki, S., S. Iwabuchi and R. Kanzaki. 2008. Representation of a mixture of pheromone and host plant odor by antennal lobe projection neurons of the silkmoth *Bombyx mori*. *J Comp Physiol A* 194: 501–15.

Namiki, S. and R. Kanzaki. 2011a. Heterogeneity in dendritic morphology of moth antennal lobe projection neurons. *J Comp Neurol* 519: 3367–86.

Namiki, S. and R. Kanzaki. 2011b. Offset response of the olfactory projection neurons in the moth antennal lobe. *Biosystems* 103: 348–54.

Namiki, S., M. Takaguchi, Y. Seki, T. Kazawa, R. Fukushima, C. Iwatsuki and R. Kanzaki. 2013. Concentric zones for pheromone components in the mushroom body calyx of the moth brain. *J Comp Neurol* 521: 1073–92.

Ochieng, S. A., K. C. Park and T. C. Baker. 2002. Host plant volatiles synergise responses of sex pheromone-specific olfactory receptor neurons in male *Helicoverpa zea*. *J Comp Physiol A* 188: 325–33.

Olsen, S. R. and R. I. Wilson. 2008. Lateral presynaptic inhibition mediates gain control in an olfactory circuit. *Nature* 452: 956–60.

Olsson, P. O., O. Anderbrandt and C. Löfstedt. 2005. Experience influences oviposition behaviour in two pyralid moths, *Ephestia cautella* and *Plodia interpunctella*. *Anim Behav* 72: 545–51.

Party, V., C. Hanot, I. Said, D. Rochat and M. Renou. 2009. Plant terpenes affect intensity and temporal parameters of pheromone detection in a moth. *Chem Senses* 34: 763–74.

Perez-Orive, J., O. Mazor, G. C. Turner, S. Cassenaer, R. I. Wilson and G. Laurent. 2002. Oscillations and sparsening of odor representations in the mushroom body. *Science* 297: 359–65.

Pregitzer, P., M. Schubert, H. Breer, B. S. Hansson, S. Sachse and J. Krieger. 2012. Plant odorants interfere with detection of sex pheromone signals by male *Heliothis virescens*. *Front Cell Neurosci* 6: 42.

Qiu, Y. T., J. J. van Loon, W. Takken, J. Meijerink and H. M. Smid. 2006. Olfactory coding in antennal neurons of the malaria mosquito, *Anopheles gambiae*. *Chem Senses* 31: 845–63.

Raguso, R. A. and M. A. Willis. 2002. Synergy between visual and olfactory cues in nectar feeding by naive hawkmoths, *Manduca sexta*. *Anim Behav* 64: 685–95.

Reisenman, C. E., T. A. Christensen and J. G. Hildebrand. 2005. Chemosensory selectivity of output neurons innervating an identified, sexually isomorphic olfactory glomerulus. *J Neurosci* 25: 8017–26.

Reisenman, C. E., A. M. Dacks and J. G. Hildebrand. 2011. Local interneuron diversity in the primary olfactory center of the moth *Manduca sexta*. *J Comp Physiol A Neuroethol Sens Neural Behav Physiol* 197: 653–65.

Riffell, J., H. Lei, T. A. Christensen and J. Hildebrand. 2009a. Characterization and coding of behaviorally significant odor mixtures. *Curr Biol* 19: 335–40.

Riffell, J., H. Lei and J. Hildebrand. 2009b. Inaugural Article: Neural correlates of behavior in the moth *Manduca sexta* in response to complex odors. *Proc Natl Acad Sci U S A* 106: 19219–26.

Root, C. M., K. Masuyama, D. S. Green, L. E. Enell, D. R. Nassel, C. H. Lee and J. W. Wang. 2008. A presynaptic gain control mechanism fine-tunes olfactory behavior. *Neuron* 59: 311–21.

Rospars, J. P. 1988. Structure and development of the insect antennodeutocerebral system. *Int J Insect Morphol Embryol* 17: 243–94.

Rospars, J. P. and J. G. Hildebrand. 1992. Anatomical identification of glomeruli in the antennal lobe of the male sphinx moth *Manduca sexta*. *Cell Tissue Res* 270: 205–27.

Rospars, J. P. and J. G. Hildebrand. 2000. Sexually dimorphic and isomorphic glomeruli in the antennal lobes of the sphinx moth *Manduca sexta*. *Chem Senses* 25: 119–29.

Rospars, J., P. Lansky, M. Chaput and P. Duchamp-Viret. 2008. Competitive and noncompetitive odorant interactions in the early neural coding of odorant mixtures. *J Neurosci* 28: 2659–66.

Rössler, W., L. P. Tolbert and J. G. Hildebrand. 1998. Early formation of sexually dimorphic glomeruli in the developing olfactory lobe of the brain of the moth *Manduca sexta*. *J Comp Neurol* 396: 415–28.

Rouyar, A., V. Party, J. Prešern, A. Blejec and M. Renou. 2011. A general odorant background affects the coding of pheromone stimulus intermittency in specialist olfactory receptor neurones. *PLoS One* 6: e26443.

Sandoz, J. C., M. H. Pham-Delegue, M. Renou and L. J. Wadhams. 2001. Asymmetrical generalisation between pheromonal and floral odours in appetitive olfactory conditioning of the honey bee (*Apis mellifera L.*). *J Comp Physiol A* 187: 559–68.

Sanes, J. R. and J. G. Hildebrand. 1976. Origin and morphogenesis of sensory neurons in an insect antenna. *Dev Biol* 51: 300–19.

Saveer, A. M., S. H. Kromann, G. Birgersson, M. Bengtsson, T. Lindblom, A. Balkenius, B. S. Hansson, P. Witzgall, P. G. Becher and R. Ignell. 2012. Floral to green: Mating switches moth olfactory coding and preference. *Proc R Soc B* 279: 2314–22.

Schmidt-Büsser, D., M. von Arx and P. M. Guerin. 2009. Host plant volatiles serve to increase the response of male European grape berry moths, *Eupoecilia ambiguella,* to their sex pheromone. *J Comp Physiol A* 195: 853–64.

Seki, Y., H. Aonuma and R. Kanzaki. 2005. Pheromone processing center in the protocerebrum of *Bombyx mori* revealed by nitric oxide-induced anti-cGMP immunocytochemistry. *J Comp Neurol* 481: 340–51.

Seki, Y. and R. Kanzaki. 2008. Comprehensive morphological identification and GABA immunocytochemistry of antennal lobe local interneurons in *Bombyx mori*. *J Comp Neurol* 506: 93–107.

Shields, V. D. C. and J. G. Hildebrand. 2001. Responses of a population of antennal olfactory receptor cells in the female moth *Manduca sexta* to plant-associated volatile organic compounds. *J Comp Physiol* 186: 1135–51.

Siju, K. P., S. R. Hill, B. S. Hansson and R. Ignell. 2010. Influence of blood meal on the responsiveness of olfactory receptor neurons in antennal sensilla trichodea of the yellow fever mosquito, *Aedes aegypti*. *J Insect Physiol* 56: 659–65.

Silbering, A. F., R. Rytz, Y. Grosjean, L. Abuin, P. Ramdya, G. S. Jefferis and R. Benton. 2011. Complementary function and integrated wiring of the evolutionarily distinct *Drosophila* olfactory subsystems. *J Neurosci* 31: 13357–75.

Spathe, A., A. Reinecke, S. B. Olsson, S. Kesavan, M. Knaden and B. S. Hansson. 2012. Plant species- and status-specific odorant blends guide oviposition choice in the moth *Manduca sexta*. *Chem Senses* 38: 147–59.

Stokl, J., J. Brodmann, A. Dafni, M. Ayasse and B. S. Hansson. 2011. Smells like aphids: Orchid flowers mimic aphid alarm pheromones to attract hoverflies for pollination. *Proc Biol Sci* 278: 1216–22.

Stokl, J., A. Strutz, A. Dafni, A. Svatos, J. Doubsky, M. Knaden, S. Sachse, B. S. Hansson and M. C. Stensmyr. 2010. A deceptive pollination system targeting drosophilids through olfactory mimicry of yeast. *Curr Biol* 20: 1846–52.

Stowe, M. K., J. H. Tumlinson and R. R. Heath. 1987. Chemical mimicry: Bolas spiders emit components of moth prey species sex pheromones. *Science* 236: 964–7.

Strausfeld, N. J., I. Sinakevitch, S. M. Brown and S. M. Farris. 2009. Ground plan of the insect mushroom body: Functional and evolutionary implications. *J Comp Neurol* 513: 265–91.

Su, C. Y., K. Menuz, J. Reisert and J. R. Carlson. 2012. Non-synaptic inhibition between grouped neurons in an olfactory circuit. *Nature* 492: 66–71.

Svensson, G. P., C. Löfstedt and N. Skals. 2004. The odour makes the difference: Male moths attracted by sex pheromones ignore the threat of predatory bats. *Oikos* 104: 91–7.

Svensson, G. P., C. Löfstedt and N. Skals. 2007. Listening in pheromone plumes: Disruption of olfactory-guided mate attraction in a moth by a bat-like ultrasound. *J Insect Sci* 7: 59.

Syed, Z. and W. S. Leal. 2007. Maxillary palps are broad spectrum odorant detectors in *Culex quinquefasciatus*. *Chem Senses* 32: 727–38.

Syed, Z. and W. S. Leal. 2008. Mosquitoes smell and avoid the insect repellent DEET. *Proc Natl Acad Sci U S A* 105: 13598–603.

Szyszka, P., M. Ditzen, A. Galkin, C. G. Galizia and R. Menzel. 2005. Sparsening and temporal sharpening of olfactory representations in the honeybee mushroom bodies. *J Neurophysiol* 94: 3303–13.

Szyszka, P., J. S. Stierle, S. Biergans and C. G. Galizia. 2012. The speed of smell: Odor-object segregation within milliseconds. *PLoS One* 7: e36096.

Takken, W. and B. G. J. Knols. 1999. Odor-mediated behavior of Afrotopical malaria mosquitoes. *Annu Rev Entomol* 44: 131–57.

Thom, C., P. G. Guerenstein, W. L. Mechaber and J. G. Hildebrand. 2004. Floral CO_2 reveals flower profitability to moths. *J Chem Ecol* 30: 1285–88.

Tootoonian, S. and G. Laurent. 2010. Electric times in olfaction. *Neuron* 67: 903–5.

Tripathy, S. J., O. J. Peters, E. M. Staudacher, F. R. Kalwar, M. N. Hatfield and K. C. Daly. 2010. Odors pulsed at wing beat frequencies are tracked by primary olfactory networks and enhance odor detection. *Front Cell Neurosci* 4: 1.

Trona, F., G. Anfora, M. Bengtsson, P. Witzgall and R. Ignell. 2010. Coding and interaction of sex pheromone and plant volatile signals in the antennal lobe of the codling moth *Cydia pomonella*. *J Exp Biol* 213: 4291–303.

Turner, G. C., M. Bazhenov and G. Laurent. 2008. Olfactory representations by *Drosophila* mushroom body neurons. *J Neurophysiol* 99: 734–46.

Van der Pers, J., G. Thomas and C. Den Otter. 1980. Interactions between plant odours and pheromone reception in small ermine moths (Lepidoptera: Yponomeutidae). *Chem Senses* 5: 367–71.

Vermeulen, A. and J.-P. Rospars. 2004. Why are insect olfactory neurons grouped into sensilla? The teachings of a model investigating the effects of the electrical interaction between neurons on the transepithelial potential and the neuronal transmembrane. *Eur Biophys J* 33: 633–43.

Vickers, N. J., T. A. Christensen and J. G. Hildebrand. 1998. Combinatorial odor discrimination in the brain: Attractive and antagonist odor blends are represented in distinct combinations of uniquely identifable glomeruli. *J Comp Neurol* 400: 35–56.

Vogt, R. G. 2005. Molecular basis of pheromone detection in insects. In *Comprehensive Insect Physiology Biochemistry Pharmacology and Molecular Biology. Volume 3: Endocrinology*, L. I. Gilbert, K. Iatro and S. Gill (eds.). London, Elsevier: 753–804.

Waldrop, B. and J. G. Hildebrand. 1989. Physiology and pharmacology of acetylcholinergic responses of interneurons in the antennal lobes of the moth *Manduca sexta*. *J Comp Physiol A Sens Neural Behav Physiol* 164: 433–41.

Wehr, M. and G. Laurent. 1996. Odour encoding by temporal sequences of firing in oscillating neural assemblies. *Nature* 384: 162–6.

Wetzel, C. H., D. Brunert and H. Hatt. 2005. Cellular mechanisms of olfactory signal transduction. *Chem Senses* 30 Suppl 1: i321–2.

Wicher, D., R. Schafer, R. Bauernfeind, M. C. Stensmyr, R. Heller, S. H. Heinemann and B. S. Hansson. 2008. *Drosophila* odorant receptors are both ligand-gated and cyclic-nucleotide-activated cation channels. *Nature* 452: 1007–11.

Willis, M. A. and T. C. Baker. 1984. Effects of intermittent and continuous pheromone stimulation on the flight behaviour of the oriental fruit moth, *Grapholita molesta*. *Physiol Entomol* 9: 341–58.

Wilson, R. I. and Z. F. Mainen. 2006. Early events in olfactory processing. *Annu Rev Neurosci* 29: 163–201.

Wilson, R. I., G. C. Turner and G. Laurent. 2004. Transformation of olfactory representations in the *Drosophila* antennal lobe. *Science* 303: 366–70.

Wright, G. A., S. M. Kottcamp and M. G. Thomson. 2008. Generalization mediates sensitivity to complex odor features in the honeybee. *PLoS One* 3: e1704.

Yang, Z. H., M. Bengtsson and P. Witzgall. 2004. Host plant volatiles synergize response to sex pheromone in codling moth, *Cydia pomonella*. *J Chem Ecol* 30: 619–29.

Zack-Strausfeld, C. and K.-E. Kaissling. 1986. Localized adaptation processes in olfactory sensilla of Saturniid moths. *Chem Senses* 11: 499–512.

Zöllner, G. E., S. J. Torr, C. Ammann and F. X. Meixner. 2004. Dispersion of carbon dioxide plumes in African woodland: Implications for host-finding by tsetse flies. *Physiol Entomol* 29: 381–94.

Zufall, F. and T. Leinders-Zufall. 1997. Identification of a long-lasting form of odor adaptation that depends on the carbon monoxide/cGMP second-messenger system. *J Neurosci* 17: 2703–12.

Zufall, F. and T. Leinders-Zufall. 2000. The cellular and molecular basis of odor adaptation. *Chem Senses* 25: 473–81.

3 First Investigation of the Semiochemistry of South African Dung Beetle Species

Barend (Ben) Victor Burger

CONTENTS

3.1   Introduction 57
3.2   Chemical Ecology of the Genus *Kheper* 59
3.3   Collection of the Abdominal Secretion 62
3.4   Sample Preparation and Analysis 65
3.5   Optimization of GC-FID/EAD Instrumentation 68
3.6   Chemical Characterization of Abdominal Secretions of the Male
      Kheper Species 72
      3.6.1   *K. lamarcki* 72
      3.6.2   *K. nigroaeneus* 73
      3.6.3   *K. subaeneus* 75
      3.6.4   *K. bonellii* 78
3.7   Long-Chain Constituents of the Abdominal Secretions 82
3.8   Composition of the Pheromone-Disseminating Carrier Material 83
3.9   Intergeneric and Interspecific Chemical Signaling in Dung Beetles 86
      3.9.1   Intergeneric Communication 86
      3.9.2   Interspecific Communication 88
3.10 Peculiar Behavior in *P. femoralis* 89
3.11 Defensive Mechanisms in *Oniticellus egregius* 93
3.12 Conclusions, Perspectives, and Prospects 94
Acknowledgments 95
References 95

3.1   INTRODUCTION

The inhabitants of cities and large towns are largely unaware of the constant battle that rural communities have to wage against flies. In rural areas, on the other hand, inhabitants do not always appreciate the crucial role that dung beetles play in controlling dung-breeding fly populations. Not only do dung beetles play an important role in the destruction of the habitat of many dung-breeding flies (Heinrich and Bartholomew 1979), but by burying and dispersing dung, the coprophagous fauna associated with mammals is also responsible for returning a large proportion of plant nutrients to the soil (e.g., Bornemissza and Williams 1970).

Serious problems have been experienced in the cattle farming areas of Australia where cattle were introduced without the associated insect fauna. The dung pats left by cattle become the habitat of dung-breeding flies, some of which are blood-feeding pests and carriers of serious diseases (Hughes 1970). This resulted in the uncontrolled increase of fly populations and the deterioration of pastures in parts of that continent. It has been estimated that cow pats are responsible for the reduction of pasture, albeit only temporarily, by about 20% per animal per year (Waterhouse 1974). During the 1950s, George Bornemissza hypothesized that the introduction of foreign dung beetle species that are able to remove and bury cattle dung would aid not only Australia's soil fertility by recycling the dung nutrients back into the ground, but would also reduce the number of pestilent flies and parasitic worms that use the dung pats as a breeding resource. In 1965, this idea culminated in the establishment of the Australian Dung Beetle Project, and eventually 43 dung beetle species were imported from Africa and other continents in order to combat the fly problem (Wikipedia August 2012). The introduction of exotic dung beetles and their subsequent establishment in Australia was a highly successful venture and largely solved the fly problem in many parts of the continent.

The emergence of problems similar to those in Australia is now also being observed in other countries. The destruction of the habitat of large mammals by urbanization and modern farming practices invariably results in dwindling numbers of dung beetles and an increase in numbers of flies during the warm summer months. The importance of this process is vividly illustrated in rural areas in Africa where visitors to game reserves with normal herbivore populations are rarely bothered by flies during the summer when dung beetle activity reaches its peak, whereas in small settlements a few kilometers outside these reserves the indigenous people are plagued by swarms of flies.

Extensive research has been devoted to the ecology, ethology, and evolution of dung beetles, notably by Bornemissza, Halffter, and their coworkers. Recently, Simmons and Ridsdill-Smith (2011) edited a wide-ranging review of the literature on the ecology and evolution of dung beetles. Although it is estimated that there are more than 4000 dung beetle species in Africa alone, the olfactory ecology of dung beetles remains a largely unexplored yet potentially very fertile research field. Little research has been carried out on the chemical aspects of the ecology of dung beetles, and practically no information was available on the existence, the modes of operation, and the chemical structures of sex attractants of dung beetles before the work discussed in this chapter commenced in 1978.

On the basis of their nesting behavior, dung beetles of the subfamily Scarabaeinae can be divided into three groups: The paracoprids construct their nests under a dung pat by excavating tunnels in which the dung is packed; the endocoprids excavate a chamber in the dung pat itself, forming brood balls within this chamber; and the telecoprids detach a portion of dung from the pat, rolling it some distance from the dung source before burying it. The majority of dung beetles species in southern

Africa are mainly crepuscular paracoprids (87%) (Halffter and Matthews 1966). The diurnal telecoprids are also numerous, but are represented by fewer species (12%) (Ferreira 1972).

Larvae and adults of the majority of species of the subfamily Scarabaeinae are coprophagous, and are morphologically adapted to feed on vertebrate excrement (Halffter and Matthews 1966). Adult dung beetles ingest only the liquid or colloidal constituents of the dung by squeezing portions of the moist dung between highly specialized membranous mandibles and ingesting the expressed juice. There are differences in the courtship, mating, and brooding behavior of the Scarabaeinae species. Telecoprids prepare a brood ball from fresh dung, roll the ball to an apparently carefully selected spot at a distance from the dung pat or dung midden, where the female deposits an egg in a chamber constructed at the top of the dung ball. The larva feeds on whole dung particles with the aid of its chewing mouthparts until eventually it pupates (Waterhouse 1974). The next generation, as well as the adult beetles that have overwintered in the soil, emerge from the soil after the first summer rains when temperatures are high enough for habitation and when the soil is moist enough for the beetles to break free from the brood chamber or from the hard soil.

Dung beetles are attracted to dung on which they feed, and after a feeding period of a few weeks the beetles that have overwintered in the soil and those that breed in their first season are ready to start breeding. Dung beetles utilize volatile compounds emitted by dung to locate a fresh source of a species' preferred dung type, although dung beetles are apparently not linked exclusively to one type of dung (Dormont et al. 2004, 2007).

In areas with intact ecology, dung beetles of many species arrive in the thousands at a dung pat or rhinoceros midden. There is fierce competition for fresh dung between the beetles. Even in the case of the large volumes of dung voided by, for example, rhinoceros, dung beetles manage to dehydrate the dung or to bury or disperse it within a few hours. A photograph of rhinoceros dung voided during the night in the Mkuzi Game Reserve, South Africa, is shown in Figure 3.1a. At about 10:00 the following morning during a sampling period of 20 minutes, 720 dung beetles were recorded arriving at this resource. The entire dung heap appeared to be in constant movement from the activity of thousands of beetles, resulting in the formation of an almost uniform mixture of dung and dung beetles. By that stage, it was already too late for the larger species to gather enough of the already half-dried-out dung for the formation of a brood ball. In the foreground in Figure 3.1a, the last beetles that had managed to form a dung balls can be seen rolling them away. By 15:00 in the afternoon most of the dung was buried underneath the dung or next to it, or rolled away, and only relatively dry plant material was left at this spot.

3.2   CHEMICAL ECOLOGY OF THE GENUS *KHEPER*

The ecology and ethology of some of the ball-rolling dung beetles of southern Africa were investigated in great detail by Tribe (1976). His observation that the male dung beetles of the genus *Kheper* produce a visible sex-attracting abdominal secretion created a unique opportunity to investigate an example of the secretion of a pheromone in nonvolatile carrier material. Visible secretions are produced by several insects,

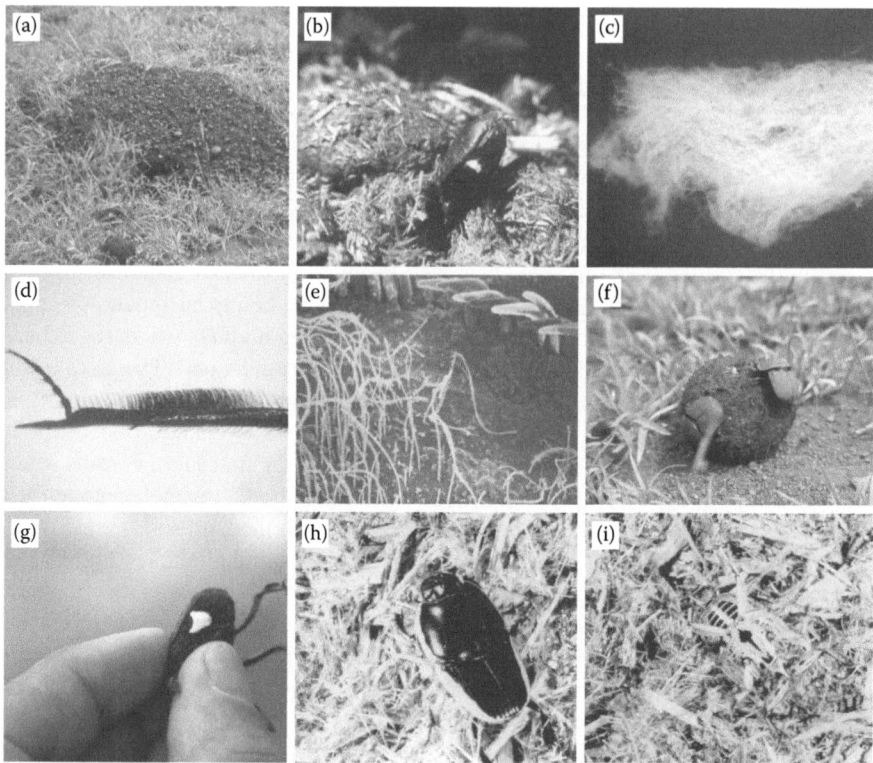

FIGURE 3.1 **(See color insert.)** (a) Dung beetles on and in rhinoceros dung that has been voided during the night. (Photograph courtesy of B. V. Burger.) (b) A *K. lamarcki* male that has lost his left hind leg in the typical secretion-producing headstand position. (Photograph courtesy of B. V. Burger.) (c) The fragile fibers of the male *K. lamarcki* secretion resemble cotton wool. (Photograph courtesy of B. V. Burger.) (d) A brush on the tibia of a *K. lamarcki* male. (Photograph courtesy of G. D. Tribe.) (e) The first row of comblike structures on the abdomen of a *K. lamarcki* male against which the secretion is broken up. (Photograph courtesy of G. D. Tribe.) (f) *K. lamarcki* male pushing a dung ball while the female clings to the ball. (Photograph courtesy of B. V. Burger.) (g) *K. lamarcki* male with secretion accumulated on the left side of the insect's abdomen and with the secretion-dispersion brush visible on the tibia of the intact hind leg. (Photograph courtesy of B. V. Burger.) (h) Dorsal (left) and (i) ventral (right) aspects of *O. egregious*. (Photographs courtesy of B. V. Burger.)

for example, by the butterfly larvae *Epipyrops anomala* (Marshall 1974), *Papilio demodocus* (Burger et al. 1978), and *Cryptoglossa verrucosa* (Hadley 1979), and the azalea lace bug, *Stehanitis pyrioides* (Nair and Braman 2012). However, these secretions are quite different in terms of function from those of dung beetles of the genus *Kheper*.

The genus *Kheper* is confined mainly to a belt stretching from the northern part of the Eastern Cape Province of South Africa, through KwaZulu-Natal, Botswana, Zimbabwe, northern Namibia, and into southern Angola. *Kheper bonelii* is an exception to this distribution; it occurs as an isolated population in the light sandy soil along the coast of the Western Cape Province. Only four of the six *Kheper* species are found in South Africa (Tribe 1976).

Tribe (1976) describes how, after arriving at a dung pat, the male *Kheper lamarcki* M'Leay carves out and then molds a dung ball or assumes the highest position on the ball or the dung pat. Supporting his body on the front two pairs of legs, the male raises his abdomen and lowers his head so that his body is at a relatively steep angle of about 45° to the dung (Figure 3.1b). A white flocculent substance resembling fine cotton wool (Figure 3.1c) is extruded from a sieve consisting of several hundred tiny openings in a depression on either side of the first abdominal sternite. The secretion is supplied by a large gland complex immediately underlying the depression in which the sieve is located. The hind legs are retracted simultaneously inward towards the sides of the body, and are then simultaneously rapidly extended. This movement results in the hardened secretion being broken up into tiny particles between brushes on the tibiae (Figure 3.1d) and comblike structures on the abdomen of the male (Figure 3.1e). Small puffs of white powder are dispersed into the air above the abdomen. After a short interval of 20–30 seconds, the legs are again withdrawn inward against the sides of the body, and again extended rapidly. The brushes on the tibiae are usually rudimentary in the females. As will be discussed below, these particles contain a complex mixture of more than 150 organic compounds, of which only a limited number are common to all *Kheper* species. This behavior continues until a female arrives, by which time both the hind legs and sides of the abdomen have become coated in the white powder.

According to Tribe (1976), normally, the male relinquishes his headstand position as soon as another beetle of the same species appears within range and then challenges the intruder with his forelegs raised. If the intruder is a male, he will respond in a corresponding way by raising his forelegs to meet the challenge. A fight then ensues, but invariably the beetle that retains the highest position on the pat, usually the one releasing the pheromone, is able to forcefully flip the other beetle off the pat with his forelegs and clypeus. An attracted female is challenged by the male; when her presence is detected the male raises his forelegs and advances on the female. Her reaction is a submissive lowering of the forelegs, followed by antennal contact and acceptance by the male. The male then approaches her, and after antennal contact, possibly involving the detection of a pheromone, the female is allowed to cling to the brood ball that the male then rolls rapidly away from the dung pat. If the male has not already made a dung ball before the female arrived, he then detaches a portion of dung from the pat with or without assistance by the female. The dung fragment is rounded by the female and the male proceeds to roll both the ball, and the female clinging to it, away from the dung pat until a suitable site has been found to bury the ball (Figure 3.1f). While the female clings to the dung ball the male removes the soil from below the ball, which results in the ball sinking gradually deeper into the soil. Once the ball is buried, copulation occurs below the soil. The male returns to the surface after four days while the female remains below to lay an egg in the ball and to brood the developing larva.

Although it has not been observed that male *K. subaeneus* produce a secretion from the top of a dung ball, this species follows basically the same breeding routine.

On arrival at a fresh dung source, the male generally vanishes into the dung, only to appear again after a while and start producing the abdominal secretion. After the male has accepted an arriving female or a female has accepted the male, as the case may be, the pair normally disappears into the dung and they mold a dung ball below the surface of the dung. Eventually the dung is broken free from the surrounding dung and is rolled away from the dung pat or midden. Such a smooth and uninterrupted procedure is only possible when the male arrives at the dung pat before a situation such as depicted in Figure 3.1a has developed, with thousands of dung beetles arriving at the dung pat, or when another male attracted by the abdominal secretion of the first male has not taken over the ball-molding process.

In contrast, *K. nigroaeneus* males first bury a brood ball before assuming the typical secretion-producing stance at the entrance to the burrow in which the ball has been buried. This also happens even when a male is not in possession of a dung ball (Edwards and Aschenborn 1988).

3.3 COLLECTION OF THE ABDOMINAL SECRETION

Permission was granted to us annually by the KwaZulu-Natal Parks Board for the collection of 250 male and 125 female *Kheper* dung beetles of each of the three species that are found in northern KwaZulu-Natal. Only a few rhinoceros dung middens were easily and safely accessible from the dirt roads in the Mkuzi Game Reserve and these species were therefore mostly trapped in 10 pitfall traps (Tribe 1976) spaced at 100-m intervals along the so-called Beacon Road and baited with fresh horse dung during the first week of November each year. Only 17 male *K. lamarcki* were collected for research during the first season of the project. Although these dung beetles can easily fly at speeds exceeding 30 km/h and can therefore cover 100 m in a few seconds, traps at specific positions were always less productive than others along that road. Increasingly, fewer dung beetles were trapped in traps positioned towards the southernmost part of this stretch of road where there were often only a few dung beetles per trap, whereas traps along the northern part of the road were often filled to capacity with dung beetles of many different species within a few hours. All collections were carried out in the morning between about 09:00 and 12:00. There was little large herbivore activity in this part of the reserve and little variation in the nature and density of the vegetation along this road. The difference in trap yields could therefore not be ascribed to the competition between the relatively small volumes of horse dung with which the traps were baited and the availability of large quantities of dung in the vicinity of the less productive traps. The largest numbers of dung beetles were always trapped in spots where the soil was somewhat sandier. Apparently dung beetles prefer, and have the ability to detect or otherwise select, places where it would be easier to bury the dung balls.

K. subaeneus is not found in Mkuzi Game Reserve; beetles of this species were collected in the Hluhluwe Game Reserve, which is situated about 50 km south of the Mkuzi Game Reserve. Here, practically no dung beetles of any species were caught in pitfall traps during the early years of the research project, possibly because the dung beetles were drawn to the large numbers of nearby dung middens in the accessible parts of the reserve. It is also possible that here the dung beetle population was depleted in accessible areas because large numbers of dung beetles were removed from those areas in the reserve for the Australian Dung Beetle Project. Fortunately, a few *K. subaeneus* could be collected from the scores of rhinoceros dung middens on the roads of the reserve. In later years, it was often possible to collect the allotted number of dung beetles during a single 3-km drive from the Monumental Entrance Gate of the reserve (coordinates 28:04:12.26S, 32:08:25:88E).

The collected beetles were transported in aluminum boxes containing moist soil by car from the game reserves and by air as hand luggage in the passenger cabin of a commercial airline from Durban to Cape Town. Once in Stellenbosch, the male and female beetles were kept together at subtropical temperatures in greenhouses (4 m × 3 m), the floors of which were covered with about 150 mm of moist sandy soil. Early on week days, the beetles were supplied with fresh horse dung. Within a few hours the dung beetles managed to spread out the dung in the greenhouses and after a week or two the soil was totally covered by a thick homogeneous layer of dry fibrous material. As soon as the first signs of any fungal growth appeared, the dried out material was removed and replaced with some fresh dung.

In an exploratory attempt at isolating and identifying the constituents of the abdominal secretion of male *Kheper* beetles, the abdominal glands of a male *K. lamarcki* were extracted with dichloromethane and the extracts concentrated for gas chromatographic-mass spectrometric (GC-MS) analysis. Such an isolation of trace quantities of a semiochemical mixture from a relatively large volume of glandular tissue is a time-consuming and often futile undertaking. In contrast, the chemical characterization of semiochemicals secreted in a carrier material or a matrix could be a quite simple exercise. In ideal cases it might even be possible to enrich volatile semiochemicals on sorptive traps during production of the secretion by the organism under investigation. Tribe (1976) has found that loss of a hind leg did not deter the male insect from producing secretion. If one of the hind legs of a calling beetle is amputated, the beetle normally disappears underground for a while, after which it reemerges and resumes its secretion-producing stance. The secretion that accumulates as a band on the incapacitated side of the beetle (Figure 3.1b) can then be collected with ophthalmic forceps.

The removal of a hind leg is the only possible way to prevent the dispersal of the secretion, which, if it is not brushed off, collects at that side of the abdomen as a tuft of cotton-wool-like material (Figure 3.1c). This behavior was also observed in captive *Kheper* males, but they generally did not reappear the same day, and very little secretion was eventually collected using this approach. Therefore, in the following and subsequent seasons, one hind leg was removed from all the males on arrival in Stellenbosch. About 5% died within the first week, but thereafter the surviving beetles appeared to behave normally and quickly learned how to roll dung balls with only one hind leg. In captivity, the production of the sex attractant was sporadic, and contrary to observations in nature, captive beetles sometimes produced secretion even when not in possession of fresh dung. They therefore had to be kept under constant observation. According to Tribe (1976), the beetles are at their most productive at temperatures around 29°C. In captivity, the beetles frequently went through the initial stages of the production of secretion, but they mostly stopped before a visible quantity of material had accumulated. In rare cases, the production of secretion continued for periods of up to 2 hours, resulting in the accumulation of substantial quantities of material. When disturbed in any way, for example by another beetle of the same or another species, male dung beetles immediately stopped secreting the attractant to investigate, and they even disappeared underground, sometimes together with a female. Males observed to assume the attractant-secreting posture were therefore fenced off from other beetles with aluminum strips. If after having secreted some material a beetle lowered its abdomen and prepared to go underground, it was taken firmly between the fingers (Figure 3.1g), and the secretion transferred to a vial with forceps or a spatula. Preferably, however, the secretion was removed periodically from the incapacitated side of the beetle with ophthalmic forceps to prevent losing the material or contaminating it with dust particles, and to avoid evaporation of the volatiles from the carrier material. Meanwhile, the vial with collected material was kept cool in an ice box.

In captivity, the secretion-producing behavior of the *Kheper* males differed from their behavior in the field. In the Hluhluwe Game Reserve, several *K. subaeneus* males were often observed simultaneously producing the secretion on rhinoceros dung middens. In captivity, the males only occasionally exhibited calling behavior. Males of this species produce their secretion as a relatively thick band about 5 mm wide. However, over a period of 5 years, only a few milligrams of the secretion was collected from about 700 males. It was also remarkable that in the wild, large numbers of *K. subaeneus* were often found in or on certain dung middens, while less than 5 m away not a single *K. subaeneus* could be found in or on a dung midden that contained equally fresh dung. As in many other insect species, the male-produced sex attractant of *K. subaeneus* apparently also acts as an efficient aggregation pheromone.

As has been mentioned, in nature male *K. nigroaeneus* first bury a brood ball before assuming the typical secretion-producing stance at the entrance to the burrow in which the ball has been buried (Edwards and Aschenborn 1988). The same behavior was observed in the captive insects. The male beetles never called from atop dung balls or when they were merely in possession of fresh dung. In captivity, often 10 or more beetles simultaneously exhibited calling behavior in one of the greenhouses. Unfortunately, the secretion of this species is extruded in an almost transparently thin and very narrow band, less than 1 mm wide, so that very little material accumulated even when a male continued producing the secretion for 30 min or longer.

K. lamarcki males produced secretion in bands of approximately the same width and thickness as *K. subaeneus*. In the wild and in captivity they produced secretion less frequently than *K. nigroaeneus*, but in captivity more frequently than *K. subaeneus*. The three *Kheper* species that were kept in greenhouses adapted quite readily to captivity. After initially trying to escape (for a few days), only to repeatedly hit the roof or sides of the greenhouse, they soon gave up trying to fly away. Surprisingly, the beetles that survived their first year in captivity appeared to have permanently lost their urge or ability to fly the next season.

Adult *K. bonellii* dung beetles were collected from cow dung pats in the vicinity of Elandsbaai on the southwest Cape coast during September each year, transported to Stellenbosch, and kept in greenhouses on horse dung. In captivity, male *K. bonellii* hardly ever produced secretion, and therefore the constituents of the sex attractant had to be extracted from the glandular material of the males.

3.4 SAMPLE PREPARATION AND ANALYSIS

Although Tribe (1976) describes the abdominal secretion as a "paraffin carrier" (i.e., a hydrocarbon wax), it does not dissolve in any of the common organic solvents. After preliminary small-scale experiments established that a complex mixture of volatile organic compounds (VOCs) could be extracted from the secretion with different solvents, such as pentane, dichloromethane and acetone, the following procedure was followed to isolate the volatile organic fraction of the secretion for GC-MS analysis and to purify the insoluble carrier material for further investigation.

The collected secretion was examined under a microscope for the presence of dust and dung particles that could have been picked up from the abdomen by the emerging secretion. Without spending too much time on the process and risking the loss of any highly volatile components present in the secretion, any visible impurities were carefully removed with ophthalmic forceps. The collected secretion (e.g., 450 µg collected from four *K. lamarcki* males) was stirred in a 1-ml Reacti-Vial with 100 µl of pentane using a winged magnetic follower to yield a suspension of the white insoluble carrier material in the solvent. To precipitate the carrier material together with any other insoluble matter, the suspension was centrifuged in the Reacti-Vial at 3000 rpm for 30 minutes. Without disturbing the precipitated carrier material, the supernatant pentane solution was carefully removed with a syringe. This extraction was repeated twice. Combination of the pentane extracts followed by slow evaporation of the solvent yielded a concentrate of the soluble VOCs for qualitative and quantitative GC-MS analyses. If only the VOCs had to be analyzed, the impurities visible under a microscope were removed as described, and the soluble organic material simply extracted with the more polar solvent dichloromethane containing just enough pentane to ensure precipitation of the carrier under centrifugation.

To purify the carrier material for further analysis, the differences in density of mixtures of pentane and dichloromethane were employed to separate the abdominal secretion from lighter dung fiber and from heavier sand and dust particles. The secretion was suspended in a mixture containing a relatively high percentage of pentane in which the dung particles remained suspended, while the carrier material and sand particles were precipitated by centrifugation at 3000 rpm. The solvent with the suspended impurities was removed with a syringe or pipette. The process was then repeated with the precipitated material using pentane/dichloromethane mixtures with increasing densities until finally the carrier material remained in suspension and the solvent plus carrier material could be removed with a pipette from the heavier dust particles. The carrier material was precipitated from the suspension by dilution with pentane and centrifuging at 3000 rpm. After removal of the solvent with a syringe and drying of the residue under reduced pressure at room temperature, the highly purified carrier material (400 µg, approximately 90% of the collected secretion) was obtained for further analysis.

As mentioned above, adult male *K. bonellii* in captivity hardly ever produced secretion. The VOCs were therefore extracted from the excised secretion-producing glands. The glands underlying the abdominal pores on both sides of the first abdominal sternite were carefully removed after the gut of the insect had been removed, the glandular tissue was extracted with dichloromethane of the best purity available (residue analysis grade or pesticide analysis grade) and the extract stored at –20°C until used for analysis.

The first four constituents isolated from the volatile organic fraction of the male abdominal secretion of *K. lamarcki*, hexadecanoic acid, 2,6-dimethyl-5-heptenoic acid, (*E*)-nerolidol and 3-methylindole (skatole), were relatively easily identified. However, exploratory GC and GC-MS analyses revealed that the secretions of *K. lamarcki, K. nigroaeneus,* and *K. subaeneus* are complex mixtures of at least 150 constituents (Figure 3.2); secretions contained hydrocarbons, alcohols, straight- and branched-chain fatty acids, and methyl and ethyl esters of these and other fatty acids with various carbon chain lengths.

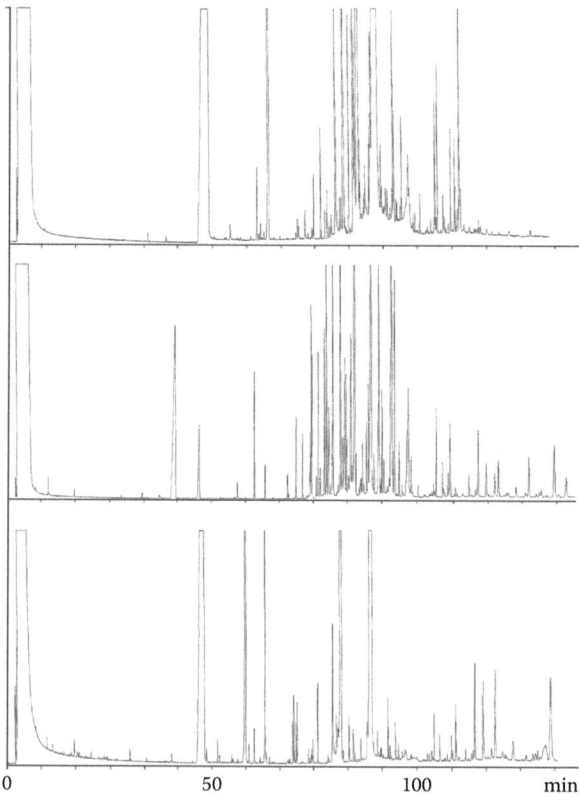

FIGURE 3.2 Gas chromatograms of extracts of the male abdominal secretions of *K. lamarcki* (top), *K. nigroaeneus* (middle), and *K. subaeneus* (bottom). (From Munro, Z. M. 1988. Semiochemiese kommunikasie: Ontwikkeling en toepassing van analitiese tegnieke vir die bepaling van die chemiese samestelling van die abdominale afskeiding van *Kheper lamarcki, K. Subaeneus* en *K. nigroaeneus* (Coleoptera: Scarabaeidae). PhD dissertation, University of Stellenbosch, South Africa.)

As shown in Table 3.1, the volatile compounds present in the secretions of these three species are similar, but only some of them are common to two or to all three of the species. Furthermore, these compounds include several chiral long-chain compounds, the structures of which cannot be readily determined even with modern technology.

TABLE 3.1
Some of the Constituents Identified in Extracts of the Male Abdominal Secretions of Dung Beetles of the Genus *Kheper*[a]

	lamarcki	*nigroaeneus*	*subaeneus*
2-Hydroxybenzaldehyde			✓
5-Methylheptanoic acid		✓	
Methyl 2,6-dimethyl-5-heptanoate	✓		✓
Glycerol	✓	✓	✓
Ethyl 2,6-dimethyl-5-heptanoate	✓		✓
Ethyl tetradecanoate		✓	✓
2,6-Dimethyl-5-heptanoic acid	✓	✓	✓
4-Dodecen-2-ol			✓
4-Tridecen-2-ol			✓
3-Methylindole (skatole)	✓	✓	✓
3,7,11-Trimethyldodeca-1,6,10-trien-3-ol (nerolidol)	✓	✓	✓
Methyl 12-methyltridecanoate		✓	✓
Methyl tetradecanoate	✓	✓	✓
Methyl 2-methyltetradecanoate	✓	✓	✓
Methyl 13-methyltetradecanoate	✓	✓	
Ethyl tetradecanoate	✓	✓	
Branched alkanone			✓
Methyl pentadecanoate	✓	✓	✓
Methyl 2-methyltetradecanoate	✓	✓	
Tetradecanoic acid	✓	✓	✓
Methyl 14-methylpentadecanoate		✓	
Ethyl pentadecanoate	✓	✓	✓
Methyl heptadecanoate	✓	✓	✓
Branched carboxylic acid		✓	
Methyl 2-methylhexadecanoate	✓	✓	
Pentadecanoic acid	✓	✓	✓
Branched carboxylic methyl ester		✓	
Ethyl hexadecanoate	✓	✓	✓
Methyl 15-methylhexadecanoate	✓	✓	
Methyl 2-methylheptadecanoate	✓	✓	
Hexadecanoic acid	✓	✓	✓
Ethyl heptadecanoate		✓	

(continued)

TABLE 3.1 (Continued)
Some of the Constituents Identified in Extracts of the Male Abdominal
Secretions of Dung Beetles of the Genus *Kheper*[a]

	lamarcki	*nigroaeneus*	*subaeneus*
Tricosane	✓	✓	✓
Tetracosane	✓		
11-Methylpentacosane			✓
13-Methylpentacosane	✓	✓	
2-Methylpentacosane	✓		
3-Methylpentacosane	✓		
Heptacosane	✓	✓	✓
11-Methylhexacosane	✓	✓	✓
13-Methylhexacosane		✓	✓
2-Methylheptacosane	✓	✓	✓
3-Methylheptacosane		✓	✓
13-Methyloctacosane		✓	
15-Methyloctacosane		✓	
11,15-Dimethylnonacosane		✓	

[a] The majority of these compounds were tentatively identified by EI- and CI-MS.

The envisaged complete chemical characterization of all the volatile constituents of the secretions was therefore suspended in favor of the identification of the constituents that could be detected by electroantennographic detection (EAD). However, the development of instrumentation for GC with flame ionization (FID) and electroantennographic detection in parallel (GC-FID/EAD) using the lamellate, clublike antennae of dung beetles proved very difficult due to the exceedingly poor signal-to-noise ratios of the EAD traces that were obtained. The excised antennae of these beetles that were often used in experiments for more than 12 hours appeared to be much more sensitive to interferences such as air turbulence, heat radiation, and fluctuations in air temperature and humidity than those of the *Lepidoptera*. It was hypothesized that, in addition to the detection of semiochemicals, dung beetle antennae are probably also used for navigation, the determination of air speed, meteorological conditions, and the detection of moisture. The influence of these parameters on EAD recordings was not investigated in any detail, but could be an interesting topic for future research by scientists interested in dung beetle ethology.

3.5   OPTIMIZATION OF GC-FID/EAD INSTRUMENTATION

In addition to the conventional instrumentation and operating parameters normally employed for GC-FID/EAD analyses, extraordinary precautions have to be considered to achieve acceptable results when dung beetle antennae are used as sensing elements.

1. *Quality, selectivity, and capacity of the GC column.* Overloading an apolar GC column with a polar compound, such as a carboxylic acid, results in the polar analyte eluting as a so-called fronting peak, which can be seen as a reversed tailing peak. In EAD analysis, fronting is probably a more serious problem than tailing because the sensilla on the antenna could become saturated by a fronting-EAD active compound before the detection threshold of the system is reached. The analytes must reach the preparation (antenna) as sharp peaks or pulses—the sharper the pulses, the better the resulting signal-to-noise ratios. Highly volatile apolar compounds are therefore mostly eluted from apolar columns as very sharp peaks with high signal-to-noise ratios. Carboxylic acids are eluted with satisfactory peak shapes from columns coated with so-called free fatty acid phase (FFAP) or an equivalent phase. If a sample contains both apolar and polar compounds with medium to long carbon chains then a column coated with the intermediate polar phase OV-1701 would be a good choice. In general, columns with thin stationary film coatings produce sharper peaks, but such columns are easily overloaded and produce broad and/or fronting peaks. Thus, if a semiochemical secretion contains analytes with widely different polarities then a column with a somewhat higher film thickness could be considered. Preferably, capillary columns coated with stationary phases of different polarities should first be tested for the best results. Sharper peaks can also be produced by employing higher temperature programming rates, provided the column's separation efficiency is not compromised.

2. *Elimination of the condensation of analytes in the column tip.* An analyte that condenses in the tip of the column at the point where the hot effluent is introduced into the cool humidified stream of air will slowly evaporate from that point, resulting in the analyte reaching the preparation as a broad band instead of a sharply eluted peak or pulse. Condensation of the analytes, especially high-boiling compounds, in the tip of the capillary column is avoided by heating the column right up to its very tip. This can best be achieved by routing the column through a heated interface consisting of an aluminum block with a sharp conical tip barely touching the glass tube conducting the humidified air to the antenna (Figure 3.3) (Burger and Petersen 1991). The tip of the column should not extend beyond the tip of the interface. In this way analyte pulses can be delivered into the air duct with a minimum of heat being conducted to the glass tube and without condensation in the tip of the capillary column. Even better results can be obtained by introducing the effluent into the humidified air duct via a retention gap connected to the column tip with a press-fit connector (Rohwer et al. 1986). A retention gap, developed by Grob and Müller (1982), consists of a length of uncoated but properly deactivated fused silica capillary that has zero retention for the analytes eluting from the column at the temperature of the interface.

3. *Elimination of condensation of high-boiling analytes in the humidified air duct.* The column effluent must be introduced into the humidified air duct at a point not too far removed from the antenna in order to avoid condensation of the heavier analytes on the inside surface of the humidified air duct,

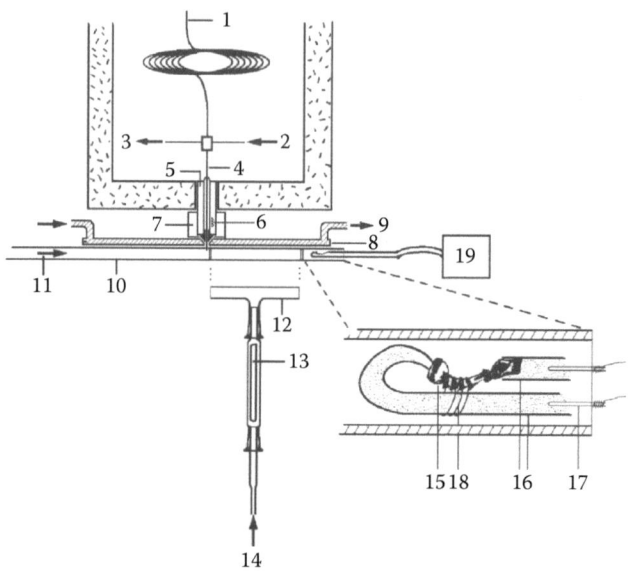

FIGURE 3.3 Schematic view of a gas chromatograph fitted with a column effluent splitter for flame ionization detection (FID) and electroantennographic detection (EAD) in parallel: 1, capillary column; 2, makeup gas at 15 ml/min; 3, FID; 4, fused silica capillary; 5, aluminum block; 6, heater; 7, insulation; 8, heat shield; 9, water at 18°C; 10, glass tube; 11, humidified air; 12, replacement T-piece for recording of EAG; 13, paper strip impregnated with sample; 14, air puffed over the paper strip and antenna; 15, antenna; 16, glass capillaries containing saline solution; 17, Ag/AgCl electrodes; 18, antenna tied to a capillary with cotton yarn; 19, amplifier and recording devices (From Burger B. V. and W. G. B. Petersen. 1991. Semiochemicals of the Scarabaeinae, III. Identification of an attractant for the dung beetle *Pachylomerus femoralis* in the fruit of the spineless monkey orange tree, *Strychnos madagascariensis*. *Zeitschrift Naturforsch.* 46c:1073. Reprinted with permission.)

but sufficiently far enough upstream to ensure proper mixing with the air. The contamination of the air duct can be elegantly eliminated by making provision for replacing a section of glass tubing with a clean section from just after the point of effluent introduction to just before the antenna. In long analyses of complex mixtures, glass sections could be exchanged several times at appropriate intervals during the experiment. To facilitate the exchange, sections of glass tubing have to be cut with a glass-cutting circular saw to exactly the same length. This part of the air duct has to be supported in a V-shaped gutter to facilitate alignment when replacing a section. For easy manipulation, each of the sections has to be fitted with a small handle. Such a setup can also be conveniently converted into an arrangement for electroantennographic (EAG) experiments by using a T-section with a B-10 ground-glass side socket for the connection of a tube containing a test sample on filter paper (Figure 3.3).

4. *Elimination of turbulence in the humidified air flowing over the antenna.* Turbulence in the humidified air duct just before it reaches the preparation and turbulent mixing of the humidified air with drier, and probably warmer, laboratory air on the surface of the antenna are probably the major causes of an unstable EAD signal and low signal-to-noise ratios. To avoid turbulence in the humidified airstream, there should not be any sharp bends in the glass tube immediately upstream from the preparation. Although it is impossible to totally eliminate turbulence, especially on the surface of the antenna, the best solution is to prepare the antenna in such a manner that it can be inserted "lengthwise" into the air duct by using electrodes that have been bent in an appropriate fashion. Dung beetles are too strong to be tethered and antennae have to be excised for EAD experiments. Fortunately, the insects' antennae can be used for periods exceeding 12 hours. The antenna can be fastened to one of the electrodes and the two electrodes can be tied together before the preparation is inserted into the air duct (Figure 3.3).

5. *Elimination of condensed water droplets in the humidified air duct or on the surface of the antenna.* The effective lifetime of an antenna in FID/ EAD analyses can be extended by cooling the humidified air to a slightly lower than ambient temperature. This is, however, not necessary in analyses with dung beetle antennae. It is also risky because it could lead to condensation of water droplets in the air duct and/or on the surface of the antenna.

6. *Elimination of pressure pulses in the humidified airsteam.* Humidifying the air by using a wide-bore tube to bubble the air through water could add to the turbulence in the air flowing over the preparation. A proper gas-washing bottle with an *inverted* sintered glass frit should be used instead.

7. *Elimination of electrostatic interference.* It is standard practice to operate the EAD setup in an earthed Faraday cage, which should preferably not be earthed to an outlet of the building's electrical supply. The cage should be connected with a thick *insulated* copper cable to a copper mesh mat buried outside the laboratory building.

8. *Elimination of heat radiation.* Finally, a further incremental improvement could possibly be achieved by shielding the preparation and the humidified air duct against heat radiation from the hot interface or other hot parts of the gas chromatograph by installing a water-cooled heat shield as shown in Figure 3.3.

9. *Recording the FID and EAD chromatograms.* Recording the chromatograms on a dual-channel integrator or on a computer using appropriate commercially available software allows for time scale manipulation in order to ensure as far as possible that FID and EAD responses that appear to be coinciding are indeed resulting from the same analyte. If two strip chart recorders are used, they should be accurately matched (synchronized). A relatively high chart speed and constant monitoring of any fine structure in the responses are also essential in order to distinguish between background noise and authentic FID and AEG responses.

3.6 CHEMICAL CHARACTERIZATION OF ABDOMINAL SECRETIONS OF THE MALE *KHEPER* SPECIES

Ideally, the EAD-active constituents of the abdominal sex attracting secretions of the male *Kheper* beetles were identified by GC-MS analysis, enantioselective GC analysis, and retention time comparison with authentic synthetic analogs of the natural compounds.

3.6.1 *K. LAMARCKI*

Initially, only four of the VOCs of the male abdominal sex attractant secretion of *K. lamarcki*, hexadecanoic acid (57%), 2,6-dimethyl-5-heptenoic acid (33%), (*E*)-nerolidol (9%) and skatole (1%), were positively identified in the extract of the secretion (Burger et al. 1983). Synthetic analogs of these compounds were used to formulate an attractant for field tests. Due to the small quantities of these compounds isolated from the secretion, and because GC methods to determine the absolute configuration of chiral compounds had not yet been developed, racemic 2,6-dimethyl-5-heptenoic acid and (*E*)-nerolidol were used in the synthetic attractant. Purified French chalk was used as inert carrier material. Confusing results were obtained. Not a single dung beetle of any species was caught in pitfall traps baited with the attractant alone, whereas large numbers of dung beetles of a variety of species were trapped in control traps baited with fresh horse dung. Combining the synthetic attractant with horse dung even seemed to reduce the attractiveness of the dung for *K. lamarcki* as well as for the other dung beetle species.

Females of the genus *Kheper* produce only one egg per year or perhaps two during a long summer, and after mating they remain underground to care for their brood. Because the tests were carried out late in the active season of *K. lamarcki*, the majority of the females had probably already gone underground, and the fact that only male *K. lamarcki* were trapped is probably not significant. Raw data obtained in bioassays carried out earlier in the next season suggested that more females were attracted to pitfall traps baited with horse dung plus the synthetic attractant than to the traps baited with the dung alone. However, rigorous statistical evaluation of the results again showed that the apparent differences in the numbers of male and females were insignificant.

The different yields of the control traps baited with dung confirmed a problem that is probably inherent in field tests with dung beetles; namely, their territorial behavior appears to be also determined by cues other than sex attractants, such as the nature of the soil—the type, texture, and moisture content. To conclude, no evidence was found that the above four VOCs are indeed components of the purported sex attractant of *K. lamarcki*.

An example of a GC-FID/EAD analysis of the VOCs extracted from the male abdominal secretion of *K. lamarcki* that was typically obtained during the early years of this project is depicted in Figure 3.4. Obviously, this analysis did not produce any useful information. Despite the incremental improvement of the GC-EAG setup over time, later GC-FID/EAD analyses of extracts of the male abdominal secretion of *K. lamarcki* did not produce appreciably better results. It is possible

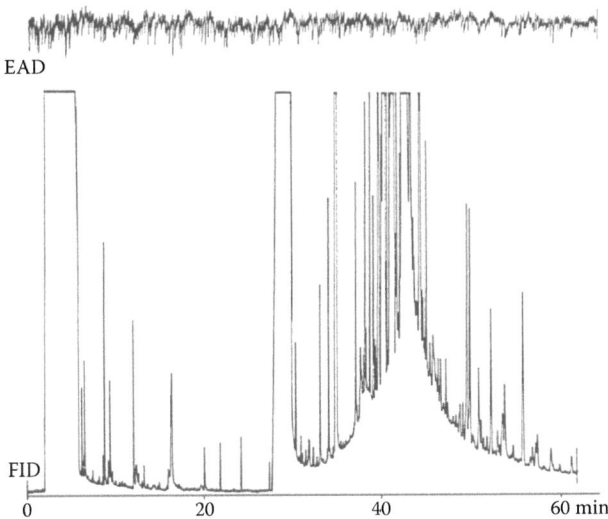

FIGURE 3.4 FID/EAD analysis of an extract of the male abdominal secretion of *K. lamarcki* using a female antenna as sensing element. (From Petersen, W. G. B. 1997. Semiochemiese kommunikasie by miskruiers van die genus Kheper (Scarabaeinae). PhD dissertation, University of Stellenbosch.)

that in the case of this species the signals generated by the antennae were simply too weak to be detected by the then-available instrumentation.

The injection of an enormous sample of an extract of the male abdominal secretion of *K. lamacki* in a GC-FID/EAD analysis gave the set of chromatograms depicted in Figure 3.5. The width of the peaks of 2,6-dimethyl-5-heptenoic acid **1** and (*E*)-nerolidol **3** is an indication of the massive overloading of the column with respect to the major constituents of the secretion. Under these conditions the antenna responded only to the leading edge of the 2,6-dimethyl-5-heptenoic acid peak. The antenna also responded to the elution of skatole **2**. Twelve constituents of the secretion eluting with retention times ranging from about 10 to 35 minutes gave weak EAD responses. Only five of these compounds were identified as ethyl butanoate, 2-hexanone, 3-hexen-2-one, 2-methylpropyl propanoate, and 3-methylbutyl propanoate. Although indications of EAD activity were found in the low retention time range of a few subsequent analyses, the responses were always quite weak and not reproducible. No further attention was given to these constituents and the possibility that they could be constituents of a male-secreted pheromone was not seriously considered. The significance of these almost forgotten results was only realized many years later in quite a different context, as will be shown in Section 3.9.1.

3.6.2 *K. NIGROAENEUS*

Better results were obtained in GC-FID/EAD analyses of the VOCs extracted from the abdominal secretion of *K. nigroaeneus* males (Burger and Petersen 2002).

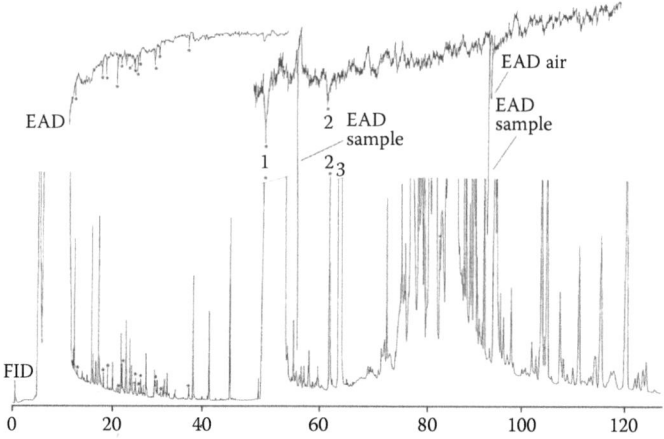

FIGURE 3.5 FID/EAD analysis of an extract of the male abdominal secretion of *K. lamarcki* using a male antenna as sensing element. Coinciding peaks and responses are indicated by black dots. (From Munro, Z. M. 1988. Semiochemiese kommunikasie: Ontwikkeling en toepassing van analitiese tegnieke vir die bepaling van die chemiese samestelling van die abdominale afskeiding van *Kheper lamarcki, K. Subaeneus* en *K. nigroaeneus* (Coleoptera: Scarabaeidae). PhD dissertation, University of Stellenbosch, South Africa.)

Figure 3.6 depicts a set of chromatograms obtained in a GC analysis of the extract with FID/EAD detection in parallel using a temperature programming rate of 4°C/min. Here a female antenna was used as an EAD sensing element. Male and female antennae mostly gave reproducible and qualitatively similar results, which is not unexpected for a sex attractant secreted by a male insect (Leal 1998).

Compound 1 gave reproducible EAD responses, but despite the relatively large samples that were used, it was not visible in the corresponding FID gas chromatograms. The low concentration in which this compound is present in the secretion and the fact that it was impossible to collect unlimited quantities of abdominal secretion precluded its identification. However, in retrospect, it might have been possible to identify this compound by using the strategy that was used to identify an approximately equally low concentration of a constituent of the abdominal secretion of *K. subaeneus* (see below).

Compound 2 was identified as 3-methylheptanoic acid by GC-MS analysis in conjunction with ^{1}H NMR analysis of a sample of the compound isolated by preparative GC. The two enantiomers of the synthetic acid are separable on a capillary column coated with a mixture of the stationary phase OV-1701-OH and heptakis(2,3,6-tri-O-methyl)-β-cyclodextrin; the *S* enantiomer eluting first from this column. Using this column, enantioselective GC analyses with FID and EAD detection in parallel confirmed the EAD activity of the second-eluting *R* enantiomer. (R)-(+)-3-Methylheptanoic acid was synthesized in high enantiomeric purity from (R)-(+)-3,7-dimethyl-6-octen-1-ol [(R)-(+)-β-citronellol] as starting material (Burger and Petersen 2002). The third EAD active compound, 3 in Figure 3.6, was identified as

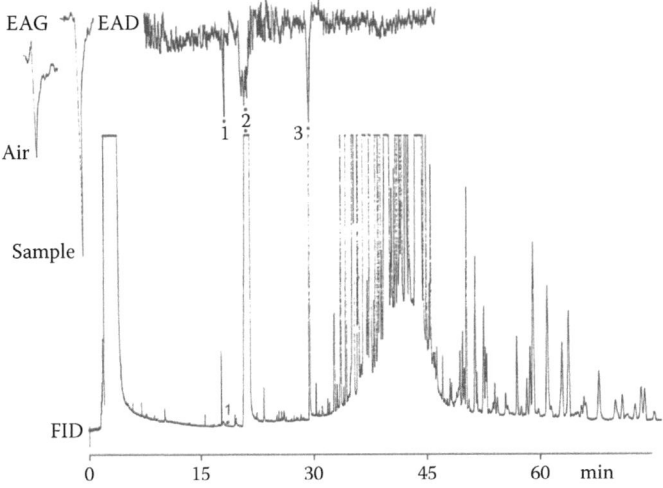

FIGURE 3.6 First part of a GC-FID/EAD analysis of an extract of the male abdominal secretion of *K. nigroaeneus* using a female antenna as a sensing element. Corresponding peaks in the FID and EAD traces are indicated by dots. The structures of the EAD active compounds are given in the text. (From Petersen, W. G. B. 1990. Aanwending van elektroantennografiese metodes in die chemiese karakterisering van semioverbindings van kewers. MSc thesis, University of Stellenbosch.)

skatole, which was also identified as a constituent of the abdominal secretions of the other dung beetles of the genus *Kheper* that were investigated.

3.6.3 *K. SUBAENEUS*

On a few occasions the male abdominal secretion of captive male *K. subaeneus* was collected in small quantities—a total of only a few milligrams over a period of 5 years from about 700 males. GC-FID-EAD analyses of extracts of the collected material gave diverse results. In some analyses several of the constituents of the extracts elicited EAD responses, but mostly different constituents produced EAD responses with varying intensities, as illustrated in Figures 3.7 and 3.8.

The major constituent of the secretion of this species, constituent 1 in the set of FID and EAD chromatograms depicted in Figure 3.7, elicited a strong EAD response in the male antenna that was used as sensing element. This compound was identified as 2,6-dimethyl-5-heptenoic acid, which had previously been identified in *K. lamarcki* (Burger et al. 1983). Racemic 2,6-dimethyl-5-heptenoic acid was resolved by GC on a stationary phase mixture of OV-1701-OH and heptakis(2,3,6-tri-O-methyl)-β-cyclodextrin, mentioned previously. The natural compound was identified as (*S*)-2,6-dimethyl-5-heptenoic acid, which coeluted with the natural compound from the enantioselective column. The synthetic *R* enantiomer did not elicit EAD responses from male or female antennae, but neither did the synthetic *S* enantiomer in a large number of male and female antennae that

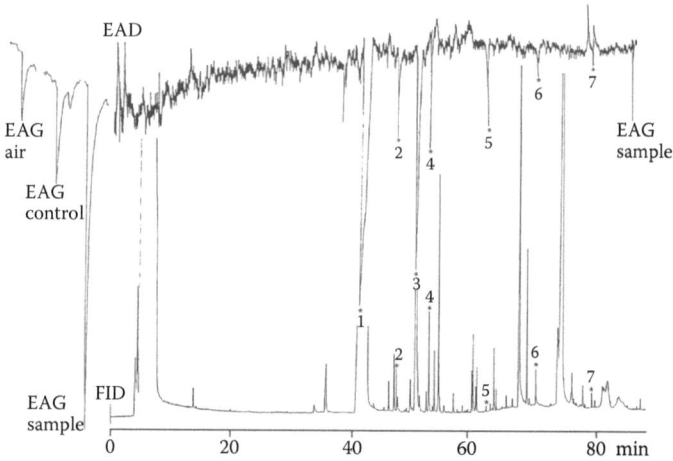

FIGURE 3.7 FID/EAD analysis of an extract of the male abdominal secretion of *K. subaeneus* using a male antenna as a sensing element. Coinciding peaks and responses are indicated by black dots. 7-Vinyldec-8-en-1-ol was used as control. The structures of the EAD active compounds are given in the text. (From Munro, Z. M. 1988. Semiochemiese kommunikasie: Ontwikkeling en toepassing van analitiese tegnieke vir die bepaling van die chemiese samestelling van die abdominale afskeiding van *Kheper lamarcki, K. subaeneus* en *K. nigroaeneus* (Coleoptera: Scarabaeidae). PhD dissertation, University of Stellenbosch, South Africa.)

were used in FID/EAD measurements. In view of the observation that the natural compound also did not reproducibly elicit EAD responses, this was not a totally unexpected result.

Constituents 2 and 3 were tentatively identified as 4-dodeken-2-ol and 4-trideken-2-ol, respectively. Constituent 4 was identified as skatole. Constituent 5 remained unidentified and constituent 6 was tentatively characterized as a branched-chain methyl ester.

Two compounds elicited exceptionally strong responses in the first part of the GC-FID/EAD analysis shown in Figure 3.8 (Burger et al. 2002). A female antenna was used as sensing element in this analysis.

Constituent 1 in the set of FID and EAD chromatograms depicted in Figure 3.8 was identified as butanoic acid (*n*-butyric acid) based on its mass spectrum and coelution with an authentic synthetic analog of the natural compound. Although the EAD response was weak, it was detected in several of the GC-EAD traces. 2,6-Dimethyl-5-heptenoic acid, the major constituent of the secretion, eluted at about 22.2 minutes, but did not elicit an EAD response in this analysis. Constituent 3 was identified as skatole.

Although barely detectable in the FID gas chromatogram, constituent 2 gave a strong and reproducible EAD response in the same antenna and also in other *K. subaeneus* antennae used as sensing elements. About 10% of the material collected during the previous five summer seasons was used for GC-FID/EAD experiments. The remaining material was used in one final attempt at obtaining some

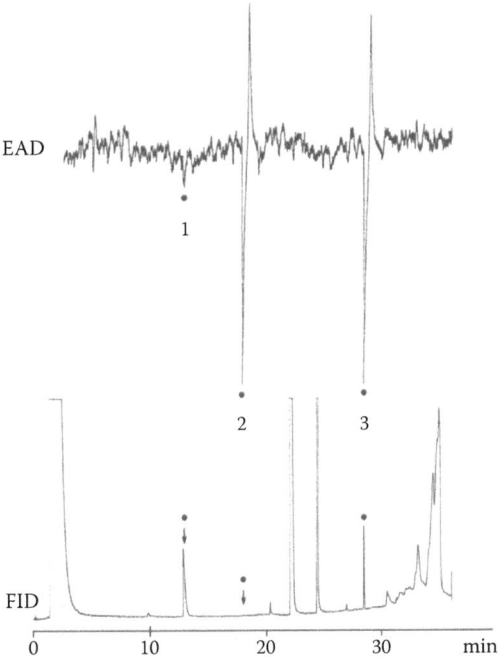

FIGURE 3.8 First part of an analysis of an extract of the male abdominal secretion of *K. subaeneus* with EAD and FID detection in parallel using a female antenna as a sensing element. The structures of the EAD-active compounds are given in the text. (From Petersen, W. G. B. 1997. Semiochemiese kommunikasie by miskruiers van die genus Kheper (Scarabaeinae). PhD dissertation, University of Stellenbosch.)

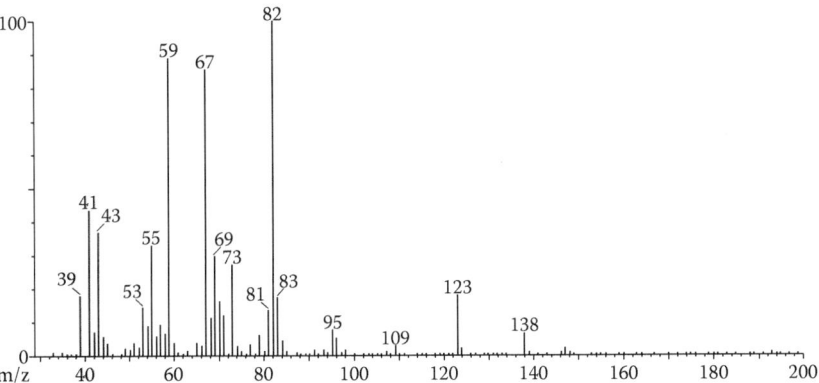

FIGURE 3.9 Mass spectrum of the EAD active constituent [(*E*)-2,6-dimethyl-6-octen-2-ol] of the male abdominal secretion of *Kheper subaeneus* eluting at 17.75 minutes in the gas chromatogram shown in Figure 3.8. (From Petersen, W. G. B. 1997. Semiochemiese kommunikasie by miskruiers van die genus Kheper (Scarabaeinae). PhD dissertation, University of Stellenbosch.)

mass spectral information on the structure of this constituent. The resulting mass spectrum (Figure 3.9) contained several ions typically found in the mass spectra of monoterpenoids. GC-MS retention time comparisons showed that constituent 2 could not be a cyclic, an acyclic monoterpene, or a primary monoterpene alcohol, but that it could possibly be a tertiary terpene alcohol other than linalool. Employing retention time data and mass spectral comparison of the unknown constituent with reference compounds synthesized in the laboratory, it was eventually identified as the new natural tertiary terpenoid alcohol (*E*)-2,6-dimethyl-6-octen-2-ol, for which the name (*E*)-subaeneol was proposed (Burger et al. 2002).

3.6.4   *K. BONELLII*

K. bonellii is the last of the four *Kheper* species of which the male abdominal secretions were investigated. This species is found in the coastal area of the Western Cape, about 200 km from Stellenbosch.

GC analyses of the extract of the abdominal glands of male *K. bonellii* with FID and EAD recording in parallel gave reproducible results; male and female antennae responded to the same constituents of the extract. An example of a typical analysis is given in Figure 3.10 (Burger et al. 2008). Constituent 1 was identified as 2-butanone,

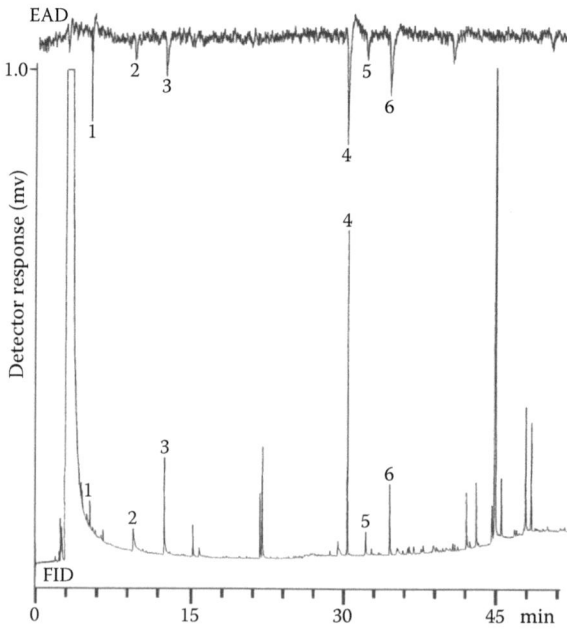

FIGURE 3.10   GC-FID/EAD analysis of an extract of the abdominal sex pheromone glands of the dung beetle *K. bonellii* using a male antenna as a sensing element. The peaks are numbered to indicate coinciding FID and EAD responses. (From Petersen, W. G. B. 1997. Semiochemiese kommunikasie by miskruiers van die genus Kheper (Scarabaeinae). PhD dissertation, University of Stellenbosch.)

a solvent impurity. Compounds 2, 3, 5, and 6 were identified as propanoic acid, buta-noic acid, indole, and skatole, respectively.

Although the presence of a reasonably abundant ion at *m/z* 74 in the mass spectrum of constituent 4 suggested the presence of a methyl ester moiety, the ions at *m/z* 87, 129, and 143 in the mass spectra of saturated methyl esters are present in uncharacteristically low abundances in this spectrum. Fortunately, extracts of the glandular abdominal tissue contained relatively high concentrations of this constituent. It was therefore possible to isolate sufficient material from about 200 beetles by preparative GC on an open-tubular wide bore capillary column for ^{1}H and ^{13}C NMR, APT, and COSY spectral analyses. This constituent was identified as methyl *cis*-2-(2-hexyl-cyclopropyl)acetate. Wilson and Prodan (1976) have identified the *trans* isomer of this ester in the essential oil of *Croton eluteria* (Euphorbiaceae), and proposed the names *cis*- and *trans*-cascarillate for the two isomers. The absolute configuration of the enantiomer present in the male abdominal secretion of *K. bonellii* was established by stereoselective synthesis, which afforded methyl *(R,R)*-2-(2-hexylcyclopropyl) acetate [methyl *(R,R)*-cascarillate] in an enantiomeric excess of 69%. Enantioselective GC-FID/EAD, using a capillary column coated with OV-1701-OH containing 10% heptakis(2,3-di-O-methyl-6-O-*tert*-butyldimethylsilyl)-β-cyclodextrin as chiral selector, showed that methyl *(R,R)*-cascarillate was the only enantiomer of racemic methyl *cis*-cascarillate that coeluted with the natural compound, and that it elicited EAD responses in the antennae of male and female *K. bonellii* (Figure 3.11) (Burger et al. 2008).

Far better GC-FID/EAD analysis results were obtained when the antennae of *K. bonellii* were used as sensing elements compared to analyses of the VOCs the abdominal secretions of the other *Kheper* species, probably because the antennae of *K. bonellii* could be used within a few hours after the insect had been captured. Analyses of the VOCs of the other species were invariably carried out late in the season and often using the antennae from insects that had spent several months in captivity under less than optimal climatic conditions.

The constituents identified in the male abdominal secretions of the four *Kheper* species are summarized in Table 3.2.

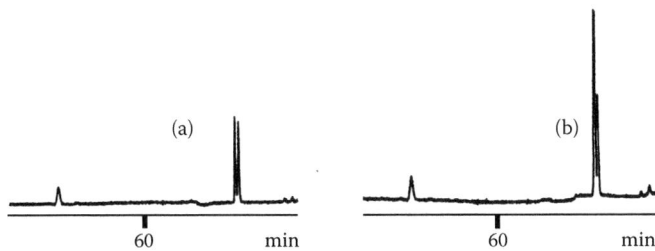

FIGURE 3.11 Enantioselective GC analysis. (a) Racemic methyl *cis*-cascarillate. (b) Synthetic racemic methyl *cis*-cascarillate spiked with the natural ester that was isolated by preparative GC from an extract of the male abdominal glands of *K. bonellii*. (From Petersen, W. G. B. 1997. Semiochemiese kommunikasie by miskruiers van die genus Kheper (Scarabaeinae). PhD dissertation, University of Stellenbosch.)

TABLE 3.2
EAD Active Compounds Identified in the Male Abdominal Secretions of *Kheper* Species[a,b]

Kheper lamarcki	*K. nigroaeneus*	*K. subaeneus*	*K. bonellii*[c]
Ethyl butanoate[d]	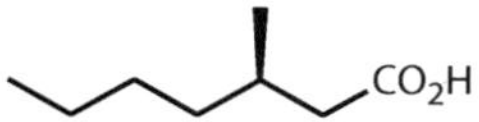	Butanoic acid	Propanoic acid
			Butanoic acid

2-Hexanone[d]

(R)-(+)-3-methylheptanoic acid

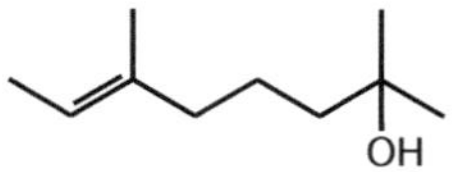

(*E*)-2,6-Dimethyl-6-octen-2-ol

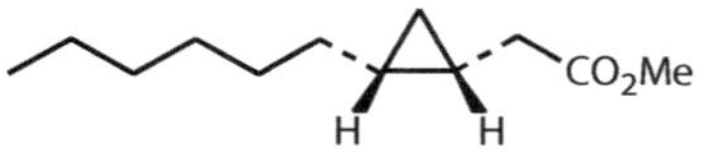

3-Hexen-2-one[d]

2-Methylpropyl propanoate[d]

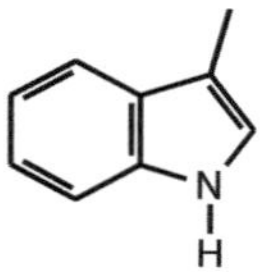

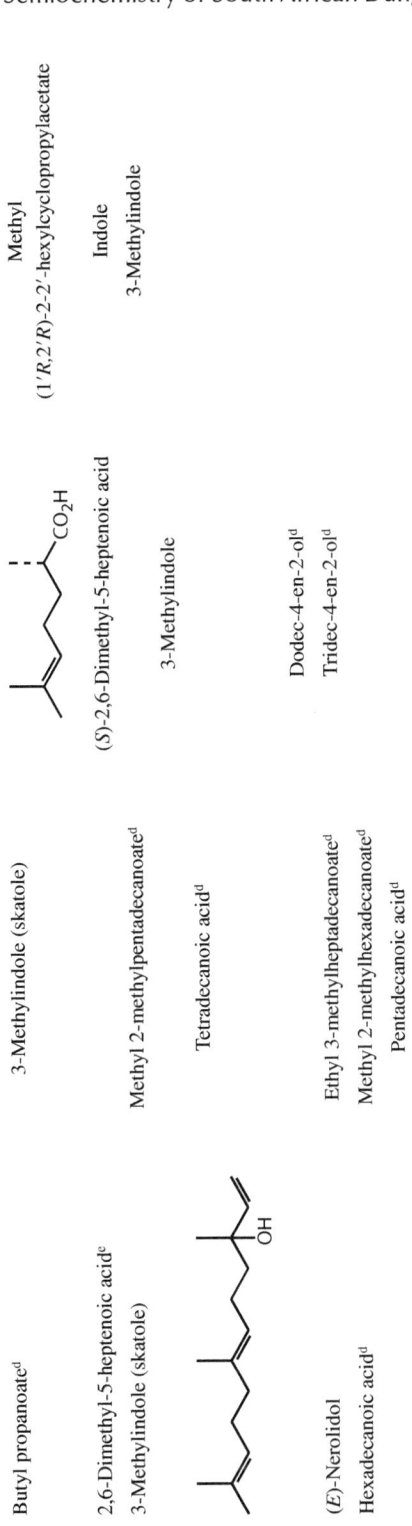

Butyl propanoate[d]

2,6-Dimethyl-5-heptenoic acid[e]

3-Methylindole (skatole)

(E)-Nerolidol

Hexadecanoic acid[d]

3-Methylindole (skatole)

Methyl 2-methylpentadecanoate[d]

Tetradecanoic acid[d]

Ethyl 3-methylheptadecanoate[d]

Methyl 2-methylhexadecanoate[d]

Pentadecanoic acid[d]

(S)-2,6-Dimethyl-5-heptenoic acid

3-Methylindole

Dodec-4-en-2-ol[d]

Tridec-4-en-2-ol[d]

Methyl
(1'R,2'R)-2-2'-hexylcyclopropylacetate

Indole

3-Methylindole

a　Compounds listed in order of elution from the gas chromatographic columns used.

b　Unless stated otherwise, the identification of compounds were substantiated by GC-MS retention time comparison with authentic synthetic compounds.

c　Compounds extracted from the glandular tissue.

d　Mass spectral identification and EAD responses not reproducible.

e　Absolute streochemical configuration not elucidated in this species.

From the research carried out with dung beetles over many years it is clear that semiochemical communication in dung beetles is extraordinarily complex. As far as the inconsistency of antennal responses in EAD analyses with the antennae of this and other *Kheper* species is concerned, there are various factors that must be considered. It is possible, for example, that there could be some natural variation in the population in terms of antennal responses. Because the research described was carried out with captured insects, it was not possible to determine the age, physiological state, and so forth of the insects that were used in these experiments—factors that could have influenced the EAD results.

The possibility that the response of an antenna to the volatile constituents of the secretion could depend on the concentration of the components flowing over the antenna was investigated. However, results do not seem to explain the inconsistent results obtained with (S)-2,6-dimethyl-5-heptenoic acid and some of the other minor constituents of the secretion. In retrospect, and bearing in mind that this compound is a carboxylic acid, the inconsistent results could possibly have resulted from insufficient attention having been paid to one or more of the extraordinary precautions mentioned earlier that have to be taken to ensure successful GC-FID/EAD analyses.

3.7 LONG-CHAIN CONSTITUENTS OF THE ABDOMINAL SECRETIONS

As far as the reproducible EAD responses are concerned, qualitatively similar results were obtained in GC-FID/EAD analyses with male and female antennae as sensing elements. However, on several occasions, examples were encountered of male and female antennae responding reproducibly to different long-chain compounds on different occasions. However, these results could not be reproduced with other antennae and the identification of these compounds could thus not be verified by EAD analyses of the synthetic compounds. These compounds were always long-chain fatty acids such as hexadecanoic acid, or the methyl or ethyl esters of long-chain acids. It was thought that the large numbers of long-chain compounds present in the abdominal secretions of *Kheper* males could contribute to the controlled release of the sex attractants of these insects. Unfortunately, no concerted attempts have yet been undertaken to confirm or refute this hypothesis. In retrospect, and in the light of remarks above concerning the importance of stringently applying appropriate GC-EAD techniques, the question arises as to whether such erratic results could perhaps be ascribed to the unfavorable elution profiles of such high molecular mass compounds, their nonoptimal introduction into the humidified air flowing over the antenna, or even to partial condensation on the walls of the air duct. The use of large sample sizes in efforts to detect small quantities of the more volatile EAD-active constituents probably even exacerbated the problem. To the best of the author's knowledge, long-chain compounds such as those present in the male abdominal secretions of the *Kheper* beetles have to date not been subjected to GC-EAD analysis. Future investigations into the function of these constituents of the male abdominal secretions of *Kheper* beetles could require the development of appropriate instrumentation for EAD analyses of these compounds.

The problems associated with analytes being eluted as fronting peaks could be solved by carrying out GC-FID/EAD analyses on a comprehensive two-dimensional gas chromatograph (GC^2 or GC × GC) in which a cryogenic modulator (Beens et al. 2001) focuses small portions of the effluent from the so-called first-dimension column into narrow concentrated bands of analytes, which are delivered into the second-dimension column for separation on a different stationary phase. Analytes are eluted from the second column of a GC^2 instrument as extremely sharp peaks. Depending on the stationary phases used in the instrument, different compound classes are grouped in different areas of the resulting comprehensive two-dimensional gas chromatogram. Thus, in addition to delivering sharp analyte pulses to the preparation, excellent separation of the constituents of complex mixtures can be achieved, and some information on the chemical properties of EAD-active compounds can even be obtained. If less complex mixtures have to be analyzed, the instrument can be operated in the one-dimensional mode by replacing the second-dimension column with a retention gap of the same internal diameter. Alternatively, a standard GC can be adapted by installing a device for cyclic cooling and heating a short section of a retention gap connected to the capillary column. Construction of an appropriate device for this purpose might require assistance from an experienced chromatographer.

3.8　COMPOSITION OF THE PHEROMONE-DISSEMINATING CARRIER MATERIAL

The minute quantities of VOCs present in the abdominal secretion of the male *Kheper* beetles are soluble in the common organic solvents and were extracted with, for example, dichloromethane, which left the insoluble white carrier material intact. The carrier material was found to be soluble in concentrated sulfuric acid and in aqueous solutions of other strong acids. When a speck of the secretion was heated in a small flame the typical odor of burning hair was detected, and it was concluded that the carrier material could be, or could contain, a polypeptide. Although the use of a synthetic polypeptide as support material in field tests of the synthetic analogs of the semiochemicals identified was not envisaged, it was evident that more specific information was required for the selection of a chemically similar and readily available protein for this purpose. As far as the general appearance and brittleness of the fibers are concerned, the males of the *Kheper* species produce a similar material, although that of *K. nigroaeneus* appears to consist of finer fibers, and that of *K. subaeneus* softer and slightly sticky fibers.

SDS polyacrylamide gel electrophoresis (SDS-PAGE) of the carrier material of three *Kheper* species (Figure 3.12) revealed the presence in each secretion of proteins, with molecular masses of approximately 17 kDa and approximately 31 kDa, with the 17-kDa protein as the major fraction.

In the absence of the reducing agent, β-mercaptoethanol, the apparent molecular mass of the 17-kDa protein decreased to 15 kDa, indicating that the naturally

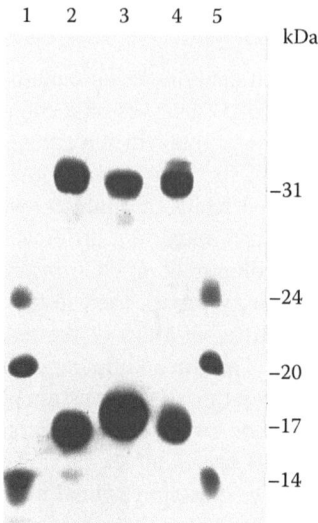

FIGURE 3.12 SDS-PAGE of the pheromone disseminating secretion of *K. lamarcki* (lane 2), *K. nigroaeneus* (lane 3), *K. subaeneus* (lane 4), and mass marker proteins (lanes 1 and 5): trypsinogen, 24 kDa; soybean trypsin inhibitor, 20.1 kDa; α-lactalbumin, 14.2 kDa. (From Munro, Z. M. 1988. Semiochemiese kommunikasie: Ontwikkeling en toepassing van analitiese tegnieke vir die bepaling van die chemiese samestelling van die abdominale afskeiding van *Kheper lamarcki*, *K. Subaeneus* en *K. nigroaeneus* (Coleoptera: Scarabaeidae). PhD dissertation, University of Stellenbosch, South Africa.)

occurring protein exists as a compactly folded protein stabilized by intramolecular disulphide linkages. On breaking these linkages with β-mercaptoethanol, the molecule unfolds to a more extended structure with lower mobility. Minor differences in the electrophoretic mobility of the proteins secreted by the three species may reflect minor differences in their overall structures. It is possible that the 31-kDa proteins are differently linked dimers of the smaller proteins. Complete amino acid sequencing was not attempted, but partial sequence analysis revealed a single sequence to be present in the 15-kDa protein isolated from the *K. nigroaeneus* carrier material, whereas two sequences appear to be present in the major constituent from the *K. lamarcki* carrier material. The molecular mass of the 15-kDa protein of the pheromone-disseminating carrier material of the two species was verified by ^{252}Cf plasma desorption MS. Average molecular masses of 15451±10 Da and 15477±10 Da were calculated for the 15-kDa proteins from the *K. lamarcki* and *K. nigroaeneus* carrier materials, respectively (Burger et al. 1990).

Due to the presence of relatively large amounts of glutamic acid and aspartic acid and smaller amounts of the basic amino acids, the proteinaceous carrier materials were expected to have low isoelectric points. The proteins also contain large proportions of hydrophobic amino acids. As most of the VOCs present in the abdominal secretions of these insects are long-chain fatty acids, these properties are not unexpected. The amino acid composition of these proteins is similar to that of serum albumin, which is known to bind a large number of hydrophobic compounds, including dodecyl sulphate and fatty acids (e.g., Noverr and Huffnagle 2004). This suggests that these proteins might have a high affinity for compounds such as the fatty acids that were identified as EAD-active constituents of the volatile organic fraction of the abdominal secretions of these insects. This supports the purported controlled-release pheromone disseminating function of the carrier material. On the other hand, in the dry state, these secretions have large surface areas, which might accelerate the evaporation of the adsorbed volatile constituents. In view of the disappointing results of the bioassays carried out with a few of the organic compounds isolated from the male abdominal secretion of *K. lamarcki*, no further investigation into the structures of these proteins was undertaken.

This could potentially be an interesting field of research for entomologists and chemists interested in applying the powerful analytical techniques that are now available for the investigation of the role of proteins in the dissemination of semiochemicals.

A number of experiments were carried out to compare the affinity of the carrier protein of *K. lamarcki* and proteins such as albumin and trypsin for the compound 2,6-dimethyl-5-heptenoic acid, which elicited EAG responses in the antennae of this insect. In these experiments, the compound was allowed to evaporate from a dispensing glass-fiber sleeve or wick, and was taken up in four receiving sleeves spaced equidistant around the dispensing wick. The wicks were coated with the carrier protein of *K. lamarcki*, bovine albumin, and bovine pancreas trypsin, respectively. The fourth glass fiber wick was left untreated, as a control. The device that was designed for this experiment is depicted in Figure 3.13. Thermal desorption and quantitative GC analysis of the 2,6-dimethyl-5-heptenoic acid retained by the receiving sleeves showed that albumin and the carrier protein retained the acid approximately equally effectively, while trypsin retained the acid about three times less effectively (Burger et al. 1990). These experiments probably constitute the earliest reported attempts at demonstrating that proteins could be involved in pheromone dissemination.

The strong retention of the acid by the carrier protein suggests that controlled release in order to retain the integrity of the message rather than rapid release of the pheromone might be the primary function of the carrier protein. This unique pheromone disseminating mechanism might be more effective under adverse climatic conditions than other release mechanisms normally employed by insects. Where related species occupy the same habitat, the integrity of their chemical signals may be of vital importance to their survival, especially when, as in the case of these *Kheper* beetles, a pair of dung beetles normally produces only one or two offspring per year. Using a solid carrier material impregnated with the semiochemicals has the advantage that a chemical message transmitted in a protein particle will retain its integrity over longer distances. As the volatile fraction of the secretion contains compounds with widely different polarities, the polypeptide carrier could also have a marked influence on the ratio in which these compounds are released into the atmosphere. This could be another interesting field of research for interested scientists.

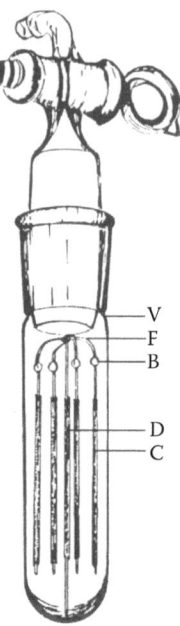

FIGURE 3.13 Device designed for the determination of the affinity of various proteins for 2,6-dimethyl-5-heptenoic acid. V, glass vial; F, five-pronged glass framework constructed from 1.0-mm glass rod and supported on the central prong; D, dispensing glass-fiber wick treated with the acid; C, adsorbing wicks (collectors); B, glass beads formed on the outer prongs to avoid contact between the collectors and the wall of the vial. (From Burger, B. V. et al. 1983. *Z. Naturforsch.* 38c:848. Reprinted with permission.)

3.9   INTERGENERIC AND INTERSPECIFIC CHEMICAL SIGNALING IN DUNG BEETLES

3.9.1   INTERGENERIC COMMUNICATION

Instead of using an antenna of *K. lamarcki* in a GC-FID/EAD analysis of an extract of the male abdominal secretion of this species, an antenna of another dung beetle, *Pachylomerus femoralis*, was on occasion inadvertently used as sensing element. The latter species is slightly larger than *K. lamarcki*. The distribution of these two dung beetles overlaps in areas with sandy soil, which appears to be a prerequisite for colonization by *P. femoralis*. A series of highly volatile constituents of the *K. lamarcki* secretion elicited strong responses in the antenna of *P. femoralis*. Remarkably sharp and strong responses were reproducibly obtained in subsequent analyses carried out under the same experimental conditions. A typical set of gas chromatograms obtained in a GC-FID/EAD analysis of an extract of the abdominal secretion of male *K. lamarcki* is depicted in Figure 3.14. The majority of the EAD-active constituents identified by GC-MS analysis and by MS and retention time comparison with authentic synthetic compounds are listed in Table 3.3 (Burger et al. 1995a).

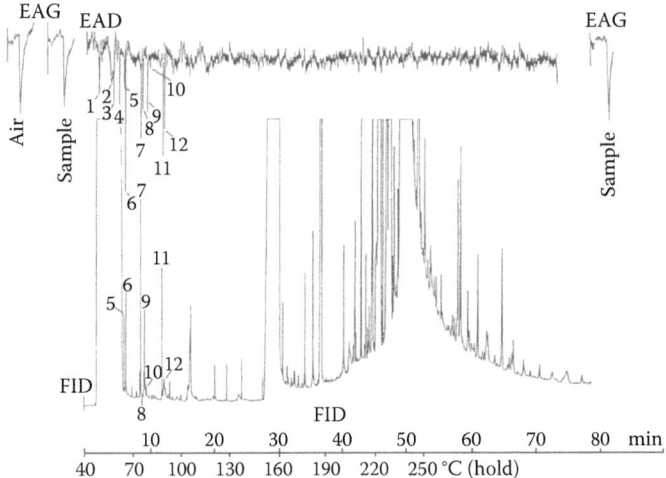

FIGURE 3.14 GC-FID/EAD analysis of the male abdominal secretion of *K. lamarcki* using an antenna of a female *P. femoralis*. Corresponding peaks in the FID and EAD traces are numbered consecutively in order of elution. For peak identities, see Table 3.3. (From Petersen, W. G. B. 1990. Aanwending van elektroantennografiese metodes in die chemiese karakterisering van semioverbindings van kewers. MSc thesis, University of Stellenbosch.)

TABLE 3.3

Constituents of the Male Abdominal Secretion of *K. lamarcki* that Elicited Antennal Responses in the Excised Antennae of *P. femoralis*

Peak Number in Figure 3.14	Constituent
1	Unidentified
2	Unidentified
3	Methyl propanoate
4	Unidentified
5	Ethyl propanoate
6	Methyl butanoate
7	Ethyl butanoate
8	Unidentified
9	Methyl 4-pentenoate
10	Methyl pentanoate
11	Ethyl 4-pentenoate
12	Ethyl pentanoate

Whereas *K. lamarcki* is an industrious producer of dung balls, *P. femoralis* has never yet been observed constructing dung balls from dung taken from dung pats or middens. However, it has been found rolling dung fragments that mostly had irregular shapes. A relatively strong EAD response was observed at a retention time between those of ethyl butanoate 7 and methyl 4-pentenoate 9 in Figure 3.14. No indication of the presence of this compound could be found in either the FID trace or the total ion current (TIC) trace obtained in a GC-MS analysis of the extract. Constituents 1, 2, 4, and 8 remained unidentified. Although not present in the sex attractant secretion of *K. lamarcki*, FID/EAD analyses were also carried out with the methyl and ethyl esters of the *E* and *Z* isomers of both 2-pentenoic acid and 3-pentenoic acid. It was found that all of these esters elicited EAD responses that were of similar strength as those elicited by ethyl and methyl 4-pentanoate in *P. femoralis* antennae.

The possibility that the identified compounds could have been adsorbed from the horse dung on which the dung beetles were fed was investigated by carrying out headspace gas analyses on samples of horse dung using solid-phase microextraction (Arthur and Pawliszyn 1990) as concentration technique in conjunction with GC-FID/EAD analysis. No indication was found that any of the EAD-active esters were present in the horse dung. In a control experiment, the dung was spiked with small amounts of methyl and ethyl pentanoate. These two esters were easily detected in the spiked samples (Burger et al. 1995a).

In field tests, approximately equal numbers of *P. femoralis* were attracted to pitfall traps baited with horse dung and traps baited with a mixture of the compounds listed in Table 3.3. In the absence of dung, methyl and ethyl (*Z*)-and (*E*)-2-butenoic acid (crotonic acid) did not attract any *P. femoralis* to pitfall traps, whereas the individual saturated esters listed in Table 3.3 attracted small numbers of this species.

Taking the results of the almost forgotten GC-FID/EAD analysis depicted in Figure 3.5 into consideration, it was realized that some or even all of the esters that elicited responses in antennae of *P. femoralis* could be constituents of the male secreted sex attractant of *K. lamarcki* or of some other pheromone of this species—unfortunately only after our research on dung beetles had largely been completed.

In the absence of sufficient information, any explanation of the response in *P. femoralis* to the presence of the short-chain esters in the male abdominal secretion of *K. lamarcki* must necessarily be of a speculative nature. However, it is possible that *P. femoralis*, not being capable of forming dung balls, employs the volatile compounds identified in the abdominal secretion of *K. lamarcki*, as a kairomone to find dung of a suitable shape, such as a dung ball, for transportation to its burrow. The calling behavior of *K. lamarcki* thus exposes it to being robbed by the aggressive *P. femoralis*. In contrast, male *K. nigroaeneus* produce abdominal secretion only after having safely buried their dung balls. In captivity in the greenhouses in Stellenbosch, secreting *K. lamarcki* males were often disturbed by *P. femoralis*, while *K. nigroaeneus* males did not seem to elicit similar interest from *P. femoralis*. Further interesting aspects of the ethology and chemical ecology of *P. femoralis* are discussed in Section 3.10.

3.9.2 INTERSPECIFIC COMMUNICATION

The discovery of the response elicited in the antennae of *P. femoralis* by certain constituents in the abdominal secretion of *K. lamarcki* led to an investigation of the possibility that, for some reason and under certain conditions, coexisting dung beetle species might be semiochemically aware of the presence of other species in their home range. Only two of the four *Kheper* species under investigation coexist in the Mkuzi Game Reserve and were used in an attempt to test this hypothesis. In GC-FID/EAD analyses, antennae of male as well as female *K. nigroaeneus* responded reproducibly to 3-methyl-2-butanone, skatole, and (*E*)-nerolidol, and nonreproducibly to 2,6-dimethyl-5-heptenoic acid (compounds that were identified in extracts of the male abdominal secretion of *K. lamarcki*). As in other analyses in which antennae of *K. lamarcki* were used as sensing elements, the VOCs of the abdominal secretion of *K. nigroaeneus* did not elicit responses in the antennae of the former species with the exception of skatole (Petersen 1990). At this stage there is no plausible explanation for these results. It is possible that many additional compounds could be involved in this type of interspecific semiochemical interaction. It is an interesting phenomenon that could lead to the discovery of interesting information, but it would require the development of an absolutely reliable EAD setup and experimental protocol.

3.10 PECULIAR BEHAVIOR IN *P. FEMORALIS*

Due to their production of a visible sex-attracting secretion, research was mainly focused on the four species of the genus *Kheper* that are found in South Africa. However, large numbers of *P. femoralis* were often found together with *K. lamarcki* in pitfall traps and sometimes on a few dung middens in the Mkuzi Game Reserve, and it was thus possible to observe the interesting nesting behavior of the latter insect. Both species of the African dung beetle genus *Pachylomerus* are found in southern Africa, but they do not overlap in distribution. *P. femoralis* inhabits wetter areas stretching from the northern parts of KwaZulu-Natal through the northern parts of South Africa and into northeastern Botswana. *P. opaca* is smaller than *P. femoralis* and inhabits the drier northwestern Cape and southern Namibia (Tribe 1976). *P. femoralis* is a large dung beetle (approximately 40 mm × 25 mm) with a highly developed prothorax and forelegs.

According to Tribe, the typical behavior of *P. femoralis* is to construct an unbranched tunnel within about 300 mm of a dung pat by digging with the foretibiae and clypeus and then turning around in the burrow and pushing out the loosened soil or sand, using the prothorax as a shovel. The excavated soil is used to build a ramp that leads to the dung pat. The burrow is provisioned with dung by the beetle making several trips to and from the dung. Irregular pieces of dung are then butted with the head or rolled to the burrow entrance and pushed inside with the head. *P. femoralis* is very aggressive, defending both a large section of the dung pat and the ramp against any intruder.

Although we never observed the construction of dung balls from dung taken from a dung pat or midden by *P. femoralis*, this species has on a few occasions been found rolling dung fragments along the roads of the reserve. The dung fragments had irregular shapes, not having been formed into balls, and were relatively small compared to the enormous dung balls that the slightly smaller *K. lamarcki* usually constructs. Another peculiarity is that, on the few occasions that this otherwise rare rolling behavior was observed, many *P. femoralis* were observed rolling dung fragments along the roads of the reserve during the late afternoon—and always when a thunderstorm was expected. This behavior was not repeated the following day in the same area after the rain. It is possible that this ball-rolling behavior could be induced by a drop in atmospheric pressure or other climatological conditions.

During field trials with synthetic analogs of the constituents of the male abdominal secretion of *K. lamarcki* in November 1989, we observed a *P. femoralis* rolling one of the seeds of a spineless monkey orange tree, *Strychnos madagascariensis* Poiret. Subsequent work revealed this apparently aberrant behavior to be quite common in this insect, at least in the Mkuzi Game Reserve. This phenomenon has not been described before, despite the intensive research that has been carried out on dung beetles in the Mkuzi and Hluhluwe Game Reserves during the Australian Dung Beetle Project.

The spineless monkey orange tree, also known as the yellow or black monkey orange tree in certain parts of South Africa, is found in large numbers in those parts of the Mkuzi Game Reserve with light sandy soil. It bears fruit with the shape, size, and color of an orange, but with an extremely hard shell. The fruit usually starts ripening towards the middle of November in the northern parts of KwaZulu-Natal. Each fruit contains 10 or more hard seeds to which a thin layer of soft flesh adheres. The seeds have irregular shapes, an average diameter of about 20–30 mm, and are interspaced with thin layers of fleshy material that has a strong pleasant flavor—almost like that of the mango. The fleshy material is consumed by animals that have strong enough jaws to fracture the shell of the fruit. Because the edible parts of the fruit form a relatively small proportion of the fruit, it has not been commercially exploited for human consumption. The dung beetle rolling the seed of this fruit must have found one that had been left intact with its fruit flesh by one of these animals. In a first test carried out immediately after this peculiar behavior had first been observed, two of 12 seeds simply put out on the ground were removed by *P. femoralis* within a few hours. In another experiment carried out in the same area, 24 *P. femoralis* beetles were caught in two pitfall traps baited with seeds of the fruit, whereas only one *P. femoralis*, in addition to several dung beetles from other species, was caught in nine traps baited with horse dung. No dung beetles were caught in traps baited with banana and pawpaw, the only fruit available commercially in the Mkuzi area early in November.

Qualitative analyses were carried out by trapping the VOCs responsible for the flavor of the fruit on an open tubular trap (Grob and Habich 1985) coated with polydimethylsiloxane (film thickness 15 μm). The trapped volatiles were desorbed from the trap and analyzed by GC-MS. The compounds that could possibly be responsible for the attraction of *P. femoralis* to the fruit flesh of the spineless monkey orange were distinguished from other inactive volatiles using FID and EAD detection in parallel. A typical set of FID/EAG gas chromatograms of the headspace volatiles from the fruit flesh of the spineless monkey orange is depicted in Figure 3.15 (Burger and Petersen 1991).

Regarding the relative sizes of the EAD signals, these analyses were remarkably reproducible. The EAD signals were sharper, and with a few exceptions, their signal-to-noise ratios were higher than those obtained in GC-FID/EAD analyses of the VOCs of the male abdominal secretion of, for example, *K. bonellii*. Male and female antennae produced identical responses. The antennae of *P. femoralis* were viable for more than 8 hours. The EAD-active compounds are listed in Table 3.4.

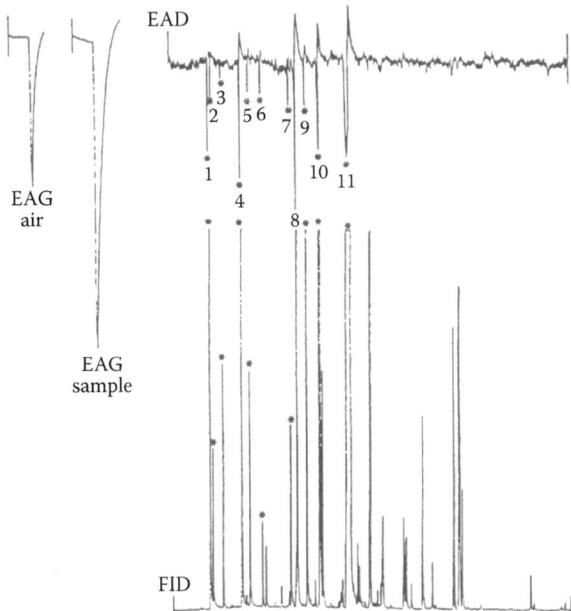

FIGURE 3.15 GC-FID/EAD analysis of the headspace gas volatiles from the fruit of *Strychnos madagascariensis* using a female *P. femoralis* antenna as a sensing element. Corresponding peaks in the FID and EAD traces are numbered consecutively in order of elution. For peak identities, see Table 3.4. (From Petersen, W. G. B. 1990. Aanwending van elektroantennografiese metodes in die chemiese karakterisering van semioverbindings van kewers. M.Sc. thesis, University of Stellenbosch.)

Many other constituents related to the EAD-active compounds, such as 1-hexanol, butanoic acid, hexanoic acid, and a large number of saturated and unsaturated esters from 2-methylpropyl ethanoate (2-methylpropyl acetate) to butyl dodecanoate that are present in the spineless monkey orange (Pretorius et al. 1987), gave no, or only very weak, EAD responses.

To obtain information on the composition of the odor plume of the fruit under field conditions, the VOCs released by the fleshy parts of a spineless monkey orange were determined in air flowing at approximately 0.16 km/h over 20 seeds and the fleshy parts of a single spineless monkey orange. The volatiles were trapped on an ultrathick film trap (film thickness 145 μm) (Burger et al. 1991). The quantitative composition of the EAD-active compounds in the odor plume of the fruit is given in Table 3.4.

Although dispensing a mixture of these EAD-active compounds from a small vial via a filter-paper wick was not expected to release the compounds in the same ratio in which they are present in the odor plume of the fruit, eight and 14 *P. femoralis* were caught within 2 hours, respectively, in two pitfall traps baited with the synthetic material, whereas five and 32 beetles were found, respectively, in two traps baited with spineless monkey orange. It was furthermore found that the individual compounds and mixtures of two or three of the compounds were also attractive to

TABLE 3.4

Constituents of the Headspace Gas of the Fruit of the Spineless Monkey Orange Tree, *Strychnos madagascariensis*, That Elicited Antennal Responses in Excised Antennae of *Pachylomerus femoralis*

Peak Number in Figure 3.15	Constituent	Quantity (µg/ml Headspace Gas)[a]
1	1-Butanol	0.16
2	Methyl butanoate	0.005
3	Ethyl 2-methylpropanoate	0.02
4	Ethyl butanoate	0.38
5	Butyl ethanoate	0.02
6	Ethyl 2-methylbutanoate	0.01
7	Propyl butanoate	0.04
8	Butyl propanoate	0.06
9	Methyl hexanoate	0.05
10	Butyl 2-methylpropanoate	0.26
11	Butyl butanoate	2.04

[a] Headspace gas samples were withdrawn from a stream of nitrogen flowing at a linear velocity of 0.16 km/h over the contents of one spineless monkey orange at a point 300 mm downstream from the fruit.

P. femoralis; three insects were found in a trap baited with a 1:1 mixture of methyl and ethyl butanoate, and eight insects in a trap baited with a 4:4:1 mixture of butyl 2-methylpropanoate, butyl butanoate, and butyl propanoate.

As in field tests with synthetic components of the abdominal secretion of male *K. lamarcki* combined with horse dung, large differences were found in the number of beetles caught in traps baited with either dung or fruit. One trap, for example, contained 38 *P. femoralis*, whereas not a single beetle of this species was found in a trap situated 100 m from the first one and baited with the same material. Due to dense vegetation, it was impossible to observe the movement of animals in that area and to determine how far a trap was situated from a fresh dung source or from the nearest spineless monkey orange tree. Under these circumstances, the resulting quantitative data were not amenable to statistical analysis. However, the results did indicate that the attractiveness of the mixture of synthetic compounds for *P. femoralis* is at least comparable to that of the fresh spineless monkey orange.

At the time that the field tests were carried out, large numbers of fruit (some of them opened by primates) were lying beneath the trees and the air was permeated with the smell of the fruit in areas with a large population of spineless monkey orange trees. No beetles were caught in traps baited with the fruit in these areas, whereas more than 30 *P. femoralis* were found in traps in the same area baited with horse dung. In contrast, *P. femoralis* beetles were attracted more strongly to the fruit than to horse dung in areas where there is much animal activity and where spineless monkey orange trees are scarce or the fruit has not yet ripened.

In spite of fierce competition for food on relatively small quantities of horse dung placed in the field, no dung ball formation or rolling of balls or fragments of dung from these dung sources by *P. femoralis* was observed. In contrast, practically all the *P. femoralis* attracted to the contents of spineless monkey oranges started frantically rolling seeds, either immediately on arriving at the fruit or after inspecting the fruit for a few seconds. A few beetles attempted to push seeds into existing holes in the ground.

The attraction to fruit or plant material is not unique to *P. femoralis*. Halffter and Matthews (1966) have reported the attraction of dung beetles to carrion and rotting fruit. Burger et al. (1991) found a few of the small golden-brown dung beetles, *Proagoderus aureiceps* d'Orbigny, in many of the traps spiked with spineless monkey orange. It is difficult to suggest a plausible explanation for this phenomenon. There are many imponderables to consider: Why do these dung beetles roll these seeds, of which the adhering fruit flesh is practically dried out by the time the seeds are buried? Could these dung beetles have learnt that rolling the seeds is a futile exercise in areas where there is an ample supply of the fruit? Are they so strongly territorial that they do not forage in neighboring areas, although such areas are not more than a few seconds apart at the speed at which they can fly? Could the confusing results of bioassays be ascribed to the attraction to either fruit or dung being overridden by the nature of the soil in the areas where the field tests were carried out?

In the light of the paucity of information currently available, an explanation of the phenomenon could be based on the attraction of *P. femoralis* by the male abdominal secretion of *K. lamarcki*, as discussed above, and the coincidental presence of identical, or at least structurally related, compounds in spineless monkey oranges. A second explanation could be based on the possibility that the attractants present in the fruit are probably excreted by the primates that consume the fruit (especially when they overindulge) and it is thus possible that *P. femoralis* could associate the aroma of the fruit with the availability of the feces of these animals.

3.11   DEFENSIVE MECHANISMS IN *ONITICELLUS EGREGIUS*

Elaborate chemical defense and associated escape behavior is found in the endocoprid dung beetle *Oniticellus egregius* Klug (Davis 1989). This species constructs its brood ovoids of dung in the soil immediately under the edge of the dropping, enveloping each ovoid in a soil shell before clearing the loose earth around the ovoids to produce a brood chamber (Davis 1989). This species has been recorded mainly from the course dung of elephant, rhinoceros, and zebra (Davis 1977).

Several protective mechanisms are utilized by *O. egregius*. The dorsal surface of this insect (9.8–15.5 mm) is metallic blue-black with a yellow border (Figure 3.1h), while the ventral surface is mottled yellow and gold. If disturbed, *O. egregius* flip themselves onto their backs and exhibit thanatosis, holding the middle and hind legs away from the body. This exposes the ventral surface that is colored similarly to the fibers of dry rhinoceros or elephant dung and renders them inconspicuous (Figure 3.1i). Another escape mechanism could also be employed: the forelegs that are held close to the body are suddenly released with sufficient force to lift the beetle as high as 60 cm into the air; it flips and becomes cryptic where it lands among the shredded dung. It simultaneously releases a brown fluid from the lateral edge of the anterior abdominal segments just posterior to the hind legs (Davis 1977). This secretion, reminiscent of oil of wintergreen, has been identified as methyl salicylate (oil of wintergreen) and 1,4-benzoquinone (Burger et al. 1995b). 1,4-Benzoquinone and its derivatives have been identified in several orders of the Diplopoda and Insecta and these compounds appear to be relatively common in the defensive secretions of the Coleoptera (e.g., Blum 1981). The esters of silacylic acid seem to be much less well represented. Methyl salicylate has been found in the secretion of the carabid beetle (Schildknecht et al. 1968). Ants and termites frequently inhabit dung pats in winter where they would encounter *O. egregius* and would likely be repelled by the wintergreen odor.

3.12   CONCLUSIONS, PERSPECTIVES, AND PROSPECTS

The identification of the constituents of the male abdominal sex-attractant secretions of the four *Kheper* species was the main focus of this first investigation into the semiochemistry of dung beetles in South Africa. The research was only partly successful, mainly due to the complexity of the volatile organic fraction of the male sex-attracting secretions, the time-consuming development of appropriate instrumentation and methodology for GC-FID/EAD analyses, and the untrustworthiness of the results of bioassays, which resulted from the fact that dung and not the sex attractant is the primary factor in the attraction of both males and females to the secretion-producing beetle. Nevertheless, some interesting information came to light.

In three of the four species, several of the constituents that reproducibly elicited EAD responses in male as well as female antennae are terpenoids or compounds with terpenoid-like structures that lack one or two carbon atoms. The role of the long-chain compounds in the male abdominal secretions was not satisfactorily elucidated. However, even though not reproducible with instrumentation available at the time, the EAD responses that some of these compounds elicited in the antennae of the insects could be seen as an indication that further investigation into the role of these compounds in chemical signaling in these species might be worthwhile.

The white abdominal pheromone-disseminating secretions of three of the four South African *Kheper* species were investigated. They were found to consist of only two proteins, with molecular masses of approximately 17 kDa and approximately 31 kDa, respectively. The proteins are similar, but not identical, in the three species. Semiochemical communication appears to be more complex in the *Coleoptera* than in other orders. Investigation of the phenomenon in dung beetles was complicated by weak EAD responses in GC-FID/EAD analyses. The unreliability of bioassays in the field was a further complicating factor. The research carried out to date has only scratched the surface of a wide and extremely interesting topic. The main benefit of this part of the research was that it demonstrated the potential that lies in this field of research for entomologists and chemists interested in semiochemical communication in the *Coleoptera*.

An investigation into the possible role of semiochemicals in the courtship behavior between male and female after arrival of the insects on a dung pat could be a difficult, but nevertheless very rewarding endeavor, particularly because of the extremely important ecological role of dung beetles. In this regard, the role of these insects in the reduction of the production of methane in the dung of herbivores has to date been largely disregarded.

In contrast to the somewhat disappointing results of the investigation of the sex attractants of the *Kheper* species, the serendipitously discovered seed-rolling behavior of *P. femoralis* produced interesting and exciting results. In GC-FID/ EAD analyses of the VOCs present in the fruit of the spineless monkey orange tree, *S. madagascariensis*, reproducibly elicited strong EAD responses in both male and female antennae of this dung beetle.

Identical and similar short-chain esters were identified in extracts of the male abdominal secretion of *K. lamarcki*. The success of this part of the research largely made up for many disappointments in the research on the *Kheper* species. Unfortunately, the limited time that was annually available for the dung beetle research precluded pursuing this research to a satisfactory conclusion. The bioassays should be repeated with the synthetic esters as well as with the fruit in areas where there are no spineless monkey orange trees or when there is no ripe fruit available in the field.

The Laboratorium vir Ekologiese Chemie Universiteit van Stellenbosch (LECUS) has now concluded its research on dung beetles. It is hoped that the research discussed in this chapter will encourage other scientists to become interested in a still largely unexplored and very lucrative field of research that holds promises of many exciting discoveries—there are so many questions still to be answered.

ACKNOWLEDGMENTS

I am indebted to Dr. Zenda Ofir (née Munro) and Dr. Warren Petersen, who devoted a total of 18 difficult years of their lives to carry out the research discussed in this chapter, to my wife, Wina, who collected the dung beetle secretions at unbearable temperatures, and to the KwaZulu-Natal Parks Board for permission to collect dung beetles and carry out behavioral experiments in the Mkuzi and Hluhluwe Game Reserves. I am grateful for the invaluable entomological expertise and assistance of Dr. Geoff Tribe, who also read the manuscript to verify its entomological content. The research has been funded by the National Research Foundation of South Africa and by Stellenbosch University.

REFERENCES

Arthur, C. L. and J. Pawliszyn. 1990. Solid phase microextraction with thermal desorption using fused silica optical fibers. *Anal. Chem.* 62:2145–2148.

Beens, J., M. Adahchour, R. J. J. Vreuls, K. V. Klaas and U. A. Th. Brinkman. 2001. Simple, non-moving modulation interface for comprehensive two-dimensional gas chromatography. *J. Chromatogr.* A 919:127–132.

Blum, M. S. 1981. *Chemical Defences of Arthropods.* Academic Press, New York.

Bornemissza, G. F. and C. H. Williams. 1970. An effect of dung beetle activity on plant yield. *Pedobiologia* 10:1–7.

Burger, B. V., M. Röth, M. le Roux, H. S. C. Spies and H. Geertsema. 1978. The chemical nature of the defensive larval secretion of the citrus swallowtail, *Papilio demodocus.* *J. Insect Physiol.* 24:803–805.

Burger, B. V., Z. Munro, M. Röth, H. S. C. Spies, V. Truter, G. D. Tribe and R. M. Crewe. 1983. Studies on the pheromones of Scarabaeinae, I. Composition of the heterogeneous sex attracting secretion of the dung beetle, *Kheper lamarcki. Zeitschrift Naturforsch.* 38c:848–855.

Burger, B. V., Z. Munro and W. F. Brandt. 1990. Pheromones of the Scarabaeinae, II. Composition of the pheromone disseminating carrier material secreted by male dung beetles of the genus *Kheper. Zeitschrift Naturforsch.* 45c:863–872.

Burger B. V. and W. G. B. Petersen. 1991. Semiochemicals of the Scarabaeinae, III. Identification of an attractant for the dung beetle *Pachylomerus femoralis* in the fruit of the spineless monkey orange tree, *Strychnos madagascariensis. Zeitschrift Naturforsch.* 46c:1073–1079.

Burger, B. V., W. G. B. Petersen and G. D. Tribe. 1995a. Semiochemicals of the Scarabaeinae, IV: Identification of an attractant for the dung beetle *Pachylomerus femoralis* in the abdominal secretion of the dung beetle *Kheper lamarcki. Zeitschrift Naturforsch.* 50c:675–860.

Burger, B. V., W. G. B. Petersen and G. D. Tribe. 1995b. Semiochemicals of the Scarabaeinae, V: Characterization of the defensive secretion of the dung beetle *Oniticellus egregius. Zeitschrift Naturforsch.* 50c:681–684.

Burger, B. V. and W. G. B. Petersen. 2002. Semiochemicals of the Scarabaeinae, VI. Identification of EAD-active constituents of abdominal secretion of male dung beetle *Kheper nigroaeneus. J. Chem. Ecol.* 28:501–513.

Burger, B. V., W. G. B. Petersen, W. G. Weber and Z. M. Munro. 2002. Semiochemicals of the Scarabaeinae, VII. Identification and synthesis of EAD-active constituents of abdominal sex attracting secretion of the male dung beetle *Kheper subaeneus. J. Chem. Ecol.* 28:2527–2539.

Burger, B. V., W. G. B. Petersen, B. T. Ewig, J. Neuhaus, G. D. Tribe, H. S. C. Spies and W. J. G. Burger. 2008. Semiochemicals of the Scarabaeinae, VIII. Identification of active constituents of the abdominal sex-attracting secretion of the male dung beetle, *Kheper bonellii*, using gas chromatography with flame ionization and electroantennographic detection in parallel. *J. Chromatogr. A* 1186:245–253.

Davis, A. L. V. 1977. The endocoprid dung beetles of southern Africa (Coleoptera: Scarabaeidae). MSc thesis, Rhodes University, Grahamstown, South Africa.

Davis, A. L. V. 1989. Nesting of Afrotropical Oniticellus (Coleoptera: Scarabaeidae) and its evolutionary trend from the soil to dung. *Ecol. Entomol.* 14:11–21.

Dormont, L., G. Epinat and J. P. Lumaret. 2004. Trophic preferences mediated by olfactory cues in dung beetles colonizing cattle and horse dung. *Environ. Entomol.* 33:370–377.

Dormont, L., S. Rapior, D. B. McKey and J. P. Lumaret. 2007. Influence of dung volatiles on the process of resource selection by coprophagous beetles. *Chemoecol.* 17:23–30.

Edwards, P. B. and H. H. Aschenborn. 1988. Male reproductive behavior of the African ball-rolling dung beetle, *Kheper nigroaeneus* (Coleoptera: Scarabaeidae). *The Coleopterists Bull.* 42:17–27.

Ferreira, M. C. 1972. Os escarabideos de África (Sul do Sáara). *Revista. Ent. Moçamb.* (1968-1969/1972) 11:1–1088.

Grob, K. and A. Habich. 1985. Headspace gas analysis: The role and the design of concentration traps specifically suitable for capillary gas chromatography. *J. Chromatogr. A* 321:45–58.

Grob, K. and R. Müller. 1982. Some technical aspects of the preparation of a retention gap in capillary gas chromatography. *J. Chromatogr. A* 244:185–196.

Hadley, N. F. 1979. Wax secretion and color phases of the desert tenebrionid beetle *Cryptoglossa verrucosa* LeConte. *Science* 203:367–369.

Halffter, G. and E. G. Matthews. 1966. The natural history of dung beetles of the subfamily Scarabaeidae. *Folia Entomol. Mexicana* 12–14:1–312.

Heinrich, B. and G. A. Bartholomew. 1979. Roles of endothermy and size in inter- and intra-specific competition for elephant dung in an African dung beetle, *Scarabaeus laevistriatus*. *Physiol. Zool.* 52:484–496.

Hughes, R. D. 1970. The bushfly. *Aust. Nat. Hist.* 16:331–334.

Leal, W. S. 1998. Chemical ecology of phytophagous scarab beetles. *Ann. Rev. Entomol.* 43:39–61.

Marshall, A. T. 1974. Paraffin tubules secreted by the cuticle of an insect, *Epipyrops anamala* (Epipyropidae: Lepidoptera). *J. Ultrastruct. Res.* 47:41–60.

Munro, Z. M. 1988. Semiochemiese kommunikasie: Ontwikkeling en toepassing van analitiese tegnieke vir die bepaling van die chemiese samestelling van die abdominale afskeiding van *Kheper lamarcki, K. Subaeneus* en *K. nigroaeneus* (Coleoptera: Scarabaeidae). PhD dissertation, University of Stellenbosch, South Africa.

Nair, S. and S. K. Braman. 2012. A scientific review on the ecology and management of the azalea lace bug *Stephanitis pyrioides* Scott (Tingidae: Hemiptera). *J. Entomol. Sci.* 47:247–263.

Noverr, M. C. and G. B. Huffnagle. 2004. Regulation of *Candida albicans* morphogenesis by fatty acid metabolites. *Infect. Immun.* 72:6206–6210.

Petersen, W. G. B. 1990. Aanwending van elektroantennografiese metodes in die chemiese karakterisering van semioverbindings van kewers. MSc thesis, University of Stellenbosch, South Africa.

Petersen, W. G. B. 1997. Semiochemiese kommunikasie by miskruiers van die genus Kheper (Scarabaeinae). PhD dissertation, University of Stellenbosch, South Africa.

Pretorius, V., E. R. Rohwer, A. Rapp and H. Mandery. 1987. Volatile constituents of the spineless monkey orange. *Dtsch. Lebensm. Rundsch.* 83:180–182.

Rohwer, E. R., V. Pretorius and P. J. Apps. 1986. Simple press-fit connectors for flexible fused silica tubing in gas-liquid chromatography. *J. High Resolut. Chromatogr.* 9:295–297.

Schildknecht, H., H. Winkler, B. Krauss and U. Maschwitz. 1968. Über Arthropoden-Abwehrstoffe, XXVIII: Über das Abwehrsekret von *Idiochroma dorsalis. Zeitschrift Naturforsch.* 23b:46–49.

Simmons, L. W. and T. J. Ridsdill-Smith. 2011. *Ecology and Evolution of Dung Beetles.* Chichester, UK, Wiley-Blackwell.

Tribe G. D. 1976. The ecology and ethology of ball-rolling dung beetles (Coleoptera: Scarabaeidae). MSc (Agric.) thesis, University of Natal, South Africa.

Waterhouse, D. F. 1974. The biological control of dung. *Sci. Am.* 230:100–109.

Wilson, S. R. and K. A. Prodan. 1976. The synthesis and stereochemistry of cascarillic acid. *Tetrahedron Lett.* 17:4231–4234.

4 Pheromone Reception in Insects
The Example of Silk Moths

Karl-Ernst Kaissling

CONTENTS

4.1 Introduction 99
4.2 Aspects of Pheromone Communication 100
4.3 Insect Antennae and Olfactory Sensilla 104
4.4 Electrophysiology 107
4.5 Sensitivity of Pheromone Receptor Neurons and Behavioral Responses..... 110
4.6 Elementary Receptor Potentials 114
4.7 Inhibition of Pheromone Receptor Neurons 117
4.8 Concentration Detectors and Flux Detectors 120
4.9 Olfactory Transduction, Extracellular 121
4.10 Diffusion on the Hairs 124
4.11 Kinetic Model 124
4.12 Five Functions of the Pheromone Binding Protein 126
4.13 Pheromone Degradation and Deactivation 128
4.14 Receptor Molecules, Ion Channels, and Sensory Neuron Membrane
     Protein.. 130
4.15 Olfactory Transduction, Intracellular 132
4.16 Temporal Coding 134
Acknowledgments.. 134
References... 134

4.1 INTRODUCTION

Pheromones, chemical signals for intraspecific communication (Karlson and Luescher 1959), are usually blends of chemical compounds in species-specific mixtures. Airborne pheromones of moths often consist of only two or three chemical components, each of which is perceived by a separate type of receptor neurons. Each of these neurons, called olfactory specialists (Boeckh et al. 1965) is tuned to one biologically significant compound; it responds to compounds other than the key compound only if presented at 10- to 10,000-fold higher stimulus concentrations. The composition of a pheromone blend is represented by the pattern of excitations across the types of specialists (Baker et al. 2004). Odor specialists are known also for compounds other than pheromones, such as plant volatiles or carbon dioxide. Less sharply tuned olfactory neurons have been called generalists. They may show varying and overlapping response spectra such as found in moths (Schneider et al. 1964) or in pine weevils (Mustaparta 1975).

Insect antennae—sense organs for various sensory modalities including the function of noses—provide simple and convenient subjects for morphological, electrophysiological, and biochemical studies. From the long hair sensilla of moths one may record the responses of individual olfactory neurons by extracellular electrodes. This review will cover various aspects of pheromone communication with an emphasis on reception and little on pheromone-controlled behavior. It will focus on two species of silk moths, *Bombyx mori* and the saturniid moth *Antheraea polyphemus*, with a few added remarks on other insects.

4.2 ASPECTS OF PHEROMONE COMMUNICATION

Olfaction in insects is described in numerous reviews and books (Blomquist and Vogt 2004; Cardé and Minks 1997; Field et al. 2000; Hansson 1999; Hildebrand and Shepherd 1997; Vogt 2005; Jacquin-Joly and Lucas 2005; Kaupp 2010; Mustaparta 2002; Pelosi 1996; Pelosi et al. 2006; Pernollet and Briandt 2004; Priesner 1979; Schneider 1984, 1992; Steinbrecht 1999; Tegoni et al. 2004). Studies of pheromone reception in insects enhance our general knowledge on chemoreception, on the control of behavior by chemical stimuli, and also provide a basis for insect pest control (Karg and Suckling 1999; Hummel and Hecker 2012).

Volatile insect pheromones are produced by a large variety of glands located at various places on the insect body. For instance, the sex attractant bombykol ((E,Z)-10, 12- hexadecadiene-1-ol) (Butenandt and Hecker 1961; Butenandt et al. 1959) is secreted by the abdominal sacculi laterales (Steinbrecht 1964) of the female silk moth *B. mori* (Figure 4.1a), together with traces of the (E,E)-isomer of the alcohol (Kasang et al. 1978a) and the analogous (E,Z)-aldehyde bombykal (Kasang et al. 1978b). Bombykol alone is able to elicit a pattern of sexual behavior of the male moth (Figure 4.1c), such as wing vibration, walking, and turning so that it is headed upwind. The excitatory effect of bombykol is partially blocked if bombykol is presented together with bombykal (Kaissling et al. 1978). In fact bombykal may elevate the threshold concentration for bombykol up to 1000-fold (Figure 4.5a later in the chapter). Extremely strong stimuli of bombykal would rapidly adapt the bombykal neuron, but elicit nerve impulses in the bombykol neuron (cf. Figures 4.8d, e, later in the chapter) and cause a behavioral response (Figure 4.5a).

The behavioral inhibition occurs by central processing of the excitatory responses of bombykol and bombykal receptor neurons. The biological function of the inhibition by bombykal is unknown. The behavioral dose-response curve obtained with the bombykol/bombykal blend released by the female increases over a range of stimulus intensities, which is wider than for bombykol alone (Figure 4.5a, dashed curve). In other species of moths the sexual behavior may be blocked by a pheromone component of a different but closely related moth species. The moth whose behavior is blocked may even possess a receptor-neuron type tuned to the behavioral inhibitor from the other species.

One example for behavioral inhibition between species is provided by the gypsy and nun moths (*Lymantria dispar* and *L. monacha*, respectively), which live

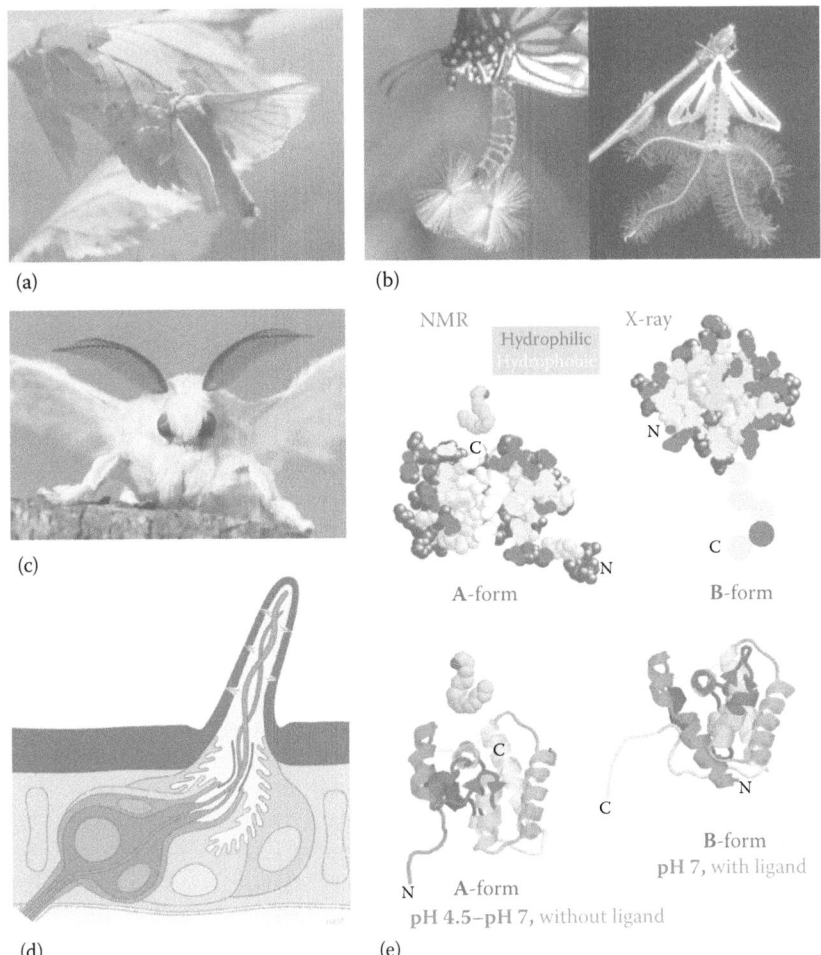

(a) (b) (c) (d) (e)

FIGURE 4.1 **(See color insert.)** (a) Female *Bombyx mori* in calling position, with everted abdominal pheromone glands. (b) Males of the milkweed or monarch butterfly *Tirumala peti-verana* (Danainae, left hand) and the Asian arctiide moth *Creatonotos gangis* with expanded androconia. (Courtesy of M. Boppré.) (c) Male *Bombyx mori* in alerted position, with combed antennae elevated. (Courtesy of R.A. Steinbrecht.) (d) Diagram of an olfactory sensillum trichodeum with two receptor cells (red) and three auxiliary cells (green). In blue: Sensillum lymph. (Courtesy of R.A. Steinbrecht.) (e) Structures of bombykol and the bombykol-binding protein, from X-ray and NMR analysis. Top: Calotte models. Bombykol molecule in grey, with red OH-group. Shape of bombykol as if bound to B-form. A-form with C-terminus occupying the inner binding cavity in white. B-form with bombykol: Five amino acids of the unfolded C-terminus, not shown by x-ray, are added in an arbitrary shape. Bottom: Backbone structures of the protein. A-form with C-terminal helix. (Made by using RASMOL, with data from Sandler B.H. et al. 2000. *Chem Biol* 7:143–151; Horst R. et al. 2001. *Proc Natl Acad Sci U S A* 98:14374–14379; Reproduced from Kaissling K.E. 2013. *J Comp Physiol A* 199:879–896. With kind permission from Springer Science+Business Media.)

sympatrically in parts of Europe and share (+)-disparlure as attractant (Hansen 1984). In addition to the attractant the female nun moth produces (–)-disparlure (2-methyl-7R,8S-epoxy-octadecane), which keeps the male gypsy moth from approaching her. The male gypsy moth has two receptor neurons, each specialized for one of the two enantiomers. The male nun moth, however, does not have receptor cells for the inhibitor and is attracted by females of both species.

Many pheromones of female moths are straight-chain unsaturated hydrocarbons with a terminal alcohol, aldehyde, or acetate function and are synthesized by abdominal glands (see database of pheromones and semiochemicals, El-Sayed 2012). When two or three components are attractive in a certain species-specific ratio, the relative amount of a synergistic component can be less than 0.1%. The same components mixed in a different ratio might represent the pheromone of a different species (Figure 4.4a later in the chapter).

The female moths of *Bombyx mori* are not able to smell their own pheromone. Although the female sensilla trichodea look similar to those of the male with neurons for bombykol and bombykal, the two neurons innervating the female sensilla are insensitive to the pheromone components. Instead one neuron is tuned to benzoic acid (De Brito Sanchez 2000; De Brito Sanchez and Kaissling 2005; Heinbockel and Kaissling 1996; Priesner 1979), the other most sensitive to the norterpene 2,6-dimethyl-5-hepten-2-ol, and tenfold less to (+)-linalool and (–)-linalool (Barrozo and Kaissling 2002; Priesner 1979). While benzoic acid is emitted from the intestinal secretion (meconium) of freshly emerged moths, the natural origin of the norterpene so far is unknown, and its significance for *Bombyx mori* females is unexplored. Behavioral responses to both of these undoubtedly very sensitively perceived compounds (Ziesmann et al. 2000) have not been found.

In other species of moths the females smell their own pheromone (Schneider et al. 1998). Both sexes of the noctuid moth *Spodoptera littoralis* perceive the pheromone; they even are able to learn—like honeybees (Vareschi and Kaissling 1970)—to extend their proboscis upon pheromone stimuli (Hartlieb et al. 1999).

Males of several lepidopteran species (e.g., of the danaine butterflies or arctiide moths) release pheromones serving as aphrodisiacs (Boppré 1986). These odors are often distributed from special scent organs called androconia (Figure 4.1b). The male monarch butterfly for instance may present to the female brushes of cuticular hairs that produce fine (3–5 μm) particles impregnated with the pheromone (Figure 4.2a–c). This "love dust" sticks to the female antennae, providing a long-lasting source of the stimulus that makes the female receptive to copulation (Boppré and Vane-Wright 1989). The aphrodisiacs of Danainae originate from plant alkaloids, which are actively sought out, taken up, and metabolized by the male (Boppré 1990). Pheromonal compounds such as danaidone, hydroxydanaidal, and danaidal, frequently used by Danainae and Arctiidae and often occurring as blends of two components, are derived from pyrrolizidine alkaloids (PAs). PAs ingested by the larvae from their host plants regulate both scent-organ morphogenesis and pheromone biosynthesis in the arctiide moth genus *Creatonotos* (Boppré and Schneider 1989; Egelhaaf et al. 1992). Males of *Bombyx mori* possess small tiltable hair pencils hidden under the front legs as detected only recently (Anderson et al. 2009), with pending discovery of an aphrodisiac pheromone.

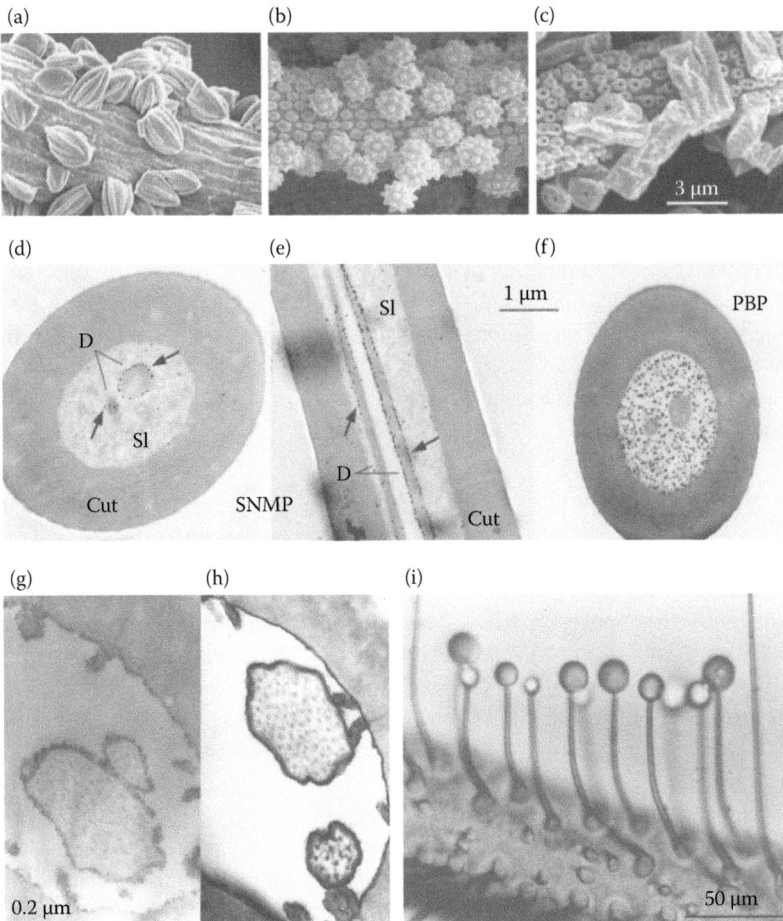

FIGURE 4.2 (a–c) "Love dust" particles on the hairs of the androconia of the male danaine butterflies *Danaus formosa, Amauris tartarea,* and *Danaus* sp. (Courtesy of M. Boppré.) (d–h) Olfactory hairs of a male moth of *Antheraea polyphemus*, electron-micrographic sections. (d, e) Sections treated with gold-labeled antibodies against SNMP. (Modified from Rogers M.E. et al. 2001a. *Cell Tissue Res* 303:433–446.) The gold particles (here intensified) are associated with the plasma membrane of the receptor neurons (D). Sl = sensillum lymph, Cut = cuticle. (f) Gold particles (here original) attached to antibodies against PBP. (Modified from Steinbrecht R.A. et al. 1995. *Cell Tissue Res* 282:203–217.) About 1% of the protein carries a gold label. (g, h) Cross section of hairs treated with cationic markers cationized ferritin (g), and ruthenium red (h). The markers indicate the presence of fixed negative charges. (Modified from Keil T.A. 1984. *Tissue Cell* 16:705–717.) (i) Antennal side branch of *A. polyphemus*. Sensillum lymph droplets ejected from cut hairs. Preparation under water-saturated paraffin oil. The size of the droplets corresponds to the volume of the basal lymph cavity (see Figure 4.1d). (Modified from Kaissling K.E. and Thorson J. 1980. Insect olfactory sensilla: Structural, chemical and electrical aspects of the functional organisation. In *Receptors for Neurotransmitters, Hormones and Pheromones in Insects*, D.B. Sattelle, L.M. Hall and J.G. Hilderbrand (eds.), 261–282. Amsterdam: Elsevier/North-Holland Biomedical Press.)

Communication by pheromones is highly developed in social insects such as honeybees, ants, or termites, which bear numerous pheromone glands on various body parts and produce a variety of chemicals. Correspondingly they possess a large number of types of specialist receptor cells (Dumpert 1972; Vareschi 1971). Pheromones of social insects are not only involved in sexual behavior, they also attract or repel conspecifics as markers of food, of trails, and of social groups, and have many other functions as well. For example, the honeybee queen (*Apis mellifera L.*) produces pheromones that attract a retinue of workers around her and drones on mating flights, that prevent workers from laying eggs and from swarming, and that regulate several other aspects of colony functioning (Keeling et al. 2003). The queen produces a synergistic, multiglandular pheromone blend of at least nine components. In termites a blend of cuticular hydrocarbons may play a key role in colony recognition (Kaib et al. 2004). Differences in the composition of cuticular hydrocarbons among colonies control variation in aggression between colonies.

4.3   INSECT ANTENNAE AND OLFACTORY SENSILLA

Insect pheromone receptor neurons innervate sensory hairs or plates, called sensilla, located on the antennae together with sensilla for stimuli of other sensory modalities. The antennae enormously differ in shape and size, often with a spectacular sexual dimorphism (Figure 4.3a). Male insects may have enlarged antennae with numerous sensilla designed for the most sensitive reception of the pheromone. Female saturniid moths are reported to attract their males via pheromones over distances of 1 km or more (Cardé and Charlton 1984; Priesner et al. 1986). The combed antennae of these moths may have an outline area of more than 1 cm^2 (Boeckh et al. 1960), a size corresponding to the nostril of a human nose.

An important tool for quantitative studies on insect olfaction was tritium-labeled pheromone. Using ^{3}H-pheromone the numbers of stimulus molecules released from the odor source and adsorbed on the antenna have been measured (Kanaujia and Kaissling 1985). From the 30% fraction of air actually passing the large combed antennae of silk moths (Kaissling 2009a; Vogel 1983), all pheromone molecules were caught. It can be calculated that a pheromone molecule—if it were reflected by the antennal surface—would hit the antenna about 100 times on its diffusional zigzag path through the lattice of the long hair sensilla (Figure 4.3b). Initially, 80% of the molecules adsorbed on the antenna were found on the hairs by measuring the radioactivity on hairs cut off immediately after stimulation with labeled pheromone (Kanaujia and Kaissling 1985; Steinbrecht and Kasang 1972). Due to the spacing of the hairs tuned to the diffusional movements of the pheromone in air and to the lipophilic surface of the hairs, the antenna serves as a kind of olfactory lens focusing the stimulus molecules to the receptor neurons. The size of a fruit fly antenna (Figure 4.3b, inset) is about 10,000-fold smaller, suggesting a much smaller absolute sensitivity to odors. Some insects with long flagellar antennae are able to recognize spatial odor patterns. Thus a topical representation of antennal areas was found in the central nervous system of cockroaches (Hoesl 1990).

In contrast to the vertebrate nasal epithelium, different types of olfactory neurons are not randomly intermingled but grouped in the sensilla, in morphologically

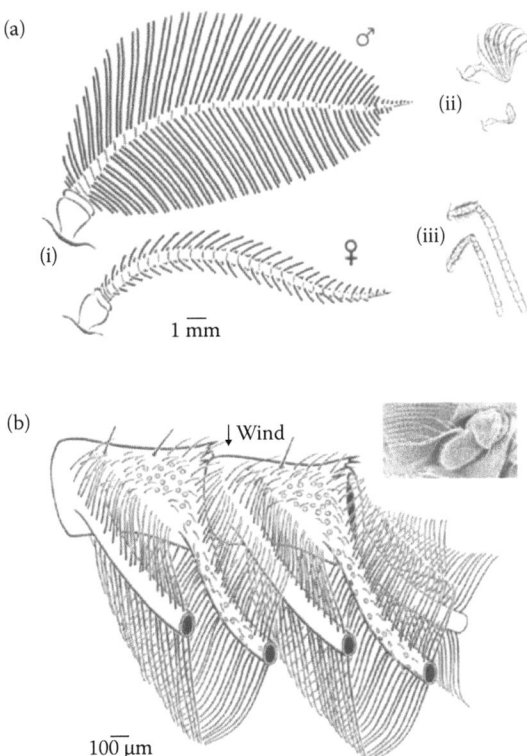

FIGURE 4.3   (a) Antennae of insects with sexual dimorphism: (i) Saturniid moth (*Antheraea pernyi*), (ii) scarabid beetle (*Rhopaea* sp.), (iii) honeybee (*Apis mellifera* L.). (Modified from Kaissling K.E. 1987. In *R.H. Wright Lectures on Insect Olfaction*, K. Colbow (ed.), 1–190. Burnaby, B.C., Canada: Simon Fraser University.) (b) Antenna of a male saturniid moth (see ai), central part of two segments with several types of sensilla. The long hairs, *sensilla trichodea,* are innervated by two or three pheromone receptor neurons. Shorter hairs and *sensilla coeloconica* (circles). Three bristles (above) for taste and mechanical stimuli. *Sensilla styloconica* for humidity and temperature (upper edge). (Modified from Kaissling K.E. 1987.) Inset: Head of *Drosophila* with one antenna, same scale. The third segment (funiculus) is covered with numerous tiny (10-μm) hair-like olfactory sensilla.

and physiologically well-defined units (Keil 1984a, 1999, 2012; Keil and Steinbrecht 1984; Steinbrecht 1973, 1984). The pheromone-sensitive sensilla of moths consist of hollow cuticular hairs up to 400 μm long and 1 to 5 μm thick, innervated by one or several olfactory receptor neurons and furnished with three auxiliary cells (Figures 4.1d and 4.2d–f). The distal processes (dendrites, 0.1 to 0.5 mm in diameter) of the receptor neurons extend into the entire hair lumen. Interestingly the hair tip can be more sensitive than the hair base (Kaissling 2009b). The axons of each neuron connect to the antennal lobe of the central nervous system (CNS) (Ai and Kanzaki 2004; Anton and Homberg 1999; Boeckh and Boeckh 1979; Hansson and Christensen 1999; Hildebrand 1996), the insect equivalent of the vertebrate olfactory

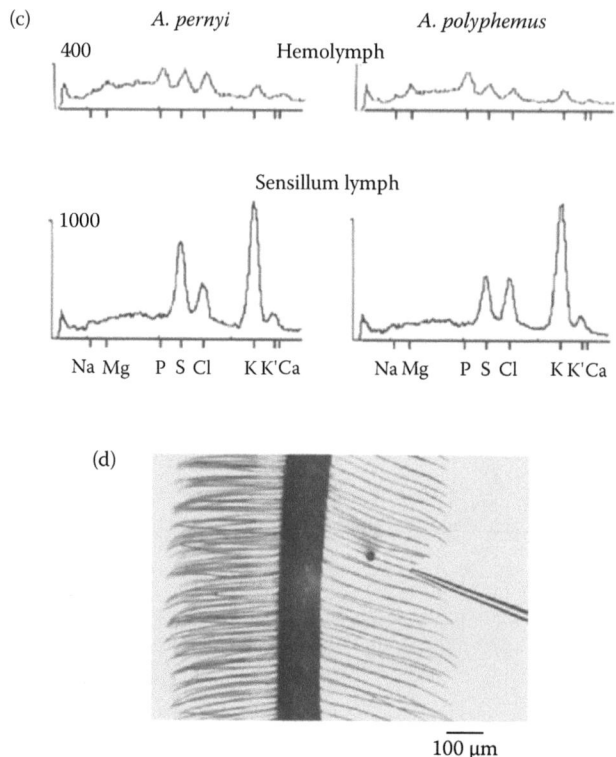

100 μm

FIGURE 4.3 *(Continued)* (c) Electron microprobe analysis. The diagrams result from single sensillum lymph droplets (Figure 4.2i), and from fine threads made from hemolymph. Abscissa: Energy of emitted x-rays (1–4 keV), characteristic for each element. Ordinate: relative numbers of counts. The relative concentrations of the elements are obtained if one raises the peak amplitudes by calibration factors (3.5 for Na, 2.4 for Mg, 1.1 for P, 1.2 for S, 0.98 for Cl, 1 for K, and 0.93 for Ca). According to flame photometry, the concentration of K^+ within the sensillum lymph was about 200 mOsmol/l, and within the hemolymph 36 mOsmol/l. The sulfur peak originates from the PBP with 14 sulfur atoms per molecule of ApolPBP1, or 15 sulfur atoms per molecule of AperPBP1 and AperPBP2. (Modified from Kaissling K.E. and Thorson J. 1980. Insect olfactory sensilla: Structural, chemical and electrical aspects of the functional organisation. In *Receptors for Neurotransmitters, Hormones and Pheromones in Insects*, D.B. Sattelle, L.M. Hall and J.G. Hildebrand (eds.), 261–282. Amsterdam: Elsevier/North-Holland Biomedical Press). (d) The glass capillary of the recording Ag/AgCl-electrode filled with sensillum lymph Ringer (Kaissling 1995) is slipped over one of several cut hairs. The reference electrode (outside the photo) supports the isolated antennal branch. Another capillary is directed to the middle of the hair for locally applying airborne stimuli. (Modified from Kaissling K.E. 1974. Sensory transduction in insect olfactory receptors. In *Biochemistry of Sensory Functions, 25. Mosbacher Colloquium der Gesellschaft für Biologische Chemie*, L. Jaenicke (ed.), 243–273. Berlin: Springer Verlag; Kaissling K.E. 2009. *J. Comp Physiol A* 195:895–922.)

bulb. During ontogeny one of the auxiliary cells produces the cuticular wall of the hairs. Later the trichogen cell withdraws from the hair lumen and secretes the sensillum lymph. Finally the neuronal dendrites grow into the hair lumen (Ernst 1972; Keil and Steiner 1991; Kumar and Keil 1996).

The dendrites have a ciliary portion about 2 μm long that separates the inner segment from the outer segment. The outer segment contains no cellular organelles except microtubules. Mysteriously, in shorter hairs it often forms numerous parallel branches (>100, Steinbrecht 1999), with at least one microtubule per branch. The (waterproof) cuticle of olfactory hairs or plates is penetrated by so-called pore tubules (10 nm in diameter) extending into the hair lumen and sometimes contacting the neuron (Ernst 1972; Steinbrecht 1997; Steinbrecht and Stankiewicz 1999). Via these structures of unknown chemical composition the usually lipophilic odorant molecules are thought to reach the hair lumen (Figure 4.2g, h).

In cases of two or three pheromone components the respective sensilla are innervated by two or three specialist neurons, one for each of the components (Meng et al. 1989). Receptor neurons for pheromone components and behavioral inhibitors can occur in the same sensillum. With several thousands of such sensilla covering the antenna, a very fine spatial resolution of the pheromone distribution in air is feasible. It has been shown experimentally that the fine-scale distribution of pheromone components or of the pheromone blend and behavioral inhibitors influences the orientation of flying males approaching an odor source (Todd and Baker 1999).

The extracellular sensillum lymph bathing the olfactory dendrites (Figure 4.2i) corresponds to the mucus covering the vertebrate olfactory epithelium. Besides pheromone binding protein (PBP) (Steinbrecht et al. 1992) and pheromone-degrading enzymes, the sensillum lymph contains an unusual ion composition (200 mM K^+, 40 mM Na^+) (Figure 4.3c). Dispersive x-ray elementary analysis of sensillum lymph microdroplets revealed a lack of anions (Kaissling and Thorson 1980; Steinbrecht and Zierold 1989); only about half of the anions are covered by chloride. This is compensated by the PBP with 23 negative and 14 positive charges and its high concentration of 10 mM (Klein 1987; Vogt and Riddiford 1986a). Incidentally, the elementary analysis also revealed a high sulfur peak (Figure 4.3c), which indicates the sulfur content of the PBP (e.g., 12 sulfur atoms per *Bmor*PBP1).

4.4 ELECTROPHYSIOLOGY

A simple method to investigate stimulus-response characteristics is the recording of the electroantennogram (EAG). The EAG represents summed fractions of receptor potentials of many olfactory sensilla located near both of the Ag/AgCl electrode capillaries inserted into tip and base of the antenna (Schneider 1957, 1992; Kaissling 1995). It is particularly useful for measuring the responses of the odor specialists, even if it does not show the contribution of each neuron type (Figure 4.7b later in the chapter). EAG recordings and also recordings from single sensilla combined with gas-chromatography have been employed to identify the effective components of blends, either of pheromones or of plant volatiles (Stranden et al. 2003).

Single sensilla allow transepithelial recording of receptor potentials and nerve impulses from two or three identified receptor neurons. Electrical contact can be obtained by slipping the recording electrode capillary over the cut hair tip (Figure 4.3d). With a special way of squeezing (Kaissling 1995), the receptor-neuron dendrites are severed but immediately sealed, a method that avoids short-circuiting the membrane potential. Between sensillum lymph and hemolymph there is a transepithelial potential of +25 mV–+50 mV produced by an electrogenic potassium pump located in the folded apical membrane of auxiliary cells (Kaissling and Thorson 1980; Küppers and Bunse 1996). The latter pump is responsible for the high potassium concentration in the sensillum lymph (Figure 4.3c).

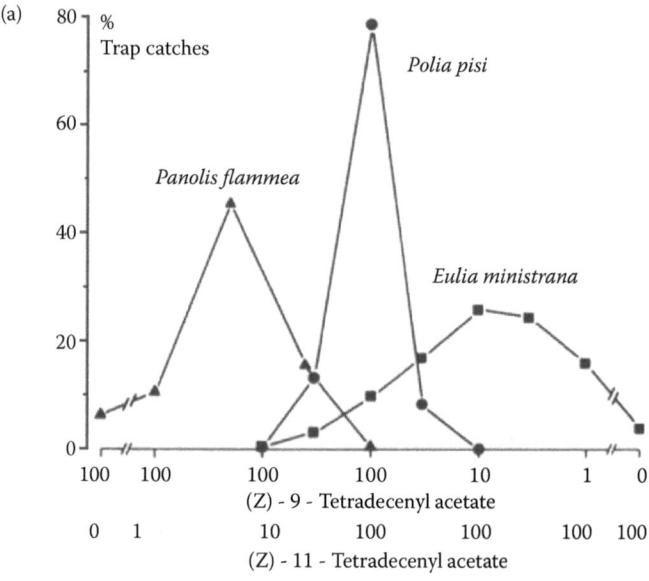

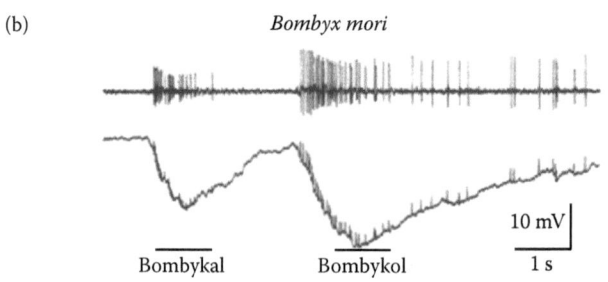

FIGURE 4.4 (a) Relative trap catches of three species of moths in response to two obligatory pheromone components. Abscissa: Ratios of the two components. The average maximal numbers of moths per trap were (from left to right) 27.0, 23.8, 40.8, respectively (six replicates for each species). Data from Priesner E. and Schroth M. 1983. *Z Naturforsch* 38c:870–873; Priesner E. 1980. *Z Naturforsch* 35c:990–994; Priesner E. 1984. *Z Naturforsch* 38c:849–852. (From Kaissling, K.E. and Kramer E. 1990. *Verh Dtsch Zool Ges* 83:109–131.) (b) Responses obtained from one hair with two receptor cells of a male moth of *Bombyx mori*. Upper trace AC-amplification, lower trace DC-amplification showing receptor potentials with superimposed nerve impulses (spikes). One cell responds to bombykal (small spikes); the other one to bombykol (large spikes). (Modified from Kaissling K.E. et al. 1978. *Naturwissenschaften* 65:382–384.)

Pheromone stimuli elicit negative deflections of the transepithelial potential (receptor potentials) up to 30 mV, reflecting changes in membrane potential of the receptor neuron. The transepithelial resistance decreases by up to 20% (Zack 1979). Nerve impulses of a few millivolts amplitude show peak frequencies up to almost 300/s (Kaissling and Thorson 1980; Zack 1979). The opposite polarities in DC recordings of the receptor potential (negative) and the nerve impulses (positive) (Figure 4.4b) suggest that the potential is initiated by depolarization of the dendritic membrane, and that the impulses are generated in the soma region of the receptor neuron (De Kramer et al. 1984; Thurm

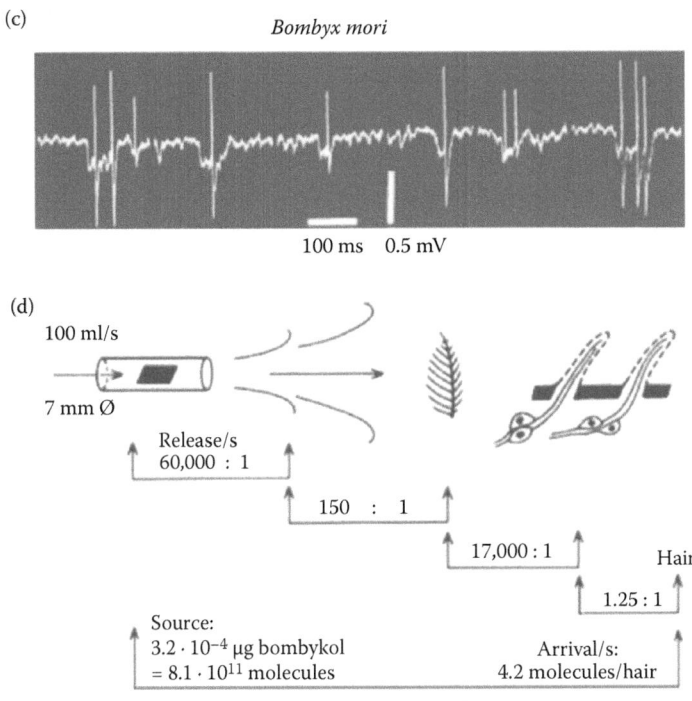

(c) *Bombyx mori*

100 ms   0.5 mV

(d)

100 ml/s

7 mm Ø

Release/s
60,000 : 1

150 : 1

17,000 : 1   Hair

1.25 : 1

Source:
$3.2 \cdot 10^{-4}$ µg bombykol
$= 8.1 \cdot 10^{11}$ molecules

Arrival/s:
4.2 molecules/hair

Response: 1 nerve impulse/neuron

FIGURE 4.4 (*Continued*) (c) *Bombyx mori*, elementary receptor potentials (ERPs). Six samples from the two receptor neurons in one hair, with large and small nerve impulses for bombykol and bombykal, respectively. (From Kaissling K.E. and Thorson J. 1980. Insect olfactory sensilla: Structural, chemical and electrical aspects of the functional organisation. In *Receptors for Neurotransmitters, Hormones and Pheromones in Insects*, D.B. Sattelle, L.M. Hall and J.G. Hildebrand (eds.), 261–282. Amsterdam: Elsevier/North-Holland Biomedical Press.) (d) Stepwise determination of the number of bombykol molecules adsorbed per one of the 17,000 olfactory hairs on the antenna of *Bombyx mori* males. The source loads were measured down to 10^{-7} µg/fp, the releases from sources with 10^{-2} up to 10^2 µg/fp. Higher loads were used for measuring the fraction of released stimulus molecules adsorbed on the antenna (1/150) and the fraction of the latter caught by the hairs (80%). The airstream velocity at the antenna was 56 cm/s. The stimulus concentration at the antenna was 40% of the concentration at the outlet of the tube. (Modified from Kaissling K.E. and Priesner E. 1970. *Naturwissenschaften* 57:23–28. With kind permission from Springer Science+Business Media.)

and Küppers 1980). This is supported by experiments with selective cooling of the long hairs of moths (Kodadová and Kaissling 1996).

Pheromone sensilla with long hairs are convenient subjects for studying the electrical organization of the sensillum circuit (De Kramer and Hemberger 1987; Kaissling and Thorson 1980; Kodadová and Kaissling 1996; Redkozubov 2000; Vermeulen and Rospars 2001). Recording from the tip of a cut hair (Figure 4.3d) provides conditions equivalent to those for loose patch-clamp recordings; in this case, the entire dendrite represents the patch of the receptor-cell membrane. Typically the transepithelial resistance is around 200 MOhm, and the sensilla are electrically well isolated from each other. The analysis revealed a high specific resistance of the dendritic membrane (3000 Ohm cm^2) providing a length constant of the dendrite large enough to conduct a distal membrane depolarization to the cell soma region with the generator region for the nerve impulses.

Recording from cut hairs allows some of the sensillum lymph to be replaced by the fluid from the capillary electrode. For instance, pheromone may be dissolved in the electrolyte and directly applied to the neuronal dendrites inside the hair (Kaissling et al. 1991; Pophof 2002, 2004; Van den Berg and Ziegelberger 1991). This method has been used for the long hairs of moths. Direct application of odorants by electrode penetration of the hair wall worked in *Drosophila* sensilla (Jones et al. 2011).

4.5 SENSITIVITY OF PHEROMONE RECEPTOR NEURONS AND BEHAVIORAL RESPONSES

Combined radiometric, electrophysiological and behavioral experiments (Kaissling 2009a; Kaissling and Priesner 1970) were employed to study the absolute sensitivity of the silk moth *Bombyx mori,* which will be briefly summarized here. An important tool was the radiolabeled pheromone bombykol (Kasang 1968). Although a high specific activity (31.7 Ci/g) was obtained by introducing one tritium atom per four bombykol molecules, the minimum amount detectable in a scintillation counter was 10^8 bombykol molecules, or 4×10^{-8} µg of bombykol. The load of the stimulus source (1 cm^2 filter paper) at the behavioral 50% threshold of the male moths was about 2×10^{-5} µg of bombykol (Figure 4.5a). The release of bombykol from this source was determined by extrapolation from the release measured with ^{3}H-bombykol at and above loads of 10^{-2} µg, assuming that the release below 10^{-2} µg was linearly proportional to the source load. This assumption was supported by the linear proportion of nerve impulses at loads below 10^{-2} µg. Loads of 10^{-3} µg, 10^{-4} µg, and 10^{-5} µg elicited 4.1, 0.31, and 0.03 nerve impulses, respectively, per neuron and per 1-s stimulus. The distribution of impulses conformed to a Poisson distribution for single random events such as expected for the arrival of single molecules (Figure 4.5b).

Stepwise determination (Figure 4.4d) revealed that about four bombykol molecules were adsorbed per hair sensillum when one nerve impulse was fired per neuron. At the behavioral 50% threshold about 24% of the antennal hairs received a molecule and about 6% of the neurons fired a nerve impulse; the antenna was exposed for 1 s to 3100 bombykol molecules per milliliter at an airstream velocity of 60 cm/s. About 52,000 molecules/ml were necessary to elicit one impulse per

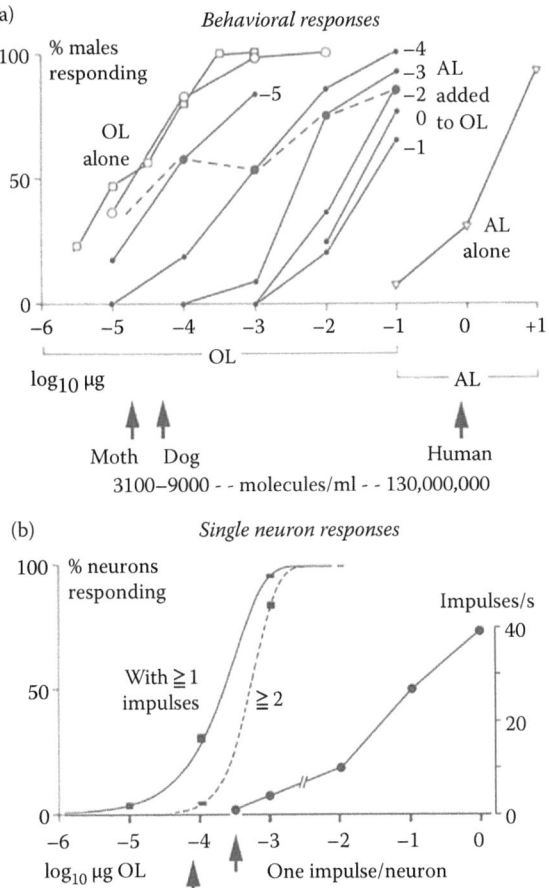

FIGURE 4.5 *Bombyx mori.* Responses of single male moths (a, c) and single neurons (b, c) under the same stimulus conditions. Abscissa (a, b): Loads of 1-cm² filter papers with bombykol (OL) or bombykal (AL). For combined stimuli in (a) two filter papers were used. (a) Dashed line indicates the response to the OL/AL ratio (10/1) of the female gland. (Modified from Kaissling K.E. et al. 1978. *Naturwissenschaften* 65:382–384.) Open squares: Data from Kaissling K.E. and Priesner E. 1970. *Naturwissenschaften* 57:23–28. Arrows indicate the behavioral 50%-threshold of the moth for bombykol, and the thresholds found in dogs for butyric acid (Neuhaus W. 1953. *Z vergl Physiol* 35:527–552) and in humans for sec. butyl mercaptan (Stuiver M. 1958. Biophysics of the sense of smell. *Thesis Rijksuniversitiet*, Groningen). (b) Left two curves: Percentages of responses with at least one nerve impulse (solid line) and at least two nerve impulses (dashed line) fired upon 1-second stimuli of bombykol, after subtraction of spontaneous firing in fresh air. The data fit the Poisson curves expected for a random distribution; 566 to 895 measurements per source load and 1070 control measurements. (From Table 2 in Kaissling K.E. and Priesner E. 1970. *Naturwissenschaften* 57:23–28.) Right curve: Average frequencies of nerve impulses. (From Table 2 in Kaissling K.E. and Priesner E. 1970. *Naturwissenschaften* 57:23–28.) Above the // sign: peak frequencies for a single neuron. (From Figure 3 in Kaissling K.E. and Priesner E. 1970. *Naturwissenschaften* 57:23–28.)

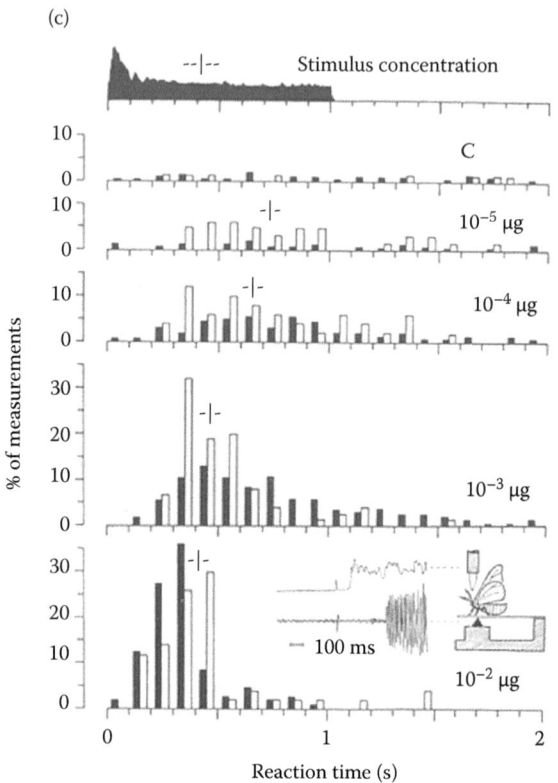

FIGURE 4.5 *(Continued) Bombyx mori.* Responses of single male moths (a, c) and single neurons (b, c) under the same stimulus conditions. (c) First nerve impulses of single bombykol neurons (black columns) and behavioral responses of single males (white columns) elicited by 1-second stimuli of bombykol. C = control with pure air. Black columns: Number of measurements (from above): 1070, 866, 895, 807, 566. White columns: Numbers of males tested (from above): 76, 64, 49, 74, 56. -l- denotes the occurrence of half of the behavioral responses. - -l- - denotes the arrival of half of the stimulus molecules. Inset shows array for measuring air stream velocity (with thermistor), and wing vibration (with Piezo device). Upper trace: Time course of the pheromone release from the filter paper, determined radiometrically. (Modified from Kaissling K.E. and Priesner E. 1970. *Naturwissenschaften* 57:23–28. With kind permission from Springer Science+Business Media.)

neuron. Our final conclusion was that one pheromone molecule is sufficient to elicit a nerve impulse.

A much larger number of nerve impulses, however, are needed in order to alert the moth (i.e., to overcome the noise of spontaneous firing of the receptor cells). Without pheromone a neuron fires a nerve impulse on average every 11.7 s, with a random distribution of intervals (Kaissling 1971). The 17,000 bombykol neurons per antenna (Schneider and Kaissling 1957; Steinbrecht 1970) permanently send an average of 1450 nerve impulses/s to the CNS. Because upon 1-s stimulation the behavioral and the impulse responses were distributed over 2 s after stimulus onset (Figure 4.5c), the noise may be calculated for 2900 spontaneous impulses; it amounts to sqrt 2900 = 54 impulses. At the behavioral 50% threshold, the 1-s bombykol stimulus induces 950 impulses fired during the 2-s time by the 17,000 neurons of one antenna (i.e., 5.6% of neurons fired an impulse). This reveals a signal-to-noise ratio of 950/54 = ~18.

The signal-to-noise ratio, however, would be smaller if the calculation is done for the time used by the CNS for integrating the nerve impulses. Half of the behavioral responses occurred within about 700 ms at weak stimuli (10^{-5} μg/fp) and within about 400 ms at stronger stimuli (10^{-2} μg/fp) (Figure 4.5c). The true integration time must be shorter than the behavioral reaction times because one has to subtract the time until a significant number of neurons fired a nerve impulse. A significant impulse signal (6% of receptor neurons firing one nerve impulse; see above) was reached at 10^{-3} μg after about 250 ms, and at 10^{-2} μg less than 200 ms after stimulus onset (Figure 4.5c). Half of the behavioral responses occurred after 450 and 400 ms, respectively. For the estimated true integration time of 200 ms the noise of spontaneous firing is sqrt 290 = 17 impulses, and the signal-to-noise ratio would be 95/17 = ~6. This value is little above the theoretical minimum signal-to-noise ratio of 3 for a significant signal. These calculations show that the moth CNS performs an astonishingly efficient signal-to-noise analysis of the input from the antennal neurons.

Olfactory receptor neurons of insects like those of vertebrates are primary sense cells. In insects they send their axons to glomeruli of the antennal lobe (Hildebrand and Shepherd 1997) where they terminate in the macroglomerular complex (MGC) of the antennal lobe, which has a subunit for every type of pheromone receptor cell (Sadek et al. 2002; Trona et al. 2013). The numerical convergence of 17,000 primary fibers onto the secondary 34 projection neurons in *B. mori* (Kanzaki et al. 2003) would be 500:1, not sufficient for integration over 17,000 neurons per antenna. Local interneurons and convergence of the projection neurons to higher neurons might be involved.

There are functional connections between the MGC and the ordinary glomeruli that receive input from receptor cells for general odors (Boeckh and Ernst 1987). The division of the antennal lobe into a region for the pheromone input and one for general odors (Hansson and Christensen 1999) resembles that of the olfactory system in many vertebrates, where the accessory and the main olfactory bulb are innervated from the vomeronasal organ and the main olfactory epithelium, respectively.

Comparing the insect antennae containing 10,000 up to 100,000 receptor neurons to the noses of vertebrates, it seems clear that the number of receptor neurons and the absolute sensitivity are not correlated. While the threshold for bombykol is at about 3000 molecules/ml (Figures 4.5a), one of the most effective odorants for humans (sec. butyl mercaptan) can be recognized at a concentration of 1.3×10^8 molecules/ml (Stuiver 1958) by the human nose with 6 million receptor neurons (Menco 1983). The dog's nose (German shepherd) may have up to 500 million olfactory neurons; it may smell butyric acid at 9×10^3 molecules/ml (Neuhaus 1953), and alpha-ionone at 4×10^5 molecules/ml of air (Moulton 1977). Factors other than the number of receptor neurons may be important for a high sensitivity, such as a high effectiveness of molecule capture and conveyance to the sensitive structures, or a low background activity of the receptor neurons.

4.6 ELEMENTARY RECEPTOR POTENTIALS

By which mechanism would a single pheromone molecule elicit a nerve impulse? Extracellular DC recordings from single sensilla at stimulus loads of 10^{-3} µg/fp and below showed elementary receptor potentials (ERPs) preceding single nerve impulses (Figure 4.4c) (Kaissling 1974; Kaissling and Thorson 1980). ERPs also may occur without being followed by a nerve impulse. The ERPs appear as single "bumps" of a few milliseconds duration, or as bursts of a few bumps with amplitudes up to 0.5 mV. The amplitudes of the bumps from the bombykol and bombykal neuron differ (Figure 4.4c), they vary strongly (Redkozubov 2000). The temporal distribution of nerve impulses generated by bombykal (Figure 4.6aiii), however, is similar to those of bombykol-induced responses (compare with the impulse distribution at 10^{-4} µg of bombykol in Figure 4.5c). The average reaction times of the ERPs and those of the nerve impulses are within the range of 0.5 s.

Minor and Kaissling (2003) assumed that each bump reflects an activation of a single receptor molecule and that a burst of bumps indicates repetitive activations of the same receptor molecule by the same pheromone molecule (assumption C in Kaissling 2001). According to an analysis of the electrical sensillum circuit, an average bump results from an increase of membrane conductance by 30 pS, which could be caused by opening of a single ion channel (Kaissling and Thorson 1980). Patch clamp experiments with extruded dendrites of receptor neurons of the saturniid moth *A. polyphemus* revealed pheromone-dependent channel openings of 56 pS, with an opening time of 1.2 ms (Zufall and Hatt 1991). These channels were observed with inside-out patches upon treatment with pheromone and cGMP (1 µM) or a membrane-permeable analog of diacylglycerol (1,2-dioctanoyl-sn-glycerol, 0.36 µM) in the presence of MgATP, but not by IP3 (1 µM) (see also Pophof and Van der Goes van Naters 2002).

Opening a single channel upon activation of a single receptor molecule would require tight functional coupling of receptor molecules and ion channels. This coupling appears to be realized in *Drosophila* and other insects (Sato et al. 2008; Wicher et al. 2008). These authors found that the olfactory receptor molecule is associated with a co-receptor molecule (Orco) forming an ion channel (Vosshall and Hansson 2011). This is supported by the recent finding that the Orco-agonist VUAA1 (found by screening over >100,000 compounds) opens channels with 22 pS conductance in outside-out membrane patches of HEK293 cells expressing AgOrco from *Anopheles gambiae* (calculated from Figure 3 of Jones et al. 2011). For the effects of VUAA1 and further Orco receptor activator molecules (OrcoRAMs), see Bohbot and Dickens 2012, Chen and Luetje 2012, and Nolte et al. 2013.

Bumps—so far observed only in *B. mori*, saturniid, and sphingid moths (Kaissling 2013)—seem to be an invention of the extremely sensitive pheromone neurons, which are able to fire one or a few nerve impulses upon the impact of a single stimulus molecule. Triggering a single nerve impulse requires a sufficiently large and long-lasting decrease of the membrane potential at the spike generator region located at the soma of the neuron (Kodadová and Kaissling 1996). The proper invention would be a receptor molecule, which—while activated by the pheromone molecule—keeps the associated ion channel open sufficiently long (for about 10 ms) to reaching a voltage change

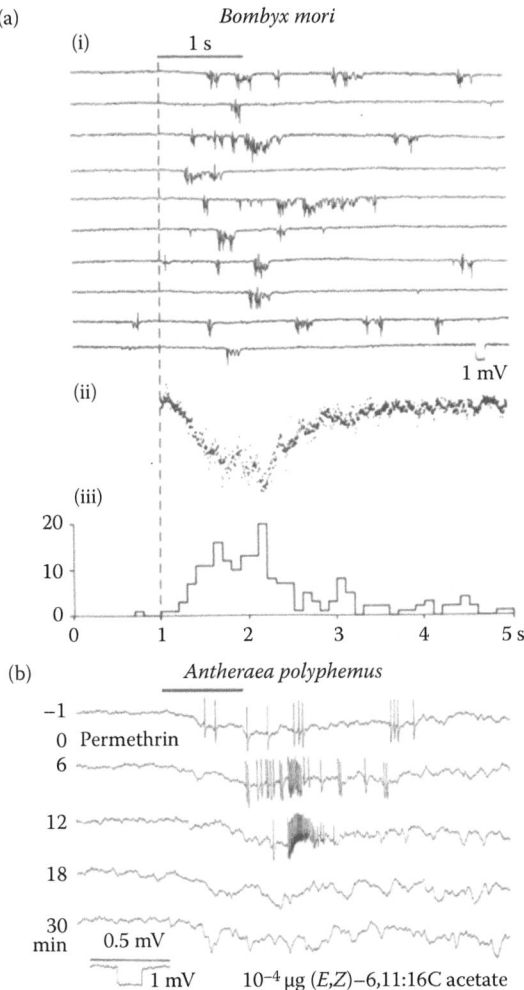

FIGURE 4.6 Extracellular DC recordings from single sensilla trichodea. (a) ERPs and nerve impulses from one sensillum. (i) 10 consecutive responses to 1-second stimuli of bombykal (1 ng/ filter paper) with 1-minute intervals between stimuli. Single and superimposed ERPs. (ii) Thirty consecutive traces as shown in (i) were added. (iii) Numbers of nerve impulses in 100-ms bins were summed from the 30 responses. The average number of nerve impulses during 2 seconds was 5.1 per stimulus, their average reaction time was around 500 ms. Same millivolt calibration for (i) and (ii); same time axis for (i), (ii), and (iii). (Modified from Kaissling K.E. 1986. *Annu Rev Neurosci* 9:21–45; Kaissling K.E. 1987. In *R.H. Wright Lectures on Insect Olfaction.* K. Colbow (ed.), 1–190. Burnaby, B.C., Canada: Simon Fraser University.) (b) Five responses from one sensillum to the major pheromone component. The insecticide permethrin applied at time zero as a droplet to a neighboring hair caused after 6-minute hypersensitivity, typical bursting, and finally, block of impulse firing. The ERPs, forming a fluctuating response, were not affected. (From Kaissling K.E. 1980. Action of chemicals, including (+)trans Permethrin and DDT, on insect olfactory receptors. In *Insect Neurobiology and Pesticide Action (Neurotox 79)*, 351–358. London: Society of Chemical Industry.)

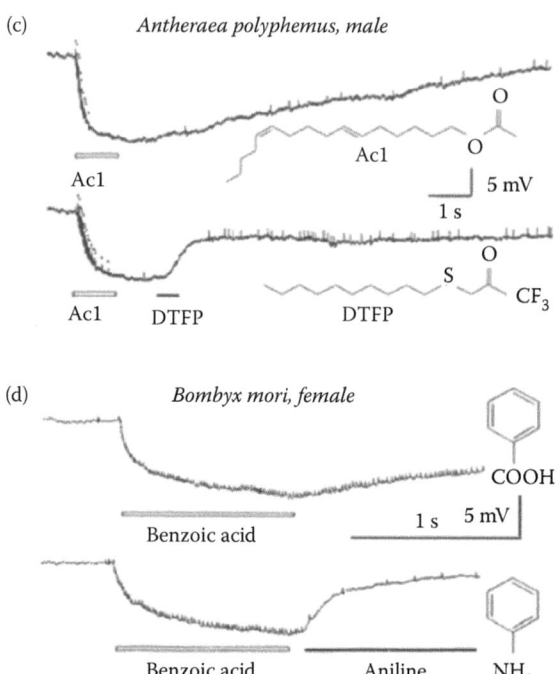

FIGURE 4.6 (*Continued*) Extracellular DC recordings from single sensilla trichodea. (c, d) Immediate reduction of the responses to key odorants by inhibitory compounds. (c) Neuron sensitive to the major pheromone component of *A. polyphemus (E,Z)*-6,11-hexadecadienyl acetate (Ac1). Initial burst of nerve impulses (highlighted by dots), phasic response of pheromone receptor neurons under strong stimulation, the impulse generator adapts. Inhibitor: decyl-thio-trifluoro propanone (DTFP). The partial repolarization caused by DTFP is followed by firing of nerve impulses (disadaptation). (Modified from Pophof B. 1998. *J Comp. Physiol A* 183:153–164.) (d) Neuron tuned to benzoic acid. This neuron shows phasic-tonic responses with little adaptation, even at strong stimulation (10 μg/filter paper). Inhibitor: aniline (10 μl/fp). (Modified from Kaissling K.E. 1987. In *R.H. Wright Lectures on Insect Olfaction*, K. Colbow (ed.), 1–190. Burnaby, B.C., Canada: Simon Fraser University. With kind permission from Springer Science+Business Media.)

large enough for eliciting an action potential. The change of the membrane potential at the cell soma induced by the dendritic 30-pS conductance increase reaches about 1 mV, as expected from the electrical circuit analysis (Kaissling 1987). It should be noted that visual neurons of locusts and flies show bumps in response to light flashes containing single or a few quanta per neuron (Kirschfeld 1966; Scholes 1965).

Less sensitive receptor neurons such as those of the female *B. mori* responding to linalool or to benzoic acid do not show bumps even though they innervate sensilla trichodea similar to those innervated by the pheromone receptor neurons of the male. The benzoic acid neuron needs impacts of more than 1000 odorant molecules per second, which could open many ion channels but with smaller conductance per channel, in order to produce an increase in impulse firing. This neuron responds to 7×10^8 molecules of benzoic acid per milliliter of air, at an air speed of

60 cm/s (Ziesmann et al. 2000), not detected by the human nose. Its apparent spontaneous impulse firing in laboratory air was found to be partially due to the previously unknown contamination with benzoic acid.

With stronger stimuli, the bumps of the pheromone neurons superimpose and form a fluctuating receptor potential like that obtained by adding to each other many responses to smaller stimuli (Figure 4.6aii). A fluctuating response can be seen after selective blocking of the nerve impulses by the insecticide permethrin (Figure 4.6b) (Kaissling 1980), a compound known to interfere with nerve impulse formation (Vijverberg et al. 1982). Further increase of stimulus strength reduces the fluctuations, presumably by opening many ion channels but with reduced conductance per channel (Kaissling 1980). This effect may be responsible for the wider range of stimulus intensities covered by the pheromone receptor neurons compared with less sensitive neurons tuned to other odorants (Kaissling 2013).

Interestingly, responses to certain less effective pheromone derivatives do not show bumps; they produce smooth receptor potentials even at weak excitation, and their dose-response curves cover a smaller range of stimulus intensities (Figure 4.7ai). Most of the derivatives show a decline of the response after stimulus end faster than observed after bombykol stimuli (Figure 4.7aii). Modeling of the receptor potential kinetics suggests that the receptor activations by these compounds do not last as long as would be necessary for the formation of visible bumps (Kaissling 2013). Reducing duration of the activated state in the model also simulates other effects such as smaller maximum amplitudes of the dose-response curves, a shift of the curves along the abscissa to stronger stimuli, and faster declines of the response.

The mechanisms controlling the conductance per channel remain to be discovered. Impaired channel opening seems to be involved also in the reduction of sensitivity (adaptation) after strong stimuli (Kaissling 2013; Kaissling et al. 1987; Zack 1979). Adaptation may be induced locally by stimulation of small sections of a hair sensillum (Zack 1979; Zack-Strausfeld and Kaissling 1986).

4.7 INHIBITION OF PHEROMONE RECEPTOR NEURONS

The mechanism producing fluctuating responses may be selectively blocked. Brief (100-ms) exposure to the vapor of osmium tetroxide strongly reduced the responses of single neurons to bombykol but not to (Z)-10-tetradecenol, one of the derivatives producing smooth responses (Figure 16 in Kaissling 1974).

Responses of pheromone receptor neurons may be inhibited by the volatile decanoyl-thio-1,1,1-trifluoropropanone (DTFP), known as an inhibitor of the sensillar esterase, the sensillar enzyme that degrades a pheromone component of *A. polyphemus* (Vogt et al. 1985). If applied directly after a pheromone stimulus, DTFP rapidly repolarizes the transepithelial receptor potential in various moth species (Figure 4.6c) (Pophof 1998; Pophof et al. 2000). It did not, however, inhibit olfactory neurons tuned to compounds other than pheromones such as the benzoic acid receptor or the linalool-sensitive receptor of the *Bombyx* female.

The number of ^{3}H-labeled DTFP molecules adsorbed per antenna necessary for inhibition was a little higher than the calculated number of pheromone receptor molecules but less than 0.1% of the number of PBPs. This protein occurs in extremely

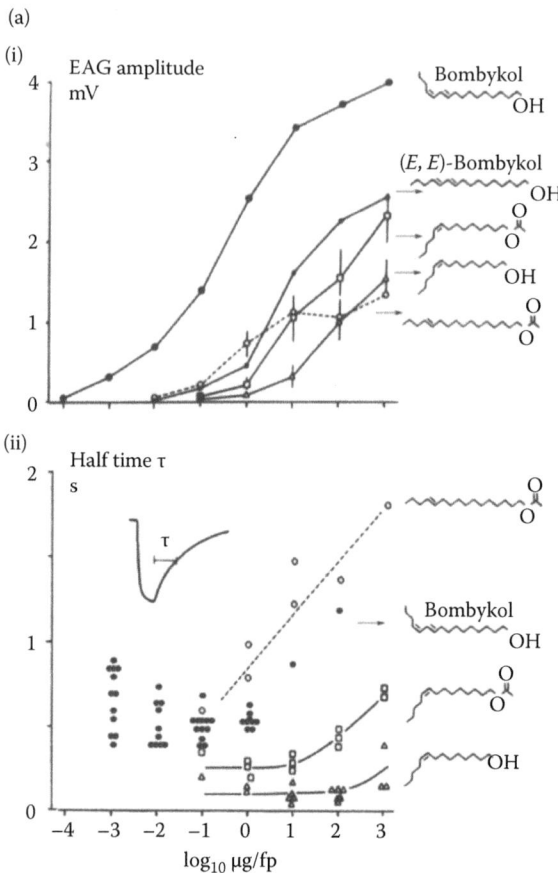

FIGURE 4.7 (a) *Bombyx mori*, EAG. Dose-response curves for bombykol ((*E,Z*)-10,12-hexadecadien-1-ol) and derivatives. Abscissae: Loads of the odor sources (filter papers). (i) EAG-amplitudes after 1-second stimuli. (ii) Half times of the declines, from the same experiments. The half times of the responses to (*E,E*)-bombykol (not shown) were indistinguishable from those to bombykol. (From Kaissling K.E. 1974. Sensory transduction in insect olfactory receptors. In *Biochemistry of Sensory Functions*, 25. *Mosbacher Colloquium der Gesellschaft für Biologische Chemie*, L. Jaenicke (ed.), 243–273. Heidelberg: Springer Verlag. With kind permission from Springer Science+Business Media. Modified from Kaissling K.E. 1977. Structures of odour molecules and multiple activities of receptor cells. In *Olfaction and Taste VI*, J. Le Magnen and P. MacLeod (eds.), 9–16. London: Inf Retrieval.) 1977. (b) *Bombyx mori*, EAG. Repetitive stimulation with pheromone and a derivative. Loads of the odor sources (filter paper). The strongest stimuli cause a delayed decline, which indicates overloading of the pheromone deactivation. The pheromone derivative shows overloading with a lower EAG-amplitude. (From Kaissling K.E. 1972. Kinetic studies of transduction in olfactory receptors of *Bombyx mori*. In *Int Symp Olfaction and Taste IV*, D. Schneider (ed.), 207–221. Stuttgart: Wiss. Verlagsgesellsch.)

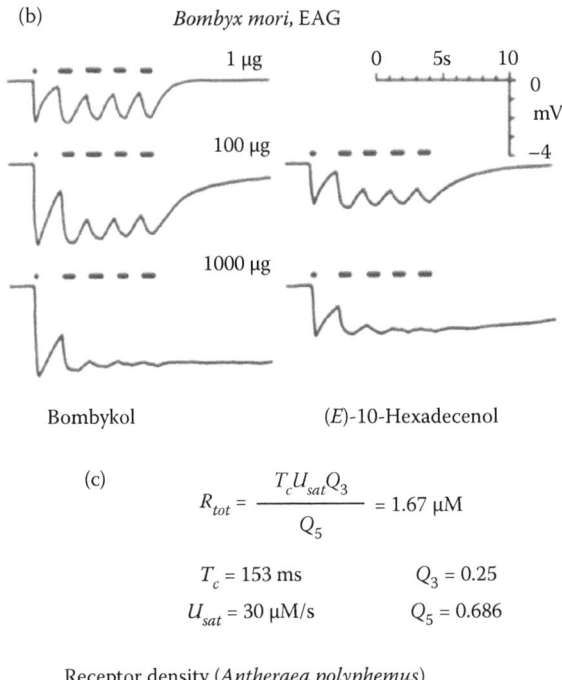

(b) *Bombyx mori*, EAG

Bombykol   (E)-10-Hexadecenol

(c)
$$R_{tot} = \frac{T_c U_{sat} Q_3}{Q_5} = 1.67 \ \mu M$$

| $T_c = 153$ ms | $Q_3 = 0.25$ |
| $U_{sat} = 30 \ \mu M/s$ | $Q_5 = 0.686$ |

Receptor density (*Antheraea polyphemus*)
= $R_{tot} \times N_{Avo} \times$ Hair volume/membrane area
= 6100 Receptor molecules/μm^2

FIGURE 4.7 (*Continued*) (c) Model calculations. Apparent density of receptor molecules within the plasma membrane of the receptor neuron. The respective density for *B. mori* would be 4300 receptor molecules/μm^2. R_{tot} = fictive concentration of the receptor molecules assuming they were distributed within the hair volume (Kaissling K.E. 2009. *J Comp Physiol A* 195:895–922, Equation 27.) $T_c = (k_6 + k_{-6})/(k_{-5} \ k_{-6})$, residence time of the ligand-PBP complex FA at the receptor molecule. $Q_5 = k_6/(k_{-5} + k_6)$. The rate constants k_6, k_{-6}, and k_{-5} were determined from the ERPs elicited by bombykal. (From Minor A.V. and Kaissling K.E. 2003. *J Comp Physiol A* 189:221–230.) U_{sat} = pheromone uptake at which the deactivation process is saturated. Q_3 = fraction of adsorbed bombykol molecules eliciting nerve impulses. (From Kaissling K.E. and Priesner E. 1970. *Naturwissenschaften* 57:23–28.) N_{Avo} = Avogadro number. Hair volume and membrane area of *A. polyphemus* = 2.6 pl and 426 μm^2, A-neuron (from Keil T.A. 1984a. *Zoomorphology* 104:147–156), of *B. mori* = 0.26 pl and 60 μm^2, B-neuron (from Steinbrecht R.A. 1973. *Z Zellforsch* 139:533–565), respectively. With kind permission from Springer Science+Business Media.

high concentrations (10 mM) in the sensillum lymph and apparently interacts—as a complex with the pheromone—with the receptor molecule (Kaissling 2013). Since DTFP strongly binds to various PBPs tested (Maida et al. 2003), it probably blocks the receptor molecule while bound to the PBP. The inhibitory effect (i.e., the repolarization of the receptor potential) is as rapid as the pheromone response, suggesting a competition between the DTFP-PBP complex and the pheromone-PBP complex formed by the preceding pheromone stimulus. A competitive inhibition might also occur between benzoic acid and the structurally related aniline (Figure 4.6d). Note that DTFP and other trifluoromethyl ketones interfere with behavioral responses to pheromone and may be suitable for insect pest control (Albajes et al. 2002; Hummel and Hecker 2012; Picimbon 2004; Karg and Suckling 1999; Levinson and Levinson 2002; Quero et al. 2004; Renou and Guerrero 2000; Renou et al. 2002).

Inhibition of pheromone receptor neurons has also been observed upon exposure to other volatiles, but it is not known whether this has a biological function. For instance, receptor potentials and the nerve impulse responses can be completely and reversibly abolished by terpenes, geraniol in *A. polyphemus* (Schneider et al. 1964), or (racemic) linalool in *B. mori* (Kaissling et al. 1989; Pophof and Van der Goes van Naters 2002). Linalool has been used to modulate long-lasting poststimulatory impulse firing of the bombykol neuron caused by (*Z,E*)-4,6-hexadecadiene (Kaissling et al. 1989). This bombykol derivative elicits anemotactic walk of *Bombyx* males, but only if the poststimulatory impulse firing is interrupted by 150-ms pulses of linalool stimuli given at a rate of 3/s (Kramer 1992). Clearly these terpenes are structurally more different than DTFP from the pheromones, whose excitatory action they inhibit. They are not, however, general inhibitors of olfactory receptor neurons, since for instance linalool effectively excites one of the *Bombyx* female neurons, as described above. Linalool does not affect the other neuron tuned to benzoic acid innervating the same sensillum.

Various compounds more generally inhibit but also irritate the cells if applied at high concentrations, including amines (Kaissling 1972, 1977). They inhibit pheromone receptor cells as well as other types of receptor cells. Often they inhibit the cell at lower concentrations and excite it at higher concentrations. Such compounds might interfere with the lipid structure of the plasma membrane so as to reduce membrane conductance at low doses. At high doses they cause increased conductance, probably destabilizing the membrane. After such stimuli recovery can be incomplete, indicating irreversible damage of the cell function. There are compounds that excite and inhibit at the same time. Often the inhibitory effect disappears more quickly than the excitatory one, leading to poststimulatory rebound effects (De Brito Sanchez 2000; De Brito Sanchez and Kaissling 2005; Pophof and Van der Goes van Naters 2002).

Inhibition and rebound was observed also after simultaneous exposure to pheromone and general anesthetics. When applied alone, general anesthetics may cause hyperpolarization and suppression of spontaneous impulse firing (Stange and Kaissling 1995). They also block the responses to pheromones or other key compounds. If applied during or directly after an excitatory stimulus, they rapidly repolarize the cell. If applied locally on the long sensilla trichodea, general anesthetics do not block the response to pheromone unless they are applied at the same locus as the pheromone. Thus, they might impair specifically the function of receptor molecules or ion channels in the receptor cell membrane either directly, or indirectly by interfering with the structure of the surrounding lipid matrix. Insecticides such as (+)-trans-Permethrin and DDT blocked the nerve impulses but not the receptor potential (Kaissling 1980).

4.8 CONCENTRATION DETECTORS AND FLUX DETECTORS

Concentration detectors and flux detectors are two types of chemoreceptors that differ in the velocity of adsorption and desorption of stimulus molecules (Kaissling

1998). In true concentration detectors the stimulus concentration at the receptor neuron equilibrates with the external stimulus concentration, limited by diffusion. Stimulus molecules would adsorb and desorb quickly, for micrometer distances to the neuron within a few milliseconds. In contrast, true flux detectors adsorb but do not desorb the stimulus molecules, as found for pheromone receptors of moths (Kanaujia and Kaissling 1985; Kasang 1971). Concentration and flux detectors might have fundamental chemical differences of the sensillum surface and the sensillum lymph composition.

During stimulation, flux detectors accumulate stimulus molecules. In order to provide the neuron at constant external stimulation with a constant concentration of stimulus molecules, and to avoid overstimulation, flux detectors need to deactivate the stimulus molecules about as fast as they are adsorbed. Stimulus deactivation must be an extracellular process, which could be relatively slow and rate-limiting for the neuronal response. Flux detectors might provide for higher sensitivity by keeping the stimulus molecule for a longer time in the neighborhood of the receptor neuron. One function of the odorant binding protein (OBP) could be to decelerate or prevent the desorption of the odorant. A rate-limiting pheromone deactivation is assumed for the quantitative model discussed below. Concentration detectors might not need odorant deactivation and odorant degradation. If these are absent other processes such as intracellular signaling may govern the response kinetics (reviewed in Gu and Rospars 2011).

Flux detectors respond to the product of stimulus concentration and the relative velocity of the medium, which reveals the adequate measure of stimulus intensity in molecules per area and per time. Concentration detectors are insensitive to airstream velocity. For a long time insect receptor neurons for carbon dioxide were the only examples known for olfactory concentration detectors, showing no response to changes in airstream velocity (Stange and Diesendorf 1973; Stange and Stowe 1999). A single CO_2-sensitive neuron of the honeybee responds well to an increase of the background concentration of 350 ppm in air CO_2 (9.4×10^{15} molecules/ml) by 5×10^{13} molecules/ml (Lacher 1964). Surprisingly true olfactory concentration detectors unaffected by airstream velocity have been found recently in *Drosophila* for three odorants (Zhou and Wilson 2012).

4.9   OLFACTORY TRANSDUCTION, EXTRACELLULAR

Extracellular transducer processes (perireceptor events) and pheromone-receptor interaction have been tentatively combined in a preliminary quantitative model (Figure 4.8a, b) (Kaissling 2001, 2009b), which has been discussed in detail recently (Kaissling 2013). Only a few aspects will be presented here. The model includes 12 chemical reactions:

- The adsorption of the pheromone (Figure 4.8a, reaction 1) and its diffusion from the adsorption site on the hair towards neuron
- The solubilization of the mostly lipophilic pheromone by binding to the PBP in the sensillum lymph (reactions 2–4)

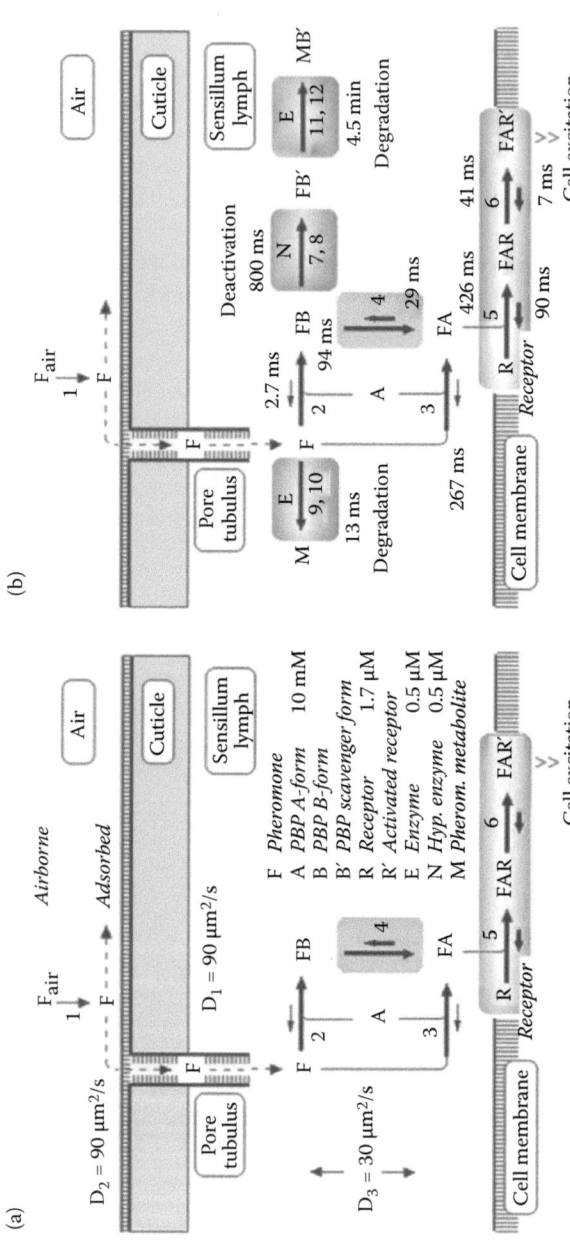

FIGURE 4.8 Model of perireceptor and receptor events (model N). (a, b) Schematic view of the perireceptor space of a sensillum, with chemical reactions. (a) Simplified scheme with coefficients for two-dimensional diffusion (D_2) on the hair surface, one-dimensional diffusion through the pore tubulus (D_1), and three-dimensional diffusion within the sensillum lymph. Reactions 1–6 leading to activation of the receptor molecules at the cell membrane. (b) Complete scheme. The ms numbers are half-lives of the reaction partner positioned near the origin of the bold arrows. They are fictive half-lives if one reaction only were to occur. The half-life of FA due to the association with the receptor was $\ln 2/(k_5\,[R_{tot}]) = 426$ ms, with $[R_{tot}] = 1.67\ \mu M$ and $k_5 = 0.974/$ (s μM). The activated complex FAR′ induces cell excitation. (Modified from Kaissling K.E. 2009b. *J Comp Physiol A* 195:895–922. With kind permission from Springer Science+Business Media.)

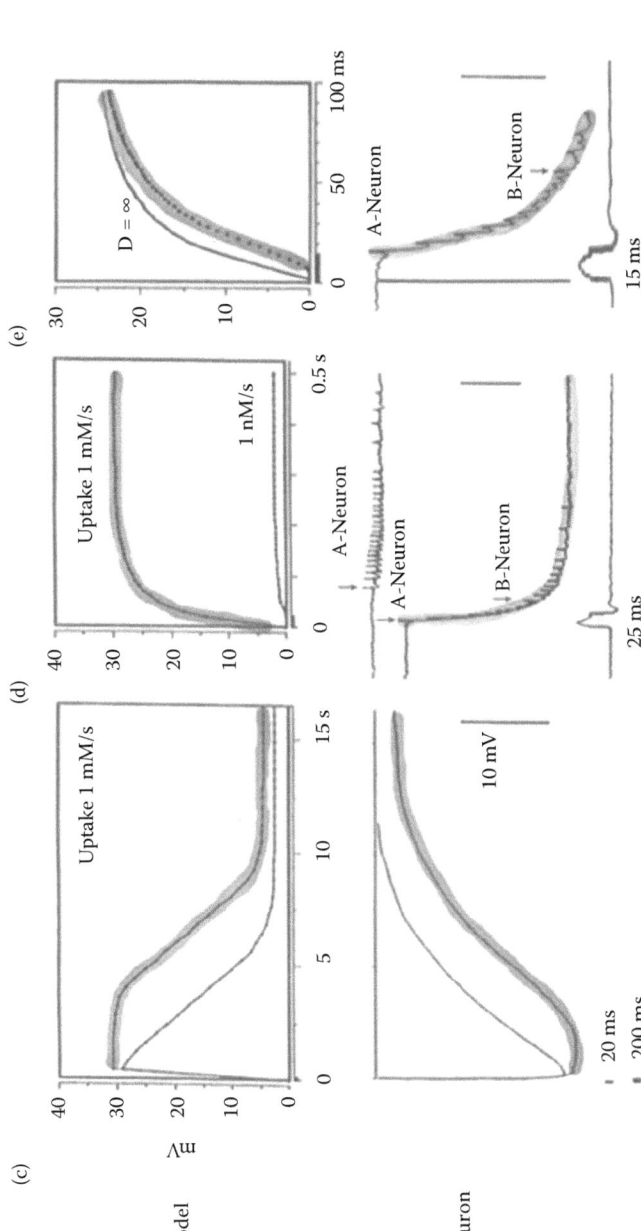

FIGURE 4.8 *(Continued)* Model of perireceptor and receptor events (model N). (c–e) DC-recordings from a single sensillum ("Neuron") of *Antheraea polyphemus* and simulated receptor potentials ("Model"), at different time scales. Very strong local stimuli of (*E,Z*)-6,11-hexadecadienyl acetate (Ac1) were given for 20 and 200 ms (see Figure 4.3d). The signal of a pressure detector (in d and e) indicates the stimulus duration (see Kaissling 1995). The model responses were obtained with an uptake of 1 mM/s or 1 nM/s. (c) The 200-ms stimulus overloaded the deactivation process and delayed the decline for about 3 seconds. (d, e) Weak response of B-neuron tuned to (*E,Z*)-6,11-hexadecadienal. (e) The 10-ms delay of the model receptor potential is absent with infinitely fast diffusion. (Modified from Kaissling K.E. 2001. *Chem Senses* 26:125–150. With kind permission from Springer Science+Business Media.)

- The activation of the receptor molecule in the plasma membrane of the receptor cell (reactions 5, 6)
- The deactivation of the pheromone within the sensillum lymph (Figure 4.8b, reactions 7, 8)
- The enzymatic degradation of the pheromone (Figure 4.8b, reactions 9–12)

4.10   DIFFUSION ON THE HAIRS

Radiolabeled pheromone was used to determine the velocity of pheromone transport to the receptor neuron. Because of the relatively small hair volume, the initial concentration of the 80% of pheromone adsorbed on the hairs is very high (cf. Figure 4.4d). Following the concentration gradient the molecules migrate from the hairs to the antennal body. By cutting hairs of male antennae at various time intervals after stimulation with ^{3}H-labeled pheromone, the velocity of this migration was determined. The coefficient for longitudinal diffusion was D = 50 μm^2/s for ^{3}H-bombykol ((E,Z)-10,12-hexadecadienol) in *B. mori* (Steinbrecht 1973; Steinbrecht and Kasang 1972) and D = 30 μm^2/s for ^{3}H-(E,Z)-6,11-hexadecadienyl acetate, the major pheromone component of the saturniid moth, *A. polyphemus* (Kanaujia and Kaissling 1985). This velocity corresponds to the range expected for diffusion within the sensillum lymph of the PBP molecule with a molecular mass of 15 kD. Since the radioactivity was shown to enter the sensillum lymph, it was concluded that the longitudinal diffusion occurs inside the hairs while the pheromone is bound to the PBP.

The velocity of the diffusion from the adsorption site on the hair surface to the pore entrance, and through the pore tubules was determined from the migration along hairs of dried antennae (Kanaujia and Kaissling 1985). The sensillum lymph was evaporated and the hairs were filled with air so that the PBP was unable to move. On these hairs the longitudinal migration of the ^{3}H-pheromone was about threefold faster than in intact antennae (D = 90 μm^2/s). In the model we used D_2 = D_1 = 90 μm^2/s for the two-dimensional diffusion along the hair surface and the one-dimensional diffusion through the tubules, and D_3 = 30 μm^2/s for the diffusion within the lymph (Figure 4.8a). The mean time between adsorption on the hair surface and the arrival at the neuron is in the 10-ms range (Kaissling 1987; Steinbrecht 1973). The diffusional delay is independent on stimulus intensity; it becomes visible at strong stimulus intensities, where the receptor potential starts after about 10 ms (Figure 4.8e). The first nerve impulse is elicited after another 5 ms. With infinitely fast diffusion the model response starts practically without a delay (Figure 4.8e) (Kaissling 2001). Note that the average reaction time of the nerve impulses at weak stimulation (in the range of half a second, Figures 4.5c, 4.6a) must be due to processes other than diffusion, as e.g. the pheromone-receptor association (Figure 4.8b, reaction 5).

4.11   KINETIC MODEL

Since the tertiary structure of the bombykol binding protein was analyzed by x-ray crystallography (Sandler et al. 2000) and by nuclear magnetic resonance (NMR) (Horst et al. 2001; Klusák et al. 2003), further insect binding proteins were studied.

The preliminary model presented here (Figure 4.8a, b) is restricted to biochemical and electrophysiological data obtained from two species of moths, *B. mori* and *A. polyphemus* by various labs. Rate constants for most reactions and the initial concentrations of the reaction partners (reviewed in Kaissling 2009b, 2013) allow to calculate fictive half-lives of the latter if one reaction only were to occur (Figure 4.8b). For instance, the half-life of F would be $\ln2/(k_2 [A]) = 2.7$ ms due to binding of F and the PBP form A alone (reaction 2, with the association rate constant $k_2 = 0.068/(s\ \mu M)$ and $[A] = 3.8$ mM). Due to the enzymatic degradation alone (reaction 9 = binding to the pheromone degrading enzyme E, and reaction 10 = catalytic step), the half-life of F would be 13 ms. The half-life of F is 2.1 ms if both binding to A and degradation are considered.

As mentioned above it is tentatively assumed that the decline of the receptor potential is governed by the pheromone deactivation (Figure 4.8b, reactions 7, 8). From the decline after stimulus end at medium stimulus intensities, and considering the dose-response relationship of the receptor potential amplitude, a half-life of 0.8 s was estimated for the activated pheromone-PBP-receptor complex FAR'. At high intensities the decline was delayed due to overloading of the deactivation process (Figures 4.7aii, b, and 4.8c). In the favored model version presented here the deactivation is catalyzed by the hypothetical enzyme N (binding of FB to N, reaction 7, and the catalytical step forming FB, reaction 8). Alternatively, it was assumed that the receptor molecules themselves could act as enzymes catalyzing the deactivation, or that this process occurs spontaneously (Kaissling 2009b).

When entering the hair lumen the pheromone F encounters one of two reaction partners dissolved in the sensillum lymph, either the PBP form A (reactions 2 and 3) or the pheromone-degrading enzyme E (reaction 9, followed by the catalytic reaction 10). Without the ligand the PBP exists in its A form independent on pH (Lautenschlager et al. 2005) with the inner binding cavity of the protein occupied by the C terminus in a helical conformation (Figure 4.1e). Within less than 3 ms, 17% of the pheromone F is metabolized to M by the enzyme E, whereas 83% of F is bound to the PBP and thereby protected from enzymatic degradation (Vogt and Riddiford 1986a).

When F binds at neutral pH to the A form, mainly the complex FB is produced (reaction 2); that is, the conformation of the PBP changes from A to B (Figure 4.1e). Now the pheromone is bound inside the inner cavity, and the C terminus forms a flexible, mainly hydrophobic tail. At low pH the binding of F and A forms the complex FA (reaction 3), with the pheromone presumably attached to the periphery of the protein. This reaction may be neglected here because the formation of FA via FB is about 10-fold faster. Upon pH changes, the complex FB may be rapidly converted into FA, and back to FB (reaction 4), with rate constants obtained from stop-flow experiments by Leal et al. (2005). The change of FB to FA is thought to occur at low pH via protonation of histidines (Horst et al. 2001; Nemoto et al. 2002; Wojtasek and Leal 1999), *in vivo* within a zone of negative charges fixed at the cell membrane of the receptor neuron (Figure 4.2g, h) (Keil 1984b). The complex FA is assumed to be the only species in vivo binding to the receptor molecule (R) (reaction 5) (see below). According to the model the association of FA and R (reaction 5) is the slowest one of the processes leading to receptor activation, with a half-life of FA of more than 400 ms (Figure 4.8b). Therefore the formation of FAR is largely responsible for the reaction times of the responses to single stimulus molecules at very low stimulus intensities (Figures 4.5c and 4.6a).

The ternary complex FAR may change one or several times to the activated state FAR′ (reaction 6), triggering a "bump" upon each change, as described above. The lumped lifetime of the states FAR and FAR′ is also considered as the residence time of the pheromone-PBP complex at the receptor molecule (T_c, Figure 4.7c).

In parallel to the reactions leading to the receptor activation, the pheromone molecules are deactivated (reactions 7, 8). Finally, while bound to the scavenger PBP form B', they are enzymatically degraded (reactions 11, 12), but 20,000-fold more slowly than the free pheromone (see below). Only $Q_3 = 25\%$ of the molecules adsorbed on the hair sensilla have the chance to activate a receptor molecule (see Figure 4.4d).

4.12 FIVE FUNCTIONS OF THE PHEROMONE BINDING PROTEIN

In vitro the free pheromone might be able to activate the receptors if applied in unphysiological concentrations. *In vivo*, however, the pheromone-PBP complex is thought to interact with the receptor molecule. The idea that the complex FA, rather than the free pheromone, interacts with the receptor molecule is supported by experimental evidence (a–c, below) and by model calculations (d, e).

a. Infusion of the sensillum lymph space with pheromone-PBP mixtures revealed that besides the pheromone also the PBP is involved in the receptor activation (Pophof 2002). Furthermore, infusion of one PBP of *A. polyphemus* (ApolPBP1) elicited nerve impulses of the bombykol receptor neuron of *B. mori* even in the absence of pheromone (Pophof 2004).

b. In *Drosophila* the PBP "Lush" activated the receptor directly via a pheromone-induced conformation of the protein, again without the pheromone (Ha and Smith 2009; Laughlin et al. 2008; Xu et al. 2005).

c. The sensitivity of HEK293 cells expressing one of the three receptors of *A. polyphemus* (ApolOR1) was about 300-fold increased for one of the three pheromone components ((E,Z)-6,11-headecadienal) when a specific PBP (ApolPBP2) was added (Forstner et al. 2009; Grosse-Wilde et al. 2006).

d. According to the model the concentration of the complex FA is—during stimulation—about 60-fold higher than that of the free pheromone F.

e. Finally the lifetime of the free pheromone is much shorter (a few milliseconds) than the time necessary for association with the receptor (several hundred milliseconds; see Figure 4.8b).

Modeling showed that the PBP serves several, at first glance contradictory, functions (Kaissling 2001, 2009b):

a. The protein binds and solubilizes the hydrophobic pheromone and serves as a transporter towards the receptor neuron.

b. Binding to the PBP protects the pheromone from enzymatic degradation (Vogt and Riddiford 1986a). Without protection, 93% of the pheromone would be degraded before binding to the receptor (Kaissling 2009b).

c. There is evidence that the PBP is involved in the pheromone-receptor interaction, as described above.

d. The PBP performs the postulated pheromone deactivation (see below).

e. The PBP serves as an organic anion compensating the partial lack of chloride ions found by electron microprobe analysis of the sensillum lymph (Figure 4.3c).

A typical PBP, 3 nm wide, has a molecular mass of 15 kD and 142 amino acids, and possesses six highly conserved cysteines forming three disulfide bridges (Leal et al. 1999; Maida et al. 2005; Scaloni et al. 1999; Vogt et al. 1999). The amino acid sequence is known for many of these proteins. Species with a larger number of pheromone components possess a diversity of PBPs with different binding specificities (Nagnan-Le Meillour and Jacquin-Joly 2003). Nonpheromone sensilla contain so-called general odorant binding proteins (GOBPs), which are related to the PBPs and share the six conserved cysteines (He et al. 2010; Pelosi and Maida 1995; Steinbrecht et al. 1995; Zhang et al. 2001; Zhou et al. 2009). Besides these OBPs further proteins of lower homology with possible chemosensory function (chemosensory proteins [CSPs]) have been found in several insect orders (Dani et al. 2011; Picimbon 2003; Tegoni et al. 2004; Vogt 2003). These are similar in size to or smaller than OBPs but differ in amino acid sequence and cysteine content. *Drosophila melanogaster* has about 40 PBP-related proteins (Graham and Davies 2002).

The PBP was detected using gel electrophoresis and ^{3}H-labeled (E,Z)-6,11-hexadecadienyl acetate (Ac1), the main pheromone component of *A. polyphemus* (Vogt and Riddiford 1981). The binding survived the electrophoresis. The dissociation constant of purified PBP and the pheromone component Ac1 from *A. polyphemus* antennae was 60 nM (Kaissling et al. 1985), from an assay where PBP dissolved in Ringer solubilized the pheromone initially bound to a glass surface. A different assay revealed 640 nM (Du et al. 1994). The values for bombykol and PBP A form and B form of *B. mori* were $K_{d3} = 1.6$ µM and $K_{d2} = 105$ nM, respectively (Leal et al. 2005). The solubilization of pheromone by PBP was also shown by means of electrophysiological recording during direct application of the *polyphemus* pheromone and the PBP to the sensillum lymph via the recording glass capillary (Van den Berg and Ziegelberger 1991). In these experiments bovine serum albumin (BSA) solubilized the pheromone equally well.

PBP binding may contribute to the specificity of the neuron response (Grosse-Wilde et al. 2006; He et al. 2010; Hooper et al. 2009; Steinbrecht 1996). For instance in *A. polyphemus* (E,Z)-6,11-hexadecadienol bound to the purified PBP 1000-fold less strongly than the pheromone (E,Z)-6,11-hexadecadienyl acetate, and was 1000-fold less effective as a stimulus for the neuron (Du and Prestwich 1995; Du et al. 1994). However, the saturated acetate bound to the PBP only 10-fold weaker than the pheromone, whereas its effect on the neuron response was 1 million times weaker than that of the pheromone (De Kramer and Hemberger 1987). The dissociation constants of (+)- and (–)-disparlure and two recombinant PBPs in the gypsy moth differed by about two- to fourfold (Plettner et al. 2000). In contrast, the sensitivities of both types of receptor neurons for the two enantiomers differed by factors of more than 100 (Hansen 1984).

The same sensillum may contain several PBPs with different binding specificities and in very different amounts. Thus the same three types of PBPs occur in the sensilla trichodea of *A. polyphemus* and *A. pernyi* together with three receptor neurons, each tuned to one of the three pheromone components (Maida et al. 2003). Each of the PBPs preferentially binds one of these components. The binding results agree with the finding of Mohanty et al. (2004) that a more bulky amino acid joins in the pheromone-binding cavity of the PBP preferring the shorter pheromone molecule.

Bombykal bound to the bombykol binding protein (BmorPBP1) similarly as bombykol (Graeter et al. 2006; He et al. 2010). However, bombykal in combination with BmorPBP1 failed to activate the bombykal neuron, but it did activate it in combination with ApolPBP1 (Pophof 2004). So far no PBP for bombykal has been found (Forstner et al. 2006). Surprisingly, BmorGOBP2 bound bombykol as well as BmorPBP1, and even discriminated it from bombykal (He et al. 2010).

Since the structure of BmorPBP became known, many other antennal PBPs of insects have been analyzed (Dani et al. 2011; Fan et al. 2011). The ApolPBP1 of the moth *A. polyphemus* is similar in secondary and tertiary structure to BmorPBP (Mohanty et al. 2004). BmorPBP and ApolPBP1 have 5 histidines. The PBPs of the cockroach *Leucophaea madera* (LmadPBP, binding a pheromone component [Lartigue et al. 2003]), the honeybee *Apis mellifera* (Amel-ASP1, binding two major pheromone components [Lartigue et al. 2004]), and the fly *D. melanogaster* (LUSH, binding short-chain n-alcohols [Kruse et al. 2003] and the pheromone (Z)-11 vaccenyl acetate [Xu et al. 2005]) show interesting differences from BmorPBP (e.g., only two histidines in LUSH, one in Amel-ASP1, and none in LmadPBP). The C-terminus of Amel-ASP1 is placed against the "body" of the protein along the wall of the internal cavity. LmadPBP has a C-terminus shortened by 24 amino acids. Thus pH-dependent changes such as found in BmorPBP are not expected for all PBPs.

Besides PBPs and the related GOBPs there are various smaller chemosensory proteins of unknown function (Dani et al. 2011). Finally it should be noted that PBP-related proteins also occur in insect taste receptors (Nagnan Le Meillour et al. 2000), possibly involved in the transport and perception of noxious taste substances (Ozaki et al. 2003).

Interestingly, the principle of a double-walled nanocapsule has been implemented at least twice in evolution. OBPs of mammals belong to the lipocalin family with a size and function similar to those of insect OBPs, also called encapsulins (Leal 2003). However lipocalins, serving as OBPs in vertebrates (Pernollet and Briand 2004), have a different structure characterized by antiparallel beta-sheet folding and in addition comprise two alpha helices near the N terminal. The sheets, held together by hydrogen bridges, form a container-like structure called the beta barrel.

4.13   PHEROMONE DEGRADATION AND DEACTIVATION

Degradation of pheromone was first found in the silk moth *B. mori* by Kasang (1971). Living antennae exposed in air for 10 seconds to ^{3}H-bombykol were subsequently eluted for 10 minutes by pentane and for another 10 minutes by a chloroform-methanol mixture, and the amounts of bombykol and its metabolites in the resulting solutions were checked by thin-layer chromatography. Bombykol had been turned into aldehyde and acid, and later into esters, with a half-life of 4 to 5 minutes. The degradation was sensitive to temperature, suggesting catalysis by an enzyme, probably a dehydrogenase. Interestingly, pheromone degradation was also found in female *Bombyx* antennae lacking pheromone receptor neurons and on other body parts such as the wings or legs of both sexes (Kasang and Kaissling 1972; Kasang et al. 1988, 1989a, 1989b). While these body parts are tightly covered with scales, the pheromone degradation also occurs on these dry cuticular structures devoid of cellular elements. Vogt and Riddiford (1986b) isolated an enzyme from body scales, an interesting case of enzymatic reactions in nonaqueous material. The degradation on the entire body surface prevents the generation of secondary pheromone sources that could interfere with the mating behavior of the males.

Pheromone-degrading enzymes (esterases, aldehyde oxidases) were isolated from moth antennae (Rybczynski et al. 1990; Vogt et al. 1985), besides enzymes belonging to the cytochrome P450 family (Maibeche-Coisne et al. 2002). Kinetic studies with the enriched pheromone esterase of *A. polyphemus* revealed a K_m of 2.2 μM, a catalytic rate constant in the range of 98/s, and an estimated concentration in vivo of 1 μM (Vogt et al. 1985). The respective values for the cloned enzyme were 1.2 μM, 127/s, and 0.5 μM (Ishida and Leal 2005). With these values—and without protection—the half-life of the pheromone *in vivo* would be 15 or 13 ms, respectively, in contrast to the 4.5 minutes found by Kasang et al. (Kasang et al. 1988, 1989a, 1989b). In fact Kasang's curves show a twofold time course, with an initial very rapid decrease of the intact pheromone by 17% (Kaissling 2009b, 2013). Modeling revealed that this fraction of the pheromone is degraded within the first few milliseconds when the pheromone enters the hair lumen, before most of it is bound to the PBP and thereby protected from the enzyme. That the degradation is not responsible for the decline of the receptor potential is supported by recordings from single antennae with very little enzyme activity but normal decline (Maida et al. 1995).

Stimulus *deactivation* was postulated by Kaissling (1972) in order to explain why the receptor potential declines within seconds after cessation of a brief external stimulus although intact pheromone molecules remain for minutes on the antenna, and for tenths of seconds on and even inside the hairs (Kanaujia and Kaissling 1985). The receptor potential, however, is not adapted and responds to a new stimulus. The process of deactivation seems to be saturable since after strong stimuli the response declines more slowly (Figure 4.7aii), and after extremely strong stimuli the response does not decline but continues for a time interval depending on the amount of odorant loaded onto the antenna (Figures 4.7b and 4.8c). Since the saturation for pheromone derivatives may occur at a submaximal excitation level (Figure 4.7b), the deactivation is likely an extracellular rather than an intracellular process.

In the modeled deactivation the PBP carrying a pheromone molecule undergoes a hypothetical structural change from B to B' (Figure 4.8b, reactions 7, 8), which blocks the release of the pheromone from the PBP and renders it "invisible" for the receptor. As a result of a discussion with F. Damberger and W. Leal, we proposed that a hypothetical enzyme N might be able to recognize the pheromone-carrying B-form by the exposed hydrophobic C-terminal tail, and to discriminate it from the empty A-form with the exposed hydrophilic N-terminal tail (Figure 4.1e). The enzyme N could block or even remove the C-terminal tail of FB, and thus prevent the stimulatory complexes FA, FAR, and finally FAR' from being formed.

The latter idea is supported by the finding that experimental removal of the C-terminus eliminated the FB → FA transformation at low pH (Figure 4.8b, reaction 4). The pheromone was irreversibly locked inside the binding cavity of the truncated PBP (Leal et al. 2005; Michel et al. 2011). Furthermore, the pheromone binding of the PBP at low pH was retained by one-point mutation of the C terminus (Xu and Leal 2008). Apparently the formation of a C-terminal alpha helix—necessary for the ejection of the pheromone from the inner binding cavity—was blocked.

The deactivated pheromone is not completely protected from degradation, although the velocity of the latter is reduced by a factor of 20,000 (Kaissling 2009b). Enzymatic degradation may have a useful function in removing traces of the adsorbed pheromone. This is important to guarantee full recovery of the receptor neurons from previous stimulation and to reduce the nerve impulse discharge to the level of spontaneous activity.

The idea of two processes—a deactivation followed by a much slower degradation—is supported by the response the bombykol neuron to the derivative (Z,E)-4,6-hexadecadiene (Kaissling et al. 1989). A 1-s stimulus by this compound produces a response, which—after stimulus offset—declines like a response to bombykol. Since the initial decline is incomplete (see tailing in Figure 4.8c), it is followed by a prolonged (15-minute) firing of nerve impulses of the bombykol receptor neuron. The initial decline could be due to deactivation whereas the poststimulatory firing may indicate that the hexadecadiene cannot be degraded by the dehydrogenase postulated by Kasang (see above).

Besides pheromones other odorants can be metabolized on antennae of moths. Thus benzoic acid, the most effective compound for the B neuron of the female sensilla trichodea was conjugated with serine to N-benzoylserine 10 s after exposure of fresh (but not of heat-treated) antennae to benzoic acid. While pheromone degradation occurs also on the body scales of the moth, no derivatization of benzoic acid was found on body parts other than the antennae (Oldenburg et al. 2001).

4.14 RECEPTOR MOLECULES, ION CHANNELS, AND SENSORY NEURON MEMBRANE PROTEIN

In *D. melanogaster* more than 60 types of candidate odorant receptor molecules have been identified, each having seven transmembrane domains activating G-proteins. Each receptor cell expresses only one type of receptor protein (Clyne et al. 1999; Dobritsa et al. 2003; Vosshall 2001). Molecules belonging to the seven-transmembrane-domain category were also identified and localized by *in situ* hybridization in antennae of the moth *Heliothis virescens* (Krieger et al. 2002, 2003). Receptor molecules for bombykol and bombykal were identified by Sakurai et al. (2004) and Krieger et al. (2005). An impressive overview and evolutionary tree of the receptor molecules of males and females of *B. mori* is given by Anderson et al. (2009). For further receptor molecules in moths see Grosse-Wilde et al. (2010).

Nakagawa et al. (2005) determined an $EC_{50} = 1.5$ µM for bombykol and the bombykol receptor molecule expressed in *Xenopus* oocytes. For the complex FA and the receptor molecule our model revealed $EC_{50\ model} = 6.8\ \mu M$ (Kaissling 2009b). Considering the different experimental conditions, no closer agreement with the results of Nakagawa et al. would be expected. Comparing these EC_{50} values with the above dissociation constants of pheromone and PBP (between 60 nM and 1.6 μM), the pheromone seems to be bound more strongly to the PBP than the pheromone-PBP complex to the receptor molecule. Using oocyte expression of the bombykol receptor Xu et al. (2012) found an EC_{50} for bombykol of 0.99 μM, but of 9.6 μM for bombykal, for the same receptor. This result does not reflect the *in vivo* situation: The bombykol neuron does not respond to bombykal unless the stimulus intensity is at least 10,000-fold higher than for bombykol (Kaissling et al. 1978).

In principle, processes such as the postulated pheromone deactivation or the binding of the odorant to extracellular binding proteins (see Section 4.12) may contribute to the response specificity. However, the receptor-neuron specificity seems mainly bound to the neuron, and is most likely determined by the interaction of stimulus molecules with receptor molecules since the specificity of other processes such as binding to PBPs or deactivation seems much less sharp than the specificity of the cell response. It should be noted that the bombykol receptor molecule was functionally expressed in an "empty" olfactory neuron of *D. melanogaster* (Syed et al. 2006) which then responded to bombykol. Similarly, a receptor for (Z)-11-hexadecenal from the diamondback moth *Plutella xylostella* was expressed in the bombykol neuron of *B. mori* males, which then responded behaviorally to the *xylostella* pheromone almost as well as to bombykol (Sakurai et al. 2011).

A preliminary calculation reveals densities of olfactory receptor molecules at the neuronal membrane of about 4000/μm^2 for *B. mori* and 6000/μm^2 for *A. polyphemus* (Figure 4.7c). The average density of repetitive structures—putative receptor molecules—found by negative staining in isolated membrane vesicles obtained from isolated sensilla in *A. polyphemus* was 10,000 units/μm^2 (maximally 30,000 units/μm^2) (Klein and Keil 1984). The number of receptor molecules estimated per neuron amounts to 260,000 for *B. mori,* and to 2.6 million for *A. polyphemus*. The large numbers of receptor molecules are required for a wide working range of stimulus intensities covered by the neuronal response. This applies especially to flux detectors like insect pheromone sensilla where the number of receptor molecules occupied by stimulus molecules linearly depends on the stimulus uptake (Kaissling 1998).

The calculated densities of olfactory receptor molecules are far higher than the presumed minimum density of ion channels in the plasma membrane of the olfactory receptor neuron. The analysis of the electrical sensillum circuit revealed that opening of 10,000 ion channels per receptor neuron of *A. polyphemus* with a conductance per channel of 30 pS would suffice for full depolarization of the receptor neuron (Kaissling 2013; Kaissling and Thorson 1980). This number would correspond to a minimum channel density of 23 per μm^2. A much higher number of channels, however, are expected if each receptor molecule together with a co-receptor molecule forms an ion channel (see Section 4.6). This would mean that the real number of ion channels is close or equal to the number of receptor molecules as determined here for model N.

Of particular interest is a membrane protein known as the sensory neuron membrane protein (SNMP), a member of the so-called CD36 protein family (Rogers et al. 1997, 2001a, 2001b). Its members are characterized by two terminal transmembrane domains and a large extracellular domain; they function as docking sites, where extracellular protein molecules can become coupled to the cell membrane. In *Drosophila* the receptor molecules are associated with the SNMP (Benton et al. 2007), and this protein is required for pheromone sensitivity (Jin et al. 2008). The density of SNMP molecules may be roughly estimated from impressive electron micrographs showing gold-labeled antibodies against SNMP associated with the neuronal cell membrane of *A. polyphemus* (Figure 4.2d, e). The density of SNMP could well be equal to that of the receptors if only a few percent of the SNMP molecules carried a gold particle. A different type of SNMP was found in the auxiliary cells of pheromone sensilla (Forstner et al. 2008).

Assuming a ternary association of receptor, co-receptor, and SNMP we arrive at 18,000 protein units per μm^2. This is a minimum estimate because the pheromone uptake U (molecules per hair volume and per s) has been related to the entire hair volume of 2.6 pl (Keil 1984a). If the pheromone molecules stay within the hair lumen only, the pheromone uptake U including U_{sat}, and, consequently, the receptor number calculated (Figure 4.7c) would be even higher. This means the protein density of the olfactory neuron is close to the density of rhodopsin in the outer disc membrane of vertebrate visual cells with 40,000 units/μm^2 (Dratz and Hargrave 1983).

4.15   OLFACTORY TRANSDUCTION, INTRACELLULAR

After the activation of receptor molecules by the odorant, a variety of intracellular signal compounds seems to be involved in the transduction process; among them are diacyl glycerol, cGMP, and Ca^{++} (Gu and Rospars 2011; Krieger and Breer 2003; Stengl et al. 1999). The function of 1,4,5 inositol trisphosphate (IP3) seems questionable (Kaissling 1994; Kaissling and Boekhoff 1993). Various constituents of intracellular pathways have been identified and immunolocalized (Jacquin-Joly et al. 2002; Laue et al. 1997; Maida et al. 2000).

Often it is thought that high sensitivity of olfactory receptor neurons requires amplification by intracellular signaling processes (Gu et al. 2009; Nakagawa and Vosshall 2009; Stengl 2010; Wicher et al. 2008). At least for the extremely sensitive pheromone receptors of moths—producing bumps eliciting nerve impulses upon a conductance increase of 30 pS—it seems clear that amplification is performed solely by the electrical organization of the sensillum. Intracellular messengers, however, may play a role as modulators and for adaptation after strong stimulation, for instance for the supposed reduction of ion channel opening (see Section 4.6).

Intracellular signaling processes are more likely rate-limiting for the neuronal response in concentration detectors, which do not need extracellular odorant degradation and deactivation (see Section 4.8). Thus olfactory response characteristics like long-lasting impulse firing upon brief stimuli, such as observed in *Drosophila* (Montague et al. 2011) and in moths (Kaissling et al. 1989), could have different origins (e.g., from intracellular or extracellular processes, respectively).

The model of perireceptor- and receptor events (Section 4.9 ff.) satisfactorily simulates the rise and fall of the receptor potential except that the fall measured at high stimulus intensities proceeds much more slowly than the simulated one. A simplified version of the model of Kaissling (2001) was combined with a model of intracellular signaling (Rospars et al. 2007). When square wave pulses of the concentration of activated receptor molecules (FAR', Figure 4.8a,b) were applied in the combined model, the intracellular processes turned out to be relatively fast and contributed relatively little to the receptor potential kinetics (Gu et al. 2009). Considering intracellular processes, however, can improve the simulation at high stimulus intensities (Gu and Rospars 2011). Adaptation phenomena such as diminished ERPs (Section 4.6) still await modeling.

More than one type of ion channel appears to contribute to the receptor potential, and further channels must be involved in the generation of nerve impulses in the soma region of the receptor cell (Stengl et al. 1999). In *A. polyphemus* the initial burst of nerve impulses observed at relatively high stimulus intensities might be induced by opening of a Ca^{++}-activated nonspecific ion (CAN) channel located in the soma region of the receptor cell (Zufall et al. 1991). This phasic response adapts very quickly, possibly because this type of channel is blocked by cGMP. The cloned cDNA of a cyclic nucleotide and voltage-activated ion channel from the antennae of the moth *Heliothis virescens* was heterologously expressed and analyzed by patch clamp recordings and *in situ* hybridization (Krieger et al. 1999). It was suggested that this channel plays a role in regulating the responsiveness of the cell via intracellular cAMP-levels, possibly controlled by the neuromodulator octopamine (Pophof 2000; Von Nickisch-Rosenegk et al. 1996).

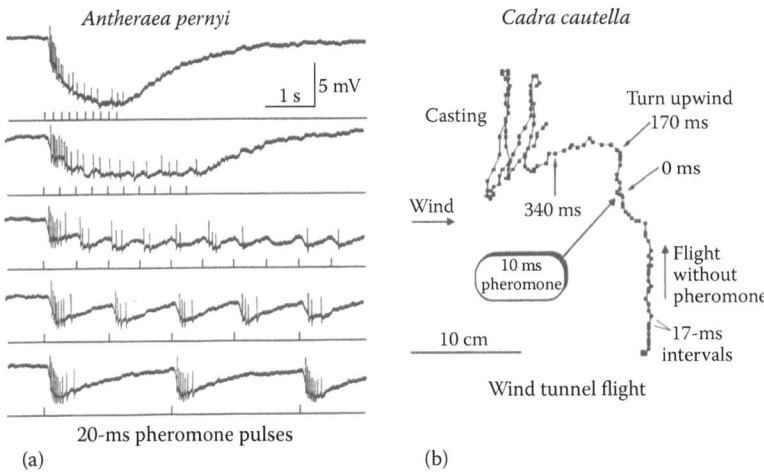

FIGURE 4.9   (a) Responses of a single pheromone receptor neuron repetitively stimulated at various frequencies by 20-ms pulses of the main pheromone component (E,Z)-6,11-hexadecadienal, at 8°C. (Modified from Kodadová B. 1996. *J Comp Physiol A* 179:301–310.) (b) Flight path of a male almond moth *Cadra cautella*, photographed every 17 ms. The moth turns upwind after a single pheromone stimulus of 10 ms. About 440 ms after "losing the pheromone plume" the moth started casting with regular flip-flop turns. (Modified from Mafra-Neto A. and Cardé R.T. 1996. *Experientia* 52:373–379. With kind permission from Springer Science+Business Media.)

4.16 TEMPORAL CODING

With stronger stimulation, the elementary receptor potentials add up to an overall receptor potential that can reach 30 mV. While the average latency of the responses to single pheromone molecules is about 0.5 seconds, at high stimulus intensities the onset of the overall receptor potential may be delayed by 10 ms only (Figure 4.8e). At high stimulus intensities, insect olfactory receptor neurons and also higher-order neurons within the antennal lobe (Christensen and Hildebrand 1988; Lei et al. 2002) resolve repetitive stimulus pulses up to frequencies of 10 pulses per second (Almaas et al. 1991; Barrozo and Kaissling 2002; Kaissling 1986; Rumbo and Kaissling 1989). The time resolution of the nerve impulse response depends on the type of receptor neuron and on temperature (Kodadová 1996) (Figure 4.9a). The astonishing resolution is restricted to higher stimulus intensities where the response latency is short. It was first shown by Kramer (1986) that the anemotactic walk of a male moth of *B. mori* near the odor source consists of several pheromone-elicited turns per second into the upwind direction (Kaissling 1997; Kaissling and Kramer 1990; Kramer 1996; Todd and Baker 1999). Each turn was elicited by a brief odor pulse such as a male encounters due to turbulence within a pheromone plume. A single upwind turn of a flying almond moth elicited by a 10-ms pheromone stimulus has been marvelously demonstrated by Mafra-Neto and Cardé (1994) (Figure 4.9b). It should be noted that the upwind orientation during flight requires visual reference to the ground. For temporal coding see also M. Renou (Chapter 2, this volume).

ACKNOWLEDGMENTS

The author thanks A. Krikellis and his team for librarian help, A.M. Biederman-Thorson for linguistic improvements, and C. Mucignat for generous editorial support.

REFERENCES

Ai H. and Kanzaki R. 2004. Modular organization of the silkmoth Antennal lobe macroglomerular complex revealed by voltage-sensitive dye imaging. *J Exp Biol* 207:633–644.

Albajes R., Konstantopoulou M., Etchepare O., Eizaguirre M., Frerot B., Sans A., Krokos F., Ameline A., and Mazomenos B. 2002. Mating disruption of the corn borer *Sesamia nonagrioides* (Lepidoptera: Noctuidae) using sprayable formulations of pheromone. *Crop Protection* 21:217–225.

Almaas T.J., Christensen T.A. and Mustaparta H. 1991. Chemical communication in Heliothine moths I. Antennal receptor neurons encode several features of intra- and interspecific odorants in the male corn earworm moth *Helicoverpa zea*. *J Comp Physiol A* 169:249–258.

Anderson A.R., Wanner K.W., Trowell S.C., Warr C.G., Jaquin-Joly E., Zagatti P., Robertson H. and Newcomb R.D. 2009. Functional analysis of female-biased odorant receptors from the silkworm, *Bombyx mori*. *Insect Biochem Molec Biol* 39:189–197.

Anton S. and Homberg U. 1999. Antennal lobe structure. In *Insect Olfaction*, B.S. Hansson, ed., 98–124. Berlin: Springer Verlag.

Baker T.C., Ochieng S.A., Cossé A.A., Lee S.G., Todd J.L., Quero C. and Vickers N.J. 2004. A comparison of responses from olfactory receptor neurons of *Heliothis subflexa* and *Heliothis virescens* to components of their sex pheromone. *J Comp Physiol A* 190:155–165.

Barrozo R.B. and Kaissling K.E. 2002. Repetitive stimulation of olfactory receptor cells in female silkmoths *Bombyx mori* L. *J Insect Physiol* 48:825–834.

Benton R., Vannice K.S. and Vosshall L. 2007. An essential role for a CD36-related receptor in pheromone detection in Drosophila. *Nature* 450:289–203.

Blomquist G.J. and Vogt R.G. (eds.) 2003. *Insect Pheromone Biochemistry and Molecular Biology*. London: Elsevier Academic Press.

Boeckh J., Kaissling K.E. and Schneider D. 1960. Sensillen und Bau der Antennengeißel von *Telea polyphemus*. *Zool Jahrb Anat Ontog* 78:559–584.

Boeckh J., Kaissling K.E. and Schneider D. 1965. Insect olfactory receptors. *Cold Spring Harbor Symp Quant Biol* 30:1263–1280.

Boeckh J. and Boeckh V. 1979. Threshold and odor specificity of pheromone-sensitive neurons in the deutocerebrum of *Antheraea pernyi* and *A. polyphemus* (Saturnidae). *J Comp Physiol* 132:235–242.

Boeckh J. and Ernst K.D. 1987. Contribution of single unit analysis in insects to an understanding of olfactory function. *J Comp Physiol* 161:549–565.

Bohbot J.D. and Dickens J.C. 2012. Odorant receptor modulation: Ternary paradigm for mode of action of insect repellents. *Neuropharmacology* 62:2086–2095.

Boppré M. 1986. Insects pharmacophageously utilizing defensive plant chemicals (pyrrolizidine alkaloids). *Naturwissenschaften* 73:17–26.

Boppré M. 1990. Lepidoptera and pyrrolizidine alkaloids, exemplification of complexity in chemical ecology. *J Chem Ecol* 16:165–185.

Boppré M. and Schneider D. 1989. The biology of *Creatonotos* (Lepidoptera: Arctiidae) with special reference to the androconial system. *Zool J Linnean Society* 96:339–356.

Boppré M. and Vane-Wright R.I. 1989. Androconial system in Danainae (Lepidoptera): Functional morphology of *Amauris*, *Danaus*, *Tirumala* and *Euploea*. *Zool J Linn Soc* 97:101–133.

Butenandt A., Beckmann R., Stamm D. and Hecker E. 1959. Über den Sexuallockstoff des Seidenspinners *Bombyx mori*. Reindarstellung und Konstitution. *Z Naturforschung* 14b:283–284.

Butenandt A. and Hecker E. 1961. Synthese des Bombykols, des Sexual-Lockstoffes des Seidenspinners, und seiner geometrischen Isomeren. *Angew Chemie* 73:349–353.

Cardé R.T. and Charlton R.E. 1984. Olfactory sexual communication in Lepidoptera: Strategy, sensitivity and selectivity. In *Insect Communication*, T. Lewis (ed.), 241–265. New York: Academic Press.

Cardé R.T. and Minks A.K. (eds.) 1997. *Insect Pheromone Research: New Directions*. New York: Chapman & Hall.

Chen S. and Luetje C.W. 2012 Identification of new agonists of the insect odorant receptor co-receptor subunit. *PLoS One* 7 (5):e36784.

Christensen T.A. and Hildebrand J.G. 1988. Frequency coding of central olfactory neurons in the sphinx moth *Manduca sexta*. *Chem Senses* 13:123–130.

Clyne P.J., Warr C.G., Freeman M.R., Lessing D., Kim J. and Carlson J.R. 1999. A novel family of divergent seven-transmembrane proteins: Candidate odorant receptors in Drosophila. *Neuron* 22:327–338.

Dani F.R., Michelucci E., Francese S., Mastrobuoni G., Cappellozza S., La Marca G., Niccolini A., Felicioli A., Moneti G. and Pelosi P. 2011. Odorant-binding proteins and chemosensory proteins in pheromone detection and release in the silkmoth *Bombyx mori*. *Chem Senses* 36: 335–347.

De Brito Sanchez M.G. 2000. Estudios electrofisiológicos de estructura-actividad de la cèlula antenal receptora del ácido benzoico de la hembra de *Bombyx mori* L. PhD thesis, University of Buenos Aires, Argentina.

De Brito Sanchez M.G. and Kaissling, K.E. 2005. Inhibitory and excitatory effects of iodobenzene on the antennal benzoic acid receptor cell of the female silk moth *Bombyx mori* L. *Chem Senses* 30:435–442.

De Kramer J.J., Kaissling K.E. and Keil T. 1984. Passive electrical properties of insect sensilla may produce the biphasic shape of spikes. *Chem Senses* 8:289–295.

De Kramer J.J. and Hemberger K. 1987. The neurobiology of pheromone reception. In *Pheromone Biochemistry,* G.D. Prestwich and G.J. Blomquist (eds.), 433–472. New York: Academic Press.

Dobritsa A.A., Van der Goes van Naters W., Warr C., Steinbrecht R.A. and Carlson J.R. 2003. Integrating the molecular and cellular basis of odor coding in the *Drosophila* antenna. *Neuron* 37:827–841.

Dratz E.A. and Hargrave P.A. 1983. The structure of rhodopsin and the rod outer segment disk membrane. *Trends Biochem Sci* 8:128–131.

Du G., Ng C.S. and Prestwich G.D.1994. Odorant binding by a pheromone binding protein: Active site mapping by photoaffinity labeling. *Biochemistry* 33:4812–4819.

Du G. and Prestwich G.D. 1995. Protein structure encodes the ligand binding specificity in pheromone binding proteins. *Biochemistry* 34:8726–8732.

Dumpert K. 1972. Alarmstoffrezeptoren auf der Antenne von *Lasius fuliginosus* (Latr.) (Hymenoptera, Formicidae). *Z Vergl Physiol* 76:403–425.

Egelhaaf A., Rick-Wagner S. and Schneider D. 1992. Development of the male scent organ of *Creatonotos transiens* (Lepidoptera, Arctiidae) during metamorphosis. *Zoomorphology* 111:125–139.

El-Sayed A. M. 2012. The Pherobase: Data base of pheromones and semiochemicals, <http://www.pherobase.com>.

Ernst K.D. 1972. Die Ontogenie der basiconischen Riechsensillen auf der Antenne von *Necrophorus* (Coleoptera). *Z Zellforsch Mikrosk Anat* 129:217–236.

Fan J., Francis F., Liu Y., Chen J.L. and Cheng D.F. 2011. An overview of odorant-binding protein functions in insect peripheral olfactory reception. *Genet Mol Res* 10:3056–3069.

Field L.M., Pickett J.A. and Wadhams L.J. 2000. Molecular studies in insect olfaction. *Insect Molec Biol* 9:545–551.

Forstner M., Gohl T., Breer H. and Krieger J. 2006. Candidate pheromone binding proteins of the silkmoth *Bombyx mori. Invert Neurosci* 6:177–187.

Forstner M., Gohl T., Gondesen I., Raming K., Breer H. and Krieger J. 2008. Differential expression of SNMP-1 and SNMP-2 proteins in pheromone-sensitive hairs of moths. *Chem Senses* 33:291–299.

Forstner M., Breer H. and Krieger J. 2009. A receptor and binding protein interplay in the detection of a distinct pheromone component in the silkmoth *Antheraea polyphemus. Int J Biol Sci* 5:745–757.

Graeter F., Xu W., Leal W. and Grubmueller H. 2006. Pheromone discrimination by the pheromone-binding protein of *Bombyx mori. Structure* 14:1577–1586.

Graham L.A. and Davies P.L. 2002. The odorant-binding proteins of *Drosophila melanogaster*: annotation and characterization of a divergent gene family. *Gene* 292:43–55.

Grosse-Wilde E., Svatos A. and Krieger J. 2006. A pheromone-binding protein mediates the bombykol-induced activation of a pheromone receptor *in vitro. Chem Senses* 31:547–555.

Grosse-Wilde E., Stieber R., Forstner M., Krieger J., Wicher D. and Hansson B. 2010. Sex-specific odorant receptors of the tobacco hornworm *Manduca sexta. Front Cell Neurosci* 4:22.

Gu Y., Lucas P. and Rospars J.-P. 2009. Computational model of the insect pheromone transduction cascade. *Plos Comput Biol* 5(3):e1000321.

Gu Y. and Rospars J.-P. 2011. Dynamical modeling of the moth pheromone-sensitive olfactory receptor neuron within its sensillar environment. *PLoS ONE* 6(3):e17422.

Ha T.S. and Smith D.P. 2009. Odorant and pheromone receptors in insects. *Front Cell Neurosci* 3:10–15.

Hansen K. 1984. Discrimination and production of disparlure enantiomers by the gypsy moth and the nun moth. *Physiol Entomol* 9:9–18.

Hansson B.S. (ed.) 1999. *Insect Olfaction.* Berlin: Springer Verlag.

Hansson B.S. and Christensen T.A. 1999. Functional characteristics of the antennal lobe. In *Insect Olfaction*, B.S. Hansson (ed.). Berlin: Springer Verlag.

Hartlieb E., Anderson P. and Hansson B.S. 1999. Appetitive learning of odours with different behavioural meaning in moths. *Physiol Behav* 67:671–677.

He X., Tzotzos G., Woodcock C., Pickett J.A., Hooper T., Field L.M. and Zhou J.J. 2010. Binding of the general odorant binding protein of *Bombyx mori* BmorGOBP2 to the moth sex pheromone components. *J Chem Ecol* 36:1293–1305.

Heinbockel, T. and Kaissling K.E. 1996. Variability of olfactory receptor neuron responses of female silkmoths (*Bombyx mori* L.) to benzoic acid and (+)-linalool. *J Insect Physiol* 42:565–578.

Hildebrand J.G. 1996. Olfactory control of behavior in moths: Central processing of odor information and the functional significance of olfactory glomeruli. *J Comp Physiol A* 178:5–19.

Hildebrand J.G. and Shepherd G.M. 1997. Mechanisms of olfactory discrimination: Converging evidence for common principles across phyla. *Annu Rev Neurosci* 20:595–631.

Hoesl M. 1990. Pheromone-sensitive neurons in the deutocerebrum of *Periplaneta americana*: Receptive fields on the antenna. *J Comp Physiol* 167:321–327.

Hooper A.M., Dufour S., He X., Muck A., Zhou J.J., Almeida R., Field L.M., Svatos A. and Pickett J.A. 2009. High-throughput ESI-MS analysis of binding between the *Bombyx mori* pheromone-binding protein BmorPBP, its pheromone components and some analogues. *Chem Commun* 2009:5725–5727.

Horst R., Damberger F., Luginbühl P., Güntert P., Peng G., Nikonova L., Leal W.S. and Wüthrich K. 2001. NMR structure reveals intramolecular regulation mechanism for pheromone binding and release. *Proc Natl Acad Sci U S A* 98:14374–14379.

Hummel H.E. and Hecker E. 2012. Half a century of pheromones—Their contributions to sustainable insect pest management (IPM). *Mitt Dtsch Ges Allg Angew Ent* 18:451–460.

Ishida Y. and Leal W.S. 2005. Rapid inactivation of a moth pheromone. *Proc Natl Acad Sci U S A* 102:14075–14079.

Jacquin-Joly E., Francois M.C., Burnet M., Lucas P., Bourrat F. and Maida R. 2002. Expression pattern in the antennae of the newly isolated lepidopteran Gq protein alpha subunit cDNA. *Eur J Biochem* 269:2133–2142.

Jacquin-Joly E. and Lucas P. 2005. Pheromone reception and transduction: Mammals and insects illustrate converging mechanisms across phyla. *Curr Top Neurochem* 4:75–105.

Jin X., Ha T.S. and Smith D.P. 2008. SNMP is a signaling component required for pheromone sensitivity in *Drosophila*. *Proc Natl Acad Sci USA* 105:10995–11000.

Jones P.L., Pask G.M., Rinker D.C. and Zwiebel L.J. 2011. Functional agonism of insect odorant receptor ion channels. *Proc Natl Acad Sci U S A* 108:8821–8825.

Kaib M., Jmhasly P., Wilfert L., Durka W., Franke S., Francke W., Leuthold R.H. and Brandl R. 2004. Cuticular hydrocarbons and aggression in the termite *Macrotermes subhyalinus*. *J Chem Ecol* 30:365–385.

Kaissling K.E. 1972. Kinetic studies of transduction in olfactory receptors of *Bombyx mori*. In *Int Symp Olfaction and Taste IV*, D. Schneider (ed.), 207–21. Stuttgart: Wiss Verlagsgesellsch.

Kaissling K.E. 1974. Sensory transduction in insect olfactory receptors. In *Biochemistry of sensory functions*, L. Jaenicke (ed.). *Mosbacher Coll Ges Biolog Chemie* 25:243–273. Berlin: Springer.

Kaissling K.E. 1977. Structures of odour molecules and multiple activities of receptor cells. In *Olfaction and Taste VI*. J. Le Magnen and P MacLeod (eds.), 9–16. London: Inf Retrieval.

Kaissling K.E. 1980. Action of chemicals, including (+)trans Permethrin and DDT, on insect olfactory receptors. In *Insect Neurobiology and Pesticide Action (Neurotox 79)*, 351–358. London: Society of Chemical Industry.

Kaissling K.E. 1986. Chemo-electrical transduction in insect olfactory receptors. *Annu Rev Neurosci* 9:21–45.

Kaissling K.E. 1987. *R.H. Wright Lectures on Insect Olfaction*, K. Colbow (ed.), 1–190. Burnaby, B.C., Canada: Simon Fraser University.

Kaissling K.E. 1994. IP3 effects in moth pheromone receptors: Calculations. In *Sensory Transduction*, Proceedings of the 22nd Göttingen Neurobiological Conference, Vol. 1, N. Elsner and H. Breer (eds.), 94. Stuttgart: Thieme Verlag.

Kaissling K.E. 1995. Single unit and electroantennogram recordings in insect olfactory organs. In *Experimental Cell Biology of Taste and Olfaction: Current Techniques and Protocols*, A.L. Spielman and J.G. Brand (eds.), 361–386. Boca Raton, FL: CRC Press.

Kaissling K.E. 1997. Pheromone-controlled anemotaxis in moths. In *Orientation and Communication in Arthropods*, M. Lehrer (ed.), 343–374. Basel: Birkhäuser.

Kaissling K.E. 1998. Flux detectors versus concentration detectors: Two types of chemoreceptors. *Chem Senses* 23:99–111.

Kaissling K.E. 2001. Olfactory perireceptor and receptor events in moths: A kinetic model. *Chem Senses* 26:125–150.

Kaissling K.E. 2009a. The sensitivity of the insect nose: The example of *Bombyx mori*. In *Biologically Inspired Signal Processing, SCI*, A. Gutiérrez and S. Marco (eds.), 188:45–52. Berlin, Heidelberg: Springer Verlag.

Kaissling K.E. 2009b. Olfactory perireceptor and receptor events in moths: A kinetic model revised. *J Comp Physiol A* 195:895–922.

Kaissling K.E. 2013. Kinetics of olfactory responses might largely depend on the odorant-receptor interaction and the odorant deactivation postulated for flux detectors. *J Comp Physiol A* 199:879–896.

Kaissling K.E. and Priesner E. 1970. Die Riechschwelle des Seidenspinners. *Naturwissenschaften* 57:23–28.

Kaissling K.E., Kasang G., Bestmann H.J., Stransky W. and Vostrowsky O. 1978. A new pheromone of the silkworm moth *Bombyx mori*. *Naturwissenschaften* 65:382–384.

Kaissling K.E. and Thorson J. 1980. Insect olfactory sensilla: Structural, chemical and electrical aspects of the functional organisation. In *Receptors for Neurotransmitters, Hormones and Pheromones in Insects*, D.B. Sattelle, L.M. Hall and J.G. Hildebrand (eds.), 261–282. Amsterdam, New York: Elsevier/North-Holland Biomedical Press.

Kaissling K.E., Klein U., de Kramer J.J., Keil T.A., Kanaujia S. and Hemberger J. 1985. Insect olfactory cells: Electrophysiological and biochemical studies. In *Molecular Basis of Nerve Activity. Proceedings of the International Symposium in Memory of David Nachmansohn*, J.P. Changeux, F. Hucho, A. Maelicke and E. Neumann (eds.), 173–183. Berlin: W. de Gruyter.

Kaissling K.E., Zack Strausfeld C. and Rumbo E.R. 1987. Adaptation processes in insect olfactory receptors. Mechanisms and behavioral significance. *Ann N Y Acad Sci* 510:104–112.

Kaissling K.E., Meng L.Z. and Bestmann H.J. 1989. Responses of bombykol receptor cells to (*Z,E*)-4,6-hexadecadiene and linalool. *J Comp Physiol A* 165:147–154.

Kaissling, K.E. and Kramer E. 1990. Sensory basis of pheromone-mediated orientation in moths. *Verh Dtsch Zool Ges* 83:109–131.

Kaissling, K.E., Keil T.A. and Williams L. 1991. Pheromone stimulation in perfused olfactory hairs of *Antheraea polyphemus*. *J Insect Physiol* 37:71–78.

Kaissling K.E. and Boekhoff I. 1993. Transduction and intracellular messengers in pheromone receptor cells of the moth *Antheraea polyphemus*. In *Arthropod Sensory Systems*, K. Wiese, F.G. Gribakin, A.V. Popov and G. Renninger (eds.), 489–502. Basel: Birkhäuser Verlag.

Kanaujia S. and Kaissling K.E. 1985. Interactions of pheromone with moth antennae: Adsorption, desorption and transport. *J Insect Physiol* 31:71–81.

Kanzaki R., Soo K., Seki Y. and Wada S. 2003. Projections to higher olfactory centres from subdivisions of the antennal lobe macroglomerular complex of the male silkmoth. *Chem Senses* 28:113–130.

Karg G. and Suckling M. 1999. Applied aspects of insect olfaction. In *Insect Olfaction*, B.S. Hansson (ed.), 351–377. Berlin: Springer Verlag.

Karlson P. and Lüscher M. 1959. "Pheromones": A new term for a class of biologically active substances. *Nature (Lond.)* 183:55–56.

Kasang G. 1968. Tritium labeling of the sex attractant Bombykol. *Z Naturforschung* 23b:1331–1335.

Kasang G. 1971. Bombykol reception and metabolism on the antennae of the silkmoth *Bombyx mori*. In *Gustation and Olfaction*, G. Ohloff and A.F. Thomas (eds.), 245–250. London/ New York: Academic Press.

Kasang G. and Kaissling K.E. 1972. Specificity of primary and secondary olfactory processes in *Bombyx* antennae. In *Intern Symp Olfaction and Taste IV*, D. Schneider (ed.), 200– 206. Stuttgart: Wissensch Verlagsgesellsch.

Kasang G., Schneider D. and Schäfer, W. 1978a. The silkworm moth *Bombyx mori*. Presence of the (*E,E*)-stereoisomer of bombykol in the female pheromone gland. *Naturwissenschaften* 65:337–338.

Kasang G., Kaissling K.E., Vostrowsky O. and Bestmann H.J. 1978b. Bombykal, a second pheromone component of the silkworm moth *Bombyx mori*. L. *Angew Chemie* 90:74– 75, or *Angew Chemie, Int ed Engl* 17:60.

Kasang G., von Proff L. and Nicholls M. 1988. Enzymatic conversion and degradation of sex pheromones in antennae of the male silkworm moth *Antheraea polyphemus*. *Z Naturforschung* 43c:275–284.

Kasang G., Nicholls M. and von Proff L. 1989a. Sex pheromone conversion and degradation in antennae of the silkworm moth *Bombyx mori* L. *Experientia* 45:81–87.

Kasang G., Nicholls M., Keil T. and Kanaujia S. 1989b. Enzymatic conversion of sex pheromones in olfactory hairs of the male silkworm moth *Antheraea polyphemus*. *Z Naturforschung* 44c:920–926.

Kaupp U.B. 2010. Olfactory signalling in vertebrates and insects: Differences and commonalities. *Nat Rev Neurosci* 11:188–200.

Keeling C.I., Slessor K.N., Higo H.A. and Winston M.L. 2003. New components of the honey bee (*Apis mellifera* L.) queen retinue pheromone. *Proc Natl Acad Sci U S A* 100:4486–4491.

Keil T.A. 1984a. Reconstruction and morphometry of silkmoth olfactory hairs: a comparative study of sensilla trichodea on the antennae of male *Antheraea polyphemus* and *Antheraea pernyi* (Insecta, Lepidoptera). *Zoomorphology* 104:147–156

Keil T.A. 1984b. Surface coats of pore tubules and olfactory sensory dendrites of a silkmoth revealed by cationic markers. *Tissue Cell* 16:705–717.

Keil T.A. 1999. Morphology and development of the peripheral olfactory organs. In *Insect Olfaction*, B.S. Hansson (ed.), 5–47. Berlin: Springer Verlag.

Keil T.A. 2012. Sensory cilia in arthropods. *Arth Struct & Dev* 41:515–534.

Keil T.A. and Steinbrecht R.A. 1984. Mechanosensitive and olfactory sensilla of insects. In *Insect Ultrastructure, Vol. 2*, R.C. King and H. Akai (eds.), 477–516. New York: Plenum Publishers.

Keil T.A. and Steiner C. 1991. Morphogenesis of the antenna of the male silkmoth, *Antheraea polyphemus* III. Development of olfactory sensilla and the properties of hair-forming cells. *Tissue Cell* 23:821–851.

Kirschfeld K. 1966. Discrete and graded receptor potentials in the compound eye of the fly *(Musca)*. Proceedings of the International Symposium on the Functional Organization of the Compound Eye, 291–307. Oxford: Pergamon Press.

Klein U. 1987. Sensillum-lymph proteins from antennal olfactory hairs of the moth *Antheraea polyphemus* (Saturniidae). *J Insect Biochem* 17:1193–1204.

Klein U. and Keil T.A. 1984. Dendritic membrane from insect olfactory hairs: Isolation method and electron microscopic observations. *Cell Mol Neurobiol* 4:385–396.

Klusák V., Havlas Z., Rulísek L., Vondrásek J. and Svatos A. 2003. Sexual attraction in the silkworm moth: Nature of binding of bombykol in pheromone binding protein—An ab initio study. *Chem Biol* 10:331–340.

Kodadová B. 1996. Resolution of pheromone pulses in receptor cells of *Antheraea polyphemus* at different temperatures. *J Comp Physiol A* 179:301–310.

Kodadová B. and Kaissling K.E. 1996. Effects of temperature on responses of silkmoth olfactory receptor neurones to pheromone can be simulated by modulation of resting cell membrane resistances. *J Comp Physiol* 179:15–27.

Kramer E. 1986. Turbulent diffusion and pheromone triggered anemotaxis. In *Mechanisms in Insect Olfaction*, T.L. Payne, M.C. Birch and C.E.J. Kennedy (eds.), 58–67. Oxford: Oxford University Press.

Kramer E. 1992. Attractivity of pheromone surpassed by time-patterned application of two nonpheromone compounds. *J Insect Behav* 5:83–97.

Kramer E. 1996. A tentative intercausal nexus and its computer model on insect orientation in windborne pheromone plumes. In *Pheromone Research: New Directions*, R.T. Cardé and A. Minks (eds.), 232–247. New York: Chapman & Hall.

Krieger J., Strobel J., Vogl A., Hanke W. and Breer H. 1999. Identification of a cyclic nucleotide and voltage-activated ion channel from insect antennae. *Insect Biochem Mol Biol* 29:255–267.

Krieger J., Raming K., Dewer Y.M.E., Bette S., Conzelmann S. and Breer H. 2002. A divergent gene family encoding candidate olfactory receptors of the moth *Heliothis virescens*. *Eur J Neurosci* 16:619–628.

Krieger J. and Breer H. 2003. Transduction mechanisms of olfactory sensory neurons. In *Insect Pheromone Biochemistry and Molecular Biology*, G.J. Blomquist and R.G. Vogt (eds.), 593–607. London: Elsevier Academic Press.

Krieger J., Klink O., Mohl C., Raming K. and Breer H. 2003. A candidate olfactory receptor subtype highly conserved across insect orders. *J Comp Physiol A* 189:519–526.

Krieger J., Grosse-Wilde E., Gohl T. and Breer H. 2005. Candidate pheromone receptors of the silkmoth *Bombyx mori*. *Eur J Neurosci* 21:2167–2176.

Kruse S.W., Zhao R., Smith D. and Jones N. 2003. Structure of a specific alcohol-binding site defined by the odorant binding protein LUSH from *Drosophila melanogaster*. *Nat Struct Biol* 10:694–700.

Kumar G.L. and Keil T.A. 1996. Pheromone stimulation induces cytoskeletal changes in olfactory dendrites of male silkmoths (Lepidoptera, Saturniidae, Bombycidae). *Naturwissenschaften* 83:476–478.

Küppers J. and Bunse I. 1996. A primary cation transport by a V-type ATPase of low specificity. *J Exp Biol* 199:1327–1334.

Lacher V. 1964. Elektrophysiologische Untersuchungen an einzelnen Rezeptoren für Geruch, Kohlendioxyd, Luftfeuchtigkeit und Temperatur auf den Antennen der Arbeitsbiene und der Drohne (*Apis mellifica* L.). *Z Vergl Physiol* 48:587–623.

Lartigue A., Gruez A., Spinelli S., Riviere S., Brossut B., Tegoni M. and Cambillau C. 2003. The crystal structure of a cockroach pheromone-binding protein suggests a new ligand binding and release mechanism. *J Biol Chem* 278:30213–30218.

Lartigue A., Gruez A., Briand L., Blon F., Bézirard V., Walsh M., Pernollet J.C., Tegoni M. and Cambillau C. 2004. Sulfur single-wavelength anomalous diffraction crystal structure of a pheromone-binding protein from the honeybee *Apis mellifera* L. *J Biol Chem* 279:4459–4464.

Laue M., Maida R. and Redkozubov A. 1997. G-protein activation, identification and immunolocalization in pheromone-sensitive sensilla trichodea of moths. *Cell Tissue Res* 288:149–158.

Laughlin J.D., Ha T.S., Jones D.N.M. and Smith D.P. 2008. Activation of pheromone-sensitive neurons is mediated by conformational activation of pheromone-binding protein. *Cell* 133:1255–1265.

Lautenschlager C., Leal W.S. and Clardy J. 2005. Coil-to-helix and ligand release of *Bombyx mori* pheromone-binding protein. *Biochem Biophysic Res Commun* 335:1044–1050.

Leal W.S. 2003. Proteins that make sense. In *Insect Pheromone Biochemistry and Molecular Biology: The Biosynthesis and Detection of Pheromones and Plant Volatiles*, G.J. Blomquist and R.G. Vogt (eds.), 447–476. London: Elsevier Academic Press.

Leal W.S., Nikonova L. and Peng G. 1999. Disulfide structure of the pheromone binding protein from the silkworm moth, *Bombyx mori*. *FEBS Lett* 464:85–90.

Leal W.S., Chen A.M., Ishida Y., Chiang V.P., Erickson M.L., Morgan T.I. and Tsuruda J.M. 2005. Kinetics and molecular properties of pheromone binding and release. *Proc Natl Acad Sci U S A* 102:5386–5391.

Lei H., Christensen T.A. and Hildebrand J.G. 2002. Local inhibition evoked synchronization of glomerulus-specific output neurons. *Nat Neurosci* 5:557–565.

Levinson H. and Levinson A. 2002. Insectistasis as a means of controlling pest populations in the storage environment. In *Encyclopedia of Pest Management*, 402–406. New York: Marcel Dekker.

Mafra-Neto A. and Cardé R.T. 1996. Dissection of the pheromone-modulated flight of moths using single-pulse response as a template. *Experientia* 52:373–379.

Maibeche-Coisne M., Jacquin-Joly E., Francois M.C. and Nagnan Le Meillour P. 2002. cDNA cloning of biotransformation enzymes belonging to the cytochrome P450 family in the antennae of the noctuid moth *Mamestra brassicae*. *Insect Mol Biol* 11:273–281.

Maida R., Ziegelberger G. and Kaissling K.E. 1995. Esterase activity in the olfactory sensilla of the silkmoth *Antheraea polyphemus*. *NeuroReport* 6:822–824.

Maida R., Redkozubov A. and Ziegelberger G. 2000. Identification of PLCß and PKC in pheromone receptor neurons of *Antheraea polyphemus*. *Neuroreport* 11:1773–1776.

Maida R., Ziegelberger G. and Kaissling K.E. 2003. Ligand binding to six recombinant pheromone-binding proteins of *Antheraea polyphemus* and *Antheraea pernyi*. *J Comp Physiol B* 173:565–573.

Maida R., Mameli M., Mueller B., Krieger J. and Steinbrecht R.A. 2005. The expression pattern of four odorant-binding proteins in male and female silk moths, *Bombyx mori*. *J Neurocytol* 34:149–163.

Menco B.P. 1983. The ultrastructure of olfactory and nasal respiratory epithelium surfaces. In *Nasal Tumors in Animals and Man, Vol 1 Anatomy, Physiology, and Epidemiology*, G. Reznik and S.F. Stinson (eds.), 45–102. Boca Raton, FL: CRC Press.

Meng L.Z., Wu C.H., Wicklein M., Kaissling K.E. and Bestmann H.J. 1989. Number and sensitivity of three types of pheromone receptor cells in *Antheraea pernyi* and *A. polyphemus*. *J Comp Physiol A* 165:139–146.

Michel E., Damberger F.F., Ishida Y., Fiorito F., Lee D., Leal W.S. and Wüthrich K. 2011. Dynamic conformational equilibria in the physiological function of the *Bombyx mori* pheromone-binding protein. *J Mol Biol* 408:922–931.

Minor A.V. and Kaissling K.E. 2003. Cell responses to single pheromone molecules may reflect the activation kinetics of olfactory receptor molecules. *J Comp Physiol A* 189:221–230.

Mohanty S., Zubkov S. and Gronenborn A.M. 2004. The solution NMR structure of *Antheraea polyphemus* PBP provides new insight into pheromone recognition by pheromone-binding proteins. *J Mol Biol* 337:443–451.

Montague S.A., Mathew D. and Carlson J.R. 2011. Similar odorants elicit different behavioral and physiological responses, some supersustained. *J Neurosci* 31:7891–7899.

Moulton D.G. 1977. Minimum odorant concentrations detectable by the dog and their implications for olfactory receptor sensitivity. In *Chemical Signals in Vertebrates*, D. Müller Schwarze and M.M. Mozell (eds.), 455–464. New York: Plenum Press.

Mustaparta H. 1975. Responses of single olfactory cells in the pine weevil *Hylobius abietes* L. (Col.: Curculionidae). *J Comp Physiol* 97:271–290.

Mustaparta H. 2002. Encoding of plant odour information in insects: Peripheral and central processing. *Entomologia Experimentalis et Applicata* 104:1–13.

Nagnan Le Meillour P., Cain A.H., Jacquin-Joly E., Francois M.C., Ramachandran S., Maida R. and Steinbrecht R.A. 2000. Chemosensory proteins from the proboscis of *Mamestra brassicae*. *Chemical Senses* 25:541–553.

Nagnan-Le Meillour P. and Jacquin-Joly E. 2003. Biochemistry and diversity of insect odorant-binding proteins. *In Insect Pheromone Biochemistry and Molecular Biology*, G.J. Blomquist and R.G. Vogt (eds.), 509–537. London: Elsevier Academic Press.

Nakagawa T., Sakurai T., Nishioka T. and Touhara K. 2005. Insect sex-pheromone signals mediated by specific combinations of olfactory receptors. *Science* 307:1638–1642.

Nakagawa T. and Vosshall L.B. 2009. Controversy and consensus: Noncanonical signaling mechanisms in the insect olfactory system. *Curr Opin Neurobiol* 19:284–292.

Nemoto T., Uebayashi M. and Komeiji Y. 2002. Flexibility of a loop in a pheromone binding protein from Bombyx mori: A molecular dynamics simulation. *Chem-Bio Info J* 2:32–37.

Neuhaus W. 1953. Über die Riechschärfe des Hundes für Fettsäuren. *Z vergl Physiol* 35:527–552.

Nolte A., Funk N.W., Mukunda L., Gawalek P., Werckenthin A. et al. 2013. *In situ* tip-recordings found no evidence for an Orco-based ionotropic mechanism of pheromone-transduction in *Manduca sexta*. *PLoS ONE* 8(5):e62648.

Oldenburg C., Kanaujia S., Spiteller D., Oldham N. J., Boland W. and Kaissling K.E. 2001. Benzoic acid, a stimulant of odorant receptors of *Bombyx mori*, is rapidly metabolized to *N*-benzoylserine on the antennae. *Chemoecology* 11:183–190.

Ozaki M., Takahara T., Kawahara Y., Wada-Katsumata A., Seno K., Amakawa T., Yamaoka R. and Nakamura T. 2003. Perception of noxious compounds by contact chemoreceptors of the blowfly, *Phormia regina*: Putative role of an odorant-binding protein. *Chem Senses* 28:349–359.

Pelosi P. 1996. Perireceptor events in olfaction. *J Neurobiol* 30:3–19.

Pelosi P. and Maida R. 1995. Odorant-binding proteins in insects. *Comp Biochem Physiol* 111B:503–514.

Pelosi P., Zhou J.J., Ban L.P. and Calvello M. 2006. Soluble proteins in insect chemical communication. *Cell Mol Life Sci* 63:1658–1676.

Pernollet J.C. and Briand L. 2004. Structural recognition between odorants, olfactory-binding proteins and olfactory receptors, primary events in odor coding. In *Flavor Perception*, A. Taylor and D. Roberts (eds.), 86–150. Oxford: Blackwell Publishing.

Picimbon J.F. 2003. Biochemistry and evolution of OBP and CSP proteins. In *Insect Pheromone Biochemistry and Molecular Biology*, G.J. Blomquist and R.G. Vogt (eds.), 539–566. London: Elsevier Academic Press.

Picimbon J.F. 2004. Synthesis of odorant reception-suppressing agents: Odorant binding proteins (OBPs) and chemosensory proteins (CSPs) as molecular targets for pest management. In *Phytoprotection*, B. Philogène, C. Regnault-Roger and C. Vincent (eds.), 245–266. Paris: Lavoisier Tech and Doc.

Plettner E., Lazar J., Prestwich E.G. and Prestwich G.D. 2000. Discrimination of pheromone enantiomers by two pheromone binding proteins from the gypsy moth *Lymantria dispar*. *Biochemistry* 39:8953–8962.

Pophof B. 1998. Inhibitors of sensillar esterase reversibly block the responses of moth pheromone receptor cells. *J Comp Physiol A* 183:153–164.

Pophof B. 2000. Octopamine modulates the sensitivity of silkmoth pheromone receptor neurons. *J Comp Physiol A* 186:307–313.

Pophof B. 2002. Moth pheromone binding proteins contribute to the excitation of olfactory receptor cells. *Naturwissenschaften* 89:515–518.

Pophof B. 2004. Pheromone-binding proteins contribute to the activation of olfactory receptor neurons in the silkmoths *Antheraea polyphemus* and *Bombyx mori*. *Chem Senses* 29:117–126.

Pophof B., Gebauer T. and Ziegelberger G. 2000. Decyl-thio-trifluoropropanone, a competitive inhibitor of moth pheromone receptors. *J Comp Physiol A* 186:315–323.

Pophof B. and Van der Goes van Naters W. 2002. Activation and inhibition of the transduction process in silkmoth olfactory receptor neurons. *Chem Senses* 27:435–443.

Priesner E. 1979. Progress in the analysis of pheromone receptor systems. *Ann Zool Ecol Anim* 11:533–546.

Priesner E. 1980. Sex attractant system in *Polia pisi* (Lepidoptera: Noctuidae). *Z Naturforsch* 35c:990–994.

Priesner E. 1984. The pheromone receptor system of male *Eulia ministrana* L., with notes on other Cnephasiini moths. *Z Naturforsch* 39c:849–852.

Priesner E. and Schroth M. 1983. Supplementary data on the sex attractant system of *Panolis flammea*. *Z Naturforsch* 38c:870–873.

Priesner E., Witzgall P. and Voerman S. 1986. Field attraction response of raspberry clearwing moths, *Pennisetia hyleiformis* Lasp. (Lepidoptera: Sesiidae), to candidate pheromone chemicals. *J Appl Ent* 102:195–210.

Quero C., Bau J., Guerrero A. and Renou M. 2004. Responses of the olfactory receptor neurons of the corn stalk borer *Sesamia nonagrioides* to components of the pheromone blend and their inhibition by trifluoromethyl ketone analogue. *Pest Manag* Sci 60:719–726.

Redkozubov A. 2000. Elementary receptor currents elicited by a single pheromone molecule exhibit quantal composition. *Pfluegers Arch: Eur J Physiol* 440:896–901.

Renou M. and Guerrero A. 2000. Insect parapheromones in olfaction research and semiochemicalbased pest control strategies. *Annu Rev Entomol* 45:605–630.

Renou M., Berthier A. and Guerrero A. 2002. Disruption of responses to pheromone by (Z)-11-hexadecenyl trifluoromethyl ketone, an analogue of the pheromone, in the cabbage armyworm *Mamestra brassicae*. *Pest Manag Sci* 58:839–844.

Rogers M., Sun M., Lerner M.R. and Vogt R.G. 1997. Snmp-1, a novel membrane protein of olfactory neurons of the silk moth *Antheraea polyphemus* with homology to the CD36 family of membrane proteins. *J Biol Chem* 272:14792–14804.

Rogers M.E., Steinbrecht R.A. and Vogt R.G. 2001a. Expression of SNMP-1in olfactory neurons and sensilla of male and female antennae of the silkmoth *Antheraea polyphemus*. *Cell Tissue Res* 303:433–446.

Rogers M.E., Krieger J. and Vogt R.G. 2001b. Antennal SNMPs (sensory neuron membrane proteins) of Lepidoptera define a unique family of invertebrate CD36-like proteins. *J Neurobiol* 49:47–61.

Rospars J.P., Lucas P. and Coppey M. 2007. Modelling the early steps of transduction in insect olfactory receptor neurons. *Biosystems* 89:101–109.

Rumbo E.R. and Kaissling K.E. 1989. Temporal resolution of odour pulses by three types of pheromone receptor cells in *Antheraea polyphemus*. *J Comp Physiol A*, 165:281–291.

Rybczynski R., Vogt R.G. and Lerner, M.R. 1990. Antennal-specific pheromone-degrading aldehyde oxidases from the moths *Antheraea polyphemus* and *Bombyx mori*. *J Biol Chem* 32:19712–19715.

Sadek M.M., Hansson B.S., Rospars J.P. and Anton S. 2002. Glomerular representation of plant volatiles and sex pheromone components in the antennal lobe of *Spodoptera littoralis*. *J Exp Biol* 205:1363–1376.

Sakurai T., Nakagawa T., Mitsuno H., Mori H., Endo Y., Tanoue S., Yasukochi Y., Touhura K. and Nishioka T. 2004. Identification and functional characterization of a sex pheromone receptor in the silkmoth *Bombyx mori*. *Proc Natl Acad Sci U S A* 101:16653–16658.

Sakurai T., Mitsuno H., Haupt S.S., Uchino K., Yokohari F., Nishioka T., Kobayashi I., Sezutsu H., Tamura T. and Kanzaki R. 2011. A single sex pheromone receptor determines chemical response specificity of sexual behavior in the silkmoth *Bombyx mori*. *PLoSGenet* 7(6):e1002115.

Sandler B.H., Nikonova L., Leal W.S. and Clardy J. 2000. Sexual attraction in the silkworm moth: structure of the pheromone-binding-protein-bombykol complex. *Chem Biol* 7:143–151.

Sato K., Pellegrino M., Nakagawa T., Vosshall L.B. and Touhara K. 2008. Insect olfactory receptors are heteromeric ligand-gated ion channels. *Nature* 452:1002–1006.

Scaloni A., Monti M., Angeli S. and Pelosi P. 1999. Structural analysis and disulfide-bridge pairing of two odorant-binding proteins from *Bombyx mori*. *Biochem Biophys Res Commun* 266:386–391.

Schneider D. 1957. Electrophysiological investigation of the antennal receptors of the silk moth during chemical and mechanical stimulation. *Experientia* 13:89.

Schneider D. 1984. Insect olfaction—Our research endeavour. In *Foundations of Sensory Science*, W.W. Dawson and J.M. Enoch (eds.), 381–481. Berlin, Heidelberg: Springer-Verlag.

Schneider D. 1992. 100 years of pheromone research, an essay on Lepidoptera. *Naturwissenschaften* 79:241–250.

Schneider D. and Kaissling K.E. 1957. Der Bau der Antenne des Seidenspinners *Bombyx mori* L. II. Sensillen, cuticulare Bildungen und innerer Bau. *Zool Jahrb Anat Ontog* 76:224–250.

Schneider D., Lacher V. and Kaissling K.E. 1964. Die Reaktionsweise und das Reaktionsspektrum von Riechzellen bei *Antheraea pernyi* (Lepidoptera, Saturniidae). *Z vergl Physiol* 48:632–662.

Schneider D., Schulz S., Priesner E., Ziesmann J. and Francke W. 1988. Autodetection and chemistry of female and male pheromone in both sexes of the tiger moth *Panaxia quadripunctaria*. *J Comp Physiol A* 182:153–161.

Scholes J. 1965. Discontinuity of the excitation process in locust visual cells. *Cold Spring Harbour Symp Quant Biol* 30:517–527.

Stange G. and Diesendorf M. 1973. The response of the honey bee antennal CO_2-receptors to N_2O and Xe. *J Comp Physiol* 86:139–158.

Stange G. and Kaissling K.E. 1995. The site of action of general anaesthetics in insect olfactory receptor neurons. *Chem Senses* 20:421–432.

Stange G. and Stowe S. 1999. Carbon-dioxide sensing structures in terrestrial arthropods. *Microsc Res Tech* 47:416–427.

Steinbrecht R.A. 1964. Feinstruktur und Histochemie der Sexualduftdrüse des Seidenspinners *Bombyx mori*. *Z Zellforsch* 64:227–261.

Steinbrecht R.A. 1970. Zur Morphometrie der Antenne des Seidenspinners, *Bombyx mori L.*: Zahl und Verteilung der Riechsensillen (Insecta, Lepidoptera). *Z Morph Tiere* 66:93–126.

Steinbrecht R.A. 1973. Der Feinbau olfaktorischer Sensillen des Seidenspinners (*Insecta, Lepidoptera*). Rezeptorfortsätze und reizleitender Apparat. *Z Zellforsch* 139:533–565.

Steinbrecht R.A. 1984. Chemo-, hygro-, and thermoreceptors. In *Biology of the Integument*, Vol. 1, J. Bereiter-Hahn, A.G. Matoltsy and K.S. Richards (eds.), 523–553. Berlin, Heidelberg: Springer Verlag.

Steinbrecht, R.A. 1996. Are odorant-binding proteins involved in odorant discrimination? *Chem Senses* 21:719–727.

Steinbrecht, R.A. 1997. Pore structures in insect olfactory sensilla—A review of data and concepts. *Int J Insect Morphol Embryol* 26:229–245.

Steinbrecht, R.A. 1999. Olfactory receptors. In *Atlas of Arthropod Sensory Receptors: Dynamic Morphology in Relation to Function*, E. Eguchi and Y. Tominaga (eds.), 155–176. Tokyo: Springer Verlag.

Steinbrecht R.A. and Kasang G. 1972. Capture and conveyance of odour molecules in an insect olfactory receptor. In *Int Symp Olfaction and Taste IV*, D. Schneider (ed.), 193–199. Stuttgart: Wiss Verlagsgesellsch.

Steinbrecht R.A. and Zierold K. 1989. Electron probe X-ray microanalysis in the silkmoth antenna—Problems with quantification in ultrathin cryosections. In *Electron Probe Microanalysis—Applications in Biology and Medicine*, K. Zierold and H.K. Hagler (eds.), 87–97. Berlin, Heidelberg: Springer Verlag.

Steinbrecht R.A., Ozaki M. and Ziegelberger G. 1992. Immunocytochemical localization of pheromone-binding protein in moth antennae. *Cell Tissue Res* 270:287–302.

Steinbrecht R.A., Laue M. and Ziegelberger G. 1995. Immunolocalization of pheromone-binding protein and general odorant-binding protein in olfactory sensilla of the silk moths *Antheraea* and *Bombyx*. *Cell Tissue Res* 282:203–217.

Steinbrecht, R.A. and Stankiewicz, B.A. 1999. Molecular composition of the wall of insect olfactory sensilla—The chitin question. *J Insect Physiol* 45:785–790.

Stengl M. 2010. Pheromone transduction in moths. *Front Cell Neurosci* 4:1–15.

Stengl M., Ziegelberger G. and Boekhoff I. 1999. Perireceptor events and transduction mechanisms in insect olfaction. In *Insect Olfaction*, B.S. Hansson (ed.), 49–66. Berlin: Springer Verlag.

Stranden M., Røstelien T., Liblikas I., Almaas T.J., Borg-Karlson A.K. and Mustaparta H. 2003. Receptor neurones in three heliothine moths responding to floral and inducible plant volatiles. *Chemoecology* 13:143–154.

Stuiver M. 1958. Biophysics of the sense of smell. Thesis Rijksuniversiteit, Groningen.

Syed Z., Ishida Y., Taylor K., Kimbrell D.A. and Leal W.S. 2006. Pheromone reception in fruit flies expressing a moth's odorant receptor. *Proc Natl Acad Sci U S A* 103:16538–16543.

Tegoni M., Campanacci V. and Cambillau C. 2004. Structural aspects of sexual attraction and chemical communication in insects. *Trends Biochem Sci* 29:257–264.

Thurm U. and Küppers J. 1980. Epithelial physiology of insect sensilla. In *Insect Biology in the Future*, M. Locke and D.S. Smith (eds.), 735–763. New York: Academic Press.

Todd J.L. and Baker T.C. 1999. Function of peripheral olfactory organs. In *Insect Olfaction*, B.S. Hansson(ed.), 67–96. Berlin: Springer Verlag.

Trona F., Anfora G., Balkenius A., Bengtsson M., Tasin M., Knight A., Janz N., Witzgall P. and Ignell R. 2013. Neural coding merges sex and habitat chemosensory signals in an insect herbivore. *Proc R Soc B* 280:1760.

Van den Berg M.J. and Ziegelberger G. 1991. On the function of the pheromone-binding protein in the olfactory hairs of *Antheraea polyphemus*. *J Insect Physiol* 37:79–85.

Vareschi E. 1971. Duftunterscheidung bei der Honigbiene—Einzelzellableitungen und Verhaltensreaktionen. *Z Vergl Physiol* 75:143–173.

Vareschi E. and Kaissling K.E. 1970. Dressur von Bienenarbeiterinnen und Drohnen auf Pheromone und andere Duftstoffe. *Z vergl Physiol* 66:22–26.

Vermeulen A. and Rospars J.P. 2001. Electrical circuitry of an insect olfactory sensillum. *Neurocomputing* 29:587–596.

Vijverberg H.P.M., van der Zalm J.M. and van den Bercken J. 1982. Similar mode of action of pyrethroids and DDT on sodium channel gating in myelinated nerves. *Nature (Lond.)* 295:601–603.

Vogel S. 1983. How much air passes through a silkmoth antenna? *J Insect Physiol* 29:597–602.

Vogt R.G. 2003. Biochemical diversity of odor detection: OBPs, ODEs and SNMPs. In *Insect Pheromone Biochemistry and Molecular Biology*, G.J. Blomquist and R.G. Vogt (eds.), 391–446. London: Elsevier Academic Press.

Vogt R.G. 2005. Molecular basis of pheromone detection in insects. In *Comprehensive Insect Physiology, Biochemistry, Pharmacology and Molecular Biology. Volume 3. Endocrinology*, L.I. Gilbert, K. Iatro and S. Gill (eds.), 753–804. London: Elsevier.

Vogt R.G. and Riddiford L.M. 1981. Pheromone binding and inactivation by moth antennae. *Nature (London)* 293:161–163.

Vogt R.G. and Riddiford L.M. 1986a. Pheromone reception: A kinetic equilibrium. In *Mechanisms in Insect Olfaction*, T.L. Payne, M. Birch, and C.E.J. Kennedy (eds.), 201–208. Oxford: Clarendon Press.

Vogt R.G. and Riddiford L.M. 1986b. Scale esterase: A pheromone degrading enzyme from the wing scales of the silk moth *Antheraea polyphemus. J Chem Ecol* 12:469–482.

Vogt R.G., Riddiford L.M. and Prestwich GD. 1985. Kinetic properties of a pheromone degrading enzyme: The sensillar esterase of *Antheraea polyphemus. Proc Natl Acad Sci U S A* 82:8827–8831.

Vogt R.G., Callahan F.E., Rogers M.E. and Dickens J.C. 1999. Odorant binding protein diversity and distribution among the insect orders, as indicated by LAP, an OBP-related protein of the true bug *Lygus lineolaris* (Hemiptera, Heteroptera). *Chem Senses* 24:481–495.

Von Nickisch-Rosenegk E., Krieger J., Kubick S., Laage R., Strobel J., Strotmann J. and Breer H. 1996. Cloning of biogenic amine receptors from moth (*Bombyx mori* and *Heliothis virescens*). *Insect Biochem Mol Biol* 26:817–827.

Vosshall L.B. 2001. The molecular logic of olfaction in *Drosophila. Chem Senses* 26:207–213.

Vosshall L.B. and Hansson S.B. 2011. A unified nomenclature system for the insect olfactory coreceptor. *Chem Senses* 36:497–498.

Wicher D., Schaefer R., Bauernfeind R., Stensmyr M.C., Heller R., Heinemann S.H. and Hansson B.S. 2008. *Drosophila* odorant receptors are both ligand-gated and cyclic-nucleotide-activated cation channels. *Nature* 452:1007–1011.

Wojtasek H. and Leal W.S. 1999. Conformational change in the pheromone-binding protein from *Bombyx mori* induced by pH and by interaction with membranes. *J Biol Chem* 274:30950–30956.

Xu P.X., Atkinson R., Jones D.N.M. and Smith D.P. 2005. Drosophila OBP LUSH is required for activity of pheromone-sensitive neurons. *Neuron* 45:193–200.

Xu P.X., Hooper A.M., Pickett J.A. and Leal W.S. 2012. Specificity determinants of the silkworm moth sex pheromone. *PLoS ONE* 7(9):e44190.

Xu W. and Leal W.S. 2008. Molecular switches for pheromone release from a moth pheromone-binding protein. *Biochem Biophys Res Commun* 372:559–564.

Zack C.W.M. 1979. Sensory adaptation in the sex pheromone receptor cells of saturniid moths. Diss Fak Biol LMU München.

Zack-Strausfeld C. and Kaissling, K.E. 1986. Localized adaptation processes in olfactory sensilla of Saturniid moths. *Chem Senses* 11:499–512.

Zhang S.G., Maida R. and Steinbrecht R.A. 2001. Immunolocalization of odorant-binding proteins in noctuid moths (Insecta, Lepidoptera). *Chem Senses* 26:885–896.

Zhou J.J., Robertson G., He X., Dufour S., Hooper A.M., Pickett J.A., Keep N.H. and Field L.M. 2009. Characterisation of *Bombyx mori* odorant-binding proteins reveals that a general odorant-binding protein discriminates between sex pheromone components. *J Mol Biol* 389:529–545.

Zhou Y. and Wilson R.I. 2012. Transduction in *Drosophila* olfactory receptor neurons is invariant to air speed. *J Neurophysiol* 108:2051–2059.

Ziesmann J., Valterova I., Haberkorn K., De Brito Sanchez M.G. and Kaissling K.E. 2000. Chemicals in laboratory room air stimulate olfactory neurons of female *Bombyx mori. Chem Senses* 25:31–37.

Zufall F. and Hatt H. 1991. Dual activation of sex pheromone-dependent ion channel from insect olfactory dendrites by protein kinase C activators and cyclic GMP. *Proc Natl Acad Sci U S A* 88:8520–8524.

Zufall F., Hatt H. and Keil T.A. 1991. A calcium-activated nonspecific cation channel from olfactory receptor neurons of the silkmoth *Antheraea polyphemus. J Exp Biol* 161:455–468.

5 Chemical Communication in the Honey Bee Society

Laura Bortolotti and Cecilia Costa

CONTENTS

5.1 Pheromones of the Honey Bee Colony 148
    5.1.1 Queen Pheromones 149
        5.1.1.1 Main Component of the Queen Signal: The Queen
                 Mandibular Pheromone... 149
        5.1.1.2 Beyond the QMP: Other Queen Pheromones 157
    5.1.2 Worker Pheromones.. 161
        5.1.2.1 Regulation of Worker Reproduction: The Mandibular
                 Gland Pheromones 162
        5.1.2.2 Regulation of Worker Activity: Ethyl Oleate.................... 163
        5.1.2.3 Orientation and Recruitment: The Nasonov Gland
                 Pheromones... 164
        5.1.2.4 Marking and Recruiting in Foraging Behavior: Tarsal
                 Glands and Other Pheromones 165
        5.1.2.5 Defensive Behavior: Alarm Pheromones........................... 165
        5.1.2.6 Nestmate Recognition: The Cuticular Hydrocarbons........ 168
    5.1.3 Drone Pheromones.. 169
        5.1.3.1 Drone Acceptance in the Colony 169
    5.1.4 Brood Pheromones.. 170
        5.1.4.1 Regulation of Brood Development and Care...................... 170
        5.1.4.2 Regulation of Worker Reproduction 171
        5.1.4.3 Regulation of Worker Behavioral Development................ 171
    5.1.5 Pheromone Complexity and Evolution of Sociality in Bees............. 172
5.2 Neurophysiology of Chemical Communication: Pheromone Processing
    in the Bee Brain... 175
    5.2.1 Reception of the Pheromonal Signal 176
        5.2.1.1 Olfactory Receptor Neurons 176
        5.2.1.2 Antennal Lobes and the Glomeruli 178
        5.2.1.3 Pheromone Processing in the Glomeruli 179
        5.2.1.4 Sexual Communication: Drone Reception of QMP in
                 Macroglomerular Complexes.. 180
        5.2.1.5 Pheromone Processing in Higher Centers 181

5.2.2   Processing and Modulation of the Pheromonal Signal.................... 182
    5.2.2.1   Modulation of the Signal: The Role of Biogenic
        Amines and Juvenile Hormones 183
    5.2.2.2   Direct Modulation of Worker Behavior: HVA Mimic
        of Dopamine 187
    5.2.2.3   Worker Attraction and Aversion: The Role of
        Pheromones on Appetitive and Defense Behavior............. 188
    5.2.2.4   Modulation of Worker Metabolism: The Effect of
        Pheromones on Nutrient Stores.. 190
5.2.3   From Signal to Behavior: Pheromones and Gene Expression 191
    5.2.3.1   Insights into the Pheromone-Mediated Genetic
        Mechanism Underlying Worker Behavioral
        Development.. 192
    5.2.3.2   Alarm Pheromone and the Expression of Immediate
        Early Genes.. 193
5.3   Concluding Remarks 193
References.. 195

5.1   PHEROMONES OF THE HONEY BEE COLONY

Together with the honey bee dance, honey bee pheromones represent one of the most advanced ways of communication among social insects.

Pheromones are chemical substances secreted by an animal's exocrine glands that elicit a behavioral or physiological response by another animal of the same species. In honey bees the targets of pheromonal messages are usually members of the same colony, but there are some exceptions in which the target can be a member of another colony (Free 1987).

The composite organization of the honey bee society, which consists of three adult castes (queen, worker, and male) and non-self-sufficient brood, provides for many coordinated activities and developmental processes and thus needs a similar elaborate way of communication among the colony members. Pheromones are the key factor in generating and maintaining this complexity, assuring a broad plasticity of functions that allow the colony to deal with unforeseen events or changing environmental conditions.

Pheromones are involved in almost every aspect of the honey bee colony life: development and reproduction (including queen mating and swarming), foraging, defense, orientation, and in general the whole integration of colony activities, from foundation to decline. Pheromones allow communication among all the honey bee castes: queen–workers, workers–workers, queen–drones, and between adult bees and brood (Trhlin and Rajchard 2011; Winston 1987).

In honey bees, as in other animals, there are two types of pheromones: primer pheromones and releaser pheromones. Primer pheromones act at a physiological level, triggering complex and long-term responses in the receiver and generating both developmental and behavioral changes. Releaser pheromones have a weaker effect, generating a simple and transitory response that influences the receiver only at the behavioral level.

Most of the pheromones known in insects are of the releaser type; they are clas-sified into several categories based on their function (e.g., sexual, aggregation, dis-persal, alarm, recruitment, trail, territorial, recognition) (Ali and Morgan 1990). Primer pheromones are especially developed in social insects, where they represent the major driving force in the evolution of social harmony and in maintaining colony homeostasis (Le Conte and Hefetz 2008). Among honey bee pheromones, the queen signal and the brood pheromones (described in detail below) are principally primer pheromones (having also some releaser functions), while most worker pheromones are to be considered releaser pheromones.

In the following paragraphs the main honey bee pheromones are described, based on the honey bee caste to which they belong and the glands responsible for their production. In the first part of the chapter the effect (or the effects) exerted by each pheromone on the receivers and on the bee colony will be illustrated, while the neuro-physiologic and molecular mechanisms of the response to the chemicals will be dis-cussed in the second part of the chapter.

5.1.1   QUEEN PHEROMONES

The honey bee queen represents the main regulating factor of the colony functions. This regulation is largely achieved by means of pheromones, which are produced by different glands and emitted as a complex chemical blend, known as the "queen signal."

The queen signal acts principally as a primer pheromone, inducing several physi-ological and behavioral modifications in the worker bees of the colony that result in maintenance of colony homeostasis through establishment of social hierarchy and preservation of the queen's reproductive supremacy. More specifically, the effects of the queen signal are maintenance of worker cohesion, suppression of queen rear-ing, inhibition of worker reproduction, and stimulation of worker activities: cleaning, building, guarding, foraging, and brood feeding (Figure 5.1). It is known that when the queen is old or sick (low pheromonal signal) or it dies (no pheromonal signal), workers are driven to rear new queens from young brood within 12–24 hours; the removal of the queen in absence of young brood soon leads to the decline of the colony: the workers stop performing their activities and start to lay unfertilized eggs that develop into male adults (drones); the colony becomes disorganized, unfit, dirty, susceptible to diseases and prey of predators; it rapidly depopulates and goes toward a certain death.

In addition to its primer effect, the queen signal exerts an attractive releaser effect: it calls workers around the queen in a retinue group, which is stimulated to feed and groom her; in young premating queens it acts as attractant for drones during the mat-ing flights; during swarming it keeps the swarm together.

5.1.1.1   Main Component of the Queen Signal: The Queen Mandibular Pheromone

The queen mandibular pheromone (QMP) is by far the most studied and well-known chemical signal in the honey bee society. Its first identification dates back to 1960,

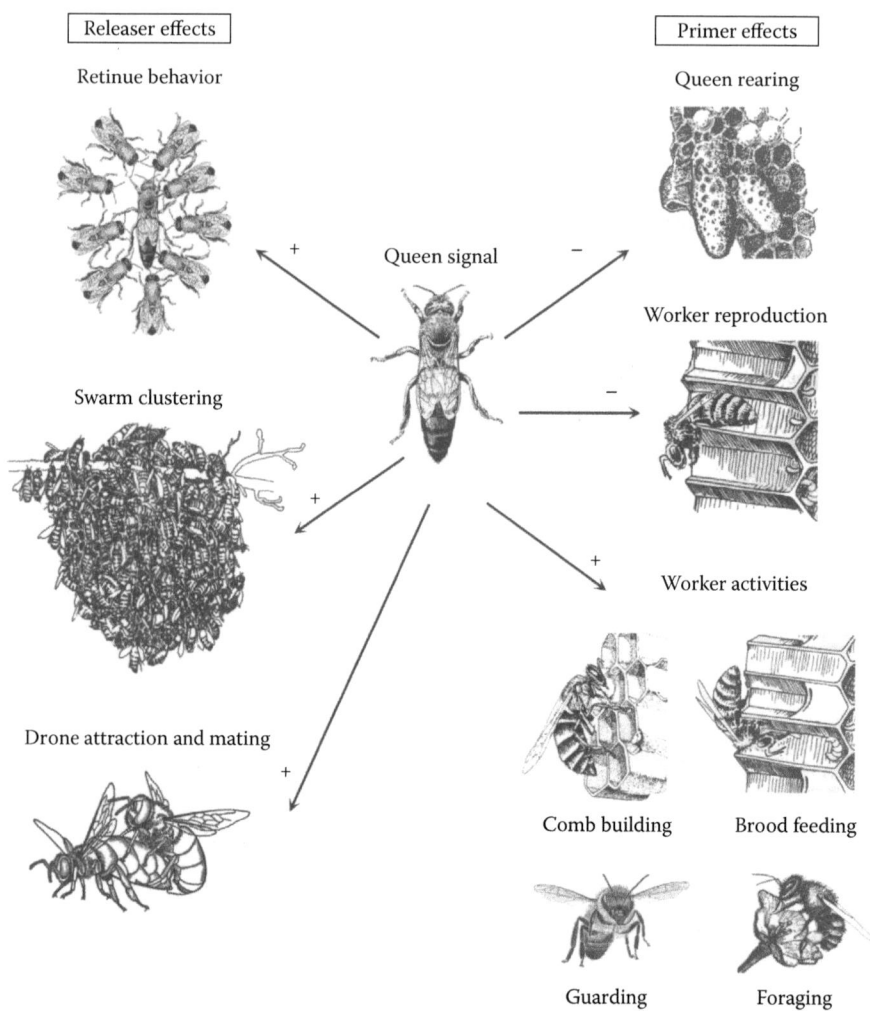

FIGURE 5.1 Releaser and primer effects of the queen signal, which regulates colony functions and development. Stimulating effects are indicated as "+" and inhibiting effects as "–." (Adapted from Winston, M.L. (1987) *The Biology of the Honey Bee.* Cambridge, MA: Harvard University Press; Frilli, F. et al. (2001) *L'ape: Forme e funzioni.* Bologna, Italy: Calderini Edagricole; Goodman, L. (2003) *Form and Function in the Honey Bee.* Cardiff, UK: IBRA–International Bee Research Association; Contessi, A. (2004) *Le api: Biologia, allevamento, prodotti.* Bologna, Italy: Edagricole.)

when (E)-9-oxodec-2-enoic-acid, more simply known as 9-ODA, was detected as the substance secreted by the queen mandibular glands (Barbier and Lederer 1960; Callow and Johnston 1960). The secreting organs are a pair of saclike glands located inside the head above the base of the mandible (Figure 5.2). The glands open through a short duct at the base of the mandible and their secretion runs along a deeper channel surrounded by hairs (Billen 1994).

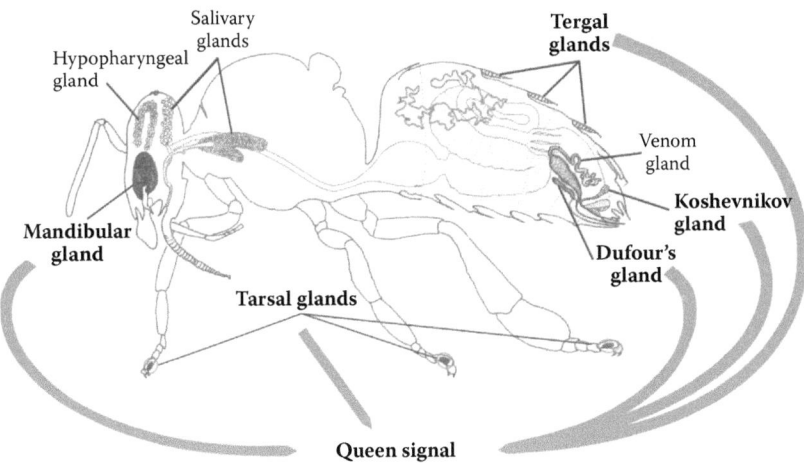

FIGURE 5.2 Exocrine glands of the honey bee queen. The pheromone-producing glands that concur to the formation of the queen signal are highlighted in bold. (Adapted from Goodman, L. (2003) *Form and Function in the Honey Bee.* Cardiff, UK: IBRA.)

In 1988, Slessor et al. discovered four other compounds secreted by the mandibular glands that act synergistically with 9-ODA: the two enantiomers of 9-hydroxydec-2-enoic acid (9-HDA), methyl p-hydroxybenzoate (HOB) and 4-hydroxy-3-methoxy-phenylethanol (homovanillyl alcohol [HVA]). The five components together were more active than any of the single substances, alone or in combination, in forming the retinue of worker bees (see Section 5.1.1.1.1). It was concluded that these five chemicals together form the base of QMP secretion, which represents the main constituent of the queen signal.

Several authors analyzed the evolution of QMP components during queen aging, from emergence until the full dominant status. Generally, the amount of volatiles was found to increase with age, but the findings concerning the different compounds and their relative amounts were inconsistent among authors, as illustrated by the studies described below.

Engels et al. (1997) identified three different ontogenetic patterns of QMP in queens: (1) virgin premating queens presented a weak signal, with oleic acid (OLA) as the main component, (2) mating queens intensified the signal, mainly consisting of 9-ODA along with OLA and small amounts of 9-HDA, and (3) postmating dominant queens exhibited a strong signal with high concentrations of 9-ODA combined with medium proportions of 9-HDA, less OLA, and small amounts of oxygenated aromates. The authors suggested that these oxygenated aromates, especially HOB and the late appearing HVA, could be the typical signal of old egg-laying and dominant queens.

Plettner et al. (1997) compared the quantities of the QMP components between 6-day-old virgins and 1-year-old mated egg-laying queens of several species; they found that mated *Apis mellifera* queens had significantly higher levels of 9-ODA, 9-HDA, HOB, and HVA, while an opposite trend was found for 10-HDA and

10-HDAA, which are typical components of worker mandibular glands (see Section 5.1.2.1) and are produced in higher amounts by virgin queens.

Slessor et al. (1990), who compared virgin with mated queens of different ages, found slightly different results, with levels of 9-ODA almost constant in the different groups, 9-HDA levels higher in mated than in virgin queens, and HOB and HVA levels higher in the oldest mated queens compared to the virgin and the young mated ones. In all cases pheromone levels were highest in mature, mated, laying queens.

On the contrary, Rhodes et al. (2007), comparing 7-day-old virgin and mated queens, found that the former had higher levels of 9-HDA, 9-ODA, and 10-HDA than the latter. In a similar comparison, Richards et al. (2007) found that the quantities of 9-ODA, 9-HDA, and HVA were all significantly lower in mated queens compared to virgins, although the extract of mated queens' mandibular glands was more attractive towards workers than the one of virgin queens.

Finally, Strauss et al. (2008), analyzing the mandibular gland compounds of virgin, drone-laying, and mated queens, found similar amounts of 9-ODA in the three groups and increasing amounts of all the other components (9-HDA, 10-HDA, 10-HDAA, and HVA) except HOB, from virgin to mated queens. The constant level of 9-ODA, which corresponds to a higher relative proportion in the virgin queen, suggests that 9-ODA plays a greater role in drone attraction of virgin queens than of retinue attraction in mated queens; on the contrary, 9-HDA, 10-HDA, 10-HDAA, and HVA are positively correlated with the reproductive potential and the ovarian activation of the queen (Strauss et al. 2008).

These contradictory results suggest that the role of single QMP components in queen signal is not fully understood and that there could be additional unknown compounds in mated queens' mandibular glands that synergize with 9-ODA and 9-HDA in the attractive function.

5.1.1.1.1 Attracting the Workers: The Retinue

The first functions of QMP to be discovered were due to its attractant properties toward the workers: the formation of the queen retinue and the constitution and maintenance of the swarm cluster (Kaminsky et al. 1990; Winston et al. 1989).

When the queen is stationary on the comb she is surrounded by a circle of workers known as "court" or "retinue" that face toward her and feed, palpate, and lick her. Usually the retinue is composed of eight or 10 workers. Several studies demonstrated that QMP and its components are accountable for the formation of the retinue (Free 1987), and this is supported by the fact that worker attraction towards the queen could be related to the modifications in the QMP pattern. The attention paid by retinue workers increases when the virgin queen becomes mated and lays eggs, and then decreases as she grows old; the degree of attractiveness of the queen is null at 0–1 day old, medium from 2 to 4 days old, and high from 5 days to 18 months old (De Hazan et al. 1989). Richards et al. (2007) tested the mandibular gland extracts of virgin and inseminated queens on worker retinue responses and found that gland extracts of the inseminated queens were more attractive than those of the virgin queens, and those of queens inseminated with more than one drone were more attractive than those of queens inseminated with a single drone. This suggests that mating is a crucial factor for the development of the chemical signal of the queen and its attractive effect on workers.

In 2003, Keeling et al. identified four additional compounds produced by the queen that act synergistically with QMP in attracting workers to form the retinue group: coniferyl alcohol (CA), methyl oleate (MO), hexadecane-1-ol (PA), and linoleic acid (LA); the first is secreted by the mandibular glands, while the others are produced in different parts of the queen's body. These substances were inactive alone, but in combination with QMP they were found to greatly increase activity of the queen's retinue. Furthermore, in a recent study, Maisonnasse et al. (2010a) showed that queens artificially deprived of mandibular glands can still attract workers in the retinue, suggesting that QMP was not the only pheromone able to attract workers and that in its absence other substances can take its role.

5.1.1.1.2 Attracting the Workers: The Swarm

Swarming is the way in which the colony reproduces itself; workers rear new queens and the first emerging one will kill the others, and after mating will become the new colony regnant, while the old queen drives the swarm toward a new nest. The presence of the queen is essential to keep the swarming bee cluster together: if the queen dies or is unable to fly, the swarm soon returns to the parental hive. The queen's attractiveness towards the swarm cluster is triggered by means of pheromonal signals, mainly the QMP. In 1989, Winston et al. compared the effects of the queen, the mandibular gland extracts, and the five-component-blend on the swarming and thus demonstrated that the component blend and the gland extract showed comparable effects, while the queen alone always had the strongest attractiveness. This suggested, as with induction of retinue behavior, that other extra mandibular components could be involved in formation of the swarm cluster.

5.1.1.1.3 Attracting the Drones: QMP as a Sexual Pheromone

Soon after its discovery it became clear that the QMP is used by virgin queens to attract drones during mating flights (Gary 1962); more specifically, by using queen dummies 9-ODA was clearly shown to attract drones (Gary and Marston 1971).

In further experiments, different combinations of 9-HDA, 10-HDA, and HOB were also found to increase the number of drones making contact with the queen dummy; 9-HDA and 10-HDA in particular seemed to be responsible for the increased mating contacts, although they were active only at short range, contrary to 9-ODA, which also acted at a higher distance (Brockmann et al. 2006; Loper et al. 1996). Comparing QMP components in virgin and mated queens it emerges that 10-HDA is more represented in the former while it is highly reduced in quantity in the latter (Plettner et al. 1997). The fact that 10-HDA is produced in large amounts by virgin queens suggests its role as a sex pheromone in mating behavior.

An increase in the frequency of mating behavior was observed also when tergal gland extracts were added to 9-ODA (Renner and Vierling 1977). This suggests that several glandular sources could cooperate in increasing the effectiveness of the pheromonal stimulus, leading to a stronger response and a more complete performance of the mating behavior sequence.

Thus, the relative contribution of different components of QMP and other glands on the sex pheromone blend is not yet completely clear.

5.1.1.1.4   Unique Queen: Suppression of Queen Rearing and Swarming

Many insect societies are monogynous, which means that a single queen is present in each colony. In small and primitive social species the maintenance of queen dominance is achieved by fight and physical competition among females; in contrast, in large monogynous colonies a physical dominance is not possible and a more efficient system has evolved for the maintenance of queen dominance that is based on pheromonal signals.

As previously stated, the removal of the queen from a colony of *A. mellifera* results in worker bees building special cells (queen cups) for rearing new queens (Winston 1992), but the precise way in which this happens is still in part unclear.

The rearing of new queens in a colony has two main scopes: reproduction of the colony through swarming and replacement of the queen when it is old or weak (this phenomenon is known as supersedure) or if it dies for some apicultural or pathological reason. QMP suppresses both queen supersedure and swarming by its dispersal throughout the colony (Winston et al. 1989). Several studies were addressed to elucidate the mechanisms of dispersal of QMP inside the colony and its transfer among workers. In 1991 Naumann et al. identified the group of retinue workers as the first actors in transferring queen pheromones toward the other workers and self-grooming as the means by which the pheromone is translocated from the mouthparts and the head to the abdomen of the workers (Naumann 1991). The distribution of QMP seems to be influenced by the size of the colony, since in populous colonies workers at the periphery gain a lower amount of pheromone than in unpopulous nests (Naumann et al. 1993). This explains why populous colonies swarm: the pheromone signal "the queen is present" tends to decline when the colony grows because pheromone dispersal is reduced and workers thus perceive a lower amount of pheromone, and the result is colony reproduction through queen rearing and colony swarming. When the queen dies or is removed, the pheromone signal disappears completely and workers are stimulated to rear new queens.

The role of QMP in the suppression of initiating queen rearing was confirmed by several studies that showed that administration of synthetic QMP to queenless colonies (i.e. colonies without a queen) suppresses the production of queen cups (Pettis et al. 1995) if the administration occurs within 24 hours from queen loss; in fact if synthetic QMP is applied 4 days after queen loss no effect is observed, indicating that QMP inhibits the initiation of queen-rearing but not the maintenance of established cells (Melathopolous et al. 1996).

5.1.1.1.5   Unique Mother: Suppression of Worker Reproduction

One of the main features of the honey bee society is the presence of two female castes (queen and workers) among which the queen is the only reproductive one. Workers are anatomically equipped with ovaries (which contain a lower number of ovarioles compared to queens) but development of the oocytes is inhibited by presence of the queen. If the queen is absent (and the colony and its workers are termed "queenless" as opposite to "queenright"), the workers' ovaries can become active and workers can thus lay eggs (Butler and Fairey 1963; de Groot and Voogd 1954; Jay 1968; Velthuis et al. 1990). However, they can only produce haploid unfertilized eggs that give rise to male offspring, since they have not mated and do not have a spermatheca. A further confirmation of the inhibiting role of queen presence on ovarian development is provided by the findings of Malka et al. (2007), who showed that queenless egg-laying workers reintroduced into queenright colonies showed a clear regression in ovary development, whereas this did not happen if they were reintroduced into queenless colonies. This indicates that ovarian regression was not due to the change of colony but to the presence of a queen and her signals.

The specific role of QMP in the suppression of worker reproduction has been hotly debated for a long time. The first hypothesized mechanism for inhibition of ovarian development is that QMP acts by lowering the titer of juvenile hormone (JH) in workers and that this low titer corresponds to a low level of ovarian development and vice versa, as demonstrated in other social insects (Hartfelder 2000; Robinson and Vargo 1997). However, different studies exploring this hypothesis in honey bees gave contrasting results. First, the correlation between JH level and ovarian development in adult hemolymph is not confirmed (Robinson et al. 1991, 1992a). Furthermore, the administration of synthetic QMP alone was able to lower the titer of JH but not able to suppress the development of worker ovaries and egg-laying (Kaatz et al. 1992; Willis et al. 1990).

Disregarding the involvement of JH, the work of Hoover et al. (2003) clearly demonstrated that synthetic QMP alone is active in suppressing ovary development in workers and that its effect is comparable to that of the whole queen extract, thus excluding the involvement of other queen-produced substances. The components of queen pheromone identified by Keeling et al. in 2003 failed in suppressing ovary development and did not improve the efficacy of QMP when applied in combination, suggesting that they play a role in retinue behavior but not in workers' ovary development.

More recently Katzav-Gozansky et al. (2006) observed that QMP alone or in combination with the secretion of Dufour's gland inhibited ovarian development, with the combination of the two pheromones being the most effective treatment, but always less effective than the presence of a live queen, redrawing attention to the hypothesis of a queen multisignal in the regulation of worker reproduction.

It thus appears that the regulation of worker ovary development is probably an even more complex process, involving both queen and brood signals, since two components of the brood pheromone, ethyl palmitate and methyl linoleate, also play a role in suppressing worker ovary development (Mohammedi et al. 1998). Possibly the queen regulation alone becomes essential when no brood is present in the colony, such as during interruption of egg-laying due to environmental conditions (in winter or in summer in southern climates).

5.1.1.1.6 *Regulation of Worker Activity and Behavioral Development*

In highly advanced insect societies there is a typical organization of the infertile caste that determines an age-dependent division of labor, called temporal polyethism, in which workers progress from tasks performed inside the nest (cleaning, building, feeding) during the first 2–3 weeks of life, to those performed outside it (ventilating, guarding, foraging) in the last 1–3 weeks (Robinson 1992). This behavioral progress seems to be driven by endogenous factors, as it is linked to the amount of JH in worker hemolymph, which increases with the increasing age (Huang et al. 1991). Nevertheless, the task allocation of each worker cohort (group of workers of the same age) can be modulated by environmental factors that modify the requirements of the colony: a loss of older workers (e.g., forager bees that die in the field) can result in faster development of young bees into foragers, while a lack of young bees (e.g., for a natural or artificial interruption of queen egg-laying) can result in a slower behavioral development or a reverse from foragers to nest bees (Huang and Robinson 1992; Robinson et al. 1992b).

This plasticity in worker task allocation is at least partly regulated by QMP, via suppression of JH titers in workers; in colonies supplemented with synthetic QMP workers showed a reduced level of JH associated with a delay in behavioral development and a reduction of foraging activity (Pankiw et al. 1998). This mechanism could have an adaptive significance, such that the presence of the queen prevents workers from developing too rapidly into foragers, thus preserving a stock of young workers in case of loss of foragers due to adverse environmental causes (Winston and Slessor 1998).

Nevertheless, it is likely that the regulation of worker behavioral development is primarily modulated by the workers themselves, since the artificial alteration of worker demography is effective in changing age polyethism development even with constant presence of the queen (Huang and Robinson 1996; Robinson and Huang 1998). It is more likely that QMP acts as an auxiliary inhibitor on the workers' division of labor rather than as a primary motor; this modulating role can be exerted by the queen's strict contact with nurse bees in the nest, which determines an augmented inhibiting effect on the behavioral development of this group of bees.

The influence of QMP has been demonstrated on the activity of single workers, such as comb building. In the presence either of a mated queen or of artificial QMP, workers are stimulated to produce a higher amount of wax for the comb than in the presence of a virgin queen or in queen absence. In the latter case, furthermore, workers tend to produce male cells, demonstrating that the presence of the mated queen or QMP inhibits the production of male brood (Ledoux et al. 2001). A small difference was observed in the effects produced by presence of a mated queen and application of synthetic QMP, suggesting that QMP is not the only queen pheromone that influences comb building. Among the QMP components, HVA and HOB seem to play a major role in comb building, confirmed by the fact that virgin queens produce very low amounts of this component in their mandibular glands and do not stimulate comb building (Ledoux et al. 2001).

QMP also has an effect on stimulating foraging behavior and brood rearing. Higo et al. (1992) observed that in newly established spring colonies treated with synthetic QMP the number of foragers and the weight of pollen loads increased. Brood rearing also increased, but not significantly. In contrast, no effects were observed on large established colonies at their summer population peak, suggesting that QMP affects foraging, but its effect is influenced by colony conditions and environmental factors.

The defensive behavior is one of the best known features of a honey bee colony and consists in recognition of predators, alerting nestmates, and enacting antipredator behavior (from threat postures to buzzing and finally stinging). The defensive behavior will be illustrated more in detail in the description of the alarm pheromones but we mention it here because the presence of the queen seems to be important for its regulation, since it was observed that queenless colonies exhibit an increased aggressive behavior compared to queenright ones, and that synthetic QMP decreases stinging response in caged honey bees (Kolmes and Njehu 1990). The effect of QMP on the colony's defensive behavior was recently confirmed by Gervan et al. (2005), who showed that the administration of synthetic QMP significantly reduces defensive behavior in both queenright and queenless colonies, with a decrease in the number of bees that respond to a simulated danger and a slight reduction of sting reaction, and by Vergoz et al. (2007) who found that QMP blocks aversive learning in young workers.

5.1.1.2   Beyond the QMP: Other Queen Pheromones

Mandibular glands are not the only source of chemicals with a role in social cohesion and colony homeostasis. For many years researchers presumed that QMP alone could account for the regulation of all colony functions. Later on, other pheromone sources were discovered that were in agreement with a multicomponent nature of the queen signal. Already in 1970, Velthuis already found that queens from which mandibular glands were removed were still able to exert some regulatory functions on workers (retinue behavior, inhibition of queen cup construction, suppression of worker ovary development). But it was not until 2010 that Maisonnasse et al. confirmed these results, showing that demandibulated queens retain their full regulatory role on the above mentioned functions (Maisonnasse et al. 2010a). The authors discovered that the levels of QMP components were similar in demandibulated and control queens, with the exception of 9-ODA, which was not detected in the former. This suggests that only 9-ODA is uniquely produced and stored in the mandibular glands, while the other substances (HOB and 9-HDA) appear to have another source of production in the queen's body. 9-ODA has always been considered the main substance acting on retinue behavior but this is conserved in demandibulated queens, suggesting that other queen substances have the potential to substitute 9-ODA in evoking this behavior.

Alternative sources of the queen signal have been identified in the tergal, tarsal, Dufour's, and Koschevnikov glands (Figure 5.2). Their secretions can either cooperate with QMP in the composition of the queen signal or be responsible for a single or few specific regulatory functions.

5.1.1.2.1   QMP Assistants: Tergal Gland Pheromones

Tergal glands, also known as Renner and Bumann glands, are located underneath the abdominal tergites and their ducts open through the cuticle in the tergites' posterior edge region (Renner and Baumann 1964). In queens, numerous big gland cells occur, mainly in tergites III to V, whereas workers only have very few and considerably reduced cells (Billen et al. 1986).

The role of tergal glands as the source of pheromones with a supporting function to QMP has been postulated by several authors, in particular in African honey bee races (Velthuis 1970, 1985). De Hazan et al. (1989) found that the exocrine secretions of both mandibular and tergal glands in *A. mellifera* contribute to the attraction of worker bees, although the first to a greater extent than the second. On the contrary Moritz and Crewe (1991) in *A. m. capensis* (a honey bee race that occurs in the Cape province of South Africa) found that tergal and mandibular secretions contribute equally to the total pheromone blend. According to Wossler and Crewe (1999a) the major compound of the tergal gland secretion in workers and virgin and mated queens (studied in *A. m. capensis* and *A. m. scutellata*) is (Z)-9-octadecenoic acid, while Espelie et al. (1990) had found that the major components in virgin honey bee queens of 3–10 days old were decyl decanoate and longer chain-length esters of decanoic acid.

Just like the QMP, the tergal gland secretion shows both primer and releaser properties. In *A. m. scutellata* Wossler and Crewe (1999b) showed that queen mandibular gland secretions were more effective than tergal gland secretions in formation of the retinue, but the two secretions together were even more efficient, indicating a releaser function of queen tergal gland pheromone in evoking the worker retinue behavior. On the other hand, the tergal gland extracts of virgin queens of both *A. m. capensis* and *A. m. scutellata* showed a significant effect in the inhibition of ovarian development in their own workers, indicating that the secretions from the tergal glands can also operate as primer pheromone (Wossler and Crewe 1999c). Furthermore, in *A. m. capensis* the secretions of the queen tergal glands are used by workers as kin recognition signals (Moritz and Crewe 1988).

In *A. m. ligustica* the glandular production was suggested to be a specific signal of the mated queen, since the production of tergal gland alkenes is stimulated by natural mating and not by instrumental insemination (Smith et al. 1993).

5.1.1.2.2   Footprint Pheromone: Tarsal Gland Pheromones

The tarsal glands (Arnhart 1923) are present in queens, workers, and drones and consist of a unicellular layer of glandular epithelium located in the sixth tarsomere of each of the six legs. The secretory products accumulate in a saclike reservoir inside the tarsus, which communicates with the exterior at the level of an articular slit located between the fifth tarsomere and the arolium (Lensky et al. 1985); these secretions are oily, colorless substances that are extruded through openings when the bee is walking, from which comes the name footprint pheromones. It is assumed that the secrete of tarsal glands can serve different purposes in the three bee castes, since some differences in the chemical composition in queens, workers, and males were observed (Lensky et al. 1984).

Lensky and Slabezki (1981) observed that tarsal gland secretions deposited by the mated queen on the comb inhibit queen cup construction by workers; this hypothesis is supported by the observation that in overcrowded colonies the queens' movements are restricted to the central parts of the comb, thus they are almost absent from the bottom edges of the combs where queen cups are usually built. A blend of the mandibular and tarsal pheromones is able to inhibit this behavior, giving an example of coregulation by these two pheromones.

5.1.1.2.3   Fertility Signal: Dufour's Gland Pheromones

Dufour's gland, first described in honey bees by Dufour in 1841, is a tubular gland associated with the sting apparatus together with the venom, sting sheath, and

Koschevnikov glands (Figure 5.3). It opens into the dorsal vaginal wall (Billen 1987) close to the setosa membrane, a hairy region of cuticle that surrounds the entire sting bulb and acts as a platform for pheromone release (Lensky et al. 1995; Martin et al. 2005). The peculiar position of Dufour's gland suggested different possible functions for its secretions, linked mainly to reproduction and egg-laying in queens (production of an egg coating or egg marking) and to defense in workers (production of a sting lubricant, neutralization of the remains of the acid secretion in the sting).

One of the main explored theories is the role of Dufour's gland secretion as a caste-specific egg-marking pheromone applied by the queen during deposition, which could allow the workers to distinguish between queen-laid or worker-laid eggs (Ratnieks 1988; Ratnieks and Visscher 1989). Egg recognition is fundamental to the mechanism of worker policing, by which workers kill eggs laid by fellow workers but leave queen-laid eggs (a small percentage of workers lays eggs even in presence of the queen) (Ratnieks 1993). The existence of an egg-discriminatory pheromone was postulated by Ratnieks in 1988, and in subsequent experiments he found evidence that the Dufour's gland secretion could be a source for this pheromone (Ratnieks 1995). However, in later studies it was found that neither the glandular secretion nor its ester or hydrocarbon constituents were able to protect worker-born eggs from policing; in fact, treated worker eggs were removed significantly faster than queen eggs, and at the same rate as nontreated worker eggs (Katzav-Gozansky et al. 2001; Martin et al. 2002). These results lead to the rejection of the hypothesis that the Dufour's gland secretion serves as an egg-marking pheromone.

An alternative hypothesis for the role of the Dufour's gland secretions is a fertility signal. This is supported by the caste-specific composition of its secretions: in workers they consist of a series of long-chain hydrocarbons, whereas in queens there is in addition a series of waxlike esters; these queen-specific esters are tightly correlated with ovarian development (Katzav-Gozansky et al. 1997, 2000). Queenless workers, which are likely to become egg layers, produce a queenlike secretion with an augmented level of the ester fraction that is correlated with worker ovarian development (Dor et al. 2005; Katzav-Gozansky et al. 2003). In workers of *A. m. capensis*, which are known to be social parasites, this mimicry of queen secretion profile by egg-laying workers is even more developed (Sole et al. 2002).

The Dufour's gland secretion seems to act similarly to the QMP in attracting workers, which form a retinue around the scented source. Bioassays reveal that the active constituents are the queen-specific esters rather than the hydrocarbons. The queenlike glandular secretions of egg-laying workers are also attractive to nestmates, although to a lesser degree than those of the queen, while the secretions of non-egg-laying workers are totally inactive (Katzav-Gozansky et al. 2002, 2003). Moreover, workers were more attracted to Dufour's gland extract from inseminated queens compared with virgins and to multidrone inseminated queens compared with single-drone inseminated ones, confirming the relation between Dufour's gland secretions and reproductive potential (Richard et al. 2011).

Finally, as previously stated, the combination of QMP and Dufour's gland secretions was effective in inhibiting ovarian development in workers, although neither was as effective as the presence of the queen (Katzav-Gozansky et al. 2006).

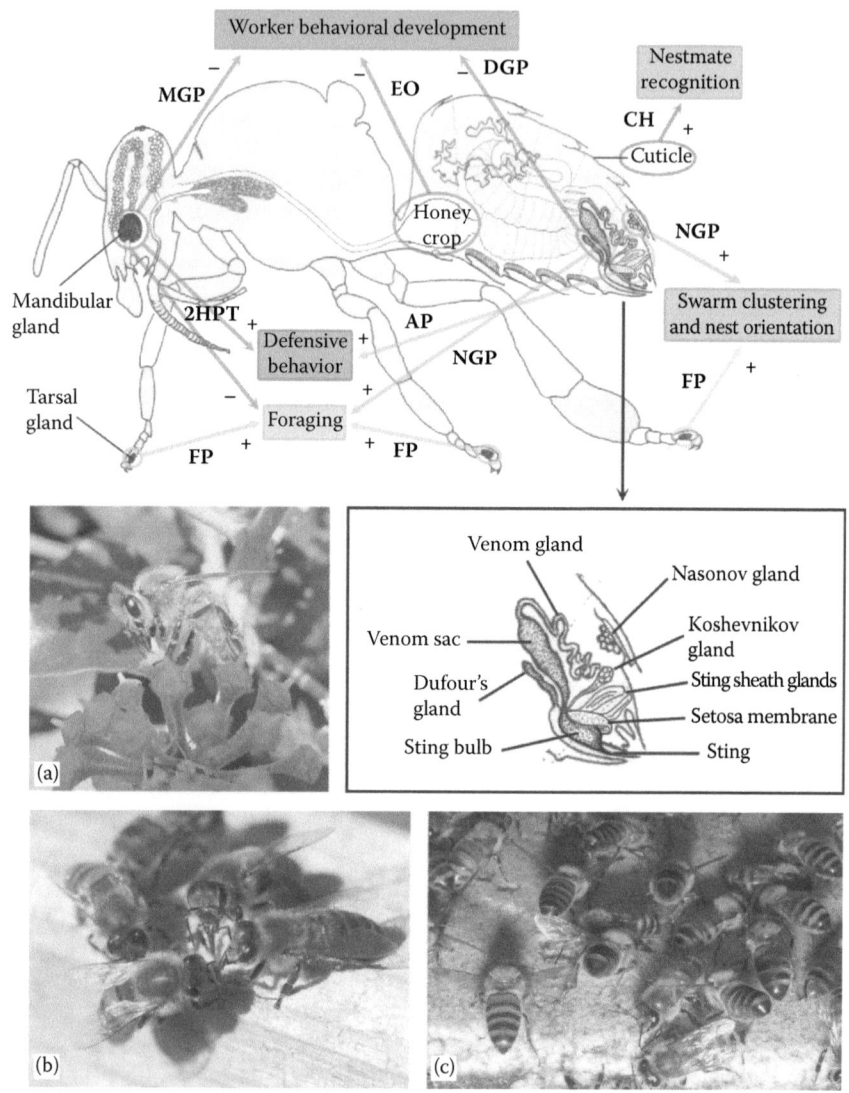

FIGURE 5.3 **(See color insert.)** Pheromone-producing glands and organs and their main products in the honey bee worker, and their effect in the different worker activities. Stimulating effects are indicated as "+" and inhibiting effects as "–." (a) Honey bee worker foraging on *Hedera helix*; during foraging workers mark flowers with the secrete of their tarsal and mandibular glands. (b) Nestmate recognition among workers in front of the hive entrance; workers use chemical cues like cuticular hydrocarbons perceived through mouth and body part contact. (c) Honey bee workers exposing the Nasonov gland in front of the hive to release the orientation pheromone. 2HPT = 2-heptanone; AP = Alarm pheromone; CH = cuticular hydrocarbons; DGP = Dufour's gland pheromone; EO = ethyl oleate; FP = Footprint pheromone; MGP = mandibular gland pheromone; NGP = Nasonov gland phero-mone. (Adapted from Goodman, L. (2003) *Form and Function in the Honey Bee.* Cardiff, UK: IBRA. Photographs by Cecilia Costa and Per Kryger.)

All this evidence suggests that Dufour's gland secretion is a component of the queen signal, both in its releaser effects, as revealed by the stimulation of retinue behavior, and in its primer effects, correlated with reproductive dominance and fertility, through the inhibition of ovarian development. The specific roles of the Dufour's gland secretion and QMP in regulating the physiological development and the behavior of workers still have to be elucidated.

5.1.1.2.4   Aging of the Queen Signal: Koschevnikov Gland Pheromones
The Koschevnikov gland is located near the sting shaft (Figure 5.3) and is composed of glandular units, each consisting of a secretory cell and a duct cell connected to the epidermis. Secretions are emitted onto the entire surface of the setosa membrane (Grandperrin and Cassier 1983), where they are released together with the alarm pheromones originating from the glandular part of the sting sheaths.

In honey bee workers the gland produces an alarm pheromone that is released when a bee stings (see Section 5.1.2.5.1, worker alarm pheromones). In queens the gland seems to play a different role, concurring to the already described queen signal. This is supported also by a caste-specific chemical composition of its extracts: 28 compounds including acids, alcohols, alkanes, and alkenes were found in queen Koschevnikov glands, which are not present in worker alarm pheromone (Lensky et al. 1991). Other studies highlighted that topical treatment of worker bees with extracts of queen Koschevnikov glands induced aggressive reaction by other workers, the so-called "balling behavior," which is usually generated by a high-concentration QMP treatment (Pettis et al. 1998).

The gland starts degenerating after the queen is 1 year of age and this contributes to the loss of signal in old queens (Grandperrin and Cassier 1983).

5.1.2   Worker Pheromones

Caste-specific secretion is an important feature of the honey bee pheromonal system. Some honey bee glands are typically developed in only one of the two female castes (e.g., the tergal glands in queens or the Nasonov gland in workers); nevertheless, most glands are developed in both queens and workers, but their secretion is caste-specific, as is the case of mandibular, Dufour's, tarsal, and Koschevnikov glands.

The glandular plasticity in honey bees is linked to two processes of caste determination: the one that results in the differences between queens and workers, and the one that results in differentiation among workers, the behavioral development referred to as temporal polyethism. The development of the glands in workers follows a temporal pattern linked to the activities connected with the gland secretions: for example, wax glands and hypopharyngeal glands (with secretions correlated to building and feeding activities) develop sooner and are more active in young bees, while the sting alarm pheromone production is low in young bees and rises as the workers become guard bees. However, gland secretion does not necessarily show a caste-specific rigidity, but it can be rather plastic and adaptive, thus supporting the changing needs of the colony (Katzav-Gozansky et al. 2002).

5.1.2.1 Regulation of Worker Reproduction: The Mandibular Gland Pheromones

Honey bee mandibular glands represent a clear model of caste-specific secretion. Queens and workers produce a caste-related blend of functionalized 8- and 10-carbon fatty acids, which match the queen's reproductive and the worker's non-reproductive roles in the colony (Plettner et al. 1996). While the queen mandibular glands produce mainly 9-ODA, 9-HDA, HOB, and HVA, in worker mandibular glands the prevailing components are 10-hydroxy-2(*E*)decenoic acid (10-HDA), 10-hydroxydecanoic acid (10-HDAA), and their respective diacids. Both castes produce the other caste's aliphatic compounds in small quantities: queens have some 10-HDA and 10-HDAA and workers have a trace of 9-HDA (Plettner et al. 1995, 1997). Occasionally, traces of 9-ODA can be found in the glands of queenless workers, more frequently in African than European subspecies (Crewe and Velthius 1980; Plettner et al. 1993).

In almost all honey bee subspecies, removal of the queen leads to the development of a certain numbers of egg-laying workers that are then able to suppress ovary development in the other workers by means of pheromones, just as the queen does; for this reason these egg-laying workers are also called false queens (Sakagami 1958; Velthuis et al. 1990) or pseudoqueens (Moritz et al. 2000). Thus, if necessary, worker mandibular glands are able to produce a set of chemicals very similar to those of queen glands and with a comparable action. In queenright colonies this secretion is suppressed by the presence of the queen by means of the complex pheromonal secretion formerly illustrated, among which the QMP plays the major role. Conversely, in queenless colonies, where the queen inhibition fails, the secretion of worker mandibular pheromone plays an important role in the regulation of reproductive dominance among workers. This phenomenon is particularly evident in *A. m. capensis* workers, in which the transformation from infertile worker to false queen is rapid and recognizable by the formation of a retinue of worker bees surrounding the false queen (Crewe and Velthuis 1980). Furthermore, workers of this subspecies show intermediate characteristics between queens and workers—they present a functional spermatheca and are able to produce female offspring by thelytokous parthenogenesis (Onions 1912). For these reasons the role of worker mandibular pheromones in the regulation of worker reproduction has been extensively studied in *A. m. capensis*, where for example, Simon et al. (2001) found that while the mandibular gland secretions of newly emerged workers are mainly composed of 10-HDA and 10-HDAA, the secretion of 4-day-old queenless workers is dominated by the queen substance 9-ODA. Tan et al. (2012), in a comparative study of *A. mellifera* and *A. cerana*, observed that 9-ODA, HOB, HVA, 9-HDA, 10-HDA, and 10-HDAA levels are higher in mandibular glands of egg-laying workers than in non-egg-laying ones.

Other studies demonstrated a direct correlation between worker ovary development and the amount of pheromone produced by worker mandibular (and occasionally tergal) glands (Velthuis et al. 1990). In *A. m. capensis* Moritz et al. (2000) observed that the amount of pheromone produced by workers represents the main selection criterion for pseudoqueens: the ones producing higher amounts of pheromones have better chances of becoming pseudoqueens since they inhibit the production of pheromones in the other workers. Studies involving the subspecies

A. m. ligustica also showed that the development of egg-laying workers appears to be mediated by mandibular pheromone production; old bees whose mandibular glands were removed were significantly less inhibitory towards young bees compared to unoperated bees (Huang et al. 1998). These results are consistent with the hypothesis that worker mandibular glands contain an inhibitor of behavioral development acting similarly to QMP (Robinson and Huang 1998).

Other authors found that not only the secretion of the mandibular gland is involved in worker reproductive development, but also that of Dufour's gland, since both glands show higher activity in egg-laying workers (Katzav-Gozansky et al. 2000). As for queens, it is likely that the Dufour's gland secretion acts as a fertility signal, since it has a very strict correlation with ovarian development (Katzav-Gozansky et al. 2004, 2006), while the mandibular gland secretion is involved in the establishment of a reproductive dominance (Plettner et al. 2003). The two signals seem to act independently: observations carried out in queenless groups established that workers that show a precocious aggressive behavior but undeveloped ovaries have a more queenlike pheromone in mandibular glands (dominance signal), while workers with late aggressive behavior and larger oocytes have a more queenlike pheromone in the Dufour's gland (fertility signal), but not in the mandibular ones (Malka et al. 2008).

Mandibular gland secretions also have additional specific functions in honey bee workers. When a worker becomes a forager its mandibular glands produce a very odorous compound, 2-heptanone, which acts as a releaser alarm pheromone. Its role will be discussed later, together with the other alarm pheromone signals.

5.1.2.2 Regulation of Worker Activity: Ethyl Oleate

When talking about the regulation of worker polyethism, it has been said that the queen is only an auxiliary factor in driving the onset of behavioral development, which is primarily modulated by the workers themselves (Huang and Robinson 1996). Indeed, Pankiw in 2004 observed that administration to young workers of substances extracted from the surface of foraging bees increased their foraging age, whereas extracts of young preforaging workers decreased it (Pankiw 2004a). This confirmed the role of substances produced by adult forager bees as primer pheromones.

The chemical substance that acts as an inhibiting factor delaying onset of foraging age was identified by Leoncini et al. (2004) as ethyl oleate. This substance was found in high concentrations on the body of adult forager bees. Further studies demonstrated that it is produced in the epithelium of the honey crop through the transformation of ethanol derived from fermented nectar, then it exudes to the esoskeleton where it is transmitted among workers as a low-volatile at close range or by physical contact, and diffused in the hive by evaporation (Castillo et al. 2012; Muenz et al. 2012).

The discovery of this new pheromone elucidates how workers are able to regulate their own task allocation: when a high number of foragers are present in the colony their secretion inhibits the development of young bees, which can devote themselves to nest occupations; when foragers grow old or are lost, the inhibition fails and young bees develop into new foragers.

Interestingly, ethyl oleate was also found as a component of the pheromone blend of queens and brood, and therefore, classified as a colony pheromone (Keeling and Slessor 2005; Slessor et al. 2005).

5.1.2.3 Orientation and Recruitment: The Nasonov Gland Pheromones

The Nasonov gland secretion is the most well known worker-exclusive pheromone in honey bees. The gland consists of a mass of cells located beneath the intersegmental membrane, between the sixth and seventh tergites (Figure 5.3). Each cell has a small duct that transports the secretion and opens through the cuticle (Cassier and Lensky 1994). The glandular secrete is released under the posterior part of the sixth tergite and the workers free it by flexing the tip of the abdomen downward; during the secretion the bee usually stands with the abdomen elevated and fans its wings to facilitate volatile dispersion (Free 1987). Thus it is very easy to recognize workers bees while they are secreting the Nasonov pheromone (Figure 5.3c).

The Nasonov secretion is composed of seven volatile compounds: geraniol, nerolic acid, geranic acid, (E)-citral, (Z)-citral, (E-E)-farnesol, and nerol (Pickett et al. 1980). It has a general attractive effect and is used by workers in several different situations, the principal being marking of hive entrance, swarm clustering, and marking of foraging sources.

Workers secrete and disperse the Nasonov pheromone by fanning at the hive entrance to orient other members of the colony toward the nest. It is also released by young workers during their first orientation flights. The pheromone release is especially evident after colony disturbance and can be elicited by nest odors such as comb, honey, pollen, propolis, and QMP (Ferguson and Free 1981).

Recent studies revealed a new possible role of the Nasonov pheromone within the hive: workers selecting young larvae to rear as new queens expose their Nasonov gland to attract other workers toward the selected larval cell. In an experimental test with queenless workers, the cells where a higher amount of pheromone was released had a higher chance of developing into queen cups, confirming the involvement of the Nasonov pheromone in recruiting workers to queen rearing (Al-Kahtani and Bienefeld 2012).

5.1.2.3.1 Swarm Clustering

Together with the QMP, the secretion of workers' Nasonov gland functions as a cohesion factor for swarm clustering. When the swarm leaves the hive it settles in a temporary location; the first bees reaching the site immediately expose their Nasonov gland to call the rest of the swarm. Here a certain number of workers, called scout bees, are in charge of finding a new suitable place to establish the nest. When a scout bee finds a potential nest site, it communicates the location to the other workers by the waggle dance, then returns to the site and starts to release the Nasonov pheromone to drive the swarm exactly to the new nest entrance (Free 1987; Seeley 2010).

The potential of Nasonov gland components in attracting clustering bees was demonstrated in behavioral assays (Free et al. 1981a,b). Among the various components, geraniol, (E)-citral, and nerolic acid were the most attractive ones. In further studies a synthetic pheromone blend was able to attract swarms to the artificial nest cavities where it was applied (Schmidt 1994, 1999).

5.1.2.3.2   Foraging Recruitment
The function of the Nasonov gland in recruiting workers toward foraging sites has been known for some time, but its precise mechanism is still debated. When a forager finds a profitable food source, it exposes its Nasonov glands to orientate nestmates and stimulates them to land on it (Free 1987). Foragers, however, were seldom observed to expose the Nasonov gland while visiting flowers, whereas this behavior was more frequently observed during water collection. In an experimental trial with artificial nectars, the release of Nasonov pheromone was stimulated only by sugar concentrations much greater than those of natural nectars and the pheromone release was limited when the colony had already abundant nectar supplies (Pflumm and Wilhelm 1982). This suggests that the Nasonov pheromone is mainly used to recruit workers toward water sources and is involved in nectar source location only when the reward is very high or the nest storages are particularly scarce. In agreement with this, Fernandez et al. (2003) in experimental tests observed that the duration of Nasonov gland exposure at the feeding place was higher when the bees exploited the highest sugar reward.

5.1.2.4   Marking and Recruiting in Foraging Behavior: Tarsal Glands and Other Pheromones

The secretion of worker tarsal glands, also known as the worker footprint pheromone, is thought to have an auxiliary role in marking the hive entrance and food sources. It is deposited on the hive entrance by landing workers and probably also on visited flowers, enhancing the attractiveness of the Nasonov pheromone (Williams et al. 1981). In this shared function, the footprint pheromone probably acts as a proximity signal, being active at short distances, while the Nasonov pheromone is a broad attractive, with its volatile compounds being effective also at higher distances (Ferguson and Free 1981).

Thom et al. (2007) found that waggle-dancing bees produce and release four cuticular hydrocarbons (two alkanes, tricosane and pentacosane, and two alkenes, Z-(9)-tricosene and Z-(9)-pentacosene) from their abdomens into the air. These compounds are produced subcutaneously and secreted on the surface of the cuticle. When they are injected into a hive, the number of foragers leaving the hive increases significantly, suggesting a pheromonal role in worker recruitment.

Another pheromone involved in foraging behavior is 2-heptanone secreted by workers' mandibular glands. It exerts a repellent effect and therefore seems to be correlated to a repellent forage marking scent; its role is discussed in Section 5.1.2.5.2.

5.1.2.5   Defensive Behavior: Alarm Pheromones

The defensive response is one of the most well known honey bee behaviors, especially after the introduction of the African honey bee *A. m. scutellata*, also named "killer bee," to Brazil and its subsequent spread through tropical and subtropical New World habitats (Breed et al. 2004).

Defensive behavior is the first out-nest task performed by workers and is thus initiated at a younger age than foraging. There are two different kinds of workers involved in the defensive behavior: the guards and the defenders. Guard bees are workers that patrol the entrance of the hive in search of bees, insects, animals, or any other object or creature that approaches the colony (Arechavaleta-Velasco and Hunt 2003). They also inspect all bees that land at the hive entrance through antennation to recognize nestmates and reject non-nestmates (Breed et al. 1992). Defenders, also called stingers, are bees that respond to a danger or a disturbance by flying out, sting-ing, and sometimes pursuing intruders (Breed et al. 1990).

Alarm pheromones integrate the defensive response, the signaling potential, or actual dangers to the members of the colony. They can be released by bees while extruding the sting without stinging, during stinging, and also from stings left in the victim. The receiver bees are activated to a defensive behavior in the form of dispersal, or more often, attack against the source of danger. Defensive response is an example of collective action based on recruitment and amplification processes (Millor et al. 1999).

The possibility of modulating this response largely depends on the peculiar attri-butes of alarm pheromones, which are represented by a multicomponent and mul-tisource pheromonal blend. Two main groups of substances with alarm effect have been identified in honey bees: the sting apparatus alarm pheromones, which have as a main component isopentyl acetate (Blum et al. 1978; Boch et al. 1962), and the man-dibular gland alarm pheromone, with its single-component 2-heptanone (Shearer and Boch 1965). Both substances elicit defensive behavior against intruders at the hive entrance. Nevertheless many studies suggest that the two substances have dif-ferent functional values even if both are capable of evoking deterrent responses in a defensive context (Balderrama et al. 2002).

5.1.2.5.1   Sting Apparatus

The alarm pheromones are mainly produced in worker honey bees by the Koschevnikov gland and by the glandular areas of the sting sheaths (Figure 5.3); the secretions are quickly volatilized on the hairs of the setosa membrane at the base of the sting bulb (Cassier et al. 1994; Lensky et al. 1994).

During defensive behavior guard bees appear at the hive entrance, raising their abdomen and exposing the sting chamber to release alarm pheromones; at the same time they fan their wings, dispersing the volatile components.

Over 40 compounds (including precursor, intermediate, and final biosynthetic products) have been identified from extracts of the worker sting apparatus, among which about 15 components stimulate one or more alarm behaviors (flying from the nest to locate the source of disturbance, pursuing, biting, and stinging) (Pankiw 2004b; Wager and Breed 2000).

Isopentyl acetate (IPA, or isoamyl acetate) is the principal active component of the alarm pheromone blend and is responsible for the majority of sting-releasing activity. Its main functional value seems related to alerting and eliciting defensive responses at the hive entrance.

It is absent in queens and young workers and increases as a worker bee ages, reaching its highest level when the worker is about 2–3 weeks old, just at the time she begins to perform guarding tasks. The amount then decreases as she becomes a forager (Allan et al. 1987; Boch and Shearer 1966).

Tests with IPA show that it is effective in alerting bees and inducing them to sting, but not as effective as the complete alarm pheromone blend (Boch et al. 1962;

Free and Simpson 1968). Hence, the sting apparatus appears to be a multicomponent gland in which at least some component is specialized for a different function or can be more usefully used in one context than another. The most effective components in alerting bees are IPA and 2-nonanol (Collins and Blum 1982), while many other components stimulate attack and stinging (Free et al. 1983). In laboratory tests, honey bee workers stimulated with an electric shock increased their threshold of responsiveness after the application of IPA (Nuñez et al. 1998).

Pickett et al. (1982) identified a less volatile component, (Z)-11-eicosenol, as another effective alarm pheromone component for inducing stinging behavior. (Z)-11-eicosenol prolongs the activity of the more volatile IPA, presumably by slowing down its evaporation; as a consequence the blend of IPA and (Z)-11-eicosenol is active for a longer time than IPA alone.

There is evidence that honey bee races which differ for the intensity of defensive behavior can show differences in the amount and composition of alarm pheromones. Collins et al. (1989) found that Africanized honey bees have higher levels for nine of 12 alarm pheromone components compared to European honey bees and twice as much IPA; they also have a much more intense defense reaction when artificial pheromones are released at the hive entrance, indicating a lower threshold of pheromone response (Collins et al. 1982). Hunt et al. (2003) analyzed the alarm pheromone components from colonies of Africanized honey bees and they found a specific unsaturated derivative of IPA (3-methyl-2-buten-1-yl acetate, 3M2BA) that was able to recruit worker bees as efficiently as IPA; the two substances act synergistically and a mixture of these two compounds recruited bees more efficiently than either of the compounds alone. These two characteristics of Africanized bees can partly account for their higher defensive behavior rates.

5.1.2.5.2 Mandibular Glands: 2-Heptanone

The role of 2-heptanone (2HPT) produced by worker mandibular glands in colony defense is less clear than IPA. Close to the nest there is a strong response to it by guard bees (Shearer and Boch 1965) but in general it shows a much lower ability to recruit bees and to induce stinging than IPA: small amounts of IPA are 20–70 times more efficient than equivalent amounts of 2HPT in eliciting alarm behavior at the hive entrance (Boch et al. 1970; Lensky and Cassier 1995). Furthermore it has been observed that the level of 2HPT in workers did not differ significantly between docile and aggressive colonies or between Africanized and European honey bees (Sakamoto et al. 1990a,b; Vallet et al. 1991). Finally, in laboratory tests on the aversive stinging extension reflex (SER) small amounts of IPA led to an increase of responsiveness to the electric shock, while the same response is achieved by large amounts of 2HPT, confirming that the two pheromones are capable of evoking deterrent responses in a defensive context, but at different concentrations (Balderrama et al. 2002). A recent study by Papachristoforou et al. (2012) showed a further role of 2HPT in defensive behavior: it acts as an anesthetic in small arthropods such as wax moth larva (WML) and Varroa mites, which are paralyzed after a honey bee bite (honey bees use their mandibles to bite invaders that are too small to sting).

From these results it can be deduced that while IPA is a true defensive substance, 2HPT can have another principal role in the colony. This is supported also by the observation that the amount of 2HPT in workers progressively increases with age, its level being higher in foragers than in guard bees (Boch and Shearer 1967; Vallet et al. 1991), which suggests that the main function of 2HPT could be associated with foraging.

The hypothesis of a correlation between 2HPT and foraging behavior has been examined in behavioral assays, which showed a repulsive effect of 2HPT when added to sucrose solution visited by workers and a temporary, repulsive effect on the visitation of flowers by foraging bees. Hence it seems to act as a repellent forage-marking pheromone that may aid honey bee foragers in quickly discarding recently visited flowers (Giurfa 1993; Vallet et al. 1991).

5.1.2.6   Nestmate Recognition: The Cuticular Hydrocarbons

Cuticular lipids are a complex mixture of compounds in which aliphatic long-chain hydrocarbons are generally the major component. They evolved primarily as a water-impermeable layer for protecting the insect from desiccation thanks to their chemical composition, which also gives them a resistance to high temperatures. In social insects the cuticular lipids show an unusual richness of branched hydrocarbons, which increase insect vulnerability to desiccation, because they have a considerably lower cuticular break point compared to straight-chain hydrocarbons. This apparently meaningless disadvantage can be justified by the supposition that in the course of evolution cuticular hydrocarbons (CHCs) shift their role from insect protection to communication, in particular in those insects showing a social behavior. The highly diversified composition of branched hydrocarbons, in fact, enables the creation of highly specific blends, which can serve as communication cues (Le Conte and Hefetz 2008).

Nestmate recognition is the capability of an individual belonging to a social group (namely a colony) to discriminate between nestmates (members of the same colony) and conspecific non-nestmates (members of other colonies). This function requires that nestmates present a uniform chemical pattern and tolerate individuals that present the same pattern, while non-nestmates, which have a different pattern, are recognized as invaders, eliciting an aggression response. In social insects, nestmate recognition is the base of defense behavior against parasites or conspecific invaders. It is mainly based on olfactory signals, and many studies have demonstrated that such chemical cues are contained within the lipid layer covering the insect cuticle.

The specific chemical profile that characterizes individuals of a same colony is achieved partly by genetic inheritance and partly by the environment (e.g., nest odors, diet, colony environment). In honey bees, heritable self-produced cues appear to be of minor importance in nestmate recognition; in fact guard bees are unable to discriminate between related and unrelated conspecifics if they have been living within the same hive during adulthood (Downs and Ratnieks 1999). Among the environmental factors, nest material—in particular, wax components—rather than food source and flower scent seem to be the most important source of recognition cues (Breed et al. 1998; Downs et al. 2000, 2001). Page et al. (1991) demonstrated that variability in hydrocarbons of individual workers is determined at least in part genetically, as they found the highest correlations of cuticular hydrocarbon extracts among closely related individuals.

Evidence of the importance of cuticular hydrocarbons as nestmate recognition pheromones derives from the common observation that their composition is less variable among nestmates than among individuals belonging to different colonies (Breed 1998). Moreover, young bees, which have fewer hydrocarbons in their cuticle, are accepted more readily into an unrelated colony, while the removal of the hydrocarbons from older bees improves acceptance (Breed et al. 2004b).

Among the aliphatic compounds that make up the honey bee cuticular hydrocarbons, alkenes and alkanes seem to be the most effective ones in nestmate recognition. In a conditioning proboscis extension reflex (PER) test, Chaline et al. (2005) observed that honey bees learn and discriminate alkenes better than alkanes, suggesting that the former may constitute the main compounds used as cues in the social recognition processes. Dani et al. (2005) found that honey bees treated with alkenes were attacked more intensively than bees treated with alkanes, and they conclude that the two different classes of compounds have a different effect on acceptance and this may correspond to a differential importance in the recognition signature.

Finally, exposure to the queen mandibular pheromone (QMP) significantly alters cuticular hydrocarbon patterns of worker bees; Fan et al. (2010) showed that QMP-treated nestmates are no longer recognized as nestmates by untreated bees, and vice versa.

It appears that nestmate recognition is a complex phenomenon triggered by several different cues and strictly connected with the already described process of colony cohesion and organization.

5.1.3 DRONE PHEROMONES

Very few pheromonal signals are known in the honey bee drones and most are linked to sexual features. This reflects the minor role of males in honey bee society, almost entirely limited to the mating function.

Lensky et al. (1985) verified the role of drone mandibular gland secretions in attracting other flying drones to congregation areas. Drone mandibular glands are much smaller than those of queens and workers and their size varies according to age. The secretory activity increases from 0–3 days old to a maximum at 7 days of age, while after 9 days the glands were no longer active.

The drone tarsal gland secretion also differs chemically from the female's, and its biological effects are still obscure (Lensky et al. 1984).

5.1.3.1 Drone Acceptance in the Colony

Similarly to workers, honey bee drones show features that determine their acceptance or refusal in the colony. This depends on whether they belong to the colony or not and on their age. According to Free (1957), in late summer young drones of 7 days of age are fed and cleaned by workers, while older, sexually mature drones (average age 23 days) are rejected and attacked. The mechanism that might be used by the workers to distinguish between drones of different ages could be the perception of chemical signals on the drone surface. Wakonigg et al. (2000) in *Apis cerana* observed that drone cuticular hydrocarbons have an age-related profile; although only small differences exist in cuticular composition among colonies, there is an evident trend in their profiles during drone aging.

Kirchner and Gadagkar (1994) observed that honey bee drones undergo a similar nestmate recognition process at the hive entrance as honey bee workers, suggesting the presence of an analogous recognition mechanism based on a cuticular phero-monal signal. The cuticular profile is sex-specific, since there are several compounds that are barely present or totally absent in the female but exist on the male cuticle. Compared with workers, drones of *A. dorsata* and *A. cerana* had higher propor-tions of branched alkanes and short-chain alkenes and lower proportions of normal alkanes and long-chain alkenes (Francis et al. 1989). This supports the hypothesis that worker bees are able to distinguish between the two sexes by chemical cues.

5.1.4 BROOD PHEROMONES

Larvae of *A. mellifera* produce a complex mixture of compounds that act both as primer and releaser pheromones, regulating worker development and colony growth. This brood pheromone (BP) is a blend of 10 fatty-acid esters: methyl palmitate, methyl oleate, methyl stearate, methyl linoleate, methyl linolenate, ethyl palmitate, ethyl oleate, ethyl stearate, ethyl linoleate, and ethyl linolenate (Le Conte et al. 1990). All together these compounds form the brood primer pheromone, but each compo-nent, alone or in combination, shows one or more releaser effects on adult bees (Le Conte et al. 2001).

5.1.4.1 Regulation of Brood Development and Care

BP is secreted by larval salivary glands, which are better known in honey bee larvae for their function in the silk secretion necessary for the construction of the pupal cocoon. The epithelial cells of the larval salivary glands secrete the fatty acids into the lumen of the glands, which act as a reservoir of the ester components of BP (Le Conte et al. 2006). The production of the different pheromone constituents varies in function of caste and larval age (Le Conte et al. 1994/1995) and this modulation of the signal is functional to guarantee the proper response to the needs of any larval age and caste by receiver nurse bees. For instance, methyl linolenate, methyl linole-ate, methyl oleate, and methyl palmitate, which are produced in large quantities by the larvae during the cell capping, were found to induce the worker bees to cap the cells (Le Conte et al. 1994). Methyl palmitate and ethyl oleate increase the activity of the workers' hypopharyngeal glands, which produce proteinaceous material (royal jelly) that is fed by nurse bees to young larvae (Mohammedi et al. 1996).

During queen rearing, methyl stearate increases the acceptance of the queen cups, methyl linoleate enhances the production and administration of royal jelly, and methyl palmitate increases the weight of the queen larvae (Le Conte et al. 1995).

Several studies demonstrated that colonies treated with BP rear more brood and more adults than controls, thus regulating colony growth. The BP treatment lead to an increased brood area and an augmented number of bees; the amount of pro-tein consumption was augmented, as well as the amount of extractable protein from hypopharyngeal glands, indicating an enriched nutritional environment (Pankiw et al. 2008). Moreover, the queen in BP-treated colonies was fed longer, was more active, and laid more eggs; the workers spent more time cleaning cells and rearing the brood, resulting in a larger brood area in BP-treated colonies compared non-treated colonies (Pankiw et al. 2004; Sagili and Pankiw 2009).

5.1.4.2 Regulation of Worker Reproduction

Beside its releaser effects on workers linked to larval development, BP components act as primer pheromones regulating, in synergy with QMP, worker ovarian development. In particular, ethyl palmitate and methyl linolenate were found to act as worker ovary development inhibitors (Mohammedi et al. 1998), probably lowering worker JH titers, since the administration of high doses of BP resulted in lower JH levels in both laboratory and field experiments (Le Conte et al. 2001). This is consistent with JH inhibitory effect on behavioral development: JH titers and rates of JH biosynthesis are low in nurse bees and high in foragers, and JH treatments cause precocious foraging (Robinson 1992a; Robinson and Vargo 1997). Moreover, BP shunts vitellogenin transport to the hypopharyngeal gland rather than to the ovaries, thus redirecting worker metabolism from reproduction to brood care (Le Conte and Hefetz 2008).

All the above described BP components are nonvolatile substances and their distribution is probably mediated by worker to worker contact. Recently a new highly volatile pheromone, E-β-ocimene, has been identified in honey bee larvae (Maisonnasse et al. 2009); it belongs to the terpene family and has an aerial transmission, being easily dispersed within the colony. Compared to BP, E-β-ocimene has a prevalent effect in the regulation of adult worker physiology and development. In particular it exerts two main effects: inhibition of worker ovaries and modulation of worker behavioral maturation (Maisonnasse et al. 2010b). It seems that BP and E-β-ocimene act in a synergistic manner to repress the activation of the workers' ovaries; this influence also has a consequence on brood care, since reproductive workers do not work as hard as sterile workers, showing a reduction in both larval care and foraging tasks.

5.1.4.3 Regulation of Worker Behavioral Development

In addition to the effects on larval development, BP induces an increase in colony growth also through a modulation of worker behavioral development. The treatment of colonies with BP caused an increased number of pollen foragers and augmented the weight of pollen load they transport (Pankiw et al. 2004). Pollen intake is also modulated by BP by altering the ratio of pollen to nonpollen foragers. Treatment with BP significantly decreased pollen forager turnaround time in the hive, increasing the ratio of pollen to non-pollen foragers entering the colony (Pankiw 2007).

Later studies demonstrated that BP acts in a dose-dependent manner in altering the demographics of colony foraging behavior: a low amount decreases the foraging age, resulting in a higher proportion of pollen foragers compared to nurse bees; high doses slow down the development of young bees from nest to foraging duties, so the foraging age increases, resulting in a lengthened nursing phase (Le Conte et al. 2001; Sagili et al. 2011). Moreover, pollen foragers exposed to a low amount of BP return to the nest with larger pollen loads compared to those treated with a higher amount (Sagili et al. 2011). This dose-dependent action is functional to the regulation of worker tasks on the basis of brood presence and age, as will be described.

As for worker ovarian development, E-β-ocimene cooperates with BP in the regulation of worker activity; in particular E-β-ocimene induces an earlier worker development toward foraging tasks, thereby optimizing food collection. Maisonnasse et al. (2010b) tried to explain how these two pheromones act synergically in maintaining colony homeostasis through the retention of a proper nurse/forager ratio and the inhibition of worker reproduction. It is known that in presence of brood, workers initiate foraging earlier compared to broodless colonies (Amdam et al. 2009; Tsuruda and Page 2009), thus assuring adequate food collection; however, an overabundance of foragers could lead to a lack of and a decline in brood care. Conversely, too many nurses cause a decrease in food collection and storage in the colony and a subsequent decline in brood nourishment. BP and E-β-ocimene are able to control this equilibrium since young and old larvae emit different types and quantities of these two pheromones: young larvae emit principally E-β-ocimene, while BP is produced in a growing amount during larval growth, reaching the highest concentrations during the capping stage.

In this way the young larvae, which have lower nursing needs, promote foraging and pollen collection by emitting a low quantity of BP and a large amount of E-β-ocimene, whereas old larvae, which have higher nursing needs, delay foraging and promote an increased brood care by producing a high quantity of BP, which also stimulates the development of worker hypopharyngeal glands and brood care tasks like cleaning, nourishment, and cell capping (Figure 5.4). Thus, young and old larvae play opposite roles in the behavioral maturation of worker bees according to their specific needs: young larvae promote foraging and old larvae promote brood care (Maisonnasse et al. 2010b).

It is clear that worker behavioral development, which leads to the typical honey bee age polyethism, is a complex and flexible process, involving more than one stimulus. The combined effect of queen signals, worker pheromones (ethyl oleate), and brood pheromones results in a plastic modulation of worker activity that is able to adapt worker response to the needs of the colony, which vary depending on colony developmental stage and environmental factors (Castillo et al. 2012).

5.1.5 PHEROMONE COMPLEXITY AND EVOLUTION OF SOCIALITY IN BEES

The deep diversification of chemical signaling in the honey bee society is strictly linked to the progression toward an increasing social complexity that evolved during the development of eusociality. In fact the success of social insect colonies lies in the capacity of all members of the society to act concertedly and in a well-organized and context-dependent manner. This ability is mainly based on the sophisticated means of communication represented by pheromones.

The study of pheromones in social bees, and in general in the superfamily Apoidea, is a viable means to understand the evolution of sociality in these insects, thanks to the gradual development of pheromonal regulation from the more simple bee societies to the highly organized *A. mellifera* (Bloch and Grozinger 2011). Similarly, the development of exocrine glands is much greater in social Apoidea compared to solitary ones (Billen and Morgan 1998). Some of these glands have a role in producing building substances (like wax glands) or digestive enzymes (like

Regulation of worker behavioral development

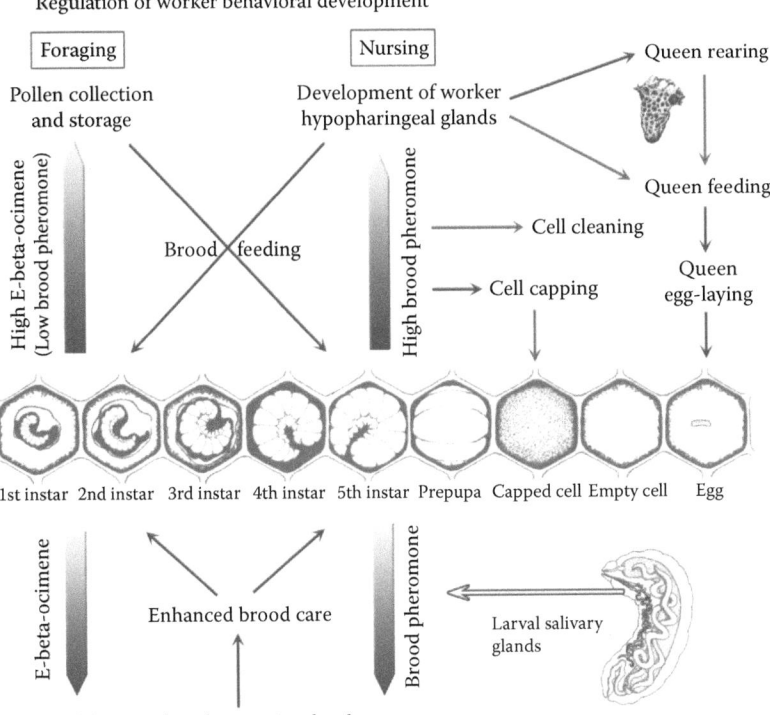

FIGURE 5.4 Effect of brood pheromones on the regulation of worker behavioral and sexual development. Large arrows indicate effect of the pheromones on development; small black arrows indicate effect of the pheromones on activities. In the lower part of the figure: both E-beta-ocimene and brood pheromone (secreted by larval salivary glands) inhibit ovarian development in workers, which in turn enhances their brood care commitment. In the upper part of the figure: high levels of E-beta-ocimene and low levels of brood pheromone, secreted by young larvae, boost worker behavioral development toward the foraging phase, stimulating pollen collection and thus supporting larval nutrition; high levels of brood pheromone, secreted by older larvae, enhance the nursing phase through the development of worker hypopharyngeal glands, which produce the royal jelly used for nutrition of young brood and the queen; in addition, brood pheromone stimulates cell capping and cell cleaning, the rearing of new queens, and queen egg-laying. (Adapted from Winston, M.L. (1987) *The Biology of the Honey Bee.* Cambridge, MA: Harvard University Press; and Crane, E. (1990) *Bees and Beekeeping: Science, Practice and World Resources.* Ithaca, NY: Cornell University Press.)

salivary glands) but most of them have a function related to the social organization of the colony, as we saw in the preceding paragraphs.

For classification purposes the different pheromones were described separately here according to the producing caste and the gland source, but they must be considered as the single components of a unique and multifaceted language of chemical communication. The most clarifying example is the queen signal with its main component QMP, whose main aim is to establish the reproductive dominance over male parentage with respect to the other females. Why should such a complex signal, with such a complex blend and multiple glandular sources, have evolved?

In support of the evolutionary theory is the finding that other *Apis* species show fewer components in their mandibular gland extract compared to *Apis mellifera*. For example, the QMP of the more primitive open-nesting species *A. dorsata*, *A. florea*, and *A. andreniformis* contain only the three acid components, 9-ODA and +/− 9-HDA, and lack the aromatic HOB and HVA compounds. *A. cerana*, the other cavity-nesting species besides *A. mellifera*, has four of the five QMP components but lacks HVA. In addition, mandibular gland secretions from open-nesting species show less differentiation between workers and queens (Plettner et al. 1997; Winston and Slessor 1998).

According to some authors, the complex blend of the queen signal may have evolved as a result of a social struggle for reproduction, an evolutionary "arms race" in which the queen signals and the workers' attempts to overcome this signal have coevolved, giving rise to the increasing chemical complexity (Katzav-Gozansky 2006). The capacity of egg-laying workers to mimic, at least partly, the queen signal, and the existence of workers able to overcome the queen signal and to reproduce even in queenright colonies (Hoover et al. 2005) give support to this theory. The end result is that the queen developed multiple pheromones, none of which is individually sufficient to obtain the full effect, but whose combination may act either additively or synergistically in establishing the complete reproductive dominance.

Another interesting feature of honey bee pheromones is the evident redundancy of signals, so that several chemical substances act in synergy or in cooperation in regulating one or more processes. This complementarity is responsible for a special chemical syntax that is probably functional to fine-tuning social regulation: more than one substance can regulate the same function, exerting similar effects but with a slightly different and specific target, location, or time lapse, thus assuring both a reinforcement and a modulation of the signal. An example is the regulation of worker behavioral development, mediated by the queen, workers, and two different brood signals. Again, why are four different pheromones produced by three separate sources needed to regulate this function? The queen, the forager workers, and the old brood produce three different pheromone blends—respectively, QMP, ethyl oleate, and BP—which slow down the behavioral development of workers toward foraging tasks, whereas young larvae produce E-β-ocimene, which has the opposite effect. A possible explanation for this multiple source signal is that the regulation of honey bee behavioral maturation, whose outcome strongly affects colony growth and development, and in ultimate analysis, colony survival, needs a multilevel feedback control network.

Some specific worker tasks appear to be influenced by multiple chemical signals, as is the case of foraging. This activity is regulated by worker-worker signals, which act both in an attractive (Nasonov and tarsal gland pheromones) and in a repulsive (2-heptanone) way, to guide and recruit nestmates to the most profitable food source.

Even from these few examples it is evident that the pheromonal system has evolved to fulfill the needs of efficiency and modulation typical of a complex insect society, acting according to cooperation, synergy, or complementarity schemes. The whole comprehensive mechanism of pheromone communication and its role in the regulation of sociality will be further elucidated in the next chapter, where we will describe the decoding process of the multiple pheromonal signals, the neurophysiological changes resulting from their processing, and the final effects on worker behavioral modules and colony functions.

5.2 NEUROPHYSIOLOGY OF CHEMICAL COMMUNICATION: PHEROMONE PROCESSING IN THE BEE BRAIN

The study of honey bee pheromones started in the 1960s and since then many advancements have been made in the knowledge of composition of pheromone blends, their glandular origin, and their colony target. But while we knew the effects of many of these pheromones, for a long time we could only speculate as to the neuronal mechanisms that mediate between pheromone and function. Only recently, with the development of molecular and genetic tools, some progress has been achieved in this direction. Thus, we are just starting to gain some awareness of the neurophysiological pathways of pheromone reception and processing in the bee brain, and only a few mechanisms have been entirely elucidated.

As previously reported, releaser and primer pheromones exert different effects on the receiver, the former being immediate and transitory and the latter delayed and long-termed. This difference suggests that two different mechanisms may exist by which pheromones influence the receiver: a direct effect on neural transmission for releaser pheromones against an effect on physiological processes (e.g., hormonal, metabolic, or genetic changes) for primer pheromones.

A common question regarding queen primer pheromones is their mode of action in regulating worker reproduction and behavioral development: is it by means of a controlling mechanism (queen pheromone as a suppressive agent) or a signaling one (queen pheromone as an "honest" signal) (Strauss et al. 2008)? In the "control" hypothesis the queen pheromones manipulate workers coercively by inhibiting their ovarian and behavioral development. In the signal hypothesis, also called the cooperation hypothesis, queen pheromones simply act to signal to workers the queen presence and its egg-laying potential, rather than to manipulate worker behavior and/or physiology. In the presence of a strong and healthy queen, workers refrain themselves from reproducing and prevent nestmate workers from reproducing (worker policing) in order to maximize colony fitness. When workers perceive a decline in the fecundity of the queen, they can activate their ovaries to produce their own male offspring (Keller and Nonacs 1993; Kocher and Grozinger 2011; Le Conte and Hefetz 2008; Strauss et al. 2008).

Both hypotheses make sense from an evolutionary point of view, and several authors tried to collect evidence to support one or the other theory, but without giving a definite and unquestionable response. Both theories could explain the richness and variety of queen pheromones, whose components increase with increasing level of sociality, such as both theories could support the variability of response given by different workers to the colony pheromones (Kocher and Grozinger 2011; Strauss et al. 2008). In either case, the way the pheromone is detected and processed in the brain of different receiver workers seems to play a crucial role in the regulation mechanism.

Different pheromones use different ways of transmission from the producer to the receiver. Volatile substances, such as the alarm and Nasonov pheromones produced by the workers, and the components of the QMP that attract drones for mating and workers for swarm clusters, use a dispersal mechanism; footprint pheromones, BP, and most components of QMP are transmitted primarily by contact, and the same is true for the esters produced in other queen glands (e.g., tergal and Dufour's), which are delivered as an integral part of the queen signal together with QMP; for this reason some authors have named them "passenger pheromones" (Keeling and Slessor 2005; Slessor et al. 2005).

Whatever the way of transmission of the pheromone, the reception process starts in the receiver olfactory system.

5.2.1 RECEPTION OF THE PHEROMONAL SIGNAL

5.2.1.1 Olfactory Receptor Neurons

In insects, the peripheral odour detection starts in the peripheral chemosensory system with the detection of the chemical signal by olfactory receptor neurons (ORNs), which express specific olfactory receptors (ORs). ORs are seven-transmembrane domain proteins coupled to G proteins; following the binding with odorant molecules cellular transduction cascades are activated, implicating the production of cAMP, leading to depolarization and action potentials.

These receptors are located mainly in the antennae, where they are organized in olfactory sensilla of various shapes; the poreplate sensilla are the most frequent sensilla in the honey bee antennae (Figure 5.5). A poreplate sensillum is formed by an oval-shaped thin cuticular plate with numerous minute pores and is innervated by 5 to 35 ORNs with their corresponding ORs. Each poreplate contains the whole range of ORs and thus represents a whole miniature system (Brockmann and Brueckner 1995; Sandoz 2011).

Odorant molecules reach the dendrites of ORNs by diffusing through an extracellular fluid, called sensillum lymph, filling the sensillum cavity. In this fluid, odorant binding proteins (OBPs) transport the odorants to the ORNs (Sandoz 2011). While OBPs bind general odorants, a specific class of OBPs, the PBPs (pheromone binding proteins) are specialized in binding sexual pheromones and are present mainly in male insect sensilla (Laughlin et al. 2008; Leal 2005). OBPs and PBPs play an essential role in the detection of general odors and pheromonal molecules and in their transduction, passing the molecules to the sensory neuron membrane protein, which then delivers it to the olfactory receptor (Pesenti et al. 2008, 2009). Another class of soluble chemosensory proteins (CSPs), which shares no sequence homology with either PBPs or general OBPs, has been described in honey bees (Danty et al. 1998). However, in honey bees only 21 genes coding for OPBs and six coding for CSPs have been found in the genome, so that the relative importance of these molecules in the process of odor perception is still unclear (Forêt and Maleska 2006). The results of a recent study using a proteomic approach show that 12 of the 21 OBPs

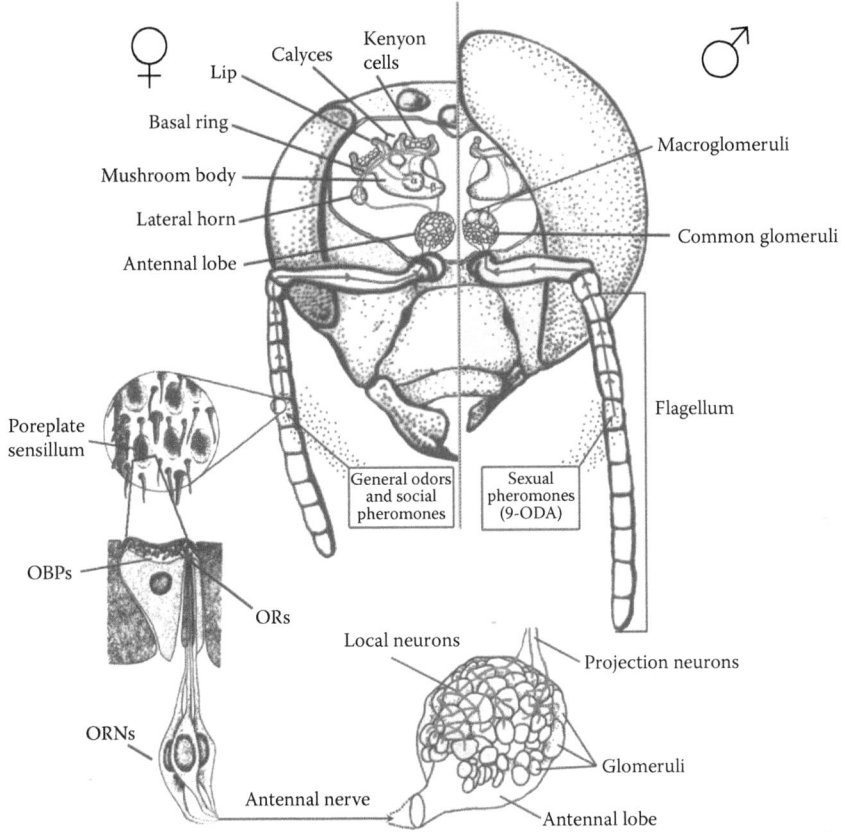

FIGURE 5.5 Schematic representation of the reception pathway for general odors and social pheromones in the worker honey bee (left) and for sexual pheromones in the honey bee drone (right) from the antenna poreplate sensilla to the antennal lobe, and the following transmission of the signal in the central nervous system. In the left and lower part of the figure the detail of odor reception from olfactory reception neurons (ORNs) and odorant-binding proteins (OBPs) to the glomeruli of the antennal lobe is highlighted. (Adapted from Winston, M.L. (1987) *The Biology of the Honey Bee*. Cambridge, MA: Harvard University Press; Morgan, S.M. et al. (1998) *Behav Brain Res* 91(1–2): 115–26; Frilli, F. et al. (2001) *L'ape: Forme e funzioni*. Bologna, Italy: Calderini Edagricole.)

and 2 of the 6 CSPs predicted in the honey bee genome are present in the foragers' antennae (Dani et al. 2010) and some OBPs are found to be more highly expressed in the mandibular glands of different honey bee castes, suggesting their involvement also in solubilization and release of semiochemicals (Iovinella et al. 2011). Three main subclasses of OBPs are defined in honey bees on the basis of antennal specific proteins (ASPs), namely ASP1, ASP2, and ASP3 (Danty et al. 1997, 1998). ASP1 is thought to be associated with QMP because of its higher abundance in drone sensilla and the ability to bind 9-ODA and 9-HDA, the most active components of the queen pheromone blend, while ASP2 and ASP3 bind general odorants (Danty et al. 1999).

One of the CSPs, called ASP3c, specifically binds brood pheromone components and not general odorants or other pheromones (Briand et al. 2002).

The wide range of pheromones described in honey bees, together with the great number of environmental odors they encounter, suggests a highly developed olfactory system that must be able to discriminate a large number of volatile substances. Indeed, the sequencing of the honey bee genome allowed the identification of an exceptionally high number of OR types (160–170), compared to the already known ORs of *Drosophila melanogaster* (62 ORs) and *Anopheles gambiae* (79 ORs) (Robertson and Wanner 2006). This high number is evidently linked to the extraordinary olfactory abilities of honey bees, whose social life requires the perception of several pheromone blends as well as kin recognition signals and numerous floral odors.

It is presumable that different ORs are differentially expressed according to caste and function; indeed, among the identified antennal ORs, the AmOR11, which is upregulated in drones, was recently demonstrated in male antennae to specifically detect 9-ODA and to respond to all the main QMP components (Wanner et al. 2007). On the contrary, a number of other receptors (OR63, OR81, OR109, OR150, OR151, OR152) are more highly expressed in worker bees than in drones and are probably linked to floral odorant reception, being differentially expressed in bees which live in different environments and thus experience diverse floral scents (Reinhard and Claudianos 2013).

5.2.1.2   Antennal Lobes and the Glomeruli

The ORNs project their axons to a specific area of the deutocerebrum called antennal lobe, which is organized in densely packed nervous structures termed glomeruli (Figure 5.5). The axons of several ORNs converge to the glomeruli through four sensory tracts (T1–T4), which define four subpopulations of glomeruli, two containing about 70 glomeruli each (T1 and T3) and two with seven glomeruli each (T2 and T4). The ORNs of an individual poreplate project to all four glomerular subpopulations and are therefore distributed across the whole antennal lobes (Brockmann and Brueckner 1995; Flanagan and Mercer 1989; Kelber et al. 2006).

The arrangement and number of glomeruli are largely species-specific and vary from about 32 in the mosquito *Aedes aegypti* to more than 1000 in locusts and social wasps. In honey bees the workers possess 166 glomeruli and the drones 103. The latter also have four large glomerular complexes exclusively committed to processing sexual pheromones, probably with a functional specialization for a specific pheromone substance in each of the four complexes (Arnold et al. 1985).

It has to be noted that the number of glomeruli in the honey bee antennal lobe is almost equal to the number of ORN types expressing a given OR in the antennae, supporting the hypothesis of a linear relationship one-receptor/one-neuron/one-glomerulus.

Within the glomeruli the ORN axons synapse with two other kinds of neurons: the local neurones and the projection neurones (Figure 5.5). The former are mainly GABAergic neurons with an inhibitory output, while the latter are cholinergic neurons that show either excitatory or inhibitory responses to odors. The local neurons can be classified into two main types: the homogeneous local neurons, which innervate most if not all glomeruli in a uniform manner, and the heterogeneous local neurons, which innervate only a small subset of glomeruli with one dominant glomerulus that is densely innervated and a few others with very sparse processes (Flanagan and Mercer 1989; Sandoz 2011). The function of local interneurons is to interconnect the glomeruli and modulate the signal coming from ORs.

Projection neurons leave the antennal lobe via a variable number of pathways called antennocerebral tracts, connecting it with different areas of the protocerebrum, mostly the calyces of the mushroom bodies and the lateral protocerebrum (Hansson and Anton 2000; Kay and Stopfer 2006). Projection neurons can also be classified into two types: uniglomerular projection neurons branch in a single glomerulus within the antennal lobe and project to the mushroom body or the lateral horn through two major antennocerebral tracts, while multiglomerular projection neurons branch in most glomeruli and their axons follow three lesser antennocerebral tracts leading to other areas of the protocerebrum surrounding the α-lobe of the mushroom body or extending toward the lateral horn (Sandoz 2011, 2013).

5.2.1.3 Pheromone Processing in the Glomeruli

Within the glomeruli the olfactory signal undergoes an important integration and encoding before being transmitted to the higher centers. Glomeruli are the anatomical and functional units of the antennal lobes and constitute sites of synaptic interaction between different neuron types. The activity patterns of antennal lobes in response to odors was studied in the honey bee by optical imaging techniques (Galizia et al. 1997, 1998). Axons of ORNs expressing the same odorant receptor or with similar odor specificities converge onto the same glomerulus. Considering that a single type of molecule interacts with a number of different ORNs, which activate a similar number of glomeruli, an odor blend is represented by the activation of a variable number of glomeruli, resulting in a spatial representation of the odor within the antennal lobe (Galizia et al. 1999; Joerges et al. 1997; Sachse and Galizia 2003; Sachse et al. 1999). This representation is variable in time and depends on the olfactory experience; therefore, odorants are represented in the antennal lobe as changing spatiotemporal patterns of glomerular activity (Sandoz et al. 2003). The early olfactory learning during young adulthood enhances glomerular activity and modifies the spatiotemporal response patterns; these changes affect neural activity until the time when bees initiate foraging activities (Arenas et al. 2009; Galizia and Vetter 2005).

Social (nonsexual) pheromones, like citral and geraniol (components of the Nasonov gland), IPA (the major component of the alarm pheromone associated with the sting apparatus), and the worker mandibular gland pheromone 2-heptanone, are coded in the antennal lobe as "general" odors since they elicit activity in the same brain region as environmental odors (Galizia et al. 1999; Joerges et al. 1997; Sachse et al. 1999). IPA elicits strong responses in several glomeruli that also exhibit strong responses to orange, clove oil, limonene, and several plant extracts (Galizia and Menzel 2001). Nevertheless, Sandoz et al. (2001) found that IPA and 2-heptanone, which share an alarm role but have a different chemical structure and source, induce a reciprocal generalization in olfactory conditioning tests, suggesting that a similarity in the neural representation of odor could rely not only on the chemical structure but also on their functional value (Sandoz et al. 2001).

Wang et al. (2008) investigated the neural activity elicited by eight components of the sting pheromone, compared with the whole bee sting apparatus, at the level of the antennal lobes of honey bee workers. They found that the sting preparation evokes a clearly distinct glomerular pattern compared to those of individual pheromone components (e.g., IPA-activated glomeruli in the medial part of the antennal lobes), whereas the stings activated the lateral dorsal part. It seems that the sting apparatus pheromone is processed in a similar way to general odors, since the main determinant of glomerular activation is its chemical structure rather than its pheromonal value. However, in contrast to the elemental strategy used for processing nonpheromonal mixtures, where the neural representation of mixtures of two to four nonpheromonal odors could be linearly predicted based on the neural representation of each component (Deisig et al. 2006), pheromonal blends do not follow such a linear representation, revealing more complex strategies for the processing of pheromonal mixtures in the honey bee antennal lobe (Wang et al. 2008).

5.2.1.4 Sexual Communication: Drone Reception of QMP in Macroglomerular Complexes

Male insects, including honey bee drones, have a specialized olfactory subsystem to detect female sexual pheromones even at long distances. This subsystem is characterized by a large number of ORs and ORNs sensitive to the components of the female pheromones. Their axons converge to hypertrophied glomerular subunits called macroglomerular complexes that are located in the antennal lobes. In honey bees the sexual dimorphism of the reception system is evident (Figure 5.5); compared with worker bees, drones have larger antennae, with a flagellum surface twice as large as that of the workers, and about seven times as many poreplate sensilla (around 18,000 versus 2700) and ORNs (around 340,000 versus 65,000) (Esslen and Kaissling 1976). In addition, the female antennal lobe is composed of isomorphic glomeruli (about 160 in workers and 150 in queens), whereas in drones there is a reduced number of these "ordinary" isomorphic glomeruli (about 100) but there are four voluminous macroglomeruli (Arnold et al. 1985).

The sexual dimorphism of the reception system corresponds to different neuronal strategies to detect and respond to the pheromone signals. Electroantennographic studies showed that while worker antennae have a very similar response to the various QMP components, suggesting that there is no antennal specialization with regard to the number of sensory neurons, in contrast, drone antennae showed marked responses to 9-ODA and to synthetic QMP compared to the other QMP components. This high antennal response is characteristic of sexual pheromones that elicit a long-distance reaction and is attributed to a much higher number of sensory neurons in the male antennae (Brockmann et al. 1998).

These results confirm that worker antennae have a kind of generalized antennal tuning with no significant differences in the number of sensory neurons for individual mandibular pheromone components, while drone antennae are specialized in the perception of one component of the mandibular pheromone, 9-ODA. This scenario is in accordance with the above described differential morphology in the olfactory system between workers and drones, and confirmed by the finding that drones of the more primitive honey bee species, *Apis florea*, have only 1200 poreplate sensilla per antenna and only two macroglomeruli in their antennal lobes, corresponding to a minor complexity of the sex pheromone mixture in this species compared to *A. mellifera* (Brockmann and Bruchner 2001).

Wanner et al. (2007) identified four candidate sex pheromone ORs from the honey bee genome based on their higher expression in drone antennae compared to worker antennae. This number coincides with the number of macroglomeruli in the drone antennal lobe, but only one of them, the already cited AmOr11, specifically responds to 9-ODA, while the other three could not be linked to any queen pheromone component.

Further analysis of drone antennal lobes led to the discovery that the ventral part carries only ordinary glomeruli while the dorsal part shows two of the four macroglomeruli, one located dorsomedially (MG1) and the other on the dorso-lateral side (MG2). Optical imaging of the antennal lobe showed that floral odors and blend mixtures induced focal responses on the ventromedial side of the antennal lobe, a region rich in ordinary glomeruli. In contrast MG2 is clearly and specifically devoted to the reception of the QMP main component 9-ODA, which does not induce signals in regions other than MG2. Among the other QMP components, HOB and HVA induced activity mostly in two ordinary glomeruli in the center of the frontal region, which showed responses also to floral odorants 1-hexanol, limonene, clove oil, and orange oil blends, while 9-HDA and 10-HDA induced only very low and diffuse signals in ordinary glomeruli that could not be measured (Sandoz 2006, 2007). The fact that HVA and HOB are detected in drones by the general olfactory system and not by the pheromonal subsystem can be explained by their different pheromonal role: in fact they are produced mainly by mated queens and not (or very little) by virgin queens, suggesting that they are used for the induction of workers' retinue behavior and not for drone attraction by virgin queens (Plettner et al. 1997). The role of 9-HDA and 10-HDA as sex attractants remains unclear.

The different organization of the olfactory system between workers and drones reflects their diverse role in the honey bee society: drones exhibit a clear olfactory specialization for the sexual pheromone 9-ODA consistent with their exclusive reproductive role in the hive, while workers show a broader range and less specific response for both pheromonal and nonpheromonal odors consistent with the use of these different signals in different behavioral contexts (Sandoz et al. 2007).

5.2.1.5 Pheromone Processing in Higher Centers

Processed olfactory information leaves the antennal lobe by the projection neurons, towards higher-order brain centers, especially the mushroom bodies and the lateral horn (Figure 5.5).

Olfactory inputs project to a specific area of the mushroom bodies, the Kenyon cells, which form two cup-shaped regions called calyces in each brain hemisphere. The calyces are anatomically and functionally subdivided into the lip, the collar, and the basal ring. The lip region and the inner half of the basal ring receive olfactory input, whereas the collar and outer half of the basal ring receive visual input. The axons of Kenyon cells project in bundles into the midbrain, forming the peduncle and the vertical and horizontal lobes, also called α and β lobes (Strausfeld 2002).

The mushroom bodies receive not only olfactory and visual signals, but also mechanosensory and gustatory inputs. They play an important role in the process of associative learning of olfactory stimuli but also act as a multisensory integration center with a feedback and modulatory function (Mercer and Erber 1983). They are also involved in higher nervous functions, such as learning, memory, and cognitive processes.

In contrast, the processing of olfactory stimuli in the lateral horn are still mostly unknown, including the topography of neurons leaving the lateral horn and the descending pathways involved in behavioral output. In *Drosophila* this region is divided in two main subregions that separately process pheromones and fruit odors (Jefferis et al. 2007); since the honey bee's lateral horn shows a specific compartmentalization with at least four subcompartments, an organization similar to that of *Drosophila* could exist in the honey bee, with a specific pheromone processing region in the lateral horn.

Given that no dedicated glomeruli have been found in workers for the processing of pheromones, Sandoz et al. (2007) hypothesized that specific recognition of pheromones, especially the social ones, may take place at higher processing levels downstream from the antennal lobes. It is conceivable that particular Kenyon cells could recognize specific combinations of activated projection neurons, which would indicate that the detected stimulus is a pheromone.

5.2.2 PROCESSING AND MODULATION OF THE PHEROMONAL SIGNAL

The reception and processing of pheromones lead to a response in the receiver that corresponds to a behavioral and physiological change. But how does this process work? The response to pheromones involves both environmental and physiological factors, since pheromones induce a behavioral plasticity in the receiver through a shift in neural response thresholds to environmental conditions.

Releaser pheromones act through a direct and unambiguous pathway in which one pheromone evokes one response in the receivers. In contrast, primer pheromones induce more deep and prolonged effects that can be modulated by the receiver to give a different behavioral response according to its physiological state. These different patterns suggest a different way of action for these two types of pheromones, but until now evidence suggests that the two pathways are partly overlapping and involve similar neuronal and physiological mechanisms.

Study of the mode of action of pheromones should first take into account that many factors affect their reception and processing. The same chemicals can be perceived and processed in a different manner according to the physiological state of the receiver, which in turn is influenced by both genetic and environmental factors correlated to the social environment and the individual developmental stage.

A well-known example is the response to QMP by workers of different ages: Pham-Delègue et al. (1993) demonstrated that there is an age-dependency and experience-dependency in the attraction effect of QMP toward workers. Furthermore, they showed that the olfactory environment experienced in the first day of adult life can strongly modify the functioning of the olfactory nervous system and thus worker behavioral responses (De Jong and Pham-Delègue 1991; Pham-Delègue et al. 1991). This was observed both for general olfactory sensitivity and for pheromonal stimuli, suggesting that age and experience induce different behavioral responses linked to the plasticity of the olfactory system at a peripheral or central level. The relationships between peripheral sensitivity, signal processing, and behavioral responses have only recently started to be elucidated.

The behavioral development from nurses to foragers is accompanied by a brain plasticity that involves in particular the antennal lobes and the mushroom bodies. This transition from life inside the hive to activities outside the hive is associated with a distinct increase in antennal lobe and mushroom body size: the volume of glomeruli changes with the shift to foraging duties, and forager bees have a larger mushroom body calyx than nurse bees of the same age (Brown et al. 2004; Farris et al. 2001; Maleszka et al. 2009). This increase is due to a growing number of neural connections, driven by the richer sensory experience of the outside life.

Another useful approach to uncover the physiological mechanism of pheromone effects exploits genetic differences in worker responses. For instance, some workers are highly responsive to QMP, while others respond poorly or not at all in laboratory bioassays (Kaminski et al. 1990; Pankiw et al. 1994, 1995). There may be genetic and physiological differences between high and low responding workers in receiving or responding to the queen pheromonal message and these differences could provide a powerful tool to dissect the neurochemical pathways of QMP effects (Winston and Slessor 1998).

Pheromones could act by modulating sensory response thresholds which affect the probability of workers performing certain behaviors, such as nursing, foraging, or defence. Besides QMP, alarm pheromones also show this modulating effect, for example on appetitive and aversive learning, which are important behaviors in forager and guard bee workers (Hunt 2007; Urlacher et al. 2010).

The different substances that are possibly involved in the neuromodulation of pheromone signals in the bee brain will be described in the following section, together with some interesting discovered cases of the pheromonal effect on specific functions.

5.2.2.1   Modulation of the Signal: The Role of Biogenic Amines and Juvenile Hormones

5.2.2.1.1   Brain Amines as Neuromodulators

In the honey bee brain several biogenic amines with potential modulatory function have been detected both in the central and peripheral nervous system. These molecules function as neurotransmitters, neuromodulators, and neurohormones, mediating a diversity of physiological and behavioral functions. In particular, dopamine (DA), serotonin (5-hydroxy-tryptamine, 5-HT) and octopamine (OA), which are all neurotransmitters and long-term brain modulators, seem to be involved in the modulation of behavior, which is functionally linked to pheromone activity (Mercer 1987; Mercer and Menzel 1982).

Biogenic amines in the honey bee brain are synthesized by a relatively small number of modulatory neurons, which often possess widespread projections. The mushroom body calyces in particular receive input from OA and DA neurons, which play an important role in associative learning (Bicker 1999).

DA neurons are present in most parts of the bee brain and in the subesophageal ganglion, representing about 0.1% of the entire neuronal population. Most are located in the mushroom bodies below the lateral calyx and in the anterior-ventral protocerebrum. DA neurons occupy large volumes of neuropil and DA fibers synapse onto the antennal lobes and the Kenyon cell bodies, suggesting a role in mediating distant rather than local neural interactions (Schaefer and Rehder 1989; Schuermann et al. 1989).

5-HT neurons are found in all areas of the brain, in particular the optic lobes, but 5-HT-immunoreactive fibers innervate the mushroom bodies outside the calyces, the antennal lobes, and almost all parts of the central body (Gauthier and Grünewald 2013). Antennal glomeruli contain 5-HT fibers restricted around the margin (Schuermann and Klemm 1984) and a large 5-HT interneuron interconnects the deutocerebral antennal and dorsal lobes with the subesophageal ganglion and descends into the ventral nerve chord (Rehder et al. 1987).

OA neurons are represented in most of the cerebral ganglion, but mainly in five brain regions: in the pars intercerebralis, mediodorsal to the antennal lobes, on both sides of the protocerebrum midline, between the lateral protocerebral lobes and the dorsal lobes, and on either side of the central body. Fine networks invade the antennal lobes, the calyces, and a small part of the α-lobes of the mushroom bodies, the protocerebrum, and all three optic ganglia (Kreissl et al. 1994). Another unpaired median cluster of OA neurons is located within the subesophageal ganglion, where the VUM neurons were identified (see Section 5.2.2.1.2).

The level of these three biogenic amines (5-HT, DA, OA) in the honey bee brain has been shown to vary during worker development, namely active foragers had significantly higher levels of amines than younger bees working in the hive. These variations are age- and task-dependent and can be correlated to the behavioral development of workers (Schulz and Robinson 1999; Taylor et al. 1992; Wagener-Hulme et al. 1999). This variability thus reflects a differential responsiveness to stimuli associated respectively with brood care or with foraging, such as optical cues (nurse bees live in the dark while foragers need light to orientate), odorant signals (flower and environmental scent), and also learning and memory, since foraging tasks demand cognitive functions for orientation, flower handling, and communication. Furthermore, high levels of DA in the honey bee brain were found to be correlated with ovarian development (Sasaki and Nagao 2001) and the dietary administration of dopamine is able to activate ovaries in queenless workers, suggesting a role of dopamine in the regulation of the reproductive status of honey bee workers (Dombroski et al. 2003).

The levels of amines can vary also independent of age: a different level of DA and 5-HT was found in the optic lobes of nectar foragers and pollen foragers, behaviors that are typically performed at similar ages (Taylor et al. 1992), and between food storers and comb builders, the former having significantly lower levels of DA (Wagener-Hulme et al. 1999). This non-age-dependent difference can be correlated to a differential development of specific brain functions correlated to the performed tasks. There is a different modulation of amine levels in the two brain regions involved in the division of labor, the optical lobes, and the mushroom bodies. In the optical lobes the amounts of DA, 5-HT and OA vary significantly with worker age, but not with task, whereas in mushroom bodies they vary significantly with worker behavior, but not with age (Schulz and Robinson 1999).

Among the three amines, OA is the one that exhibits the most robust association with behavior: foragers had significantly higher brain levels of OA compared to bees performing in-hive tasks, such as nursing or food storing, independent of age (Schulz and Robinson 1999; Wagener-Hulme et al. 1999). The strong correlation between OA concentration in the antennal lobes and worker task suggests that it plays a causal role in the regulation of honey bee behavioral development. In particular, its increase in the antennal lobes seems to be involved in the release and maintenance of the foraging state since the administration of OA to workers at the foraging age results in an earlier onset of foraging, but when administered to younger workers it produces no effects (Schulz and Robinson 2001; Schulz et al. 2002a).

The influence of OA on foraging behavior probably acts through the regulation of response to foraging-related stimuli that involve learning and memory. This is supported both by anatomical and experimental findings: OA fibers were found in all neuropils that contain pathways for proboscis extension learning (Kreissl et al. 1994); OA administration enhances worker responsiveness to unconditioned olfactory stimuli, probably producing a central excitatory state in which the effectiveness of sensory stimuli is improved (Mercer and Menzel 1982); furthermore, while both DA and 5-HT injected into the bee brain reduce the response to a conditioned olfactory stimulus, OA-treated bees do not have a reduced response. The application of DA in the mushroom body causes a reduction of potentials after antennal stimulation that can account for the reduced response (Mercer and Erber 1983). Further studies confirmed the role of OA in appetitive olfactory learning in bees: injections of this amine in the honey bee brain provide a substitute for sucrose reward and induce olfactory learning (Hammer and Menzel 1998); last, blocking OA receptors disrupts olfactory conditioning (Farooqui et al. 2003). Recent research has examined in depth the role of DA neurons in aversive learning and of OA neurons in appetitive learning (see Sections 5.2.2.3.1 and 5.2.2.3.2).

5.2.2.1.2   Brain Amines and Pheromones

It is known that queen pheromones act as typical tranquillizer signals, suppressing perception and stabilizing emotional agitation especially of young worker bees (Lipinski 2006). For instance, workers in queenless colonies tend to be agitated, nervous, and aggressive; it seems that queen pheromones act on workers as a sort of social peacemaker. This effect is achieved through different physiological and hormonal mechanisms. In queenright colonies young workers have significantly lower levels of all three main biogenic amines and JH titers compared to queenless colonies: the calming effect is probably exerted by lowering the level of neurotransmitters and by decreasing the excitation of corpora allata, which results in a reduced arousal to external stimuli. A similar calming effect is exerted by brood pheromones and by mandibular pheromones of older workers (Lipinski 2006).

To understand the role of brain amines in the modulation of pheromonal signals, the relationship between their level in the worker brain and the worker response to pheromones was investigated in several studies. For example, Harris and Woodring (1999) found that in honey bees the ingestion of 5-hydroxytryptophan, a precursor of 5-HT, causes a reduction of the worker response to IPA, measured as buzzing response. On the contrary, the ingestion of L-DOPA, precursor of DA, has no effect on the buzzing response stimulated by IPA, suggesting that response to alarm pheromone in honey bees is regulated only by 5-HT metabolism, while it is known that DA and 5-HT are both involved in the neuromodulation of aggressive behavior in many vertebrates and invertebrates (Hunt 2007).

OA has been shown to be quite strictly involved in the response to pheromones linked to behavioral development, which we know to be regulated by the demographic composition of the colony and by the presence of brood, through worker and brood pheromones. Barron et al. (2002) showed that OA is able to enhance worker responsiveness to brood pheromones and to decrease responsiveness to social inhibition exerted by adult bees. OA thus acts as a modulator of pheromonal communication by regulating the response thresholds to worker and brood pheromones. However, the modulation of brood pheromone response is selective for the foraging stimuli, since other functions regulated by this pheromone are not enhanced by OA, like capping behavior (Barron and Robinson 2005). Furthermore, OA does not enhance the response to other pheromonal signals, like retinue response to QMP. The specific mechanism by which OA achieves these results is not yet clear; it may act by modulating ORNs in the antennal lobes or by modulating the neuronal circuits involved in the processing of the olfactory stimulus within the mushroom bodies (Schulz et al. 2002a). Neurons of the octopaminergic VUM family may be involved in this modulating function: the VUM mx1 neuron projects from the subesophageal ganglion, where it gets gustatory input from sucrose receptors, to the brain, meeting the olfactory pathway in three areas: the antennal lobes, the mushroom bodies calyces, and the lateral horn; thus it may act by combining olfactory and gustatory stimuli with higher functions (Hammer 1993; Schröter et al. 2006).

5.2.2.1.3 Juvenile Hormone and Pheromones

Similar to brain amines, the level of JH is functionally correlated to worker behavioral development: JH levels are higher in foragers compared to nurses, and treatment with JH or JH analogues results in precocious foraging (Huang et al. 1991; Robinson 1987a). It has been demonstrated that QMP is able to reduce the titer of JH in workers (Kaatz et al. 1992; Pankiw et al. 1998), which results in the lower level of JH in nurse bees, which are in strict contact with the queen and thus with QMP, compared to foragers.

There is a strict relation between the level of OA in the honey bee brain and the level of JH in the hemolymph: OA stimulates production of JH in vitro (Kaatz et al. 1994) and treatment with the JH analog methoprene results in increased forager-like levels of OA in the antennal lobes of preforager workers (Schulz et al. 2002b). The regulation of foraging behavior probably passes through an increase in OA levels in the brain, since allatectomized bees (no JH production) can still initiate foraging after an OA treatment. The timing of OA and JH presence is consistent with the hypothesis that JH acts earlier in the process of forager development as a trigger factor, while OA acts later but more rapidly as a releaser factor of foraging behavior

(Schulz et al. 2002b). These findings suggest that the variability in JH and OA levels between workers of different age and task are a key factor in modulating the worker behavioral response to pheromones, but it is not fully established whether JH and OA act through the same or different neural pathways.

The hypothesis that JH influences age-dependent olfaction was tested by examining the effect of the JH analog methoprene on alarm pheromone perception (Robinson 1987b). Worker sensitivity to alarm pheromone increases with age (Collins 1980) and with increasing group size (Southwick and Moritz 1985), indicating a strong influence of the social context on pheromone processing. Methoprene strengthens the behavioral response to alarm pheromone at every age, but is strongest between 5 and 8 days of age. Contrary to behavioral assays, the electroantennographic response to alarm pheromone did not increase in workers after day 5 and was not affected by methoprene: this shows that the honey bee peripheral olfactory system reaches full maturity 4 days after adult emergence and suggests that hormonal modulating effects on pheromone perception occur in the central nervous system (Masson and Arnold 1984; Robinson 1987b).

5.2.2.2   Direct Modulation of Worker Behavior: HVA Mimic of Dopamine

Another interesting cue in the study of pheromone processing in the bee brain came from the observation that one of the components of QMP, HVA, has a similar structure to DA, one of the biogenic amines that plays a role in honey bee behavioral regulation (Beggs et al. 2007). The presence of this compound within the pheromonal blend suggested that exposure to HVA might affect DA function, modulating dopaminergic neural pathways.

Three DA receptor genes *Amdop1*, *Amdop2*, and *Amdop3* were identified in the honey bee mushroom bodies; the receptor density, their gene transcript, and levels of gene expression have been found to change during the lifetime of the adult worker bee (Humphries et al. 2003; Kokay et al. 1999; Kurshan et al. 2003; Mustard et al. 2003). Beggs and Mercer (2009) demonstrated that HVA selectively activates the D2-like DA receptor *Amdop3*.

The application of QMP to worker honey bees alters DA receptor gene expression, mainly lowering *Amdop1* transcript levels; consistently, the DA-evoked response, measured as intracellular cAMP level, is lower in mushroom bodies of workers exposed to QMP or HVA (Beggs et al. 2007). This finding is in agreement with the hypothesis that HVA plays a direct role on the modulation of DA levels in the brain. Further confirmation came from an experiment in which workers exposed to a mated queen (which produces higher levels of HVA) showed significantly lower brain DA levels than workers of the same age exposed to a virgin queen (low or absent HVA production); HOB, the other QMP component produced by mated and virgin queens, showed no effect on DA levels of worker brain. Finally, activity levels in bees treated with QMP are reduced, but this effect can be reversed by a treatment with L-dopa, a precursor of DA (Beggs et al. 2007). Taken together, all these results confirm that HVA alone is able to mimic the effects of QMP on DA levels in the honey bee brain and that DA pathways are not affected by other components of the QMP blend.

Another possible role of HVA in the QMP blend focuses on the inhibition of ovarian reproduction in workers: since the treatment of queenless workers with dopamine enhances ovarian development in workers (Dombroski et al. 2003), HVA may inhibit ovarian activation by acting agonistically on the dopamine pathway. However, a direct effect of HVA on ovarian development has not yet been confirmed.

5.2.2.3   Worker Attraction and Aversion: The Role of Pheromones on Appetitive and Defense Behavior

The appetitive learning conditioning in honey bees is a well-known experimental technique in which bees rewarded with sucrose on particular stimuli become able to respond to the same stimulus or to a similar one even without sucrose reward; the response is typical and measurable, consisting in the proboscis extension reflex (PER) (Giurfa 2007). On the contrary, the aversive learning conditioning consists in training bees to a defensive response, namely the SER, in response to potentially noxious stimuli. This is achieved through a modified protocol for the PER, in which the stimulus is not associated with a sucrose reward, but to a mild electric shock (Carcaud et al. 2009). The PER and SER tests were used to reveal the modulating role of some pheromones on worker appetitive and aversive learning.

5.2.2.3.1   QMP and Queen Attraction

Vergoz et al. (2007a) demonstrated that while OA mediates appetitive learning, as already shown by other authors, aversive learning in honey bees is mediated by DA; in fact it is suppressed by blocking of DA, but not OA, receptors. Since it has been demonstrated that HVA can mimic DA function, Vergoz et al. (2007a) postulated that QMP, through its component HVA, is responsible for blocking aversive learning in young workers. This hypothesis was proved in a further study (Vergoz et al. 2007b), which showed that QMP does block aversive learning in young bees while leaving appetitive learning intact. The authors postulate that QMP production by the mated queen gives her an advantage by preventing young workers, which are in close contact with her and on which she depends for feeding, form an aversion to her pheromonal bouquet.

During their studies on appetitive and aversive behavior, Vergoz et al. (2009) observed that worker responsiveness to QMP is strongly age-dependent, since 2-day-old workers are more strongly attracted to QMP than 6-day-old ones, while foragers are even repelled by QMP. They also showed that this behavior is likely to be modulated by receptors in honey bee antennae: those of 2-day-old workers strongly attracted by QMP have a higher expression level of OA receptor *Amoa1* and of DA receptor *Amdop3* compared to 6-day-old workers; the level of *Amdop3* transcript decreases during the first week of adult life, together with the attraction towards QMP. However, this pattern is true only for bees that have been exposed to QMP since adult emergence, while young bees that have not been exposed previously to QMP are not attracted to it and show a higher expression level of the DA receptor *Amdop1*. Thus it seems that the queen possesses several ways to modulate worker behavior through QMP at the level of the antennal sensory neurons: by suppressing avoidance behavior (by blocking the DA signal) and by enhancing the attractiveness of her pheromone (by increasing the OA signal). This is supported by the fact that high expression levels of OA receptor gene *Amoa1* and DA receptor *Amdop3* in the antennae augment the attractive qualities of QMP, while suppression of DA receptor *Amdop1* also enhances attraction to QMP by reducing worker sensitivity to unattractive components of the pheromone.

Similar results were found by McQuillan et al. (2012), who analyzed OA and DA antennal receptors in workers of different age and task commitment. The expression levels of the receptors *Amoa1* and *Amdop2* show an increase with age, being higher in older workers, while the opposite trend is shown for *Amdop3* expression levels, which clearly decrease with age. Furthermore, expression levels of *Amoa1* are higher in same-age pollen foragers than in nurses, consistently with the higher OA brain level in foragers (Schulz and Robinson 1999; Wagener-Hulme et al. 1999). Although the physiological significance of this variability in receptor gene expression has not been fully determined, the dynamics of gene expression in the antennae are indicative of a functional role of the periphery in the behavioral changes of honey bee workers.

5.2.2.3.2 Alarm Pheromones and Defense Behavior

Several studies have shown that alarm pheromones, besides their important role in triggering bee defense behavior, can act as modulators of the sensitivity to environmental stimuli.

Stress-induced analgesia is a mechanism that increases the threshold of responsiveness to external stimuli that elicit innate defensive responses by activating endogenous opioid pathways. In honey bees the threshold of stinging response (the main defense behavior) was artificially increased with injection of morphine, and this effect was antagonized by naloxone, demonstrating the presence of an endogenous opioid system in the honey bee and its involvement in the modulation of the stinging response (Nuñez et al. 1983). The exposure of workers to IPA causes a reduction in the responsiveness to a nociceptive stimulus (electrical stimulation) by increasing the threshold of responsiveness. This effect is antagonized by naloxone, indicating the involvement of an opioid system, as a typical opioid analgesia is induced. The social meaning of this analgesic effect is to increase the worker defensive efficiency by reducing the probability of withdrawal when facing an enemy (Nuñez et al. 1998).

In the experiments of Balderrama et al. (2002) IPA exposure led to a decrease in responsiveness to sucrose and an increase of responsiveness to a noxious stimulus (i.e., an electric shock). In a followup study the exposure to alarm pheromones or IPA showed a strong effect on appetitive learning by decreasing the learning success of exposed bees (Urlacher et al. 2010). These effects are not in contrast with the main hypothesized role of alarm pheromones, as the depression of foraging activity, through the decrease in sucrose responsiveness and the appetitive learning, allows workers to freely contrast a potential danger or enemies signaled by the release of alarm pheromones. This can strengthen worker's commitment to their role in colony guarding and defense.

The physiological mechanism subtending this modulating effect could involve biogenic amines, which are known to regulate aversive and appetitive learning, respectively, through DA and OA pathways (Giurfa 2007; Vergoz et al. 2007a). Alternatively, the activation of an opioid-like system, which was shown to be affected by this pheromone, could lead to a general learning impairment for its analgesic effects (Nuñez et al. 1998).

5.2.2.4 Modulation of Worker Metabolism: The Effect of Pheromones on Nutrient Stores

We saw that two pheromones have a prevalent role in the regulation of worker development by slowing the worker transition from nurse to foragers: QMP and brood pheromone. In the previous sections we showed that QMP acts through a central or peripheral modulation of brain amines, which influences the subsequent behavioral and physiological pathways, including the reception level of the pheromone itself. Moreover, OA modulates worker responsiveness to brood pheromone by regulating worker response thresholds.

Another way of action of these two pheromones seems to be the regulation of worker metabolism. Nurse bees have higher lipid stores than foragers and isolated worker bees have lower lipid levels than bees kept in a colony, regardless of food availability (Toth and Robinson 2005; Toth et al. 2005); thus pheromones may partly exert their effects by regulating workers' nutrient storage. Moreover, among worker proteins, vitellogenin (Vg), an egg yolk protein, is produced in higher levels by the fat bodies of nurse bees than forager bees (Fluri et al. 1982) and thus can serve as a molecular marker for the nurselike physiological state.

In an experiment by Fischer and Grozinger (2008), the administration of QMP on caged workers increased protein and Vg level in the fat bodies. According to the authors, this effect could be achieved by behavioral, physiological, or molecular mechanisms: QMP modulates feeding behavior, inducing treated bees to consume more food or to reduce activity; it decreases level of JH, which is known to increase metabolism (Sullivan et al. 2003), and this reduction in turn increases Vg levels and potential lipid storage; finally, it can modulate metabolic pathways through regulation of the genes involved in the insulin signaling pathway, which is associated with nutrient storage (Fischer and Grozinger 2008).

A confirmation of the role of pheromones in regulating worker metabolism comes from the researches of Smedal et al. (2009), which demonstrated that BP also regulates Vg accumulation in the fat body. Beside its role in oogenesis Vg is utilized by workers for food production and is involved in the regulation of foraging behavior and the enhancing of worker lifespan, possibly by scavenging free radicals and enhancing honey bee immunity (Amdam et al. 2003, 2004, 2005; Nelson et al. 2007; Seehuus et al. 2006). Exposure to synthetic BP blend causes an increase in the amount of Vg in the fat bodies of young bees (3–4 days old) and a decrease in older workers (23–24 days old). This is consistent with the results of Pankiw et al. (2008), who showed that brood pheromone stimulates pollen consumption, leading to an increase of protein content in hypopharyngeal glands, but also showed that workers of different ages are affected in an opposite manner by the pheromone, confirming the differential perception of pheromones according to worker age and task. In this case brood pheromone acts on young workers by enhancing their capacity to produce brood food and to store a surplus from Vg synthesis, and on older workers by inhibiting an extensive Vg storage, ensuring that more protein remains free in the hemolymph to be converted into brood food (Smedal et al. 2009).

5.2.3 FROM SIGNAL TO BEHAVIOR: PHEROMONES AND GENE EXPRESSION

Pheromone processing in the bee brain leads to neurophysiological changes that result in the production of a specific behavior or to changes in sensory thresholds that result in altered behavior under different contexts. In either case, the molecular mechanisms by which pheromones are transduced in the brain to influence behavior are only beginning to be understood. A great breakthrough was made with the completion of the honey bee genome (Honey Bee Genome Sequencing Consortium 2006) and the development of a genome-wide honey bee microarray, which enabled to search for differences associated with variation in responsiveness to pheromones.

A number of authors found that worker division of labor is based, in addition to the already mentioned age and environmental factors, on genetic differences among workers performing different tasks. Thus the probability of performing a particular task within a specific age caste would be determined not only by the endogenous and exogenous environment, but also by the genotype of the worker (Calderone and Page 1988; Frumhoff and Baker 1988). These genetic differences could influence, for example, the probability of a worker to become a guard, a nectar forager, a pollen forager, or a nest-site scout (Robinson and Page 1988, 1989).

The natural variation in honey bee pheromone response, observed by several authors (Pankiw et al. 1994) may be potentially adaptive, because it creates variability in task performance that supports colony plasticity and thus productivity. Kocher et al. (2010) found variability in worker attraction to QMP and consequently in the retinue response of adult workers, which appears to be associated with brain gene expression patterns and linked to the reproductive potential in honey bees. The authors found 960 gene transcripts that are differentially expressed between high and low responder workers, and a negative correlation between individual retinue response and ovariole number, a trait strongly linked to reproductive potential (Makert et al. 2006). This indicates that workers with the highest reproductive potential (e.g., the greatest number of ovarioles) avoid the queen, while those with lower reproductive potential are attracted to her. Under queenless condition workers with high reproductive potential would activate their ovaries, whereas the ones with low reproductive potential would be in charge of rearing a new queen (Kocher et al. 2010). This would confirm the observations by Moritz et al. (2002) that in *A.m. capensis*, workers that are likely to become reproductively active are indeed more likely to avoid the queen.

One way that a pheromone can influence behavior is by orchestrating large-scale changes in brain gene expression. In recent years several authors demonstrated that a differential gene expression exists between workers performing different tasks (Whitfield et al. 2003, 2006) and that exposure of honey bee workers to pheromones causes changes in brain gene expression that are associated with downstream changes in behavior. Therefore, it should be possible to investigate the mode of action of pheromones by correlating the changes in gene expression and the resulting behavioral expression. The first attempt in this direction was made by Grozinger et al. (2003) with QMP.

5.2.3.1 Insights into the Pheromone-Mediated Genetic Mechanism Underlying Worker Behavioral Development

We know that QMP has a delaying effect on the transition from hive tasks to foraging in workers. Several genes have been identified as correlated to nursing or foraging conditions (Whitfield et al. 2003) and the exposure of young honey bee workers to QMP was found to activate genes associated with nursing and to repress genes associated with foragers. In the study by Grozinger et al. (2003) the gene that was more robustly and chronically regulated was found to be an ortholog of the *Drosophila* transcription factor *krüppel homolog 1* (*Kr-h1*). This gene encodes for a zinc finger transcription factor that plays an important role in orchestrating development and cell differentiation. Although the different components of QMP taken individually were thought to elicit limited responses, two of them, 9-HDA and 9-ODA, were both able to produce a strong QMP-like gene activation. In particular, they were able to downregulate expression of *Kr-h1*, suggesting that 9-ODA and 9-HDA are the QMP components that influence the timing of the transition from hive work to foraging (Grozinger et al. 2007).

From all the reported observations about the role of pheromones in modulating worker behavior, it is interesting to investigate the functional relation between QMP, which regulates the transition from nurse to forager, and OA and JH, which have levels with strong correlation to these behavioral stages. Grozinger and Robinson (2007) studied the effects of these three factors on the modulation of the gene *Kr-h1*. JH analog, methoprene, or OA are unable, alone, to modulate *Kr-h1* expression, demonstrating that these molecules do not have a direct influence on the gene expression. Conversely, methoprene, but not OA, significant reduces the effect of QMP on *Kr-h1* brain expression in young bees, suggesting that high JH titers, typical of foragers, prevent downregulation of *Kr-h1* expression by QMP in older bees (Grozinger and Robinson 2007). The authors' interpretation is that QMP affects workers' transition to foragers partly via JH regulation, since the pheromone is able to lower JH levels, and JH levels in turn modulate pheromone response, but other mechanisms must be involved, since a JH analog is not able to affect gene (namely *Kr-h1*) expression.

Together with QMP, BP is responsible for the regulation of worker behavioral development, delaying the transition of workers from nurses to foragers. Its way of action seems to be even more complex than QMP, since it has a dose- and age-dependent effect, and in addition to a primer effect on behavioral maturation, it acts as a releaser, stimulating the foraging activity of older bees that are competent to forage (Pankiw 2004c; Pankiw and Page 2001).

Alaux et al. (2009) showed that BP effect on foraging ontogeny is linked to a variation in gene expression, since BP treatment upregulates brain genes that are highly expressed in workers specialized in brood care, and downregulates genes that are highly expressed in foragers. According to its age-related effect, the exposure to BP for 5 days caused a brain gene expression profile similar to the profile of nurse bees, while this similarity to nurse bees was absent in bees exposed to BP for 15 days. In fact, although there was a significant overlap between the gene sets controlled by BP in young and old bees, many were regulated in opposite directions. For example, the gene *malvolio* (*mvl*), which is activated in precocious foragers (Ben-Shahar et al. 2004), was upregulated by BP in 15-day-old bees, but not in 5-day-old bees, suggesting that *mvl* represents a key component of the regulation of foraging behavior by BP. This differential effect on brain gene expression of 15-day-old bees is consistent with the role of BP as releaser pheromone, triggering foraging behavior in older bees (Alaux et al. 2009).

Comparing these results with those obtained by Grozinger et al. (2003) with QMP, which exerts a similar effect on behavioral development, it emerges that some genes are regulated by both BP and QMP, probably because of the different chemical composition of the two pheromones, which are also found to use different peripheral receptors (Robertson and Wanner 2006; Wanner et al. 2007).

5.2.3.2 Alarm Pheromone and the Expression of Immediate Early Genes

It has been demonstrated that primer pheromones exert their effects partly by causing changes in brain gene expression (Alaux et al. 2009; Grozinger et al. 2003). Releaser pheromones, which cause immediate and short-term responses, are thought to act through more direct neurophysiologic modulation systems. Today, the study of the mode of action of these two kinds of pheromones has changed this rigid distinction. For example, the exposure of honey bee colonies to IPA, originally classified as a releaser pheromone, caused a significant increase in expression of the gene *c-Jun* (Alaux and Robinson 2007). *C-Jun* belongs to the group of immediate early genes (IEGs), which are activated transiently and rapidly in response to a wide variety of stimuli. They are activated at the transcription level before any new proteins are synthesized and are known as early regulators of cell growth and differentiation signals, but are also involved in synaptic plasticity. The correlation between IPA exposure and *c-Jun* expression in honey bees blurs the long-standing distinction between primer and releaser pheromones and highlights the importance of brain gene expression in social regulation (Robinson et al. 2005).

5.3 CONCLUDING REMARKS

If one compares the very high number of pheromonal substances identified in the honey bee colony with the relative scarcity of uncovered physiological mechanisms subtending pheromonal effects, it clearly emerges that there is still much study required to fully understand the pathways from pheromone production to pheromonal output.

Pheromone processing starts in the peripheral sensory system where the chemical signal is transduced, and partially elaborated in the glomeruli of the antennal lobes (Figure 5.6).

The pheromonal signal is probably elaborated in the mushroom bodies and in the lateral horn during its transmission from the neuronal fibers leaving the antennal lobes to the central nervous system. At this level, the outcome of the signal is modulated by biogenic amines, which act as neurotransmitters and neuromodulators in several neuronal functions, and whose fibers are well represented in these two

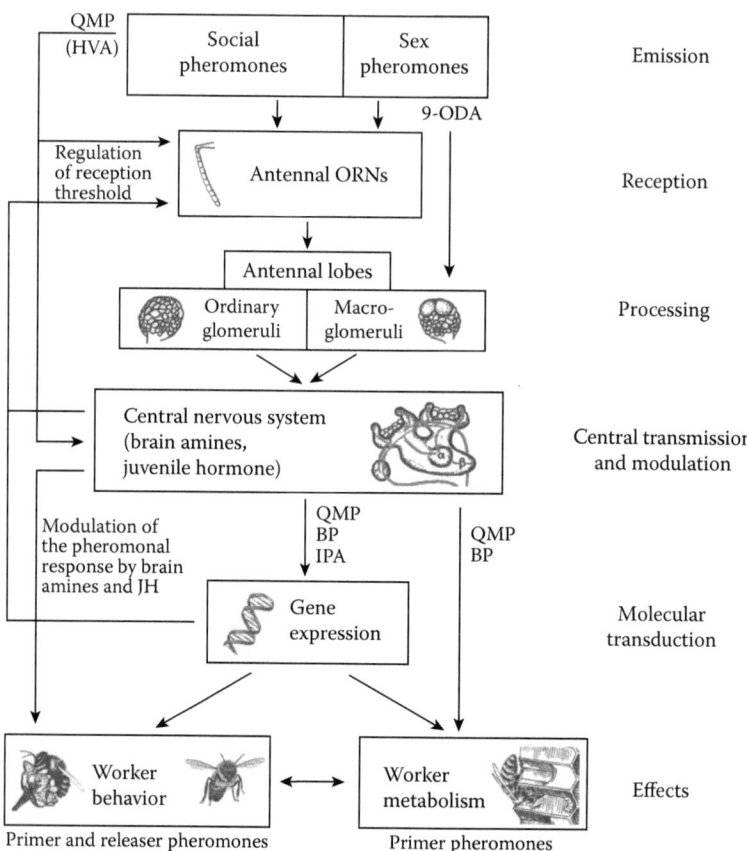

FIGURE 5.6 Schematic representation of the pheromone processing pathway from its reception to its behavioral and physiological effects. Pheromones are transduced by the antennal odorant reception neurons (ORNs) and processed in the antennal lobes, respectively, in the ordinary glomeruli for social pheromones and in the macroglomeruli for sexual pheromones (9-ODA). The signal is transmitted to the central nervous system, mainly to the mushroom bodies and the lateral horns, where it influences the level of brain amines (especially octopamine [OA]) and juvenile hormone (JH). In turn, the levels of brain amines and JH modulate pheromonal response by regulating the reception threshold at peripheral and central level. Homovanillyl alcohol (HVA), one of the components of QMP, is able to mimic dopamine and thus acts directly on the dopaminergic pathways in the central nervous system, but also at the peripheral level on DA and OA antennal receptors. At the molecular level some pheromones were found to act on gene expression (QMP, BP, and IPA), others on worker metabolism (primer pheromones like QMP and BP), while for all of them the final effect is the regulation of worker behavior.

neuropils. The precise role of biogenic amines in the transmission of pheromonal signals has not been clearly elucidated, but it is quite certain that they are involved in the process of worker behavioral development, which is triggered by the queen pheromones. This is confirmed by the fact that one component of QMP, HVA, which shares a similar chemical structure with DA, is able to skip the reception step in antennal lobes and directly affect worker behavior by modulating DA neuronal pathway and DA levels in the brain (Figure 5.6).

The recent advances in genomics have strongly contributed to understanding the mechanisms that regulate pheromone communication by showing a direct correlation between pheromonal signal and expression of genes involved in worker behavior. This was shown in particular for two primer pheromones, QMP and BP, but also for a releaser one, the sting alarm pheromone, thus questioning the old distinction between primer and releaser pheromones, at least for their operating mechanism (Figure 5.6).

In 1998, Winston and Slessor, in their article "Honey bee primer pheromones and colony organization: gaps in our knowledge," stated that "we know a considerable amount about the functions in which these pheromones are active, but the modes of action, physiological effects and precise chemical nature of specific pheromonal activities remain as subjects for future research". Today, after 15 years of research, we can say that several of these gaps have been filled, but still our understanding of neuronal and molecular mechanisms in pheromone processing is represented by separate pieces of an extremely more complex puzzle. A complete comprehension of the mechanism of pheromone communication in honey bees needs to put all the pieces together in an organic and all-inclusive picture that is able to display the complex routes and the multiple connections of this sophisticated chemical communication system.

REFERENCES

Alaux, C. and Robinson, G.E. (2007) Alarm pheromone induces immediate–Early gene expression and slow behavioral response in honey bees. *J Chem Ecol* 33: 1346–50.

Alaux, C., Le Conte, Y., Adams, H.A., Rodriguez-Zas, S., Grozinger, C.M., Sinha S. and Robinson, G.E. (2009) Regulation of brain gene expression in honey bees by brood pheromone. *Gene Brain Behav* 8: 309–19.

Ali, M.F. and Morgan, E.D. (1990) Chemical communication in insect communities: A guide to insect pheromones with special emphasis on social insects. *Biol Rev Camb Philos* 65(3): 227–47.

Al-Kahtani, S.N. and Bienefeld, K. (2012) The Nasonov gland pheromone is involved in recruiting honey bee workers for individual larvae to be reared as queens. *J Insect Behav* 25: 392–400.

Allan, S.A., Slessor, K.N., Winston, M.L. and King, G.G.S. (1987) The influence of age and task specialization on the production and perception of honey bee pheromones. *J Insect Physiol* 33: 917–22.

Amdam, G.V., Aase, A.L.T.O., Seehuus, S.C., Norberg, K., Hartfelder, K. and Fondrk, M.K. (2005) Social reversal of immunosenescence in honey bee workers. *Exp Gerontol* 40: 939–47.

Amdam, G.V., Norberg, K., Hagen, A. and Omholt, S.W. (2003) Social exploitation of vitellogenin. *Proc Natl Acad Sci U S A* 100: 1799–802.

Amdam, G.V., Rueppell, O., Fondrk, M.K., Page, R.E. and Nelson, C.M. (2009) The nurse's load: Early-life exposure to brood-rearing affects behavior and lifespan in honey bees (*Apis mellifera*). *Exp Gerontol* 44: 467–71.

Amdam, G.V., Simões, Z.L.P., Hagen, A., Norberg, K., Schrøder, K., Mikkelsen, O., Kirkwood, T.B.L. and Omholt, S.W. (2004) Hormonal control of the yolk precursor vitellogenin regulates immune function and longevity in honey bees. *Exp Gerontol* 39: 767–73.

Arechavaleta-Velasco, M.E. and Hunt, G.J. (2003) Genotypic variation in the expression of guarding behavior and the role of guards in the defensive response of honey bee colonies. *Apidologie* 34: 439–47.

Arenas, A., Giurfa, M., Farina, W.M. and Sandoz, J.-C. (2009) Early olfactory experience modifies neural activity in the antennal lobe of a social insect at the adult stage. *Eur J Neurosci* 30: 1498–508.

Arnhart, L. (1923) Das krallenglied der honigbiene. *Arch Bienenk* 5: 37–86.

Arnold, G., Masson, C. and Budharugsa, S. (1985) Comparative study of the antennal lobes and their afferent pathways in the worker bee and the drone (*Apis mellifera*). *Cell Tissue Res* 242: 593–605.

Balderrama, N., Nuñez, J., Guerrieri, F. and Giurfa, M. (2002) Different functions of two alarm substances in the honey bee. *J Comp Physiol A* 188: 485–91.

Barbier, J. and Lederer, E. (1960) Structure chimique de la substance royale de la reine d'abeille (*Apis mellifera* L.). *C R Acad Sci Ser III Sci Vie* 251: 1131–35.

Barron, A.B. and Robinson, G.E. (2005) Selective modulation of task performance by octopamine in honey bee (*Apis mellifera*) division of labour. *J Comp Physiol A* 191: 659–68.

Barron, A.B., D.J. Schulz and G.E. Robinson (2002) Octopamine modulates responsiveness to foraging-related stimuli in honey bees (*Apis mellifera*). *J Comp Physiol A* 188: 603–10.

Beggs, K.T. and Mercer, A.R. (2009) Dopamine receptor activation by honey bee queen pheromone. *Curr Biol* 19: 1206–9.

Beggs, K.T., Glendining, K.A., Marechal, N.M., Vergoz, V., Nakamura, I., Slessor, K.N. and Mercer, A.R. (2007) Queen pheromone modulates brain dopamine function in worker honey bees. *Proc Natl Acad Sci U S A* 104: 2460–64.

Ben-Shahar, Y., Dudek, N.L. and Robinson, G.E. (2004) Phenotypic deconstruction reveals involvement of manganese transporter *malvolio* in honey bee division of labor. *J Exp Biol* 207: 3281–88.

Bicker, G. (1999) Biogenic amines in the brain of the honey bee: Cellular distribution, development, and behavioral functions. *Microsc Res Techniq* 44(2–3): 166–78.

Billen, J.P.J. (1987) New structural aspects of the Dufour's and venom glands in social insects. *Naturwissenschaften* 74: 340–41.

Billen, J.P.J. (1994) Morphology of exocrine glands in social insects: An update 100 years after Ch. Janet. In: *Les Insectes Sociaux* (A. Lenoir, G. Arnold and M. Lepage, eds.) Paris: Publication Université de Paris Nord.

Billen, J.P.J. and Morgan, E.D. (1998) Pheromones communication in social insects: Sources and secretions. In: *Pheromone Communication in Social Insects: Ants, Wasps, Bees, and Termites* (R.K. Vander Meer, M.D. Breed, K.E. Espelie and M.L. Winston, eds.), pp. 3–33. Boulder, CO: Westview Press.

Billen, J.P.J., Dumortier, K.T.M. and Velthuis, H.H.W. (1986) Plasticity of honey bee castes: Occurrence of tergal glands in workers. *Naturwissenschaften* 73: 332–33.

Bloch, G. and Grozinger C.M. (2011) Social molecular pathways and the evolution of bee societies. *Philos T Roy Soc B* 366: 2155–70.

Blum, M.S., Fales, H.M., Tucker, K.W. and Collins, A.M. (1978) Chemistry of the sting apparatus of the worker honey bee. *J Apicult Res* 17: 218–21.

Boch, R. and Shearer, D.A. (1966) Iso-pentyl acetate in stings of honey bees of different ages. *J Apicult Res* 5: 65–70.

Boch, R. and Shearer, D.A. (1967) 2-heptanone and 10-hydroxy-trans-dec-2-enoic acid in the mandibular glands of worker honey bees of different ages. *Z Vergl Physiol* 54: 1–11.

Boch, R., Shearer, D.A. and Petrasovits, A. (1970) Efficacies of two alarm substances of the honey bee. *J Insect Physiol* 16: 17–24.

Boch, R., Shearer, D.A. and Stone, B.C. (1962) Identification of iso-amyl acetate as an active component in the sting pheromone of the honey bees. *Nature* 195: 1018–20.

Breed, M.D. (1998) Recognition pheromones of the honey bee. *Bioscience* 48: 463–70.

Breed, M.D., Guzman-Novoa, E. and Hunt, G.J. (2004a) Defensive behavior of honey bees: Organization, genetics, and comparisons with other bees. *Annu Rev Entomol* 49: 271–98.

Breed, M.D., Leger, E.A., Pearce, A.N. and Wang, Y.J. (1998) Comb effects on the ontogeny of honey bee nestmate recognition. *Anim Behav* 55: 13–20.

Breed, M.D., Perry, S. and Bjostad, L.B. (2004b) Testing the blank slate hypothesis: Why honey bee colonies accept young bees. *Insect Soc* 51: 12–16.

Breed, M.D., Robinson, G.E. and Page, R.E. (1990) Division of labor during honey bee colony defense. *Behav Ecol Sociobiol* 27: 395–401.

Breed, M.D., Smith, T.A. and Torres, A. (1992) Role of guard honey bees (Hymenoptera: Apidae) in nestmate discrimination and replacement of removed guards. *Ann Entomol Soc Am* 85: 633–37.

Briand, L., Swasdipan, N., Nespoulous, C., Bézirard, V., Blon, F., Huet, J.-C., Ebert, P. and Pernollet, J.-C. (2002) Characterization of a chemosensory protein (ASP3c) from honey bee (*Apis mellifera* L.) as a brood pheromone carrier. *Eur J Biochem* 269: 4586–96.

Brockmann, A. and Brueckner, D. (1995) Projection pattern of poreplate sensory neurones in honey bee worker, *Apis mellifera* L. (Hymenoptera: Apidae). *Int J Insect Morphol* 24(4): 405–11.

Brockmann, A. and Brueckner, D. (2001) Structural differences in the drone olfactory system of two phylogenetically distant *Apis* species, *A. florea* and *A. mellifera*. *Naturwissenschaften* 88: 78–81.

Brockmann, A., Brückner, D. and Crewe, R. (1998) The EAG response spectra of workers and drones to queen honey bee mandibular gland components: The evolution of a social signal. *Naturwissenschaften* 85: 283–85.

Brockmann, A., Dietz, D., Spaethe, J. and Tautz, J. (2006) Beyond 9-ODA: Sex pheromone communication in the European honey bee *Apis mellifera* L. *J Chem Ecol* 32(3): 657–67.

Brown, S.M., Napper, R.M. and Mercer, A.R. (2004) Foraging experience, glomerulus volume, and synapse number: A stereological study of the honey bee antennal lobe. *J Neurobiol* 60(1): 40–50.

Butler, C.G. and Fairey, E.M. (1963) The role of the queen in preventing oogenesis in worker honey bees. *J Apicult Res* 2: 14–18.

Calderone, N.W. and Page, R.E. (1988) Genotype variability in age polyethism and task specialization in the honey bee, *Apis mellifera* (Hymenoptera: Apidae). *Behav Ecol Sociobiol* 22: 17–25.

Callow, R.K. and Johnston, N.C. (1960) The chemical constitution and synthesis of queen substance of honey bees (*Apis mellifera*). *Bee World* 41: 152–53.

Carcaud, J., Roussel, E., Giurfa, M. and Sandoz, J.-C. (2009) Odour aversion after olfactory conditioning of the sting extension reflex in honey bees. *J Exp Biol* 212: 620–26.

Cassier, P. and Lensky, Y. (1994) The Nassanov gland of the workers of the honey bee (*Apis mellifera* L.): Ultrastructure and behavioural function of the terpenoid and protein component. *J Insect Physiol* 40: 577–84.

Cassier, P., Tel-Zur, D. and Lensky, Y. (1994) The sting sheaths of honey bee workers (*Apis mellifera* L.): Structure and alarm pheromone secretion. *J Insect Physiol* 40: 23–32.

Castillo, C., Chen, H., Graves, C., Maisonnasse, A., Le Conte, Y. and Plettner, E. (2012) Biosynthesis of ethyl oleate, a primer pheromone, in the honey bee (*Apis mellifera* L.). *Insect Biochem Molec* 42: 404–16.

Chaline, N., Sandoz, J.-C., Martin, S.J., Ratnieks, F.L.V. and Jones, G.R. (2005) Learning and discrimination of individual cuticular hydrocarbons by honey bees (*Apis mellifera*). *Chem Senses* 30: 327–35.

Collins, A.M. (1980) Effect of age on the response to alarm pheromones by caged honey bees. *Ann Entomol Soc Am* 73: 307–9.

Collins, A.M. and Blum, M.S. (1982) Bioassay of compounds derived from the honey bee sting. *J Chem Ecol* 8(2): 463–70.

Collins, A.M., Rinderer, T.E., Daly, H.V., Harbo, J.R. and Pesante, D. (1989) Alarm phero-
  mone production by two honey bee (*Apis mellifera*) types. *J Chem Ecol* 15: 1747–56.
Collins, A.M., Rinderer, T.E., Harbo, J.R. and Bolten, A.B. (1982) Colony defense by
  Africanized and European honey bees. *Science* 218: 72–74.
Contessi, A. (2004) *Le api: Biologia, allevamento, prodotti*. Bologna, Italy: Edagricole.
Crane, E. (1990) *Bees and Beekeeping: Science, Practice and World Resources*. Ithaca, NY:
  Cornell University Press.
Crewe, R.M. and Velthuis, H.H.W. (1980) False queens: A consequence of mandibular gland
  signals in worker honey bees. *Naturwissenschaften* 67: 467–69.
Dani, F.R., Iovinella, I., Felicioli, A., Niccolini, A., Calvello, M.A., Carucci, M.G., Qiao, H.,
  Pieraccini, G., Turillazzi, S., Moneti, G. and Pelosi, P. (2010) Mapping the expression of
  soluble olfactory proteins in the honey bee. *J Proteome Res* 9: 1822–33.
Dani, F.R., Jones, G.R., Corsi, S., Beard, R., Pradella, D. and Turillazzi, S. (2005) Nestmate
  recognition cues in the honey bee: Differential importance of cuticular alkanes and
  alkenes. *Chem Senses* 30: 477–89.
Danty, E., Arnold, G., Huet, J.-C., Huet, D., Masson, C. and Pernollet, J.-C. (1998) Separation,
  characterization and sexual heterogeneity of multiple putative odorant-binding proteins
  in the honey bee *Apis mellifera* L. (Hymenoptera: Apidea). *Chem Senses* 23: 83–91.
Danty, E., Briand, L., Michard-Vanhée, C., Perez, V., Arnold, G., Gaudemer, O., Huet, D.,
  Huet, J.-C., Ouali, C., Masson, C. and Pernollet, J.-C. (1999) Cloning and expression
  of a queen pheromone-binding protein in the honey bee: An olfactory-specific, develop-
  mentally regulated protein. *J Neurosci* 19: 7468–75.
Danty, E., Michard-Vanhée, C., Huet, J.-C., Genecque, E., Pernollet, J.-C. and Masson, C.
  (1997) Biochemical characterization, molecular cloning and localization of a putative
  odorant-binding protein in the honey bee *Apis mellifera* L. (Hymenoptera: Apidea).
  FEBS Lett 414: 595–98.
de Groot, A.P. and Voogd, S. (1954) On the ovary development in queenless worker bees (*Apis
  mellifera* L.). *Experientia* 10: 384–85.
De Hazan, M., Lensky, Y. and Cassier, P. (1989) Effect of queen honey bee (*Apis mellifera*)
  ageing on her attractiveness to workers. *Comp Biochem Physiol* 93A: 777–83.
De Jong, R. and Pham-Delègue, M.H. (1991) Electroantennogram responses related to olfactory conditioning in the honey bee (*Apis mellifera ligustica*). *J Insect Physiol* 37: 319–24.
Deisig, N., Giurfa, M., Lachnit, H. and Sandoz, J.-C. (2006) Neural representation of olfactory
  mixtures in the honey bee antennal lobe. *Eur J Neurosci* 24: 1161–74.
Dombroski, T., Simões, Z. and Bitondi, M. (2003) Dietary dopamine causes ovary activation
  in queenless *Apis mellifera* workers. *Apidologie* 34: 281–89.
Dor, R., Katzav-Gozansky, T. and Hefetz, A. (2005) Dufour's gland pheromone as a reliable
  fertility signal among honey bee (*Apis mellifera*) workers. *Behav Ecol Sociobiol* 58:
  270–76.
Downs, S.G. and Ratnieks, F.L.W. (1999) Recognition of conspecifics by honey bee guards
  (*Apis mellifera*) uses non-heritable cues applied in the adult stage. *Anim Behav* 58:
  643–48.
Downs, S.G., Ratnieks, F.L., Badcock, N. and Mynott, A. (2001) Honey bee guards do not
  use food derived odours to recognise non-nestmates: A test of the odour convergence
  hypothesis. *Behav Ecol* 12: 47–50.
Downs, S.G., Ratnieks, F.L., Jefferies, S.L. and Rigby, H.E. (2000) The role of floral oils in the
  nestmate recognition system of honey bees (*Apis mellifera* L.). *Apidologie* 31: 357–65.
Engels, W., Rosenkranz, P., Adler, A., Taghizadeh, T., Lübke, G. and Francke, W. (1997)
  Mandibular gland volatiles and their ontogenetic patterns in queen honey bees, *Apis
  mellifera carnica*. *J Insect Physiol* 43(4): 307–13.
Espelie, K.E., Butz, V.M. and Dietz, A. (1990) Decyl decanoate: a major component of the
  tergite glands of honey bee queens (*Apis mellifera* L.). *J Apicult Res* 29: 15–19.

Esslen, J. and Kaissling, K.E. (1976) Zahl und verteilung antennaler sensillen bei der honigbiene (*Apis mellifera* L.). *Zoomorphologie* 83: 227–51.

Fan, Y., Richard, F.-J., Rouf, N. and Grozinger, C.M. (2010) Effects of queen mandibular pheromone on nestmate recognition in worker honey bees, *Apis mellifera. Anim Behav* 79: 649–56.

Farooqui, T., Robinson, K., Vaessin, H. and Smith, B.H. (2003) Modulation of early olfactory processing by an octopaminergic reinforcement pathway in the honey bee. *J Neurosci* 23: 5370–80.

Farris, S.M., Robinson, G.E. and Fahrbach, S.E. (2001) Experience- and age-related outgrowth of intrinsic neurons in the mushroom bodies of the adult worker honey bee. *J Neurosci* 21(16): 6395–404.

Ferguson, A.W. and Free, J.B. (1981) Factors determining the release of Nasonov pheromone by honey bees at the hive entrance. *Physiol Entomol* 6(1): 15–19.

Fernandez, P.C., Gil, M. and Farina W.M. (2003) Reward rate and forager activation in honey bees: Recruiting mechanisms and temporal distribution of arrivals. *Behav Ecol Sociobiol* 54: 80–87.

Fischer, P. and Grozinger, C.M. (2008) Pheromonal regulation of starvation resistance in honey bee workers (*Apis mellifera*). *Naturwissenschaften* 95: 723–29.

Flanagan, D. and Mercer, A.R. (1989) An atlas and 3-D reconstruction of the antennal lobes in the worker honey bee, *Apis mellifera* L. (Hymenoptera: Apidae). *Int J Insect Morphol* 18: 145–59.

Fluri, P., Luscher, M., Wille, H. and Gerig, L. (1982) Changes in the weight of the pharyngeal gland and haemolymph titres of juvenile hormone, protein and vitellogenin in worker honey bees. *J Insect Physiol* 28: 61–68.

Forêt, S. and Maleska, R. (2006) Function and evolution of a gene family encoding odorant binding-like proteins in a social insect, the honey bee (*Apis mellifera*). *Genome Res* 16: 1404–13.

Francis, B.R., Blanton, W.E., Littlefield, J.L. and Nunamaker, R.A. (1989) Hydrocarbons of the cuticle and hemolymph of the adult honey bee (Hymenoptera: Apidae). *Ann Entomol Soc Am* 82(4): 486–94.

Free, J.B. (1957) The food of adult drone honey bees (*Apis mellifera*). *J Anim Behav* 5: 7–11.

Free, J.B. (1987) *Pheromones of Social Bees*. London: Chapman and Hall.

Free, J.B. and Simpson, J. (1968) The alerting pheromones of the honey bee. *Z Vergl Physiol* 61: 361–65.

Free, J.B., Ferguson, A.W. and Pickett, J.A. (1981a) Evaluation of the various components of the Nasonov pheromone used by clustering honey bees. *Physiol Entomol* 6(3): 263–68.

Free, J.B., Pickett, J.A., Ferguson, A.W. and Smith, M.C. (1981b) Synthetic pheromones to attract honey bee (Apismellifera) swarms. *J Agr Sci* 97: 427–31.

Free, J.B., Ferguson, A.W., Simpkins, J.R. and Al Sa'ad, B.N. (1983) Effect of honey bee Nasonov and alarm pheromone components on behaviour at the nest entrance. *J Apicult Res* 22(4): 214–23.

Frilli, F., Barbattini, R. and Milani, N. (2001) *L'ape: Forme e funzioni*. Bologna, Italy: Calderini, Edagricole.

Frumhoff, P.C. and Baker, J. (1988) A genetic component to division of labour within honey bee colonies. *Nature* 33: 358–61.

Galizia, C.G. and Menzel, R. (2001) The role of glomeruli in the neural representation of odours: Results from optical recording studies. *J Insect Physiol* 47: 115–30.

Galizia, C.G. and Vetter, R. (2005) Optical methods for analyzing odor-evoked activity in the insect brain. In: *Advances in Insect Sensory Neuroscience* (T.A. Christensen, ed.), pp. 349–92. Boca Raton, FL: CRC Press.

Galizia, C.G., Joerges, J., Kuttner, A., Faber, T. and Menzel, R. (1997) A semi-in-vivo preparation for optical recording of the insect brain. *J Neurosci Meth* 76: 61–69.

Galizia, C.G., Nägler, K., Hölldobler, B. and Menzel, R. (1998) Odour coding is bilaterally symmetrical in the antennal lobes of honey bees (*Apis mellifera*). *Eur J Neurosci* 10: 2964–74.

Galizia, C.G., Sachse, S., Rappert, A. and Menzel, R. (1999) The glomerular code for odor representation is species-specific in the honey bee *Apis mellifera*. *Nat Neurosci* 2: 473–78.

Gary, N.E. (1962) Chemical mating attractants in the queen honey bee. *Science* 136: 773–74.

Gary, N.E. and Marston, J. (1971) Mating behavior of drone honey bees with queen models (*Apis mellifera* L.). *Anim Behav* 19: 299–304.

Gauthier, M. and Grünewald, B. (2013) Neurotransmitter systems in the honey bee brain: Functions in learning and memory. In: *Honeybee Neurobiology and Behavior. A Tribute to Randolf Menzel* (C.G. Galizia, D. Eisenhardt and M. Giurfa, eds.), pp. 155–69. Dordrecht, Netherlands: Springer.

Gervan, N., Winston, M., Higo, H. and Hoover, S. (2005) The effects of honey bee (*Apis mellifera*) queen mandibular pheromone on colony defensive behaviour. *J Apicult Res* 44: 175–79.

Giurfa, M. (1993) The repellent scent mark of the honey bee *Apis mellifera ligustica* and its role as communication cue during foraging. *Insect Soc* 40: 59–67.

Giurfa, M. (2007) Behavioral and neural analysis of associative learning in the honey bee: A taste from the magic well. *J Comp Physiol A* 193: 801–24.

Goodman, L. (2003) *Form and Function in the Honey Bee*. Cardiff, UK: IBRA.

Grandperrin, D. and Cassier, P. (1983) Anatomy and ultrastructure of the Koschewnikow's gland of the honey bee, *Apis mellifera* L. (Hymenoptera: Apidae). *Int J Insect Morphol* 12(1): 25–42.

Grozinger, C. and Robinson, G. (2007) Endocrine modulation of a pheromone-responsive gene in the honey bee brain. *J Comp Physiol A* 193(4): 461–70.

Grozinger, C.M., Fischer, P. and Hampton, J.E. (2007) Uncoupling primer and releaser responses to pheromone in honey bees. *Naturwissenschaften* 94(5): 375–79.

Grozinger, C.M., Sharabash, N.M., Whitfield, C.W. and Robinson, G.E. (2003) Pheromone-mediated gene expression in the honey bee brain. *Proc Natl Acad Sci U S A* 100(2): 14519–25.

Hammer, M. (1993) An identified neurone mediates the unconditioned stimulus in associative olfactory learning in honey bees. *Nature* 366: 59–63.

Hammer, M. and Menzel, R. (1998) Multiple sites of associative odor learning as revealed by local brain microinjections of octopamine in honey bees. *Learn Mem* 5: 146–56.

Hansson, B.S. and Anton, S. (2000) Function and morphology of the antennal lobe: New developments. *Annu Rev Entomol* 45: 203–31.

Harris, J. and Woodring, J. (1999) Effects of dietary precursors to biogenic amines on the behavioural response from groups of caged worker honey bees (*Apis mellifera*) to the alarm pheromone component isopentyl acetate. *Physiol Entomol* 24: 285–91.

Hartfelder, K. (2000) Insect juvenile hormone: From "status quo" to high society. *Braz J Med Biol Res* 33(2): 157–77.

Higo, H.A., Colley, S.J., Winston, M.L. and Slessor, K.N. (1992) Effects of honey bee (*Apis mellifera*) queen mandibular gland pheromone on foraging and brood rearing. *Can Entomol* 124: 409–18.

Honey Bee Genome Sequencing Consortium (2006) Insights into social insects from the genome of the honey bee *Apis mellifera*. *Nature* 443: 931–49.

Hoover, S.E.R., Keeling, C.I., Winston, M.L. and Slessor, K.N. (2003) The effect of queen pheromones on worker honey bee ovary development. *Naturwissenschaften* 90: 477–80.

Hoover, S.E.R., Winston, M.L. and Oldroyd, B.P. (2005) Retinue attraction and ovary activation: Responses of wild type and anarchistic honey bees (*Apis mellifera*) to queen and brood pheromones. *Behav Ecol Sociobiol* 59: 278–84.

Huang, Z.-Y. and Robinson, G.E. (1992) Honey bee colony integration: Worker-worker interactions mediate hormonally regulated plasticity in division of labor. *Proc Natl Acad Sci U S A* 89: 11726–29.

Huang, Z.-Y. and Robinson, G.E. (1996) Regulation of honey bee division of labor by colony age demography. *Behav Ecol Sociobiol* 39: 147–58.

Huang, Z.-Y., Plettner, E. and Robinson, G.E. (1998) Effect of social environment and mandibular gland removal on division of labor in worker honey bees. *J Comp Physiol A* 183: 143–52.

Huang, Z.-Y., Robinson, G.E., Tobe, S.S., Yagi, K.J., Strambi, C., Strambi, A. and Stay, B. (1991) Hormonal regulation of behavioural development in the honey bee is based on changes in the rate of juvenile hormone biosynthesis. *J Insect Physiol* 37: 733–41.

Humphries, M.A., Mustard, J.A., Hunter, S.J., Mercer, A., Ward, V. and Ebert, P.R. (2003) Invertebrate D2 type dopamine receptor exhibits age-based plasticity of expression in the mushroom bodies of the honey bee brain. *J Neurobiol* 55: 315–30.

Hunt, G.J. (2007) Flight and fight: A comparative view of the neurophysiology and genetics of honey bee defensive behavior. *J Insect Physiol* 53: 399–410.

Hunt, G.J., Wood, K.V., Guzmán-Novoa, E., Lee, H.D., Rothwell, A.P. and Bonham, C.C. (2003) Discovery of 3-methyl-2-buten-1-yl acetate, a new alarm component in the sting apparatus of Africanized honey bees. *J Chem Ecol* 29: 451–61.

Iovinella, I., Dani, F.R., Niccolini, A., Sagona, S., Michelucci, E., Gazzano, A., Turillazzi, S., Felicioli, A. and Pelosi, P. (2011) Differential expression of odorant-binding proteins in the mandibular glands of the honey bee according to caste and age. *J Proteome Res* 10: 3439–49.

Jay, S.C. (1968) Factors influencing ovary development of worker honey bees under natural conditions. *Can J Zool* 46: 345–47.

Jefferis, G.S., Potter, C.J., Chan, A.M., Marin, E.C., Rohlfing, T., Maurer, C.R. Jr. and Luo, L. (2007) Comprehensive maps of *Drosophila* higher olfactory centers: Spatially segregated fruit and pheromone representation. *Cell* 128: 1187–203.

Joerges, J., Küttner, A., Galizia, C.G. and Menzel, R. (1997) Representations of odours and odour mixtures visualized in the honey bee brain. *Nature* 387: 285–88.

Kaatz, H.H., Eichmueller, S. and Kreissl, S. (1994) Stimulatory effect of octopamine on juvenile hormone biosynthesis in honey bees (*Apis mellifera*): Physiological and immunocytochemical evidence. *J Insect Physiol* 40: 865–72.

Kaatz, H.H., Hildebrandt, H. and Engels, W. (1992) Primer effect of queen pheromone on juvenile hormone biosynthesis in adult worker bees. *J Comp Physiol* 162: 588–92.

Kaminski, L.-A., Slessor, K.N., Winston, M.L., Hay, N.W. and Borden, J.H. (1990) Honey bee response to queen mandibular pheromone in laboratory bioassays. *J Chem Ecol* 16: 841–50.

Katzav-Gozansky, T. (2006) The evolution of honey bee multiple queen-pheromones–A consequence of a queen-worker arm race? *Braz J Morphol Sci* 23(3–4): 287–94.

Katzav-Gozansky, T., Boulay, R., Soroker, V. and Hefetz, A. (2004) Queen-signal modulation of worker pheromonal composition in honey bees. *Proc R Soc Lond B Bio* 271: 2065–69.

Katzav-Gozansky, T., Boulay, R., Soroker, V. and Hefetz, A. (2006) Queen pheromones affecting the production of queen-like secretion in workers. *J Comp Physiol A* 192: 737–42.

Katzav-Gozansky, T., Soroker, V. and Hefetz, A. (2000) Plasticity in caste related exocrine secretion biosynthesis in the honey bee (*Apis mellifera*). *J Insect Physiol* 46: 993–98.

Katzav-Gozansky, T., Soroker, V. and Hefetz, A. (2002) Honey bees Dufour's gland: Idiosyncrasy of a new queen signal. *Apidologie* 33: 525–37.

Katzav-Gozansky, T., Soroker, V. and Hefetz, A. (2003) Honey bee egglaying workers mimic a queen signal. *Insect Soc* 50: 20–23.

Katzav-Gozansky, T., Soroker, V., Hefetz, A., Cojocaru, M., Erdmann, D.H. and Francke, W. (1997) Plasticity of caste-specific Dufour's gland secretion in the honey bee (*Apis mellifera* L.). *Naturwissenschaften* 84: 238–41.

Katzav-Gozansky, T., Soroker, V., Ibarra, F., Francke, W. and Hefetz, A. (2001) Dufour's gland secretion of the queen honey bee (*Apis mellifera*): An egg discriminator pheromone or a queen signal? *Behav Ecol Sociobiol* 51: 76–86.

Kay, L.M. and Stopfer, M. (2006) Information processing in the olfactory systems of insects and vertebrates. *Sem Cell Devel Biol* 17(4): 433–42.

Keeling, C.I. and Slessor, K.N. (2005) A scientific note on aliphatic esters in queen honey bees. *Apidologie* 36: 559–60.

Keeling, C.I., Slessor, K.N., Higo, H.A. and Winston, M.L. (2003) New components of the honey bee (*Apis mellifera* L.) queen retinue pheromone. *Proc Natl Acad Sci U S A* 100: 4486–91.

Kelber, C., Roessler, W. and Kleineidam, C.J. (2006) Multiple olfactory receptor neurons and their axonal projections in the antennal lobe of the honey bee *Apis mellifera*. *J Comp Neurol* 496: 395–405.

Keller, L. and Nonacs, P. (1993) The role of queen pheromones in social insects: Queen control or queen signal? *Anim Behav* 45: 787–94.

Kirchner, W.H. and Gadagkar, R. (1994) Discrimination of nestmate workers and drones in honey bees. *Insect Soc* 41(3): 335–38.

Kocher, S.D. and Grozinger, C.M. (2011) Cooperation, conflict, and the evolution of queen pheromones. *J Chem Ecol* 37: 1263–75.

Kokay, I.C., Ebert, P.R., Kirchhof, B.S. and Mercer, A.R. (1999) Distribution of dopamine receptors and dopamine receptor homologs in the brain of the honey bee, *Apis mellifera* L. *Microsc Res Technol* 44: 179–89.

Kolmes, S.A. and Njehu, N. (1990) Effect of queen mandibular pheromones on *Apis mellifera* worker stinging behavior (Hymenoptera: Apidae). *J New York Entomol S* 98(4): 495–98.

Kreissl, S., Eichmüller, S., Bicker, G., Rapus, J. and Eckert, M. (1994) Octopamine-like immunoreactivity in the brain and subesophageal ganglion of the honey bee. *J Comp Neurol* 348(4): 583–95.

Kurshan, P.T., Hamilton, I.S., Mustard, J.A. and Mercer, A.R. (2003) Developmental changes in expression patterns of two dopamine receptor genes in mushroom bodies of the honey bee, *Apis mellifera*. *J Comp Neurol* 466: 91–103.

Laughlin, J.D., Ha, T.S., Jones, D.N. and Smith, D.P. (2008) Activation of pheromone-sensitive neurons is mediated by conformational activation of pheromone binding protein. *Cell* 133: 1255–65.

Le Conte, Y. and Hefetz, A. (2008) Primer pheromones in social Hymenoptera. *Annu Rev Entomol* 53: 523–42.

Le Conte, Y., Arnold, G., Trouiller, J. and Masson, C. (1990) Identification of a brood pheromone in honey bees. *Naturwissenschaften* 77: 334–36.

Le Conte, Y., Becard, J.M., Costagliola, G., de Vaublanc, G., El Maataoui, D.C., Plettner, E. and Slessor, K.N. (2006) Larval salivary glands are a source of primer and releaser pheromone in honey bee (*Apis mellifera* L.). *Naturwissenschaften* 93: 237–41.

Le Conte, Y., Mohammedi, A. and Robinson, G.E. (2001) Primer effects of a brood pheromone on honey bee behavioural development. *Proc R Soc Lond B Bio* 268: 163–68.

Le Conte, Y., Sreng, L. and Poitout, S.H. (1995) Brood pheromone can modulate the feeding behavior of *Apis mellifera* workers (Hymenoptera: Apidae). *J Econ Entomol* 88: 798–804.

Le Conte, Y., Sreng, L. and Trouiller, J. (1994) The recognition of larvae by worker honey bees. *Naturwissenschaften* 81: 462–65.

Le Conte, Y., Sreng, L., Sacher, N., Trouiller, J., Dusticier, G. and Poitout, S.H. (1994/1995) Chemical recognition of queen cells by honey bee workers *Apis mellifera* (Hymenoptera: Apidae). *Chemoecology* 5/6: 6–12.

Leal, W.S. (2005) Pheromone reception. *Top Curr Chem* 240: 1–36.

Ledoux, M.N., Winston, M.L., Higo, H., Keeling, C.I., Slessor, K.N. and Le Conte, Y. (2001) Queen and pheromonal factors influencing comb construction by simulated honey bee (*Apis mellifera* L.) swarms. *Insect Soc* 48(1): 14–20.

Lensky, Y. and Cassier, P. (1995) The alarm pheromones of queen and worker honey bees. *Bee World* 76: 119–29.

Lensky, Y. and Slabezki, Y. (1981) The inhibiting effect of the queen bee (*Apis mellifera* L.) foot-print pheromone on the construction of swarming queen cups. *J Insect Physiol* 27(5): 313–23.

Lensky, Y., Cassier, P., Finkel, A., Delorme-Joulie, C. and Levinsohn, M. (1985) The fine structure of the tarsal glands of the honey bee *Apis mellifera* L. (Hymenoptera). *Cell Tissue Res* 240: 153–58.

Lensky, Y., Cassier, P., Finkel, A., Teeslshee, A., Shelensinger, R., Delorme-Joulie, C. and Levinsohn, N. (1984) Le glandes tarsales de l'abeille mellifique (*Apis mellifera* L.): Reines, ouvriéres et faux bourdons (Hymenoptera, Apidae). II. Role biologique. *Ann Sci Nat Zool Biol Anim* 6: 167–75.

Lensky, Y., Cassier, P., Rosa, S. and Grandperrin, D. (1991) Induction of balling in worker honey bees (*Apis mellifera* L.) by "stress" pheromone from Koschewnikow glands of queen bees: Behavioural, structural and chemical study. *Comp Biochem Physiol* 100A: 585–94.

Lensky, Y., Cassier, P. and Tel-Zur, D. (1995) The setaceous membrane of honey bee (*Apis mellifera* L.) workers' sting apparatus: Structure and alarm pheromone distribution. *J Insect Physiol* 41: 589–95.

Leoncini, I., Le Conte, Y., Costagliola, G., Plettner, E., Toth, A.L., Wang, M., Huang, Z., Bécard, J., Crauser, D., Slessor, K.N. and Robinson, G.E. (2004) Regulation of behavioral maturation by a primer pheromone produced by adult worker honey bees. *Proc Natl Acad Sci U S A* 101: 17559–64.

Lipínski, Z. (2006) The calming properties of the honey bee queen, young brood and older bees. *J Apic Sci* 50(1): 63–70.

Loper, G.M., Taylor, O.R. Jr., Foster, L.J. and Kochansky, J. (1996) Relative attractiveness of queen mandibular pheromone components to honey bee (*Apis mellifera* L.) drones. *J Apic Res* 35: 122–23.

Maisonnasse, A., Alaux, C., Beslay, D., Crauser, D., Gines, C., Plettner, E. and Le Conte, Y. (2010a) New insights into honey bee (*Apis mellifera*) pheromone communication. Is the queen mandibular pheromone alone in colony regulation? *Front Zool* 7: 18–25.

Maisonnasse, A., Lenoir, J.C., Beslay, D., Crauser, D. and Le Conte Y (2010b) E-β-ocimene, a volatile brood pheromone involved in social regulation in the honey bee colony (*Apis mellifera*). *PLoS ONE* 5(10): e13531.

Maisonnasse, A., Lenoir, J.C., Costagliola, G., Beslay, D., Choteau, F., Crauser, D., Becard, J.M., Plettner, E. and Le Conte, Y. (2009) A scientific note on E-β-ocimene, a new volatile primer pheromone that inhibits worker ovary development in honey bees. *Apidologie* 40: 562–64.

Makert, G.R., Paxton, R.J. and Hartfelder K. (2006) Ovariole number—A predictor of differential reproductive success among worker subfamilies in queenless honey bee (*Apis mellifera* L.) colonies. *Behav Ecol Sociobiol* 60: 815–25.

Maleszka, J., Barron, A., Helliwell, P. and Maleszka, R. (2009) Effect of age, behaviour and social environment on honey bee brain plasticity. *J Comp Physiol A* 195(8): 733–40.

Malka, O., Shnieor, S., Hefetz, A. and Katzav-Gozansky, T. (2007) Reversible royalty in worker honey bees (*Apis mellifera*) under the queen influence. *Behav Ecol Sociobiol* 61: 465–73.

Malka, O., Shnieor, S., Katzav-Gozansky, T. and Hefetz, A. (2008) Aggressive reproductive competition among hopelessly queenless honey bee workers triggered by pheromone signalling. *Naturwissenschaften* 95: 553–59.

Martin, S.J., Dils, V. and Billen, J. (2005) Morphology of the Dufour gland within the honey bee sting gland complex. *Apidologie* 36: 543–46.

Martin, S.J., Jones, G.R., Châline, N., Middleton, H. and Ratnieks, F.L.W. (2002) Reassessing the role of the honey bee (*Apis mellifera*) Dufour's glands in egg marking. *Naturwissenschaften* 89: 528–32.

Masson, C. and Arnold, G. (1984) Ontogeny, maturation and plasticity of the olfactory system in the worker bee. *J Insect Physiol* 30: 7–14.

McQuillan, H.J., Barron, A.B. and Mercer, A.R. (2012) Age- and behaviour-related changes in the expression of biogenic amine receptor genes in the antennae of honey bees (*Apis mellifera*). *J Comp Physiol A* 198: 753–61.

Melathopoulos, A.P., Winston, M.L., Pettis, J.S. and Pankiw, T. (1996) Effect of queen mandibular pheromone on initiation and maintenance of queen cells in the honey bee (*Apis mellifera* L.). *Can Entomol* 128: 263–72.

Menzel, R., Heyne, A., Kinzel, C., Gerber, B. and Fiala, A. (1999) Pharmacological dissociation between the reinforcing, sensitizing, and response-releasing functions of reward in honey bee classical conditioning. *Behav Neurosci* 113: 744–54.

Mercer, A. (1987) Biogenic amines and the bee brain. In: *Neurobiology and Behaviour of Honey bees* (R. Menzel and A. Mercer, eds.), pp. 245–52. Berlin: Springer-Verlag.

Mercer, A.R. and Erber, J. (1983) The effects of amines on evoked potentials recorded in the mushroom bodies of the bee brain. *J Comp Physiol* 151: 469–76.

Mercer, A.R. and Menzel, R. (1982) The effects of biogenic amines on conditioned and unconditioned responses to olfactory stimuli in the honey bee, *Apis mellifera*. *J Comp Physiol* 145: 363–68.

Millor, J., Pham-Delègue, M., Deneubourg, J.L. and Camazine, S. (1999) Self-organized defensive behavior in honey bees. *Proc Natl Acad Sci U S A* 96: 12611–15.

Mohammedi, A., Crauser, D., Paris, A. and Le Conte, Y. (1996) Effect of a brood pheromone on honey bee hypopharyngeal glands. *C R Acad Sci* 319: 769–72.

Mohammedi, A., Paris, A., Crauser, D. and Le Conte, Y. (1998) Effect of aliphatic esters on ovary development of queenless bees (*Apis mellifera* L.). *Naturwissenschaften* 85: 455–58.

Morgan, S.M., Butz Huryn, V.M., Downes, S.R. and Mercer, A.R. (1998) The effects of queenlessness on the maturation of the honey bee olfactory system. *Behav Brain Res* 91(1–2): 115–26.

Moritz, R.F.A. and Crewe, R.M. (1988) Chemical signals of queens in kin recognition of honey bees, *Apis mellifera* L. *J Comp Physiol A* 164: 83–89.

Moritz, R.F.A. and Crewe, R.M. (1991) The volatile emission of honey bee queens (*Apis mellifera* L.). *Apidologie* 22: 205–12.

Moritz, R.F.A., Crewe, R.M. and Hepburn, H.R. (2002) Queen avoidance and mandibular gland secretion of honey bee workers (*Apis mellifera* L.). *Insectes Soc* 49: 86–91.

Moritz, R.F.A., Simon, U.E. and Crewe, R.M. (2000) Pheromonal contest between honey bee workers (*Apis mellifera capensis*). *Naturwissenschaften* 87: 395–97.

Muenz, T.S., Maisonnasse, A., Plettner, E., Le Conte, Y. and Rössler, W. (2012) Sensory reception of the primer pheromone ethyl oleate. *Naturwissenschaften* 99: 421–25.

Mustard, J.A., Blenau, W., Hamilton, I.S., Ward, V.K., Ebert, P.R. and Mercer, A.R. (2003) Analysis of two D1-like dopamine receptors from the honey bee, *Apis mellifera*, reveals agonist-independent activity. *Mol Brain Res* 113: 67–77.

Naumann, K. (1991) Grooming behaviors and the translocation of queen mandibular gland pheromone on worker honey bees (*Apis mellifera* L.). *Apidologie* 22: 523–31.

Naumann, K., Winston, M.L. and Slessor, K.N. (1993) Movement of honey bee (*Apis mellifera* L.) queen mandibular gland pheromone in populous and unpopulous colonies. *J Insect Behav* 6: 211–23.

Naumann, K., Winston, M.L., Slessor, K.N., Prestwich, G.D. and Webster, F.X. (1991) The production and transfer of honey bee (*Apis mellifera* L) queen mandibular gland pheromone. *Behav Ecol Sociobiol* 29: 321–32.

Nelson, C.M., Ihle, K., Amdam, G.V., Fondrk, M.K. and Page, R.E. (2007) The gene vitellogenin has multiple coordinating effects on social organization. *PLoS Biol* 5: 673–77.

Nuñez, J.A., Almeida, L., Balderrama, N. and Giurfa, M. (1998) Alarm pheromone induces stress analgesia via an opioid system in the honey bee. *Physiol Behav* 63: 75–80.

Nuñez, J.A., Maldonado, H., Miralto, A. and Balderrama, N. (1983) The stinging response of the honey bee: Effects of morphine, naloxone and some opioid peptides. *Pharm Biochem Behav* 19: 921–24.

Onions, G.W. (1912) South African "fertile" worker bees. *Agric J S Afr* 7: 44–46.

Page, R.E. Jr., Metcalf, R.A., Metcalf, R.L., Erickson, E.H. Jr. and Lampman, R.L. (1991) Extractable hydrocarbons and kin recognition in honey bee (*Apis mellifera* L.). *J Chem Ecol* 17(4): 745–56.

Pankiw, T. (2004a) Worker honey bee pheromone regulation of foraging ontogeny. *Naturwissenschaften* 91: 178–81.

Pankiw, T. (2004b) Cued in: Honey bee pheromones as information flow and collective decision-making. *Apidologie* 35: 217–26.

Pankiw, T. (2004c) Brood pheromone regulates foraging activity of honey bees (Hymenoptera: Apidae). *J Econ Entomol* 97: 748–51.

Pankiw, T. (2007) Brood pheromone modulation of pollen forager turnaround time in the honey bee (*Apis mellifera* L.). *J Insect Behav* 20: 173–80.

Pankiw, T. and Page, R.E. (2001) Brood pheromone modulates honey bee (*Apis mellifera* L.) sucrose response thresholds. *Behav Ecol Sociobiol* 49: 206–13.

Pankiw, T., Huang, Z., Winston, M.L. and Robinson, G.E. (1998) Queen mandibular gland pheromone influences worker honey bee (*Apis mellifera* L.) foraging ontogeny and juvenile hormone titers. *J Insect Physiol* 44(7–8): 685–92.

Pankiw, T., Roman, R., Sagili, R.R. and Metz, B.N. (2008) Brood pheromone effects on colony protein supplement consumption and growth in the honey bee (Hymenoptera: Apidae) in a subtropical winter climate. *J Econ Entomol* 101(6): 1749–55.

Pankiw, T., Roman, R., Sagili, R.R. and Zhu-Salzman, K. (2004) Pheromone-modulated behavioral suites influence colony growth in the honey bee (*Apis mellifera*). *Naturwissenschaften* 91: 575–78.

Pankiw, T., Winston, M.L. and Slessor, K.N. (1994) Variation in worker response to honey bee (*Apis mellifera* L.) queen mandibular pheromone. *J Insect Behav* 7: 1–15.

Pankiw, T., Winston, M.L. and Slessor, K.N. (1995) Queen attendance behavior of worker honey bees (*Apis mellifera* L.) that are high and low responding to queen mandibular pheromone. *Insect Soc* 41: 371–78.

Papachristoforou, A., Kagiava, A., Papaefthimiou, C., Termentzi, A., Fokialakis, N., Skaltsounis, A.-L., Watkins, M., Arnold, G. and Theophilidis, G. (2012) The bite of the honey bee: 2-Heptanone secreted from honey bee mandibles during a bite acts as a local anaesthetic in insects and mammals. *PLoS ONE* 7(10): e47432.

Pesenti, M.E., Spinelli, S., Bezirard, V., Briand, L., Pernollet, J.-C., Tegoni, M. and Cambillau, C. (2008) Structural basis of the honey bee PBP pheromone and pH-induced conformational change. *J Mol Biol* 380: 158–69.

Pesenti, M.E., Spinelli, S., Bezirard, V., Briand, L., Pernollet, J.-C., Campanacci, V., Tegoni, M. and Cambillau, C. (2009) Queen bee pheromone binding protein pH-Induced domain swapping favors pheromone release. *J Mol Biol* 390: 981–90.

Pettis, J.S., Westcott, L.C. and Winston, M.L. (1998) Balling behaviour in the honey bee in response to exogenous queen mandibular gland pheromone. *J Apicult Res* 37(2): 125–31.

Pettis, J.S., Winston, M.L. and Slessor K.N. (1995) Behaviour of queen and worker honey bees (Hymenoptera: Apidae) in response to exogenous queen mandibular gland pheromone. *Ann Entomol Soc Am* 88(4): 580–88.

Pflumm, W. and Wilhelm, K. (1982) Olfactory feedback in the scent-marking behaviour of foraging honey bees at the food source? *Physiol Entomol* 7: 203–7.

Pham-Delègue, M.H., Trouiller, J., Bakchine, E., Roger, B. and Masson, C. (1991) Age dependency of worker bee response to queen pheromone in a four-armed olfactometer. *Insect Soc* 38: 283–92.

Pham-Delègue, M.H., Trouiller, J., Caillaud, C.M., Roger, B. and Masson, C. (1993) Effect of queen pheromone on worker bees of different ages: Behavioral and electrophysiological responses. *Apidologie* 24: 267–81.

Pickett, J.A., Williams, I.H. and Martin, A.P. (1982) (Z)-11-Eicosen-1-ol, an important new pheromonal component from the sting of the honey bee, *Apis mellifera* L. (Hymenoptera, Apidae). *J Chem Ecol* 8(1): 163–75.

Pickett, J.A., Williams, I.H., Martin, A.P. and Smith, M.C. (1980) Nasonov pheromone of the honey bee *Apis mellifera* L. (Hymenoptera:Apidae). I. Chemical characterization. *J Chem Ecol* 6: 425–34.

Plettner, E., Otis, G.W., Wimalaratne, P.D.C., Winston, M.L., Slessor, K.N, Pankiw, T. and Punchihewa, P.W.K. (1997) Species- and caste-determined mandibular gland signals in honey bees (Apis). *J Chem Ecol* 23: 363–77.

Plettner, E., Slessor, K.N., Winston, M.L. and Oliver, J.E. (1996) Caste-selective pheromone biosynthesis in honey bees. *Science* 271: 1851–53.

Plettner, E., Slessor, K.N., Winston, M.L., Robinson, G.E. and Page, R.E. (1993) Mandibular gland components and ovarian development as measures of caste differentiation in the honey bee (*Apis mellifera* L.). *J Insect Physiol* 39: 235–40.

Plettner, E., Sutherland, G.R.J., Slessor, K.N. and Winston, M.L. (1995) Why not be a queen? Regioselectivity in mandibular secretions of honey bee castes. *J Chem Ecol* 21: 1017–29.

Ratnieks, F.L.W. (1988) Reproductive harmony via mutual policing by workers in eusocial Hymenoptera. *Am Nat* 132: 217–36.

Ratnieks, F.L.W. (1993) Egg-laying, egg-removal, and ovary development by workers in queenright honey bee colonies. *Behav Ecol Sociobiol* 32: 191–98.

Ratnieks, F.L.W. (1995) Evidence for a queen-produced egg-marking pheromone and its use in worker policing in the honey bee. *J Apic Res* 34: 31–37.

Ratnieks, F.L.W. and Visscher, P.K. (1989) Worker policing in honey bees. *Nature* 342: 796–97.

Rehder, V., Bicker, G. and Hammer, M. (1987) Serotonin-immunoreactive neurons in the antennal lobes and suboesophageal ganglion of the honey-bee. *Cell Tissue Res* 247: 59–96.

Reinhard, J. and Claudianos, C. (2013) Molecular insights into honey bee brain plasticity. In: *Honey bee Neurobiology and Behavior. A Tribute to Randolf Menzel* (C.G. Galizia, D. Eisenhardt and M. Giurfa, eds.), pp. 359–72. Dordrecht: Springer.

Renner, M. and Baumann, M. (1964) Über Komplexe von subepidermalen Drüsenzellen (Duftdrüsen?) der Bienenkönigin. *Naturwissenschaften* 3: 68–69.

Renner, M. and Vierling, G. (1977) Die Rolle des Taschendruesenpheromons beim Hochzeitsflug der Bienenkoenigin. (The secretion of the tergite glands and the attractiveness of the queen honey bee to drones in the mating flight). *Behav Ecol Sociobiol* 2: 329–38.

Rhodes, J.W., Lacey, M.J. and Harden, S. (2007) Changes with age in queen honey bee (*Apis mellifera*) head chemical constituents (Hymenoptera: Apidae). *Sociobiology* 50: 11–22.

Richard, F.-J., Schal, C., Tarpy, D.R. and Grozinger, C.M. (2011) Effects of instrumental insemination and insemination quantity on Dufour's gland chemical profiles and vitellogenin expression in honey bee queens (*Apis mellifera*). *J Chem Ecol* 37: 1027–36.

Richard, F.-J., Tarpy, D.R. and Grozinger, C.M. (2007) Effects of insemination quantity on honey bee queen physiology. *PLoS ONE* 2(10): e980.

Robertson, H.M. and Wanner, K.W. (2006) The chemoreceptor superfamily in the honey bee, *Apis mellifera*: Expansion of the odorant, but not gustatory, receptor family. *Genome Res* 16: 1395–403.

Robinson, G.E. (1987a) Hormonal regulation of age polyethism in the honey bee, *Apis mellifera*. In: *Neurobiology and Behaviour of Honey bees* (R. Menzel and A. Mercer, eds.), pp. 266–79. Berlin: Springer-Verlag.

Robinson, G.E. (1987b) Modulation of alarm pheromone perception in the honey bee: Evidence for division of labor based on hormonally regulated response thresholds. *J Comp Physiol A* 160: 613–19.

Robinson, G.E. (1992) Regulation of division of labor in insect societies. *A Rev Entomol* 37: 637–65.

Robinson, G.E. and Huang, Z.-Y. (1998) Colony integration in honey bees: Genetic, endocrine and social control of division of labor. *Apidologie* 29: 159–70.

Robinson, G.E. and Page, R.E. (1988) Genetic determination of guarding and undertaking in honey bee colonies. *Nature* 333: 356–58.

Robinson, G.E. and Page, R.E. (1989) Genetic determination of nectar foraging, pollen foraging and nestsite scouting in honey bee colonies. *Behav Ecol Sociobiol* 24: 317–23.

Robinson, G.E. and Vargo, E.L. (1997) Juvenile hormone in adult eusocial Hymenoptera: Gonadotropin and behavioral pacemaker. *Arch Insect Biochem Physiol* 35: 559–83.

Robinson, G.E., Grozinger, C.M. and Whitfield, C.W. (2005) Sociogenomics: social life in molecular terms. *Nat Rev Genet* 6(4): 257–70.

Robinson, G.E., Page, R.E., Strambi, C. and Strambi, A. (1992b) Colony integration in honey bee: Mechanism of behavioural reverse. *Ethology* 90(4): 336–48.

Robinson, G.E., Strambi, C., Strambi, A. and Feldlaufer, M.F. (1991) Comparison of juvenile hormone and ecdysteroid haemolymph titres in adult worker and queen honey bees (*Apis mellifera*). *J Insect Physiol* 37(12): 929–35.

Robinson, G.E., Strambi, C., Strambi, A. and Huang, Z.-Y. (1992a) Reproduction in worker honey bees is associated with low juvenile hormone titers and rates of biosynthesis. *Gen Comp Endocr* 87: 471–80.

Sachse, S. and Galizia, C.G. (2003) The coding of odour-intensity in the honey bee antennal lobe: Local computation optimizes odour representation. *Eur J Neurosci* 18: 2119–32.

Sachse, S., Rappert, A. and Galizia, C.G. (1999) The spatial representation of chemical structures in the AL of honey bees: Steps towards the olfactory code. *Eur J Neurosci* 11: 3970–82.

Sagili, R.R. and Pankiw, T. (2009) Effects of brood pheromone modulated brood rearing behaviors on honey bee (*Apis mellifera* L.) colony growth. *J Insect Behav* 22: 339–49.

Sagili, R.R., Pankiw, T. and Metz, B.N. (2011) Division of labor associated with brood rearing in the honey bee: How does it translate to colony fitness? *PLoS ONE* 6(2): e16785.

Sakagami, S.F. (1958) The false queen: fourth adjustive response in dequeened colonies. *Behaviour* 8: 280–96.

Sakamoto, H.C., Soares, A.E.E. and Lopes, J.N.C. (1990a) Relationship between 2-heptanone and some biological and environmental variables in *Apis mellifera* L. *J Apicult Res* 29(4): 194–98.

Sakamoto, H.C., Soares, A.E.E. and Lopes, J.N.C. (1990b) A comparison of 2-heptanone production in Africanised and European strains of the honey bee, *Apis mellifera* L. *J Apicult Res* 29(4): 199–205.

Sandoz, J.-C. (2006) Odour-evoked responses to queen pheromone components and to plant odours using optical imaging in the antennal lobe of the honey bee drone *Apis mellifera* L. *J Exp Biol* 209: 3587–98.

Sandoz, J.-C. (2011) Behavioral and neurophysiological study of olfactory perception and learning in honey bees. *Front Syst Neurosci* 5: 98.

Sandoz, J.-C. (2013) Olfaction in honey bees: From molecules to behavior. In: *Honey bee Neurobiology and Behavior. A Tribute to Randolf Menzel* (C.G. Galizia, D. Eisenhardt and M. Giurfa, eds.), pp. 235–52. Dordrecht: Springer.

Sandoz, J.-C., Deisig, N., de Brito Sanchez, M.G. and Giurfa, M. (2007) Understanding the logics of pheromone processing in the honey bee brain: From labeled-lines to across-fiber patterns. *Front Behav Neurosci* 1: 5.

Sandoz, J.-C., Galizia, C.G. and Menzel, R. (2003) Side-specific olfactory conditioning leads to more specific odor representation between sides but not within sides in the honey bee antennal lobes. *Neuroscience* 120: 1137–48.

Sandoz, J.-C., Pham-Delègue, M.H., Renou, M. and Wadhams, L.J. (2001) Asymmetrical generalisation between pheromonal and floral odours in appetitive olfactory conditioning of the honey bee (*Apis mellifera* L.). *J Comp Physiol A* 187: 559–68.

Sasaki, K. and Nagao, T. (2001) Distribution and levels of dopamine and its metabolites in brains of reproductive workers in honey bees. *J Insect Physiol* 47: 1205–16.

Schaefer, S. and Rehder, V. (1989) Dopamine-like immunoreactivity in the brain and suboesophageal ganglion of the honey bee. *J Comp Neurol* 280: 43–58.

Schmidt, J.O. (1994) Attraction of reproductive honey bee swarms to artificial nests by Nasonov pheromone. *J Chem Ecol* 20: 1053–56.

Schmidt, J.O. (1999) Attractant or pheromone: The case of Nasonov secretion and honey bee swarms. *J Chem Ecol* 25: 2051–56.

Schröter, U., Malun, D. and Menzel, R. (2006) Innervation pattern of suboesophageal ventral unpaired median neurones in the honey bee brain. *Cell Tissue Res* 327: 647–67.

Schuermann, F.W. and Klemm, N. (1984) Serotonin-immunoreactive neurons in the brain of the honey bee. *J Comp Neurol* 225: 570–80.

Schuermann, F.W., Elekes, K. and Geffard, M. (1989) Dopamine-like immunoreactivity in the bee brain. *Cell Tissue Res* 256: 399–410.

Schulz, D.J. and Robinson, G.E. (1999) Biogenic amines and division of labor in honey bee colonies: Behaviorally related changes in the antennal lobes and age related changes in the mushroom bodies. *J Comp Physiol A* 184: 481–88.

Schulz, D.J. and Robinson, G.E. (2001) Octopamine influences division of labor in honey bee colonies. *J Comp Physiol A* 187: 53–61.

Schulz, D.J., Barron, A.B. and Robinson, G.E. (2002a) A role for octopamine in honey bee division of labor. *Brain Behav Evolut* 60: 350–59.

Schulz, D.J., Sullivan, J.P. and Robinson, G.E. (2002b) Juvenile hormone and octopamine in the regulation of division of labour in honey bee colonies. *Horm Behav* 42: 222–31.

Seehuus, S.C., Norberg, K., Gimsa, U., Krekling, T. and Amdam, G.V. (2006) Reproductive protein protects sterile honey bee workers from oxidative stress. *Proc Natl Acad Sci U S A* 103: 962–67.

Seeley, T. (2010) *Honey bee Democracy*. Princeton, NJ: Princeton University Press.

Shearer, D.A. and Boch, R. (1965) 2-Heptanone in the mandibular gland secretion of the honey bee. *Nature* 206: 530.

Simon, U.E., Moritz, R.F. and Crewe, R.M. (2001) The ontogenetic pattern of mandibular gland components in queenless worker bees (*Apis mellifera capensis* Esch.). *J Insect Physiol* 47(7): 735–38.

Slessor, K.N., Kaminski, L.A., King, G.G.S., Borden, J.H. and Winston, M.L. (1988) Semiochemical basis for retinue response to queen honey bees. *Nature* 332: 354–56.

Slessor, K.N., Kaminski, L.A., King, G.G.S. and Winston, M.L. (1990) Semiochemicals of the honey bee queen mandibular glands. *J Chem Ecol* 16: 851–60.

Slessor, K.N., Winston, M.L. and Le Conte, Y. (2005) Pheromone communication in the honey bee (*Apis mellifera* L.). *J Chem Ecol* 31(11): 2731–45.

Smedal, B., Brynem, M., Kreibich, C.D. and Amdam, G.V. (2009) Brood pheromone suppresses physiology of extreme longevity in honey bees (*Apis mellifera*). *J Exp Biol* 212: 3795–801.

Smith, R.K., Spivak, M., Taylor, O.R. Jr., Bennett, C. and Smith, M.L. (1993) Maturation of tergal gland alkene profiles in European honey bee queens, *Apis mellifera* L. *J Chem Ecol* 19: 133–42.

Sole, C.L., Kryger, P., Hefetz, A., Katzav-Gozansky, T. and Crewe, R.M. (2002) Mimicry of queen Dufour's gland secretions by workers of *Apis mellifera scutellata* and *A. m. capensis. Naturwissenschaften* 89: 561–64.

Southwick, E.E. and Moritz, R.F.A. (1985) Metabolic response to alarm pheromone in honey bees. *J Insect Physiol* 31: 389–92.

Strausfeld, N.J. (2002) Organization of the honey bee mushroom body: Representation of the calyx within the vertical and gamma lobes. *J Comp Neurol* 450: 4–33.

Strauss, K., Scharpenberg, H., Crewe, R.M., Glahn, F., Foth, H. and Moritz, R.F.A. (2008) The role of the queen mandibular gland pheromone in honey bees (*Apis mellifera*): Honest signal or suppressive agent? *Behav Ecol Sociobiol* 62: 1523–31.

Sullivan, J.P., Fahrbach, S.E., Harrison, J.F., Capaldi, E.A., Fewell, J.H. and Robinson, G.E. (2003) Juvenile hormone and division of labor in honey bee colonies: Effects of allatectomy on flight behavior and metabolism. *J Exp Biol* 206: 2287–96.

Tan, K., Yang, M., Wang, Z., Radloff, S.E. and Pirk, C.W.W. (2012) The pheromones of laying workers in two honey bee sister species: *Apis cerana* and *Apis mellifera. J Comp Physiol A* 198: 319–23.

Taylor, D.J., Robinson, G.E., Logan, B.J., Laverty, R. and Mercer, A.R. (1992) Changes in brain amine levels associated with the morphological and behavioural development of the worker honey bee. *J Comp Physiol A* 170: 715–21.

Thom, C., Gilley, D.C., Hooper, J. and Esch, H.E. (2012) The scent of the waggle dance. *PLoS Biol* 5(9): e228.

Toth, A.L. and Robinson, G.E. (2005) Worker nutrition and division of labour in honey bees. *Anim Behav* 69(2): 427–35.

Toth, A.L., Kantarovich, S., Meisel, A.F. and Robinson, G.E. (2005) Nutritional status influences socially regulated foraging ontogeny in honey bees. *J Exp Biol* 208: 4641–49.

Trhlin, M. and Rajchard, J. (2011) Chemical communication in the honey bee (*Apis mellifera* L.): A review. *Vet Med-Czech* 56(6): 265–73.

Tsuruda, J.M. and Page, R.E. (2009) The effects of young brood on the foraging behavior of two strains of honey bees (*Apis mellifera*). *Behav Ecol Sociobiol* 64: 161–67.

Urlacher, E., Francés, B., Giurfa, M. and Devaud, J.M. (2010) An alarm pheromone modulates appetitive olfactory learning in the honey bee (*Apis mellifera*). *Front Behav Neurosci* 4: 157.

Vallet, A., Cassier, P. and Lensky, Y. (1991) Ontogeny of the fine structure of the mandibular glands of the honey bee (*Apis mellifera* L.) workers and the pheromonal activity of 2-heptanone. *J Insect Physiol* 37: 789–804.

Velthuis, H.H.W. (1970) Queen substances from the abdomen of the honey bee queen. *Z Vergl Physiol* 70: 210–22.

Velthuis, H.H.W. (1985) The honey bee queen and the social organization of her colony. In: *Experimental Behavioural Ecology and Sociobiology: In Memoriam Karl von Frisch 1886–1982* (B. Hölldobler and M. Lindauer, eds), pp. 343–57. Stuttgart, Germany: Gustav Fischer Verlag.

Velthuis, H.H.W., Ruttner, F. and Crewe, R.M. (1990) Differentiation in reproductive physiology and behaviour during the development of laying worker honey bees. In: *Social Insects* (W. Engels, ed.), pp. 231–43. Berlin: Springer-Verlag.

Vergoz, V., McQuillan, H.J., Geddes, L.H., Pullar, K., Nicholson, B.J., Paulin, M.G. and Mercer, A.R. (2009) Peripheral modulation of worker bee responses to queen mandibular pheromone. *Proc Natl Acad Sci U S A* 106: 20930–35.

Vergoz, V., Roussel, E., Sandoz, J.-C. and Giurfa, M. (2007a) Aversive learning in honey bees revealed by the olfactory conditioning of the sting extension reflex. *PLoS ONE* 2: e288.

Vergoz, V., Schreurs, H.A. and Mercer, A.R. (2007b) Queen pheromone blocks aversive learning in young worker bees. *Science* 317: 384–86.

Wagener-Hulme, C., Schulz, D.J., Kuehn, J.C. and Robinson, G.E. (1999) Biogenic amines and division of labor in honey bee colonies. *J Comp Physiol A* 184: 471–79.

Wager, B.R. and Breed, M.D. (2000) Does honey bee sting alarm pheromone give orientation information to defensive bees? *Ann Entomol Soc Am* 93: 1329–32.

Wakonig, G., Eveleigh, L., Arnoldj, G., and Crailsheim, K. (2000) Cuticular hydrocarbon profiles reveal age-related changes in honey bee drones (*Apis mellifera carnica*). *J Apicult Res* 39(3–4): 137–41.

Wang, S., Sato, K., Giurfa, M. and Zhang, S. (2008) Processing of sting pheromone and its components in the antennal lobe of the worker honey bee. *J Insect Physiol* 54: 833–41.

Wanner, K.W., Nichols, A.S., Walden, K.K., Brockmann, A., Luetje, C.W. and Robertson, H.M. (2007) A honey bee odorant receptor for the queen substance 9-oxo-2-decenoic acid. *Proc Natl Acad Sci U S A* 104: 14383–88.

Whitfield, C.W., Ben-Shahar, Y., Brillet, C., Leoncini, I., Crauser D., Le Conte, Y., Rodrigues-Zas, S. and Robinson, G.E. (2006) Genomic dissection of behavioral maturation in the honey bee. *Proc Natl Acad Sci U S A* 103: 16068–75.

Whitfield, C.W., Cziko, A.M. and Robinson, G.E. (2003) Gene expression profiles in the brain predict behavior in individual honey bees. *Science* 302: 296–99.

Williams, I.H., Martin, A.P. and Pickett, J.A. (1981) The Nasonov pheromone of the honey bee *Apis mellifera* L. (Hymenoptera, Apidae). II. Bioassay of the components using foragers. *J Chem Ecol* 7 (2): 225–37.

Willis, L.G., Winston, M.L. and Slessor, K.N. (1990) Queen honey bee mandibular pheromone does not affect worker ovary development. *Can Entomol* 122: 1093–99.

Winston, M.L. (1987) *The Biology of the Honey Bee*. Cambridge, MA: Harvard University Press.

Winston, M.L. (1992) Semiochemicals and insect sociality. In: *Insect Chemical Ecology: An Evolutionary Approach* (B.D. Roitberg and M.B. Isman, eds.), pp. 315–33. New York: Routledge, Chapman & Hall.

Winston, M.L. and Slessor, K.N. (1998) Honey bee primer pheromones and colony organization: Gaps in our knowledge. *Apidologie* 29: 81–95.

Winston, M.L., Slessor, K.N., Willis, L.G., Naumann, K., Higo, H.A., Wyborn, M.H. and Kaminski, L.-A. (1989) The influence of queen mandibular pheromones on worker attraction to swarm clusters and inhibition of queen rearing in the honey bee (*Apis mellifera* L.). *Insect Soc* 36: 15–27.

Wossler, T.C. and Crewe, R.M. (1999a) Mass spectral identification of the tergal gland secretions of female castes of two African honey bee races (*Apis mellifera*). *J Apicult Res* 38(3–4): 137–48.

Wossler, T.C. and Crewe, R.M. (1999b) The releaser effects of the tergal gland secretion of queens honey bees (*Apis mellifera* L.). *J Insect Behav* 12(3): 343–51.

Wossler, T.C. and Crewe, R.M. (1999c) Honey bee queen tergal gland secretion affects ovarian development in caged workers. *Apidologie* 30: 311–20.

6 Drosophila Pheromones From Reception to Perception

Wynand van der Goes van Naters

CONTENTS

6.1 Introduction 211
6.2 Olfactory Sensilla 212
6.3 Responses of Trichoid ORNs in *Drosophila* to Fly Odors 213
6.4 Molecular Basis of Odor Response in Trichoid Sensilla 215
6.5 *Drosophila* Pheromones in Courtship and Aggression 219
References... 223

6.1 INTRODUCTION

Olfaction in insects has been intensely studied to address fundamental questions in sensory biology. It has yielded surprising answers and insights into reception of sensory stimuli, sensory coding, and the architecture and function of the neural circuits that drive behavior. *Drosophila melanogaster* is an experimentally tractable model organism that can teach us much about how insects use their senses of smell and taste to communicate, to find the hosts from which they feed, and to find sites for laying eggs. *Drosophila* is a member of the Diptera. The Diptera include flies and mosquitoes that transmit some of the most serious communicable diseases such as malaria, dengue fever, leishmaniasis, and African trypanosomiasis (sleeping sickness in humans). An understanding of how Diptera use olfactory cues to seek their hosts and to communicate with mating partners may have medical and economic impact if this knowledge can be translated to develop new technologies, or improve existing technologies, for pest control. In addition, courtship in *Drosophila* has emerged as a model for how chemicals are received in the sense organs and then perceived through processing in the neural circuits of the brain. Superficially simple, the interaction of flies in courtship and other behaviors—including aggression—turns out to be remarkably complex and modifiable by experience and learning. Pheromones are chemicals released by an organism to which a conspecific responds either behaviorally or physiologically. Hydrocarbons in the cuticle may be important in interindividual recognition of species and gender in many insects, and in social insects they additionally may act in recognition of colony membership, nestmates, caste, and kin (Howard and Blomquist 2005). Due to their low volatility, it is generally assumed that cuticular hydrocarbons act on contact and are perceived through the taste organs, and recent studies have confirmed the involvement of taste in pheromone detection (Miyamoto and Amrein 2008; Montell 2009). However, using a stimulus method that mimics the proximity of two interacting insects, it was shown in *Drosophila* that extracts of flies stimulate the olfactory system at close range (Van der Goes van Naters and Carlson 2007). This chapter focuses on how candidate volatile pheromones in *Drosophila* are received by the sense organs and discusses recent advances in our understanding of how olfactory information modulates the flies' behavior towards each other.

6.2 OLFACTORY SENSILLA

The sense of smell in adult insects is mediated by bipolar sensory neurons, the olfactory receptor neurons (ORNs), located in cuticular structures called sensilla. In *Drosophila*, olfactory sensilla are found in the antennae and maxillary palps (Su et al. 2011). As in many flies, the third segment of the *Drosophila* antenna is enlarged. This funiculus bears all the antennal olfactory sensilla. The ultrastructural features of olfactory sensilla in *Drosophila* have been investigated in remarkable detail (Shanbhag et al. 1999, 2000). In each sensillum of the *Drosophila* adult there are 1–4 ORNs, whose function is supported by three auxiliary cells; these are the thecogen cell, which closely envelopes part of the neurons in a glia-like way, and the trichogen and tormogen cells. The structure of a sensillum in insects can be understood by investigating its development (Keil 1997). A general model for sensillogenesis is that a single precursor cell in the epidermis undergoes differential mitoses leading to a side-by-side arrangement of the tormogen, trichogen, sensory, and thecogen cells. The sensory cell or cells migrate below the plane of the epithelium, extending a process basally that becomes the axon and a process, which will be the sensory dendrite, apically through the thecogen cell. These are then surrounded by the trichogen and tormogen cells, which ultimately form the shaft and socket of the sensillum and act in the secretion of a lymph that fills the cavity inside it. In *Drosophila*, while the basic structure of the sensillum is similar, the cells that generate the olfactory sensilla may not be clonally related. Rather, selection of a founder cell in the epidermis leads to recruitment of several neighbors to form a presensillum cluster (Reddy et al. 1997; Ray and Rodrigues 1995), which then leads to the formation of a sensillum with 1–4 ORNs surrounded by the auxiliary cells. There are three morphological types of sensilla in *Drosophila*: coeloconic, basiconic, and trichoid. The shaft of the sensillum in the case of basiconic and trichoid sensilla is a hair- or fingerlike protrusion from the cuticle holding the dendrites within it. Pores permeate the wall of the shaft and are thought to be the entry point for odor molecules that have adsorbed to the outside of the hair. The shaft of coeloconic sensilla is formed by a number of apposed fingerlike protrusions forming a palisade around the dendrites in the lymph. Slits or radial spoke channels may provide openings that allow odor molecules to pass through the palisade. In cross section coeloconic sensilla look double-walled, whereas basiconic and trichoid sensilla are single-walled. Coeloconic and basiconic sensilla contain neurons that respond to food, including fruit, and to other general odors (De Bruyne et al. 2001; Yao et al. 2005). Trichoid sensilla in a number of insects house neurons sensitive to pheromones.

Reception of odors at the dendritic membrane is mediated by proteins encoded by the *Olfactory receptor* (*Or*) (Clyne et al. 1999; Robertson et al. 2003; Vosshall et al. 1999) and *Ionotropic receptor* (*Ir*) (Benton et al. 2009) gene families although several other membrane proteins, such as sensory neuron membrane proteins (SNMPs) (Benton et al. 2007; Jin et al. 2008), may also interact with odor molecules. *D. melanogaster* has a family of 60 *Or* genes, with different genes expressed in different subsets of sensory neurons. The proteins encoded by the *Or* gene family act in association with a noncanonical member of the family, the Or coreceptor Orco (previously called Or83b), to form a ligand-gated ion channel but the heterodimer may also signal through G proteins (Sato et al. 2008; Wicher et al. 2008). When expressed in *Xenopus* oocytes, *Or* genes conferred responses to odors thereby confirming their identity as odor receptor genes (Wetzel et al. 2001). *Or* genes have been identified in other insect species. Importantly, function of the receptors for sex pheromones in the silk moth *Bombyx mori* has been investigated by heterologous expression analysis (Nakagawa et al. 2005; Sakurai et al. 2004). BmOr1- and BmOr3-mediated specific responses to the sex pheromone components (E10, Z12)-hexadecadienol and (E10, Z12)-hexadecadienal in *Xenopus* oocytes. The ligand specificity of many other odor receptors has been analyzed in an *in vivo* ectopic expression system that may recapitulate their endogenous context better than provided by heterologous systems such as *Xenopus* oocytes. This expression system, sometimes called the empty neuron, is based on a mutant fly in which the endogenous receptor genes of one of the ORN classes (ab3A) are deleted; the promoter of the endogenous receptor gene *Or22a* drives expression of another *Or* in the neuron via the *UAS-GAL4* system (Dobritsa et al. 2003). The fidelity of the empty neuron has been validated and the response properties of *Drosophila* receptors (Dobritsa et al. 2003; Hallem and Carlson 2004, 2006; Kreher et al. 2005) and *Anopheles* receptors (Carey et al. 2010) have been characterized in detail using this system.

6.3 RESPONSES OF TRICHOID ORNs IN *DROSOPHILA* TO FLY ODORS

While ORNs in basiconic and coeloconic sensilla mediate responses to food odors and other odors from the environment, recordings from ORNs in trichoid sensilla suggest they respond to odors produced by the flies themselves (Van der Goes van Naters and Carlson 2007). The trichoid sensilla are found in both males and females. Under the light microscope they appear as a forest of sharp-tipped, light brown fluid-filled hairs covering a large ventrolateral area of the anterior and posterior funicular surfaces. The hairs have a thickening at the base, the socket, or basal drum, and their shafts are straight or slightly curved. In several other *Drosophila* species the distal tip of the trichoids are strongly curved, giving them a hooked or walking-stick-like appearance. Shanbhag et al. (1999) have counted 166 ± 7 trichoid sensilla in males of *D. melanogaster* and 144 ± 8 in females under the light microscope, and by EM they counted 117 and 113, respectively. There are approximately 400 sensilla in each of the two funiculi of the antennae, so the trichoid sensilla represent more than a quarter of the total number. Trichoid sensilla are approximately 20 µm long and 1.8 µm wide at the base of the hair, and the drum is 3 µm wide and high. They can be subdivided by the number of ORNs that innervate them into sensilla with one (T1 sensilla), two (T2 sensilla), or three (T3 sensilla) ORNs. The trichoids with three ORNs are found most distolaterally in the antenna, while the T1 sensilla are most medially located. T2 trichoids are in a zone between the T1 and T3 populations.

The ORN in the T1 sensilla responds to *cis*-vaccenyl acetate (Clyne et al. 1997; Ha and Smith 2006; Kurtovic et al. 2007; Van der Goes van Naters and Carlson 2007). *Cis*-vaccenyl acetate (*cis*-11-octadecenyl acetate) is an ester exclusively produced by males that has complex roles in communication, and are discussed in detail below. When air passed over a group of males is directed at the antenna from a distance, the T1 ORNs indeed respond robustly. Air passed over males and virgin females was subsequently used to screen for responses in the T2 and T3 sensilla, but no strong responses were elicited to fly odors presented in this way. During courtship flies approach each other closely, and it may be that some ORNs are adapted for reception at short range. To simulate the proximity of two interacting flies, a different stimulus method was developed. Chemicals were applied to the tip of a microcapillary, and the tip was approached to the antenna from which recordings of trichoid sensilla were made. Presented in this way, approach of a stimulus of *cis*-vaccenyl acetate caused the action potential activity of T1 neurons to ramp up rapidly (Figure 6.1). The end point of approach is when the tip carrying the stimulus is brought into contact with the sensillum. On contact, ORNs in T1 sensilla respond to a 0.5-pg load on the tip with approximately 50 impulses/s during the first 0.5 second. Withdrawal of the stimulus leads to a rapid return to prestimulus activity. The stimulus could be made to contact the sensillum repeatedly with no discernible change in responsiveness. The rate at which the action potential activity of ORNs ramped up during approach of a stimulus was very variable, while the response on contact between the sensillum and the tip carrying the stimulus appeared to be more reproducible. Responses were therefore quantified after contact of the stimulus tip with the sensillum. Doses of 0.05 pg to 0.5 ng *cis*-vaccenyl acetate in decadic steps

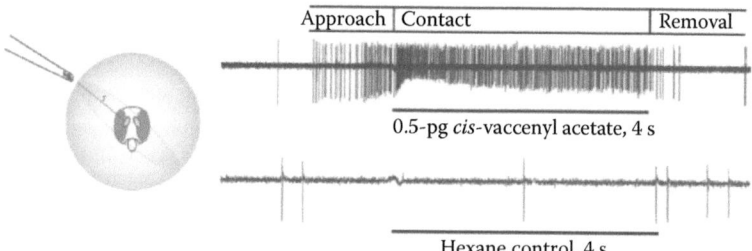

FIGURE 6.1 Recording responses to a stimulus approaching the antenna on a capillary tip. Left: schematic of stimulus presentation showing a capillary tip with material; r indicates the radius at which the stimulus rate exceeds the spontaneous rate by >20 impulses/s. Capillary tip and fly head not drawn to scale. Top right: T1 neuron responds to the approach, contact, and removal of an intermediate dose of *cis*-vaccenyl acetate. Bottom right: control stimulus. (Taken from Van der Goes van Naters W and Carlson JR (2007) *Curr Biol* 17(7):606–612. With permission.)

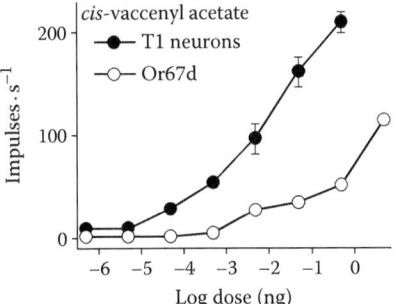

FIGURE 6.2   Responses to *cis*-vaccenyl acetate in T1 neurons (black circles) and responses mediated by Or67d in the expression system (open circles). N = 10–12, error bars indicate SEM. (Taken from Van der Goes van Naters W and Carlson JR (2007) *Curr Biol* 17(7):606–612. With permission.)

elicited increasing responses from T1 ORNs (Figure 6.2). The range of doses applied covered a large part of the dynamic range of the ORN; saturation was not achieved, suggesting that the doses were within the physiological operating range of the ORN. Using the short-range stimulus method, a systematic analysis of the trichoid population with *cis*-vaccenyl acetate and with extracts from male and virgin female flies showed that all trichoid sensilla hold ORNs that respond to fly derived chemicals. All trichoid sensilla tested appeared to have one or more ORNs that respond to male extract. Virgin female extract elicited strong responses from one or more ORNs in T2 and T3 sensilla. *Cis*-vaccenyl acetate, in addition to exciting T1 ORNs, also was a ligand for an ORN in distolateral T3 sensilla. In extracellular recordings of trichoid sensilla, it was not feasible to reliably distinguish the activity of individual ORNs. It was therefore not possible to determine whether *cis*-vaccenyl acetate and the male and female extracts acted on the same or different ORNs in the T2 and T3 sensilla. However, this analysis did show that ORNs in trichoid sensilla mediate responses to fly odors, and that the odor of females and males is represented differently in the activities of trichoid ORNs.

6.4   MOLECULAR BASIS OF ODOR RESPONSE IN TRICHOID SENSILLA

Of the 60 *Or* genes in *Drosophila*, 12 are reported to map to ORNs in trichoid sensilla: *Or2a*, *Or19a*, *Or19b*, *Or23a*, *Or43a*, *Or47b*, *Or65a*, *Or65b*, *Or65c*, *Or67d*, *Or83c*, and *Or88a* (Couto et al. 2005). The function of these odor receptors was analyzed in the empty neuron system, and four receptors mediated responses to fly chemicals in this system (Van der Goes van Naters and Carlson 2007). Or47b and Or88a were sensitive to both male and female extracts, and Or88a additionally responded to a rubbing from the genital region of males but it did not respond to *cis*-vaccenyl acetate. Or67d responded to *cis*-vaccenyl acetate as did Or65a, but Or67d was more sensitive and mediated stronger responses than Or65a. To date we know very little about the function of the other eight *Or* genes. The molecular mechanisms underlying the response to *cis*-vaccenyl acetate by T1 trichoids are being intensely investigated, and this system has become a model for pheromone sensing in *Drosophila*.

Expressed in the empty neuron, the responses to *cis*-vaccenyl acetate mediated by Or67d differ from the response in its endogenous context in T1 sensilla in several ways. With a threshold at 5 pg on the stimulus microcapillary, the ectopically expressed receptor is two orders of magnitude less sensitive than the T1 ORN. Moreover, the basiconic ORN with the ectopically expressed receptor showed several features characteristic of an adapted sensory neuron. Spontaneous activity was higher, on- and off-kinetics were slower, and the magnitudes of the responses were reduced. How does the T1 achieve its low basal activity level, high sensitivity, and fast temporal characteristics? In addition to the odor receptor there are several other molecular components that may contribute to the pheromone response in T1 sensilla, including OBPs and SNMP (Figure 6.3).

Odorant binding proteins (OBPs) are water-soluble proteins characterized by six highly conserved cysteine residues and six alpha helices that appear to provide a pocket for binding hydrophobic ligands. Insect OBPs are not related to the vertebrate lipocalin-like OBPs, which have a beta-barrel structure, even though they may have similar functions. The *Drosophila* genome carries 51 OBP genes (Hekmat-Scafe et al. 2002). The expression pattern of five OBPs (OS-E, OS-F, LUSH, PBPRP2, and PBPRP5) has been localized by immunogold labeling at the EM level (Shanbhag

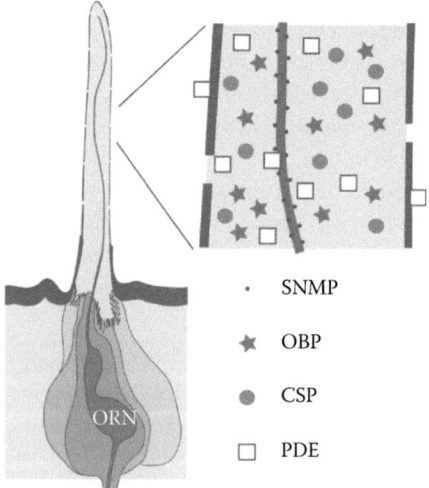

FIGURE 6.3 Schematic of a trichoid sensillum. Olfactory receptor neuron (ORN) sends a dendrite up the shaft of the sensillum; an axon projects basally toward the CNS. Inset enlarged schematic: several molecular components that have been identified in trichoid sensilla and that may act, together with the receptor (not shown) at the membrane, in the response to pheromones. Dendrite: dark gray process studded with sensory neuron membrane protein (SNMP). OBP = odorant binding protein, which are small soluble proteins characterized by three disulphide bridges (carrier function hypothesized); CSP = chemosensory protein, which are small soluble proteins characterized by two disulphide bridges with unknown function; PDE = esterases, cytochrome P450, and other enzymes: involved in ligand degradation.

et al. 2001). Intriguingly, antibodies against three OBPs (LUSH, OS-E, and OS-F) out of five tested showed strong labeling in all trichoid sensilla. A direct numerical extrapolation would suggest each sensillum could hold up to approximately 30 OBPs, but the real number may be considerably smaller as the five OBPs whose localization was studied were not a random sample from the OBP gene family. The five OBPs were originally identified by their association with the olfactory system; other family members may have functions outside the olfactory organs. However, it is clear that trichoid sensilla may harbor a lymph of remarkable complexity.

LUSH is an OBP that appears to be necessary for high sensitivity of T1 neurons to *cis*-vaccenyl acetate. It was originally implicated in olfactory function because flies homozygous for a null mutation of the gene failed to avoid strong concentrations of certain alcohols, including ethanol (Kim et al. 1998). LUSH undergoes a conformational change upon binding of *cis*-vaccenyl acetate; structural rearrangement was previously shown for the binding protein that interacts with the pheromone component Bombykol of the domesticated silkmoth *Bombyx mori* (Horst et al. 2001; Sandler et al. 2000). Experiments based on infusion into sensilla of recombinant LUSH with amino acid substitutions that reduce or mimic the conformational shift caused by pheromone binding suggested a model in which the OBP rather than the pheromone is the ligand for Or67d (Laughlin et al. 2008). In this model LUSH acts as an extracellular receptor for *cis*-vaccenyl acetate, the pheromone changes the conformation of the OBP to an activated form, and it is the activated form of the OBP that then is the ligand for Or67d at the membrane of the ORN. Particularly compelling was the observation that recombinant LUSH^{D118A} produced a dominant-active conformation that appeared to stimulate T1 neurons in complete absence of the pheromone. However, recent data with identical constructs expressed transgenically is not consistent with this hypothesis (Gomez-Diaz et al. 2013). *In vivo* expression of LUSH^{D118A} did not enhance activity of T1 neurons, and even appeared to reduce spontaneous activity relative to an appropriate control line. Moreover, it was shown that *cis*-vaccenyl acetate can activate the T1 neurons in a *lush* mutant that lacks endogenous LUSH. These results suggest that, while LUSH is important for high sensitivity of the T1 sensilla, the pheromone may directly activate the receptor Or67d. In Chapter 7, Sengupta and Smith describe a new model that combines aspects of both interpretations. In this new model SNMP acts to stabilize an inactive state of the receptor Or67d. The OBP-pheromone complex removes the inhibition by SNMP, allowing pheromone released from the complex to activate the receptor.

SNMP is a CD36-like protein first purified from an antennal preparation of the wild silk moth *Antheraea polyphemus* (Rogers et al. 1997). The class B scavenger receptor CD36 functions in binding at the cell membrane and translocation of diverse ligands, including lipids. Intriguingly, SNMP is found to be highly expressed in ORNs that respond to fatty-acid-derived pheromone components. SNMP in *Drosophila* is present in the dendrites of trichoid ORNs and only at very low levels in soma and axons; it is also expressed in support cells throughout the antenna. In mutants with a deletion of 5 bp approximately in the middle of the coding region of *Snmp* (Jin et al. 2008), and in null-mutants generated by gene targeting (Benton et al. 2007), T1 neurons have elevated spontaneous activity and show no responses to *cis*-vaccenyl acetate. Both spontaneous activity and pheromone responses could

be rescued by supplying a wild-type copy of *Snmp*. Moreover, function of the pheromone component receptor HR13 of *Heliothis virescens* (Grosse-Wilde et al. 2007)
expressed in T1 sensilla was almost completely abolished in *snmp* mutants and was
restored by transgenic rescue of Snmp. This result confirms the importance of Snmp
as a component for normal responses in trichoid ORNs to *cis*-vaccenyl acetate and
to some chemically similar pheromone components. Is SNMP required for function
of pheromone receptors in basiconic sensilla? Benton et al. (2007) found a requirement for SNMP to obtain robust responses from Or67d when it is expressed in the
empty neuron. However, using the close-range stimulus paradigm, it was shown
that Or67d expressed in the empty neuron did mediate responses to *cis*-vaccenyl
acetate, although the threshold for response was two orders of magnitude higher
than in the endogenous context (Van der Goes van Naters and Carlson 2007). The
dose required to elicit responses in the empty neuron was less than the dose eliciting
half-maximal responses in T1 neurons (Figure 6.2), and thereby can be regarded as
being within physiological range. Similarly, the pheromone receptor of the domesticated silk moth *B. mori*, BmorOr1, was functional in the empty neuron although
its response could be enhanced by coexpression with its binding protein BmorPBP1
(Syed et al. 2006).

In summary, for realizing the sensitivity and rapid on- and off-kinetics displayed
by ORNs in trichoid sensilla, several molecular components in addition to the odor
receptor are required. These are the OBPs and SNMP, but may also include other
chemosensory proteins and the diverse pheromone degrading enzymes (Leal 2013).
The sequence of events leading to a response appears to be solubilization of the
hydrophobic pheromone at the pores by an OBP, diffusion to the membrane, interaction with SNMP at the membrane, and then transfer of the pheromone to the
receptor.

Are any other models for how SNMP might function consistent with the data?
One possibility is that SNMP is involved in translocation of background levels of
pheromone and other fatty acid (derived) compounds. The transport mechanism of
hydrophobic molecules from the pores to the dendrite, probably mediated by OBPs,
is likely to have evolved to be rapid and very efficient in trichoid sensilla. In moths
even at very low levels of stimulus, 25% of stimulus molecules adsorbed on the sensillum may reach the dendrite to elicit a nerve impulse, while 75% is lost to degradation and deactivation (Kaissling 2013; see also Chapter 4). A consequence of this
may be an accumulation of pheromones and lipid-like foreign molecules against and
in the dendritic membranes. Pheromones, including *cis*-vaccenyl acetate, are produced by the flies themselves, and it is therefore remarkable that flies can respond
sensitively to approaching stimuli against a constitutively present background of
pheromone on their antennae. While much research in communication focuses on
how the signal is transmitted, the critical measure for information transfer is the
signal-to-noise ratio. To improve this ratio, it is also effective to reduce the noise.
Could SNMP act to scavenge background levels of ligand and lipid debris from the
membrane? Without a way to remove background we would expect the ORNs to
be continuously activated, an expectation consistent with the elevated spontaneous
activity in the *snmp* mutants. SNMP is related to CD36, a scavenger receptor serving
a variety of functions in association with different partners. Binding to the cognate

ligand initiates a signaling cascade that results in the internalization of the complex (Collins et al. 2009). One of the functions of CD36 is in the uptake of lipids and fatty acids, and it is possible that SNMP acts similarly to provide a continuous removal of background noise.

Detailed investigation of the mechanism of Bombykol binding and release by BmorPBP1 has yielded the most comprehensive understanding of how a specific binding protein acts. The mechanisms of transport, protection from degradation, and inactivation of the odorant following receptor activation that are served by binding proteins in moths have been analyzed in a kinetic model (Kaissling 2013). BmorPBP1 has a high affinity for Bombykol at the bulk pH of the sensillum lymph, and encapsulates its ligand within it (Sandler et al. 2000). Close to the membrane BmorPBP1 is thought to undergo a conformational change, whereby the C terminus thrusts into the binding pocket and ejects the ligand (Horst et al. 2001; Michel et al. 2011). A low pH favors the low-affinity conformation of the binding protein. The observation that there is negative charge on the surface of dendrites (Keil 1984) and the hypothesis that a local accumulation of protons occurs as a consequence (Leal 2005) provides a mechanism to explain the release of the pheromone from the binding protein at the receptor. How LUSH acts to bind and release the pheromone component is not yet clear, but a mechanism similar to the BmorPBP1-Bombykol system is one of the possibilities. Further research will also hopefully be able to explain the functional significance of the two or more other OBPs in each trichoid sensillum. Specifically, why is there more than one OBP present in T1 sensilla if there is only one ORN and one key ligand? Could the additional OBPs function as scavengers for molecules that may interfere with the pheromone response? Moreover, why is LUSH also expressed in T2 sensilla, which do not respond to *cis*-vaccenyl acetate?

6.5 *DROSOPHILA* PHEROMONES IN COURTSHIP AND AGGRESSION

Mating is the outcome of successful courtship, and courtship depends on communication between males and females. Courtship in *Drosophila* resembles a dance with an innate choreography that has been best studied in the male although females may subtly determine the outcome (Greenspan and Ferveur 2000). In courtship, males and females exchange signals that should enable them to evaluate whether the partner is suitable for reproduction. Signals are from several modalities. From a distance vision is probably most important in bringing partners together. In the dark, the sound of movement of females may affect the arousal state of males at a distance (Ejima and Griffith 2008). Males delay to initiate courtship when the female is immobilized, and courtship initiation is restored by playing a recording of the sound of movement or of white noise. Male abdominal shaking causes substrate-borne vibrations that are sensed by the female and induce her to become receptive and stop walking (Fabre et al. 2012). Sound in the form of a song generated by vibration of an extended wing is an important component of early male courtship. As the male approaches, the exchange of signals becomes more complex and more intense. At close range olfaction plays a major role. Touch in the form of tapping also involves the taste hairs on the legs. Finally, licking of the female genital region is often the immediate precursor to attempted copulation by the male.

Aggression is behavior associated with attack or threat, and may have evolved in interactions with others over limited resources (Zwarts et al. 2012). In *Drosophila* male-to-male aggression is observed during competition over food or a female, and female-to-female aggression can be observed especially when a food source with yeast is available. Males do not attack females. Male high-intensity agonistic behavior includes the lunge, boxing, and tussling. Females do head-butts and shove each other. Related behaviors are the establishment of dominance in males, where males remember and are affected by the outcome of earlier encounters with other males, and the defense of a territory. Dominance is established when males compete for a resource, and intriguingly, males appear to be able to recognize individual conspecifics from previous encounters (Yurkovic et al. 2006). As in mating, several sensory modalities play important roles in aggression. The role of hydrocarbons and behavior in eliciting aggression has been investigated by separately reversing these gender-specific characteristics in flies (Fernandez et al. 2010). Expression of a transgene (*traIR*) for generating dsRNA against the sex determination factor *transformer* in oenocytes, abdominal cells in which cuticular hydrocarbons are synthesized, masculinizes the hydrocarbon profile of a female. Expression of *traIR* in the nervous system masculinized the behavior of females. Male-to-female aggression could be elicited when either the cuticular hydrocarbon profile or the behavior of females is masculinized. Females with masculine hydrocarbons but normal female behavior were attacked by their male partners immediately after copulation, showing the effect of conflicting information. Females behaving like males also triggered aggression. When males were feminized by expression of *traF* in oenocytes and in the nervous system, other males courted them. Males therefore use both behavioral and pheromonal cues to recognize the sex of a conspecific.

Cis-vaccenyl acetate was identified in male extracts more than 40 years ago (Brieger and Butterworth 1970). Produced in the ejaculatory bulb, it has multiple roles in communication among flies:

a. *Cis*-vaccenyl acetate is transferred during copulation to the female where it acts acutely as an antiaphrodisiac to suppress courtship by males in subsequent encounters (Jallon et al. 1981).

b. It also has more chronic effects in courtship suppression and acts as the pheromonal cue in generalization of courtship learning (Ejima et al. 2007). Courtship learning is a modification of reproductive behavior whereby males exposed to a mated mature female subsequently show suppressed courtship to all females.

c. *Cis*-vaccenyl acetate acts as an aphrodisiac to virgin females by enhancing their receptivity to courtship by a male (Kurtovic et al. 2007).

d. Males and especially mated females deposit *cis*-vaccenyl acetate on food, where it appears to act as an aggregation pheromone attracting flies to a breeding substrate (Bartelt et al. 1985; Symonds and Wertheim 2005).

e. Among males, *cis*-vaccenyl acetate acutely promotes aggression (Wang and Anderson 2010), although it is not a necessary cue for aggression between a pair of males (Wang et al. 2011).

f. By contrast, chronic exposure to *cis*-vaccenyl acetate reduces aggression among males (Liu et al. 2011). *Cis*-vaccenyl acetate is detected by the flies through at least two receptors, Or67d and Or65a (Van der Goes van Naters and Carlson 2007), which are expressed in T1 and T3 sensilla, respectively (Couto et al. 2005).

Intriguingly, acute responses to *cis*-vaccenyl acetate appear to be mediated through Or67d, while chronic responses (b and f above) to *cis*-vaccenyl acetate appear to be mediated through Or65a.

Although much recent research has focused on *cis*-vaccenyl acetate as a pheromone, the chemical language of *Drosophila* is far richer and may rely predominantly on the hydrocarbons in the cuticle. In addition to a function in communication, the hydrocarbons restrict water loss (Gibbs et al. 2003) and may function as a barrier to infection (Howard and Blomquist 2005). The C21 to C35 hydrocarbons are the best-studied cuticular fraction and form a complex mixture of straight-chain and branched, saturated, and unsaturated compounds of very limited volatility (Everaerts et al. 2010). In addition to these hydrocarbons, a major group of oxygenated hydrocarbons has recently been identified on the surface of intact fruit flies by ultraviolet laser desorption/ionization orthogonal time-of-flight mass spectrometry (UV-LDI-o-TOF MS) (Yew et al. 2009). One of these, 3-O-acetyl-1,3-dihydroxyoctacosa-11,19-diene, is associated with the male genitalia. Similarly to *cis*-vaccenyl acetate, this chemical is transferred to females during copulation and then acts as an antiaphrodisiac, suppressing courtship to the female in subsequent encounters with males. The Δ9-desaturase encoded by *desat1* functions in the synthesis of key pheromone components in *D. melanogaster*, and it also may be involved in the response to pheromones as it is expressed in several brain centers associated with processing of pheromone information (Bousquet and Ferveur 2012). The hydrocarbon mixture is sexually dimorphic in *Drosophila melanogaster*. The principal hydrocarbons in females are the dienes (Z,Z)-7,11-heptacosadiene and (Z,Z)-7,11-nonacosadiene; in males they are the monoenes (Z)-7-tricosene and (Z)-7-pentacosene. The presence of more of the female hydrocarbon 7,11-heptacosadiene increases duration and frequency of mating (Antony and Jallon 1982; Marcillac and Ferveur 2004). If cuticular hydrocarbons are eliminated by ablating the oenocytes that synthesize them and *cis*-vaccenyl acetate is applied to oenocyteless females to suppress attention from males, the females' attractiveness can be restored by perfuming them with 7,11-heptacosadiene (Billeter et al. 2009). Oenocyteless females are also very attractive to males from sibling species, but application of 7,11-heptacosadiene to the females can restore the species barrier. In males, increased levels of 7-tricosene cause females to mate faster and more often (Grillet et al. 2006). Males depleted for unsaturated hydrocarbons by an insertion in the *desat1* gene were perfumed with variable quantities of 7-tricosene, and females with intact antennae could distinguish among these males while antennectomized females could not. In male-to-male interactions, courtship is suppressed by 7-tricosene (Ferveur and Sureau 1996; Wang et al. 2011) and, while *cis*-vaccenyl acetate promotes aggression among males but is not necessary for it, 7-tricosene is essential for normal levels of aggression among males (Wang et al. 2011). Receptors for dienes or the oxygenated hydrocarbons are still elusive. 7-tricosene appears to be sensed through gustatory receptors, although olfactory receptors may also be involved. The *Gr* genes that underlie taste of sugars and bitter chemicals, and possibly pheromones, encode a large family of highly diverse gustatory receptors (Clyne et al. 2000; Dunipace et al. 2001; Robertson et al. 2003; Scott et al. 2001). A knockout of *Gr32a*, which encodes a protein with similarity in amino acid sequence to the receptor Gr68a that has previously been implicated in pheromone communication (Bray and Amrein 2003), showed that Gr32a is essential for suppression of male courtship by 7-tricosene (Miyamoto and Amrein 2008). Gr32a is also required for normal aggression among males. From electrophysiological studies of taste reception on the proboscis, it was concluded that 7-tricosene and caffeine elicit responses from the same taste neurons (Lacaille et al. 2007). Indeed, caffeine applied to flies could inhibit courtship while 7-tricosene could inhibit extension of the proboscis to sucrose, suggesting a taste that is perceived as aversive. Based on expression analysis of *Gr-GAL4* drivers, individual bitter sensitive taste neurons express distinct subsets of *Gr* genes (Weiss et al. 2011). Gr32a, together with four other Grs, is expressed in all bitter sensitive neurons. Some of these shared Grs may function as a coreceptor, analogous to Orco in the olfactory system, as has been suggested for Gr33a (Moon et al. 2009). Other molecular components that function in pheromone reception are ion channel genes belonging to the degenerin/epithelial sodium channel/pickpocket (*ppk*) family, which are expressed in *fruitless* positive sensilla on the legs (Liu et al. 2012; Starostina et al. 2009, 2012; Thistle et al. 2012) and members of the *CheB* family (Toda et al. 2012). While it is not yet known where in the signaling cascade the Ppk proteins operate, they are a necessary component for the suppression of courtship to males and enhancing courtship to females.

Fifty-eight hydrocarbons can be detected in mature flies by gas chromatography; 19 are specific to virgin females, 3 are male-specific and 36 are found in both sexes (Everaerts et al. 2010). Physical parameters such as temperature influence the hydrocarbon profile displayed by a fly (Ferveur 2005). In addition, individual flies are part of the environment of other flies, and have measurable indirect genetic effects on the hydrocarbon display (Kent et al. 2008; Krupp et al. 2008). The social context in males alters the display of monoenes and methyl-branched alkanes, but not of n-alkanes. Early studies showed pheromones are essential to induce courtship in males and indicated a role of the female dienes in this (Jallon 1984). However, in flies generated to lack the dienes, females are still highly attractive to males, suggesting they produce potent additional aphrodisiac pheromones (Savarit et al. 1999). The presence of these cryptic "Ur" pheromones together with the ability to learn the odor profile of age classes (Ejima et al. 2007), and the influence of social context on the display of the hydrocarbon profile suggest *Drosophila* employs a highly complex and subtle chemical language.

A fascinating frontier in sensory biology is the investigation of how sensory information from the periphery is processed in the neural circuits of the brain to drive changes in behavior (Pavlou and Goodwin 2013). Detailed neuroanatomical maps are a prerequisite, and recent studies show sexual dimorphisms in wiring, especially in the lateral horn, which may provide an anatomical basis for opposing responses to pheromones in males and females (Cachero et al. 2010). A detailed neural map of *fruitless* expressing neurons that extends from sensory input to motor output has been assembled and it suggests that, while input and output circuits are largely similar between males and females, sex-specific behavior may arise by dimorphic connections in the brain that couple sensory input to motor output (Yu et al. 2010). Through stimulation of *fru* neurons by activating thermosensitive dTrpA1 channels expressed in them, courtship acts can be induced in solitary males without a partner. Using mosaic analysis with a repressible cell marker (MARCM) to reduce the number of neurons expressing the thermosensitive channel, specific interneurons in the brain have been identified that when activated initiate courtship in *Drosophila* (Kohatsu et al. 2011).

REFERENCES

Antony C and Jallon J (1982) The chemical basis for sex recognition in *Drosophila melanogaster*. *J Insect Physiol* 28:873–880.

Bartelt R, Schaner A and Jackson L (1985) Cis-vaccenyl acetate as an aggregation pheromone in *Drosophila melanogaster*. *J Chem Ecol* 11:1747–1756.

Benton R, Vannice KS, Gomez-Diaz C and Vosshall LB (2009) Variant ionotropic glutamate receptors as chemosensory receptors in *Drosophila*. *Cell* 136(1):149–162.

Benton R, Vannice KS and Vosshall LB (2007) An essential role for a CD36-related receptor in pheromone detection in *Drosophila*. *Nature* 450(7167):289–293.

Billeter JC, Atallah J, Krupp JJ, Millar JG and Levine JD (2009) Specialized cells tag sexual and species identity in *Drosophila melanogaster*. *Nature* 461(7266):987–991.

Bousquet F and Ferveur JF (2012) desat1: A Swiss army knife for pheromonal communication and reproduction? *Fly* 6(2):102–107.

Bray S and Amrein H (2003) A putative *Drosophila* pheromone receptor expressed in male-specific taste neurons is required for efficient courtship. *Neuron* 39(6):1019–1029.

Brieger G and Butterworth FM (1970) *Drosophila melanogaster*: Identity of male lipid in reproductive system. *Science* 167(3922):1262.

Cachero S, Ostrovsky AD, Yu JY, Dickson BJ and Jefferis GS (2010) Sexual dimorphism in the fly brain. *Curr Biol* 20(18):1589–1601.

Carey AF, Wang G, Su CY, Zwiebel LJ and Carlson JR (2010) Odorant reception in the malaria mosquito *Anopheles gambiae*. *Nature* 464(7285):66–71.

Clyne P, Grant A, O'Connell R and Carlson JR (1997) Odorant response of individual sensilla on the *Drosophila* antenna. *Invert Neurosci* 3(2–3):127–135.

Clyne PJ et al. (1999) A novel family of divergent seven-transmembrane proteins: Candidate odorant receptors in *Drosophila*. *Neuron* 22(2):327–338.

Clyne PJ, Warr CG and Carlson JR (2000) Candidate taste receptors in *Drosophila*. *Science* 287(5459):1830–1834.

Collins RF et al. (2009) Uptake of oxidized low density lipoprotein by CD36 occurs by an actin-dependent pathway distinct from macropinocytosis. *J Biol Chem* 284(44):30288–30297.

Couto A, Alenius M, and Dickson B (2005) Molecular, anatomical, and functional organization of the *Drosophila* olfactory system. *Curr Biol* 15:1535–1547.

De Bruyne M, Foster K and Carlson JR (2001) Odor coding in the *Drosophila* antenna. *Neuron* 30(2):537–552.

Dobritsa AA, Van der Goes van Naters W, Warr CG, Steinbrecht RA and Carlson JR (2003) Integrating the molecular and cellular basis of odor coding in the *Drosophila* antenna. *Neuron* 37(5):827–841.

Dunipace L, Meister S, McNealy C and Amrein H (2001) Spatially restricted expression of candidate taste receptors in the *Drosophila* gustatory system. *Curr Biol* 11(11):822–835.

Ejima A and Griffith LC (2008) Courtship initiation is stimulated by acoustic signals in *Drosophila melanogaster*. *PLoS One* 3(9):e3246.

Ejima A et al. (2007) Generalization of courtship learning in *Drosophila* is mediated by cis-vaccenyl acetate. *Curr Biol* 17(7):599–605.

Everaerts C, Farine JP, Cobb M and Ferveur JF (2010) *Drosophila* cuticular hydrocarbons revisited: mating status alters cuticular profiles. *PLoS One* 5(3):e9607.

Fabre CC et al. (2012) Substrate-borne vibratory communication during courtship in *Drosophila melanogaster*. *Curr Biol* 22(22):2180–2185.

Fernández MP et al. (2010) Pheromonal and behavioral cues trigger male-to-female aggression in *Drosophila*. *PLoS Biology* 8(11):541–541.

Ferveur JF (2005) Cuticular hydrocarbons: Their evolution and roles in *Drosophila* pheromonal communication. *Behav Genet* 35(3):279–295.

Ferveur JF and Sureau G (1996) Simultaneous influence on male courtship of stimulatory and inhibitory pheromones produced by live sex-mosaic *Drosophila melanogaster*. *Proc Biol Sci* 263(1373):967–973.

Gibbs AG, Fukuzato F and Matzkin LM (2003) Evolution of water conservation mechanisms in *Drosophila*. *J Exp Biol* 206(Pt 7):1183–1192.

Gomez-Diaz C, Reina JH, Cambillau C and Benton R (2013) Ligands for pheromone-sensing neurons are not conformationally activated odorant binding proteins. *PLoS Biology* 11(4):e1001546.

Greenspan RJ and Ferveur JF (2000) Courtship in *Drosophila*. *Annu Rev Genet* 34:205–232.

Grillet M, Dartevelle L and Ferveur JF (2006) A *Drosophila* male pheromone affects female sexual receptivity. *Proc Biol Sci* 273(1584):315–323.

Grosse-Wilde E, Gohl T, Bouche E, Breer H and Krieger J (2007) Candidate pheromone receptors provide the basis for the response of distinct antennal neurons to pheromonal compounds. *Eur J Neurosci* 25(8):2364–2373.

Ha TS and Smith DP (2006) A pheromone receptor mediates 11-cis-vaccenyl acetate-induced responses in *Drosophila*. *J Neurosci* 26(34):8727–8733.

Hallem EA and Carlson JR (2006) Coding of odors by a receptor repertoire. *Cell* 125(1):143–160.

Hallem EA, Ho MG and Carlson JR (2004) The molecular basis of odor coding in the *Drosophila* antenna. *Cell* 117:965–979.

Hekmat-Scafe DS, Scafe CR, McKinney AJ and Tanouye MA (2002) Genome-wide analysis of the odorant-binding protein gene family in *Drosophila melanogaster*. *Genome Res* 12(9):1357–1369.

Horst R et al. (2001) NMR structure reveals intramolecular regulation mechanism for pheromone binding and release. *Proc Natl Acad Sci U S A* 98:14374–14379.

Howard RW and Blomquist GJ (2005) Ecological, behavioral, and biochemical aspects of insect hydrocarbons. *Annu Rev Entomol* 50:371–393.

Jallon JM (1984) A few chemical words exchanged by *Drosophila* during courtship and mating. *Behav Genet* 14(5):441–478.

Jallon J, Antony C and Benamar O (1981) Un anti-aphrodisiac produit par les mâles de *Drosophila melanogaster* et transféré aux femelles lors de la copulation. *C R Acad Sci Paris* 292D:1147–1149.

Jin X, Ha TS and Smith DP (2008) SNMP is a signaling component required for pheromone sensitivity in *Drosophila*. *Proc Natl Acad Sci U S A* 105(31):10996–11001.

Kaissling KE (2013) Kinetics of olfactory responses might largely depend on the odorant-receptor interaction and the odorant deactivation postulated for flux detectors. *J Comp Physiol A* 199:879–896.

Keil TA (1984) Surface coats of pore tubules and olfactory sensory dendrites of a silkmoth revealed by cationic markers. *Tissue Cell* 16(5):705–717.

Keil TA (1997) Comparative morphogenesis of sensilla: A review. *Int J Insect Morphol* 26(3–4):151–160.

Kent C, Azanchi R, Smith B, Formosa A and Levine JD (2008) Social context influences chemical communication in *D. melanogaster* males. *Curr Biol* 18(18):1384–1389.

Kim MS, Repp A and Smith DP (1998) LUSH odorant-binding protein mediates chemosensory responses to alcohols in *Drosophila melanogaster*. *Genetics* 150(2):711–721.

Kohatsu S, Koganezawa M and Yamamoto D (2011) Female contact activates male-specific interneurons that trigger stereotypic courtship behavior in *Drosophila*. *Neuron* 69(3):498–508.

Kreher SA, Kwon JY and Carlson JR (2005) The molecular basis of odor coding in the *Drosophila* larva. *Neuron* 46(3):445–456.

Krupp JJ et al. (2008) Social experience modifies pheromone expression and mating behavior in male *Drosophila melanogaster*. *Curr Biol* 18(18):1373–1383.

Kurtovic A, Widmer A and Dickson BJ (2007) A single class of olfactory neurons mediates behavioural responses to a *Drosophila* sex pheromone. *Nature* 446(7135):542–546.

Lacaille F et al. (2007) An inhibitory sex pheromone tastes bitter for *Drosophila* males. *PLoS One* 2(7):e661.

Laughlin JD, Ha TS, Jones DN and Smith DP (2008) Activation of pheromone-sensitive neurons is mediated by conformational activation of pheromone-binding protein. *Cell* 133(7):1255–1265.

Leal WS (2005) Pheromone reception. *Top Curr Chem* 240:1–36.

Leal WS (2013) Odorant reception in insects: Roles of receptors, binding proteins, and degrading enzymes. *Annu Rev Entomol* 58:373–391.

Liu T, Starostina E, Vijayan V and Pikielny CW (2012) Two *Drosophila* DEG/ENaC channel subunits have distinct functions in gustatory neurons that activate male courtship. *J Neurosci* 32(34):11879–11889.

Liu W et al. (2011) Social regulation of aggression by pheromonal activation of Or65a olfactory neurons in *Drosophila*. *Nat Neurosci* 14(7):896–902.

Marcillac F and Ferveur JF (2004) A set of female pheromones affects reproduction before, during and after mating in *Drosophila*. *J Exp Biol* 207(22):3927–3933.

Michel E et al. (2011) Dynamic conformational equilibria in the physiological function of the *Bombyx mori* pheromone-binding protein. *J Mol Biol* 408(5):922–931.

Miyamoto T and Amrein H (2008) Suppression of male courtship by a *Drosophila* pheromone receptor. *Nat Neurosci* 11(8):874–876.

Montell C (2009) A taste of the *Drosophila* gustatory receptors. *Curr Opin Neurobiol* 19(4):345–353.

Moon SJ, Lee Y, Jiao Y and Montell C (2009) A *Drosophila* gustatory receptor essential for aversive taste and inhibiting male-to-male courtship. *Curr Biol* 19(19):1623–1627.

Nakagawa T, Sakurai T, Nishioka T and Touhara K (2005) Insect sex-pheromone signals mediated by specific combinations of olfactory receptors. *Science* 307(5715):1638–1642.

Pavlou HJ and Goodwin SF (2013) Courtship behavior in *Drosophila melanogaster*: Towards a "courtship connectome". *Curr Opin Neurobiol* 23(1):76–83.

Ray K and Rodrigues V (1995) Cellular events during development of the olfactory sense organs in *Drosophila melanogaster*. *Dev Biol* 167(2):426–438.

Reddy GV, Gupta B, Ray K and Rodrigues V (1997) Development of the *Drosophila* olfactory sense organs utilizes cell-cell interactions as well as lineage. *Development* 124(3):703–712.

Robertson HM, Warr CG and Carlson JR (2003) Molecular evolution of the insect chemoreceptor gene superfamily in *Drosophila melanogaster*. *Proc Natl Acad Sci U S A* 100(Suppl 2):14537–14542.

Rogers ME, Sun M, Lerner MR and Vogt RG (1997) Snmp-1, a novel membrane protein of olfactory neurons of the silk moth *Antheraea polyphemus* with homology to the CD36 family of membrane proteins. *J Biol Chem* 272(23):14792–14799.

Sakurai T et al. (2004) Identification and functional characterization of a sex pheromone receptor in the silkmoth *Bombyx mori*. *Proc Natl Acad Sci U S A* 101(47):16653–16658.

Sandler BH, Nikonova L, Leal WS and Clardy J (2000) Sexual attraction in the silkworm moth: Structure of the pheromone-binding-protein-bombykol complex. *Chem Biol* 7(2):143–151.

Sato K, Pellegrino M, Nakagawa T, Vosshall LB and Touhara K (2008) Insect olfactory receptors are heteromeric ligand-gated ion channels. *Nature* 452(7190):1002–1006.

Savarit F, Sureau G, Cobb M and Ferveur JF (1999) Genetic elimination of known pheromones reveals the fundamental chemical bases of mating and isolation in *Drosophila*. *Proc Natl Acad Sci U S A* 96(16):9015–9020.

Scott K et al. (2001) A chemosensory gene family encoding candidate gustatory and olfactory receptors in *Drosophila*. *Cell* 104:661–673.

Shanbhag SR, Müller B, and Steinbrecht RA (1999) Atlas of olfactory organs of *Drosophila melanogaster*. 1. Types, external organization, innervation and distribution of olfactory sensilla. *Int J Insect Morphol* 28:377–397.

Shanbhag SR, Müller B and Steinbrecht RA (2000) Atlas of olfactory organs of *Drosophila melanogaster*. 2. Internal organization and cellular architecture of olfactory sensilla. *Arthropod Struct Dev* 29(3):211–229.

Shanbhag SR et al. (2001) Expression mosaic of odorant-binding proteins in *Drosophila* olfactory organs. *Microsc Res Tech* 55(5):297–306.

Starostina E, Xu A, Lin H and Pikielny CW (2009) A *Drosophila* protein family implicated in pheromone perception is related to Tay-Sachs GM2-activator protein. *J Biol Chem* 284(1):585–594.

Starostina E et al. (2012) A *Drosophila* DEG/ENaC subunit functions specifically in gustatory neurons required for male courtship behavior. *J Neurosci* 32(13):4665–4674.

Su C-Y, Menuz K and Carlson JR (2011) Olfactory perception: Receptors, cells, and circuits. *Cell* 139(1):45–49.

Syed Z, Ishida Y, Taylor K, Kimbrell DA and Leal WS (2006) Pheromone reception in fruit flies expressing a moth's odorant receptor. *Proc Natl Acad Sci U S A* 103(44):16538–16543.

Symonds MR and Wertheim B (2005) The mode of evolution of aggregation pheromones in *Drosophila* species. *J Evol Biol* 18(5):1253–1263.

Thistle R, Cameron P, Ghorayshi A, Dennison L and Scott K (2012) Contact chemoreceptors mediate male-male repulsion and male-female attraction during *Drosophila* courtship. *Cell* 149(5):1140–1151.

Toda H, Zhao X and Dickson BJ (2012) The *Drosophila* female aphrodisiac pheromone activates *ppk23+* sensory neurons to elicit male courtship behavior. *Cell Reports* 1(6):599–607.

Van der Goes van Naters W and Carlson JR (2007) Receptors and neurons for fly odors in *Drosophila*. *Curr Biol* 17(7):606–612.

Vosshall LB, Amrein H, Morosov PS, Rzhetsky A and Axel R (1999) A spatial map of olfactory receptor expression in the *Drosophila* antenna. *Cell* 96:725–736.

Wang L and Anderson DJ (2010) Identification of an aggression-promoting pheromone and its receptor neurons in *Drosophila*. *Nature* 463(7278):227–231.

Wang L et al. (2011) Hierarchical chemosensory regulation of male-male social interactions in *Drosophila*. *Nat Neurosci* 14(6):757–762.

Weiss LA, Dahanukar A, Kwon JY, Banerjee D and Carlson JR (2011) The molecular and cellular basis of bitter taste in *Drosophila*. *Neuron* 69(2):258–272.

Wetzel CH et al. (2001) Functional expression and characterization of a *Drosophila* odorant receptor in a heterologous cell system. *Proc Natl Acad Sci U S A* 98:9377–9380.

Wicher D et al. (2008) *Drosophila* odorant receptors are both ligand-gated and cyclic-nucleotide-activated cation channels. *Nature* 452(7190):1007–1011.

Yao CA, Ignell R and Carlson JR (2005) Chemosensory coding by neurons in the coeloconic sensilla of the *Drosophila* antenna. *J Neurosci* 25(37):8359–8367.

Yew JY et al. (2009) A new male sex pheromone and novel cuticular cues for chemical communication in *Drosophila*. *Curr Biol* 19(15):1245–1254.

Yu JY, Kanai MI, Demir E, Jefferis GS and Dickson BJ (2010) Cellular organization of the neural circuit that drives *Drosophila* courtship behavior. *Curr Biol* 20(18):1602–1614.

Yurkovic A, Wang O, Basu AC and Kravitz EA (2006) Learning and memory associated with aggression in *Drosophila melanogaster*. *Proc Natl Acad Sci U S A* 103(46):17519–17524.

Zwarts L, Versteven M and Callaerts P (2012) Genetics and neurobiology of aggression in *Drosophila*. *Fly* 6(1):35–48.

7 How *Drosophila* Detect Volatile Pheromones
Signaling, Circuits, and Behavior

Samarpita Sengupta and Dean P. Smith

CONTENTS

7.1 Introduction 230
    7.1.1 *Drosophila melanogaster* as a Volatile Pheromone
        Model System..230
    7.1.2 *Drosophila* Olfactory Anatomy..232
    7.1.3 *Drosophila* Odorant Receptor (Or) Family 235
    7.1.4 *Drosophila* Odorant Binding Proteins..235
    7.1.5 Gene Products Underlying cVA Pheromone Detection
        in *Drosophila*..236
    7.1.6 LUSH OBP 237
    7.1.7 SNMP 241
7.2 cVA-Induced Behaviors 243
    7.2.1 Courtship 243
    7.2.2 Aggregation 245
    7.2.3 Aggression 245
7.3 Other *Drosophila* Pheromones 246
7.4 Learning Modulates Hard-Wired Pheromone Behaviors............................247
7.5 Contact Chemoreception 248
7.6 Future Prospects 249
References...250

Pheromones and the mechanisms to detect them have evolved to transmit biologically relevant information from one member of a species to another, often with miniscule amounts of chemical. In *Drosophila*, the fatty acid pheromone 11-*cis*-vaccenyl acetate (cVA) is a male-specific pheromone that functions as a courtship cue to ensure an appropriate partner is selected for mating. However, cVA also underlies other behaviors including aggression and aggregation. A specialized population of sensory neurons tuned solely to cVA mediates the detection of cVA pheromone. These neurons express a unique set of signal transduction machinery essential for detection of low levels of cVA in air. Stimulation of these neurons activates a labeled-line circuit to higher processing centers in the brain. These circuits trigger behavior outputs that are hardwired, but that can be modulated by learning. In this chapter we review the current state of our understanding of cVA pheromone biology in *Drosophila*, with emphasis on recent findings in pheromone detection mechanisms and the circuits underlying pheromone-induced behaviors.

7.1 INTRODUCTION

Pheromones are chemicals released from one individual to influence the behavior of another animal of the same species (Karlson and Luscher 1959). Detection of these pheromones can produce broad developmental or endocrine changes (priming phero-mones) or elicit specific behaviors (releaser pheromones). Releaser pheromones elicit innate behaviors in the receiving individual, and are widely used in the animal king-dom, often to guide mating behavior toward appropriate partners. Pheromones are used to guide social interactions in both vertebrate and invertebrate animals. The social insects (ants, bees, and termites) have taken great advantage of pheromone sig-naling to create a chemical language that guides an array of behaviors and develop-mental programs essential for the overall functioning of the colony (reviewed in Alaux et al. 2010). Therefore, understanding how pheromones are detected and how this information is ultimately converted into specific behaviors is of great interest. In this chapter we focus on volatile insect pheromone detection and processing. Insects are well known to have exquisite sensitivity to pheromones. For example, sex pheromones released from female moths attract male mating partners over great distances (Carde and Willis 2008; Fabre 1916), and males can detect single molecules (Kaissling and Priesner 1970). How this remarkable sensitivity is achieved remains poorly understood.

Studies utilizing *Drosophila melanogaster* have been instrumental in elucidating the molecular mechanisms underlying volatile pheromone transduction (reviewed in Ha and Smith 2009; Ronderos and Smith 2009; Smith 2012; Vosshall 2008). Here we review recent progress in understanding the detection and neuronal circuitry underlying behaviors elicited by the *Drosophila* releaser pheromone 11-*cis*-vaccenyl acetate (cVA). In addition, the neuronal circuits activated by cVA are beginning to be worked out. We discuss recent findings suggesting that the mechanisms for detection of contact (taste) pheromones (including cVA) are distinct from those used for vola-tile pheromones. Finally, to put these findings in the larger context, lessons learned in the fruit fly are likely to be relevant to other insect pheromone systems, and may reveal general principles underlying pheromone-induced behaviors in all animals. This information will provide the basis for novel approaches that are more selec-tive than chemical pesticides to control insect species that cause human disease and inflict crop damage.

7.1.1 *Drosophila melanogaster* as a Volatile Pheromone Model System

Nocturnal insect species like moths have developed extremely sensitive olfactory mechanisms to signal reproductive availability to appropriate mates in the absence of visual cues. Indeed, moths remain an attractive model system to study pheromone detection due to their relatively large size and amazing sensitivity to sex pheromones. However, the lack of potent genetic tools limits the use of these animals to dissect the molecular basis for pheromone transduction. *Drosophila melanogaster* has proven to be a valuable system to explore the molecular mechanisms mediating insect phero-mone detection and to elucidate of the neuronal circuits underlying pheromone-induced behaviors. The tools developed from over a century of fly research allow us to produce single gene mutants and evaluate the contribution of that single gene on pheromone detection *in vivo*. Using flies, it is possible to identify gene products responsible for virtually any scorable phenotype. Genetic screens for mutants that are defective for pheromone responses have revealed mutations in several genes criti-cal for this process (Jin et al. 2008). These mutations are mapped to the fly genome and fine mapping and sequence analysis identify the genes required for pheromone sensitivity. The advantage of such a forward genetic approach to find pheromone sensitivity factors is that it is unbiased. Any gene product required for any aspect of pheromone detection will be recovered, even if it encodes an unexpected factor. One major challenge for a forward genetic approach for pheromone-defective mutants is that lethal genes will not be recovered in a typical screen. Since we expect most or all genes involved exclusively in pheromone detection to be dispensable for life, this should not be a major problem. A second major challenge for a genetic screen is to elucidate the biological role for these important gene products in the pheromone detection process once they are identified. In addition to genes encoding factors that directly mediate signaling, a genetic screen will also recover mutations in transcrip-tion factors and other biosynthetic proteins required for proper production of signal-ing factors, since these will also cause pheromone insensitivity.

Flies offer additional advantages as a model to study pheromone biology. The fly genome is easily manipulated. Transgenic animals can be produced with single-copy transgenes, allowing the expression of modified proteins in the mutant background, essentially replacing the wild type protein with the mutant variant. This allows the evaluation of any defects under the most physiological relevant conditions possible.

The ability to manipulate the genome is also useful if one takes a reverse genetic approach. In this case, interesting genes may be identified in the genome or have an interesting homolog in another species. Using one of a plethora of tools, one can engineer mutants in virtually any *Drosophila* gene. Homologous recombination can be used to target individual genes and produce null mutants (Rong et al. 2002). In addition, the gene disruption project has produced thousands of fly stocks, each car-rying a single transposable element that often disrupts expression of one or more genes at the integrations site (Bellen et al. 2004). Some elements encode flippase recognition target (FRT) sites, the recombination site for flippase (FLP) recombinase enzyme (Golic and Lindquist 1989). One can produce small deletions to remove a gene of interest by crossing two FRT strains with elements integrated on either side of the gene of interest, and inducing FLP. FLP will mediate mitotic recombination between the FRT sites on the two chromosomes, resulting in a chromosome lacking the DNA between the FRT sites. These approaches are the gold standard for analyz-ing the function of *Drosophila* genes, because they completely eliminate expres-sion of the gene product of interest. Alternatively, transgenic RNA interference (RNAi) approaches are useful, and transgenic lines carrying RNAi constructs are available that target every gene in the genome. RNAi works well in *Drosophila*, with >95% knockdown of protein products achieved routinely (Kalidas and Smith 2002). However, like RNAi in vertebrate cells, knockdown is never complete and there may be enough protein produced to confer pheromone sensitivity. In such cases, a protein essential for pheromone detection might be missed in a leaky RNAi line.

Olfaction in flies can be probed using electrophysiology, behavior, cell biology, and biochemistry. While small, flies are amenable to electrophysiology. For olfaction, we perform single sensillum recordings (SSRs) that are extracellular recordings of the olfactory neurons within a single sensillum. Single sensillum electrophysiology allows monitoring of real-time neuronal activity in pheromone-sensitive neurons. Optogenetic approaches continue to improve and be applied to *Drosophila* pheromone biology. For example, transgenic expression of calcium-activated fluorescent proteins like GcAMP3 (Tian et al. 2009) emit light when they bind the calcium that floods the cytoplasm when neurons are activated. These reporters lack the exquisite time resolution of electrophysiology but have the advantage that the activity of many neurons can be monitored simultaneously. Finally, techniques are available in flies that allow us to trace the neuronal circuits underlying pheromone behaviors. Mosaic analysis with a repressible cell marker (MARCM) (Lee and Luo 2001) was developed by Liqun Luo and coworkers and utilizes FLP-mediated expression of a green fluorescent protein (GFP) reporter in small subsets of cells, producing an effect analogous to a fluorescent golgi stain (Lee and Luo 2001). Photoactivatable GFP is a GFP variant that fluoresces only after exposure to high-intensity 710-nm light (Patterson and Lippincott-Schwartz 2002). Thus, if the axonal target for a neuron in the pheromone detection circuit is known, the target can be illuminated, allowing the next order neuron in the circuit to be labeled and traced (Datta et al. 2008). These techniques allow us to follow the information flow from olfactory neurons activated by pheromone to the downstream circuits and determine how that information is delivered and processed by the central nervous system (see below) (Jefferis et al. 2007; Ruta et al. 2010).

7.1.2 *Drosophila* Olfactory Anatomy

Flies have two paired olfactory organs located on the head; the antennae and the maxillary palps (Figure 7.1a, b). These organs are covered with hollow, hairlike cuticle structures called sensilla, each of which contains the dendrites of 1–4 olfactory neurons (Figure 7.1b, c). Thus the olfactory neurons are partitioned into discrete isolated units. Similar to human hair cells in the ear, the olfactory neuron dendrites within the sensilla are bathed in a potassium-rich fluid called the sensillum lymph. In *Drosophila*, there are four morphologically distinct classes of sensilla, the food-odorant detecting basiconic sensilla, the coeloconic sensilla, the trichoid sensilla, and the intermediate sensilla (Stocker 1994). The latter are so named because they are intermediate in morphology between the trichoid and basiconic sensilla. In most insects, pheromones are detected by neurons located within the trichoid sensilla. The olfactory neurons project dendrites into the hollow sensilla (Figure 7.1c), and their axons travel to the antennal lobes in the fly central nervous system where they synapse with projection neurons in neuronal clusters called glomeruli (Figure 7.1d).

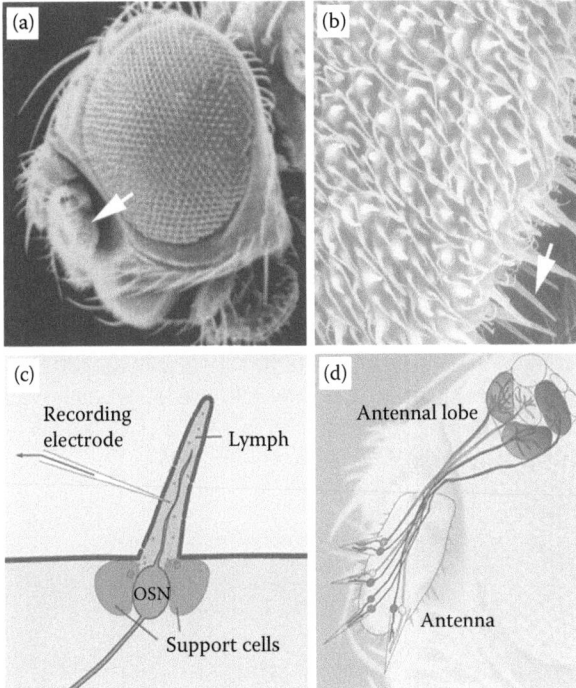

FIGURE 7.1 **(See color insert.)** Anatomy of the *Drosophila* olfactory system. (a) SEM of a *Drosophila* head. Arrow indicates the third segment of the antenna that contains the olfactory neurons. (b) Higher magnification of the antenna surface showing individual hairs or sensilla. Arrow depicts one of the long, needle-shaped trichoid sensillum that detect pheromones. (c) Drawing of a single sensillum impaled by a glass recording pipette. The dendrites of the olfactory neurons (OSN) project into the sensillum lymph-filled cavity within the sensillum shaft. Odorant binding proteins (red dots) are secreted by the support cells into the lymph. (d) Olfactory neurons project to the antennal lobe glomeruli. Neurons expressing the same receptor innervate the same glomerulus. (From Ronderos, D. S. and D. P. Smith (2009). *Fly (Austin)* 3(4): 290–297.)

There are many parallels between insect and vertebrate olfactory processing (Hildebrand and Shepherd 1997). Like vertebrates, the olfactory neurons in insects tend to express a single tuning odorant receptor gene, and neurons expressing the same receptor innervate the same glomerulus in the brain (Figure 7.1d). Therefore, the odorant code established in the brain results from the pattern of glomeruli activated by an odor, which in turn, depends on the odorant sensitivity of individual odorant receptors. Neuronal activity in each glomerulus is relayed to higher processing centers by projection neurons (analogous to mitral cells in vertebrates). A total of 20 to 50 olfactory neurons expressing the same odorant receptor converge at the glomerulus and innervate a small number of dedicated projection neurons. Therefore, there is signal amplification and noise reduction occurring through convergence at the glomerulus (Bhandawat et al. 2007). In *Drosophila*, the projection neurons innervating most glomeruli activated by food odorants innervate distinct targets from the projection neurons activated by pheromones (Jefferis et al. 2007). Indeed, when analyzed *in vivo*, exposure of females to male fly volatiles activates a single glomerulus, DA1, corresponding to activation of the cVA-sensitive neurons (Masuyama et al. 2012). In contrast to food odorants that usually activate multiple odorant receptor types, cVA detection activates a single labeled line (Schlief and Wilson 2007).

In most insects, pheromone detection occurs in the trichoid sensilla neurons. *Drosophila* has three classes of trichoid sensilla, the at1, at3, and at4 classes (Figure 7.2). The at2 sensillum class, previously thought to be a member of the trichoid class (Couto et al. 2005) is actually an intermediate sensillum class we renamed ai2 (Ronderos and Smith, in press). The 50 or so at1 sensilla are specialized for cVA detection, and the single neuron located within these sensilla uniquely express the odorant receptor Or67d (Ha and Smith 2006; Kurtovic et al. 2007). The at3 and at4 sensilla each contain three neurons, but little is known about their odorant sensitivity or function. The projections of all seven classes of neurons have been mapped to the antennal lobe glomeruli (Figure 7.2) (Couto et al. 2005).

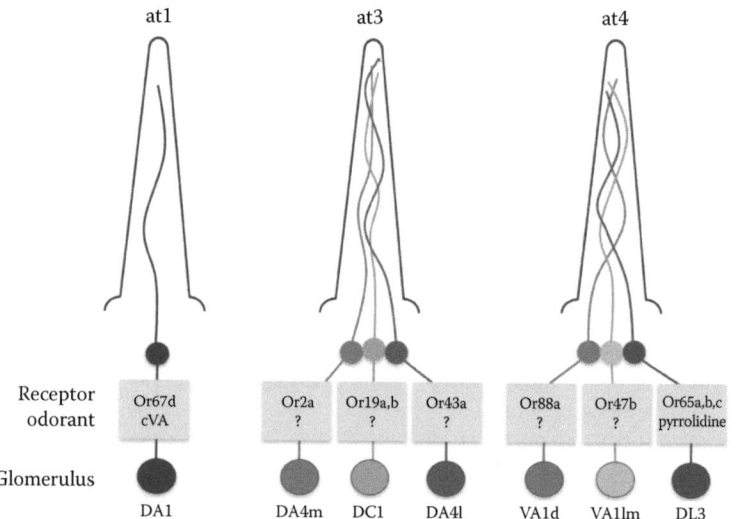

FIGURE 7.2   (See color insert.) Drawing depicting the *Drosophila* trichoid sensilla. Three classes are present on the antenna, at1, at3, and at4. at1 sensilla contain a single neuron that expresses Or67d and is exquisitely tuned to cVA pheromone. at3 sensilla contain three neurons expressing Or2a, Or19a and b and Or43a. Or19a and b are recent gene duplications that encode identical proteins. Or43a has been reported to respond to benzaldehyde and similar odorants when misexpressed in the empty neuron system (Stortkuhl 2001). However, other reports find no sensitivity to benzaldehyde in the same system (Hallem and Carlson 2006), so the tuning of this receptor remains controversial. at4 sensilla contain three olfactory neurons expressing Or88a, Or47b and Or65a, b, and c in the aT4A, B, and C neurons, respectively. Or65-expressing neurons are the only *Drosophila* olfactory neurons expressing three distinct receptors, perhaps to broaden the tuning of this neuron. Unlike the food-detecting odorant receptors expressed in basiconic sensilla, the receptors depicted here do not respond well to odorants when misexpressed in the empty neuron system (Hallem and Carlson 2006). Note: at2 sensilla, previously classified as trichoid sensilla (Couto et al. 2005) are actually intermediate sensilla (Ronderos and Smith, submitted).

7.1.3 *Drosophila* Odorant Receptor (Or) Family

The *Drosophila* Or family consists of 62 seven-transmembrane odorant receptors derived from 60 genes. These receptors have seven-transmembrane domains but are not G-protein-coupled receptors (GPCRs), as their topology is inverted in the membrane relative to classical GPCRs (Benton et al. 2006; Lundin et al. 2007). In contrast to vertebrate odorant receptors, the insect Ors are ligand-gated ion channels (Sato et al. 2008; Wicher et al. 2008). One putative receptor, Orco, is expressed in most olfactory neurons and functions as a coreceptor and forms a heterodimer with the tuning members of the Or family that are responsible for providing the odorant specificity (Larsson et al. 2004). Nearly all tuning odorant receptors have been mapped to specific functional classes of olfactory neurons (Couto et al. 2005; Vosshall et al. 2000). The tuning olfactory receptors appear necessary and sufficient to confer odorant sensitivity on basiconic neurons, and the tuning receptors present in basiconic sensilla are activated directly by food odorants. Mis-expression of an odorant receptor in a basiconic neuron lacking its normal tuning receptors converts the odorant specificity of that host neuron to that of the donated receptor (Hallem et al. 2004). By contrast, receptors normally found in trichoid sensillum neurons remain silent when expressed in basiconic neurons (Laughlin et al. 2008). Yet, mis-expressing Or67d, normally expressed exclusively in the cVA sensing neurons, in other trichoid neurons confers robust cVA sensitivity that is nearly identical to that of at1 neurons. This suggests there are additional trichoid factors important for cVA detection that are not present in basiconic neurons. Therefore, pheromone signal transduction is distinct from food odorant sensing, requiring additional signaling factors (see below).

7.1.4 *Drosophila* Odorant Binding Proteins

Odorant binding proteins (OBPs) are small (~14 kD) proteins that are secreted into the sensillum lymph, bathing the olfactory neuron dendrites by nonneuronal support cells. In contrast to vertebrate OBPs (that are members of the lipocalin protein family), insects have evolved a novel OBP family. Fifty OBP genes are present in the *Drosophila* genome and share three features. All members studied to date are low-molecular weight (~14 kD), have six conserved cysteines, a signal sequence for secretion, and are expressed in chemosensory organs (Galindo and Smith 2001; Hekmat-Scafe et al. 2002; Robertson et al. 2003). The OBPs are expressed in stereotypic subsets of chemosensory sensilla, both in the olfactory and gustatory organs, and share little sequence homology, consistent with odorant-specific roles (Galindo and Smith 2001).

The first insect OBP was identified by Vogt and Riddiford (1981). These workers identified a pheromone-binding protein in the sensillum lymph of the pheromone-sensitive sensilla in male moths. Later, other OBP members were identified in moths that were expressed in food-sensing sensilla, and were called general odorant binding proteins (GOBPs). The first *Drosophila* OBPs were discovered in John Carlson's lab and Mike Rosbash's lab using molecular screens for genes specifically expressed in the antenna (McKenna et al. 1994; Pikielny et al. 1994). The completed *Drosophila* genome sequence revealed the large size of this gene family that is rivaled only by the chemoreceptor families (Galindo and Smith 2001; Hekmat-Scafe et al. 2002). The function of OBPs remains controversial. Because they are expressed by the support cells and not neurons and are secreted into the sensillum lymph, they must function upstream of any neuronal function in the odorant detection process. The role for one OBP, LUSH, is discussed below.

7.1.5 Gene Products Underlying cVA Pheromone Detection in *Drosophila*

Pheromone signal transduction mechanisms are distinct from those mediating detection of general odorants in both vertebrate and insect species (Chamero et al. 2012; Dulac and Axel 1995; Ha and Smith 2009; Ronderos and Smith 2009). cVA pheromone detection requires a set of gene products not required for olfaction to food odorants, yet does share a requirement for Orco. Orco, the Or coreceptor that makes up part of the odorant-gated ion channels, is required for cVA responses, as $Orco^2$ mutants lack cVA sensitivity (Figure 7.3). The requirement for Orco for cVA sensitivity logically implicated one of the 62 tuning receptors was likely required for cVA sensitivity as well. Indeed, one receptor, Or67d, is expressed exclusively in at1 neurons and mediates cVA responses (Ha and Smith 2006; Kurtovic et al. 2007).

Or67d was discovered by Hugh Robertson, scanning for odorant receptors in the *Drosophila* genome (Robertson et al. 2003). This receptor was first demonstrated to be the cVA tuning receptor for at1 neurons by Ha et al. (Ha and Smith 2006). Ha undertook a genetic screen for mutants defective for cVA detection (Jin et al. 2008). One gene identified was a transcription factor called Rotund. Mutants lacking *rotund* have defects specifying the cell fate of various olfactory neuron classes (Li et al. 2013). One consequence of this is that the at1 sensilla are transformed into at4 sensilla that normally do not express *rotund*. Therefore, there are no at1 sensilla on the antenna

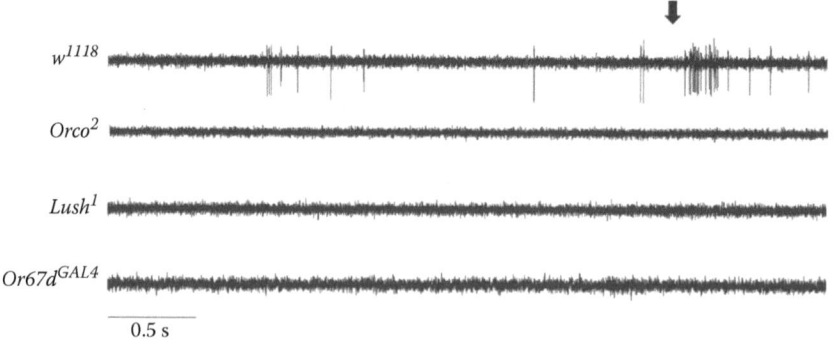

FIGURE 7.3 LUSH, Orco, and Or67d are required for cVA responses and normal spontaneous activity. Sample traces from wild type (w^{1118}), $Orco^2$ mutants, *lush^1* mutants, and *Or67d^{GAL4}* mutants. Arrow indicates application of 1% cVA for each trace.

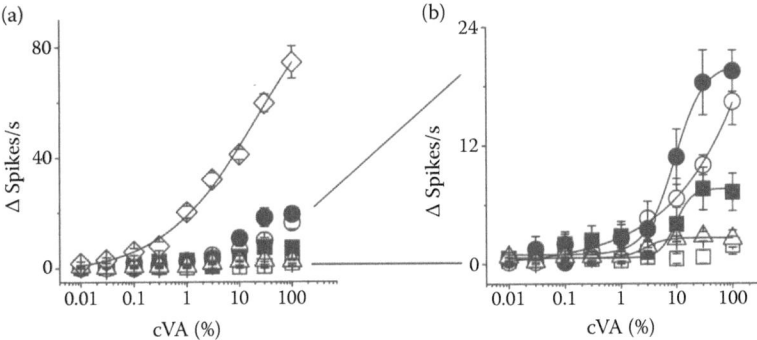

FIGURE 7.4 Expression of the known cVA sensitivity factors, LUSH, Or67d, and SNMP in the empty basiconic neuron system fails to restore full cVA responsiveness. (a) Single sensillum electrophysiological recordings using the empty neuron system expressing cVA sensitivity factors. Δab3A, lacking endogenous receptors but expressing Orco, fails to respond to any concentration of cVA (open squares). Diamonds represent wild type T1 neuronal responses for comparison. Misexpression of Or67d in the Δab3A neurons (open triangles) fails to sensitize to cVA. However, infusion of wild type LUSH into these sensilla provides minimal sensitization to high cVA application (black squares, $p < 0.01$ for 30% cVA and $p > 0.02$ for 100% cVA). Coexpression of Or67d with SNMP (open circles) weakly sensitizes the neurons to cVA. However, the presence of LUSH (black circles) fails to further sensitize Or67d/SNMP expressing neurons to cVA. (b) Expanded view of the Y-axis to better compare the low-level activation of these neurons expressing various cVA sensitizing factors. Data represents mean responses ± SEM. ($n = 6$–12). (From Laughlin et al. (2008). *Cell* 133(7): 1255–1265.)

of *rotund* mutants. Ha then performed RT-PCR experiments looking for Or members expressed in wild type antenna, but not in rotund mutants. Or67d was absent in *rotund* mutant antenna but present in wild type by RT-PCR experiments. To prove Or67d actually confers cVA sensitivity, he misexpressed this receptor in all neurons using the pan-neuronal driver pELAV. Recordings from at4 sensilla containing neurons that are normally insensitive to cVA now responded robustly to cVA. cVA sensitivity in these at4 neurons also required LUSH, as cVA sensitivity was lost to basiconic neurons when the *lush* mutant was engineered into this stock (Ha and Smith 2006). However, Or67d, Orco and LUSH are not sufficient to confer cVA sensitivity to basiconic neurons (Figure 7.4) (Laughlin et al. 2008). Therefore there must be additional cVA sensitivity factors present in trichoid sensilla that are missing in basiconic sensilla.

7.1.6 LUSH OBP

LUSH is an OBP expressed exclusively in all trichoid sensilla in the antenna, but is also found in a subset of tarsal chemoreceptors on the forelegs (see below). The *lush* was identified in a transposable element-based enhancer-trapping screen designed to identify genes expressed in the antenna (Kim et al. 1998). A transposable element carrying the LacZ gene was mobilized and allowed to randomly integrate throughout the genome (Callahan and Thomas 1994). Among several thousand insertion lines, one expressed β-galactosidase exclusively in the ventral-lateral surface of each antenna (Figure 7.5a). Cloning the integration site of the transposon insertion from this line revealed it had integrated just downstream of a gene predicted to encode a new member of the invertebrate OBP family we called *lush* (Kim et al. 1998) that was abundantly expressed in the antenna but absent from heads and bodies. Immunofluorescence studies using antiserum against the recombinant LUSH protein revealed this OBP is expressed exclusively in trichoid sensilla in the same pattern as the LacZ in the enhancer-trap line (Kim et al. 1998). The next question was, what does LUSH do?

The role of OBPs in chemosensation was unknown at that time, and remains controversial. The work of Vogt and Riddiford indicated members of this family bind directly to odorant ligands (Vogt and Riddiford 1981). This led to two prominent models. In the first, the OBP was postulated to play a role in removing stray odorant molecules from the sensillum lymph so the animal is ready for the next odor plume. Indeed, to follow a concentration gradient, there must be some way to rapidly clear stray pheromone molecules. The second model suggested OBPs transport olfactory ligands through the sensillum lymph to the dendrites. We called this the carrier model. No mutants in any OBP gene had been previously described in any species to provide additional insight into OBP function to distinguish among these models. Having a transposable element integrated near the *lush* gene provided an exciting opportunity to generate the first OBP mutant. When excising, transposons can produce local deletions. Remobilizing the transposon from the integration site near the *lush* gene produced several hundred excision lines. One line lacked LUSH protein expression, confirmed by western blots from antennal extracts (Kim et al. 1998). Sequence analysis of the genomic region from this line revealed a 1569 base pair deletion removing the entire *lush* coding sequence. This mutant is called *lush¹*.

Initially, we tested general odorant detection in *lush¹* mutants. Screening olfactory behavior in *lush¹* mutants revealed these flies are more likely to enter traps

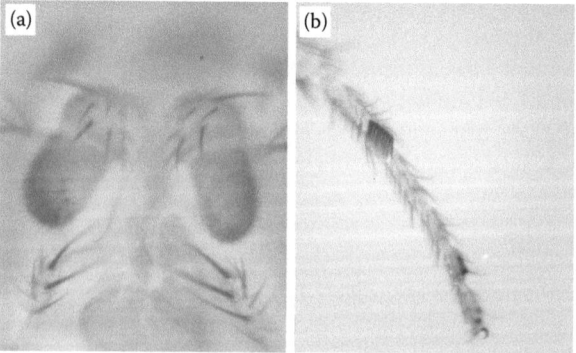

FIGURE 7.5 (**See color insert.**) LacZ expression under control of the *lush* promoter. (a) Head of a transgenic fly expressing nuclear-localized β-galactosidase showing expression restricted to the support cells of the trichoid sensilla. (b) β-Galactosidase expression is also observed in the tarsi in support cells of chemosensory sensilla.

containing high concentrations of short-chain alcohols, including ethanol. Since they showed affinity for high alcohol environments, the mutant was named *lush* (Kim et al. 1998). However, this attraction behavior (or perhaps lack of avoidance) only occurs at extremely high alcohol concentrations that are unlikely to be encountered in nature, so the significance of the alcohol phenotype is unknown. Structural studies with LUSH confirmed that LUSH protein binds directly to ethanol, and defined the first ethanol-binding site defined in a protein (Kruse et al. 2003). One possibility is that LUSH blocks access of ethanol molecules to neurons that mediate avoidance behavior to alcohol, but future work will be required to understand the behavior of *lush¹* mutants to alcohol. What became clear is that *lush¹* mutants are completely anosmic to physiological concentrations of the male-specific pheromone, cVA (Xu et al. 2005). This very dramatic effect on cVA detection reveals LUSH is required to facilitate activation of the at1 neurons in the presence of cVA pheromone. Since LUSH is required for activation of the cVA sensitive neurons, this eliminates the pheromone removal hypothesis in which the sole function of LUSH is to remove odorant molecules from the sensillum lymph.

A second phenotype that was entirely unexpected was that *lush¹* mutants are associated with a profound loss of spontaneous activity (action potentials in the absence of cVA) in the at1 neurons. Normally, in the absence of cVA, at1 neurons spike at one action potential per second. The spontaneous activity rate of the at1 neurons in *lush¹* mutants in the absence of cVA averages only one spike every 400 seconds. If LUSH functioned as a simple pheromone transporter, why would its absence alter the spontaneous activity rate of the at1 neurons in the absence of cVA? The effect on spontaneous activity is specific to at1 neurons and there is no impact on other trichoid neurons that suffer the same loss of LUSH from the sensillum lymph. Therefore it is not likely this defect results from an osmotic effect secondary to the loss of this abundant LUSH protein from the lymph. This finding indicates LUSH has a more intimate relationship with cVA signaling, perhaps as a partial agonist at the at1 neuronal receptors.

For LUSH to function as a partial agonist in the absence of cVA, the LUSH protein itself must interact with neuronal receptors expressed in at1 neurons. One possibility is that LUSH can isomerize from an inactive state to an activated conformation capable of activating at1 neuronal receptors, and the role of the pheromone is to stabilize or induce the activated conformation in LUSH. If true, this model makes two testable predictions. First, there should be a cVA-dependent conformational change in LUSH that is distinct from the conformation induced by other ligands (such as ethanol) that can bind LUSH but do not mediate at1 activation. Second, it should be possible to engineer dominantly active mutant forms of LUSH that activate at1 neurons in the absence of cVA.

To explore the nature of the LUSH-cVA complex, the x-ray crystal structure of LUSH with and without cVA bound were solved (Laughlin et al. 2008). cVA is completely encapsulated within the LUSH-cVA structure. cVA binding induces a conformational change in LUSH, ejecting phenyalanine 121 out of the cVA binding pocket, leading to disruption of the salt bridge between aspartate 118 and lysine 87. This causes the C-terminal loop of LUSH to flip outward. Confirming that the conformational change is important for activation of at1 neurons, mutating the F121 to alanine should reduce the ability to induce the activated conformational shift. Indeed, LUSH^{F121A}, when infused through the recording pipette into *lush^1* mutants, reduces cVA sensitivity 50-fold compared to infusing recombinant wild type LUSH. This mutation has no effect on cVA binding (Laughlin et al. 2008). Conversely, mutating phenylalanine 121 to the bulkier tryptophan (LUSH^{F121W}) makes at1 neurons neurons five times more sensitive to cVA when this protein is infused into the sensillum lymph! These findings support the idea that the conformational change is an important component of the at1 activating ability of the LUSH-cVA complex. Disruption of the salt bridge between K87 and D118 (LUSH^{D118A}) that helps maintain LUSH in an inactive state in the absence of cVA results in a protein that spontaneously activates at1 neurons when infused into these sensilla. at1 neuron activity is activated after a few minutes of LUSH^{D118A} infusion and peaks after approximately 20 minutes at an average of 20 spikes/sec (Laughlin et al. 2008). Infusion of LUSH^{D118A} into other sensilla has no effect on neuronal activity, confirming LUSH^{D118A} specifically activates receptors present on the at1 neurons but not in other olfactory neurons. Indeed, the crystal structure of LUSH^{D118A} is almost identical to that of cVA-bound LUSH, beautifully explaining the mechanism for the dominant activation by LUSH^{D118A}. While LUSH^{D118A} selectively activates at1 neurons in the absence of cVA, it does not completely mimic full activation achieved by high concentrations of cVA (20 spikes per second vs >50 spikes per second with cVA).

When transgenic *lush^1* flies express *lush^{D118A}* under control of the *lush* promoter, the at1 activation effect is much weaker than when recombinant protein is infused directly into the sensillum lymph. There is only a twofold increase in spontaneous activity compared to wild type flies (Ronderos and Smith 2010). Despite this modest increase in activity of the at1 circuit, the transgenic flies act as if there is cVA present. Male flies exhibit delayed courtship, while females expressing *lush^{D118A}* exhibit enhanced courtship (Ronderos and Smith 2010). This is consistent with the sexually dimorphic behavioral effects of cVA (Kurtovic et al. 2007; Ronderos and Smith 2010). Can doubling the spontaneous activity rate affect behavior? It is known that even small increases in at1 neuronal activity are amplified because of convergence. Multiple at1 neurons innervate the DA1 glomerulus, increasing the likelihood of activating the projection neurons (Bhandawat et al. 2007). Therefore, small increases in at1 firing are sufficient to alter behavior. The relatively weak activation of the at1 neurons by transgenic *lush^{D118A}* was surprising, given that infusion of bacterially expressed LUSH^{D118A} activates at1 neurons strongly. One possibility is that the olfactory neurons desensitize to the presence of activated LUSH^{D118A} when expressed chronically as a transgene. Alternatively, there may be chaperone-like factors present in the secretory system of the support cells that prevent secretion of LUSH in the active conformation. Consistent with this notion, when LUSH^{D118A} is expressed as a transgene, it rescues cVA sensitivity (Ronderos and Smith 2010), while direct infusion does not (Laughlin et al. 2008). The fact that *lush^1* mutants show defective spontaneous activity supports the idea that the binding protein is a partial agonist that activates at1 neuronal receptors in the absence of cVA. This is further supported by the mutational studies described above that functionally dissociate cVA from the conformational shifts in the LUSH protein structure that activate at1 neurons. Is the conformational activation model the whole story? Since dominant LUSH does not fully activate the at1 neurons, other protein interactions or perhaps cVA itself may be important for full at1 activation. Although the complete picture remains to be unraveled, what is clear at this point is that conformational activation of LUSH is part of the at1 activation mechanism.

Extremely high cVA levels can directly activate at1 neurons in *lush¹* mutants, demonstrating a LUSH-independent at1 activation mechanism. This is consistent with the idea that pheromone detection evolved from a system in which cVA originally activated at1 receptors directly. The LUSH cVA pheromone binding protein may have evolved initially as a cVA carrier or capturing factor to increase sensitivity and was later incorporated into the at1 signaling mechanism. It will be interesting to determine if mutations in the pheromone-binding OBPs from other species have a similar defect as is observed in *lush¹* mutants.

7.1.7 SNMP

SNMP was first identified in moths as an abundant 67-kD protein expressed on the dendrites of a subset of olfactory neurons (Rogers et al. 1997). However, the function of this two-transmembrane-spanning protein was not clear. SNMP is a member of the CD36 protein family. Many CD36 members are scavenger receptors or play a role in the transmembrane transport of lipoprotein complexes (reviewed in Vogt et al. 2009). For example, in humans, CD36 is required for uptake of oxidized cholesterol by macrophages resulting in conversion of macrophages into foam cells—an important step in the formation of atherosclerotic plaques in arteries (Collot-Teixeira et al. 2007). In *Drosophila*, other CD36 homologs are important for recognition and removal of dead cells (Franc et al. 1996) and bacteria (Philips et al. 2005), absorption of vitamin A from the gut (Gu et al. 2004; Kiefer et al. 2002), and transfer from the hemolymph into the retina (Wang et al. 2007). Mutants defective for the *Drosophila* homolog of moth SNMP were independently produced and studied in the Smith lab (Jin et al. 2008) and the Vosshall lab (Benton et al. 2007). The Vosshall group produced an SNMP mutant using homologous recombination, following up on the findings of Rogers et al. showing SNMP is restricted to dendrites of the pheromone sensitive neurons (Rogers et al. 1997; Rogers et al. 2001a; Rogers et al. 2001b). Tal Soo Ha and Xin Jin identified SNMP mutants as one of several mutants defective for cVA detection recovered in a genetic screen (Jin et al. 2008). Mutants lacking SNMP are insensitive to cVA (Benton et al. 2007; Jin et al. 2008). However, unlike virtually all other mutants defective for cVA detection studied to date, *Snmp* mutants show increased, not decreased, spontaneous activity in the at1 neurons. The increased spontaneous firing rate, around 20 spikes per second, is similar to that observed when dominant *lush^{D118A}* is infused into the lymph of at1 sensilla (Laughlin et al. 2008). SNMP protein is expressed in subsets of olfactory neurons and support cells, but cVA sensitivity specifically requires SNMP expression in the at1 neurons (Benton et al. 2007; Jin et al. 2008). SNMP functions on the surface of the at1 neuron dendrites, the site of pheromone signal transduction, as infusion of anti-SNMP antiserum into the sensillum lymph mimics the mutant phenotype (Jin et al. 2008). The Vosshall group also showed fluorescence energy transfer (FRET) between SNMP and Orco, indicating these proteins are in close proximity to each other (Benton et al. 2007). This data is consistent with SNMP acting as a component of the neuronal cVA-activated receptors.

One model for SNMP function in pheromone detection is that it mediates the transfer of pheromone, directly or from the binding protein, to the Or67d/Orco complex (Benton et al. 2007; Rogers et al. 2001). While this model would explain the loss of cVA sensitivity at low concentrations, several lines of evidence are not consistent with this model. First, the loss of a pheromone transfer protein would not be expected to affect the spontaneous activity in the at1 neurons. *Snmp* mutants have higher than normal spontaneous activity. Second, if SNMP transfers cVA, it should be possible to activate the at1 receptors with high cVA concentrations in the absence of SNMP. No amount of cVA is capable of activating at1 neurons in *Snmp* mutants (Benton et al. 2007; Jin et al. 2008). Third, SNMP appears to be required for dominant LUSH[D118A] to activate at1 neurons. While infusion of recombinant LUSH[D118A] into the sensillum lymph of wild type at1 neurons activates at1 neurons, LUSH[D118A] fails to activate at1 neurons from *Snmp* mutants (Jin et al. 2008). Either cVA and

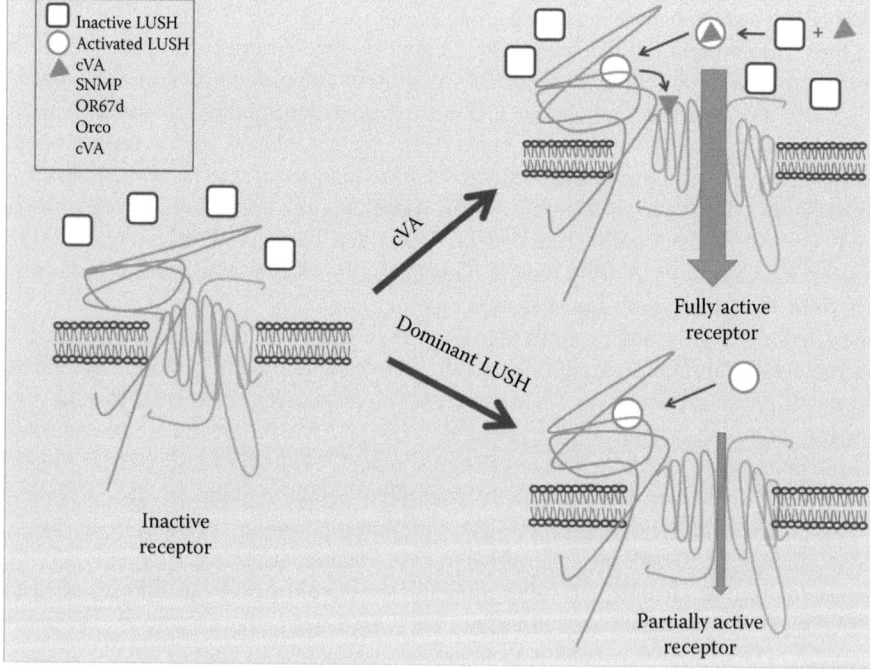

FIGURE 7.6 **(See color insert.)** Model for cVA detection in *Drosophila* at1 neurons. Left, inactive LUSH (open squares) does not activate at1 neuronal receptors, composed of Or67d (blue), Orco (orange), and SNMP (purple). SNMP is postulated to maintain the at1 receptors in the off state. cVA pheromone (red triangles) interacts with LUSH (upper right), inducing the activated conformation (red circles). Activated LUSH binds to SNMP, relieving inhibition on the Or67d/Orco ion channel. Interaction with SNMP may release cVA from LUSH, producing full activation of the receptor. Dominant LUSH (lower right, red circles) is active without cVA bound, but activates at1 receptors submaximally.

LUSH activate atl neurons through SNMP, or the neurons lacking SNMP are simply not capable of further activation over their higher endogenous spontaneous rate. The latter seems less likely given that wild type atl neurons can fire at over 50 spikes per second when stimulated by high levels of cVA. Therefore, it seems most likely that SNMP functions as a negative regulator of Or67d/Orco complex and its inhibition is relived by interactions with activated LUSH accounting for partial activation. Since dominant LUSH (and *Snmp* mutants) only produce partial activation, there may be role for cVA itself in full activation. This would also be consistent with cVA acting as a weak agonist in the absence of LUSH. Figure 7.6 shows a current model for cVA signal transduction in atl neurons.

7.2 cVA-INDUCED BEHAVIORS

7.2.1 COURTSHIP

In *Drosophila*, like all animals, behaviors are elicited in response to sensory inputs. Courtship is a well-characterized set of stereotypical behavior patterns that occurs prior to mating. Males actively court females through a series of ritual behaviors that ultimately conclude in copulation (reviewed in Dickson 2008; Greenspan and Ferveur 2000; Hall 1994; Manoli et al. 2006; Pan et al. 2011; Vosshall 2008). Progression through courtship requires interactions between partners that are mediated through the olfactory, visual, auditory, tactile, and gustatory senses to insure mating is directed toward an appropriate partner (Amrein 2004; Dickson 2008). cVA activates atl neurons equally in both male and female fruit flies. This pheromone plays a role in courtship behavior in both sexes but induces different behaviors in each sex. What is the role of cVA in courtship, and how can cVA trigger different behaviors in males and females?

The first clue that cVA is important for courtship came from the work of Jallon who noted an antiaphrodisiac effect of cVA (Jallon et al. 1981). The effect of cVA on courtship became clearer when Or67d mutants were studied. Males lacking Or67d display increased courtship toward other males, compared to wild type controls (Kurtovic et al. 2007). Females lacking this receptor show prolonged latency to copulation (Kurtovic et al. 2007). The opposite behaviors are observed for each sex when atl neurons are activated by LUSH[D118A], indicating cVA is an aphrodisiac for females but an antiaphrodisiac for males (Ronderos and Smith 2010). Are atl neurons the sole source of cVA detection? Studies reported that the odorant receptor Or65a, normally expressed by neurons in the at4 sensilla (Couto et al. 2005) could be activated by high cVA concentrations when this receptor was misexpressed in basiconic neurons (van der Goes van Naters and Carlson 2007). Furthermore, expressing tetanus toxin under control of an Or65a promoter, which blocks synaptic transmission in these neurons, disrupts cVA-induced suppression of courtship, while tetanus toxin had no effect when expressed by Or67d promoter (Ejima et al. 2007). However, the Or65a promoter used in this work also drives tetanus toxin in other brain neurons, which could affect behavior independently of activity in the Or65a neurons (Ejima et al. 2007). Regardless, it is possible that activation of multiple neuronal circuits, triggered through different odorant receptors, underlies cVA-induced behavior.

To address whether cVA acts through Or67d-expressing at1 neurons alone or through additional circuits to modulate courtship behaviors, David Ronderos studied transgenic flies expressing the dominant LUSH[D118A] in trichoid sensilla in *lush[1]* mutants. By expressing LUSH[D118A] as a transgene under control of the *lush* promoter, he can specifically activate at1 neurons in the absence of cVA (Ronderos and Smith 2010), since dominant LUSH[D118A] only activates at1 neurons (Laughlin et al. 2008). Dominant LUSH[D118A] induced suppression of courtship in males and increased receptivity to courtship advances in females (Ronderos and Smith 2010) (Figure 7.7). Importantly, the LUSH[D118A] behavioral effects on both sexes was absent when Or67d was removed, confirming these effects are mediated through at1 neurons. These findings are consistent with the receptor mutant studies by Kurtovic et al. (2007) and indicate at1 neurons are necessary and sufficient for cVA-induced courtship behavior.

How can cVA cause different behavior in males and females? As previously discussed, there is no difference in cVA sensitivity or responses at the at1 neuron level between males and females. Therefore the differences must be downstream in the circuit. Using photoactivatable GFP, Datta et al. illuminated the DA1 glomerulus in flies broadly expressing the GFP construct, resulting in activation of GFP only in

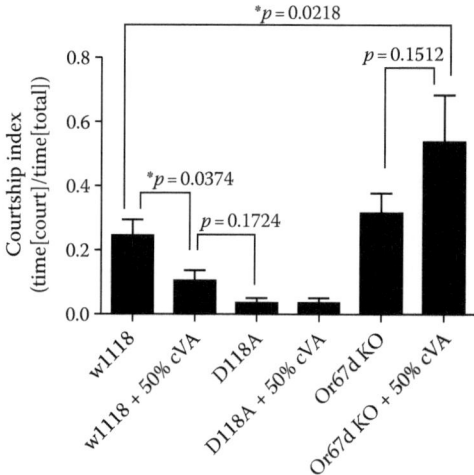

FIGURE 7.7 Sexually dimorphic cVA-triggered mating behaviors are mediated specifically by T1 neurons. Wild type male and female pairings result in a courtship index of ~30% (left). cVA exposure suppresses male courtship behavior in wild type males to a level indistinguishable from male flies expressing *lush[D118A]* in the absence of cVA ($p > 0.1$). cVA exposure has no further suppression of courtship by *lush[D118A]* males ($p > 0.1$). cVA fails to suppress courtship of *Or67d[GAL4]* mutant males, and when crossed to wild type females courtship index is enhanced due to the aphrodisiac effect of cVA on the female. This enhancement by cVA is lost when both males and females lack *Or67d*. Error bars represent SEM. Significant differences from the wild type controls calculated using Student's *t*-test. For all bars, $n \geq 6$. *$p < 0.05$, **$p < 0.01$. (Reproduced from Ronderos, D. S. and D. P. Smith (2010). *J Neurosci* 30(7): 2595–2599. With permission.)

the illuminated neurons (Datta et al. 2008). They were able to trace the DA1 projection neurons (PNs), whose dendrites were labeled in this fashion, and examine the synaptic connections these cells made. Remarkably, when they looked at the structure of these PNs, they noted a clear difference in males and females. Males have a unique axonal branch in the lateral horn that innervated cells that are not innervated in females, and this branch was observed in all males examined (Datta et al. 2008). We do not know if this axon branch is responsible for any of the sexually dimorphic behavior induced by cVA, but future studies may find ways to cleave this branch and evaluate the effects on courtship. Richard Axel's group has now extended the cVA circuit to four neurons connected by three synapses using photoactivatable GFP and illuminating the axonal targets of each group of neurons (Ruta et al. 2010). The last neuron innervates the ganglia in the ventral nerve cord. Interestingly, these downstream neurons are also sexually dimorphic and express the sex determination transcription factor fruitless (Fru) (Yu et al. 2010). Understanding precisely how these circuits operate to produce behavior is still unknown, but a wiring diagram of the circuit is certainly a good roadmap. Perhaps computer modeling of the circuit will lead to new insight.

7.2.2 AGGREGATION

In addition to effects on courtship, cVA also modulates aggregation and aggression behavior. The combination of cVA and food odorants provide a potent cue that fruit flies find extremely attractive (Bartelt et al. 1985). cVA is also transferred to females upon mating, both by contact with male cuticle and within the seminal fluid (Butterworth 1969). The latter results in cVA deposition on the food substrate when eggs are laid. Both males and females are attracted to this odorant cocktail, making this a nice mechanism to aggregate both sexes to the same location where they will be close enough to promote courtship. Additionally, a member of the ionotropic odorant receptor family, IR84a, is activated by the fruit odorants phenylacetic acid and phenylacetaldehyde, and is expressed in olfactory neurons that activate a FRU-positive circuit that enhances male courtship behaviors (Grosjean et al. 2011). These mechanisms promote reproductive behavior at appropriate sites for feeding and egg laying.

7.2.3 AGGRESSION

Male aggression is a social behavior that is manifested in almost all species of the animal kingdom. Males display aggressive behaviors to establish dominance, to compete for a limited food resource, and for the chance to mate with receptive females. In *Drosophila*, aggressive behavior was first observed in 1915 by Sturtevant and has since been characterized in terms of behavior traits. Flies exhibit characteristic aggressive behaviors during male–male interactions (Dahanukar and Ray 2011; Dickson 2008). These behaviors include (1) approaching, when a male fly will lower his body and advance towards the second fly, (2) wing threats, when one fly will quickly raise its wing towards another, (3) lunging, when the aggressive fly will throw himself on the other fly, (4) boxing, when flies will raise up on their hind legs and hit each other using their forelegs, and finally, (5) tussling, when both flies fall over each other, holding, kicking, and chasing each other (Chen et al. 2002; Dow and von Schilcher 1975; Skrzipek et al. 1979; Zwarts et al. 2012). Aggression assays are usually measured by placing pairs of flies in a chamber with a resource, either food or a female, and videotaping behavior and counting the number of times the aggressor threatens or lunges, or boxing episodes (Certel and Kravitz 2012).

Fruitless expression was shown to underlie dominance behavior in male fruit flies (Vrontou et al. 2006) and the cVA circuit contains several Fru-expressing neurons (Vrontou et al. 2006). In a series of elegant experiments cVA was shown to directly promote aggression between two male flies when present in the observation chambers (Liu et al. 2011; Wang and Anderson 2010; Wang et al. 2011). Activating at1 neurons by expressing an activated cation channel under the *Or67d* promoter that causes chronic depolarization of at1 can recapitulate the aggression phenotype in the absence of cVA (Wang and Anderson 2010). The effect of cVA on aggression is dose-dependent (Wang and Anderson 2010). At lower concentrations of cVA, the pheromone promotes aggregation (Bartelt et al. 1985; Wang and Anderson 2010). When the number of flies increases, the cVA concentration increases and may produce increased male–male aggression that disperses the flies. Thus, cVA may regulate population density at food resources. Whether this model is correct and how the relative concentration of cVA causes a change in behavior is unknown. However, this could easily be tested using cVA-insensitive mutants.

Interestingly, cVA fails to promote aggression in Gr32a null males (Wang et al. 2011). Gr32 is a taste receptor expressed in the legs and is thought to be stimulated by the male-specific cuticle hydrocarbon, 7-(z)-tricosene (Wang et al. 2011). Perhaps dual inputs from both the cVA and Gr32 circuits are required to promote aggressive behavior. This could account for aggressive behaviors being restricted to males and only when they are in direct contact (Wang et al. 2011). It will be interesting to determine how and where the Gr32a and Or67d circuits interact to promote aggression.

Serotonin, octopamine, and the product of the *white* gene have been implicated in modulating aggression (Certel et al. 2010; Dierick and Greenspan 2007; Hoyer et al. 2008; McDermott et al. 2009; Zhou and Rao 2008). A cytochrome P450 homolog, Cyp6a20, has also been suggested to have a role in modulating aggression in *Drosophila* (Wang et al. 2008). *Cyp6a20* is expressed by support cells in the pheromone-sensitive trichoid sensilla (van der Goes van Naters and Carlson 2007). These support cells also express LUSH (Kim et al. 1998). Since Cyp6a20 is not secreted into the sensillum lymph, it remains to be determined exactly how this factor influences aggressive behavior.

7.3 OTHER *DROSOPHILA* PHEROMONES

There are a number of cuticular hydrocarbons present in *Drosophila* that affect behavior (Ferveur 2005; Siwicki et al. 2005). Linking these effects with specific olfactory receptors has been challenging. Neurons expressing Or67d and Or47b express the sex determination transcription factor, fruitless (Fru) (Yu et al. 2010). Neurons expressing Fru are likely to have sexually dimorphic functions. Activation of Or67d neurons by cVA is well established and produces sexually dimorphic mating behavior (Kurtovic et al. 2007; Ronderos and Smith 2010). What activates Or47b neurons? Using a fusion between the calcium-activated transcription factor NFAT and the yeast transcription factor LexA to drive a LexA-dependent GFP reporter, Jing Wang's group showed one can specifically label activated neurons in living flies. When flies expressing this reporter system were exposed to the odors of male flies, the DA1 glomerulus, the target of Or67d-expressing olfactory neurons, was specifically labeled. When male flies expressing this reporter system were exposed to virgin female flies, the VA1lm (also called VA1v) glomerulus was labeled. This glomerulus is the target of Or47b-expressing neurons, and indicates there is a volatile pheromone detected by Or47b. To date, the volatile ligand that activates Or47b remains unknown. However, mutants in Or47b show no defects in mating latency (Wang and Anderson 2010). Indeed the only defect observed in this mutant is that the high male courtship toward males lacking 7D and 7T is suppressed in males lacking Or47b (Wang and Anderson 2010). Thus, there may be a cuticle lipid in both males and females that promotes mating behavior in males, but this response is normally suppressed in the presence of male cuticle hydrocarbons. Perhaps Or47b acts to assure that a male is courting a female of the correct species.

Additional pheromone candidates have been identified as minor hydrocarbon fractions present on the cuticle (Ferveur 2005; Yew et al. 2008). Using single-fly gas chromatography/mass spectroscopy direct analysis in real time (DART), Ed Kravitz and coworkers identified several potential candidate pheromones that differ between males and females (Yew et al. 2008). One, CH305 (3-O-acetyl-1,3-dihydroxyoctacosa-11, 19-diene) acted similar to cVA, in that female flies perfumed with this male-specific hydrocarbon showed a dose-dependent reduction in courtship initiation by males (Yew et al. 2009). However, unlike cVA, which has a short half-life of a few hours, CH305 remained on the cuticle for up to 10 days. Thus, CH305 and cVA may both act to reduce courtship in males subsequent to previous mating.

7.4 LEARNING MODULATES HARD-WIRED PHEROMONE BEHAVIORS

Survival of all species depends on adequate production of progeny. *Drosophila* males try to maximize mating to produce maximum viable progeny. *Drosophila* females are less receptive toward mating if they have mated previously (Siegel and Hall 1979). This stems from sex peptide, a peptide transferred to the female by the male during mating. Sex peptide activates a specific receptor and triggers rejection behavior in females—primarily kicking and running away when a male attempt to court (Yapici et al. 2008). Virgin females, on the other hand, are receptive toward mates and constitute a higher chance of producing viable offspring for males of the species. Therefore, it is advantageous for a male to discriminate between virgin and mated females and focus its mating efforts towards receptive females.

Mated females differ from virgin females in the profile of their cuticular lipids (Ejima et al. 2007). Mated females contain trace amounts of cVA on their cuticles left behind during mating. cVA exposure reduces courtship in males (Jallon et al. 1981;

Ronderos and Smith 2010). Thus, cVA could be responsible for mediating learning between mated and virgin females.

In a series of nice experiments, the Dickson group showed that courtship rejection training dramatically enhances the sensitivity of males for cVA (Keleman et al. 2012). Naïve males will court previously mated and immature females equally well. In contrast, while males that have previously attempted to mate with mated females and been rejected do not court mated females as vigorously in the future. Thus, males learn to discriminate between mature and immature females. Mutants lacking Or67d court mated and virgin females equally, and do not benefit from training. This indicates that cVA is the salient cue used for this learning process. In a set of simple but elegant experiments, the Dickson laboratory asked whether cVA detection, the lack of courtship success, or an association between the two is essential for learning to distinguish mated and virgin females. Females mated with males lacking sex peptide (pseudovirgin females) have cVA transferred to the cuticle yet remain receptive to courtship despite recent mating. Pseudomated females are transgenic flies that overexpress sex peptide, and thus constitutively perform courtship rejection behaviors, but have no cVA. Surprisingly, pseudomated females were just as effective as genuinely mated females to train the naive males to distinguish mated from virgin females. By contrast, pseudovirgin females, exposed to cVA during mating but that receive no sex peptide were not. However, pseudovirgin females, but not pseudomated females, were as effective as mated females when used as tester females following training. This suggested that the salient feature of training is simply the lack of courtship success, not its association with cVA, and somehow training alters the sensitivity of the male to cVA. Indeed, by spiking pseudomated females with various doses of cVA, the trained males avoided courtship with these flies at much lower doses than normally required to inhibit courtship by naïve males. (Keleman et al. 2012).

Taking this a step further, the Dickson group showed that *fru*+ dopaminergic neurons are required for the formation of the courtship memory. By permitting dopaminergic signaling in specific subsets of dopaminergic neurons (by manipulating DopR1 receptor expression), they were able to pinpoint which neurons are required. The aSP13 class of dopaminergic neurons conveys a learning signal to MBγ neurons via the dopamine receptor DopR1. This signal causes a change in the circuit that processes the cVA signal that discriminates mated from virgin females (Keleman et al. 2012). This suggests a mating strategy for males in which they are initially promiscuous but become more selective if a mating attempt fails, and this involves an effect of the aSP13 dopaminergic neurons on the mushroom body that induces lasting changes in cVA circuits that mediates discrimination of mated and virgin females.

7.5   CONTACT CHEMORECEPTION

Claudio Pikielny, Kristen Scott, and coworkers have shown that some gustatory neurons present on the tarsi (legs) are required for courtship behavior and are selective for the detection of male or female cuticle lipids, including cVA (Liu et al. 2012; Thistle et al. 2012). The Scott group has focused on paired Fru-positive neurons in the legs that coexpress the epithelial sodium channel family pickpocket (ppk) isoforms ppk23 and ppk29 (Thistle et al. 2012), while the Pikielny group focused on neurons expressing ppk25 and Nope (Liu et al. 2012). Both groups showed these taste neurons are distinct from those detecting water, sugar, and bitter taste neurons. The ppk23 and ppk29 neuron projections are sexually dimorphic (Thistle et al. 2012). Distinct populations of ppk23, ppk29-expressing neurons detect male and female hydrocarbon cues, either to the male lipids 7-tricosene (7T) and cVA or female factors (7,11-heptacosadiene (7,11-HD) and 7,11-nonacosadiene (7,11-ND). The ppk25/Nope receptors are expressed in only one of the two Fru-positive neurons and mark the female-pheromone detecting cell, as they respond to 7,11-HD and 7,11-ND but not to 7T, 7P or cVA (Starostina et al. 2012). Flies lacking any of these 4-ppk subunits have defects in courtship behaviors, indicating all are important (Lin et al. 2005; Liu et al. 2012; Starostina et al. 2012; Thistle et al. 2012). Mate selection and courtship initiation behaviors were the most affected, suggesting a role for these ion channels in contact chemosensation for partner discrimination. Expressing ppk23 in ppk29 null cells did not rescue the defective phenotype and vice versa, suggesting that these two channels do not have redundant roles, perhaps functioning as obligate heteromultimers.

Using GCaMP as a reporter for cell activation, mutant neurons lacking ppk23 and ppk29 are not activated by 7-tricosene, cVA, 7,11-heptacosadiene, or 7,11-nonacosadiene, while the wild type neurons are (Thistle et al. 2012). This suggests that ppk23 and ppk29 are involved in the signaling pathway that detects and activates sensory neurons in the presence of both male and female cuticle hydrocarbons that regulate mating behavior. These channels are probably not gated by pheromones directly, as misexpression of ppk23 and ppk 29 in water-sensing gustatory neurons fails to confer pheromone sensitivity (Thistle et al. 2012). Alternatively, ppk23 and ppk 29 may be activated by a signaling pathway initiated by unknown hydrocarbon receptors or may be gated directly by the hydrocarbons, but require additional ppk subunits. Given that ppk channels are thought to form trimers (Jasti et al. 2007), it will be interesting to find out how these ppk subunits interact to contribute to pheromone sensitivity and how they are gated.

Interestingly, ppk23 and ppk 29 channels are not required for cVA responses in the at1 neurons in the antenna, as mutants lacking these channels have normal antennal cVA responses (Sengupta and Smith 2013, unpublished observation). Therefore, there are differences in the signaling mechanisms between tarsal and antennal cVA sensitive neurons. LUSH is expressed in both antennal sensilla sensitive to cVA and in several tarsal sensilla (Figure 7.5b). Perhaps LUSH is required for cVA sensitivity in both the gustatory and olfactory detection of cVA. It will be interesting to determine if LUSH is expressed in the same sensilla as ppk23 and ppk 29, and if so, whether the initial steps of cVA detection are similar in tarsal and at1 neurons, or if there are distinct receptors in these organs.

7.6 FUTURE PROSPECTS

While much has been uncovered about the molecular signaling mechanisms activated downstream of the cVA pheromone, as well as the cellular circuits and their roles in behavior, much remains unknown. Clearly, additional factors remain to be discovered that are required for cVA detection in the at1 neurons, and cVA detection by tarsal neurons may be entirely different or share a subset of signaling factors. How the various labeled lines initiated in the antenna and other sites of phero-mone detection converge and are integrated remains a mystery. Finally, how these circuits trigger specific behaviors also remains an enigma. However, the problem is finally yielding thanks to new approaches and new techniques. What we learn in *Drosophila* will surely serve as a guidepost to better understand and manipu-late pheromone biology in other insects that plague human populations around the world.

REFERENCES

Alaux, C., A. Maisonnasse et al. (2010). Pheromones in a superorganism: From gene to social regulation. *Vitam Horm* 83: 401–423.

Amrein, H. (2004). Pheromone perception and behavior in *Drosophila*. *Curr Opin Neurobiol* 14(4): 435–442.

Bartelt, R. J., A. M. Schaner et al. (1985). *cis*-Vaccenyl acetate as an aggregation pheromone in *Drosophila melanogaster*. *J Chem Ecol* 11: 1747–1756.

Bellen, H., R. W. Levis et al. (2004). The BDGP gene disruption project: Single transposon insertions associated with 40% of *Drosophila* genes. *Genetics* 188: 731–743.

Benton, R., S. Sachse et al. (2006). Atypical membrane topology and heteromeric function of *Drosophila* odorant receptors in vivo. *PLoS Biol* 4(2): e20.

Benton, R., K. S. Vannice et al. (2007). An essential role for a CD36-related receptor in phero-mone detection in *Drosophila*. *Nature* 450(7167): 289–293.

Bhandawat, V., S. R. Olsen et al. (2007). Sensory processing in the *Drosophila* antennal lobe increases reliability and separability of ensemble odor representations. *Nat Neurosci* 10(11): 1474–1482.

Butterworth, F. M. (1969). Lipids of *Drosophila*: A newly detected lipid in the male. *Science* 163: 1356–1357.

Callahan, C. A. and J. B. Thomas (1994). Tau-β-galactosidase, an axon-targeted fusion pro-tein. *Proc Natl Acad Sci U S A* 91: 5972–5976.

Carde, R. T. and M. A. Willis (2008). Navigational strategies used by insects to find distant, wind-borne sources of odor. *J Chem Ecol* 34(7): 854–866.

Certel, S. J. and E. A. Kravitz (2012). Scoring and analyzing aggression in *Drosophila*. *Cold Spring Harb Protoc* 2012(3): 319–325.

Certel, S. J., A. Leung et al. (2010). Octopamine neuromodulatory effects on a social behavior decision-making network in *Drosophila* males. *PLoS One* 5(10): e13248.

Chamero, P., T. Leinders-Zufall et al. (2012). From genes to social communication: molecular sensing by the vomeronasal organ. *Trends Neurosci* 35(10): 597–606.

Chen, S., A. Y. Lee et al. (2002). Fighting fruit flies: a model system for the study of aggres-sion. *Proc Natl Acad Sci U S A* 99(8): 5664–5668.

Collot-Teixeira, S., J. Martin et al. (2007). CD36 and macrophages in atherosclerosis. *Cardiovasc Res* 75(3): 468–477.

Couto, A., M. Alenius et al. (2005). Molecular, anatomical, and functional organization of the *Drosophila* olfactory system. *Curr Biol* 15(17): 1535–1547.

Dahanukar, A. and A. Ray (2011). Courtship, aggression and avoidance: Pheromones, recep-tors and neurons for social behaviors in *Drosophila*. *Fly (Austin)* 5(1): 58–63.

Datta, S. R., M. L. Vasconcelos et al. (2008). The *Drosophila* pheromone cVA activates a sexu-ally dimorphic neural circuit. *Nature* 452(7186): 473–477.

Dickson, B. J. (2008). Wired for sex: The neurobiology of *Drosophila* mating decisions. *Science* 322(5903): 904–909.

Dierick, H. A. and R. J. Greenspan (2007). Serotonin and neuropeptide F have opposite modulatory effects on fly aggression. *Nat Genet* 39(5): 678–682.

Dow, M. A. and F. von Schilcher (1975). Aggression and mating success in *Drosophila melanogaster. Nature* 254(5500): 511–512.

Dulac, C. and R. Axel (1995). A novel family of genes encoding putative pheromone receptors in mammals. *Cell* 83: 195–206.

Ejima, A., B. P. Smith et al. (2007). Generalization of courtship learning in *Drosophila* is mediated by cis-vaccenyl acetate. *Curr Biol* 17(7): 599–605.

Fabre, J.-H. (1916). *Life of a Caterpillar.* New York: Dodd, Mead and Company.

Ferveur, J. F. (2005). Cuticular hydrocarbons: Their evolution and roles in *Drosophila* pheromonal communication. *Behav Genet* 35(3): 279–295.

Franc, N. C., J. L. Dimarcq et al. (1996). Croquemort, a novel *Drosophila* hemocyte/macrophage receptor that recognizes apoptotic cells. *Immunity* 4(5): 431–443.

Galindo, K. and D. P. Smith (2001). A large family of divergent odorant-binding proteins expressed in gustatory and olfactory sensilla. *Genetics* 159: 1059–1072.

Golic, K. G. and S. Lindquist (1989). The FLP recombinase of yeast catalyzes site-specific recombination in the *Drosophila* genome. *Cell* 59: 499–509.

Greenspan, R. J. and J. F. Ferveur (2000). Courtship in *Drosophila. Annu Rev Genet* 34: 205–232.

Grosjean, Y., R. Rytz et al. (2011). An olfactory receptor for food-derived odours promotes male courtship in *Drosophila. Nature* 478(7368): 236–240.

Gu, G., J. Yang et al. (2004). *Drosophila* ninaB and ninaD act outside of retina to produce rhodopsin chromophore. *J Biol Chem* 279(18): 18608–18613.

Ha, T. S. and D. P. Smith (2006). A pheromone receptor mediates 11-cis-vaccenyl acetate-induced responses in *Drosophila. J Neurosci* 26(34): 8727–8733.

Ha, T. S. and D. P. Smith (2009). Odorant and pheromone receptors in insects. *Front Cell Neurosci* 3: 10.

Hall, J. C. (1994). The mating of a fly. *Science* 264(5166): 1702–1714.

Hallem, E. A. and J. R. Carlson (2006). Coding of odors by a receptor repertoire. *Cell* 125(1): 143–160.

Hallem, E. A., M. G. Ho et al. (2004). The molecular basis of odor coding in the *Drosophila* antenna. *Cell* 117(7): 965–979.

Hekmat-Scafe, D. S., C. R. Scafe et al. (2002). Genome-wide analysis of the odorant-binding protein gene family in *Drosophila melanogaster. Genome Research* 12: 1357–1369.

Hildebrand, J. G. and G. M. Shepherd (1997). Mechanisms of olfactory discrimination: Converging evidence for common principles across phyla. *Annu Rev Neurosci* 20: 595–631.

Hoyer, S. C., A. Eckart et al. (2008). Octopamine in male aggression of *Drosophila. Curr Biol* 18(3): 159–167.

Jallon, J.-M., C. Antony et al. (1981). Un anti-aphrodisiaque produit par les mâles de *Drosophila melanogaster* et tranféré aux femelles lors de la copulation. *C. R. Acad Sci Paris* 292: 1147–1149.

Jasti, J., H. Furukawa et al. (2007). Structure of acid-sensing ion channel 1 at 1.9 A resolution and low pH. *Nature* 449(7160): 316–323.

Jefferis, G. S., C. J. Potter et al. (2007). Comprehensive maps of *Drosophila* higher olfactory centers: Spatially segregated fruit and pheromone representation. *Cell* 128(6): 1187–1203.

Jin, X., T. S. Ha et al. (2008). SNMP is a signaling component required for pheromone sensitivity in *Drosophila. Proc Natl Acad Sci U S A* 105(31): 10996–11001.

Kaissling, K.-E. and E. Priesner (1970). Die Riechschwelle des Seidenspinners. *Naturwiss* 57: 23–28.

Kalidas, S. and D. P. Smith (2002). Novel genomic cDNA hybrids produce effective RNA interference in adult *Drosophila*. *Neuron* 33: 177–184.

Karlson, P. and M. Luscher (1959). Pheromones: A new term for a class of biologically active substances. *Nature* 183(4653): 55–56.

Keleman, K., E. Vrontou et al. (2012). Dopamine neurons modulate pheromone responses in *Drosophila* courtship learning. *Nature* 489(7414): 145–149.

Kiefer, C., E. Sumser et al. (2002). A class B scavenger receptor mediates the cellular uptake of carotenoids in *Drosophila*. *Proc Natl Acad Sci U S A* 99(16): 10581–10586.

Kim, M.-S., A. Repp et al. (1998). LUSH odorant binding protein mediates chemosensory responses to alcohols in *Drosophila melanogaster*. *Genetics* 150: 711–721.

Kruse, S. W., R. Zhao et al. (2003). Structure of a specific alcohol-binding site defined by the odorant binding protein LUSH from *Drosophila melanogaster*. *Nat Struct Biol* 10: 694–700.

Kurtovic, A., A. Widmer et al. (2007). A single class of olfactory neurons mediates behavioural responses to a *Drosophila* sex pheromone. *Nature* 446(7135): 542–546.

Larsson, M. C., A. I. Domingos et al. (2004). Or83b encodes a broadly expressed odorant receptor essential for *Drosophila* olfaction. *Neuron* 43(5): 703–714.

Laughlin, J. D., T. S. Ha et al. (2008). Activation of pheromone-sensitive neurons is mediated by conformational activation of pheromone-binding protein. *Cell* 133(7): 1255–1265.

Lee, T. and L. Luo (2001). Mosaic analysis with a repressible cell marker (MARCM) for *Drosophila* neural development. *Trends Neurosci* 24(5): 251–254.

Li, Q., T.-S. Ha et al. (2013). Combinatorial rules of precursor specification underlying olfactory neuron diversity. *Current Biology* (in press).

Lin, H., K. J. Mann et al. (2005). A *Drosophila* DEG/ENaC channel subunit is required for male response to female pheromones. *Proc Natl Acad Sci U S A* 102(36): 12831–12836.

Liu, T., E. Starostina et al. (2012). Two *Drosophila* DEG/ENaC channel subunits have distinct functions in gustatory neurons that activate male courtship. *J Neurosci* 32(34): 11879–11889.

Liu, W., X. Liang et al. (2011). Social regulation of aggression by pheromonal activation of Or65a olfactory neurons in *Drosophila*. *Nat Neurosci* 14(7): 896–902.

Lundin, C., L. Kall et al. (2007). Membrane topology of the *Drosophila* OR83b odorant receptor. *FEBS Lett* 581(29): 5601–5604.

Manoli, D. S., G. W. Meissner et al. (2006). Blueprints for behavior: Genetic specification of neural circuitry for innate behaviors. *Trends Neurosci* 29(8): 444–451.

Masuyama, K., Y. Zhang et al. (2012). Mapping neural circuits with activity-dependent nuclear import of a transcription factor. *J Neurogenet* 26(1): 89–102.

McDermott, R., D. Tingley et al. (2009). Monoamine oxidase A gene (MAOA) predicts behavioral aggression following provocation. *Proc Natl Acad Sci U S A* 106(7): 2118–2123.

McKenna, M. P., D. S. Hekmat-Scafe et al. (1994). Putative pheromone-binding proteins expressed in a subregion of the olfactory system. *J Biol Chem* 269: 1–8.

Pan, Y., C. C. Robinett et al. (2011). Turning males on: Activation of male courtship behavior in *Drosophila melanogaster*. *PLoS One* 6(6): e21144.

Patterson, G. H. and J. Lippincott-Schwartz (2002). A photoactivatable GFP for selective photolabeling of proteins and cells. *Science* 297(5588): 1873–1877.

Philips, J. A., E. J. Rubin et al. (2005). *Drosophila* RNAi screen reveals CD36 family member required for mycobacterial infection. *Science* 309(5738): 1251–1253.

Pikielny, C. W., G. Hasan et al. (1994). Members of a family of *Drosophila* putative odorant-binding proteins are expressed in different subsets of olfactory hairs. *Neuron* 12: 35–49.

Robertson, H. M., C. G. Warr et al. (2003). Molecular evolution of the insect chemoreceptor gene superfamily in *Drosophila melanogaster*. *Proc Natl Acad Sci U S A* 100 Suppl 2: 14537–14542.

Rogers, M. E., J. Krieger et al. (2001a). Antennal SNMPs (sensory neuron membrane proteins) of Lepidoptera define a unique family of invertebrate CD36-like proteins. *J Neurobiol* 49(1): 47–61.

Rogers, M. E., R. A. Steinbrecht et al. (2001b). Expression of SNMP-1 in olfactory neurons and sensilla of male and female antennae of the silkmoth *Antheraea polyphemus*. *Cell Tissue Res* 303(3): 433–446.

Rogers, M. E., M. Sun et al. (1997). Snmp-1, a novel membrane protein of olfactory neurons of the silk moth *Antherea polyphemus* with homology to the CD36 family of membrane proteins. *J Biol Chem* 272: 14792–14799.

Ronderos, D. S. and D. P. Smith (2009). Diverse signaling mechanisms mediate volatile odorant detection in *Drosophila*. *Fly (Austin)* 3(4): 290–297.

Ronderos, D. S. and D. P. Smith (2010). Activation of the T1 neuronal circuit is necessary and sufficient to induce sexually dimorphic mating behavior in *Drosophila melanogaster*. *J Neurosci* 30(7): 2595–2599.

Rong, Y. S., S. W. Titen et al. (2002). Targeted mutagenesis by homologous recombination in *D. melanogaster*. *Genes and Development* 16: 1568–1581.

Ruta, V., S. R. Datta et al. (2010). A dimorphic pheromone circuit in *Drosophila* from sensory input to descending output. *Nature* 468(7324): 686–690.

Sato, K., M. Pellegrino et al. (2008). Insect olfactory receptors are heteromeric ligand-gated ion channels. *Nature* 452(7190): 1002–1006.

Schlief, M. L. and R. I. Wilson (2007). Olfactory processing and behavior downstream from highly selective receptor neurons. *Nat Neurosci* 10(5): 623–630.

Siegel, R. G. and J. C. Hall (1979). Conditioned responses in courtship behavior of normal and mutant *Drosophila*. *Proc Natl Acad Sci U S A* 76: 3430–3434.

Siwicki, K. K., P. Riccio et al. (2005). The role of cuticular pheromones in courtship conditioning of *Drosophila* males. *Learn Mem* 12(6): 636–645.

Skrzipek, K., B. Kroner et al. (1979). Inter-male aggression in *Drosophila melanogaster*— Laboratory study. *Z Tierpsych-J Comp Ethol* 43: 107–120.

Smith, D. P. (2012). Volatile pheromone signaling in *Drosophila*. *Physiol Entomol* 37: 19–24.

Starostina, E., T. Liu et al. (2012). A *Drosophila* DEG/ENaC subunit functions specifically in gustatory neurons required for male courtship behavior. *J Neurosci* 32(13): 4665–4674.

Stocker, R. F. (1994). The organization of the chemosensory system in *Drosophila melanogaster*: A review. *Cell Tissue Res* 275: 3–26.

Stortkuhl, K. F. (2001). Functional analysis of an olfactory receptor in *Drosophila melanogaster*. *Proc Natl Acad Sci U S A* 98: 9381–9387.

Sturtevant, A. (1915). Experiments on sex recognition and the problem of sexual selection in Drosophilia. *J Anim Behav* 5: 351–366.

Thistle, R., P. Cameron et al. (2012). Contact chemoreceptors mediate male-male repulsion and male-female attraction during *Drosophila* courtship. *Cell* 149(5): 1140–1151.

Tian, L., S. A. Hires et al. (2009). Imaging neural activity in worms, flies and mice with improved GCaMP calcium indicators. *Nat Methods* 6(12): 875–881.

van der Goes van Naters, W. and J. R. Carlson (2007). Receptors and neurons for fly odors in *Drosophila*. *Curr Biol* 17(7): 606–612.

Vogt, R. G., N. E. Miller et al. (2009). The insect SNMP gene family. *Insect Biochem Mol Biol* 39(7): 448–456.

Vogt, R. G. and L. M. Riddiford (1981). Pheromone binding and inactivation by moth antennae. *Nature* 293: 161–163.

Vosshall, L. B. (2008). Scent of a fly. *Neuron* 59(5): 685–689.

Vosshall, L. B., A. M. Wong et al. (2000). An olfactory sensory map in the fly brain. *Cell* 102: 147–159.

Vrontou, E., S. P. Nilsen et al. (2006). *Fruitless* regulates aggression and dominance in *Drosophila*. *Nat Neurosci* 9(12): 1469–1471.

Wang, L. and D. J. Anderson (2010). Identification of an aggression-promoting pheromone and its receptor neurons in *Drosophila*. *Nature* 463(7278): 227–231.

Wang, L., H. Dankert et al. (2008). A common genetic target for environmental and heritable influences on aggressiveness in *Drosophila*. *Proc Natl Acad Sci U S A* 105(15): 5657–5663.

Wang, L., X. Han et al. (2011). Hierarchical chemosensory regulation of male-male social interactions in *Drosophila*. *Nat Neurosci* 6: 757–762.

Wang, T., Y. Jiao et al. (2007). Dissection of the pathway required for generation of vitamin A and for *Drosophila* phototransduction. *J Cell Biol* 177(2): 305–316.

Wicher, D., R. Schafer et al. (2008). *Drosophila* odorant receptors are both ligand-gated and cyclic-nucleotide-activated cation channels. *Nature* 452(7190): 1007–1011.

Xu, P., R. Atkinson et al. (2005). *Drosophila* OBP LUSH is required for activity of pheromone-sensitive neurons. *Neuron* 45(2): 193–200.

Yapici, N., Y. J. Kim et al. (2008). A receptor that mediates the post-mating switch in *Drosophila* reproductive behaviour. *Nature* 451(7174): 33–37.

Yew, J. Y., R. B. Cody et al. (2008). Cuticular hydrocarbon analysis of an awake behaving fly using direct analysis in real-time time-of-flight mass spectrometry. *Proc Natl Acad Sci U S A* 105(20): 7135–7140.

Yew, J. Y., K. Dreisewerd et al. (2009). A new male sex pheromone and novel cuticular cues for chemical communication in *Drosophila*. *Curr Biol* 19(15): 1245–1254.

Yu, J. Y., M. I. Kanai et al. (2010). Cellular organization of the neural circuit that drives *Drosophila* courtship behavior. *Curr Biol* 20(18): 1602–1614.

Zhou, C. and Y. Rao (2008). A subset of octopaminergic neurons are important for *Drosophila* aggression. *Nat Neurosci* 11(9): 1059–1067.

Zwarts, L., M. Versteven et al. (2012). Genetics and neurobiology of aggression in *Drosophila*. *Fly (Austin)* 6(1): 35–48.

FIGURE 3.1 (a) Dung beetles on and in rhinoceros dung that has been voided during the night. (Photograph courtesy of B. V. Burger.) (b) A *K. lamarcki* male that has lost his left hind leg in the typical secretion-producing headstand position. (Photograph courtesy of B. V. Burger.) (c) The fragile fibers of the male *K. lamarcki* secretion resemble cotton wool. (Photograph courtesy of B. V. Burger.) (d) A brush on the tibia of a *K. lamarcki* male. (Photograph courtesy of G. D. Tribe.) (e) The first row of comblike structures on the abdomen of a *K. lamarcki* male against which the secretion is broken up. (Photograph courtesy of G. D. Tribe.) (f) *K. lamarcki* male pushing a dung ball while the female clings to the ball. (Photograph courtesy of B. V. Burger.) (g) *K. lamarcki* male with secretion accumulated on the left side of the insect's abdomen and with the secretion-dispersion brush visible on the tibia of the intact hind leg. (Photograph courtesy of B. V. Burger.) (h) Dorsal (left) and (i) ventral (right) aspects of *O. egregious*. (Photographs courtesy of B. V. Burger.)

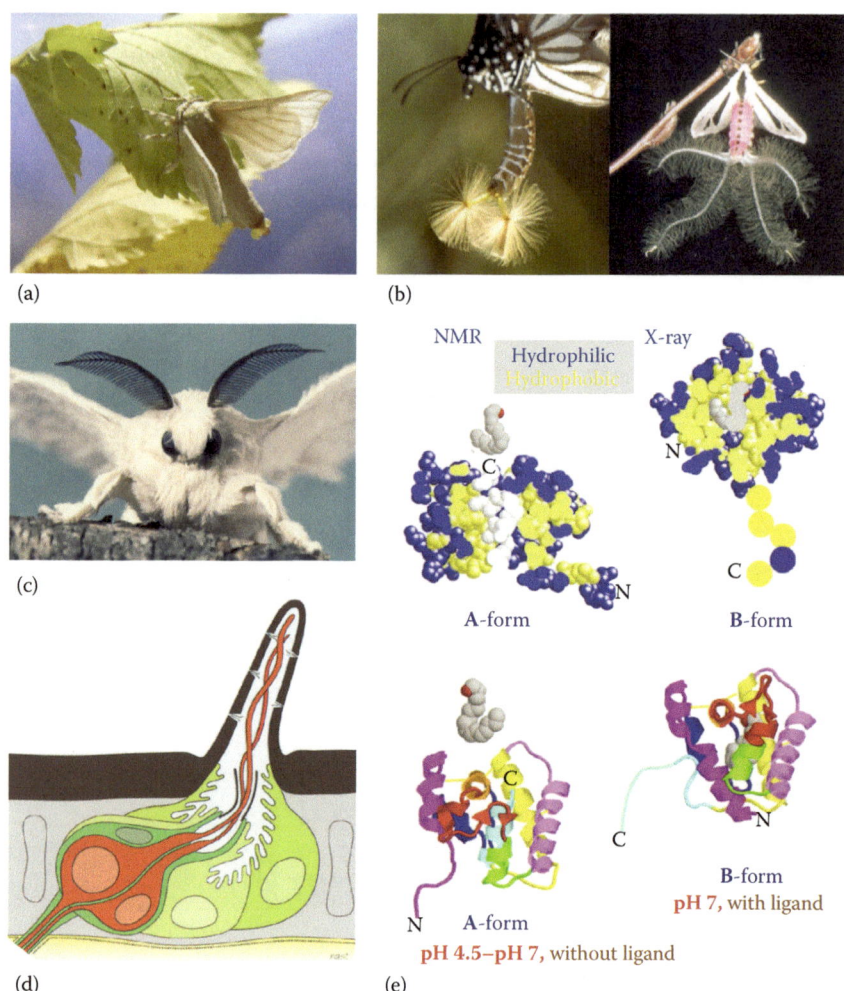

FIGURE 4.1 (a) Female *Bombyx mori* in calling position, with everted abdominal pheromone glands. (b) Males of the milkweed or monarch butterfly *Tirumala petiverana* (Danainae, left hand) and the Asian arctiide moth *Creatonotos gangis* with expanded androconia. (Courtesy of M. Boppré.) (c) Male *Bombyx mori* in alerted position, with combed antennae elevated. (Courtesy of R.A. Steinbrecht). (d) Diagram of an olfactory sensillum trichodeum with two receptor cells (red) and three auxiliary cells (green). In blue: Sensillum lymph. (Courtesy of R.A. Steinbrecht.) (e) Structures of bombykol and the bombykol-binding protein, from x-ray and NMR analysis. Top: Calotte models. Bombykol molecule in grey, with red OH-group. Shape of bombykol as if bound to B-form. A-form with C-terminus occupying the inner binding cavity in white. B-form with bombykol: Five amino acids of the unfolded C-terminus, not shown by x-ray, are added in an arbitrary shape. Bottom: Backbone structures of the protein. A-form with C-terminal helix. (Made by using RASMOL, with data from Sandler B.H. et al. 2000. *Chem Biol* 7:143–151; Horst R. et al. 2001. *Proc Natl Acad Sci U S A* 98:14374–14379; Reproduced from Kaissling K.E. 2013. *J Comp Physiol A* 199:879–896. With kind permission from Springer Science+Business Media.

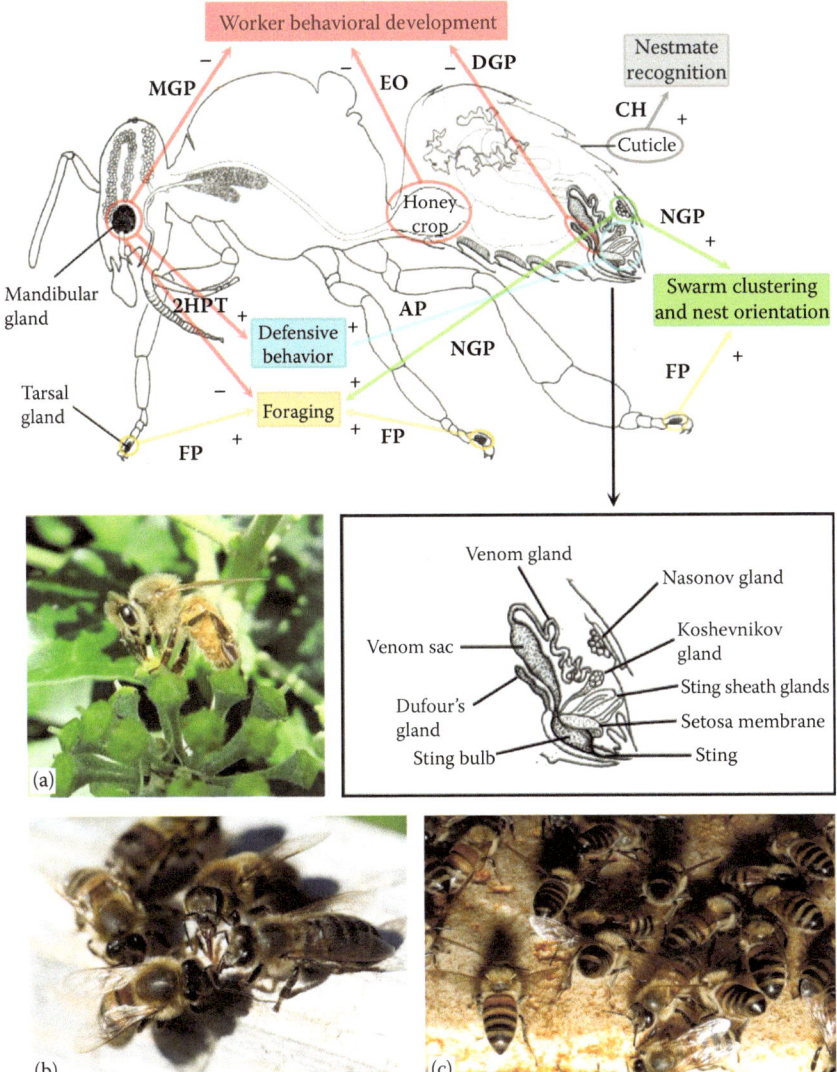

FIGURE 5.3 Pheromone-producing glands and organs and their main products in the honey bee worker, and their effect in the different worker activities. Stimulating effects are indicated as "+" and inhibiting effects as "–." (a) Honey bee worker foraging on *Hedera helix*; during foraging workers mark flowers with the secrete of their tarsal and mandibular glands. (b) Nestmate recognition among workers in front of the hive entrance; workers use chemical cues like cuticular hydrocarbons perceived through mouth and body part contact. (c) Honey bee workers exposing the Nasonov gland in front of the hive to release the orientation pheromone. 2HPT = 2-heptanone; AP = Alarm pheromone; CH = cuticular hydrocarbons; DGP = Dufour's gland pheromone; EO = ethyl oleate; FP = Footprint pheromone; MGP = mandibular gland pheromone; NGP = Nasonov gland pheromone. (Adapted from Goodman, L. (2003) *Form and Function in the Honey Bee.* Cardiff, UK: IBRA. Photographs by Cecilia Costa and Per Kryger.)

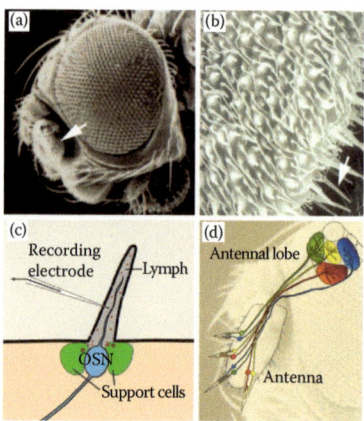

FIGURE 7.1 Anatomy of the *Drosophila* olfactory system. (a) SEM of a *Drosophila* head. Arrow indicates the third segment of the antenna that contains the olfactory neurons. (b) Higher magnification of the antenna surface showing individual hairs or sensilla. Arrow depicts one of the long, needle-shaped trichoid sensillum that detect pheromones. (c) Drawing of a single sensillum impaled by a glass recording pipette. The dendrites of the olfactory neurons (OSN) project into the sensillum lymph-filled cavity within the sensillum shaft. Odorant binding proteins (red dots) are secreted by the support cells into the lymph. (d) Olfactory neurons project to the antennal lobe glomeruli. Neurons expressing the same receptor innervate the same glomerulus. (From Ronderos, D. S. and D. P. Smith (2009). *Fly (Austin)* 3(4): 290–297.)

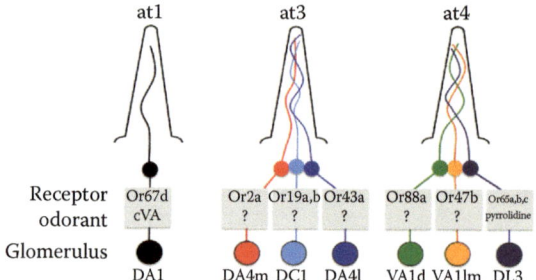

FIGURE 7.2 Drawing depicting the *Drosophila* trichoid sensilla. Three classes are present on the antenna, at1, at3, and at4. at1 sensilla contain a single neuron that expresses Or67d and is exquisitely tuned to cVA pheromone. at3 sensilla contain three neurons expressing Or2a, Or19a and b and Or43a. Or19a and b are recent gene duplications that encode identical proteins. Or43a has been reported to respond to benzaldehyde and similar odorants when misexpressed in the empty neuron system (Stortkuhl 2001). However, other reports find no sensitivity to benzaldehyde in the same system (Hallem and Carlson 2006), so the tuning of this receptor remains controversial. at4 sensilla contain three olfactory neurons expressing Or88a, Or47b and Or65a, b, and c in the aT4A, B, and C neurons, respectively. Or65-expressing neurons are the only *Drosophila* olfactory neurons expressing three distinct receptors, perhaps to broaden the tuning of this neuron. Unlike the food-detecting odorant receptors expressed in basiconic sensilla, the receptors depicted here do not respond well to odorants when misexpressed in the empty neuron system (Hallem and Carlson 2006). Note: at2 sensilla, previously classified as trichoid sensilla (Couto et al. 2005) are actually intermediate sensilla (Ronderos and Smith, submitted).

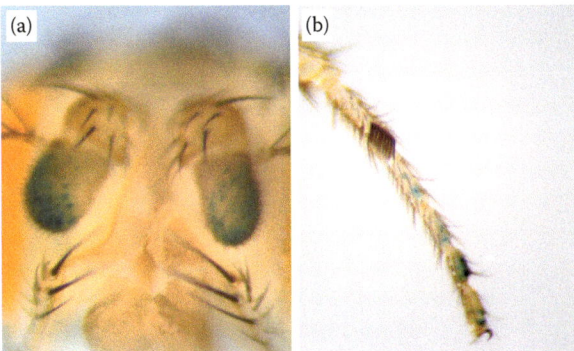

FIGURE 7.5 LacZ expression under control of the *lush* promoter. (a) Head of a transgenic fly expressing nuclear-localized β-galactosidase showing expression restricted to the support cells of the trichoid sensilla. (b) β-Galactosidase expression is also observed in the tarsi in support cells of chemosensory sensilla.

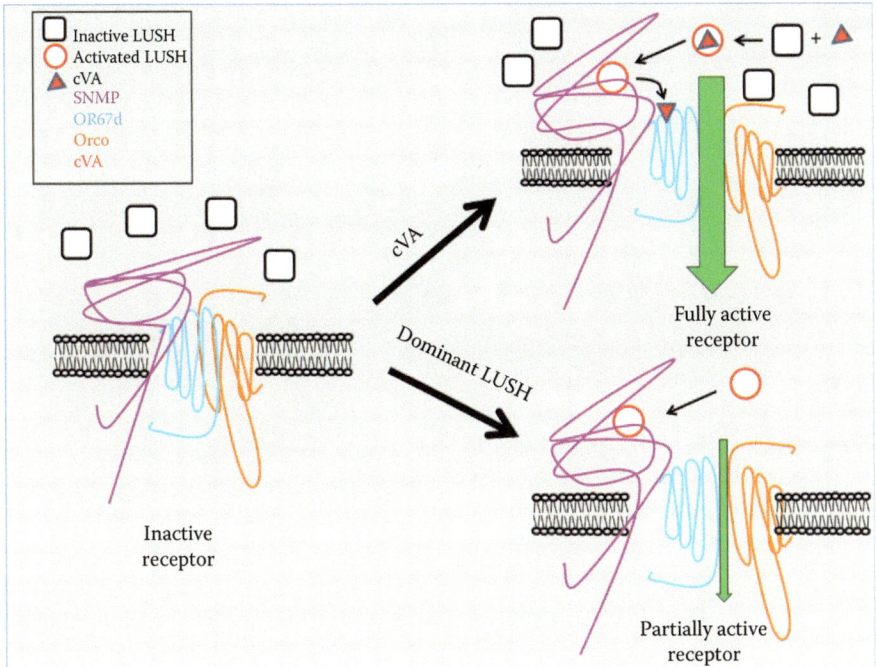

FIGURE 7.6 Model for cVA detection in *Drosophila* at1 neurons. Left, inactive LUSH (open squares) does not activate at1 neuronal receptors, composed of Or67d (blue), Orco (orange), and SNMP (purple). SNMP is postulated to maintain the at1 receptors in the off state. cVA pheromone (red triangles) interacts with LUSH (upper right), inducing the activated conformation (red circles). Activated LUSH binds to SNMP, relieving inhibition on the Or67d/Orco ion channel. Interaction with SNMP may release cVA from LUSH, producing full activation of the receptor. Dominant LUSH (lower right, red circles) is active without cVA bound, but activates at1 receptors submaximally.

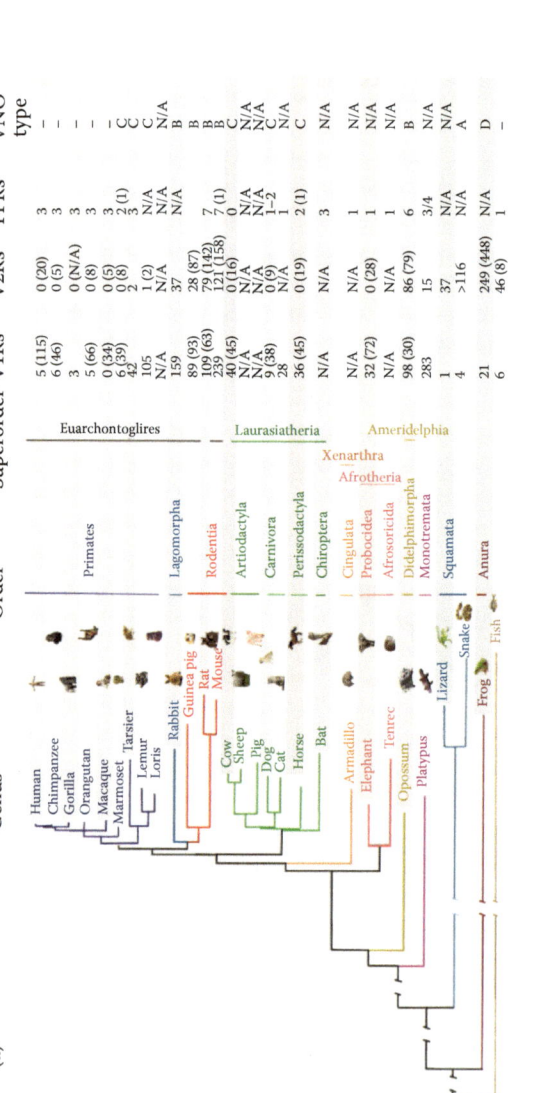

Genus	Order	Superorder	V1Rs	V2Rs	FPRs	VNO type
Human	Primates	Euarchontoglires	5 (115)	0 (20)	3	—
Chimpanzee			6 (46)	0 (5)	3	—
Gorilla			3	0 (N/A)	3	—
Orangutan			5 (66)	0 (8)	3	—
Macaque			0 (34)	0 (5)	3	C
Marmoset			6 (39)	0 (8)	2 (1)	C
Tarsier			42		3	C
Lemur			105	1 (2)	N/A	N/A
Loris			N/A	N/A	N/A	B
Rabbit	Lagomorpha		159	37		B
Guinea pig	Rodentia		89 (93)	28 (87)		B
Rat			109 (63)	79 (142)	7	B
Mouse			239	121 (158)	7 (1)	C
Cow	Artiodactyla	Laurasiatheria	40 (45)	0 (16)	0	N/A
Sheep			N/A	N/A	N/A	N/A
Pig			N/A	N/A	N/A	C
Dog	Carnivora		9 (38)	0 (9)	1–2	N/A
Cat			28	N/A	1	
Horse	Perissodactyla		36 (45)	0 (19)	2 (1)	C
Bat	Chiroptera		N/A	N/A	3	N/A
Armadillo	Cingulata	Xenarthra	N/A	N/A		
Elephant	Proboscidea	Afrotheria	32 (72)	0 (28)	1	N/A
Tenrec	Afrosoricida		N/A	N/A	1	N/A
Opossum	Didelphimorpha	Ameridelphia	98 (30)	86 (79)	6	B
Platypus	Monotremata		283	15	3/4	N/A
Lizard	Squamata		1	37	N/A	N/A
Snake			4	>116	N/A	A
Frog	Anura		21	249 (448)	N/A	D
Fish			6	46 (8)	1	—

(a)

FIGURE 10.2 (a) The phylogenetic tree of mammals. Mammals comprise about 5400 species that are distributed in six superorders including Euarchontoglires (Rodentia, Lagomorpha, primates), Laurasiatheria (Artiodactyla, Carnivora, Perissodactyla, and Chiroptera), Xenarthra (Cingulata) and Afroteria (Proboscidea, Afrosoricida, Afrosoricida) and Ameridelphia (Didelphimorpha) (Churakov et al. 2010; Huchon et al. 2007; Michaux et al. 2001; Nikolaev et al. 2007). Primates include prosimians (loris, lemur, tarsier), New World monkeys (marmoset), Old World monkeys (macaque), apes (orangutan, gorilla, chimpanzee), and humans. Doubts remain on the location of the Rodentia order in the phylogenetic tree of mammals. The most supported hypothesis (shown here) suggests that the Rodentia order is closely related to primates and Laurasiatheria (Luo et al. 2012). The number of intact genes for V1Rs, V2Rs, FPRs and pseudogenes (in brackets) are shown (Dong et al. 2012; Grus et al. 2007; Liberles et al. 2009; Nei et al. 2008; Yang & Shi 2010; Young et al. 2010). The anatomical complexity of the adult VNO (from A to D) is indicated according to Takami et al. (2002). Type A VNO belongs to the ophidian species (water snakes); Type B is present in rodents and lagomorphs. Type C is found in cows, sheep, dogs, prosimian primates, and in New World monkeys. Amphibians developed type D VNO (Takami 2002). Lizard and snake (Squamata), frog (Anura), and fish have been taken as outgroups.

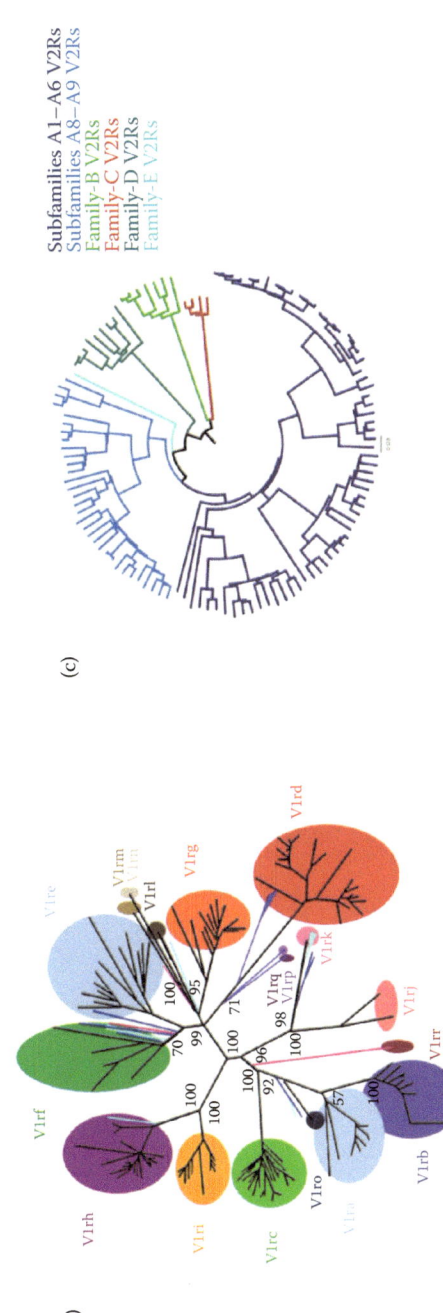

Subfamilies A1–A6 V2Rs
Subfamilies A8–A9 V2Rs
Family-B V2Rs
Family-C V2Rs
Family-D V2Rs
Family-E V2Rs

(c)

(b)

FIGURE 10.2 (*Continued*) (b) Phylogeny of V1Rs. Unrooted phylogenetic tree of functional V1Rs in rat, mouse, cow, dog, and human. The mouse V1Rs repertoire consist of 250 members organized in 12 phylogenetically isolated subfamilies (a to k). Species-specific V1R families of rat (V1rm, V1rn, V1ro), cow (V1rp and V1rq), and human (V1rr) are shown. Mouse and rat branches are in black, cow branches are in blue, dog branches are in light blue, and human branches are in purple. The bootstrap percentage supports the monophyly of all families with the exception of subfamily V1r1a. The tree was reconstructed using the neighbor-joining method with Poissoon-corrected protein distances. (c) Phylogeny of V2Rs in mouse. Family ABDE full-length genes are shown. Family E corresponds to the former subfamily A10 (Vmn2r18).

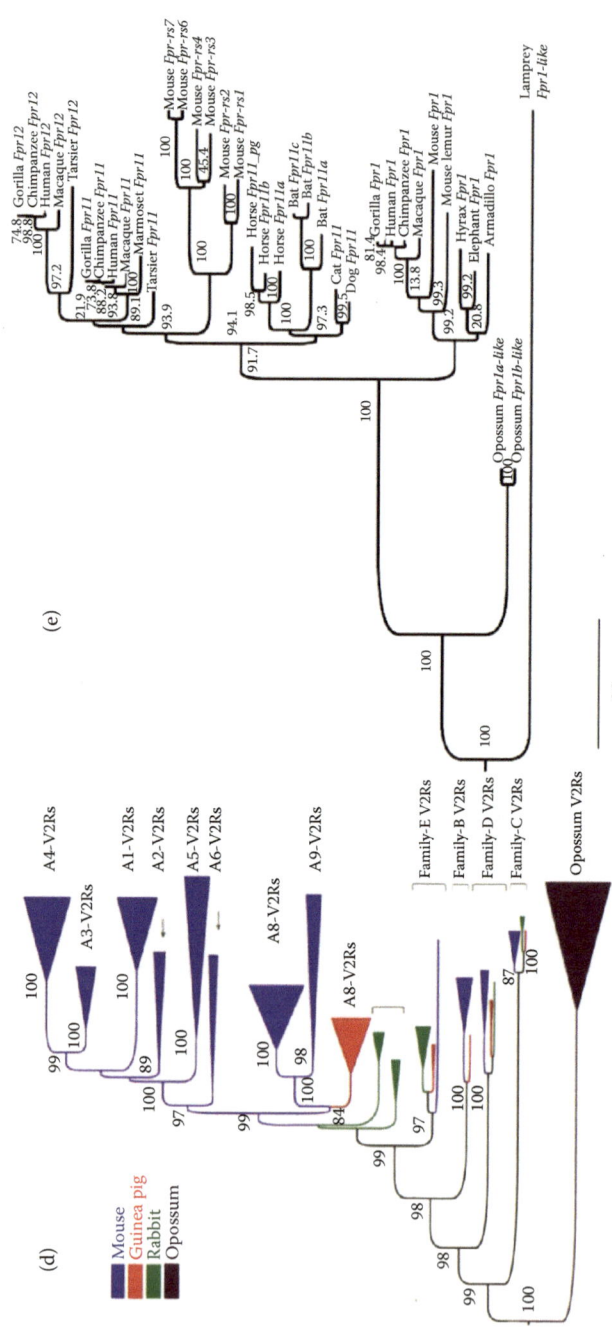

FIGURE 10.2 *(Continued)* (d) Phylogeny of V2Rs in mammals. V2Rs coding sequences of Rodentia (mouse, guinea pig), Lagomorpha (rabbit), and Didelphimorpha (opossum) are used to create a neighbor-joining phylogenetic tree. The bootstrap percentage is shown at each node. (e) Phylogenetic tree of FPR genes in mammals. FPR sequences were identified in whole genome databases and used to create a phylogenetic tree performed by maximum likelihood (ML) analysis. The ML bootstrap percentage is shown in each node. The horse fprl1 is a pseudogene (Fprl1_pg). Opossum and lamprey FPR-like genes were considered as outgroups.

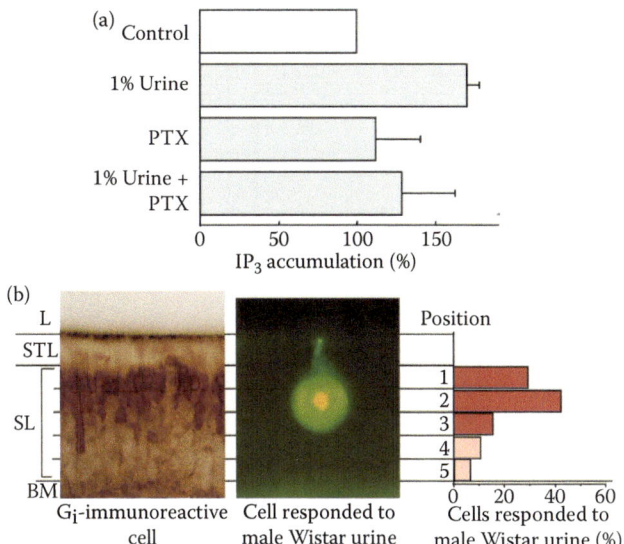

FIGURE 12.5 (a) Inhibition of IP$_3$ accumulation in response to 1% male Wistar urine induced by the treatment with 50 mg/ml Pertussis toxin (PTX). (Modified from Sasaki et al. (1999) *Brain Res* 823: 161–168.) (b) Laminar distribution of neurons responding to various urine preparations. (Modified from Inamura et al. (1999b) *J Physiol (Lond)* 517: 731–739.) Left, sensory neurons immunoreactive to anti-G$_{i2\alpha}$ in the rat vomeronasal sensory epithelium. Center, vomeronasal sensory neuron dialyzed with 1% Lucifer yellow. Right, graphical representation of the lamina distribution of neurons responding to male Wistar rat urine. Abscissa indicates percentage of neurons responding to male Wistar urine. L = lumen of the vomeronasal organ, STL = layer of supporting cells, SL = layer of receptor neurons, BM = basal membrane.

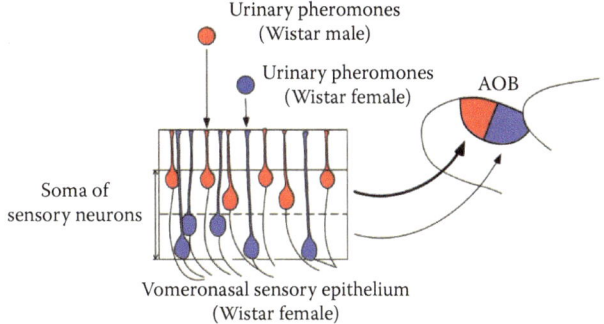

FIGURE 12.6 Schematic representation of the transmission pathway of male Wistar urinary pheromone from the vomeronasal sensory neurons at the apical layer of the epithelium to the rostral region of the accessory olfactory bulb. The vomeronasal pathway in rodents. Pheromonal information of male and female Wistar rats are received by vomeronasal sensory neurons at upper and under layers of female Wistar rats, respectively. (From Inamura et al. (1999b) *J Physiol (Lond)* 517: 731–739.) Pheromonal information received by sensory neurons having Gi and G1Rs are transmitted to the rostral region of the AOB. (From Inamura et al. (1999a) *Eur J Neurosci* 11: 2254–2260.)

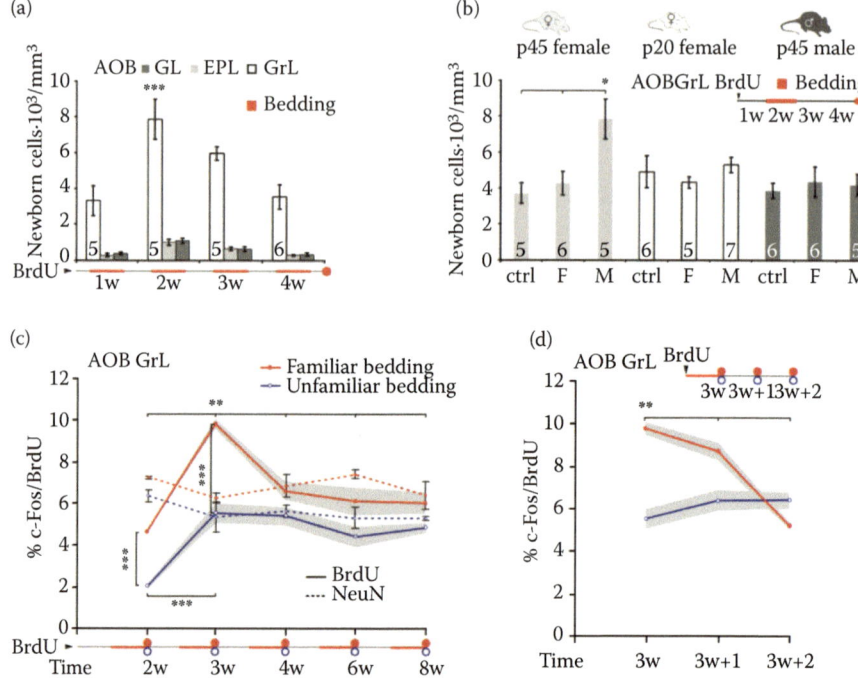

FIGURE 13.2 Sensory-driven integration and function of newborn neurons in the AOB of adult female mice. (a) Quantification of BrdU-labeled cells at 28 dpi in multiple AOB layers of p45 female mice after familiarization with male bedding performed in four different weeks. Male bedding exposure enhances integration of newborn neurons aged between 7 and 14 days. (b) Density of newborn granule cells in adult females, prepubertal females, and adult males evaluated at 28 dpi of BrdU and after 1-week exposure to female (F) or male (M) bedding (from 7th to 14th dpi of BrdU). (c, d) Percentage of c-Fos/BrdU (continuous lines) and c-Fos/NeuN-positive cells (dotted lines) induced by familiar (red) and unfamiliar (blue) bedding/pheromones in AOB GrL. Notably, AOB newborn granule neurons, but not preexisting ones (NeuN-positive, mature neurons), are preferentially activated (c-Fos-positive) by experienced pheromones. This activity is transient and picks around the third week of age of newborn cells. Data is means ± sem (represented by gray areas in c, d) and the numbers in the graph bars indicate the amount of animals used. *$P < 0.05$, **$P < 0.01$, ***$P < 0.001$. (Modified from Oboti, L. et al. 2011. *Front. Neurosci.* 5:113.)

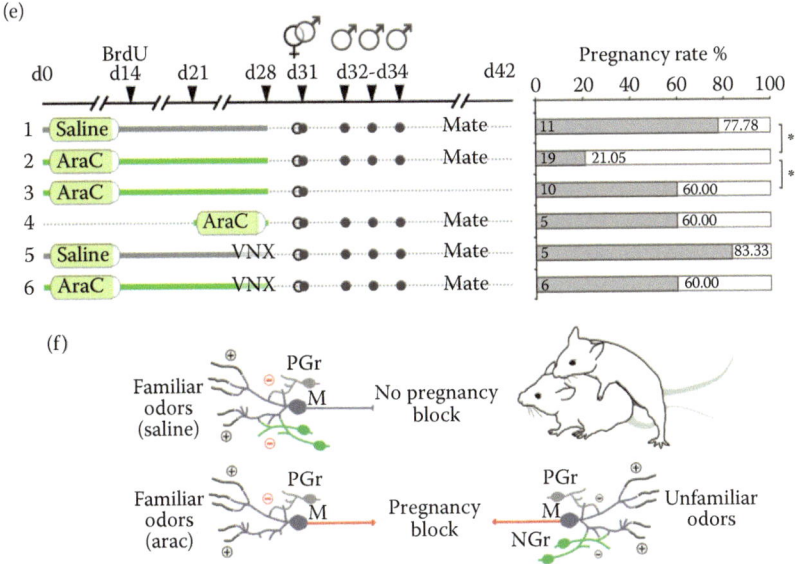

FIGURE 13.2 (*Continued*) Sensory-driven integration and function of newborn neurons in the AOB of adult female mice. (e) Matings and male stimulations on female mice after different protocols of Ara-C/saline treatment in normal and vomeronasal nerve-lesioned mice (VNX). Shown in the graph are the pregnancy rates (in percentage) as a function of the different treatment conditions evaluated 11 days after mating. (f) Schematic diagram illustrating the role of AOB newborn granule cells (NGr) in the modulation of a mate's familiar signals (left side) and unfamiliar ones (right side): granule cells are preferentially involved in the detection of male individual odors once integrated into preexisting circuits. When highly responsive newborn granule cells (NGr) are eliminated after Ara-C treatment (left side, bottom), preexisting granule cells (PGr) are not sufficient to prevent pregnancy block by a mate's familiar odors (red arrows). *$P < 0.05$. (Modified from Oboti, L. et al. 2011. *Front. Neurosci.* 5:113.)

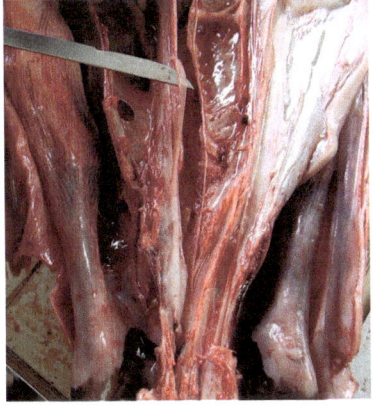

FIGURE 16.2 Location of the vomeronasal organ in the buffalo.

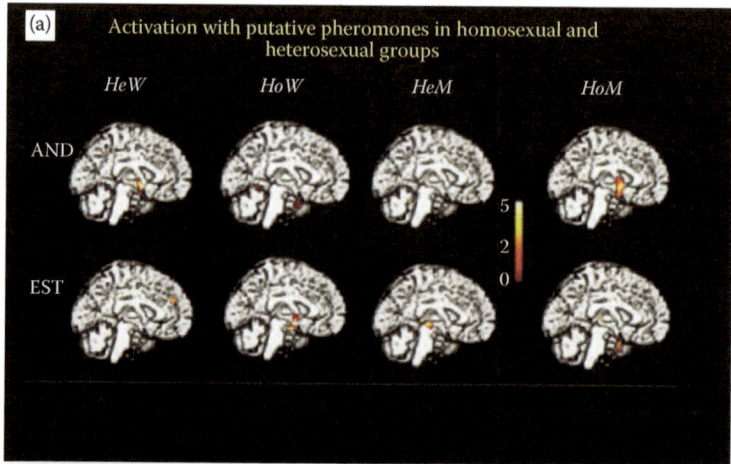

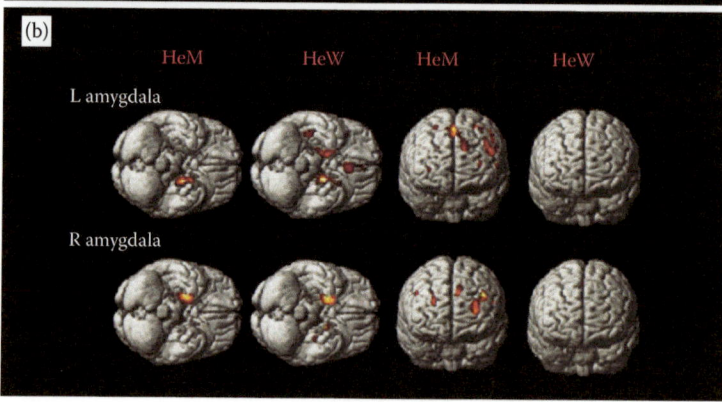

FIGURE 18.1 (a) Activation with AND and EST. Significant activations illustration of group-specific activations with putative pheromones during smelling of AND and EST in relation to smelling of air (which was the so-called baseline condition). Clusters of activated regions are superimposed on the standard brain. HeM = heterosexual men, HeW = heterosexual women, HoM = homosexual men, HoW = homosexual women, MRI = midsagittal plane. (Remodeled from Figure 1 from Savic I, Lindström P. (2008) *Proc Natl Acad Sci U S A* 105: 9403–8. doi: 10.1073. With permission.) (b) Functional connectivity from the right (R) and left (L) amygdala. (Reprinted from Figure 1 from Savic I, Lindström P. (2008) *Proc Natl Acad Sci U S A* 105: 9403–8. doi: 10.1073. With permission.)

8 Chemical Signaling in Amphibians

Sarah K. Woodley

CONTENTS

8.1 Introduction 256
8.2 Production of Amphibian Chemosignals...257
    8.2.1 Skin Glands..257
        8.2.1.1 Urodele Skin Glands..257
        8.2.1.2 Anuran Skin Glands 259
    8.2.2 Cloacal Glands...259
    8.2.3 Fecal Excretions...260
    8.2.4 Oviduct 260
    8.2.5 Summary and Conclusions 260
8.3 Amphibian Chemosignals Related to Mating..260
    8.3.1 Male Responses to Female Chemosignals....................................261
        8.3.1.1 Urodeles: Mate Location, Recognition, and Choice 261
        8.3.1.2 Anurans: Multimodal Signaling 262
    8.3.2 Female Responses to Male Chemosignals....................................262
        8.3.2.1 Urodeles: Mate Location...262
        8.3.2.2 Urodeles: Mate Recognition 262
        8.3.2.3 Urodeles: Mate Choice..263
        8.3.2.4 Anurans: Mate Recognition 265
    8.3.3 Summary and Conclusions 266
8.4 Amphibian Chemosignals Related to Territoriality and Male–Male
Interactions 267
    8.4.1 Territorial Advertisement 267
    8.4.2 Male–Male Interactions..268
    8.4.3 Summary and Conclusions 268
8.5 Chemosignal Detection...269
    8.5.1 Anatomy of the Amphibian Nasal Cavity 269
    8.5.2 Central Olfactory Projections...270
    8.5.3 Chemosensory Receptors 271
        8.5.3.1 Trace-Amine Associated Receptors 271
        8.5.3.2 ORs 271
        8.5.3.3 VRs 272
        8.5.3.4 Accessory Proteins...273

8.5.4   Responses of Chemosensory Epithelia 273
8.5.5   Primer Effects 274
8.5.6   Summary and Conclusions 275
8.6   Future Directions and Concluding Remarks 275
References 276

8.1   INTRODUCTION

The term pheromone has been used to describe chemical signaling in most inverte-brate and vertebrate groups, including amphibians. Pheromones are typically defined as chemical substances (e.g., a single molecule or a blend of a few molecules) that elicit an innate stereotyped behavior or developmental change in another individual of the same species (reviewed in Wyatt 2010). Pheromones may be water-soluble, volatile, or nonvolatile. In tetrapod vertebrates, they are detected by chemosensory neurons of the olfactory system, both the main olfactory system and the accessory (vomeronasal) olfactory system (Baum and Kelliher 2009). Pheromones are usu-ally categorized according to function as releasers, primers, modulators, and alarm pheromones. Despite years of study, it is still unclear whether pheromones operate differently from nonpheromones, especially in mammals (Doty 2010). Nonetheless, many favor the use of the term pheromone, arguing that the term has intrinsic heuris-tic value if restricted to chemical emissions shaped by evolution to have a signaling function within a species (Wyatt 2010). Some of the controversy regarding phero-mones may arise from the difficulty of translating a concept originally developed in insects to behaviorally and cognitively sophisticated mammals.

Study of amphibian chemical signaling can offer insight to the controversy over the nature of pheromones as well as provide general insights to vertebrate chemical communication. Compared to mammals, amphibians have relatively simple nervous systems and express stereotyped behaviors. A number of amphibian pheromones have been characterized and can be used to evaluate concepts related to phero-mones. Further, as modern representatives of basal tetrapods, amphibians can pro-vide insight to the evolution of aspects of chemical signaling, such as the evolution of the vomeronasal system. Finally, due to their biphasic life cycle, amphibians are a good model for understanding how chemical signaling functions in aquatic versus terrestrial environments.

Chemical signaling is widespread in amphibians. Extant amphibians descended from a common ancestor about 250 million years ago (San Mauro et al. 2005) and consist of anurans (frogs and toads), urodeles (salamanders), and gymnophionans (caecilians). Chemical communication has been well studied in urodeles, for which chemical communication is a dominant sensory modality. Chemical communication has been less studied in anurans but may be more prevalent than realized (Belanger and Corkum 2009; Lee and Waldman 2002; Waldman and Bishop 2004). Compared to the other amphibian orders, little is known about caecilian chemical communica-tion (but see Eisthen and Polese 2006; Reiss and Eisthen 2008).

The goals of this chapter are to review chemical signaling in amphibians and to highlight how study of amphibians can offer insight to vertebrate chemical com-munication. In this review, I will use the term chemosignal to describe chemical information conveyed among members of the same species that has signaling properties. I will discuss the sources of amphibian chemosignals, the function and chemical identity of chemosignals, and the sensory detection of amphibian chemosignals. The review is not meant to be comprehensive, and readers are referred elsewhere for more information (Dawley 1998; Eisthen and Polese 2006; Houck 2009; Kikuyama et al. 2005; Reiss and Eisthen 2008; Woodley 2010).

8.2 PRODUCTION OF AMPHIBIAN CHEMOSIGNALS

8.2.1 Skin Glands

Amphibian skin is unique among vertebrate classes because it is embedded with multicellular exocrine glands (Duellman and Trueb 1986). Glandular secretions serve a variety of functions, including chemical signaling (Table 8.1) (Apponyi et al. 2004; Toledo and Jared 1995). The primary dermal glands are mucous and granular glands, and release of glandular contents may be episodic or tonic (Quagliata et al. 2006). Despite a similar basic structure in all amphibians, skin glands and their secretions are diverse across species. Although many glandular secretions are nonvolatile, many are volatile. In fact, amphibians can have distinct odors that humans can smell (Smith et al. 2004).

8.2.1.1 Urodele Skin Glands

Skin glands in urodeles release chemosignals that function in courtship and mating interactions (Houck and Sever 1994; Sever 2003; Sever and Staub 2010). In plethodontid salamanders, courtship and mating occurs on land and requires coordination of a male and female for successful sperm transfer (Houck and Arnold 2003). Males possess a specialized mental gland, which is a cluster of modified mucous glands that develops on the chin during the mating season (Brizzi et al. 2002; Woodley 1994). During courtship, a male rubs its mental gland onto the skin and/or nose of a female. A mental gland is found in males of almost all of the 250+ species of plethodontid salamanders, implying that the mental gland plays a vital role in reproductive success in this salamander group (Houck and Sever 1994). The structure of the mental gland is highly specialized, and varies across species in a manner related to mode of delivery of the mental gland secretions to the female (Houck and Sever 1994). The glands are directly innervated (Brizzi et al. 2002) and produce glycoproteins that enhance female receptivity during courtship (Houck 1998). In addition to the mental gland, males of some species of plethodontid salamanders have prominent glands on the dorsal surface of the tail, apparently derived from granular glands (Sever 1976, 1989; Sever and Staub 2010). Chemosignals from the dorsal tail gland are believed to be transmitted to females during a highly conserved courtship behavior called tail-straddling walk.

Specialized skin glands may function in courtship in salamandrid salamanders as well. Male newts of the genus *Notophthalmus* have glands in their cheeks (genial) that develop during the breeding season under the influence of hormones (Pool and Dent 1977). During courtship, a male clasps the female in amplexus and rubs his cheek against the female's nose, suggesting a signaling function for genial gland secretions (Houck and Arnold 2003).

TABLE 8.1
Properties of Glands Associated with Chemosignal Production

Taxon	Gland	Location of Gland	Sex with Gland	Signaling Function	Chemosignal	Chemical Nature
Plethodontid salamanders	Mental gland	Mandible	Male	Yes	PRF, PMF, SPF	Protein
Notophthalmus salamanders	Genial gland	Cheek	Male	Not confirmed	?	?
Some plethodontid salamanders	Dorsal tail gland	Dorsal tail skin	Male	Not confirmed	?	?
Plethodon salamanders	Postcloacal glands	Ventral tail skin	Male	Not confirmed	?	?
Salamandrid salamanders	Dorsal gland/vent gland	Cloaca	Male	Yes	Sodefrin, Silefrin	Peptide
Most urodeles	Ventral gland	Cloaca	Female	Not confirmed	?	?
Hynobius leechi	?	Oviduct	Female	Yes	15-epi-PF2α	Prostaglandin
Mantellid frogs	Femoral glands	Thigh	Male	Yes	Phoracantholide J, 8-methyl-2-nonanol	Volatile: macrolide and alchohol
Rana frogs	Dorsal skin glands	Dorsal skin	Male	Not confirmed		
Litoria splendida	Parotoid and rostral glands	Near head	Male, female	Yes	Splendipherin	Peptide
Leptodactylus fallax	?	?	Male, female	Yes	LASP	Peptide

Note: LASP = *Leptodactylus* aggression-stimulating peptide; PMF = plethodon modulating factor; PRF = plethodon receptivity factor; SPF = sodefrin precursor-like factor.

Skin glands may also function in territoriality (Mathis et al. 1995). Plethodontid salamanders in the genus *Plethodon* scent mark by pressing the ventral surface of the tail onto the substrate (Jaeger 1984; Largen and Woodley 2008). The ventral surface of the tail contains postcloacal granular glands that empty during scent marking (Hecker et al. 2003; Largen and Woodley 2008; Simons et al. 1999; Staub and Paladin 1997).

In addition to the specialized glands described above, generalized granular gland secretions may also have signaling functions. Many salamanders release noxious skin secretions that function in predator defense. In red-legged salamanders (*Plethodon shermani*), antipredator secretions elicited behavioral responses from conspecifics and also activated sensory neurons in the vomeronasal organ (Schubert et al. 2008).

8.2.1.2 Anuran Skin Glands

A large number of skin glands have been linked to breeding in anurans. These glands develop during the breeding season and likely have diverse functions, including chemosignal production (Brizzi et al. 2003). Recent studies discovered chemosignals in the femoral glands of several species of mantelline frogs (Poth et al. 2012). Femoral glands are specialized macroglands that cluster on the ventral shanks of males but not females (Vences et al. 2007). Another potential source of chemosignals is the dorsal skin glands of frogs (*Lithobates*, formerly *Rana*). These glands are dispersed throughout the dorsal skin, are male-specific and androgen-dependent, and are speculated to produce chemosignals (Brizzi et al. 2002; Thomas and Licht 1993; Thomas et al. 1993).

In addition to the breeding skin glands, anuran skin contains granular glands that secrete defensive compounds. Most studies of anuran granular glands focus on finding bioactive compounds with therapeutic potential (Erspamer 1994). In the process, compounds with chemosignaling properties have been discovered. For example, the rostral and parotoid glands of male tree frogs (*Litoria splendida*) produced a peptide that was only present during the breeding season. Subsequent study found that the peptide attracted females (Wabnitz et al. 1999, 2000). The production of a male-specific pheromone by the parotoid and rostral glands was surprising because the glands are present in both males and females and produce host-defense peptides. As another example, extraction of glandular secretions from the mountain chicken frog (*Leptodactylus fallax*) revealed a peptide produced only by males and that promoted male–male aggression (King et al. 2005).

8.2.2 CLOACAL GLANDS

Cloacal glands are exocrine glands that empty into the cloacal chamber (Sever 2003). The dorsal cloacal gland (sometimes called the abdominal gland) is specialized for pheromone production in Japanese and European newts (Cedrini and Fasolo 1971; Kikuyama et al. 1995; Malacarne and Vellano 1987). These newts mate in the water, and courtship is characterized by displays in which males fan their tails, presumably to direct dorsal cloacal gland emissions towards females (Halliday 1990). The dorsal cloacal gland enlarged in response to testosterone and prolactin (Kikuyama et al. 1975). During the height of the breeding season, the dorsal cloacal gland represented about 5% of a male's total mass (Verrell et al. 1986). Secretion of the gland was triggered by arginine vasotocin (Toyoda et al. 2003). *Ambystoma*, another group of aquatic-breeding salamanders, also have well-developed dorsal cloacal glands (also called vent glands) and tail fanning behavior. It is hypothesized that pheromone production is the ancestral condition for the dorsal/vent cloacal gland (Sever 2003). Similar cloacal glands are also present in Plethodontid salamanders, but they are not as well developed and their role in chemical communication has not been studied (Brizzi et al. 2002).

Cloacal glands may be important for female pheromone production. Ventral cloacal glands are an ancestral trait in female salamanders. In the plethodontid salamander, *Eurycea bislineata*, the ventral gland actively secreted only during the mating season, suggesting that the gland may secrete a chemosignal important for mating (Sever 1988).

8.2.3 FECAL EXCRETIONS

In red-backed salamanders (*Plethodon cinereus*), a terrestrial salamander, several lines of evidence indicate that male fecal pellets contain chemosignals of interest to sexually receptive females (reviewed in Jaeger and Forester 1993; Jaeger and Wise 1991). Females were more interested in fecal pellets from males that fed on high-quality diets, suggesting that the fecal pellet is an honest signal of a male's ability to garner resources. It is not clear whether the fecal pellet chemosignals are signals per se, or simply cues of the male's diet. Similar to red-backed salamanders, archaic frogs (*Leiopelma hamiltoni*) inhabit home ranges and individuals distinguished between their own feces versus feces from conspecifics (Lee and Waldman 2002).

8.2.4 OVIDUCT

In red-bellied newts (*Cynops pyrrhogaster*), water that had previously contained females attracted males, but not if the oviduct had been removed (Toyoda et al. 1994). The development of the oviduct and the male-attracting properties of females required prolactin and estrogen (Kikuyama et al. 1986; Toyoda et al. 1994).

8.2.5 SUMMARY AND CONCLUSIONS

Chemosignals emanate from a variety of sources in amphibians (Table 8.1). Many specialized skin glands produce chemosignals, both in urodeles and anurans. Cloacal glands are also important sources of chemosignals and are hypothesized to have an ancient function in pheromonal communication, at least in urodeles. There are also nonglandular sources of chemosignals, such as fecal excretions. Compared to males, very little is known about glands producing chemosignals in female amphibians, but the oviduct may be important.

The highly specialized mental glands of plethodontid salamanders and the femoral glands of mantelline frogs suggest that the glands evolved specifically for the production of pheromonal chemosignals. However, glands specialized for other functions have also been shown to produce secretions with pheromonal effects. For example, the parotoid/rostral glands of magnificent tree frogs function to produce host defense peptides, yet pheromones were also secreted (Wabnitz et al. 2000). In addition, nonspecialized sources of chemosignals, like the oviduct, or feces, appear have signaling properties in amphibians. In these latter examples, it is unknown if the chemosignals have been shaped by evolution for a signaling function or are merely by-products of other physiological processes like reproduction or digestion.

8.3 AMPHIBIAN CHEMOSIGNALS RELATED TO MATING

Social behavior in adult amphibians generally centers on reproduction, and pheromones have been identified largely in the contexts of mating. Individuals are usually solitary during the nonbreeding season but aggregate during the breeding season in order to find suitable mating partners. In some species, the mating period is very brief (days), but in others it is prolonged (several weeks to months).

Chemical signals are used to locate and recognize suitable mating partners and may signal species, sex, reproductive status, and mate quality (Johansson and Jones 2007). In urodeles, chemosignals are particularly important for mate recognition because vocalizations are rare and mating occurs under low light. In anurans, acoustic communication is very important, but chemosignals may also play a role, especially in voiceless species or when acoustic communication is inefficient. In addition to facilitating mate localization and recognition, chemosignals may also function to coordinate mating partners and to modulate sexual motivation during courtship and mating. Coordination of mating partners is very important in amphibians. In anurans, gamete release occurs during amplexus, and partners must release gametes at the same time to ensure fertilization. In most urodeles, sperm transfer is external but fertilization is internal. Sperm is deposited on a substrate in the form of a spermatophore, and a female must be "persuaded" during courtship to pick up the spermatophore into its cloacal vent for internal sperm storage.

8.3.1 Male Responses to Female Chemosignals

8.3.1.1 Urodeles: Mate Location, Recognition, and Choice

There is abundant behavioral evidence that male urodeles locate and recognize suitable mating partners via female chemical signals. In *Plethodon* salamanders, a group of terrestrial salamanders, males preferred both volatile and substrate-borne chemosignals of conspecific females over those of males or blank substrates. Males did not prefer chemosignals of females from closely related species (Dawley 1984a,b, 1986; Palmer and Houck 2005). Males seemed particularly attuned to female chemosignals because males investigated chemosignals more than did females, and investigation was increased by androgen treatment (Schubert et al. 2006, 2008).

Chemosignals may also provide information about female reproductive condition. For example, in several species of salamandrids, males were attracted to water in which reproductively active or hormonally primed females had previously been housed (Dawley 1984b; Thompson et al. 1999; Toyoda et al. 1994; Twitty 1955). Males were not attracted to water containing males or nonreproductive females. Males also were not attracted to water from females whose oviducts had been removed, suggesting that the source of the chemosignal was the oviduct (Toyoda et al. 1994). Reproductive hormones may have been co-opted to cue female reproductive status to help males identify ovulating females. Hynobiid salamanders are a group of aquatic salamanders for which fertilization is external. In the Korean salamander (*Hynobius leechi*), water that held ovulating females attracted males. The water contained high levels of F-series prostaglandins, which mediate ovulation. In particular, 15-epi-prostaglandin F2α attracted males and activated sensory neurons of the main olfactory system (Eom et al. 2009).

Finally, female chemosignals may contribute to male mate choice. In amphibians, female body size is usually correlated with the number of follicles, which is an indicator of female fecundity. Males preferred chemosignals derived from larger, more fecund females over those from smaller, less fecund females in several species of salamandrid and plethodontid salamanders (Marco et al. 1998; Sullivan et al. 1995).

To summarize, females produce chemical substances that convey information about species, sex, reproductive status, and female fecundity, thereby permitting mate recognition, assessment, and even choice. Little is known about the identity of these female chemosignals, and whether they evolved for specific signaling functions or are merely by-products of female reproductive processes.

8.3.1.2 Anurans: Multimodal Signaling

In anurans, chemical and acoustic signaling may work in concert for species and mate recognition. Although calling behavior attracts females from a distance and signals male quality (Sullivan et al. 1995), calling is energetically costly and cannot be sustained for long periods. Chemical emissions may be important, especially at short ranges. In terrestrial toadlets, males were attracted to substrates containing female-derived chemosignals but not to blank substrates or those containing male scents. Furthermore, exposure to odors from females caused males to increase their rate of advertisement calling (Byrne and Keough 2007). The authors proposed that males use female-derived chemosignals at short ranges to modulate their calling. In this way, males can use chemosensory information to adjust their calling rates to conserve energy while still maintaining high mating success.

8.3.2 FEMALE RESPONSES TO MALE CHEMOSIGNALS

8.3.2.1 Urodeles: Mate Location

Females do not appear to use long-range male chemosignals to locate suitable mating partners. Female newts were equally attracted to waterborne chemosignals from females and males (Secondi et al. 2005). Female *Plethodon* salamanders had no consistent preferences for substrate-borne chemosignals of males or females, or from conspecifics or heterospecifics (Dawley 1986; Palmer and Houck 2005). It is possible that females limit mate searching in order to divert resources towards egg production, which is energetically costly (e.g., Fitzpatrick 1973).

8.3.2.2 Urodeles: Mate Recognition

Although females do not appear to be preferentially attracted to long-range male-derived chemosignals, females respond strongly to short-range emissions from males. In many species of salamandrid newts, males fan or vibrate their tails during courtship in order to direct a stream of chemosignals from the male dorsal cloacal gland towards females (Halliday 1990). A decapeptide called sodefrin was isolated from the dorsal cloacal gland of male red-bellied newts (*C. pyrrhogaster*) that attracted reproductively active conspecific females but not males or heterospecific females (Kikuyama et al. 1995). The dorsal cloacal gland of the sword-tailed newt (*C. ensicauda*) produced a decapeptide called silefrin, which differed from sodefrin by two amino acids (Yamamoto et al. 2000). Slight changes in amino acid sequence of sodefrin and silefrin resulted in a failure to be recognized by conspecific females, suggesting the pheromones function in species recognition. Dorsal cloacal gland pheromones are likely to operate in salamandrid newts from Europe as well (Osikowski et al. 2008).

Mate recognition pheromones may work in concert with other sensory modalities. In European newts (*Lissotriton*, formerly *Triturus*), breeding males have large crests on the tail, decorated with conspicuous patterns. During courtship, males express frequent bouts of tail displays and the more a male displays, the greater the probably that the female will be inseminated (Green 1991; Teyssedre and Halliday 1986). Tail displays are energetically costly and females probably assess male quality by the intensity of his visual displays as well as by the amount of pheromone delivered by the tail movements. It is possible that tail fanning and a large tail crest evolved initially to direct pheromones to females, and then the visual appearance of the tail evolved later to create a multimodal signal.

8.3.2.3 Urodeles: Mate Choice

The role of male chemosignals in mate choice has been well studied in plethodontid salamanders. During a long and stereotyped courtship ritual, males apply secretions from a submandibular mental gland to females. Experimental application of mental gland secretions to females during courtship reduced the duration of courtship preceding sperm transfer (i.e., increased female receptivity) (Houck and Reagan 1990; Houck et al. 1998, 2008a,b; Rollmann et al. 1999). The increase in female receptivity was due to a direct augmentation of female sexual responsiveness to olfactory stimuli rather than to an increase in general arousal (Vaccaro et al. 2010). Due to the behavioral effects of mental gland secretions on female receptivity, mental gland secretions are often called courtship pheromones.

Courtship pheromones do not function in species recognition, sex recognition, or mate attraction, because they are applied to receptive females during the later stages of courtship after potential mates have already been identified. Instead, courtship pheromones are hypothesized to function in sexual selection by female mate choice. Courtship pheromones have been proposed to be aphrodisiacs that prime the female to increase the probability that she stays with the male until successful sperm transfer occurs (Arnold 1977; Noble 1927). This sort of manipulation of female behavior may not necessarily benefit the female, potentially leading to a conflict between male and female interests (Gershman and Verrell 2002). Alternatively, courtship pheromones may convey meaningful and honest information about a male's quality or genetic compatibility, information that would increase female fitness either directly or indirectly (Johansson and Jones 2007). Regardless of the ultimate function of courtship pheromones, the proximate result is that courtship pheromones increase female receptivity to mating.

Courtship pheromones are delivered to the nares (olfactory mode) or to the skin (transdermal mode) of females, depending on the species (Figure 8.1) (Houck and Sever 1994). The transdermal mode of courtship pheromone delivery is the ancestral condition for plethodontid salamanders, evolving more than 100 million years ago. Males in species that use transdermal delivery possess hypertrophied premaxillary teeth that they use to scrape the back of the female so that the skin of the female is abraded. Due to the highly vascularized skin of plethodontids, mental gland secretions are thereby injected into the circulation (Arnold 1977; Organ 1961). The olfactory mode of delivery evolved about 20 million years ago and is restricted to species in the *Plethodon glutinosus* species complex. In this group, a male repeatedly slaps or rubs its mental gland against the nares of a female during courtship. Application of mental gland secretions to the nares of a female activated the females's vomeronasal

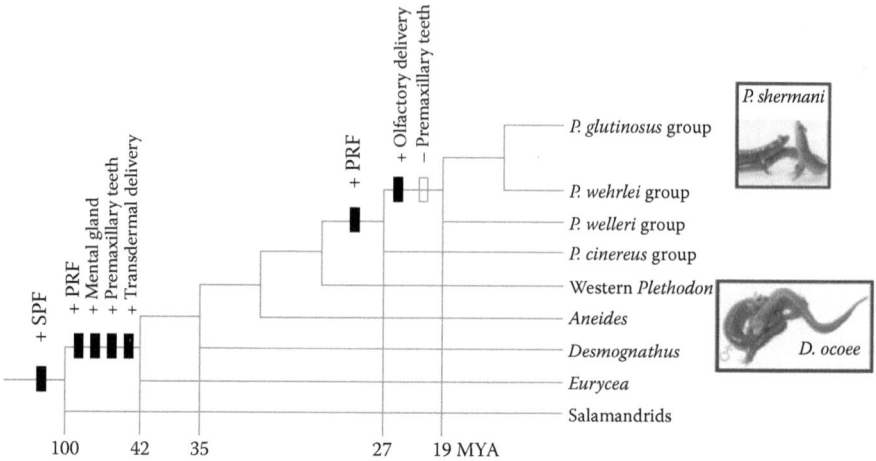

FIGURE 8.1 Cladogram showing the evolution of courtship pheromones and associated traits in plethodontid salamanders. Solid bars indicate the presence of particular traits and empty bars indicate the loss of traits. The family Salamandridae is an out group. Photographs to the left show pheromone delivery by the red-legged salamander (*P. shermani*), a species utilizing olfactory delivery of pheromones, and pheromone delivery by the Ocoee salamander (*D. ocoee*), a species utilizing transdermal delivery of pheromones. (Photographs used with permission from Stevan J. Arnold. With kind permission from Springer Science+Business Media: *J. Comp. Physiol. A Neuroethol. Sens. Neural. Behav. Physiol.,* Pheromonal communication in amphibians, 196, 2010, 713–727, Woodley, S. K.)

chemosensory epithelium of the nasal cavity and altered brain activity (Laberge et al. 2008; Wirsig-Wiechmann et al. 2002, 2006).

Three courtship pheromones have been characterized from mental gland secretions. In the red-legged salamander (*P. shermani*), a species that uses the olfactory mode of delivery, most of the mental gland secretions consisted of two courtship pheromones. These were called plethodon receptivity factor (PRF) and plethodon modulating factor (PMF) (Chouinard et al. 2013; Palmer et al. 2007a; Rollmann et al. 1999) because they were shown, in purified preparations or in recombinant form, to affect female receptivity (Houck et al. 2007, 2008b; Rollmann et al. 1999, 2003). PRF is a 22-kDa protein related to the Il-6 family of cytokines. PMF is a 7 kDa-protein related to the three-finger protein superfamily. In the Ocoee salamander (*Desmognathus ocoee*), a species with transdermal delivery of courtship pheromones, mental gland secretions were more complex than those of the red-legged salamander. About 30 compounds were found, of which nine were purified (Leichty 2012). Twenty-five percent of the secretions were a 20-kDa protein called sodefrin precursor-like factor (SPF) because of the sequence similarity to the precursor of sodefrin, the pheromone found in newts (see earlier). Like PMF, SPF is related to the three-finger protein superfamily (Palmer et al. 2007a). A preparation of mental gland extract that was enriched with SPF was experimentally applied to the dorsum of females and found to increase female receptivity (Houck et al. 2008b). Other mental gland secretions may also be important chemosignals. Several proteins in addition to SPF have been purified, some of which had sequence similarity to relaxins and serine protease inhibitors. It is speculated that some of the differences in mental gland secretions between lineages may be related to whether the mode of pheromone delivery is olfactory or transdermal.

Courtship pheromones are represented as variable isoforms. In red-legged salamanders (*P. shermani*), three main isoforms of PRF, and about 30 isoforms of PMF were found. Individual males varied in the relative proportions of the PRF isoforms, PMF isoforms, as well as unidentified proteins that were present at low levels (Chouinard et al. 2013; Feldhoff et al. 1999; Rollmann et al. 2000). In contrast, a single isoform made up the majority of SPF in Ocoee salamanders, suggesting that the SPF chemosignal is similar among individual males within a species. However, as described above, many additional compounds were found in Ocoee salamander mental glands that have not yet been characterized. Individual variation in courtship pheromones and other mental gland secretions may influence female choice, perhaps by maximizing sensory stimulation of a female so that she mates more readily, or by signaling male quality.

Courtship pheromones and other mental gland secretions have experienced rapid positive selection. Mental gland cDNA corresponding to PRF, PMF, and SPF has been sequenced in dozens of Plethodontid species (Kiemnec-Tyburczy et al. 2009; Leichty 2012; Palmer et al. 2005, 2007a,b, 2010; Watts et al. 2004). SPF and PMF are found in all species sequenced to date but PRF is restricted to *Plethodon* salamanders found in the eastern United States (Figure 8.1). High levels of genetic diversity exist for SPF, PMF, and PRF, both within and between species, although the amount depends on the species. Analysis of predicted amino acid sites indicates that those sites implicated in receptor binding have been subjected to strong and rapid positive selection, perhaps driven by an ongoing molecular tango or coevolution between chemosignal and receptor. Rapid positive selection appears to have slowed down for SPF in eastern *Plethodon* species, suggesting an evolutionary replacement of SPF by PRF (Palmer et al. 2007a). In contrast, PMF continues to be shaped by positive selection in eastern *Plethodon*, suggesting that PMF and PRF work synergistically to affect female behavior (Palmer et al. 2010).

8.3.2.4   Anurans: Mate Recognition

There are several chemically identified chemosignals produced by male anurans that influence female behavior at close range. In the male magnificent tree frog (*L. splendida*), the rostral and parotoid glands produce a peptide called splendipherin during the breeding season. Both natural and synthetic splendipherin, when placed on a pad located in water, attracted females at short range but not males. Females from another species were not attracted (Wabnitz et al. 1999, 2000). These results indicate that splendipherin signals both species and sex. Splendipherin is not water-soluble but has a hydrophobic region that allows it to travel along the surface of the water much faster than by diffusion through the water (Perriman et al. 2008). Little is known about the natural history or mating habits of this species, so it is unclear how splendipherin functions in the wild.

As another example of an anuran chemosignal, postaxillary glands present in male dwarf African clawed frogs (*Hymenochirus* sp.) produced chemosignals that attracted females (Pearl et al. 2000). In an aquatic Y-maze, females approached chemical signals derived from males as well as to postaxillary gland homogenates. Females were not attracted to chemosignals from males whose postaxillary glands had been removed. Because little is known about the ecology of this species, it is unclear how gland secretion functions in mating, but it is speculated that the glandular sections attract females and perhaps induce female ovulation (Pearl et al. 2000).

In terrestrial toadlets, females were equally interested in chemosignals from conspecific males and females, but clearly responded to a male scent trail in the presence of a calling male. The authors suggest that females use male chemosignals in combination with acoustic signals to better locate suitable mating partners (Byrne and Keough 2007).

8.3.3 SUMMARY AND CONCLUSIONS

A large number of behavioral studies indicate that chemosignals are used in male–female interactions for locating, recognizing, and choosing mates. The most well-characterized chemosignals are found in urodeles, for which there is a good understanding of the ecological context in which the signals operate. Although little studied, anuran chemical communication is likely to be a productive area of research, especially for study of multimodal communication. Although male-derived chemosignals have been well studied, female chemosignals are understudied, despite compelling behavioral evidence that males exploit female chemosignals to locate mates.

In several cases, chemosignals have been chemically characterized and found to elicit clear behavioral responses. Chemically characterized molecules with behavioral effects have been called pheromones (Table 8.2). Amphibian pheromones may be water-soluble peptides (sodefrin), water-soluble prostaglandins, water surface-active peptides (splendipherin), nonvolatile proteins (PRF, PMF, SPF), and volatiles (alcohols and macrolides) (Table 8.1). In some cases, the behavioral responses are simple and straightforward, such as mate recognition in newts. In other cases, behavioral responses are more complex, such as the role of courtship pheromones in changing the pace of mating in plethodontid salamanders. Pheromones functioning in mate recognition (e.g., sodefrin) are invariant within a species, whereas those involved in mate choice (e.g., PRF, PMF) are individually variable.

Many of the amphibian chemosignals that have been called pheromones do not fully meet strict definitions of pheromones (Wyatt 2010). A prime example is the courtship pheromones produced by the mental gland of male plethodontid salamanders. Courtship pheromones do not elicit simple stereotyped behavioral responses such as mate attraction or recognition. Instead, courtship pheromones function in female choice and have relatively subtle effects on female behavior, altering the probability that a female will remain with a male until successful sperm transfer occurs. In addition, the courtship pheromones PRF and PMF are not invariant within a species, but are highly variable among individuals of the same species. Another interesting feature of plethodontid courtship pheromones is that they are part of complex blends. For example, in Ocoee salamanders, the courtship pheromone SPF represents only 25% of the mental gland secretions, and there

TABLE 8.2
Properties of Amphibian Pheromones

Pheromone	Gland Specialized to Make Pheromones?	Part of a Blend?	Invariant within Species?	Olfactory Modality?	Function	Triggers Innate Response?
Sodefrin/silefrin	Yes	No	Yes	Yes	Mate recognition	?
Splendipherin	No	No	Yes	?	Mate recognition	?
15(epi)-PF2α	No	?	Yes	Yes	Mate recognition	?
Phoracantholide J 8-methyl-2-nonanol	Yes	Yes	Yes	?	?	?
PRF	Yes	Yes	No	Yes	Mate choice	?
PMF	Yes	Yes	No	Yes	Mate choice	?
SPF	Yes	Yes	No	No	Mate choice	?

Note: PMF = plethodon modulating factor; PRF = plethodon receptivity factor; SPF = sodefrin precursor-like factor.

are likely to be other important compounds at play. Finally, in vertebrates, pheromones are usually detected by chemosensory neurons in the nasal cavity. However, in the majority of plethodontid salamander species, courtship pheromones are applied to the skin of the back of females. This transdermal mode of chemosignal delivery is very unusual, and the signal-receptor mechanism by which mental gland secretions alter female behavior is unknown.

As a final note, pheromones are often assumed to elicit innate responses that do not require learning. To my knowledge, studies of amphibian pheromones and other chemosignals have used sexually experienced animals. In some cases subjects were collected from the wild as adults, or they were given pretrials to ensure sexually receptivity before testing. Hence, it is unknown whether documented behavioral responses to pheromones were innate or involved learning.

8.4 AMPHIBIAN CHEMOSIGNALS RELATED TO TERRITORIALITY AND MALE–MALE INTERACTIONS

8.4.1 TERRITORIAL ADVERTISEMENT

Territoriality has been well studied in plethodontid salamanders, and is particularly well developed in terrestrial species such as the red-backed salamander (*P. cinereus*) (reviewed in Mathis et al. 1995). Throughout the activity period that lasts many months, male and female red-backed salamanders occupy and defend relatively small home ranges. The territories permit increased foraging success during dry periods and may also increase access to mates during the mating season.

Chemosignals play important roles in the territoriality of red-backed salamanders (reviewed in Jaeger and Forester 1993). Chemosignals repelled conspecific as well as heterospecific intruders (Jaeger and Gergits 1979). Red-backed salamanders were also able to recognize familiar territorial neighbors (dear enemy effect) and familiar mating partners on the basis of chemosignals (Dantzer and Jaeger 2007; Horne and Jaeger 1988; Jaeger 1981; Jaeger and Gergits 1979; Jaeger et al. 1986).

Chemical signals used to advertise territories of red-backed salamanders include volatile and nonvolatile signals (Dantzer and Jaeger 2007). Nonvolatile signals are long-lasting but not very directional. Volatile signals may be especially effective when salamanders live in high densities. The identities of chemosignals used in territorial interactions are unknown, but probably emanate from postcloacal skin glands, ventral tail glands, and/or fecal pellets as described earlier (Jaeger and Wise 1991; Simons et al. 1994, 1999).

8.4.2 MALE–MALE INTERACTIONS

Chemosignals are used in male–male interactions in both salamanders and anurans. Male newts were repelled by chemosignals from reproductively active males (Park and Propper 2001). These responses may allow males to avoid aggressive interactions while searching for females. In anurans, compounds isolated from male skin altered the behavior of other males. *Leptodactylus* aggression-stimulating peptide (LASP) was purified from norepinephrine-stimulated skin secretions of male mountain chicken frogs (*Leptodactylus fallax*) (King et al. 2005). When presented with LASP on a sponge, males quickly approached the sponge and exhibited aggressive-like behaviors like leaping. LASP was not observed in female skin secretions and had no effect on females. It is not clear how LASP functions in the wild, but males of this species aggressively compete for sites to build foam nests.

A group of mantellid frogs from Madagascar have male-specific femoral glands that secrete mixtures of species-specific volatile compounds (Poth et al. 2012). Two volatiles (phoracantholide J and 8-methyl-2-nonanol) were shown to increase motor activity in male *Mantidactylus multiplicatus*. It is unknown whether the effects of the volatile components are similar to the effects of the entire gland. It is difficult to interpret changes in motor activity in response to the chemosignals because little is known about the social behavior in mantellid frogs. Also, there was a nonsignificant trend for females to respond to the volatile compounds, suggesting a role in male–female interactions.

8.4.3 SUMMARY AND CONCLUSIONS

Many amphibians inhabit and defend territories that are used for foraging or attracting mates. In salamanders, there is much behavioral evidence that chemosignals contribute to social interactions involved in territoriality, but the nature of the chemosignals has not been determined. In anurans, there are two examples of pheromones that affect male–male interactions, and more studies are necessary to understand the ecological and social contexts of chemical signaling in these species.

8.5 CHEMOSIGNAL DETECTION

Vertebrates possess multiple systems for the detection of chemical information (Breer et al. 2006; Eisthen and Polese 2006). In tetrapods, chemical information is detected by at least two anatomically separate chemosensory neuroepithelia: the main olfactory epithelium (MOE) and the accessory (vomeronasal [VNO]) system. These two systems are anatomically similar in many ways, and the functional differences between the two systems are unclear. At one time, it was hypothesized that the VNO was specialized to detect pheromones and the MOE was specialized in detection of environmental odorants. It is now generally accepted that pheromones and other chemosignals can be detected by both the MOE and VNO (Baum and Kelliher 2009).

The MOE and VNO express genes representing several families: the olfactory receptor (OR) gene family, the vomeronasal receptor gene families (V1R and V2R), the trace-amine-associated receptor (TAAR) gene family, and the formyl peptide receptor-like gene family (Ji et al. 2009; Nei et al. 2008; Riviere et al. 2009). OR genes are present in chordates (Grus and Zhang 2009; Niimura and Nei 2005), suggesting that the main olfactory system is very ancient. Although an anatomically separate vomeronasal organ arose during the evolution of amphibians, elements of the vomeronasal system are present in teleosts, cartilaginous fish, and jawless fish (Grus and Zhang 2009; Hansen et al. 2004). Compared to ORs and VRs, less is known about the phylogenetic distribution of TAARs or the formyl peptide receptor-like proteins.

OR and VR gene families are large, with lineage-specific diversification. It has been proposed that much of the diversity is related to adaptations to different environments; in particular, terrestrial versus aquatic environments. As modern representatives of basal tetrapods, study of amphibians can indicate how olfactory and vomeronasal systems evolved. Also, many amphibians alternate between aquatic and terrestrial environments throughout their lives, allowing study of how olfactory systems function in both aquatic and terrestrial environments within a single species. In the following sections, I review the anatomy of the amphibian olfactory systems, amphibian chemoreceptor molecules, and functional evidence that the MOE and VNO detect amphibian chemosignals.

8.5.1 ANATOMY OF THE AMPHIBIAN NASAL CAVITY

There are several excellent descriptions of the anatomy of amphibian nasal cavities that I will summarize here (Benzekri and Reiss 2012; Dawley 1998; Manzini and Schild 2010; Reiss and Eisthen 2008). Amphibians possess paired nasal sacs that open to the exterior via the external nares and to the roof of the mouth via the internal nares. The morphology of the nasal cavities varies between larval and adult forms, and between anurans and urodeles. In tadpoles, the nasal cavity consists of a chamber lined with olfactory sensory epithelium and an out-pocket containing a VNO. During metamorphosis, the larval olfactory cavity transforms into a large, principal cavity and a new middle chamber forms. The VNO is retained in the adult anuran, and the principal cavity, middle cavity, and VNO are interconnected. The principal and middle cavities are sometimes called the medial diverticulum and the lateral diverticulum, respectively.

In most anurans, the principal cavity is lined with ciliated olfactory sensory neurons and the middle cavity is lined with nonsensory neurons. The VNO consists of microvillar sensory neurons lining an internal cavity, with a groove extending to the internal naris. However, in clawed frogs (*Xenopus*) and other Pipid frogs, the middle cavity is lined with both ciliated and microvillar chemosensory neurons. The African clawed frog (*X. laevis*) is completely aquatic but samples chemical signals both in the water and at the water's surface. A valve directs water to the middle cavity (sometimes called a water-nose) and air to the principal cavity (sometimes called an air-nose). The VNO is filled with water throughout life (Altner, 1962, as cited by Manzini and Schild 2010).

Compared to anuran amphibians, salamanders have relatively simple nasal cavities. In both larval and adult salamanders, the nasal cavity consists of a dorsoventrally flattened cavity lined with both ciliated and microvillar olfactory sensory neurons, often with a recess or diverticulum in which are found vomeronasal sensory neurons. Access to the VNO may be via the main nasal cavity, and/or via the oral cavity. An additional access to the VNO is found in plethodontid salamanders, in which a nasolabial groove leading from the lip to the external naris directs nonvolatile chemosignals to the VNO (Dawley and Bass 1989).

Because an anatomically sequestered VNO first appeared in tetrapods, the VNO was once believed to be an adaptation to a terrestrial lifestyle, perhaps adapted for the detection of nonvolatile substances. However, several lines of evidence indicate that the VNO is not an adaptation to a terrestrial lifestyle (Eisthen 1997, 2000). The VNO is present in larval amphibians and is found in most species of fully aquatic amphibians, early tetrapods were aquatic, and the last common ancestor of tetrapods was most likely aquatic. Thus, the VNO did not initially evolve for the detection of nonvolatile substances, although lineage-specific adaptations may have occurred later in other taxonomic groups.

8.5.2 Central Olfactory Projections

In amphibians, sensory neurons of the MOE project to the main olfactory bulb (MOB), which is located on the anterior end of the telencephalon. Sensory neurons of the VNO project to the accessory olfactory bulb (AOB), which is a bulge on the lateral telencephalon caudal to the MOB. In *X. laevis*, different areas of the olfactory epithelia projected to different areas of the MOB, with sensory neurons from the principal cavity projecting to the dorsal main olfactory bulb, and sensory neurons from the middle cavity projecting to the ventral MOB (Saito and Taniguchi 2000). Furthermore, middle cavity sensory neurons expressing $G\alpha_{olf}$ projected to the rostromedial region of the ventral MOB, while those expressing $G\alpha_o$ projected to the caudolateral region of the ventral MOB (Nakamuta et al. 2011). In salamanders, the

VNO projected to the AOB, but species varied in the nature of the VNO projections, with some species having multiple terminal fields in the AOB and other species having only one or a few terminal fields (Schmidt and Roth 1990).

Projections from the olfactory bulb to the telencephalon have been examined in several amphibians (reviewed in Eisthen and Polese 2006). Projection neurons in the MOB terminate in the medial pallium, the postolfactory eminence, and the lateral and medial septum (via the medial olfactory tract), and in the lateral pallium, dorsal striatum, and the lateral amygdala (via the lateral olfactory tract). Projection neurons of the AOB reach the medial amygdala via the accessory olfactory tract. There is also an extrabulbar olfactory pathway in which sensory neurons from the main olfactory pathway skip the olfactory bulb and terminate in the preoptic area and hypothalamus.

An important area that receives olfactory information is the amygdala, which is important for emotional processing and learning. In mammals, the medial amygdala receives input from the AOB and the cortical amygdala receives input from the MOB. Recent work has found evidence that the amygdaloid complex in both urodeles and anurans shares features in common with amniote vertebrates. Based on developmental and neurochemical data as well as patterns of connectivity, there are areas of the amphibian amygdala that correspond to the mammalian medial and cortical amygdala (Laberge et al. 2006; Laberge and Roth 2005, 2007; Moreno and Gonzalez 2003, 2004, 2006; Moreno et al. 2005).

The dual olfactory hypothesis of functional and anatomical separation of the main olfactory system and the accessory olfactory system is no longer supported in mammals (Kang et al. 2009; Licht and Meredith 1987; Scalia and Winans 1975). Similarly, the dual olfactory hypothesis is not supported in amphibians (Roth and Laberge 2011). As shown in the salamander (*P. shermani*), there is a very high degree of convergence of input from the MOE and VNO in the brain. Responses of single cells in the telencephalon and amygdala were recorded after stimulation of either the olfactory or the vomeronasal nerve (Roth and Laberge 2011). Cells of the lateral pallium, which receives input from the MOB, responded to stimulation of both the olfactory and vomeronasal nerves. In the same way, many cells in the vomeronasal amygdala fired after stimulation of both nerves. Analysis of response latencies indicated that many individual cells received direct monosynaptic input from the MOB and AOB. It was speculated that the integration of olfactory and vomeronasal information in the salamander brain may function in associative learning (Roth and Laberge 2011).

8.5.3 CHEMOSENSORY RECEPTORS

8.5.3.1 Trace-Amine Associated Receptors

In the African clawed frog (*X. laevis*), TAARs were expressed in the larval olfactory organ and the principal and middle cavities of the MOE of adults. Ligands for these receptors are believed to be amines (Gliem et al. 2009).

8.5.3.2 ORs

Amphibian chemosensory receptor expression is often studied in order to gain insight to the function of different receptor molecules. Based in part on work in

African clawed frogs (*X. laevis*), OR genes were at one time divided into two classes (Freitag et al. 1995, 1998). Class I OR genes consisted of ORs with a large third extracellular loop. Class I OR genes had sequence similarity to fish ORs and were transcribed in the middle cavity (water-nose) of African clawed frogs. Class II OR genes had sequence similarity mammals and that were transcribed in the principle cavity (air-nose) of African clawed frogs. Class I ORs were proposed to detect water-soluble substances and class II ORs were proposed to detect airborne substances. Support for the functional distinction between class I and II ORs, however, was weakened by the finding that mammals express many class I ORs (Niimura and Nei 2005).

A more recent phylogenetic analysis of vertebrate ORs divided the family into nine groups: α, β, γ, δ, ε, ζ, η, θ, and κ (Nei et al. 2008; Niimura and Nei 2005). Groups δ, ε, ζ, and η (some of which had been formerly lumped into class I ORs) were found in fish and clawed frogs (*Xenopus*), suggesting detection of water-soluble substances. Groups α (formerly included in class I ORs) and γ (formerly class II ORs) were found in tetrapods, including clawed frogs, suggesting a role in detecting airborne stimuli. Some amphibian studies supported this functional distinction. For example, ORs in terrestrial phase tiger salamanders (*A. tigrinum*) aligned with group γ (Marchand et al. 2004). However, not all studies in amphibians support the notion that OR groups α and γ detect airborne stimuli exclusively. For example, group α transcripts were found in the middle cavity (water-nose) of western clawed frogs (*X. tropicalis*) (Amano and Gascuel 2012) and group γ genes were expressed in the main olfactory epithelium of larval *X. laevis* (Mezler et al. 1999). Also, ORs of mudpuppies (*Necturus maculosus*), a fully aquatic salamander, aligned with group γ ORs (Zhou et al. 1997).

8.5.3.3 VRs

Many VRs have been identified in amphibians. *Xenopus* has at least 21 intact V1Rs and 249 intact V2Rs, and the red-legged salamander (*P. shermani*) has at least 34 V2Rs (Kiemnec-Tyburczy et al. 2012; Nei et al. 2008). V1Rs fall into three main clades (Grus and Zhang 2009; Shi and Zhang 2007, 2009). *Xenopus* V1Rs share sequence similarity with mammalian V1Rs (clade 1) and fish V1rs (clade 2) with a lineage-specific expansion in clade 2. V2Rs fall into two clades, and *Xenopus* and red-legged salamander V2Rs are found in both clades, in some cases grouping with sequences from mammals (Grus and Zhang 2009; Kiemnec-Tyburczy et al. 2012; Shi and Zhang 2007, 2009).

Although the VNO is anatomically separate from the MOE in amphibians, expression of VRs is not necessarily restricted to the VNO. In particular, *Xenopus* V1Rs transcripts were not found in the VNO, but were found in the main olfactory system (both the principal and middle cavities) (Date-Ito et al. 2008). Also, although V2Rs transcripts were abundant the VNO, low levels of V2R transcripts were also found in the middle cavity of the main olfactory epithelium (Date-Ito et al. 2008; Hagino-Yamagishi et al. 2004). In red-legged salamanders, however, expression of V2Rs was restricted to the VNO (Kiemnec-Tyburczy et al. 2012).

There is some support for the hypothesis that V1Rs detect volatiles and V2Rs detect nonvolatiles, as has been shown in mammals (Shi and Zhang 2007). For example,

V2Rs transcripts were found in the VNO of red-legged salamanders (*P. shermani*), a terrestrial salamander for which the VNO is specialized to detect nonvolatile chemosignals (Dawley and Bass 1989). Also, in African clawed frogs (*X. laevis*), V1Rs were present in the air-filled principal cavity of the main olfactory system, where volatiles are detected. V1rs were also present in the middle cavity of the main olfactory system, which may explain the finding that the middle cavity responded to both volatile and water-soluble substances (Iida and Kashiwayanagi 1999).

8.5.3.4 Accessory Proteins

ORs and VRs were associated with G proteins and Trpc2 channel proteins, matching the patterns seen in mammals. For example, amphibian OR expression was associated with a special G-protein called $G\alpha_{olf}$ (Mezler et al. 2001). Coexpression of ORs and $G\alpha_{olf}$ was found in the MOE of adult *X. laevis*, and $G\alpha_{olf}$ was also found in the MOE of adult red-legged salamanders and in the olfactory epithelium of *Rhinella arenarum* tadpoles (Date-Ito et al. 2008; Jungblut et al. 2009; Kiemnec-Tyburczy et al. 2012). Also, *X. laevis* V1R expression was associated with $G\alpha_{i2}$, and V2R expression was associated with $G\alpha_o$ (Date-Ito et al. 2008; Hagino-Yamagishi et al. 2004). Transcripts for Trpc2, a channel protein associated with VR signal transduction, were abundant in the VNO of red-legged salamanders, where V2Rs transcripts were found (Kiemnec-Tyburczy et al. 2012).

Although the MOE and VNO are anatomically separate in amphibians, expression of $G\alpha_{olf}$ is not always restricted to the MOE. $G\alpha_{olf}$ transcripts were present in the VNO of Japanese toads (*Bufo japonicus*) and red-legged salamanders (*P. shermani*), both of which are terrestrial (Hagino-Yamagishi and Nakazawa 2011; Kiemnec-Tyburczy et al. 2012). It is unknown whether the presence of $G\alpha_{olf}$ in the VNO is associated with ORs.

8.5.4 RESPONSES OF CHEMOSENSORY EPITHELIA

The tiger salamander (*A. tigrinum*) has been important model for understanding how olfactory neurons respond to general odorants (Kauer 2002). Here, I focus on studies that have measured physiological responses to chemosignals. Specifically, I will address what types of substances activated olfactory sensory neurons and which olfactory system was activated by the substances. Unfortunately, strong conclusions about the chemoreceptors involved in potential responses cannot be made because ligands have not been identified for any of the chemoreceptors.

Study of physiological responses to chemosignals has focused on urodeles because they have anatomically simple nasal cavities for measuring olfactory neuron responsiveness and because well-characterized chemosignals are available for use. Studies have used electrophysiological or immunohistochemical methods. In red-bellied salamanders (*C. pyrrhogaster*), the peptide pheromone sodefrin elicited electrical field potentials throughout the main and accessory olfactory epithelia of females. However, responses were largest in an area of the VNO (Toyoda and Kikuyama 2000). These results confirmed results of an earlier study in which exposure of female crested newts (*Lissotriton* (formerly *Triturus*) *cristatus carnifex*) to water that had previously held males or to extracts of the dorsal cloacal gland triggered spike potentials in the olfactory bulb (Cedrini and Fasolo 1971). In males that were exposed to water holding males or females, very little olfactory bulb responsiveness was noticed.

In Korean salamanders (*H. leechi*), which are fully aquatic salamanders, a prostaglandin (15(epi)-PF2α) is produced by females and attracts males. Treatment with 15(epi)-PGF2α activated sensory neurons of the MOE in both males and females (Eom et al. 2009). The study did not measure VNO responses because the VNO was difficult to access.

In the terrestrial red-legged salamander (*P. shermani*), treatment with an extract from the mental gland, a source of proteinaceous courtship pheromones, activated VNO sensory neurons in both females and males, as shown using immunohisto-chemical methods (Schubert et al. 2006, 2008; Wirsig-Wiechmann et al. 2002). There was no evidence of activation of the MOE. Treatment with the proteinaceous courtship pheromones PRF and PMF also activated VNO sensory neurons (Wirsig-Wiechmann et al. 2006). To determine areas of the brain that were activated by court-ship pheromones, fos-immunoreactivity was measured in females exposed to mental gland extract (Laberge et al. 2008). Fos-immunoreactivity was greatest in areas of the amygdala shown to receive vomeronasal projections, as well as the preoptic area and ventromedial hypothalamus, areas involved in reproduction. Activation was greatest after application of mental gland extract compared to PRF or PMF alone.

Other studies have examined responses of chemosensory epithelia to broader mixes of chemosignals such as body rinses or general skin secretions. In the axolotl (*A. mexicanum*), a fully aquatic salamander, both the VNO and MOE responded to conspecific body rinses, with males being most responsive to body rinses derived from female and vice versa (Park et al. 2004). Interestingly, the VNO of females responded very strongly to methionine, a neutral substance. In red-legged salaman-ders (*P. shermani*), female skin secretions activated VNO neurons in both males and females (Schubert et al. 2008).

Responses of chemosensory epithelia to chemosignals may be modulated by hor-mones and neuropeptides. In red-bellied newts (*C. pyrrhogaster*), injection of newts with prolactin and estradiol prior to testing increased VNO responsiveness to sode-frin in both females and castrated males (Toyoda and Kikuyama 2000). In contrast with the newt, circulating sex hormone levels did not influence VNO sensory respon-siveness in male or female red-legged salamanders (*P. shermani*) (Schubert et al. 2006, 2008). However, treatment of the nasal cavity of red-legged salamanders with gonadotropin-releasing hormone increased responsiveness of VNO sensory neurons to male mental gland extract, indicating that gonadotropin-releasing hormone func-tions as a neuromodulator of olfactory processing (Wirsig-Wiechmann et al. 2012).

8.5.5 PRIMER EFFECTS

There is only one example of a chemosignal affecting the endocrine system in amphibians. Application of male mental gland extract to red-legged salamanders resulted in an increase in plasma corticosterone, a metabolic and stress hormone (Schubert et al. 2009). There were no changes in plasma testosterone or estradiol. The response was found in males but not in females. Treatment with saline controls or chemosignals derived from female skin secretions did not trigger elevated plasma corticosterone. Because mental gland pheromones are believed to function in male–female courtship interaction, it is difficult to explain the priming effect in males. Regardless, transient increases in corticosterone increased metabolic rate in red-legged salamanders, suggesting a change in arousal (Wack et al. 2012).

8.5.6 SUMMARY AND CONCLUSIONS

A relatively large number of chemoreceptor genes were identified in the genome of the western clawed frog (*X. tropicalis*). Chemoreceptors were cloned in additional amphibian species and were shown to be expressed in the chemosensory epithelia of clawed frogs and red-legged salamanders. Although the VNO and MOE are anatomically separate in amphibians, there were cases in which VRs were expressed in both the MOE and VNO, and $G\alpha_{olf}$, typically associated with OR expression in mammals was expressed in both the MOE and VNO. Input from olfactory and vomeronasal systems converged on single cells in the telencephalon and amygdala, indicating a high degree of integration of olfactory and vomeronasal information in the brain.

In aquatic and aquatic-phase salamanders (*H. leechi*, *C. pyrrhogaster*, *A. mexicanum*), both the VNO and MOE detected water-soluble chemosignals. In a fully terrestrial salamander (*P. shermani*), the VNO detected nonvolatile chemosignals. It is unknown whether responses were transduced via ORs or VRs because VR and OR expression may not be restricted to the VNO and MOE, respectively. Also, until more pheromonal ligands and receptors are matched, it is very difficult to make functional conclusions about the different chemosensory receptor genes.

Despite the wealth of information about chemosensory receptor genes and the well-described neuroanatomy of the olfactory systems in anurans, physiological responses to chemosignals have not been examined. This omission could be due to the historical emphasis on acoustic communication over chemical communication in anurans. It could also be due to the complicated nasal anatomy of anurans, which make electrophysiology studies more difficult. Regardless, given the large number of VRs identified in the *Xenopus* genome, more studies of the physiological responses to chemosignals using anurans are warranted.

8.6 FUTURE DIRECTIONS AND CONCLUDING REMARKS

With this review, I have tried to capture the rich diversity of amphibian chemosignals and the contexts in which they operate. Recent advances in noninvasive sampling methods of glands will hopefully stimulate more studies of signaling properties of skin gland secretions. The existence of several well-characterized amphibian chemosignals should be useful tools for determining the chemoreceptor genes that transduce the chemosignals. It will be interesting to determine whether the recently discovered formyl peptide receptor-like chemoreceptor genes are also expressed in amphibian olfactory epithelia. Given the large number of VRs in anurans, it seems particularly important to study behavioral and physiological responses to potential chemosignals in anurans and to determine the receptor genes mediating such responses. Once chemosignals are matched to chemoreceptor genes, hypotheses about the function of chemoreceptors at different stages of development or in different habitats can be tested.

Those amphibian chemosignals that have been chemically identified and shown to have some behavioral response have been called pheromones. Pheromones like sodefrin and silefrin fulfill common definitions of pheromones (Wyatt 2010) in that the pheromones are single molecules, are produced by specialized glands, are invariant within a species, have stereotyped behavioral responses, and are detected by chemosensory epithelia in the nasal cavity. In contrast, courtship pheromones of plethodontid salamander challenge common definitions. Although courtship pheromones are evolved signals produced by specialized glands, they are individually variable, are part of complex blends, and they have relatively subtle effects on behavior. Also, in many species of plethodontid salamanders, courtship pheromones are not detected by the olfactory system in the nose, but are applied to the skin. Identification of the receptors that transduce courtship pheromones will further provide insight to the pheromone concept as it applies to vertebrates.

Amphibians are a diverse group, representing a range of developmental processes and life histories (metamorphosis, paedomorphosis, direct development, secondarily aquatic) and living in a variety of habitats. Yet out of the more than 6500 species of amphibians, chemical signaling has been studied in only a handful of species. Although much is known about urodeles, almost nothing is known about caecilian chemical communication and little is known about chemical communication in anurans other than *Xenopus* frogs. It is hoped that this review stimulates additional interest in amphibian chemical communication. Further study of amphibian chemical communication will allow better testing of hypotheses about the evolution of vertebrate chemical communication and the function of different olfactory systems and chemosensory receptors.

REFERENCES

Amano, T. and J. Gascuel. 2012. Expression of odorant receptor family, type 2 OR in the aquatic olfactory cavity of amphibian frog *Xenopus tropicalis*. *PLoS One* 7:e33922.

Apponyi, M. A., T. L. Pukala, C. S. Brinkworth et al. 2004. Host-defence peptides of Australian anurans: Structure, mechanism of action and evolutionary significance. *Peptides* 25:1035–54.

Arnold, S. J. 1977. The evolution of courtship behavior in new world salamanders with some comments on old world salamandrids. In *The Reproductive Biology of Amphibians*, edited by D. H. Taylor and S. I. Guttman. New York: Plenum Press.

Baum, M. J. and K. R. Kelliher. 2009. Complementary roles of the main and accessory olfactory systems in mammalian mate recognition. *Annu. Rev. Physiol.* 71:141–60.

Belanger, R. M. and L. D. Corkum. 2009. Review of aquatic sex pheromones and chemical communication in anurans. *J. Herp.* 43:184–91.

Benzekri, N. A. and J. O. Reiss. 2012. Olfactory metamorphosis in the coastal tailed frog *Ascaphus truei* (Amphibia, Anura, Leiopelmatidae). *J. Morphol.* 273:68–87.

Breer, H., J. Fleischer and J. Strotmann. 2006. The sense of smell: Multiple olfactory subsystems. *Cell. Mol. Life Sci.* 63:1465–75.

Brizzi, R., G. Delfino and S. Jantra. 2003. An overview of breeding glands. In *Reproductive Biology and Phylogeny of Anura*, edited by B. G. M. Jamieson. Enfield, NH: Science Publishers.

Brizzi, R., G. Delfino and R. Pellegrini. 2002. Specialized mucous glands and their possible adaptive role in the males of some species of *Rana* (Amphibian, Anura). *J. Morphol.* 254:328–41.

Byrne, P. G. and J. S. Keough. 2007. Terrestrial toadlets use chemosignals to recognize conspecifics, locate mates and strategically adjust calling behaviour. *Anim. Behav.* 74:1155–62.

Cedrini, L. and A. Fasolo. 1971. Olfactory attractants in sex recognition of the crested newt, an electrophysiological research. *Monitore Zool. Ital.* 5:223–9.

Chouinard, A. J., D. B. Wilburn, L. D. Houck and R. C. Feldhoff. 2013. Individual variation in pheromone isoform ratios of the red-legged salamander, *Plethodon shermani*. In *Chemical Signals in Vertebrates 12*, edited by M. L. East and M. Dehnard. New York: Springer.

Dantzer, B. J. and R. G. Jaeger. 2007. Detection of the sexual identity of conspecifics through volatile chemical signals in a territorial salamander. *Ethology* 113:214–222.

Date-Ito, A., H. Ohara, M. Ichikawa, Y. Mori and K. Hagino-Yamagishi. 2008. *Xenopus* V1R vomeronasal receptor family is expressed in the main olfactory system. *Chem. Senses* 33:339–46.

Dawley, E. M. 1984a. Identification of sex through odors by male red-spotted newts, *Notophthalamus viridescens*. *Herpetologica* 40:101–5.

Dawley, E. M. 1984b. Recognition of individual, sex and species odours by salamanders of the *Plethodon glutinosus-P. jordani* complex. *Anim. Behav.* 32:353–61.

Dawley, E. M. 1986. Behavioral isolating mechanisms in sympatric terrestrial salamanders. *Herpetologica* 42:156–64.

Dawley, E. M. 1998. Olfaction. In *Amphibian Biology*, edited by H. Heatwole. Chipping Norton, New South Wales, Australia: Surrey Beatty & Sons.

Dawley, E. M. and A. H. Bass. 1989. Chemical access to the vomeronasal organ of a plethodontid salamander. *J. Morphol.* 200:163–74.

Doty, R. L. 2010. *The Great Pheromone Myth*. Baltimore, MD: Johns Hopkins University Press.

Duellman, W. E. and L. Trueb. 1986. *Biology of Amphibians*. New York: McGraw-Hill.

Eisthen, H. L. 1997. Evolution of vertebrate olfactory systems. *Brain Behav. Evol.* 50:222–33.

Eisthen, H. L. 2000. Presence of the vomeronasal system in aquatic salamanders. *Philos. Trans. R. Soc. Lond. B Biol. Sci.* 355:1209–13.

Eisthen, H. L. and G. Polese. 2006. Evolution of vertebrate olfactory subsystems. In *Evolution of Nervous Systems*, edited by J. H. Kaas. Oxford: Academic Press.

Eom, J., Y. R. Jung and D. Park. 2009. F-series prostaglandins function as sex pheromones in the Korean salamander, *Hynobius leechii*. *Comp. Biochem. Physiol. A Mol. Integr. Physiol.* 154:61–9.

Erspamer, V. 1994. Bioactive secretions of the amphibian integument. In *Amphibian Biology, The Integument*, edited by H. Heatwole and G. T. Barthalmus. Chipping Norton, New South Wales, Australia: Surrey Beatty & Sons.

Feldhoff, R. C., S. M. Rollmann and L. D. Houck. 1999. Chemical analysis of courtship pheromones in a plethodontid salamander. In *Advances in Chemical Signals in Vertebrates*, edited by R. E. Johnston, D. Muller-Schwartz and P. Sorenson. New York: Kluwer Academic/Plenum.

Fitzpatrick, L. C. 1973. Energy allocation in the Allegheny mountain salamander, *Desmognathus ochrophaeus*. *Ecological Monographs* 43:43–58.

Freitag, J., J. Krieger, J. Strotmann and H. Breer. 1995. Two classes of olfactory receptors in *Xenopus laevis*. *Neuron* 15:1383–92.

Freitag, J., G. Ludwig, I. Andreini, P. Rossler and H. Breer. 1998. Olfactory receptors in aquatic and terrestrial vertebrates. *J. Comp. Physiol. A* 183:635–50.

Gershman, S. A. and P. A. Verrell. 2002. To persuade or be persuaded: Which sex controls mating in a plethodontid salamander? *Behaviour* 139:447–62.

Gliem, S., D. Schild and I. Manzini. 2009. Highly specific responses to amine odorants of individual olfactory receptor neurons in situ. *Eur. J. Neurosci.* 29:2315–26.

Green, A. 1991. Competition and energetic constraints in the courting great crested newt, *Triturus cristatus* (Amphibia: Salamandridae). *Ethology* 87:66–78.

Grus, W. E. and J. Zhang. 2009. Origin of the genetic components of the vomeronasal system in the common ancestor of all extant vertebrates. *Mol. Biol. Evol.* 26:407–19.

Hagino-Yamagishi, K. and H. Nakazawa. 2011. Involvement of Gα(olf)-expressing neurons in the vomeronasal system of *Bufo japonicus. J. Comp. Neurol.* 519:3189–201.

Hagino-Yamagishi, K., K. Moriya H. Kubo et al. 2004. Expression of vomeronasal receptor genes in *Xenopus laevis. J. Comp. Neurol.* 472:246–56.

Halliday, T. R. 1990. The evolution of courtship behavior in newts and salamanders. *Adv. Study Behav.* 19:137–69.

Hansen, A., K. T. Anderson and T. E. Finger. 2004. Differential distribution of olfactory receptor neurons in goldfish: Structural and molecular correlates. *J. Comp. Neurol.* 477:347–59.

Hecker, L., D. M. Madison, R. W. Dapson and V. Holzherr. 2003. Presence of modified serous glands in the caudal integument of the red-backed salamander (*Plethodon cinereus*). *J. Herpetol.* 37:732–6.

Horne, E. A. and R. G. Jaeger. 1988. Territorial pheromones of female red-backed salamanders. *Ethology* 78:143–52.

Houck, L. D. 1998. Integrative studies of amphibians: From molecules to mating. *Am. Zool.* 38:108–17.

Houck, L. D. 2009. Pheromone communication in amphibians and reptiles. *Annu. Rev. Physiol.* 71:161–76.

Houck, L. D. and S. J. Arnold. 2003. Courtship and mating. In *Phylogeny and Reproductive Biology of Urodela (Amphibia)*, edited by D. M. Sever. Enfield, NH: Science Publishers.

Houck, L. D. and N. L. Reagan. 1990. Male courtship pheromones increase female receptivity in a plethodontid salamander. *Anim. Behav.* 39:729–34.

Houck, L. D. and D. M. Sever. 1994. Role of the skin in reproduction and behaviour. In *Amphibian Biology*, edited by H. Heatwole and G. T. Barthalmus. Chipping Norton, New South Wales, Australia: Surrey Beatty & Sons.

Houck, L. D., A. M. Bell, N. L. Reagan-Wallin and R. C. Feldhoff. 1998. Effects of experimental delivery of male courtship pheromones on the timing of courtship in a terrestrial salamander, *Plethodon jordani* (Caudata : Plethodontidae). *Copeia* 1998:214–19.

Houck, L. D., C. A. Palmer, R. A. Watts et al. 2007. A new vertebrate courtship pheromone, PMF, affects female receptivity in a terrestrial salamander. *Anim. Behav.* 73:315–20.

Houck, L. D., R. A. Watts, S. J. Arnold et al. 2008a. A recombinant courtship pheromone affects sexual receptivity in a plethodontid salamander. *Chem. Senses* 33:623–31.

Houck, L. D., R. A. Watts, L. M. Mead et al. 2008b. A candidate vertebrate pheromone, SPF, increases female receptivity in a salamander. In *Chemical Signals in Vertebrates*, edited by J. L. Hurst. New York: Springer.

Iida, A. and M. Kashiwayanagi. 1999. Responses of *Xenopus laevis* water nose to water-soluble and volatile odorants. *J. Gen. Physiol.* 114:85–92.

Jaeger, R. G. 1981. Dear enemy recognition and the costs of aggression between salamanders. *Am. Nat.* 117:962–74.

Jaeger, R. G. 1984. Agonistic behavior of the red-backed salamander. *Copeia* 1984:309–14.

Jaeger, R. G. and D. C. Forester. 1993. Social behavior of plethodontid salamanders. *Herpetelogica* 49:163–75.

Jaeger, R. G. and W. F. Gergits. 1979. Intra- and interspecific communication in salamanders through chemical signals on the substrate. *Anim. Behav.* 27:150–6.

Jaeger, R. G. and S. E. Wise. 1991. A reexamination of the male salamander "sexy faeces hypothesis." *J. Herp.* 25:370–3.

Jaeger, R. G., J. M. Goy, M. Tarver and C. E. Marquez. 1986. Salamander territoriality: Pheromonal markers as advertisement by males. *Anim. Behav.* 34:860–4.

Ji, Y., Z. Zhang and Y. Hu. 2009. The repertoire of G-protein-coupled receptors in *Xenopus tropicalis. BMC Genomics* 10:263.

Johansson, B. G. and T. M. Jones. 2007. The role of chemical communication in mate choice. *Biol. Rev. Camb. Philos. Soc.* 82:265–89.

Jungblut, L. D., D. A. Paz, J. J. Lopez-Costa and A. G. Pozzi. 2009. Heterogeneous distribution of G protein a subunits in the main olfactory and vomeronasal systems of *Rhinella (Bufo) arenarum* tadpoles. *Zoolog. Sci.* 26:722–8.

Kang, N., M. J. Baum and J. A. Cherry. 2009. A direct main olfactory bulb projection to the 'vomeronasal' amygdala in female mice selectively responds to volatile pheromones from males. *Eur. J. Neurosci.* 29:624–34.

Kauer, J. S. 2002. On the scents of smell in the salamander. *Nature* 417:336–42.

Kiemnec-Tyburczy, K. M., R. A. Watts, R. G. Gregg, von Borstal and S. J. Arnold. 2009. Evolutionary shifts in courtship pheromone composition revealed by EST analysis of plethodontid salamander mental glands. *Gene* 432:75–81.

Kiemnec-Tyburczy, K. M., S. K. Woodley, R. A. Watts, S. J. Arnold and L. D. Houck. 2012. Expression of vomeronasal receptors and related signaling molecules in the nasal cavity of a caudate amphibian (*Plethodon shermani*). *Chem. Senses* 37:335–46.

Kikuyama, S., H. Seshimo, K. Shirama, T. Kato and T. Noumara. 1986. Interaction of prolactin with sex steroid in oviduct and tail of newts, *Cynops pyrrhogaster. Zoolog. Sci.* 3:131–8.

Kikuyama, S., T. Nakada, F. Toyoda et al. 2005. Amphibian pheromones and endocrine control of their secretion. *Ann. N. Y. Acad. Sci.* 1040:123–30.

Kikuyama, S., R. Nakano and I. Yasumasu. 1975. Synergistic action of prolactin and androgen on the cloacal glands of the newt. *Comp. Biochem. Physiol. A Comp. Physiol.* 51:823–6.

Kikuyama, S., F. Toyoda, Y. Ohmiya et al. 1995. Sodefrin: A female-attracting peptide pheromone in newt cloacal glands. *Science* 267:1643–5.

King, J. D., L. A. Rollins-Smith, P. F. Nielsen, A. John and J. M. Conlon. 2005. Characterization of a peptide from skin secretions of male specimens of the frog, *Leptodactylus fallax* that stimulates aggression in male frogs. *Peptides* 26:597–601.

Laberge, F. and G. Roth. 2005. Connectivity and cytoarchitecture of the ventral telencephalon in the salamander *Plethodon shermani. J. Comp. Neurol.* 482:176–200.

Laberge, F. and G. Roth. 2007. Is there a structure equivalent to the mammalian basolateral amygdaloid complex in amphibians? *J. Anat.* 211:830; author reply 830–1.

Laberge, F., R. C. Feldhoff, P. W. Feldhoff and L. D. Houck. 2008. Courtship pheromone-induced c-Fos-like immunolabeling in the female salamander brain. *Neuroscience* 151:329–39.

Laberge, F., S. Muhlenbrock-Lenter, W. Grunwald and G. Roth. 2006. Evolution of the amygdala: New insights from studies in amphibians. *Brain Behav. Evol.* 67:177–87.

Largen, W. and S. K. Woodley. 2008. Cutaneous tail glands, noxious skin secretions, and scent marking in a terrestrial salamander (*Plethodon shermani*). *Herpetologica* 64:270–80.

Lee, J. S. F. and B. Waldman. 2002. Communication by fecal chemosignals in an archaic frog, *Leiopelma hamiltoni. Copeia* 2002:679–86.

Leichty, K. A. 2012. Co-option and adaptation of novel gene duplications for pheromone activity in a dusky salamander. MS thesis, MS, Biochemistry and Molecular Biology, University of Louisville, Louisville, KY.

Licht, G. and M. Meredith. 1987. Convergence of main and accessory olfactory pathways onto single neurons in the hamster amygdala. *Exp. Brain Res.* 69:7–18.

Malacarne, G. and C. Vellano. 1987. Behavioral evidence of a courtship pheromone in the crested newt, *Triturus cristatus carnifex* Laurenti. *Copeia* 1987:245–7.

Manzini, I. and D. Schild. 2010. Olfactory coding in larvae of the African clawed frog *Xenopus laevis.* In *Neurobiology of Olfaction,* edited by A. Menini. Boca Raton, FL: CRC Press.

Marchand, J. E., X. Yang, D. Chikaraishi et al. 2004. Olfactory receptor gene expression in tiger salamander olfactory epithelium. *J. Comp. Neurol.* 474:453–67.

Marco, A., D. P. Chivers, J. M. Kiesecker and A. R. Blaustein. 1998. Mate choice by chemical cues in western redback (*Plethodon vehiculum*) and Dunn's (*P. dunni*) salamanders. *Ethology* 104:781–8.

Mathis, A., R. G. Jaeger, W. H. Keen et al. 1995. Aggression and territoriality by salamanders and a comparison with the territorial behaviour of frogs. In *Amphibian Biology*, edited by H. Heatwole and B. K. Sullivan. Chipping Norton, New South Wales, Australia: Surrey Beatty & Sons.

Mezler, M., J. Fleischer, S. Conzelmann et al. 2001. Identification of a nonmammalian Golf subtype: functional role in olfactory signaling of airborne odorants in *Xenopus laevis*. *J. Comp. Neurol.* 439:400–10.

Mezler, M., S. Konzelmann, J. Freitag, P. Rossler and H. Breer. 1999. Expression of olfactory receptors during development in *Xenopus laevis*. *J. Exp. Biol.* 202:365–76.

Moreno, N. and A. Gonzalez. 2003. Hodological characterization of the medial amygdala in anuran amphibians. *J. Comp. Neurol.* 466:389–408.

Moreno, N. and A. Gonzalez. 2004. Localization and connectivity of the lateral amygdala in anuran amphibians. *J. Comp. Neurol.* 479:130–48.

Moreno, N. and A. Gonzalez. 2006. The common organization of the amygdaloid complex in tetrapods: New concepts based on developmental, hodological and neurochemical data in anuran amphibians. *Prog. Neurobiol.* 78:61–90.

Moreno, N., R. Morona, J. M. Lopez, M. Munoz and A. Gonzalez. 2005. Lateral and medial amygdala of anuran amphibians and their relation to olfactory and vomeronasal information. *Brain Res. Bull.* 66:332–6.

Nakamuta, S., N. Nakamuta and K. Taniguchi. 2011. Distinct axonal projections from two types of olfactory receptor neurons in the middle chamber epithelium of *Xenopus laevis*. *Cell Tissue Res.* 346:27–33.

Nei, M., Y. Niimura and M. Nozawa. 2008. The evolution of animal chemosensory receptor gene repertoires: Roles of chance and necessity. *Nat. Rev. Genet.* 9:951–63.

Niimura, Y. and M. Nei. 2005. Evolutionary dynamics of olfactory receptor genes in fishes and tetrapods. *Proc. Natl. Acad. Sci. U S A* 102:6039–44.

Noble, G. K. 1927. The plethodontid salamanders; some aspects of their evolution. *Am. Mus. Novit.* 249:1–26.

Organ, J. A. 1961. Studies of the local distribution, life history, and population dynamics of the salamander genus *Desmognathus* in Virginia. *Ecol. Monogr.* 31:189–220.

Osikowski, A., W. Babik, P. Grzmil and J. M. Szymura. 2008. Multiple sex pheromone genes are expressed in the abdominal glands of the smooth newt (*Lissotriton vulgaris*) and Montandon's newt (*L. montandoni*) (Salamandridae). *Zoolog. Sci.* 25:587–92.

Palmer, C. A. and L. D. Houck. 2005. Responses to sex- and species-specific chemical signals in allopatric and sympatric salamander species. In *Chemical Signals in Vertebrates*, edited by R. T. Mason, M. P. LeMaster and D. Muller-Schwartz. New York: Kluwer Academic/Plenum Publishers.

Palmer, C. A., D. M. Hollis, R. A. Watts et al. 2007a. Plethodontid modulating factor, a hypervariable salamander courtship pheromone in the three-finger protein superfamily. *FEBS J.* 274:2300–10.

Palmer, C. A., R. A. Watts, R. G. Gregg et al. 2005. Lineage-specific differences in evolutionary mode in a salamander courtship pheromone. *Mol. Biol. Evol.* 22:2243–56.

Palmer, C. A., R. A. Watts, A. P. Hastings, L. D. Houck and S. J. Arnold. 2010. Rapid evolution of plethodontid modulating factor, a hypervariable salamander courtship pheromone, is driven by positive selection. *J. Mol. Evol.* 70:427–40.

Palmer, C. A., R. A. Watts, L. D. Houck, A. L. Picard and S. J. Arnold. 2007b. Evolutionary replacement of components in a salamander pheromone signaling complex: More evidence for phenotypic-molecular decoupling. *Evolution* 61:202–15.

Park, D. and C. R. Propper. 2001. Repellent function of male pheromones in the red-spotted newt. *J. Exp. Zool.* 289:404–8.

Park, D., J. M. McGuire, A. L. Majchrzak, J. M. Ziobro and H. L. Eisthen. 2004. Discrimination of conspecific sex and reproductive condition using chemical cues in axolotls (*Ambystoma mexicanum*). *J. Comp. Physiol. A Neuroethol. Sens. Neural Behav. Physiol.* 190:415–27.

Pearl, C. A., M. Cervantes, M. Chan et al. 2000. Evidence for a mate-attracting chemosignal in the dwarf African clawed frog *Hymenochirus. Horm. Behav.* 38:67–74.

Perriman, A. W., M. A. Apponyi, M. A. Buntine et al. 2008. Surface movement in water of splendipherin, the aquatic male sex pheromone of the tree frog *Litoria splendida. FEBS J.* 275:3362–74.

Pool, T. B. and J. N. Dent. 1977. The ultrastucture and hormonal control of product synthesis in the hedonic glands of the red-spotted newt *Notopthalmus viridescens. J. Exp. Zool.* 201:177–202.

Poth, D., K. C. Wollenberg, M. Vences and S. Schulz. 2012. Volatile amphibian pheromones: Macrolides from mantellid frogs from Madagascar. *Angew. Chem. Int. Ed. Engl.* 51:2187–90.

Quagliata, S., C. Malentachhi, C. Delfino, A. M. A. Brunasso and G. Delfino. 2006. Adaptive evolution of secretory cell lines in vertebrate skin. *Caryologia* 59:187–206.

Reiss, J. O. and H. L. Eisthen. 2008. Comparative anatomy and physiology of chemical senses in amphibians. In *Sensory Evolution on the Threshold: Adaptations in Secondarily Aquatic Vertebrates*, edited by J. G. M. Thewissen and S. Nummela. Berkeley, CA: University of California Press.

Riviere, S., L. Challet, D. Fluegge, M. Spehr and I. Rodriguez. 2009. Formyl peptide receptor-like proteins are a novel family of vomeronasal chemosensors. *Nature* 459:574–7.

Rollmann, S. M., L. D. Houck and R. C. Feldhoff. 1999. Proteinaceous pheromone affecting female receptivity in a terrestrial salamander. *Science* 285:1907–9.

Rollmann, S. M., L. D. Houck and R. C. Feldhoff. 2000. Population variation in salamander courtship pheromones. *J. Chem. Ecol.* 26:2713–24.

Rollmann, S. M., L. D. Houck and R. C. Feldhoff. 2003. Conspecific and heterospecific pheromone effects on female receptivity. *Anim. Behav.* 66:857–61.

Roth, F. C. and F. Laberge. 2011. High convergence of olfactory and vomeronasal influence in the telencephalon of the terrestrial salamander Plethodon shermani. *Neuroscience* 177:148–58.

Saito, S. and K. Taniguchi. 2000. Expression patterns of glycoconjugates in the three distinctive olfactory pathways of the clawed frog, *Xenopus laevis. J. Vet. Med. Sci.* 62:153–9.

San Mauro, D., M. Vences, M. Alcobendas, R. Zardoya and A. Meyer. 2005. Initial diversification of living amphibians predated the breakup of Pangaea. *Am. Nat.* 165:590–9.

Scalia, F. and S. S. Winans. 1975. The differential projections of the olfactory bulb and accessory olfactory bulb in mammals. *J. Comp. Neurol.* 161:31–55.

Schmidt, A. and G. Roth. 1990. Central olfactory and vomeronasal pathways in salamanders. *J. Hirnforsch* 31:543–53.

Schubert, S. N., L. D. Houck, P. W. Feldhoff, R. C. Feldhoff and S. K. Woodley. 2006. Effects of androgens on behavioral and vomeronasal responses to chemosensory cues in male terrestrial salamanders (*Plethodon shermani*). *Horm. Behav.* 50:469–76.

Schubert, S. N., L. D. Houck, P. W. Feldhoff, R. C. Feldhoff and S. K. Woodley. 2008. The effects of sex on chemosensory communication in a terrestrial salamander (*Plethodon shermani*). *Horm. Behav.* 54:270–7.

Schubert, S. N., C. L. Wack, L. D. Houck et al. 2009. Exposure to pheromones increases plasma corticosterone concentrations in a terrestrial salamander. *Gen. Comp. Endocrinol.* 161:271–5.

Secondi, J., W. Haerty and T. Lode. 2005. Female attraction to conspecific cohemical cues in the palmate newt *Triturus helveticus*. *Ethology* 111:726–35.

Sever, D. M. 1976. Induction of secondary sexual characters in *Eurycea quadridigitata*. *Copeia* 1976:830–3.

Sever, D. M. 1988. The ventral gland in female salamander *Eurycea bislineata* (Amphibia: Plethodontidae). *Copeia* 1988:572–9.

Sever, D. M. 1989. Caudal hedonic glands in salamanders of the *Eurycea bislineata* complex (Amphibia: Plethodontidae). *Herpetologica* 45:322–9.

Sever, D. M. 2003. Courtship and mating glands. In *Reproductive Biology and Phylogeny of Urodela (Amphibia)*, edited by D. M. Sever. Enfield, NH: Science Publishers.

Sever, D. M. and N. L. Staub. 2010. Hormones, sex accessory structures and secondary sexual characters in amphibians. In *Hormones and Reproduction of Vertebrates*, edited by D. O. Norris and K. Lopez. San Diego, CA: Elsevier.

Shi, P. and J. Zhang. 2007. Comparative genomic analysis identifies an evolutionary shift of vomeronasal receptor gene repertoires in the vertebrate transition from water to land. *Genome Res.* 17:166–74.

Shi, P. and J. Zhang. 2009. Extraordinary diversity of chemosensory receptor gene repertoires among vertebrates. *Results Probl. Cell Differ.* 47:1–23.

Simons, R. R., B. E. Felgenhauer and R. G. Jaeger. 1994. Salamander scent marks: site of production and their role in territorial defense. *Anim. Behav.* 48:97–103.

Simons, R. R., B. E. Felgenhauer and T. Thompson. 1999. Description of the postcloacal glands of *Plethodon cinereus*, the red-backed salamander, during bouts of scent marking. *J. Morphol.* 242:257–69.

Smith, B. P., C. R. Williams, M. J. Tyler and B. D. Williams. 2004. A survey of frog odorous secretions, their possible functions, and phylogenetic significance. *Appl. Herp.* 2:47–82.

Staub, N. L. and J. Paladin. 1997. The presence of modified granular glands in male and female *Aneides lugubris* (Amphibia: Plethodontidae). *Herpetologica* 53:339–44.

Sullivan, B. K., M. J. Ryan and P. A. Verrell. 1995. Female choice and mating system structure. In *Amphibian Biology, Vol. 2 Social Behaviour*, edited by H. Heatwole and B. K. Sullivan. New South Wales, Australia: Surrey Beatty & Sons.

Teyssedre, C. and T. R. Halliday. 1986. Cumulative effect of male's dsiplays in the sexual behavior of the smooth newt *Triturus vulgaris* (Urodela, Salamandridae). *Ethology* 71:89–102.

Thomas, E. O. and P. Licht. 1993. Testicular and androgen dependence of skin gland morphology in the anurans, *Xenopus laevis* and *Rana pipiens*. *J. Morphol.* 215:195–200.

Thomas, E. O., L. Tsang and P. Licht. 1993. Comparative histochemistry of the sexually dimorphic skin glands of anuran amphibians. *Copeia* 1993:133–43.

Thompson, R. R., Z. Tokar, D. Pistohl and F. L. Moore. 1999. Behavioral evidence for a sex pheromone in female roughskin newts, *Taricha granulosa*. In *Advances in Chemical Signals in Vertebrates*, edited by R. E. Johnston, D. Muller-Schwartz and P. Sorenson. New York: Plenum Press.

Toledo, R. C. and C. Jared. 1995. Cutaneous granular glands and amphibian venoms. *Comp. Biochem. Phy. A* 111:1–29.

Toyoda, F. and S. Kikuyama. 2000. Hormonal influence on the olfactory response to a female-attracting pheromone, sodefrin, in the newt, *Cynops pyrrhogaster*. *Comp. Biochem. Phys. B* 126:239–45.

Toyoda, F., S. Tanaka, K. Matsuda and S. Kikuyama. 1994. Hormonal control of response to and secretion of sex attractants in Japanese newts. *Physiol. Behav.* 55:569–76.

Toyoda, F., K. Yamamoto, Y. Ito et al. 2003. Involvement of arginine vasotocin in reproductive events in the male newt *Cynops pyrrhogaster*. *Horm. Behav.* 44:346–53.

Twitty, V. C. 1955. Field experiments on the biology and genetic relationships of the Californian species of *Triturus*. *J. Exp. Zool.* 129:129–47.

Vaccaro, E. A., P. W. Feldhoff, R. C. Feldhoff and L. D. Houck. 2010. A pheromone mechanism for swaying female mate choice: Enhanced affinity for a sexual stimulus in a woodland salamander. *Anim. Behav.* 80:983–9.

Vences, M., G. Walh-Boos, S. Hoegg et al. 2007. Molecular systematics of mantelline frogs from Madagascar and the evolution of their femoral glands. *Bio. J. Linn. Soc.* 92:529–39.

Verrell, P. A., T. R. Halliday and L. Griffiths. 1986. The annual reproductive cycle of the smooth newt (*Triturus vulgaris*) in England. *J. Zool. Lond.* 210:101–19.

Wabnitz, P. A., J. H. Bowie, M. J. Tyler, J. C. Wallace and B. P. Smith. 1999. Aquatic sex pheromone from a male tree frog. *Nature* 401:444–5.

Wabnitz, P. A., J. H. Bowie, M. J. Tyler, J. C. Wallace and B. P. Smith. 2000. Differences in the skin peptides of the male and female Australian tree frog *Litoria splendida*. *Eur. J. Biochem.* 267:269–75.

Wack, C. L., S. E. DuRant, C. D. Hopkins et al. 2012. Elevation of plasma corticosterone increases metabolic rate in a terrestrial salamander. *Comp. Biochem. Physiol. A* 161:153–8.

Waldman, B. and P. J. Bishop. 2004. Chemical communication in an archaic anuran amphibian. *Behav. Ecol.* 15:88–93.

Watts, R. A., C. A. Palmer, R. C. Feldhoff et al. 2004. Stabilizing selection on behavior and morphology masks positive selection on the signal in a salamander pheromone signaling complex. *Mol. Biol. Evol.* 21:1032–41.

Wirsig-Wiechmann, C. R., J. Colvard, C. E. Aston et al. 2012. Gonadotropin-releasing hormone modulates vomeronasal neuron response to male salamander pheromone. *J. Exp. Neurosci.* 6:1–10.

Wirsig-Wiechmann, C. R., L. D. Houck, P. W. Feldhoff and R. C. Feldhoff. 2002. Pheromonal activation of vomeronasal neurons in plethodontid salamanders. *Brain Res.* 952:335–44.

Wirsig-Wiechmann, C. R., L. D. Houck, J. M. Wood, P. W. Feldhoff and R. C. Feldhoff. 2006. Male pheromone protein components activate female vomeronasal neurons in the salamander *Plethodon shermani*. *BMC Neurosci.* 7:26.

Woodley, S. K. 1994. Plasma androgen levels, spermatogenesis, and secondary sexual characteristics in two species of plethodontid salamanders with dissociated reproductive patterns. *Gen. Comp. Endocrinol.* 96:206–14.

Woodley, S. K. 2010. Pheromonal communication in amphibians. *J. Comp. Physiol. A Neuroethol. Sens. Neural. Behav. Physiol.* 196:713–27.

Wyatt, T. D. 2010. Pheromones and signature mixtures: Defining species-wide signals and variable cues for identity in both invertebrates and vertebrates. *J. Comp. Physiol. A Neuroethol. Sens. Neural. Behav. Physiol.* 196:685–700.

Yamamoto, K., Y. Kawai, T. Hayashi et al. 2000. Silefrin, a sodefrin-like pheromone in the abdominal gland of the sword-tailed newt, *Cynops ensicauda*. *FEBS Lett.* 472:267–70.

Zhou, Q., G. Hinkle, M. L. Sogin, and V. E. Dionne. 1997. Phylogenetic analysis of olfactory receptor genes from Mudpuppy (*Necturus maculosus*). *Biol. Bullet.* 193:248–50.

9 Vomeronasal Organ
A Short History of Discovery and an Account of Development and Morphology in the Mouse

Carlo Zancanaro

CONTENTS

9.1 Historical Background..285
    9.1.1 Discovery and Early Findings 285
    9.1.2 Twentieth Century 286
    9.1.3 Twenty-First Century...287
        9.1.3.1 Primates 287
9.2 Essentials of Vomeronasal Organ Anatomy and Comparative Anatomy.....288
    9.2.1 General Structure..288
    9.2.2 Main Variations in Vertebrates..289
    9.2.3 Developmental Issues 289
9.3 Morphology of the Mouse Vomeronasal Organ 290
    9.3.1 Development..290
    9.3.2 Adult Mouse VNO...292
References..293

9.1 HISTORICAL BACKGROUND

9.1.1 DISCOVERY AND EARLY FINDINGS

It is believed that the first illustration of the vomeronasal organ (VNO) was provided by Frederic Ruysh, a Dutch anatomist, in a drawing of the nasal septum of a young man (Ruysh 1703). About one century later the human VNO was also drawn by Samuel Semmering (Semmering 1809). Only the VNO of nonhuman animals will be considered in the following.

The discovery of the VNO is credited to Ludvig Levis Jacobson (1783–1843), a Danish physician and anatomist. Jacobson's seminal paper describing the new organ was published in Danish in 1813. An English translation of the paper based on both the original Danish text and the French translation made by Jacobson himself is available (Jacobson et al. 1998). It should be underlined that Jacobson described the VNO in a large number of mammals including some rodents and ruminants, the horse, the pig, the dog, and the cat.

Jacobson's original description of the VNO already encompassed most of the key macroscopic features known by the today's scientist: the cartilaginous capsule placed in the nasal septum, the internal receptacle (VNO lumen) showing a crescent shape in coronal section, the associated secretory apparatus (VNO glands and cavernous tissue), the exit duct (VNO duct), the rich vascular supply, the innervation from the trigeminal nerve, and the nervous connection with the olfactory bulb. Jacobson suggested a secretory role for the VNO; however, he also put forward the hypothesis of a sensory function, the secretion of the VNO glands being instrumental to sensation.

Later in the nineteenth century the availability of the Golgi technique allowed to show the similar morphology of neurones in the olfactory and vomeronasal epithelia (Retzius 1894) of the snake, thereby establishing the sensory function of the VNO. A review of the nineteenth century's knowledge on VNO is found in von Mihalkovics (1899).

9.1.2 TWENTIETH CENTURY

In the early twentieth century, investigations showed that the VNO is present in amphibians (where it has been proposed the organ first evolved [Eisthen 1992]) and absent in fish and birds. Results are reviewed in the extensive work of Pearlman (1934). The hypothesis that air is pumped in and out the VNO lumen by the alternate empting and filling of the blood sinuses placed at the posterior end of the VNO was proposed by Hamlin in 1929 (Hamlin 1929); earlier investigators (Broman 1920; Seydel 1895) also proposed a pumping role for that vascular structure, but they thought that fluid and not air was moved.

In the middle of the century, Negus summarized his investigation of the VNO in several species as follows (Negus 1956): "On anatomical grounds it appear probable that air and not fluid enters the tubule, and that it conveys odours or flavours connected with the diet of the animal. ... The most likely function of the organ would seem to be related to the after-smell of food rather than to its immediate detection." For a useful list of speculations on the function of the VNO up to 1979, refer to Johns (1986).

A decisive step in understanding the function of the VNO took place in the 1970s and 1980s when the relationship between the VNO and the sexual behavior became clear (review in Wysocki 1979; Halpern 1987). In 1975 Powers and Winans proposed the involvement of the VNO in the reproductive behavior of hamsters (Power and Winans 1975). The first nonbehavioral effect associated with the VNO (i.e., the urine-induced reflex ovulation of anovulatory rats) was shown soon after (Johns et al. 1978). The ability of substances present in the urine of mice to affect reproductive physiology had been known long before: female urine delayed the onset of puberty in prepuberal females (Lee and Boot 1955), male urine induced estrus in anestrous females (Whitten 1956), prevented pregnancy in recently mated females

(Bruce 1960), and accelerated the onset of puberty (Vandenbergh 1969). It was then hypothesized that the VNO is able to detect chemical cues in the ambient, collectively denominated "pheromones" after the definition of a class of biologically active substances secreted by insects (Karlson and Lüscher 1959).

At the end of the century, several papers authoritatively reviewed current knowledge of the VNO in general terms (Døving and Trotier 1998; Keverne 1999), or focusing on evolution patterns (Esthien 1992) and functional characteristics in mammals (Meredith 1998). By that time, the following features were recognized: (1) VNO is used by many animals to gain direct and specific contact with nonvolatile chemical cues (pheromones) released in the ambient by congeners, (2) VNO possesses two distinct types of seven-transmembrane receptors coupled to GTP-binding protein activating the 1,4,5-triphospate signaling, which differ from each other (and from the large family of olfactory receptors), (3) VNO signaling leads to activation of the hypothalamus by the way of the accessory olfactory bulb and amygdala thereby contributing to the regulation of reproductive, defensive, and ingestive behavior, and neuroendocrine secretion, (4) whether VNO serves nonpheromonal functions is unknown, but that is not an impossibility, and conversely, not everything that acts as a pheromone need stimulate the VNO, and (5) analysis of central olfactory projections show that VNO belongs to a fourth olfactory subsystem arisen in tetrapods in addition to the three olfactory subsystems broadly present in vertebrates (lateral, medial, and extrabulbar).

9.1.3 TWENTY-FIRST CENTURY

In the last 15 years a number of broad-ranging papers dealing with VNO and related issues have been published (Baxi et al. 2006; Bigiani et al. 2005; Bohem 2006; Brennan 2010; Brennan and Zufall 2006; Garrosa et al. 1998; Halpern and Martinez-Marcos 2003; Houck 2009; Isogai et al. 2011; Ma 2010; Mucignat-Caretta 2010; Rodriguez and Boehm 2009; Salazar and Sanchez Quintero 2009; Swaney and Keverne 2009; Takami 2002; Tirindelli et al. 2009; Witt and Wozniak 2006; Woodley 2010; Zufall and Leinders-Zufall 2007). The reader is referred to those papers for further information. The aim of this section is to update information therein to the most recent findings, with special attention to morphological issues.

9.1.3.1 Primates

It has been shown that in some taxa of primates VNO is similar to that of rodents (Schilling 1987) whereas in others the organ is completely lacking (Bhatnagar and Melsami 1998; Evans 2006; Smith et al. 2001). Moreover, in primates the VNO function is possibly not superimposable to that of rodents (Aujard 1997; Barret et al. 1993). In the last years investigations were conducted in several primates' species (Dennis et al. 2004; Smith et al. 2010, 2011) expanding on previous knowledge. Cells expressing neuron-specific β-tubulin, protein gene product 9.5, and olfactory marker protein were demonstrated by immunohistochemistry in the vomeronasal epithelium of infant *Eulemur mongoz* and adult *Otolemur crassicaudatus* (strepsirrhine), adult *Callithrix jacchus*, and 4-month-old *Leontopithecus rosalia* (haplorhine), thereby suggesting the presence of mature olfactory neurones therein, a prerequisite for function (Dennis et al. 2004). In contrast, only a few cells expressing the neuronal markers were found in *Saguinus geoffroyi*. Histologically, these authors confirmed that, in comparison with rodents, the VNO of primates shares similarities with that of sheep (Cohen-Tannoudji et al. 1989; Kratzing 1971; Salazar et al. 2003), ferrets (Kelliher 2001; Weiler et al. 1999), swine (Dorries et al. 1997; Salazar et al. 1997), and goats (Takigami et al. 2000, 2004).

In the lesser lemur (*Microcebus murinus*), a prosimian primate possessing a functional VNO, its removal was associated with maintenance of successful mating and normal testosterone plasma levels, but both sexual behavior and intermale aggressive behavior were much reduced (Aujard 1997). According to the author, this was due to functional disturbances of central nervous areas connected to the vomeronasal system and not to a chemosensory deficit per se, but direct evidence for that was not provided.

9.2 ESSENTIALS OF VOMERONASAL ORGAN ANATOMY AND COMPARATIVE ANATOMY

9.2.1 GENERAL STRUCTURE

The VNO (Figure 9.1) is located on both sides of the lower part of the nasal septum. The essential structure therein is the vomeronasal sensory epithelium (VNSE) containing vomeronasal receptor neurones (VRNs) whose axons leave the VNSE, coalesce along the nasal septum to form the vomeronasal nerve entering the cranial cavity, and synapse in a distinct part of the olfactory bulb. The VNO is enclosed in a bony or cartilaginous capsule and its lumen is connected to the nasal cavity by a duct (vomeronasal duct). The VNSE is a thick, crescent-shaped pseudostratified epithelium comprised of VRNs and supporting and basal cells. On the opposite side of the vomeronasal lumen a nonsensory epithelium (NSE) is usually found resembling the respiratory epithelium. Beyond the NSE, connective (cavernous) tissue with large blood vessels and sinuses are found whose contraction/dilation depending on autonomic sympathetic innervation produce a pumping action for stimulus access to the lumen. The lumen itself contains the secretion of associated vomeronasal glands (VNGs). The VNSE presents similarities and differences with the sensory epithelium of the main olfactory system (olfactory epithelium [OE] proper), which is located in the upper region of the nasal cavity. Similarities include (1) pseudostratified columnar epithelium containing sensory, supporting, and basal cells with nuclei of supporting cells in the luminal portion of the epithelium and nuclei of sensory cells more deeply located, and (2) an apical region containing dendrites of sensory cells and the supranuclear portion of supporting cells. Differences include (1) the presence of olfactory knobs in sensory cell dendrites, bearing olfactory hair in OE, (2) the presence of a basal cell layer in OE, which is absent in VNSE (of rodents), basal cells mainly being found at the border with the NSE (see e.g., Giacobini et al. 2000), and (3) the presence of ducts of olfactory glands (Bowman's) across the neuroepithelium in OE.

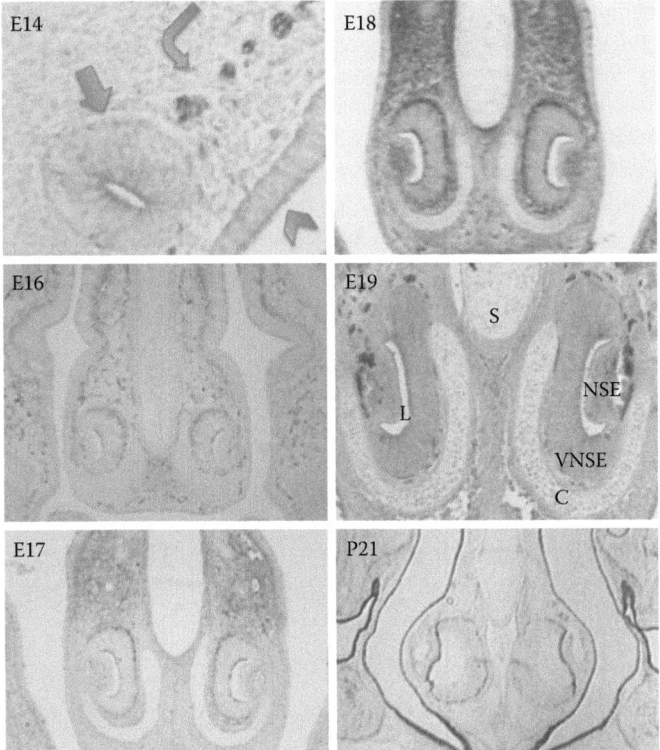

FIGURE 9.1 Pre- and postnatal developmental stages of the mouse vomeronasal organ. C = vomeronasal cartilage; E = embryonal day; L = vomeronasal lumen; NSE = nonsensory epithelium; P = postnatal day; S = nasal septum; VNSE = vomeronasal sensory epithelium; straight arrow = vomeronasal organ; curved arrow = vomeronasal nerve; arrowhead = respiratory epithelium. For a detailed description, see text.

9.2.2 Main Variations in Vertebrates

Several variations in VNO morphology are found across vertebrates. A connection with the oral cavity through the nasopalatine duct can be found in carnivores and ungulates. In amphibians (and turtles) the VNO is a part of the nasal cavity and the NSE is not present. In ophidians, the VNSE is especially thick and it is subdivided in columns by connective septa containing blood vessels, and a cavernous tissue is absent. In rodents, intraepithelial capillary loops are frequent as well as in lagomorphs but has not been found in marsupials. An interesting attempt at classifying the diversity of VNO morphology has been proposed by Takami (2002).

9.2.3 Developmental Issues

VRNs originate in the anterior end of the neural plate where a thickening of the ectodermal layer appears (olfactory placode) to invaginate and give rise to gonadotropin releasing hormone cells, receptor neurones of the olfactory and vomeronasal epithelium, and olfactory ensheathing cells giving growth and guidance support to receptor neurones. Recent findings in the mouse (Forni et al. 2011) showed that olfactory ensheathing cells as well as subpopulations of gonadotropin-releasing hormone (GnRH), olfactory, and vomeronasal cells originate from neural crest cells placed at the border of the olfactory placode, which intermix with ectodermal cells at early stages of development. A critical role for the transcription factor Bcl11b/Ctip2 in vomeronasal system development has been shown by Enomoto et al. (2011) in the mouse. In the absence of Bcl11b VRNs showed impaired axonal projections, reduction in the expression of vomeronasal receptor genes, and increased apoptosis and defective differentiation; moreover, loss of Bcl11b associated with an increased number of vomeronasal receptor $1/G\alpha_{i2}$-positive and reduced number of vomeronasal receptor $2/G\alpha_0$-positive neurons, suggesting that Bcl11b contributes regulating the fate choice between these two VRNs types.

9.3 MORPHOLOGY OF THE MOUSE VOMERONASAL ORGAN

Given the diverse morphology of VNO in different species, the following description is especially referred to the mouse VNO; it should be considered that most functional investigations of VNO have been conducted in this species.

9.3.1 DEVELOPMENT

A schematic pattern of development of VNO in rodents is found in Garrosa et al. (1992). Garrosa et al. proposed a six-stage developmental pattern of the VNO, three of which (anlage, early morphogenesis, late morphogensis) take place prenatally and the remaining three (initiation of secretory activity, cytoarchitectural maturity, complete histogenesis) postnatally.

The Anlage is found in the mouse at embryonal (E) day 11 (Cuschieri and Bannister 1975). At this stage, cells immunoreactive for luteinizing hormone releasing homone (Schwanzel-Fukuda and Pfaff 1989) as well as glutamate decarboxylase (the key enzyme for the synthesis of gamma-aminobutirric acid) (Wray et al. 1989) are present. Representative pictures of the mouse developing VNO are presented in Figure 9.1.

During early morphogenesis (E12-E14), the vomeronasal lumen appears as well as nerve projections from the developing VNSE to the brain and the vomeronasal cartilage. The precursor of the VNSE is now thicker than the NSE and the two epithelia become medially and laterally placed, respectively, and the vomeronasal lumen is crescent-shaped. In the VNSE cells labeled with the general marker of neuronal cells protein gene product (PGP) 9.5 appear (Zancanaro et al. 2002) and neuroblasts showing dendritic and axonal projections are recognizable together with supporting cells precursors and a large number of basal cells. The proteins stathmin and SCG10, which are involved in neuronal differentiation/proliferation, are found in the VNSE at E12 (Camoletto et al. 2001). PGP 9.5 labeling is also found in the vomeronasal nerve. In the mesenchyme lateral to the NSE, a group of blood vessels is developing; here, sparse substance P fibers (putatively sensitive) have been found (Nagahara et al. 1995), a finding not confirmed at any developmental stage by others (Zancanaro et al. 2002). Another marker of sensory nerve, the calcitonin gene-related peptide (CGRP), as well as atrial natriuretic peptide (ANP), a vasoactive factor, were similarly absent in the developing VNO at any stages (Zancanaro et al. 2002).

During late morphogenesis (E15 to birth) the mouse VNO reaches its adult shape by becoming increasingly crescent-shaped and elongating to cigar-shaped while its cartilagineous capsule comes to completion. The VNSE gets thicker and the NSE thinner, a distinct cavernous tissue is present beneath the latter, and the VNGs appear dorsally. Nitric oxide synthase type I (NOS I), the constitutive enzyme producing the gaseous neurotransmitter nitric oxide, is expressed in the vomeronasal nerve by E15 and in the area of cavernous tissue development by E16; vasoactive intestinal peptide immunoreactivity appears in the latter by E18 (Zancanaro et al. 2002). During late morphogenesis, the neuronal intermediate filament proteins internexin, nestin, and neurofilament M(NF-M) are sequentially expressed in the VNSE (Merigo et al. 2005); further, molecules are expressed in the VNSE, which are involved in water/ions handling or antioxidant mechanisms; that is, the cAMP-activated chloride channel (CFTR) and several aquaporins (AQP2, AQP3, AQP4, AQP5), and the Clara cell secretory protein CC26 (Merigo et al. 2011). During the last week of gestation the specific marker olfactory marker protein (OMP) is expressed in the VNSE (Tarozzo et al. 1998). At E18 fluorescent beads injected in the amniotic fluid enter the nasal cavity but not the vomeronasal lumen, suggesting that the vomeronasal duct is not open yet (Coppola and O'Connell 1989).

At birth all the key structural component of the VNO are present; however, postnatal development is required for the organ being morphologically mature and fully functional.

During postnatal development (from birth to 2 months of age) NOS I is increasingly expressed in extrinsic nerves of the cavernous tissue and around blood vessels under the VNSE by postnatal (P) day 1, and around VNGs by P8 (Zancanaro et al. 1999a) up to sexual maturity. Although the expression of NOS I in the adult mouse VNO was denied by Kishimoto et al. (1993), it was confirmed by Matsuda et al. (1996) and by Kulkarni et al. (1994) in the adult rat. Matsuda et al. (1996) found NOS immunoreactive fibers distributed in the receptor-free epithelium, around the blood vessels, and around glands. No NOS fibers were seen in the receptor area. Double immunofluorescence labeling showed that a part of the NOS immunoreactive nerve fibers in the cavernous tissue contained substance P and that all the NOS immunoreactive nerve fibers—immunoreactive nerve fibers around the blood vessels and glands in the cavernous tissue—contained vasoactive intestinal polypeptide (VIP). Immunohistochemical investigation of substance P expression in the mouse VNO during postnatal development (Zancanaro et al. 1999b) showed no immunoreactivity in the VNSE. The NSE and blood vessels of the VNSE showed CGRP immunoreactivity from P8 and with substance P immunoreactivity from P21 onwards; ANP and neuropeptide Y were found in the cavernous tissue from P8 and P21 onward. The above results suggest that VNO may be to some extent functional during postnatal development. The vomeronasal duct becomes patent sometime after the first day of life; however, during the normal prepubertal period it remains in an immature condition being characterized by an internal surface that is rapidly desquamating (Coppola et al. 1993). Accordingly, when Horowitz et al. (1999) selectively expressed barley lectin in sensory neurons in the OE and VNO of transgenic mice by using an olfactory-specific promoter, transneuronal transfer of the lectin was detected prenatally in the odor-sensing pathway but only postnatally in the pheromone-sensing pathway, suggesting that odors, but not pheromones, may be sensed *in utero*.

9.3.2 ADULT MOUSE VNO

In the adult mouse, the VNSE is a pseudostratified epithelium of the columnar type with a maximum height of about 180 μm (Mendoza 1993). From the surface downward, an apical region containing the microvilli bearing dendrites of VRNs, the microvilli, and the supranuclear portions of the supporting cells, supporting cell nuclear region, and a very thick receptor cell nuclear region are visible by light microscopy. Intraepithelial capillary loops are present coming from the connective lamina propria, which is devoid of glands.

Dendrites of receptor cells end with a short projection branching into several microvilli; instead, microvilli of supporting cells do not branch. Many desmosomes and thigh junctions connecting receptor and supporting cells are found in the apical region of VNSE. Supporting cells connect with each other by thin projections provided with desmosomes, thereby isolating dendrites from one another. The large dendrites of VRNs are rather electron-lucent by electron microscopy with a number of mitochondria and smooth endoplasmic reticulum formations. The supranuclear cytoplasm of supporting cells is rather electron-dense, containing granules and small lipid droplets.

In the supporting cell nuclear region a Golgi apparatus, mitochondria, some profiles of rough endoplasmic reticulum, and free ribosomes are found.

The VRNs nuclear region contains some rough endoplasmic reticulum, abundant smooth endoplasmic reticulum, multiple Golgi apparatus, mitochondria, and lipofuscin granules.

In the lower part of the VNSE, the infranuclear portion of supporting cells reaching the basal lamina contains bundles of tonofilaments, and axons of VNRs incompletely surrounded by supporting cells run to leave the epithelium. At the periphery of the VNSE where it joints the NSE a limited number of poorly differentiated basal cells is found, which represent a reservoir for replacing dying neurons throughout life.

Beneath the basal lamina, bundles of unmyelinated axons forming fascicles of the vomeronasal nerve run in a connective lamina propria containing blood vessels.

The serous VNG complex elongates parallel to the VNO long axis, and its duct ends at the border between VNRE and NSE. The secretory cells of VNG contain the typical organelle complement of serous cells (i.e., abundant rough endoplasmic reticulum), a well-developed Golgi apparatus, and apical secretory granules. Besides tight junctions and desmosomes, these cells are also connected by gap junctions. Axonal terminals are found between secretory cells containing both clear and electron-dense granules (Mendoza 1980).

Unlike development, aging has been not investigated in detail in the VNO. Cell proliferation, as determined by labeling with tritiated thymidine, was found to sharply decrease in the mouse by 1.5 years of age in comparison with P19 (Dodson and Bannister 1980); in the rat, proliferation density (cells/mm, BrdU labeling) at P333 was one-fourth less than P1 (Weiler and Farbman 1997). Wilson and Raisman (1980) found that the estimated total number of neurosensory cells decreased by 21% from month 4 to 8 of age in the mouse. No further decrease was present at month 18, suggesting longer life of neuronal cells in older mice.

REFERENCES

Aujard F. 1997. Effect of vomeronasal organ removal on male socio-sexual responses to female in a prosimian primate (*Microcebus murinus*). *Physiol. Behav.* 62:1003–1008.

Barrett J., Abbott D.H. and George L.M. 1993. Sensory cues and the suppression of reproduction in subordinate female marmoset monkeys, *Callithrix jacchus*. *J. Reprod. Fertil.* 97: 301–310.

Baxi K.N., Dorries K.M. and Eisthen H.L. 2006. Is the vomeronasal system really specialized for detecting pheromones? *Trends Neurosci.* 29:1–7.

Bhatnagar K.P. and Meisami E. 1998. Vomeronasal organ in bats and primates: Extremes of structural variability and its phylogenetic implications. *Microsc. Res. Tech.* 43:465–475.

Bigiani A., Mucignat-Caretta C., Montani G. and Tirindelli R. 2005. Pheromone reception in mammals. *Rev. Physiol. Biochem. Pharmacol.* 154:1–35.

Boehm U. 2006. The vomeronasal system in mice: From the nose to the hypothalamus—and back! *Semin. Cell Dev. Biol.* 17:471–479.

Brennan P.A. 2010. Pheromones and mammalian behavior. In *The Neurobiology of Olfaction*. A. Menini, editor. Boca Raton, FL: CRC Press.

Brennan P.A. and Zufall F. 2006. Pheromonal communication in vertebrates. *Nature* 444:308–315.

Broman I. 1920. Das Organon Vomero-nasale Jacobsoni—ein Wassergeruchorgane. *Anat. Hefte* 58:187–191.

Bruce H.M. 1960. A block to pregnancy in the mouse caused by proximity of strange males. *J. Reprod. Fertil.* 1:96–103.

Camoletto P., Colesanti A., Ozon S., Sobel A. and Fasolo A. 2001. Expression of stathmin and SCG10 proteins in the olfactory neurogenesis during development and after lesion in the adulthood. *Brain Res. Bull.* 54:19–28.

Cohen-Tannoudji J., Lavenet C., Locatelli A., Tillet Y. and Signoret J.P. 1989. Non-involvement of the accessory olfactory system in the LH response of anoestrous ewes to male odour. *J. Reprod. Fertil.* 86:135–144.

Coppola D.M. and O'Connell R.J. 1989. Stimulus access to olfactory and vomeronasal receptors in utero. *Neurosci Lett.* 106:241–248.

Coppola D.M., Budde J. and Millar L. 1993. The vomeronasal duct has a protracted postnatal development in the mouse. *J. Morphol.* 218:59–64.

Cuschieri A. and Bannister L.H. 1975. The development of the olfactory mucosa in the mouse: Light microscopy. *J. Anat.* 119:277–286.

Dennis J.C., Smith T.D., Bhatnagar K.P. et al. 2004. Expression of neuron-specific markers by the vomeronasal neuroepithelium in six species of primates. *Anat. Rec. A Discov. Mol. Cell. Evol. Biol.* 281:1190–1200.

Dodson H.C. and Bannister L.H. 1980. Structural aspects of ageing in the olfactory and vomeronasal epithelia in mice. In *Olfaction and Taste VII*, H. Van der Starre, editor. Oxford: IRL, pp. 151–154.

Dorries K.M., Adkins-Regan E. and Halpern B.P. 1997. Sensitivity and behavioral responses to the pheromone androstenone are not mediated by the vomeronasal organ in domestic pigs. *Brain Behav. Evol.* 49:53–62.

Døving K.B. and Trotier D. 1998. Structure and function of the vomeronasal organ. *J. Exp. Biol.* 201:2913–2925.

Eisthen H.L. 1992. Phylogeny of the vomeronasal system and of receptor cell types in the olfactory and vomeronasal epithelia of vertebrates. *Microsc. Res. Tech.* 23:1–21.

Enomoto T., Ohmoto M., Iwata T., Uno A., Saitou M., Yamaguchi T., Kominami R., Matsumoto I. and Hirota J. 2011. Bcl11b/Ctip2 controls the differentiation of vomeronasal sensory neurons in mice. *J. Neurosci.* 31:10159–10173.

Evans C.S. 2006. Accessory chemosignaling mechanisms in primates. *Am. J. Primatol.* 68:525–544.

Forni P.E., Taylor-Burds C., Melvin V.S., Williams T. and Wray S. 2011. Neural crest and ectodermal cells intermix in the nasal placode to give rise to GnRH-1 neurons, sensory neurons, and olfactory ensheathing cells. *J. Neurosci.* 31:6915–6927.

Garrosa M., Gayoso M.J. and Esteban F.J. 1998. Prenatal development of the mammalian vomeronasal organ. *Microsc. Res. Tech.* 41:456–470.

Garrosa M., Iñiguez C., Fernandez J.M. and Gayoso M.J. 1992. Developmental stages of the vomeronasal organ in the rat: A light and electron microscopic study. *J. Hirnforsch.* 33:123–132.

Giacobini P., Benedetto A., Tirindelli R. and Fasolo A. 2000. Proliferation and migration of receptor neurons in the vomeronasal organ of the adult mouse. *Brain Res. Dev. Brain. Res.* 123:33–40.

Halpern M. 1987. The organization and function of the vomeronasal system. *Annu. Rev Neurosci.* 10:325–362.

Halpern M. and Martínez-Marcos A. 2003. Structure and function of the vomeronasal system: An update. *Prog. Neurobiol.* 7:245–318.

Hamlin H.E. 1929. Working mechanism for the liquid and gaseous intake and output of Jacobson's organ. *Am. J. Physiol.* 91:201–205.

Houck L.D. 2009. Pheromone communication in amphibians and reptiles. *Annu. Rev. Physiol.* 71:161–176.

Isogai Y., Si S., Pont-Lezica L., Tan T., Kapoor V., Murthy V.N. and Dulac C. 2011. Molecular organization of vomeronasal chemoreception. *Nature* 478:241–245.

Jacobson L., Trotier D. and Døving K.B. 1998. Anatomical description of a new organ in the nose of domesticated animals by Ludvig Jacobson (1813). *Chem. Senses* 23:743–754.

Johns M.A. 1986. The role of the vomeronasal organ in behavioral control of reproduction. *Ann. N. Y. Acad. Sci.* 474:148–157.

Johns M.A., Feder H.H., Komisaruk B.R. and Mayer A.D. 1978. Urine-induced reflex ovulation in anovulatory rats may be a vomeronasal effect. *Nature* 272:446–448.

Karlson P. and Lüscher M. 1959. "Pheromones": A new term for a class of biologically active substances. *Nature* 183:55–56.

Kelliher K.R., Baum M.J. and Meredith M. 2001. The ferret's vomeronasal organ and accessory olfactory bulb: Effect of hormone manipulation in adult males and females. *Anat. Rec.* 263:280–288.

Keverne E.B. 1999. The vomeronasal organ. *Science* 286:716–720.

Kishimoto J., Keverne E.B., Hardwick J. and Emson P.C. 1993. Localization of nitric oxide synthase in the mouse olfactory and vomeronasal system: A histochemical, immunological and in situ hybridization study. *Eur. J. Neurosci.* 5:1684–1694.

Kratzing J. 1971. The structure of the vomeronasal organ in the sheep. *J. Anat.* 108:247–260.

Kulkarni A.P., Getchell T.V. and Getchell M.L. 1994. Neuronal nitric oxide synthase is localized in extrinsic nerves regulating perireceptor processes in the chemosensory nasal mucosae of rats and humans. *J. Comp. Neurol.* 345:125–138.

Lee S. and Boot L.M. 1955. Spontaneous pseudopregnancy in mice. *Acta Physiol. Pharmacol. Neurol.* 4:422–443.

Ma M. 2010. Multiple olfactory subsystems convey various sensory signals. In: *The Neurobiology of Olfaction*, A. Menini, editor. Boca Raton, FL: CRC Press.

Matsuda H., Kusakabe T., Kawakami T., Takenaka T., Sawada H. and Tsukuda M. 1996. Coexistence of nitric oxide synthase and neuropeptides in the mouse vomeronasal organ demonstrated by a combination of double immunofluorescence labeling and a multiple dye filter. *Brain Res.* 712:35–39.

Mendoza A.S. 1980. The mouse vomeronasal glands: A light and electron microscopy study. *Chem. Senses* 12:541–555.

Mendoza A.S. 1993. Morphological studies on the rodent main and accessory olfactory systems: The regio olfactoria and vomeronasal organ. *Ann. Anat.* 175:425–446.

Meredith M. 1998. Vomeronasal function. *Chem. Senses* 23:463–466.

Merigo F., Mucignat-Caretta C., Cristofoletti M. and Zancanaro C. 2011. Epithelial membrane transporters expression in the developing to adult mouse vomeronasal organ and olfactory mucosa. *Dev. Neurobiol.* 71:854–869.

Merigo F., Mucignat-Caretta C. and Zancanaro C. 2005. Timing of neuronal intermediate filament proteins expression in the mouse vomeronasal organ during pre- and postnatal development. An immunohistochemical study. *Chem. Senses.* 30:707–717.

Mucignat-Caretta C. 2010. The rodent accessory olfactory system. *J. Comp. Physiol. A Neuroethol. Sens. Neural Behav. Physiol.* 196:767–777.

Nagahara T., Matsuda H., Kadota T. and Kishida R. 1995. Development of substance P immunoreactivity in the mouse vomeronasal organ. *Anat. Embryol. (Berl.)* 192:107–115.

Negus V.E. 1956. The organ of Jacobson. *J. Anat.* 90:515–519.

Pearlman S.M. 1934. Jacobson's organ: Its anatomy, gross, microscopic and comparative, with some observations as well on its function. *Ann. Otol. Rhinol. Lar.* 43:739–768.

Powers J.B. and Winans S.S. 1975. Vomeronasal organ: Critical role in mediating sexual behavior of the male hamster. *Science* 187:961–963.

Retzius G. 1894. Die Riechzellen der Ophidier in der riechschleimhaut und im Jacobson' sches Organ. *Biol. Untersusch. Neue Folge* 6:48–51.

Rodriguez I. and Boehm U. 2009. Pheromone sensing in mice. *Results Probl Cell Differ.* 47:77–96.

Ruysh F. 1703. *Thesaurus Anatomicus III*. Table IV, Figure V, p. 49. Amsterdam: Wolters.

Salazar I. and Sanchez Quintero S. 2009. The risk of extrapolation in neuroanatomy: The case of the Mammalian vomeronasal system. *Front. Neuroanat.* 3:22.

Salazar I., Lombardero M., Alemañ N. and Sánchez Quinteiro P. 2003. Development of the vomeronasal receptor epithelium and the accessory olfactory bulb in sheep. *Microsc. Res. Tech.* 61:438–447.

Salazar I., Quinteiro P.S. and Cifuentes J.M. 1997. The soft-tissue components of the vomeronasal organ in pigs, cows and horses. *Anat. Histol. Embryol.* 26:179–186.

Schilling A. 1987. L'organe voméronasal des Mammifères. *J. Psychol. Norm. Pathol. (Paris)* 3:221–278.

Schwanzel-Fukuda M. and Pfaff D.W. 1989. Origin of luteinizing hormone-releasing hormone neurons. *Nature* 338:161–164.

Semmering S.T. 1809. Abbildungen der menschlichen Organe des Geruches. Frankfurt: Varrentrap und Wenner.

Seydel O. 1895. Uber die NasenhShle und das Jacobson'sche Organ der Amphibien: Eine vergleichendanatomische Untersuchung, Morphol. *Jahrb.* 23:453–543.

Smith T.D., Dennis J.C., Bhatnagar K.P. et al. 2010. Olfactory marker protein expression in the vomeronasal neuroepithelium of tamarins (*Saguinus spp*). *Brain Res.* 23:1375: 7–18.

Smith T.D., Garrett E.C., Bhatnagar K.P. et al. 2011. The vomeronasal organ of New World monkeys (platyrrhini). *Anat. Rec.* 294:2158–2178.

Smith T.D., Siegel M.I. and Bhatnagar K.P. 2001. Reappraisal of the vomeronasal system of catarrhine primates: Ontogeny, morphology, functionality, and persisting questions. *Anat. Rec.* 2001. 265:176–192.

Swaney W.T. and Keverne E.B. 2009. The evolution of pheromonal communication. *Behav. Brain Res.* 200:239–247.

Takami S. 2002. Recent progress in the neurobiology of the vomeronasal organ. *Microsc. Res. Tech.* 58:228–250.

Takigami S., Mori Y. and Ichikawa M. 2000. Projection pattern of vomeronasal neurons to the accessory olfactory bulb in goats. *Chem. Senses* 25:387–393.

Takigami S., Mori Y., Tanioka Y. and Ichikawa M. 2004. Morphological evidence for two types of mammalian vomeronasal system. *Chem Senses* 29:301–310.

Tarozzo G., Cappello P., De Andrea M., Walters E., Margolis F.L., Oestreicher B. and Fasolo A. 1998. Prenatal differentiation of mouse vomeronasal neurones. *Eur. J. Neurosci.* 10:392–396.

Tirindelli R., Dibattista M., Pifferi S. and Menini A. 2009. From pheromones to behavior. *Physiol. Rev.* 89:921–956.

Vandenbergh J.G. 1969. Male odor accelerates female sexual maturation in mice. *Endocrinology* 84:658–660.

von Mihalkovics V. 1899. Nasenhole und Jacobson'sches Organ. *Anat. Embryol. (Berl.)* 11:1–108.

Weiler E. and Farbman A.I. 1997. Proliferation in the rat olfactory epithelium: age-dependent changes. *J. Neurosci.* 17:3610–3622.

Weiler E., Apfelbach R. and Farbman A.I. 1999. The vomeronasal organ of the male ferret. *Chem. Senses* 24:127–136.

Whitten W.K. 1956. Modification of the oestrous cycle of the mouse by external stimuli associated with the male. *J. Endocrinol.* 13:399–404.

Wilson K.C. and Raisman G. 1980. Age-related changes in the neurosensory epithelium of the mouse vomeronasal organ: Extended period of postnatal growth in size and evidence for rapid cell turnover in the adult. *Brain Res.* 185:103–113.

Witt M. and Woźniak W. 2006. Structure and function of the vomeronasal organ. *Adv. Otorhinolaryngol.* 63:70–83.

Woodley S.K. 2010. Pheromonal communication in amphibians. *J. Comp. Physiol. A Neuroethol. Sens. Neural Behav. Physiol.* 196:713–727.

Wray S., Nieburgs A. and Elkabes S. 1989. Spatiotemporal cell expression of luteinizing hormone-releasing hormone in the prenatal mouse: Evidence for an embryonic origin in the olfactory placode. *Brain Res. Dev. Brain. Res.* 46:309–318.

Wysocki C.J. 1979. Neurobehavioral evidence for the involvement of the vomeronasal system in mammalian reproduction. *Neurosci. Biobehav. Rev.* 3:301–341.

Zancanaro C., Merigo F., Mucignat-Caretta C. and Cavaggioni A. 2002. Neuronal nitric oxide synthase expression in the mouse vomeronasal organ during prenatal development. *Eur. J. Neurosci.* 16:659–664.

Zancanaro C., Mucignat-Caretta C., Merigo F. and Cavaggioni A. 1999a. Immunohistochemical investigation of the vomeronasal organ: Nitric oxide synthase expression in the mouse during postnatal development. *Neurosci. Lett.* 269:5–8.

Zancanaro C., Mucignat-Caretta C., Merigo F. and Osculati F. 1999b. Neuropeptide expression in the mouse vomeronasal organ during postnatal development. *Neuroreport* 10:2023–2027.

Zufall F. and Leinders-Zufall T. 2007. Mammalian pheromone sensing. *Curr. Opin. Neurobiol.* 17:483–489.

10 Vomeronasal Receptors and Signal Transduction in the Vomeronasal Organ of Mammals

Simona Francia, Simone Pifferi,
Anna Menini, and Roberto Tirindelli

CONTENTS

10.1 Introduction 297
10.2 Pheromone Receptors 302
    10.2.1 V1Rs 302
    10.2.2 V2Rs 306
    10.2.3 FPRs...308
10.3 Signal Transduction Mechanisms.. 310
    10.3.1 G Proteins 311
    10.3.2 Phospholipase C.. 312
    10.3.3 Ion Channels 313
        10.3.3.1 TRPC2 313
        10.3.3.2 Ca^{2+}-Activated Cl^- Channels 314
        10.3.3.3 Other Ion Channels 315
References.. 316

10.1 INTRODUCTION

In most species, there are two chemosensory systems, both located in the nasal cavity but physiologically and anatomically distinct (Figure 10.1). The main olfactory epithelium (MOE) is principally involved in the airborne odor perception, whereas the vomeronasal organ (VNO) of Jacobson in the detection of pheromones that are chemical compounds secreted or excreted by individuals of the same species (conspecifics) (Dulac & Torello 2003; Tirindelli et al. 2009). Deficits in social and reproductive behavior are marked in animals underlying the surgical removal of the VNO but also the chemical inactivation of the MOE (Kiyokawa et al. 2007; Kolunie &

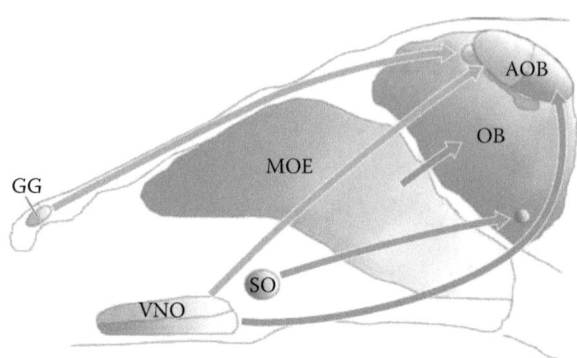

FIGURE 10.1 Chemosensory systems sensing pheromones. The main olfactory system (MOE) and the vomeronasal system (VNO) together with the septal organ of Masera (SO) and the Grueneberg ganglion (GG) are implicated in pheromone detection. Projections of sensory neurons to the main olfactory bulb (OB) and accessory olfactory bulb (AOB) are shown. (Modified from Tirindelli, R. et al. (2009) *Physiol Rev*, 89, 921–956.)

Stern 1995; Leypold et al. 2002; Wysocki & Lepri 1991). Furthermore, there is evidence that two olfactory structures, namely the septal organ of Masera and the Grueneberg ganglion also contribute to pheromone detection (Ma 2007; Roppolo et al. 2006; Tirindelli et al. 2009) (Figure 10.1).

Primary sensory VNO neurons (VSNs) send axons to mitral cells in the glomerular region of the accessory olfactory bulb (AOB), whereas olfactory neurons (OSNs) to the mitral cells of the main olfactory bulb (MOB) (Baum & Kelliher 2009; Clapham & Neer 1997) (Figure 10.1). The chemosensory neurons of the VNO and MOE are equipped with specific receptors that activate distinct molecular cascades and mediate sociosexual behaviors (Del Punta et al. 2002; Stowers et al. 2002).

The anatomical components of the VNO firstly developed in a tetrapod ancestor and led to the appearance of a rudimentary structure in amphibians that became highly organized in Squamata and in many mammalian orders as Didelphimorphia, Rodentia, and in primates (prosimians and New World monkeys). In contrast, VNO is virtually absent in birds, bats, Old World monkeys, apes, and humans (Dennis et al. 2004; Grus et al. 2005; Shi & Zhang 2007; Smith et al. 2005; Zhao et al. 2011) (Figure 10.2a).

In Didelphimorphia, Lagomorpha, and Rodentia, the VNO chemosensory epithelium presents two separate neuronal layers (apical and basal) characterized by the expression of different G protein subunits and receptors (Young & Trask 2007). Apical and basal neurons project to two distinct regions (anterior and posterior) of the AOB. From the AOB, the majority of the centripetal projections of the apical and basal neurons remain segregated in the amygdala and hypothalamus (Martinez-Marcos 2009; Yoon et al. 2005).

(a)

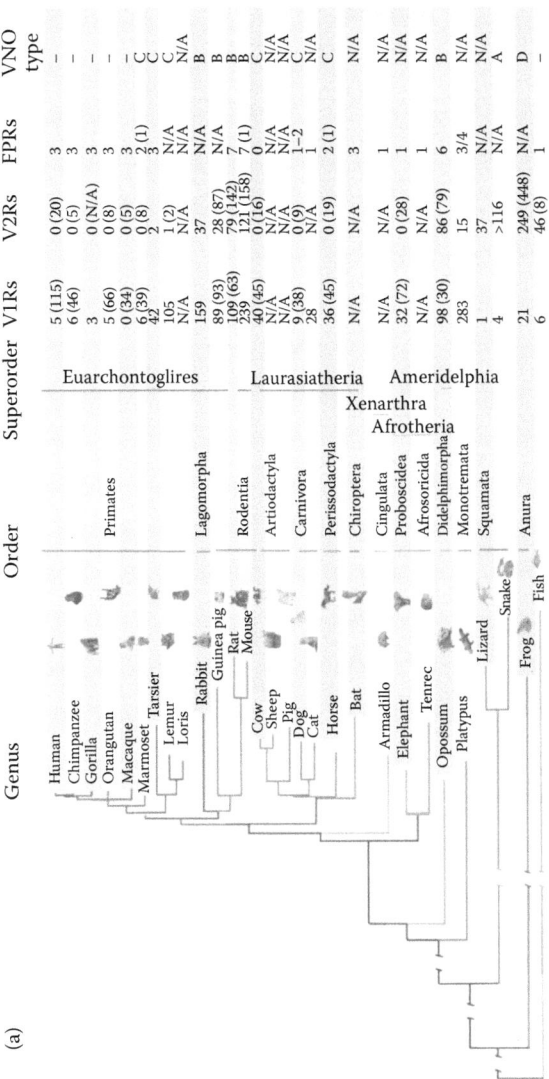

Genus	Order	Superorder	V1Rs	V2Rs	FPRs	VNO type
Human	Primates	Euarchontoglires	5 (115)	0 (20)	3	–
Chimpanzee			6 (46)	0 (5)	3	–
Gorilla			3	0 (N/A)	3	–
Orangutan			5 (66)	0 (8)	3	–
Macaque			0 (34)	0 (5)	3	C
Marmoset			6 (39)	0 (8)	2 (1)	C
Tarsier			42	2	3	C
Lemur			105	1 (2)	N/A	C
Loris			N/A	N/A	N/A	C
Rabbit	Lagomorpha		159	37	N/A	B
Guinea pig	Rodentia		89 (93)	28 (87)	7 (1)	B
Rat			109 (63)	79 (142)	7 (1)	B
Mouse			239	121 (158)	0	B
Cow	Artiodactyla	Laurasiatheria	40 (45)	0 (16)	N/A	C
Sheep			N/A	N/A	N/A	C
Pig			N/A	N/A	1–2	N/A
Dog	Carnivora		9 (38)	0 (9)	1	C
Cat			28	0	2 (1)	N/A
Horse	Perissodactyla		36 (45)	0 (19)	1	C
Bat	Chiroptera		N/A	N/A	3	N/A
Armadillo	Cingulata	Xenarthra	N/A	N/A	1	N/A
Elephant	Proboscidea	Afrotheria	32 (72)	0 (28)	1	N/A
Tenrec	Afrosoricida		N/A	N/A	1	N/A
Opossum	Didelphimorpha	Ameridelphia	98 (30)	86 (79)	6	B
Platypus	Monotremata		283	15	3/4	N/A
Lizard	Squamata		1	37	N/A	N/A
Snake			4	>116	N/A	A
Frog	Anura		21	249 (448)	N/A	D
Fish			6	46 (8)	1	–

FIGURE 10.2 (**See color insert.**) (a) The phylogenetic tree of mammals. Mammals comprise about 5400 species that are distributed in six superorders including Euarchontoglires (Rodentia, Lagomorpha, primates), Laurasiatheria (Artiodactyla, Carnivora, Perissodactyla, and Chiroptera), Xenarthra (Cingulata) and Afroteria (Proboscidea, Afrosoricida) and Ameridelphia (Didelphimorpha) (Churakov et al. 2010; Huchon et al. 2007; Michaux et al. 2007; Nikolaev et al. 2007). Primates include prosimians (loris, lemur, tarsier), New World monkeys (marmoset), Old World monkeys (macaque), apes (orangutan, gorilla, chimpanzee), and humans. Doubts remain on the location of the Rodentia order in the phylogenetic tree of mammals. The most supported hypothesis (shown here) suggests that the Rodentia order is closely related to primates and Laurasiatheria (Luo et al. 2012). The number of intact genes for V1Rs, V2Rs, FPRs and pseudogenes (in brackets) are shown (Dong et al. 2012; Grus et al. 2007; Liberles et al. 2009; Nei et al. 2008; Yang & Shi 2010; Young et al. 2010). The anatomical complexity of the adult VNO (from A to D) is indicated according to Takami et al. (2002). Type C is found in cows, sheep, dogs, prosimian primates, and in New World monkeys. Type B is present in rodents and lagomorphs. Type A VNO belongs to the ophidian species (water snakes); Amphibians developed type D VNO (Takami 2002). Lizard and snake (Squamata), frog (Anura), and fish have been taken as outgroups.

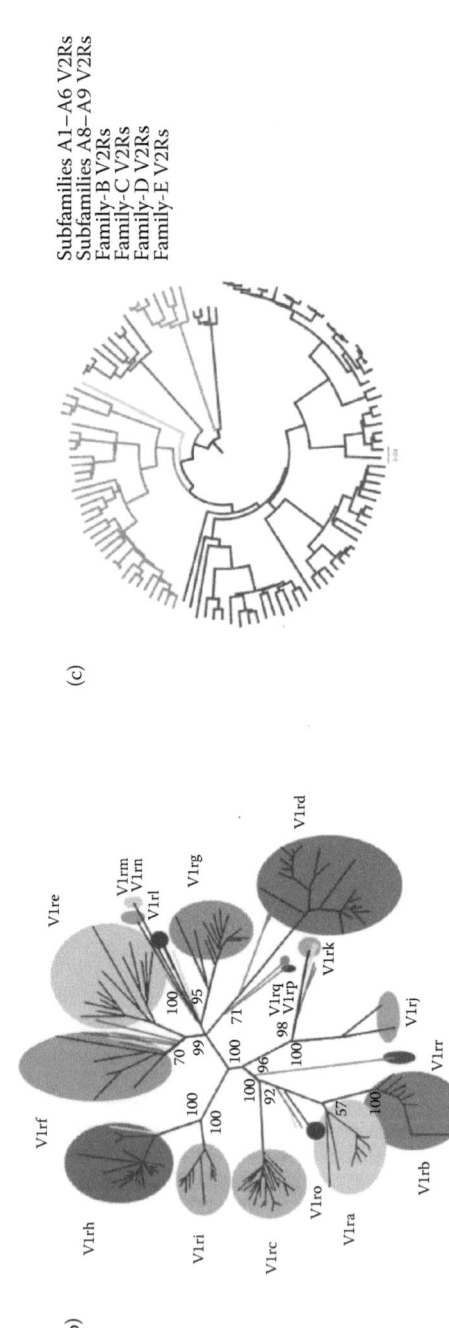

Subfamilies A1–A6 V2Rs
Subfamilies A8– A9 V2Rs
Family-B V2Rs
Family-C V2Rs
Family-D V2Rs
Family-E V2Rs

(c)

(b)

FIGURE 10.2 (*Continued*) (b) Phylogeny of V1Rs. Unrooted phylogenetic tree of functional V1Rs in rat, mouse, cow, dog, and human. The mouse V1Rs repertoire consist of 250 members organized in 12 phylogenetically isolated subfamilies (a to k). Species-specific V1R families of rat (V1rm, V1rn, V1ro), cow (V1rp and V1rq), and human (V1rr) are shown. Mouse and rat branches are in black, cow branches are in blue, dog branches are in light blue, and human branches are in purple. The bootstrap percentage supports the monophyly of all families with the exception of subfamily V1r1a. The tree was reconstructed using the neighbor-joining method with Poisoon-corrected protein distances. (c) Phylogeny of V2Rs in mouse. Family ABDE full-length genes are shown. Family E corresponds to the former subfamily A10 (Vmn2r18).

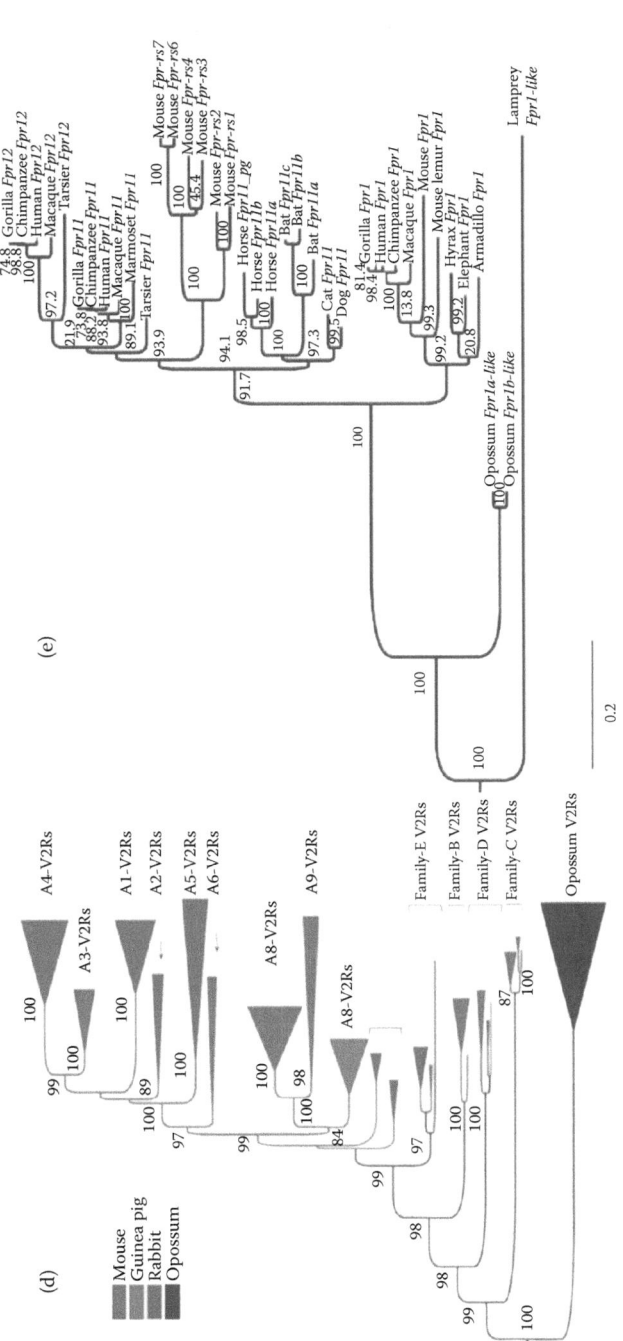

FIGURE 10.2 (*Continued*) (d) Phylogeny of V2Rs in mammals. V2Rs coding sequences of Rodentia (mouse, guinea pig), Lagomorpha (rabbit), and Didelphimorpha (opossum) are used to create a neighbor-joining phylogenetic tree. The bootstrap percentage is shown at each node. (e) Phylogenetic tree of FPR genes in mammals. FPR sequences were identified in whole genome databases and used to create a phylogenetic tree performed by maximum likelihood (ML) analysis. The ML bootstrap percentage is shown in each node. The horse fprl1 is a pseudogene (Fprl1_pg). Opossum and lamprey FPR-like genes were considered as outgroups.

In other mammalian orders (Carnivora) the dichotomic organization of the vomeronasal projections is not well defined or even absent (Artiodactyla) (Dennis et al. 2003).

10.2 PHEROMONE RECEPTORS

The molecular organization of the rodent vomeronasal neuroepithelium in apical and basal neurons is based on the specific expression pattern of two transduction molecules, G protein $G\alpha_{i2}$ and $G\alpha_o$, and two distinct superfamilies of pheromone receptors, type-1 vomeronasal receptors (V1Rs) (Dulac & Axel 1995) and type-2 vomeronasal receptors (V2Rs) (Herrada & Dulac 1997; Matsunami & Buck 1997; Ryba & Tirindelli 1997).

$G\alpha_{i2}$ is expressed in the apical neurons and colocalizes with V1Rs, whereas $G\alpha_o$ coexpresses with V2Rs in the basal neurons (Rodriguez et al. 2002).

Although both are G protein coupled receptors, V2Rs differ from V1Rs for the presence of introns in the genes and for the long N-terminal extracellular region that reflects the chemical properties of their ligands (Boschat et al. 2002; Emes et al. 2004; Ishii & Mombaerts 2008; Isogai et al. 2011; Kimoto et al. 2005). Furthermore, V1Rs and V2Rs display a differential evolutionary expansion in terrestrial vertebrates (Brykczynska et al. 2013; Dong et al. 2012; Grus et al. 2007; Liberles et al. 2009; Ryu et al. 2013; Yang & Shi 2010) (Figure 10.2a, b, and d).

However, although only V1Rs and V2Rs have been proved to respond to pheromones, the vomeronasal expression of a third class of putative pheromone receptors, formyl-peptide receptors (VNO-FPRs), was recently demonstrated (Boschat et al. 2002; Del Punta et al. 2002; Dulac & Torello 2003). FPRs are expressed in both the apical and basal regions of the VNOs of some murine rodents and may represent a link between the G_i and G_o subsystems (Liberles et al. 2009; Riviere et al. 2009).

Additionally, basal VNO neurons express a family of nonclassical molecules of the major histocompatibility complex (H2-Mv) (Ishii & Mombaerts 2008; Leinders-Zufall et al. 2009; Loconto et al. 2003). The role of the H2-Mv complex in the VNO is still debated and whether these molecules are to be considered as pheromone receptors remains to be demonstrated (Ishii & Mombaerts 2008; Loconto et al. 2003).

10.2.1   V1Rs

V1Rs are already present in lower vertebrate (fish and frog) genomes (Date-Ito et al. 2008; Grus & Zhang 2009; Hashiguchi et al. 2008; Saraiva & Korsching 2007). In mammals, V1Rs expansion occurred independently from a small number of ancestral genes that pseudogenized or completely disappeared in the Squamata order (Young et al. 2010). Remarkable gene duplications occurred in the Rodentia order with the generation of several subfamilies (Grus & Zhang 2004; Rodriguez et al. 2002) (Figure 10.2b).

The V1R repertoire is characterized by a rapid gene turnover, significant variation of the size among species, positive selection, and lineage-specific clustering (Emes et al. 2004; Shi et al. 2005; Young et al. 2005).

V1R genes diverge among mammals following species-specific duplications that resulted in the formation of distinct clades with the occurrence of very few orthologs only present in closely related species (Park et al. 2011).

The high rate of duplication led to a large degree of pseudogenization also in species that present a considerable number of intact V1R genes and a well-developed VNO. In total, of about 7000 V1R-like sequences that have been identified in 37 species, only 1809 represent functional genes (Young et al. 2010).

In general, V1R genes expanded in mammals that possess an highly organized VNO, such as platypus (283 genes), opossum (98), mouse (239), rat (109), rabbit (159), and mouse lemur (105), whereas they are absent or pseudogenized in Old World monkeys, apes, and humans that possess a vestigial and presumably non-functional VNO (Brykczynska et al. 2013; Grus & Zhang 2004; Grus et al. 2007; Hohenbrink et al. 2013; Rodriguez et al. 2002; Young et al. 2010) (Figure 10.2a). Surprisingly, the dog (Carnivora) genome displays a very small repertoire of intact V1R genes (9) in contrast to a much larger number of pseudogenes (38); a deterioration that does not seem to be correlated with the selective pressure that occurred during domestication, since the wolf genome presents homologous copies of the dog pseudogenes (Young et al. 2010). Interestingly, and in contrast to what was previously described, ruminant species such as cow, sheep, and goat evolved a significant number of V1R orthologs that are expressed in both the MOE and VNO. In cow, 70% of V1R genes has orthologs in the cross-species ruminant counterpart in contrast to the 9% of the mouse V1R orthologs identified in rat (mouse-rat are phylogenetically much closer species than cow-sheep) (Grus et al. 2005; Ohara et al. 2009) (Figure 10.2b). These unusual findings might suggest a possible involvement of both VNO and MOE in the detection of V1R ligands. Moreover, in these ruminant species (herbivores) V1Rs may also subserve to other functions rather than purely pheromonal. For instance, ruminant V1R orthologs could be involved in the detection of the same or closely related chemical cues as heterospecific odorants (allomones) (Ohara et al. 2009; Takigami et al. 2000; Wakabayashi et al. 2002).

In rodents, there are indications that V1Rs originated from an ancestor that probably possessed very few V1R genes. The majority of these genes duplicated independently generating a species-specific lineage. Lineage specificity appears so rigorous that a limited degree of orthology is only detectable in very closely related species such as *Mus spretus* and *Mus musculus* that diverged 1 million years ago (Grus & Zhang 2004; Park et al. 2011; Young et al. 2010).

The very large human V1R repertoire shows an almost complete pseudogenization (about 200 pseudogenes and only four intact genes) that occurred before the separation of hominoids (apes and humans) from Old World monkeys and was probably linked to the complete loss of the VNO (Zhang & Webb 2004).

Mouse V1Rs represent the first superfamily of pheromone receptors that have been discovered, and as with olfactory receptors, they show a monogenic expression that means that each VNO neuron expresses only one receptor type. In this murine species, V1Rs account for about 250 functionally receptors expressed in the apical VNO neurons. V1Rs are organized in 12 phylogenetically isolated subfamilies (a to k) with an identity ranging from 40% to 99% within subfamilies and from 15%

to 40% between subfamilies (DelPunta et al. 2000; Rodriguez et al. 2002) (Figure 10.2b).

Positive selection (a ratio between nonsynonymous and synonymous amino acid substitution greater than 1) that is an index of a fast gene evolution has been reported acting on the V1R repertoire (Grus & Zhang 2004; Hohenbrink et al. 2012, 2013; Mundy & Cook 2003; Wyckoff et al. 2000). Thus, the rapid diversification of the mouse V1Rs developed in association with an increasingly high predominance of the pheromone-mediated communication. As a consequence, if a pheromone blend evolved rapidly and accumulated chemical changes, the V1Rs recognition system might have been subjected to a strong selective pressure (Shi et al. 2005).

In particular, the analysis of the sequences corresponding to the transmembrane domain 2 and the extracellular loops I and II of these receptors has shown the highest variability within species in combination with a strong positive selection, suggesting that these regions are implicated in pheromone binding (Rodriguez 2004).

Investigations on V1R expressing neurons (apical) led to the identification of four candidate V1R ligands, 2-hepatanone, 6-hydroxyl-6-methyl-3-heptanone, n-pentyl-acetate and isobutylamine (Table 10.1). These molecules are detectable in the mouse urine and have shown to activate, in a very specific fashion, distinct V1Rs (Boschat et al. 2002).

In contrast, other V1Rs appear to respond to multiple chemical cues, some of which are also emitted by heterospecific animals (Isogai et al. 2011) (Table 10.1).

From these studies, it is possible to envisage that distinct subpopulations of apical neurons, presumably expressing different V1R subtypes, are tuned to detect either individual or multiple pheromonal molecules (Boschat et al. 2002; Isogai et al. 2011).

A single behavioral study on deficient mice for V1R genes has provided clues on the function of some of these receptors. Mice bearing the complete deletion of the V1Ra and V1Rb gene clusters display a significant reduction of male-to-male and maternal aggressiveness compared to controls. In addition, these mutant mice do not show electrophysiological responses to 6-hydroxyl-6-methyl-3-heptanone, n-pentyl-acetate and isobutylamine (Del Punta et al. 2002).

In conclusion, a reliable correlation between the size of the V1R repertoire and the habits, the ecological system, and the sociosexual behavior of each species is difficult to establish. There appears to be a weak association between the size of intact V1R genes in terricolous mammals and two ecological factors: spatial (nest-living and open-living behavior) and rhythms (diurnal and nocturnal) activity (Wang et al. 2010). With few but significant exceptions (tenrec, sloth, squirrel, and pika), nest-living and nocturnal mammals (mouse, rat, kangaroo rat, rabbit, and lemur) seem to require more functional genes than open-living and diurnal mammals (among which horse, cow, and apes) perhaps in order to explore restricted and dark environments (Grus et al. 2005, 2007; Wang et al. 2010).

TABLE 10.1
V1R and V2R Ligands and Their Suggested Function in
Pheromonal Communication

Mouse	Receptor Gene	VR Ligands	Ligand Function
V1Rs	V1Rb2	n-pentylacetate; isobutylamine; 2-heptanone; 6-hydroxyl-6-methyl-3-heptanone	Produced in mouse male urine. They are involved in estrus acceleration and extension and in puberty induction (Jemiolo et al. 1989; Novotny et al. 1999; Tirindelli et al. 2009).
	V1re2	Corticosteroids (Q1570, Q3910, Q2525)	Present in female urine. Unknown pheromonal fuction (Isogai et al. 2011; Nodari et al. 2008).
	V1re6	Glucocorticoid: (Q1570); corticosteroids: (Q1570, Q3910, Q2525)	Present in female urine. Unknown pheromonal fuction (Isogai et al. 2011; Nodari et al. 2008).
	V1rf3	Estrogen: (E1050); estradiols: (E0893, E0588, E1100); estriol: (E2734)	Present in female urine. Unknown pheromonal fuction (Isogai et al. 2011).
	V1rj2	Androgen: (A7864); estradiols: (E0893, E0588, E1100); estrogen: (E1050)	Present in female urine. Unknown pheromonal fuction (Isogai et al. 2011).
V2Rs	V2Rp5	Exocrine-gland-secreting peptide 1 (ESP1)	Released into male tear fluids, ESP1 enhances lordosis in females (Kimoto et al. 2005, 2007; Haga et al. 2010).
	N/A	Major urinary proteins (MUPs)	Synthesized in the liver and excreted in male mouse urine, promote aggressiveness in male and accelerate puberty in females (Mucignat-Caretta et al. 1995; Logan et al. 2008; Roberts et al. 2010).
	N/A	Aphrodisin	Present in vaginal smears induces copulatory behavior in male hamsters (Singer et al. 1986).
	V2r1b	Major histocompatibility complex (MHC) peptides	Female pregnancy block (Bruce effect) following the exposure of male scent (Bruce 1959; Thompson et al. 2007).
FPRs	Fpr-rs1,3	CRAMP;	Unknown pheromonal fuction (Liberles et al. 2009; Migeotte et al. 2006; Riviere et al. 2009).
	Fpr-rs4	fMLF, lipoxin A4;	
	Fpr-rs6	CRAMP; fMLF;	
	Fpr-rs7	CRAMP; fMLF	

Note: CRAMP = cathelin-related antimicrobial peptide, fMLF = N-formyl-methionyl-leucyl-phenylalanine.

10.2:2   V2Rs

V2R genes have already appeared in fish (46 intact genes) and frog (about 300 intact genes and 450 pseudogenes) that possesses the largest repertoire among all vertebrates (Ji et al. 2009; Shi & Zhang 2007; Young & Trask 2007). A striking variation of V2R genes occurred in reptiles and mammals since (potentially) intact genes have only been reported in Squamata (lizard and snake), Didelphimorpha, (opossum), Rodentia, and Lagomorpha (rabbit). No functional V2Rs have been identified in Carnivora (dog), Artiodactyla (cow), and primates (macaque, chimpanzee, gorilla, and human) with the exception of prosimians (loris, lemur, and tarsier) (Dong et al. 2012; Hohenbrink et al. 2012; Ishii & Mombaerts 2011; Shi & Zhang 2007; Yang et al. 2005; Young & Trask 2007) (Figure 10.2a and d).

Mammalian V2Rs can be classified in five families A to E (Figure 10.2c). Families BCDE V2Rs are phylogenetically ancient and present orthologs in species of several orders. It is likely that all mammalian V2Rs originated from 4–5 ancestral genes (probably precursors of each family) that undertook a different degree of expansion according to each species.

Receptors of family A typically represent the majority of the V2Rs (95% in the mouse) in all species and show a strong lineage specificity so that orthologs can be exclusively found in related species, in which, however, they tend to form small but independent clades (Silvotti et al. 2007).

In murids, family A has further expanded originating two distinct groups including subfamilies A1–A6 (phylogenetically more recent and largely represented in mouse and rat) and subfamilies A8–A9 (with orthologs in Cricetidae and Caviidae) (Silvotti et al. 2011).

Already present in bony fish, family C is considered the most ancient among all other V2R families from which it considerably diverges (Silvotti et al. 2005). Family C-V2Rs are typically represented by a single gene in each species with the exception of Muroidea superfamily (Brykczynska et al. 2013; Rodriguez 2004; Shi & Zhang 2007). Intact genes are found in prosimians whereas Old World monkeys, apes, and humans possess only pseudogenes (Hohenbrink et al. 2013; Martini et al. 2001).

In mouse and rat, family C expanded by gene duplication and inversion originating seven and four homologue receptors, respectively; thus, this family is split into two subfamilies, C1 and C2. This expansion probably originated before the separation of the last common ancestor of Muridae and Caviidae (guinea pig) families and appears to parallel the establishment of subfamilies A1–A6 (Silvotti et al. 2011).

In the VNO, families-ABDE V2Rs are expressed monogenically; however, the basal neurons show a multigenic expression of V2Rs that seem to represent an exception of the one-neuron-one receptor rule (Belluscio et al. 1999; Bozza et al. 2002; Rodriguez et al. 1999; Serizawa et al. 2000). In fact, families-ABDE V2Rs are coexpressed with family-C V2Rs in a specific fashion (Martini et al. 2001; Silvotti et al. 2011). In the rat and mouse, the expansion of family C and family A defined two populations of basal neurons. One population expresses phylogenetically ancient V2Rs (subfamilies A8–A9, families BDE, and subfamily-C1 V2Rs), while the other population expresses multiple combinations of more recent V2R subfamilies (subfamilies A1–A6 and subfamily-C2 V2Rs) (Silvotti et al. 2011).

As for V1Rs, the multiple lineage-specific expansion of V2Rs with the concurrent sequence diversification is probably correlated with the detection of different ligands, resulting in a selective advantage. In particular, the sequence analysis of the extracellular N-terminal domain of V2Rs shows a positive selection that is absent in other region of the receptor. This observation strongly suggests that the extracellular region is involved in pheromone binding (Yang et al. 2005).

Evidence indicates that urinary proteins (MUPs), secretory peptides, and "self" molecules (MHC) are some of the V2R ligands (Ishii & Mombaerts 2008; Kimoto et al. 2005; Leinders-Zufall et al. 2009; Loconto et al. 2003; Papes et al. 2010; Sturm et al. 2013). Recently, the lacrimal peptide (ESP1) has been reported to bind and activate V2Rp5, a member of subfamily A3. ESP1 is secreted by the extraorbital gland of the male mouse and induces a specific sexual behavior (lordosis) in females. A family-B V2R member, V2rlb, is probably responsible for the detection of MHC peptides, some of which are quantifiable in urine and are believed to be involved in the genotype discrimination (Haga et al. 2010; Leinders-Zufall et al. 2004, 2009; Sturm et al. 2013). In addition, MUPs have been demonstrated to activate neurons expressing not yet characterized V2Rs. Since MUPs are involved in male-to-male aggressiveness it is likely that V2Rs are also involved in individual and sexual recognition (Chamero et al. 2007; Ishii & Mombaerts 2008; Leinders-Zufall et al. 2009) (Table 10.1). In light of these findings, V2Rs seem to respond to proteinaceous molecules mainly deriving from bodily secretions. In contrast to V1Rs, each V2R appears to be tuned to recognized one or very few pheromonal molecules (Isogai et al. 2011).

In summary, V1Rs and V2Rs are extremely varied and expanded in different lineages of terrestrial vertebrates, and, in particular, they show a rapid turnover in rodent species (Grus et al. 2005; Young et al. 2010). Interestingly, the appearance of the VNO does not seem to correlate with the appearance of V1R and V2R genes (Yang et al. 2005) (Figure 10.2a,d). Thus, it is possible that V1Rs and V2Rs expression was initially established in other organs like, for example, the main olfactory system where these receptors are indeed reportedly expressed (Pfister et al. 2007; Rodriguez & Mombaerts 2002; Syed et al. 2013).

The ancient evolutionary origin of V1Rs and V2Rs is supported by the recent identification of genes encoding these receptors in zebra fish (Hashiguchi et al. 2008; Grus & Zhang 2009; Saraiva & Korsching 2007). The tetrapod ancestor probably possessed a very small repertoire of V1Rs and V2Rs (probably 3–4 genes). V2Rs underwent a large expansion in amphibians reptiles (snakes and lizards) and in some mammalian orders such as Didelphimorpha (opossum), and Rodentia (Brykczynska et al. 2013; Dong et al. 2012; Ishii & Mombaerts 2011; Ji et al. 2009; Nei et al. 2008; Shi & Zhang 2007; Young & Trask 2007). In contrast, V1Rs scarcely duplicated in amphibians (21 genes) and reptiles (four genes). Interestingly, both V1Rs and V2Rs significantly expanded in family Muridae and reduced (V1Rs) or disappeared (no intact genes) (V2Rs) in Laurasiatheria. Thus, the rapid evolution of both V1R and V2R superfamilies in some genome lineages is interchanged with the loss of these receptors in some other species, an observation confirmed by the high number of pseudogenes (Figure 10.2a, b, and d).

The ratio between the number of V1Rs and V2Rs in vertebrates may also address some questions about the functions of these receptors. In fact, the evolutionary transition of early tetrapods from water to land corresponds to the positive inversion of the ratio between V1R and V2R genes.

This would seem to suggest the prevalence of a chemical communication mediated by airborne pheromones in mammals in terricolous species (Shi & Zhang 2007).

10.2.3 FPRs

FPRs belong to the seven-transmembrane domain G protein coupled receptor superfamily and are present in tissues and cell lines involved in the immune response as lymphocytes, monocytes, macrophages, and microglia of most mammalian genera (Le et al. 2002; Migeotte et al. 2006); however, recent findings indicate that a murine subclass of FPRs are specifically expressed in the VNO (Liberles et al. 2009; Riviere et al. 2009) (Table 10.2).

The phylogenetic tree of FPRs suggests a very complex distribution of these receptors. FPRs represent a monophyletic group divided into three clades, FPR1, FPRL1, and FPRL2 (Yang & Shi 2010).

FPR1s are phylogenetically the most ancient as a related gene is present in the lamprey (Petromyzontiformes) genome (Figure 10.2e). In placental mammals FPR1 orthologs are present in many species irrespective of the development of a functional VNO. Interestingly, Laurasiatheria, which include Carnivora (cat and dog), Perissodactyla (horse), Artiodactyla (cow, pig, and sheep), and Chiroptera (bat), have probably lost the fpr1 genes (and pseudogenes) (Figure 10.2e). Didelphimorpha (opossum) possess fpr-like genes with 40% identity with respect to placental mammals.

The second FPR clade, FPRL1, which is phylogenetically more recent, includes receptors already present in Laurasiatheria that forms a lineage-specific gene cluster that was erroneously regarded as VNO specific, since family Muridae predominantly express these receptors in the VNO.

The third FPR clade, FPRL2, which presumably arose from the duplication of the FPR1 gene, appears to be expanded in primates, forming a lineage-specific clade with a homology of 85%–99% between species (Figure 10.2e).

In mice, FPRL1s represent a rodent specific branch that is characterized by seven protein coding genes located on chromosome 17: *Fpr-rs1*, *Fpr-rs2*, *Fpr-rs3*, *Fpr-rs4*, *Fpr-rs6*, and *Fpr-rs7* (*Fpr-rs5* is a pseudogene). These genes show an identity that ranges from 67% to 96%, and with the exclusion of *Fpr-rs2*, they appear to be exclusively expressed in the VNO. In rat, a similar expansion occurred with the presence of seven functional genes that show a one-to-one ortology with the mouse genes. Moreover, VNO-FPRs appear to be already established in other rodent species of different genera as rat and gerbil (Liberles et al. 2009; Riviere et al. 2009). Thus, it is likely that VNO-FPRs were already present in a Muridae ancestor (Figure 10.2a,e).

FPR-rs3, FPR-rs4, FPR-rs6, and FPR-rs7 are exclusively or predominantly expressed in the apical neurons and coexpressed with $G\alpha_{i2}$. In contrast, FPR-rs1 is located more basally and is coexpressed with $G\alpha_o$ (Liberles et al. 2009). The FPR expression pattern appears punctuate and therefore similar to that observed for V1Rs and for most members of V2Rs, thus suggesting a monogenic transcription for these

TABLE 10.2
Expression of FPRs in Humans and Mice

Species	Receptor	Cell Expression	Tissue Expression
Mouse	FPR1	Dendritic cells, microglia, neutrophils	Liver, lung, spleen
	FPR-rs1 (LXA4R)	Leucocytes, neutrophils, VNO neurons	Heart, liver, spleen, VNO
	FPR-rs2 (FPR2)	Dendritic cells, microglia, neutrophils	Lung, spleen
	FPR-rs3	Skeletal muscle, VNO neurons	Cerebellum, kidney, intestine, lung, muscle, spleen, thymus, mouse, rat, and gerbil VNO
	FPR-rs4	VNO neurons	VNO
	FPR-rs6	VNO neurons	Brain, testis, skeletal muscle, spleen, VNO
	FPR-rs7	VNO neurons	Heart, liver, lung, pancreas, smooth muscle, spleen, VNO
Human	FPR	Astrocytes, endotelial cells, fibroblasts, hepatocytes, immature dendritic cells, microglia, monocytes/macrophages, neutrophils, vascular smooth muscle cells	Adrenal, autonomic nervous system, bone marrow, central nervous system, colon, eye, heart, kidney, liver, lung, lung alveolar carcinoma cells, neuroblastoma, ovary, placenta, stomach, spleen, thyroid
	FPRL1 (LXA4R)	Astrocytes, endotelial cells, epitelial cells, fibroblasts, hepatocytes, immature dendritic cells, microglia, monocytes/macrophages, T cells/B cells	Bone marrow, brain, lung, neuroblastoma, placenta, spleen, testis
	FPRL2	Dendritic cells, monocytes/macrophages, neutrophils	Adrenal, lymph nodes, liver, lung, placenta, small intestine, spleen, trachea

Source: Liberles, S.D., Horowitz, L.F., Kuang, D., Contos, J.J., Wilson, K.L., Siltberg-Liberles, J., Liberles, D.A. & Buck, L.B. (2009) *Proc Natl Acad Sci U S A*, **106**, 9842–9847; Migeotte, I., Communi, D. & Parmentier, M. (2006) *Cytokine Growth Factor Rev*, **17**, 501–519; Riviere, S., Challet, L., Fluegge, D., Spehr, M. & Rodriguez, I. (2009) *Nature*, **459**, 574–577.

Note: Entries in bold represent mouse VNO-FPRs. VNO = vomeronasal organ.

genes (Dulac & Axel 1995; Herrada & Dulac 1997; Ryba & Tirindelli 1997). Since there is no evidence of a coexpression of FRPs with V1Rs or V2Rs, it is likely that specific subsets of VNO neurons exclusively express FPRs (Liberles et al. 2009; Riviere et al. 2009; Yang & Shi 2010). The functional role of both non-VNO and VNO-FPRs is still unclear although a conspicuous number of FPRs agonists and antagonists has been identified (Seifert & Wenzel-Seifert 2003; Thoren et al. 2010).

Human and mouse FPRs are activated by viral and bacterial products (Bae et al. 2003a,b). The most effective ligands of almost all FPRs is the bacterial *N*-formyl-methionyl-leucyl-phenylalanine (fMLF) that triggers several biological immunological functions in neutrophils, monocytes, and macrophages (Boulay et al. 1990; Cevik-Aras et al. 2012; Le et al. 2002). In particular, the lack of FPR1 gene causes an increase of *Listeria monocytogenes* sensitivity in mouse (Gao et al. 1999).

As for VNO-FPRs, FPR-rs1 was reportedly identified to be specific to lipoxin A4 (it was in fact named as LXA4R9), a lipid mediator generated at the site of inflammation that both regulates the transport of intermediates between platelets, neutrophils, and natural killer cells and lymphocytes activation in the chronic inflammations (Migeotte et al. 2006; Takano et al. 1997). In contrast, cloned FPR-rs1 and 3 predominantly respond to the cathelin-related antimicrobial peptide (CRAMP) that acts as chemotactic factor *in vivo*. FPR-rs6 senses fMLF whereas FPR-rs4 and 7 detect both ligands. Moreover, FPR-rs4 also responds to lipoxin A4 (Riviere et al. 2009) (Table 10.1).

Similarly to all other FPRs, each VNO-FPR probably responds to a large array of ligands, most of which have not yet been characterized. Moreover, VNO-FPRs responding to ligands do not show particular selectivity for these receptors, being effective at a similar concentration on several FPR subtypes of different species. This observation appears to contrast with the most recent phylogenetic studies on FPR genes that aim at concluding that the selective pressure exclusively acting on VNO-FPRs rather than all other FPRs is possibly linked to the functional transition from host defense (FPRL1) to vomeronasal chemosensation (VNO-FPRs) (Liberles et al. 2009).

In light of these findings it appears very difficult to define a pheromonal role for these receptors. One possibility is that VNO-FPRs may exploit the specific function of detecting compounds specifically synthesized by the bacterial flora inhabiting the secretions of conspecific or heterospecific animals, thus allowing individual identification (Riviere et al. 2009).

As a summary of vomeronasal receptors, it is possible to envisage that V1Rs recognize volatile molecules also produced by sympatric species whereas V2Rs respond to excretory proteins and peptides exclusively expressed by individuals of the same species. In contrast, FPRs sense microbial peptides that, as suggested above, may also be indirectly involved in both species or individual recognition (Del Punta et al. 2002; Chamero et al. 2007; Isogai et al. 2011; Kimoto et al. 2005; Leinders-Zufall et al. 2000, 2004; Liberles et al. 2009; Riviere et al. 2009).

10.3   SIGNAL TRANSDUCTION MECHANISMS

Binding of pheromones to vomeronasal receptors initiates a transduction cascade that produces an excitatory response. Pheromonal stimuli cause membrane depolarization and increase the action potential firing rate and the intracellular calcium concentration in VSNs (Celsi et al. 2012; Holy et al. 2000; Inamura & Kashiwayanagi 2000; Leinders-Zufall et al. 2000; Nodari et al. 2008; Shimazaki et al. 2006; Spehr et al. 2002). Despite several similarities in response to stimuli are shared between vomeronasal and olfactory sensory neurons, the transduction mechanisms used by these chemosensory cells are distinct (Berghard & Buck 1996; Wu et al. 1996).

10.3.1   G PROTEINS

As previously described, the two subsets of apical and basal VSNs expressing V1Rs or V2Rs and FPRs also express different G proteins (Figure 10.3). Berghard and Buck identified two types of G protein α subunits in basal, $G\alpha_o$, and apical, $G\alpha_{i2}$, mouse VSNs (Berghard & Buck 1996). Ultrastructural studies showed that both $G\alpha_o$ and $G\alpha_{i2}$ are located in the microvilli where the vomeronasal transduction occurs (Menco et al. 2001). The activation of GPRCs causes the release of the $\beta\gamma$ complex of the heterotrimeric G protein G_o and G_{i2} responsible for phospholipase C (PLC) activation,

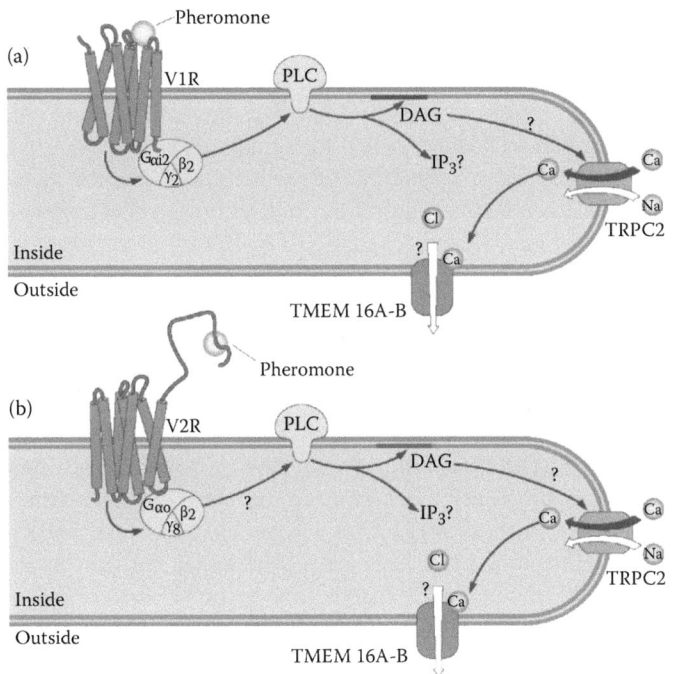

FIGURE 10.3   Signaling molecules involved in the transduction of pheromonal stimuli. (a) The binding of pheromones to V1R receptors in the microvilli activates PLC through the G protein $Go\beta_2\gamma_8$. PLC produces the second messengers DAG and IP_3. The gating of the TRPC2 channel by DAG allows a depolarizing influx of Na^+ and Ca^{2+}. TRPC2 appears to be the principal transduction channel in V1R-expressing neurons. Intracellular calcium can gate a Ca^{2+}-activated Cl^- current mediated by TMEM16A and TMEM16B. The direction of the flux of chloride ion is still unknown. (b) The binding of pheromones to V2R receptors in the microvilli activates PLC through $G\alpha_{i2}\beta_2\gamma_2$ producing DAG and IP3. In these VSN the role of TRPC2 is less clear and it has been proposed that other cation channels can be activated by PLC metabolites. Ca^{2+}-activated Cl^- channels can contribute to either the depolarization or termination of the response. (Modified from Tirindelli, R. et al. (2009) *Physiol Rev*, 89, 921–956.)

indicating that β and γ subunits play a pivotal role in vomeronasal sensory transduction (Clapham & Neer 1997). Runnenburger and colleagues reported that one type of β subunit, β_2, is expressed in the vomeronasal sensory epithelium and that an antibody against β_2 caused a reduction of IP_3 production in microvilli extracts stimulated with urine (Runnenburger et al. 2002). Moreover, two γ subunits, γ_2 and γ_8, are expressed in VSNs (Runnenburger et al. 2002; Ryba & Tirindelli 1995; Tirindelli & Ryba 1996). Interestingly, the expression of the γ_2 subunit is confined to apical neurons, whereas the γ_8 subunit is preferentially expressed in the basal neuronal population. Treatment of microvilli extract with specific antibodies against γ_2 and γ_8 reduced the production of IP_3 upon urine stimulation, suggesting that both subunits play a role in sensory transduction (Runnenburger et al. 2002).

The functional role of G proteins expressed in the VNO has been investigated in knockout mice. $G\alpha_{i2}$-deficient animals produced a severe postnatal neurodegeneration of apical VSNs, causing the death of about 50% of VSNs and a consequent reduction of about 50% of the glomerular layer of the anterior AOB (Norlin et al. 2003). Using c-Fos as a marker for neuronal activity, it has been reported that the lack of $G\alpha_{i2}$ caused a fivefold reduction of c-Fos expression in the anterior AOB of male mice exposed to bedding soiled by female mice. Behavioral studies have revealed that male mice display normal sexual behaviors whereas females present a strong reduction of maternal aggression, indicating that $G\alpha_{i2}$-expressing VSNs mediated these processes (Norlin et al. 2003). Direct measurements of responses to stimuli of VSNs lacking $G\alpha_{i2}$ would help clarify its role in sensory transduction, but at present they are not available.

The role of G proteins in survival of VSNs was confirmed also in conditional and general knockout mice for $G\alpha_o$: a severe postnatal neurodegeneration of basal VSNs caused the death of about 50% VSNs (Chamero et al. 2011; Tanaka et al. 1999). Chamero and colleagues (2011) measured the responses of VSNs with electrovomeronasogram (EVG) recordings and Ca^{2+} imaging assay. These authors reported that VSNs from $G\alpha_o$ knockout mice did not respond to stimulation with known V2R ligands, such as MHC class I peptides, ESPs, and MUPs, and with FPR-rs1 agonists. In contrast, $G\alpha_o$-deficient VSNs present normal responses to V1R ligands such as 2-heptanone and isobutylamine and to FPR-rs2-5 (Chamero et al. 2011). Furthermore, behavioral studies showed a decreased male-to-male aggressiveness mediated by MHC class I peptides and MUPs, and surprisingly, a reduction of maternal aggression, a behavior that is also mediated by V2R-$G\alpha_{i2}$-expressing neurons (Chamero et al. 2011).

10.3.2 Phospholipase C

While the molecular identity of the phospholipase C (PLC) subtypes expressed in the VNO is not identified, functional studies have indicated that some responses of VSNs to stimuli are mediated by PLC activation. Holy and colleagues (2000) measured the electrical activity of mouse VSNs in response to urine using multielectrode array recordings and showed that the PLC inhibitor U73122 blocked spiking responses to urine, whereas the inactive structural analogous U73343 did not produce a measurable effect. Similar results were obtained by Spehr and colleagues (2002) monitoring the activity of rat VSNs upon urine stimulation using Ca^{2+}-imaging. PLC hydrolizes PIP_2 into IP_3 and diacylglycerol (DAG) (Figure 10.3). Additional evidence of the involvement of PLC in the transduction cascade comes from biochemical studies showing that stimulation of VSNs with urine induced the production of IP_3 in several species (Krieger et al. 1999; Kroner et al. 1996; Sasaki et al. 1999; Thompson et al. 2004). However, the role of IP_3 in signal transduction is not clear. Indeed, urine-stimulated Ca^{2+} increase in rat VSNs was not affected by the block of IP_3 receptors (Spehr et al. 2002), although the presence of IP_3 in the patch pipette induced a nonselective cation current in about 50% of the recorded rat VSNs (Inamura et al. 1997).

10.3.3 ION CHANNELS

10.3.3.1 TRPC2

Liman and colleagues (1999) first showed that the transient receptor potential canonical 2 (TRPC2) is expressed in VSNs. TRPC2 localizes in the microvillar region of the VNO (Dibattista et al. 2012; Liman et al. 1999; Menco et al. 2001; Stowers et al. 2002). This molecule belongs to the TRPC subfamily, a class of cation channels that are characterized by PLC-dependent Ca^{2+} permeability. Drosophila homologues of mammalian TRPC, TRP and TRPL, are activated during the PLC-mediated phototransduction cascade (Yau & Hardie 2009). In Old World monkeys and in humans the TRPC2 gene has pseudogenized along with the loss of a functional VNO (Liman & Innan 2003; Zhang & Webb 2003), suggesting that the evolution of the trichromatic vision releases the selective pressure of signaling mediated by pheromones (Liman & Innan 2003; Zhang & Webb 2003).

The involvement of TRPC2 in vomeronasal sensory transduction has been clearly demonstrated using TRPC2 knockout mice. The lack of TRPC2 caused a strong postnatal degeneration of VSNs, with the loss of about 75% V2R-expressing neurons and 50% V1R-expressing neurons at 2 months of age (Stowers et al. 2002). This phenotype mirrors the effects of the deletion of the G protein α subunits (Chamero et al. 2011; Norlin et al. 2003; Tanaka et al. 1999), confirming the importance of a functional transduction machinery for survival of VSNs.

Functional properties of VSNs in TRPC2 knockout mice showed discrepancies among different studies (Kelliher et al. 2006; Kim et al. 2011; Leypold et al. 2002; Stowers et al. 2002). Some authors reported a strong reduction of the response to urine stimulation measured with EVG recordings (Leypold et al. 2002), whereas others showed the complete loss of neuronal activation by urine using multielectrode array (Stowers et al. 2002). Moreover, Kelliher and colleagues (2006) found that the EVG response to MHC class I peptides as well as the Bruce effect (pregnancy failure induced by exposure to an unrelated male (Bruce 1959)) that is mediated by this class of molecules was unaffected in TRPC2-deficient mice, indicating that TRPC2 is not involved in the mediation of some V2R-dependent processes (see Section 10.2.2). TRPC2-deficient males also show a strong decrease in the male-to-male and maternal aggressive (towards males) behavior, which can be explained as the inability of these mice to discriminate the gender (Leypold et al. 2002; Stowers et al. 2002).

Some studies indicate that TRPC2 activation involved the PLC-mediated DAG production. Indeed, experiments using inside-out excised patches from dendritic knob of mouse VSNs showed the presence of a nonselective cation current activated by DAG analogues and absent in TRPC2 deficient VSNs (Lucas et al. 2003). Moreover, whole-cell patch-clamp recordings have revealed that urine was able to induce a similar nonselective cation current blocked by the TRP channel blocker, A2B (Lucas et al. 2003). Finally, the pharmacological block of DAG degradation through DAG kinase induced the activation of TRPC2 current by PLC basal activity (Lucas et al. 2003). At present, it is unclear whether DAG directly or indirectly activates TRPC2. It is noteworthy that in Drosophila, TRP/TRPL channels are gated by protons produced by the hydrolysis of phosphoinositides to DAG (Huang et al. 2010).

10.3.3.2 Ca^{2+}-Activated Cl^- Channels

Recent studies indicate that Ca^{2+}-activated Cl^- channels are expressed in VSNs (Dibattista et al. 2012; Kim et al. 2011; Pifferi et al. 2012; Yang & Delay 2010). Yang and Delay (2010) reported that 80% of the urine-activated current, measured with perforated patch recording on dissociated mouse VSNs, is carried by Ca^{2+}-activated Cl^- channels and that a Ca^{2+} influx is necessary to activate the Cl^- channels.

Kim and colleagues (2011) confirmed that a Ca^{2+}-activated Cl^- current contributes to responses to urine, but they also suggested that this current can be activated both by Ca^{2+} influx through the TRPC2 channel and by Ca^{2+} release from internal stores mediated by IP_3 in a TRPC2-independent manner. Interestingly, Kim and colleagues (2011) obtained extracellular recordings in VNO slices and showed that both spontaneous and urine-evoked firing activities were abolished by NFA and SITS, two commonly used blockers of Ca^{2+}-activated Cl^- channels, indicating that the activation of these channels is required for neuronal activity. The presence of Ca^{2+}-activated Cl^- channels in the apical portion of VSNs was confirmed by local photorelease of Ca^{2+} from caged Ca^{2+} (Dibattista et al. 2012).

RT-PCR experiments showed that some members of the TMEM16/anoctamin family are expressed in the VNO (Billig et al. 2011; Dibattista et al. 2012; Yang & Delay 2010). TMEM16A/anoctamin1 and TMEM16B/anoctamin2 have been shown to be Ca^{2+}-activated Cl^- channels (Caputo et al. 2008; Pifferi et al. 2009; Schroeder et al. 2008; Scudieri et al. 2012; Stephan et al. 2009; Stöhr et al. 2009; Yang et al. 2008). Immunohistological data has established that both TMEM16A and TMEM16B are expressed in the apical region of the VNO (Billig et al. 2011; Dauner et al. 2012; Dibattista et al. 2012; Rasche et al. 2010). Dibattista et al. (2012) demonstrated that TMEM16A and TMEM16B largely colocalize with TRPC2 at the apical surface of the vomeronasal epithelium, and that TMEM16A and TMEM16B are coexpressed in microvilli of both apical and basal isolated vomeronasal sensory neurons, therefore suggesting that these two anion channels are likely to be involved in vomeronasal transduction (Figure 10.3). However, as an estimate of the Cl^- concentration in fluid filling the VNO lumen and inside VSNs is not currently available, the direction of the flux of Cl^- ions cannot be determined. Thus, depending on the Cl^- equilibrium potential, Ca^{2+}-activated Cl^- channels may contribute to VSNs depolarization or hyperpolarization.

10.3.3.3 Other Ion Channels

Several other ion channels are expressed in VSNs and have been proposed to be involved in signal transduction.

Kim and colleagues (2012) reported that calcium-activated and G-protein-activated inwardly rectifying potassium channels SK3 and GIRK1 contribute to the urine-mediated response in mouse VSNs. Interestingly, the depolarizing effect of these channels *in vivo* was due to an unusual high K^+ concentration (about 66 mM) in the fluid filling the lumen of VNO, resembling the well-known high K^+ concentration in the cochlear endolymph. Moreover, a decreased aggressiveness and an impairment of sociosexual behavior is associated with the deletion of SK3 and GIRK1genes (Kim et al. 2012).

Zhang and colleagues (2008) reported that the blockage of a large-conductance calcium-activated potassium channel (BK) by iberiotoxin (Ibx) induced an increase of the transduction current amplitude and of the firing rates in mouse VSNs stimulated with urine, suggesting that BK channels are also involved in vomeronasal signal transduction (Zhang et al. 2008). This data could be interpreted in light of the recent finding about the high extracellular K^+ concentration in the VNO lumen, suggesting an inhibitory role of BK channels.

An interesting open question is the possible role of arachidonic acid (AA) in sensory transduction. Spehr and colleagues (2002) first showed that AA was able to produce an increase in intracellular Ca^{2+} concentration in rat VSNs. Zhang and colleagues (2010) reported that AA was able to activate a Ca^{2+}-dependent Ca^{2+} permeable channel in inside-out patches excised from dendritic knob of mouse VSNs that is independent on TRPC2. Moreover, pharmacological blockage of DAG lipase, an enzyme necessary to the production of AA upon PLC activation, induced a reduction of urine-induced current (Zhang et al. 2010), although another report failed to detect any effect of DAG lipase inhibition (Lucas et al. 2003).

Another Ca^{2+}-dependent conductance was studied by Liman (2003) in the knob/microvilli of hamster VNSs. This author reported the presence of a Ca^{2+}-activated nonselective cation channel with a single-channel conductance of about 25 pS and with an EC_{50} for Ca^{2+} of 0.51 mM (at −80 mV). Moreover, the activity of this channel was inhibited by cAMP with an IC_{50} of 43 µM (at −80 mV). It has been suggested that TRPM4 is responsible for this conductance but a definitive proof is still missing (Liman 2003). Interestingly, Spehr and colleagues (2009) reported that a Ca^{2+}-activated nonselective cation channel was also present in the knob/microvilli of mouse VSNs with sensitivity to Ca^{2+} in micromolar range. In contrast to these results, the local photo-release of intracellular Ca^{2+} from caged compounds did not elicit any cation current in mouse VSNs (Dibattista et al. 2012). It is possible that these discrepancies could be partially due to species-specificity or to the presence of different populations of VSNs using different transduction mechanisms.

The involvement of cAMP or cGMP in the mediation of pheromonal sensation is still under debate. It has been reported that some pheromones induce a reduction of cAMP concentration (Rossler et al. 2000; Zhou & Moss 1997). However, the decrease in cAMP has a slow kinetics compared with sensory stimulation, suggesting that cAMP could have a long-lasting effect in modulation of VSNs responses.

Some components of the cAMP signaling cascade have been identified in VSNs as adenylyl cyclase types 2, 4, 5, and 6 (Berghard & Buck 1996; Lee et al. 2008; Rossler et al. 2000), the subunit CNGA4 of the olfactory cyclic nucleotide-gated channel (Berghard & Buck 1996; Lee et al. 2008), and phosphodiesterase types 1 and 4 (Cherry & Pho 2002; Lau & Cherry 2000). Possible targets for modulatory actions by cAMP include the Ca^{2+}-activated cation channel (Liman & Innan 2003) and the hyperpolarization-activated cyclic nucleotide-gated (HCN) cation channels that contribute to setting the resting membrane potential and to increasing excitability at stimulus threshold (Dibattista et al. 2012).

REFERENCES

Bae, Y.S., Song, J.Y., Kim, Y., He, R., Ye, R.D., Kwak, J.Y., Suh, P.G. & Ryu, S.H. (2003a) Differential activation of formyl peptide receptor signaling by peptide ligands. *Mol Pharmacol*, **64**, 841–847.

Bae, Y.S., Yi, H.J., Lee, H.Y., Jo, E.J., Kim, J.I., Lee, T.G., Ye, R.D., Kwak, J.Y. & Ryu, S.H. (2003b) Differential activation of formyl peptide receptor-like 1 by peptide ligands. *J Immunol*, **171**, 6807–6813.

Baum, M.J. & Kelliher, K.R. (2009) Complementary roles of the main and accessory olfactory systems in mammalian mate recognition. *Annu Rev Physiol*, **71**, 141–160.

Belluscio, L., Koentges, G., Axel, R. & Dulac, C. (1999) A map of pheromone receptor activation in the mammalian brain. *Cell*, **97**, 209–220.

Berghard, A. & Buck, L.B. (1996) Sensory transduction in vomeronasal neurons: Evidence for G alpha o, G alpha i2, and adenylyl cyclase II as major components of a pheromone signaling cascade. *J Neurosci*, **16**, 909–918.

Billig, G.M., Pal, B., Fidzinski, P. & Jentsch, T.J. (2011) Ca^{2+}-activated Cl^- currents are dispensable for olfaction. *Nat Neurosci*, **14**, 763–769.

Boschat, C., Pelofi, C., Randin, O., Roppolo, D., Luscher, C., Broillet, M.C. & Rodriguez, I. (2002) Pheromone detection mediated by a V1r vomeronasal receptor. *Nat Neurosci*, **5**, 1261–1262.

Boulay, F., Tardif, M., Brouchon, L. & Vignais, P. (1990) Synthesis and use of a novel N-formyl peptide derivative to isolate a human N-formyl peptide receptor cDNA. *Biochem Biophys Res Commun*, **168**, 1103–1109.

Bozza, T., Feinstein, P., Zheng, C. & Mombaerts, P. (2002) Odorant receptor expression defines functional units in the mouse olfactory system. *J Neurosci*, **22**, 3033–3043.

Bruce, H.M. (1959) An exteroceptive block to pregnancy in the mouse. *Nature*, **184**, 105.

Brykczynska, U., Tzika, A.C., Rodriguez, I. & Milinkovitch, M.C. (2013) Contrasted evolution of the vomeronasal receptor repertoires in mammals and squamate reptiles. *Genome Biol Evol*, **5**, 389–401.

Caputo, A., Caci, E., Ferrera, L., Pedemonte, N., Barsanti, C., Sondo, E., Pfeffer, U., Ravazzolo, R., Zegarra-Moran, O. & Galietta, L.J.V. (2008) TMEM16A, a membrane protein associated with calcium-dependent chloride channel activity. *Science*, **322**, 590–594.

Celsi, F., D'Errico, A. & Menini, A. (2012) Responses to sulfated steroids of female mouse vomeronasal sensory neurons. *Chem Senses*, **37**, 849–858.

Cevik-Aras, H., Kalderen, C., Jenmalm Jensen, A., Oprea, T., Dahlgren, C. & Forsman, H. (2012) A non-peptide receptor inhibitor with selectivity for one of the neutrophil formyl peptide receptors, FPR 1. *Biochem Pharmacol*, **83**, 1655–1662.

Chamero, P., Katsoulidou, V., Hendrix, P., Bufe, B., Roberts, R., Matsunami, H., Abramowitz, J., Birnbaumer, L., Zufall, F. & Leinders-Zufall, T. (2011) G protein G(alpha)o is essential for vomeronasal function and aggressive behavior in mice. *Proceedings of the National Academy of Sciences of the United States of America*, **108**, 12898–12903.

Chamero, P., Marton, T.F., Logan, D.W., Flanagan, K., Cruz, J.R., Saghatelian, A., Cravatt, B.F. & Stowers, L. (2007) Identification of protein pheromones that promote aggressive behaviour. *Nature*, **450**, 899–902.

Cherry, J.A. & Pho, V. (2002) Characterization of cAMP degradation by phosphodiesterases in the accessory olfactory system. *Chem Senses*, **27**, 643–652.

Churakov, G., Sadasivuni, M.K., Rosenbloom, K.R., Huchon, D., Brosius, J. & Schmitz, J. (2010) Rodent evolution: Back to the root. *Mol Biol Evol*, **27**, 1315–1326.

Clapham, D.E. & Neer, E.J. (1997) G protein beta gamma subunits. *Annu Rev Pharmacol Toxicol*, **37**, 167–203.

Date-Ito, A., Ohara, H., Ichikawa, M., Mori, Y. & Hagino-Yamagishi, K. (2008) Xenopus V1R vomeronasal receptor family is expressed in the main olfactory system. *Chem Senses*, **33**, 339–346.

Dauner, K., Lissmann, J., Jeridi, S., Frings, S. & Mohrlen, F. (2012) Expression patterns of anoctamin 1 and anoctamin 2 chloride channels in the mammalian nose. *Cell Tissue Res*, **347**, 327–341.

Del Punta, K., Leinders-Zufall, T., Rodriguez, I., Jukam, D., Wysocki, C.J., Ogawa, S., Zufall, F. & Mombaerts, P. (2002) Deficient pheromone responses in mice lacking a cluster of vomeronasal receptor genes. *Nature*, **419**, 70–74.

Del Punta, K., Rothman, A., Rodriguez, I. & Mombaerts, P. (2000) Sequence diversity and genomic organization of vomeronasal receptor genes in the mouse. *Genome Res*, **10**, 1958–1967.

Dennis, J.C., Allgier, J.G., Desouza, L.S., Eward, W.C. & Morrison, E.E. (2003) Immunohistochemistry of the canine vomeronasal organ. *J Anat*, **202**, 515–524.

Dennis, J.C., Smith, T.D., Bhatnagar, K.P., Bonar, C.J., Burrows, A.M. & Morrison, E.E. (2004) Expression of neuron-specific markers by the vomeronasal neuroepithelium in six species of primates. *Anat Rec A Discov Mol Cell Evol Biol*, **281**, 1190–1200.

Dibattista, M., Amjad, A., Maurya, D.K., Sagheddu, C., Montani, G., Tirindelli, R. & Menini, A. (2012) Calcium-activated chloride channels in the apical region of mouse vomeronasal sensory neurons. *J Gen Physiol*, **140**, 3–15.

Dong, D., Jin, K., Wu, X. & Zhong, Y. (2012) CRDB: Database of chemosensory receptor gene families in vertebrate. *PLoS One*, **7**, e31540.

Dulac, C. & Axel, R. (1995) A novel family of genes encoding putative pheromone receptors in mammals. *Cell*, **83**, 195–206.

Dulac, C. & Torello, A.T. (2003) Molecular detection of pheromone signals in mammals: From genes to behaviour. *Nat Rev Neurosci*, **4**, 551–562.

Emes, R.D., Beatson, S.A., Ponting, C.P. & Goodstadt, L. (2004) Evolution and comparative genomics of odorant- and pheromone-associated genes in rodents. *Genome Res*, **14**, 591–602.

Gao, J.L., Lee, E.J. & Murphy, P.M. (1999) Impaired antibacterial host defense in mice lacking the N-formylpeptide receptor. *J Exp Med*, **189**, 657–662.

Grus, W.E. & Zhang, J. (2004) Rapid turnover and species-specificity of vomeronasal pheromone receptor genes in mice and rats. *Gene*, **340**, 303–312.

Grus, W.E. & Zhang, J. (2009) Origin of the genetic components of the vomeronasal system in the common ancestor of all extant vertebrates. *Mol Biol Evol*, **26**, 407–419.

Grus, W.E., Shi, P. & Zhang, J. (2007) Largest vertebrate vomeronasal type 1 receptor gene repertoire in the semiaquatic platypus. *Mol Biol Evol*, **24**, 2153–2157.

Grus, W.E., Shi, P., Zhang, Y.P. & Zhang, J. (2005) Dramatic variation of the vomeronasal pheromone receptor gene repertoire among five orders of placental and marsupial mammals. *Proc Natl Acad Sci U S A*, **102**, 5767–5772.

Haga, S., Hattori, T., Sato, T., Sato, K., Matsuda, S., Kobayakawa, R., Sakano, H., Yoshihara, Y., Kikusui, T. & Touhara, K. (2010) The male mouse pheromone ESP1 enhances female sexual receptive behaviour through a specific vomeronasal receptor. *Nature*, **466**, 118–122.

Hashiguchi, Y., Furuta, Y. & Nishida, M. (2008) Evolutionary patterns and selective pressures of odorant/pheromone receptor gene families in teleost fishes. *PloS One*, **3**, e4083.

Herrada, G. & Dulac, C. (1997) A novel family of putative pheromone receptors in mammals with a topographically organized and sexually dimorphic distribution. *Cell*, **90**, 763–773.

Hohenbrink, P., Mundy, N.I., Zimmermann, E. & Radespiel, U. (2013) First evidence for functional vomeronasal 2 receptor genes in primates. *Biol Lett*, **9**, 20121006.

Hohenbrink, P., Radespiel, U. & Mundy, N.I. (2012) Pervasive and ongoing positive selection in the vomeronasal-1 receptor (V1R) repertoire of mouse lemurs. *Mol Biol Evol*, **29**, 3807–3816.

Holy, T.E., Dulac, C. & Meister, M. (2000) Responses of vomeronasal neurons to natural stimuli. *Science*, **289**, 1569–1572.

Huang, J., Liu, C.-H., Hughes, S.A., Postma, M., Schwiening, C.J. & Hardie, R.C. (2010) Activation of TRP channels by protons and phosphoinositide depletion in Drosophila photoreceptors. *Curr Biol.*, **20**, 189–197.

Huchon, D., Chevret, P., Jordan, U., Kilpatrick, C.W., Ranwez, V., Jenkins, P.D., Brosius, J. & Schmitz, J. (2007) Multiple molecular evidences for a living mammalian fossil. *Proc Natl Acad Sci U S A*, **104**, 7495–7499.

Inamura, K. & Kashiwayanagi, M. (2000) Inward current responses to urinary substances in rat vomeronasal sensory neurons. *Eur J Neurosci*, **12**, 3529–3536.

Inamura, K., Kashiwayanagi, M. & Kurihara, K. (1997) Inositol-1,4,5-trisphosphate induces responses in receptor neurons in rat vomeronasal sensory slices. *Chem Senses*, **22**, 93–103.

Ishii, T. & Mombaerts, P. (2008) Expression of nonclassical class I major histocompatibility genes defines a tripartite organization of the mouse vomeronasal system. *J Neurosci*, **28**, 2332–2341.

Ishii, T. & Mombaerts, P. (2011) Coordinated coexpression of two vomeronasal receptor V2R genes per neuron in the mouse. *Mol Cell Neurosci*, **46**, 397–408.

Isogai, Y., Si, S., Pont-Lezica, L., Tan, T., Kapoor, V., Murthy, V.N. & Dulac, C. (2011) Molecular organization of vomeronasal chemoreception. *Nature*, **478**, 241–245.

Jemiolo, B., Andreolini, F., Xie, T.M., Wiesler, D. & Novotny, M. (1989) Puberty-affecting synthetic analogs of urinary chemosignals in the house mouse, *Mus domesticus*. *Physiol Behav*, **46**, 293–298.

Ji, Y., Zhang, Z. & Hu, Y. (2009) The repertoire of G-protein-coupled receptors in *Xenopus tropicalis*. *BMC Genomics*, **10**, 263.

Kelliher, K.R., Spehr, M., Li, X.H., Zufall, F. & Leinders-Zufall, T. (2006) Pheromonal recognition memory induced by TRPC2-independent vomeronasal sensing. *Eur J Neurosci*, **23**, 3385–3390.

Kim, S., Ma, L., Jensen, K.L., Kim, M.M., Bond, C.T., Adelman, J.P. & Yu, C.R. (2012) Paradoxical contribution of SK3 and GIRK channels to the activation of mouse vomeronasal organ. *Nat Neurosci*, **15**, 1236–1244.

Kim, S., Ma, L. & Yu, C.R. (2011) Requirement of calcium-activated chloride channels in the activation of mouse vomeronasal neurons. *Nat Commun*, **2**, 365.

Kimoto, H., Haga, S., Sato, K. & Touhara, K. (2005) Sex-specific peptides from exocrine glands stimulate mouse vomeronasal sensory neurons. *Nature*, **437**, 898–901.

Kimoto, H., Sato, K., Nodari, F., Haga, S., Holy, T.E. & Touhara, K. (2007) Sex- and strain-specific expression and vomeronasal activity of mouse ESP family peptides. *Curr Biol*, **17**, 1879–1884.

Kiyokawa, Y., Kikusui, T., Takeuchi, Y. & Mori, Y. (2007) Removal of the vomeronasal organ blocks the stress-induced hyperthermia response to alarm pheromone in male rats. *Chem Senses*, **32**, 57–64.

Kolunie, J.M. & Stern, J.M. (1995) Maternal aggression in rats: Effects of olfactory bulbectomy, ZnSO4-induced anosmia, and vomeronasal organ removal. *Horm Behav*, **29**, 492–518.

Krieger, J., Schmitt, A., Lobel, D., Gudermann, T., Schultz, G., Breer, H. & Boekhoff, I. (1999) Selective activation of G protein subtypes in the vomeronasal organ upon stimulation with urine-derived compounds. *J Biol Chem*, **274**, 4655–4662.

Kroner, C., Breer, H., Singer, A.G. & O'Connell, R.J. (1996) Pheromone-induced second messenger signaling in the hamster vomeronasal organ. *Neuroreport*, **7**, 2989–2992.

Lau, Y.E. & Cherry, J.A. (2000) Distribution of PDE4A and G(o) alpha immunoreactivity in the accessory olfactory system of the mouse. *Neuroreport*, **11**, 27–32.

Le, Y., Murphy, P.M. & Wang, J.M. (2002) Formyl-peptide receptors revisited. *Trends Immunol*, **23**, 541–548.

Lee, S.J., Mammen, A., Kim, E.J., Kim, S.Y., Park, Y.J., Park, M., Han, H.S., Bae, Y.C., Ronnett, G.V. & Moon, C. (2008) The vomeronasal organ and adjacent glands express components of signaling cascades found in sensory neurons in the main olfactory system. *Mol Cells*, **26**, 503–513.

Leinders-Zufall, T., Brennan, P., Widmayer, P., S, P.C., Maul-Pavicic, A., Jager, M., Li, X.H., Breer, H., Zufall, F. & Boehm, T. (2004) MHC class I peptides as chemosensory signals in the vomeronasal organ. *Science*, **306**, 1033–1037.

Leinders-Zufall, T., Ishii, T., Mombaerts, P., Zufall, F. & Boehm, T. (2009) Structural requirements for the activation of vomeronasal sensory neurons by MHC peptides. *Nat Neurosci*, **12**, 1551–1558.

Leinders-Zufall, T., Lane, A.P., Puche, A.C., Ma, W., Novotny, M.V., Shipley, M.T. & Zufall, F. (2000) Ultrasensitive pheromone detection by mammalian vomeronasal neurons. *Nature*, **405**, 792–796.

Leypold, B.G., Yu, C.R., Leinders-Zufall, T., Kim, M.M., Zufall, F. & Axel, R. (2002) Altered sexual and social behaviors in trp2 mutant mice. *Proc Natl Acad Sci U S A*, **99**, 6376–6381.

Liberles, S.D., Horowitz, L.F., Kuang, D., Contos, J.J., Wilson, K.L., Siltberg-Liberles, J., Liberles, D.A. & Buck, L.B. (2009) Formyl peptide receptors are candidate chemosensory receptors in the vomeronasal organ. *Proc Natl Acad Sci U S A*, **106**, 9842–9847.

Liman, E.R. (2003) Regulation by voltage and adenine nucleotides of a Ca^{2+}-activated cation channel from hamster vomeronasal sensory neurons. *J Physiol*, **548**, 777–787.

Liman, E.R. & Innan, H. (2003) Relaxed selective pressure on an essential component of pheromone transduction in primate evolution. *Proc Natl Acad Sci U S A*, **100**, 3328–3332.

Liman, E.R., Corey, D.P. & Dulac, C. (1999) TRP2: A candidate transduction channel for mammalian pheromone sensory signaling. *Proc Natl Acad Sci U S A*, **96**, 5791–5796.

Loconto, J., Papes, F., Chang, E., Stowers, L., Jones, E.P., Takada, T., Kumanovics, A., Fischer Lindahl, K. & Dulac, C. (2003) Functional expression of murine V2R pheromone receptors involves selective association with the M10 and M1 families of MHC class Ib molecules. *Cell*, **112**, 607–618.

Logan, D.W., Marton, T.F. & Stowers, L. (2008) Species specificity in major urinary proteins by parallel evolution. *PloS One*, **3**, e3280.

Lucas, P., Ukhanov, K., Leinders-Zufall, T. & Zufall, F. (2003) A diacylglycerol-gated cation channel in vomeronasal neuron dendrites is impaired in TRPC2 mutant mice: Mechanism of pheromone transduction. *Neuron*, **40**, 551–561.

Luo, H., Arndt, W., Zhang, Y., Shi, G., Alekseyev, M.A., Tang, J., Hughes, A.L. & Friedman, R. (2012) Phylogenetic analysis of genome rearrangements among five mammalian orders. *Mol Phylogenet Evol*, **65**, 871–882.

Ma, M. (2007) Encoding olfactory signals via multiple chemosensory systems. *Crit Rev Biochem Mol Biol*, **42**, 463–480.

Martinez-Marcos, A. (2009) On the organization of olfactory and vomeronasal cortices. *Prog Neurobiol*, **87**, 21–30.

Martini, S., Silvotti, L., Shirazi, A., Ryba, N.J. & Tirindelli, R. (2001) Co-expression of putative pheromone receptors in the sensory neurons of the vomeronasal organ. *J Neurosci*, **21**, 843–848.

Matsunami, H. & Buck, L.B. (1997) A multigene family encoding a diverse array of putative pheromone receptors in mammals. *Cell*, **90**, 775–784.

Menco, B.P., Carr, V.M., Ezeh, P.I., Liman, E.R. & Yankova, M.P. (2001) Ultrastructural localization of G-proteins and the channel protein TRP2 to microvilli of rat vomeronasal receptor cells. *J. Comp Neurol*, **438**, 468–489.

Michaux, J., Reyes, A. & Catzeflis, F. (2001) Evolutionary history of the most speciose mammals: Molecular phylogeny of muroid rodents. *Mol Biol Evol*, **18**, 2017–2031.

Migeotte, I., Communi, D. & Parmentier, M. (2006) Formyl peptide receptors: A promiscuous subfamily of G protein-coupled receptors controlling immune responses. *Cytokine Growth Factor Rev*, **17**, 501–519.

Mucignat-Caretta, C., Caretta, A. & Cavaggioni, A. (1995) Acceleration of puberty onset in female mice by male urinary proteins. *J Physiol*, **486 (Pt 2)**, 517–522.

Mundy, N.I. & Cook, S. (2003) Positive selection during the diversification of class I vomeronasal receptor-like (V1RL) genes, putative pheromone receptor genes, in human and primate evolution. *Mol Biol Evol*, **20**, 1805–1810.

Nei, M., Niimura, Y. & Nozawa, M. (2008) The evolution of animal chemosensory receptor gene repertoires: Roles of chance and necessity. *Nat Rev Genet*, **9**, 951–963.

Nikolaev, S., Montoya-Burgos, J.I., Margulies, E.H., Rougemont, J., Nyffeler, B. & Antonarakis, S.E. (2007) Early history of mammals is elucidated with the ENCODE multiple species sequencing data. *PLoS Genet*, **3**, e2.

Nodari, F., Hsu, F.F., Fu, X., Holekamp, T.F., Kao, L.F., Turk, J. & Holy, T.E. (2008) Sulfated steroids as natural ligands of mouse pheromone-sensing neurons. *J Neurosci*, **28**, 6407–6418.

Norlin, E.M., Gussing, F. & Berghard, A. (2003) Vomeronasal phenotype and behavioral alterations in G alpha i2 mutant mice. *Curr Biol*, **13**, 1214–1219.

Novotny, M.V., Jemiolo, B., Wiesler, D., Ma, W., Harvey, S., Xu, F., Xie, T.M. & Carmack, M. (1999) A unique urinary constituent, 6-hydroxy-6-methyl-3-heptanone, is a pheromone that accelerates puberty in female mice. *Chem Biol*, **6**, 377–383.

Ohara, H., Nikaido, M., Date-Ito, A., Mogi, K., Okamura, H., Okada, N., Takeuchi, Y., Mori, Y. & Hagino-Yamagishi, K. (2009) Conserved repertoire of orthologous vomeronasal type 1 receptor genes in ruminant species. *BMC Evol Biol*, **9**, 233.

Papes, F., Logan, D.W. & Stowers, L. (2010) The vomeronasal organ mediates interspecies defensive behaviors through detection of protein pheromone homologs. *Cell*, **141**, 692–703.

Park, S.H., Podlaha, O., Grus, W.E. & Zhang, J. (2011) The microevolution of V1r vomeronasal receptor genes in mice. *Genome Biol Evol*, **3**, 401–412.

Pfister, P., Randall, J., Montoya-Burgos, J.I. & Rodriguez, I. (2007) Divergent evolution among teleost V1r receptor genes. *PLoS One*, **2**, e379.

Pifferi, S., Cenedese, V. & Menini, A. (2012) Anoctamin 2/TMEM16B: A calcium-activated chloride channel in olfactory transduction. *Exp Physiol*, **97**, 193–199.

Pifferi, S., Dibattista, M. & Menini, A. (2009) TMEM16B induces chloride currents activated by calcium in mammalian cells. *Pflugers Arch.*, **458**, 1023–1038.

Rasche, S., Toetter, B., Adler, J., Tschapek, A., Doerner, J.F., Kurtenbach, S., Hatt, H., Meyer, H., Warscheid, B. & Neuhaus, E.M. (2010) Tmem16b is specifically expressed in the cilia of olfactory sensory neurons. *Chem Senses*, **35**, 239–245.

Riviere, S., Challet, L., Fluegge, D., Spehr, M. & Rodriguez, I. (2009) Formyl peptide receptor-like proteins are a novel family of vomeronasal chemosensors. *Nature*, **459**, 574–577.

Roberts, S.A., Simpson, D.M., Armstrong, S.D., Davidson, A.J., Robertson, D.H., McLean, L., Beynon, R.J. & Hurst, J.L. (2010) Darcin: A male pheromone that stimulates female memory and sexual attraction to an individual male's odour. *BMC Biol*, **8**, 75.

Rodriguez, I. (2004) Pheromone receptors in mammals. *Horm Behav*, **46**, 219–230.

Rodriguez, I. & Mombaerts, P. (2002) Novel human vomeronasal receptor-like genes reveal species-specific families. *Curr Biol*, **12**, R409–411.

Rodriguez, I., Del Punta, K., Rothman, A., Ishii, T. & Mombaerts, P. (2002) Multiple new and isolated families within the mouse superfamily of V1r vomeronasal receptors. *Nat Neurosci*, **5**, 134–140.

Rodriguez, I., Feinstein, P. & Mombaerts, P. (1999) Variable patterns of axonal projections of sensory neurons in the mouse vomeronasal system. *Cell*, **97**, 199–208.

Roppolo, D., Ribaud, V., Jungo, V.P., Luscher, C. & Rodriguez, I. (2006) Projection of the Gruneberg ganglion to the mouse olfactory bulb. *Eur J Neurosci*, **23**, 2887–2894.

Rossler, P., Kroner, C., Krieger, J., Lobel, D., Breer, H. & Boekhoff, I. (2000) Cyclic adenosine monophosphate signaling in the rat vomeronasal organ: Role of an adenylyl cyclase type VI. *Chem Senses*, **25**, 313–322.

Runnenburger, K., Breer, H. & Boekhoff, I. (2002) Selective G protein beta gamma-subunit compositions mediate phospholipase C activation in the vomeronasal organ. *Eur J Cell Biol*, **81**, 539–547.

Ryba, N.J. & Tirindelli, R. (1995) A novel GTP-binding protein gamma-subunit, G gamma 8, is expressed during neurogenesis in the olfactory and vomeronasal neuroepithelia. *J Biol Chem*, **270**, 6757–6767.

Ryba, N.J. & Tirindelli, R. (1997) A new multigene family of putative pheromone receptors. *Neuron*, **19**, 371–379.

Ryu, S.H., Kwak, M.J. & Hwang, U.W. (2013) Complete mitochondrial genome of the Eurasian flying squirrel *Pteromys volans* (Sciuromorpha, Sciuridae) and revision of rodent phylogeny. *Mol Biol Rep*, **40**, 1917–1926.

Saraiva, L.R. & Korsching, S.I. (2007) A novel olfactory receptor gene family in teleost fish. *Genome Res*, **17**, 1448–1457.

Sasaki, K., Okamoto, K., Inamura, K., Tokumitsu, Y. & Kashiwayanagi, M. (1999) Inositol-1,4,5-trisphosphate accumulation induced by urinary pheromones in female rat vomeronasal epithelium. *Brain Res*, **823**, 161–168.

Schroeder, B.C., Cheng, T., Jan, Y.N. & Jan, L.Y. (2008) Expression cloning of TMEM16A as a calcium-activated chloride channel subunit. *Cell*, **134**, 1019–1029.

Scudieri, P., Sondo, E., Ferrera, L. & Galietta, L.J.V. (2012) The anoctamin family: TMEM16A and TMEM16B as calcium-activated chloride channels. *Exp Physiol*, **97**, 177–183.

Seifert, R. & Wenzel-Seifert, K. (2003) The human formyl peptide receptor as model system for constitutively active G-protein-coupled receptors. *Life Sci*, **73**, 2263–2280.

Serizawa, S., Ishii, T., Nakatani, H., Tsuboi, A., Nagawa, F., Asano, M., Sudo, K., Sakagami, J., Sakano, H., Ijiri, T., Matsuda, Y., Suzuki, M., Yamamori, T. & Iwakura, Y. (2000) Mutually exclusive expression of odorant receptor transgenes. *Nat Neurosci*, **3**, 687–693.

Shi, P. & Zhang, J. (2007) Comparative genomic analysis identifies an evolutionary shift of vomeronasal receptor gene repertoires in the vertebrate transition from water to land. *Genome Res*, **17**, 166–174.

Shi, P., Bielawski, J.P., Yang, H. & Zhang, Y.P. (2005) Adaptive diversification of vomeronasal receptor 1 genes in rodents. *J Mol Evol*, **60**, 566–576.

Shimazaki, R., Boccaccio, A., Mazzatenta, A., Pinato, G., Migliore, M. & Menini, A. (2006) Electrophysiological properties and modeling of murine vomeronasal sensory neurons in acute slice preparations. *Chem Senses*, **31**, 425–435.

Silvotti, L., Cavalca, E., Gatti, R., Percudani, R. & Tirindelli, R. (2011) A recent class of che-mosensory neurons developed in mouse and rat. *PloS One*, **6**, e24462.

Silvotti, L., Giannini, G. & Tirindelli, R. (2005) The vomeronasal receptor V2R2 does not require escort molecules for expression in heterologous systems. *Chem Senses*, **30**, 1–8.

Silvotti, L., Moiani, A., Gatti, R. & Tirindelli, R. (2007) Combinatorial co-expression of pher-omone receptors, V2Rs. *J Neurochem*, **103**, 1753–1763.

Singer, A.G., Macrides, F., Clancy, A.N. & Agosta, W.C. (1986) Purification and analysis of a proteinaceous aphrodisiac pheromone from hamster vaginal discharge. *J Biol Chem*, **261**, 13323–13326.

Smith, T.D., Bhatnagar, K.P., Burrows, A.M., Shimp, K.L., Dennis, J.C., Smith, M.A., Maico-Tan, L. & Morrison, E.E. (2005) The vomeronasal organ of greater bushbabies (*Otolemur* spp.): Species, sex, and age differences. *J Neurocytol*, **34**, 135–147.

Spehr, M., Hatt, H. & Wetzel, C.H. (2002) Arachidonic acid plays a role in rat vomeronasal signal transduction. *J Neurosci*, **22**, 8429–8437.

Stephan, A.B., Shum, E.Y., Hirsh, S., Cygnar, K.D., Reisert, J. & Zhao, H. (2009) ANO2 is the cilial calcium-activated chloride channel that may mediate olfactory amplification. *Proc Natl Acad Sci U S A*, **106**, 11776–11781.

Stöhr, H., Heisig, J.B., Benz, P.M., Schöberl, S., Milenkovic, V.M., Strauss, O., Aartsen, W.M., Wijnholds, J., Weber, B.H.F. & Schulz, H.L. (2009) TMEM16B, a novel protein with calcium-dependent chloride channel activity, associates with a presynaptic protein com-plex in photoreceptor terminals. *J Neurosci*, **29**, 6809–6818.

Stowers, L., Holy, T.E., Meister, M., Dulac, C. & Koentges, G. (2002) Loss of sex discrimina-tion and male-male aggression in mice deficient for TRP2. *Science*, **295**, 1493–1500.

Sturm, T., Leinders-Zufall, T., Macek, B., Walzer, M., Jung, S., Pommerl, B., Stevanovic, S., Zufall, F., Overath, P. & Rammensee, H.G. (2013) Mouse urinary peptides provide a molecular basis for genotype discrimination by nasal sensory neurons. *Nat Commun*, **4**, 1616.

Syed, A.S., Sansone, A., Nadler, W., Manzini, I. & Korsching, S.I. (2013) Ancestral amphib-ian v2rs are expressed in the main olfactory epithelium. *Proc Natl Acad Sci U S A*, **110**, 7714–7719.

Takami, S. (2002) Recent progress in the neurobiology of the vomeronasal organ. *Microsc Res Tech*, **58**, 228–250.

Takano, T., Fiore, S., Maddox, J.F., Brady, H.R., Petasis, N.A. & Serhan, C.N. (1997) Aspirin-triggered 15-epi-lipoxin A4 (LXA4) and LXA4 stable analogues are potent inhibitors of acute inflammation: Evidence for anti-inflammatory receptors. *J Exp Med*, **185**, 1693–1704.

Takigami, S., Mori, Y. & Ichikawa, M. (2000) Projection pattern of vomeronasal neurons to the accessory olfactory bulb in goats. *Chem Senses*, **25**, 387–393.

Tanaka, M., Treloar, H., Kalb, R.G., Greer, C.A. & Strittmatter, S.M. (1999) G(o) protein-dependent survival of primary accessory olfactory neurons. *Proc Natl Acad Sci U S A*, **96**, 14106–14111.

Thompson, R.N., McMillon, R., Napier, A. & Wekesa, K.S. (2007) Pregnancy block by MHC class I peptides is mediated via the production of inositol 1,4,5-trisphosphate in the mouse vomeronasal organ. *J Exp Biol*, **210**, 1406–1412.

Thompson, R.N., Robertson, B.K., Napier, A. & Wekesa, K.S. (2004) Sex-specific responses to urinary chemicals by the mouse vomeronasal organ. *Chem Senses*, **29**, 749–754.

Thoren, F.B., Karlsson, J., Dahlgren, C. & Forsman, H. (2010) The anionic amphiphile SDS is an antagonist for the human neutrophil formyl peptide receptor 1. *Biochem Pharmacol*, **80**, 389–395.

Tirindelli, R. & Ryba, N.J. (1996) The G-protein gamma-subunit G gamma 8 is expressed in the developing axons of olfactory and vomeronasal neurons. *Eur J Neurosci*, **8**, 2388–2398.

Tirindelli, R., Dibattista, M., Pifferi, S. & Menini, A. (2009) From pheromones to behavior. *Physiol Rev*, **89**, 921–956.

Wakabayashi, Y., Mori, Y., Ichikawa, M., Yazaki, K. & Hagino-Yamagishi, K. (2002) A putative pheromone receptor gene is expressed in two distinct olfactory organs in goats. *Chem Senses*, **27**, 207–213.

Wang, G., Shi, P., Zhu, Z. & Zhang, Y.P. (2010) More functional V1R genes occur in nest-living and nocturnal terricolous mammals. *Genome Biol Evol*, **2**, 277–283.

Wu, Y., Tirindelli, R. & Ryba, N.J. (1996) Evidence for different chemosensory signal transduction pathways in olfactory and vomeronasal neurons. *Biochem Biophys Res Commun*, **220**, 900–904.

Wysocki, C.J. & Lepri, J.J. (1991) Consequences of removing the vomeronasal organ. *J Steroid Biochem Mol Biol*, **39**, 661–669.

Wyckoff, G.J., Wang, W. & Wu, C.I. (2000) Rapid evolution of male reproductive genes in the descent of man. *Nature*, **403**, 304–309.

Yang, C. & Delay, R.J. (2010) Calcium-activated chloride current amplifies the response to urine in mouse vomeronasal sensory neurons. *J Gen Physiol*, **135**, 3–13.

Yang, H. & Shi, P. (2010) Molecular and evolutionary analyses of formyl peptide receptors suggest the absence of VNO-specific FPRs in primates. *J Genet Genomics*, **37**, 771–778.

Yang, H., Shi, P., Zhang, Y.P. & Zhang, J. (2005) Composition and evolution of the V2r vomeronasal receptor gene repertoire in mice and rats. *Genomics*, **86**, 306–315.

Yang, Y.D., Cho, H., Koo, J.Y., Tak, M.H., Cho, Y., Shim, W.-S., Park, S.P., Lee, J., Lee, B., Kim, B.-M., Raouf, R., Shin, Y.K. & Oh, U. (2008) TMEM16A confers receptor-activated calcium-dependent chloride conductance. *Nature*, **455**, 1210–1215.

Yau, K.W. & Hardie, R.C. (2009) Phototransduction motifs and variations. *Cell*, **139**, 246–264.

Yoon, H., Enquist, L.W. & Dulac, C. (2005) Olfactory inputs to hypothalamic neurons controlling reproduction and fertility. *Cell*, **123**, 669–682.

Young, J.M. & Trask, B.J. (2007) V2R gene families degenerated in primates, dog and cow, but expanded in opossum. *Trends Genet*, **23**, 212–215.

Young, J.M., Kambere, M., Trask, B.J. & Lane, R.P. (2005) Divergent V1R repertoires in five species: Amplification in rodents, decimation in primates, and a surprisingly small repertoire in dogs. *Genome Res*, **15**, 231–240.

Young, J.M., Massa, H.F., Hsu, L. & Trask, B.J. (2010) Extreme variability among mammalian V1R gene families. *Genome Res*, **20**, 10–18.

Zhang, J. & Webb, D.M. (2003) Evolutionary deterioration of the vomeronasal pheromone transduction pathway in catarrhine primates. *Proc Natl Acad Sci U S A*, **100**, 8337–8341.

Zhang, J. & Webb, D.M. (2004) Rapid evolution of primate antiviral enzyme APOBEC3G. *Hum Mol Genet*, **13**, 1785–1791.

Zhang, P., Yang, C. & Delay, R.J. (2008) Urine stimulation activates BK channels in mouse vomeronasal neurons. *J Neurophysiol*, **100**, 1824–1834.

Zhang, P., Yang, C. & Delay, R.J. (2010) Odors activate dual pathways, a TRPC2 and a AA-dependent pathway, in mouse vomeronasal neurons. *Am J Physiol*, **298**, C1253–1264.

Zhao, H., Xu, D., Zhang, S. & Zhang, J. (2011) Widespread losses of vomeronasal signal transduction in bats. *Mol Biol Evol*, **28**, 7–12.

Zhou, A. & Moss, R.L. (1997) Effect of urine-derived compounds on cAMP accumulation in mouse vomeronasal cells. *Neuroreport*, **8**, 2173–2177.

11 Central Processing of Intraspecific Chemical Signals in Mice

Carla Mucignat-Caretta

CONTENTS

11.1 Introduction 325
11.2 Chemosignals Modulate Various Neurohormonal Domains in Mice 326
    11.2.1 Modulation of Reproductive State 326
    11.2.2 Modulation of Adult Behavior 327
11.3 Chemosensory Systems Mediating Detection of Intraspecific Signals 328
    11.3.1 Accessory Olfactory Bulb 329
    11.3.2 Higher-Level Projection Areas 330
    11.3.3 Crosstalk between the Main and Accessory Olfactory Inputs 331
    11.3.4 Cell Activity in the Accessory Olfactory Bulb 331
    11.3.5 Convergence and Functional Integration of Main and Accessory
           Olfactory Information for the Perception of Intraspecific
           Chemosignals 332
11.4 Specific Circuits Involved in Different Effects 333
    11.4.1 Chemosignaling during Development 333
    11.4.2 Changes in AOB Chemosignals Processing after Parturition 334
    11.4.3 Chemosignaling, Memory, and the Reward System 334
    11.4.4 Memory for the Mate 335
    11.4.5 Role of Hormones and Hormone-Related Proteins in
           Intraspecific Chemical Communication 336
11.5 Concluding Remarks 337
Acknowledgments 338
References 338

11.1 INTRODUCTION

The previous chapters explored the peripheral anatomy and physiology of sensory organs responsible for detection of chemical signals shared among members of the same species. This chapter deals with the central processing of intraspecific chemosignals in the mouse; in particular, it will present data on the stimulus processing in the olfactory bulb, the amygdala, and the olfactory cortex, mainly in relation to neuroendocrine and behavioral modulation.

Communication is a primary function of living systems, from unicellular to complex organisms. It is necessary to mediate the responses of organisms to environmental changes and also to organize the actions of multiple members of the same species, to increase the fitness of both message sender and receiver, or either. Communication via chemical signals allows an honest and sensitive transmission of information, which requires the attribution of a specific link between a molecule (or blend of molecules) and a given response. Hence, central processing of intraspecific chemical signals allows stimulus detection and feature encoding, and involves different associative areas of the brain in mediating the responses to chemosignals.

Signaling molecules (not referred to here as "pheromones" due to the lack of consensus on the use of this term) are released in the environment together with biologic fluids, including tears, saliva, and genital secretions. Most of the data available for the mouse was obtained using as stimuli the molecules emitted with urine, which is excreted in larger amounts. Several sensory organs detect the relevant molecules and send information to various centers in the brain, using different, yet interconnected, pathways.

This chapter reviews what is known about signal processing in key chemosensory areas of the brain to show how chemical cues from conspecifics may influence both the behavior (including exploration, social, and sexual behavior) and the physiology of the organism that sends or receives the chemosignals.

11.2 CHEMOSIGNALS MODULATE VARIOUS NEUROHORMONAL DOMAINS IN MICE

11.2.1 MODULATION OF REPRODUCTIVE STATE

Among mice, communication by means of chemical signals represents a convenient way to share data about gender, social, and health status in order to modulate social behavior and reproduction-related actions. This complex set of information is used for affecting behavior in the receiver mouse, with respect to the environment, for example exploration or marking posts, and to other mice, for example by modulating sexual or aggressive intercourses (Wolff 2007). In addition, intraspecific chemical signals strongly impact the reproductive physiology of the receiver mouse by modulating its neuroendocrine axis. Mouse chemical signals can affect mice reactions on a short-time scale, mainly by modulating behavioral responses mediated by neuronal activation or fast hormone release, as in the case of aggression. On the other hand, chemosignals may trigger long-lasting responses, mainly mediated by activation of the neuroendocrine axis. Both short- and long-term responses imply a high degree of integrative activity involving various brain areas. In order to describe the areas involved in chemosensory processing, some of the most relevant effects induced by chemical signals in mice will be briefly introduced before turning to their neurobiological bases.

The discovery of mouse chemosignals started with the observation that interactions within and between genders heavily impacted on the hormonal status, mainly in female mice, and this impact was due to the physical presence of other mice or their urine. Female mice living in crowded groups display modifications in the length of estrus cycles up to suppression (van der Lee and Boot 1955); female urine, collected at different stages of estrus cycle or pregnancy may affect differentially cycling as well (Hoover and Drickamer 1979). Also male mice urine may influence female cycling by inducing its synchronization (Bronson and Whitten 1968) and accelerating the onset of female puberty (Vandenbergh 1969). These first-described effects may be explained in terms of variation of circulating sexual hormones in the sender mouse that modify active substance excretion, and the subsequent effect in the sex hormones setting of the receiver: in general, female chemosignals tend to suppress cycles in other females, while males tend to predispose females for mating by facilitating estrus cycles. However, this could be a problem if pregnancy has to be initiated, so the stud male chemosignals must be switched off after mating (Bruce 1960), an event that includes more cognitive and subtle processing (see below), while chemosignals from other males are processed to return females to cycling, thus preventing pregnancy. A similar block of pregnancy was induced also by urine of females kept in group, possibly indicating that at high population density, where competition for resources is high, pregnancy does not ensue (Drickamer 1999).

Urine has been proved to contain the active substances for all the above effects, thus it has been considered for decades as the main source of intraspecific chemosignals, in part for its easy collection, albeit other sources like tears or gland secretions have been explored. Secretions and their synthetic analogues have then been used to test responses evoked in various centers of the brain.

11.2.2 MODULATION OF ADULT BEHAVIOR

Urinary chemosignals have a strong impact on the behavior of adult mice and induce immediate cardiovascular responses (Lee and Wilson 2012). Male and female urines contain chemosignals that promote and inhibit aggression, respectively (Mugford and Nowell 1970, 1971). However, the effect of male urine on aggression in males is more dependent on the context than originally thought: when painted on adults of either sex, male urine strongly induce aggression, but when painted on pups strongly inhibits it; however, male urine painted on food does not influence eating, strongly supporting the view that effects of chemosignals depend on decoding the socio-environmental context (Mucignat-Caretta et al. 2004). Lacrimal gland secretions may also induce aggression, for which they act on the vomeronasal organ (VNO), like urine (Thompson et al. 2007).

A second class of behavioral effects includes alarm reactions that may be induced by the exposure to urine of mice that have experienced a threatening or stressing situation (Rottman and Snowdown 1972; Zalaquett and Thiessen 1991), and the effects on subsequent associative learning (Bredy and Barad 2008).

Other behaviors are modified by female odors that evoke ultrasound vocalization in male mice (Whitney et al. 1974). Male chemosignals may also favor long-term spatial memory in females (Roberts et al. 2012).

11.3 CHEMOSENSORY SYSTEMS MEDIATING DETECTION OF INTRASPECIFIC SIGNALS

Different chemosensory systems are involved in sensing intraspecific chemicals in mice; they are located in the nasobuccal cavity, where molecules arrive borne by the air or by direct contact of the sensors with the secretions. Although mostly related to food evaluation, it cannot be excluded that taste receptors participate in intraspecific signaling, because grooming and licking may bring secretions to the mouth. The best-known receptors for intraspecific chemosignals are located in the nose, three of which are considered minor (Breer et al. 2006): the Grünenberg ganglion, the septal (or Masera's) organ, and the free terminals of trigeminal nerve. The other two play a major role: the main olfactory epithelium, and the vomeronasal (or Jacobson's) organ. Their function is to detect chemicals that may be dangerous, to indicate food for retrieval, and to modulate social interactions (Ma 2007).

The Grünenberg ganglion consists of a cluster of olfactory neurons located at the entrance of the nostrils that express either olfactory or vomeronasal receptor proteins and send their axons to the caudal main olfactory bulb. Its function is possibly linked to suckling or alarm detection (Brechbühl et al. 2008; Fleischer et al. 2006; Roppolo et al. 2006). The septal organ consists of a cluster of chemosensory neurons located in the caudal part of the basal septum; they express some odorant receptor proteins, one of which has broadly tuned responses (Grosmaitre et al. 2009; Kaluza et al. 2004; Tian and Ma 2008), and project to the caudal ventral aspects of the main olfactory bulb (Lèvai and Strotmann 2003).

The main olfactory epithelium is located within the dorsomedial parts of the nose, and is connected to the main olfactory bulb (MOB) that projects to the lateral amygdala and the olfactory cortex (Figure 11.1). Its role in detecting the chemical characteristics of the environment is fundamental for detection of food and threatening events for survival. The VNO (see Chapters 9 and 10) is connected to the accessory olfactory bulb (AOB), the medial amygdala, and medial hypothalamus. Note that the VNO expresses also some main olfactory receptors in addition to V1R and V2Rs (Lèvai et al. 2006; Chapter 10). Electrophysiological responses in the olfactory bulb and lateral and preoptic hypothalamus indicated that no neuron was selective for male urine, and only a few could differentiate between male and castrated urine (Scott and Pfaff 1970). The VNO is stimulated by molecules that also activate the main olfactory epithelium, but also by larger molecules that may be sucked up by a vascular pumping mechanism (see Chapter 9). Since it senses molecules that may be similar to or different from those sensed by the main olfactory system and process them differently, its role is to complement olfaction (Mucignat-Caretta 2010). Vomeronasal deafferentation results in modification of reproductive-related behaviors from courtship to sexual, territorial, and aggressive behavior in male and female mice, only partially mitigated by previous sexual experience (Wysocki and Lepri 1991).

The main and accessory olfactory systems can detect environmental chemicals and intraspecific chemosignals and their central projections are greatly intermingled (Pro-Sistiaga et al. 2007). Nevertheless, the two systems show a different selectivity for intraspecific stimuli, like those present in the whole urine: activation of AOB

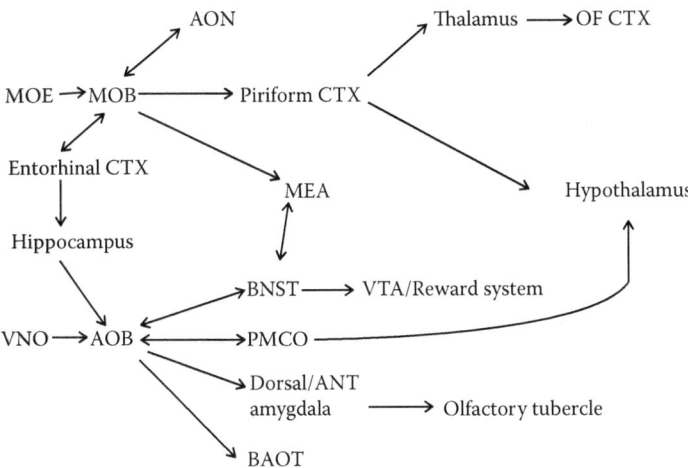

FIGURE 11.1 A schematic representation of the most important central connections of the main and accessory olfactory systems of the mouse. The centrifugal inputs from the brainstem are not indicated. ANT: anterior; AOB: accessory olfactory bulb; AON: anterior olfactory nucleus; BAOT: bed nuclei of the accessory olfactory tract; BNST: bed nuclei of the stria terminalis; CTX: cortex; MEA: medial amygdala; MOB: main olfactory bulb; MOE: main olfactory epithelium; OF: orbitofrontal; PMCO: posteromedial amygdala nuclei; VNO: vomeronasal organ; VTA: ventral tegmental area.

appears larger than in the MOB, where activation is confined to some regions (Xu et al. 2005).

Both the MOB and AOB are highly plastic regions, being targets for the newly differentiated neurons coming from the subventricular zone through the rostral migratory stream. Male chemosignals may modulate brain plasticity in the adult female by increasing neurogenesis (Mak et al. 2007; for details see Chapter 13).

11.3.1 ACCESSORY OLFACTORY BULB

In the description of the VNO, Jacobson suggested that its function could be the processing of sensory inputs in parallel with the main olfactory system in order to detect specific features that may complement the olfactory information (Jacobson et al. 1998).

The vomeronasal receptors send their axons to the ipsilateral AOB that lies in the dorsal part of the MOB and also receives modulating inputs from the brainstem regulatory nuclei and other higher chemosensory centers. In contrast to the main olfactory system, in which all of the neurons expressing the same olfactory receptor converge onto one glomerulus, the vomeronasal axons project to various glomeruli (Belluscio et al. 1999; Rodriguez et al. 1999). These axons branch and synapse with AOB mitral/tufted (MT) cell dendrites, several of which collect the signals from glomeruli that are innervated mainly by the same V1R or V2R-expressing neurons (Del Punta et al. 2002). The VNO input signal is processed in dendrites of MT cells in concurrence with periglomerular cells activity. However, each glomerulus may process inputs from different receptors (Wagner et al. 2006), hence MT cells are involved in integrative processing. Actually, AOB MT cells can integrate signals from one receptor or different types of receptors (Meeks et al. 2010). Reciprocal synapses between MT cells and inhibitory interneurons are detected in primary dendrites at variance with the MOB, in which these synapses are present on secondary branchings (Moriya-Ito et al. 2013).

Each MT cell integrates signals from up to 1000 receptor cells and is connected to interneurons that modulate the firing of neighboring MT cells, supporting a role for lateral modulation of MT cells activity (Larriva-Sahd 2008). Vomeronasal cells expressing the class 1 vomeronasal receptors (V1Rs) project to the anterior AOB, while those expressing class 2 receptors (V2Rs) project to the caudal AOB (see Chapter 10); in addition, the expression of class I major histocompatibility genes reveals a third population of cells projecting to AOB (Ishii and Mombaerts 2008).

The activity of AOB has been linked to intraspecific chemical signaling, since exposure to urine may activate c-Fos transcription in the AOB (Guo et al. 1997; Martel and Baum 2007) with same- and opposite sex chemicals activating the caudal and rostral subdivision of AOB, respectively (Kumar et al. 1999).

11.3.2 HIGHER-LEVEL PROJECTION AREAS

The axons of MT cells leave the AOB directed to the posteromedial cortical nuclei of amygdala, to the bed nucleus of the stria terminalis, and bed nucleus of accessory olfactory tract (de Olmos et al. 1978; Scalia and Winans 1975; Winans and Scalia 1970). The AOB receives centrifugal projection from the amygdala through the stria terminalis, directed to the AOB granule and MT cells (Raisman 1972); these projection neurons express estrogen receptors, thus linking their activity to hormonal modulation (Fan and Luo 2009). Amygdala projections to AOB granule cells are glutamatergic, while the bed nucleus of the stria terminalis sends GABAergic afferents to the AOB MT cells (Fan and Luo 2009).

Neurons in the amygdala receive from the main and accessory olfactory bulbs (see below), indicating a participation of both systems in the modulation of a variety of social actions, including sexual, parental, and aggressive behavior (Lanuza et al. 2008). While responses in the AOB are long-lasting but disappear within minutes after stimulus presentation, rodent amygdala activation in response to intraspecific chemosignals is scaled on a longer time frame, suggesting a possible link to long-term modulation of hypothalamic neuroendocrine activation (Mucignat-Caretta et al. 2006).

In turn, amygdala nuclei project to the preoptic, ventromedial, and premammillary hypothalamic nuclei, all involved in the modulation of gonadotrophin release, and also to various cortical areas that receive information from the main and accessory olfactory systems (Martinez-Marcos 2009; von Campenhausen and Mori 2000; Yokosuka et al. 1999). Medial amygdala projections to the ventromedial hypothalamus are glutamatergic and arise from cells similar to pyramidal neurons in the deep piriform cortex; these amygdala cells are excited from AOB afferents and inhibited by local GABAergic interneurons (Bian et al. 2008).

The connections between the posterior medial amygdala and ventromedial hypothalamus appear anatomically segregated according to their influence on defensive (ventral region of posterior amygdala) or reproductive behaviors (dorsal posterior amygdala) and use different neurotransmitters for gating reproduction in the presence of threatening stimuli (Choi et al. 2005).

Dopamine plays a crucial role in central processing of responses to chemosignals, since it links motivation to sensory elements that indicate the presence of food and potential partners. The lack of D2 receptors reduces the growth hormone output and in turn the excretion of chemosignals in urine, thereby affecting social dominance and competition for resources (Noain et al. 2013).

11.3.3 CROSSTALK BETWEEN THE MAIN AND ACCESSORY OLFACTORY INPUTS

The main and accessory olfactory systems cooperate to achieve successful interactions with conspecifics (Keller et al. 2009; Restrepo et al. 2004). The first long-distance approach between mice is apparently mediated by the main olfactory system that detects airborne stimuli and triggers active exploration and subsequent sampling for VNO access (Keverne 2004). The main and accessory olfactory systems host vasopressin neurons, which promote social odor recognition (Wacker and Ludwig 2012). Neurons of both systems express steroid receptors for linking their activity to hypothalamic releasing factors and sex hormones levels to promote reproductive-related behaviors, including territory defense, mating, and pup care (Dluzen and Ramirez 1983, Guillamon and Segovia 1997) that are different in the two sexes. In addition, male odor-induced activation in the bed nucleus of the stria terminalis and medial preoptic area is sexually dimorphic (Halem et al. 1999).

11.3.4 CELL ACTIVITY IN THE ACCESSORY OLFACTORY BULB

When compared to the main olfactory system, the accessory olfactory system displays some differences in its activity, starting from the VNO up to higher centers, mainly related to the temporal dynamics of cell responses.

The MOB is involved in intraspecific chemosignaling: single urinary volatiles activate a subset of MT cells, and a certain number of MT cells can respond selectively to male mouse urine. Therefore, MOB mitral cells act as selective feature detectors insensitive to other components of the blend (Lin et al. 2005).

Vomeronasal neurons hardly adapt; in fact they respond for long time periods to stimuli for which they are selective, for example female or male urine (Holy et al. 2000).

When exposed to potential mates and their chemosignals, rapid-onset responses are recorded from the MOB, while the AOB responds slowly, also with inhibitory activity that stems from integration processes in the AOB local circuits (Luo et al. 2003). Mitral cell responses in the MOB are transient, while the AOB MT responses are sustained and mediated by calcium-driven nonselective cationic current provoked by brief but strong activity (Shpak et al. 2012).

The AOB responses consist of gradual increasing of the firing sustained over many seconds even after the stimulus ends (Luo and Katz 2004), suggesting that accuracy of the response is more important than speed for AOB activity. Intraspecific chemosignals appear to activate anterior and posterior AOB, while heterospecific signals stimulate only the anterior portion; this pattern of activation is also present in the medial amygdala (Samuelsen and Meredith 2009).

The AOB responses to urinary chemosignals vary according to the gender of the emitter and of the receiver, being larger for opposite-sex chemosignals (Guo et al. 1997). These molecules may also activate separate groups of glomeruli in the MOB (Baum 2012; Martel and Baum 2007). The activation of AOB is not necessary to discriminate male from female urine or to elicit male sexual behavior, but it is important for coding the incentive value of opposite-sex chemosignals (Jakupovic et al. 2008) and for suppressing male-specific behavior, due to sexually dimorphic connections with the amygdala and hypothalamus (Baum 2009).

The responses of AOB output cells may be classified into three classes, and apparently act for enhancing responses to long-lasting stimuli and reducing those to brief signals (Zibman et al. 2011). Multisite recordings revealed that AOB neurons encode sex and genetic background in urine and saliva, while other neurons respond to predator cues (Ben-Shaul et al. 2010).

The flux of information from AOB to the subsequent areas is modulated by reciprocal dendrodendritic inhibitory synapses between granule and MT cells, with the glutamate receptor mGluR2 acting to reduce dendrodendritic inhibition (Taniguchi et al. 2013).

11.3.5 CONVERGENCE AND FUNCTIONAL INTEGRATION OF MAIN AND ACCESSORY OLFACTORY INFORMATION FOR THE PERCEPTION OF INTRASPECIFIC CHEMOSIGNALS

The processing of main and accessory olfactory information starts in separate epithelia that project to the main and accessory olfactory bulbs (Barber and Raisman 1974). However, the secondary projections to the basal telencephalon and then to the ventral striatum are intermingled (Martinez-Marcos 2009). The main olfactory bulb is connected to the anterior olfactory nucleus, where the responses to mixtures exceed the responses to single odorants (Lei et al. 2006) and projects diffusely to the piriform cortex so that no spatial map can be recognized there for odorant coding (Choi et al. 2011; Miyamichi et al. 2011; Sosulki et al. 2011). The olfactory cortex processes stimuli from multiple glomeruli to code for combinations of chemical features (Davison and Ehlers 2011). Recent evidence suggests that the olfactory cortex receives multisensory inputs that possibly modulate olfactory processing (Varga and Wesson 2013).

The dichotomy of the vomeronasal connections, with V1Rs projecting to the anterior AOB and V2Rs projecting to the posterior AOB, is also partially maintained in AOB projection areas: the anterior AOB sends axons to the bed nucleus of the stria terminalis, while the posterior AOB is connected to the dorsal anterior amygdala; the amygdala is in turn connected to the Islands of Calleja and olfactory tubercle and acts to mediate fear, anxiety, and aggressive and reproductive behaviors (Lanuza et al. 2008; Martínez-García et al. 2008). The bed nucleus of the stria terminalis also modulates anxiety-prone moods such as exploratory drive or novelty seeking (Kim et al. 2013) because it is connected to the ventral tegmental area and the reward system

(Jennings et al. 2013). This may explain the wide influence of chemosignals in mice behavior. Efferents of the accessory olfactory system also remain partially segregated in the hypothalamus (Mohedano-Moriano et al. 2008), while no segregation is present in the posteromedial nucleus of amygdala, bed nuclei of the stria terminalis, and bed nuclei of the accessory olfactory tract (von Campenhausen and Mori 2000).

Since the rostral AOB receives efferent fibers from the main olfactory bulb, the main and accessory olfactory systems already interact in the bulb (Larriva-Sahd 2008). The medial amygdala comprises heterogeneous populations of neurons (Niimi et al. 2012); it receives inputs from the MOB that drive responses to the opposite sex chemosignals, in addition to inputs from AOB (Kang et al. 2009). As a consequence, chemosensory activation of medial amygdala is also present in mice lacking vomeronasal Trpc2 channel in which AOB activation is absent (Hasen and Gammie 2011). While AOB MT cells terminate into medial and posterolateral cortical amygdala, MOB efferents are more branched to the piriform cortex and medial and anterior amygdala (Kang et al. 2011).

In conclusion, the two systems overlap in areas like the anterior medial amygdala and bed nucleus of the stria terminalis, as well as in classical olfactory areas, including the nucleus of the lateral olfactory tract, and in classical vomeronasal areas, including anteroventral amygdala and the bed nucleus of the accessory olfactory tract. On the other hand, the two systems remain separated in the posteromedial and posterolateral cortical nuclei of amygdala, suggesting the existence of both specific and mixed chemosensory areas (Pro-Sistiaga et al. 2007).

The processing of chemosensory information in the AOB is modulated by centrifugal pathways from both vomeronasal and olfactory amygdala, and from the bed nucleus of the stria terminalis (Fan and Luo 2009; Martel and Baum 2009; Mohedano-Moriano et al. 2012). In addition, the AOB receives modulatory inputs from the hippocampus that are modulated by the entorhinal cortex and hence from the main olfactory bulb (de la Rosa-Prieto et al. 2009), implying a crosstalk between main and accessory olfactory processing and memory-related areas.

11.4 SPECIFIC CIRCUITS INVOLVED IN DIFFERENT EFFECTS

The above-mentioned features of central processing in main and accessory olfactory systems have been explored in different experimental contexts. The neurobiological processing of some stimuli and their effects on intraspecific communication were studied in greater detail: the following section will deal with a few examples of what has been discovered.

11.4.1 CHEMOSIGNALING DURING DEVELOPMENT

In the early phases of postnatal life, altricial rodent pups collect information on the rest of the world using two main senses, the chemical and the somatosensory. These allow survival by favoring crowding at the nest and finding a nipple for milk. During this period, the main and accessory olfactory systems may be functionally heterogeneous. In the newborn rat, which has been studied in greater detail than the mouse, the dam's chemosignals are perceived through the main olfactory system (Greer et al. 1982;

Teicher et al. 1980). Olfactory cognitive functions are well developed at birth; immediately after birth, olfactory learning is facilitated in order to create associations between stimuli necessary to survive (Miller and Spear 2008) and depends on serotoninergic afferents to MT cells (McLean et al. 1995). However, learning is different in pups compared to adults, because facilitation for olfactory associations is complemented by difficulty in learning aversions during the first postnatal week (Moriceau et al. 2006).

The stratification of AOB, morphological differentiation of cells, and glomerular wiring are established between postnatal day 1 and 5 (Salazar et al. 2006), while the electrical activity progressively matures from an immature period up to postnatal day 11, to a transition stage up to day 17, and eventually to the mature stage (Sugai et al. 2005). Note that both general odors and male-urine exposure may affect olfactory bulb maturation; in fact during development the olfactory systems interactions with the environment may affect cell migration in the forebrain (Harvey and Cowley 1984; Suarez et al. 2012). An analogous transition in activation has been described in the piriform cortex (Illig 2007). Behavioral maturation results from olfactory processing and parallel neuroendocrine development, so that the same intraspecific chemical stimuli change their value before and after puberty (Kennedy and Brown 1970; Mucignat-Caretta et al. 1995, 1998).

Apparently, the main olfactory system plays a major role during the first postnatal days even if the VNO is quite mature at birth and already active *in utero* (Coppola and O'Connell 1989; see also Chapter 9). The VNO may release neurotransmitter to the AOB right after parturition, the precise connectivity to AOB being refined by early exposure to chemosignals (Hovis et al. 2012).

11.4.2 Changes in AOB Chemosignals Processing after Parturition

Not only the pup, but also the dam uses its chemosensory ability to discriminate its own pup, most probably via changes in the main olfactory system that drive female behavior from aversion to parental care (Levy and Keller 2009). Pups' chemosignals activate the accessory olfactory system in the adult male and drive the transition from infanticidal to parental behavior (Tachikawa et al. 2013). Moreover, juvenile chemosignals inhibit sexual behavior via the accessory olfactory system (Ferrero et al. 2013).

Apparently, maternal behavior benefits from both the main and the accessory olfactory systems, as the mother's VNO detects dodecyl propionate from rat pups (Brouette-Lalou et al. 1999). Moreover, at parturition a large modification in gene expression takes place in the AOB (Canavan et al. 2011), while the deletion of Trpc2, the vomeronasal-organ-specific ion channel, results in the absence of posterior AOB and severely affects maternal behavior directed to the pup or to intruders (Hasen and Gammie 2009). However maternal detection of pups' cues is also impaired in mice deficient for the olfactory-specific adenylyl cyclase 3 (Wang and Storm 2011). Hence, both the main and accessory olfactory systems appear involved in mother–pup interactions.

11.4.3 Chemosignaling, Memory, and the Reward System

Attractiveness of the male chemosignals for the adult female mouse is a major determinant of sexual approach. High molecular weight urinary substances act as rewarding stimuli to condition responses to urinary volatiles that by themselves do not attract inexperienced females (Moncho-Bogani et al. 2002). They act through the accessory olfactory system and subsequent activation of the basolateral amygdala. This kind of associative learning, which does not require the involvement of the ventral tegmental area and the reward system (Martínez-Hernández et al. 2006; Martínez-Ricós 2008; Moncho-Bogani et al. 2005), takes place in the basolateral amygdala, which is connected to the ventral striatopallidum and hence to the limbic system (Novejarque et al. 2011). The attractiveness of male chemosignals appears dependent on efficient central nitrergic neuromodulation (Augustin-Pavon et al. 2009).

A differential effect of the volatile and nonvolatile urinary chemicals is also present for individual social recognition, short-term memory being dependent on nonvolatiles, while long-term memory requires activation of both the main and accessory olfactory bulbs with both volatiles and nonvolatiles (Noack et al. 2010).

11.4.4 MEMORY FOR THE MATE

By securing ovulation, urinary male chemosignals are sufficient to prime the female neuroendocrine system for reproduction. However, once mating has occurred, the female hormonal milieu must dramatically change in order to allow the onset of pregnancy. How can this be achieved, if the stud male is present and its chemosignals tend to drive the female hormones to estrus again? Actually, dramatic modifications in the female central processing of male chemosignals take place after mating. While after mating the stud male can stay in contact with the female without stopping pregnancy, strange males or their chemosignals that come in contact with the female before the implantation of the fertilized ova in the uterus can return the female to estrus, thus preventing pregnancy. This implies that the chemosignals from the stud are processed differently from those of stranger males: mates' signals must be discriminated and remembered. This specific form of memory is established at mating and involves modification in the processing of studs' chemosignals in the olfactory bulbs. The block of pregnancy is induced by stimulation of the VNO with new male chemosignals (Bellringer et al. 1980) and involves dopaminergic circuits acting on prolactin release (Marchlewska-Koj and Jemiolo 1978) possibly modulated by cholecystokinin (Li et al. 1992). Chemical stimuli from the male activate both the main and accessory olfactory systems. The main olfactory system is responsible for cognitive processing, like discrimination, which is not necessary for pregnancy block, which depends on accessory olfactory processing (Lloyd-Thomas and Keverne 1982). Both MOB and AOB receive noradrenergic modulatory fibers from the brainstem through the medial olfactory striae: noradrenaline increase due to cervical stimulation at mating drives the establishment of memory for the stud in the subsequent hours, while this input is not relevant for retrieving the stud's memory (Kaba and Keverne 1988; Rosser and Keverne 1985). The role of noradrenaline from locus coeruleus is to enhance inhibition of MT cells by increasing GABA release from granule cells via alpha-1 adrenergic receptors (Araneda and Firestein 2006), a mechanism common to AOB and MOB (Zimnik et al. 2013). The memory for the stud is established by plastic modulation of the strength between GABA-ergic granule to MT cells synapses in the AOB (Brennan et al. 1990; Kaba et al. 1989, 1994), possibly involving also cholinergic modulation (Takahashi and Kaba 2010) and lateral inhibition (Hendrickson et al. 2008). Accordingly, in the AOB, memory induced at mating involves the expression of genes associated with the synaptic function (Upadhya et al. 2011) and changes the phenotype of the microglia, which monitor active synapses, in the external plexiform layer of anterior AOB (Okere 2012). Eventually, processing of the signal from the stud male is prevented to go on, so that medial amygdala neurons respond more strongly to strange male chemosignals (Binns and Brennan 2005).

An involvement of the main olfactory system can also be devised in the pregnancy block by stranger males because after mating a dopamine surge in the main olfactory bulb depresses the perception of social odors (Serguera et al. 2008).

A critical role in establishing memory for the mate is played by newly differentiating neurons that reach the bulb via the rostral migratory stream (for details, see Chapter 13).

An interesting issue about the involvement of the olfactory system concerns the steroids contained in male urine, including unconjugated estradiol, that may enter via the olfactory system (Guzzo et al. 2012), suggesting that chemosensory and hormonal cues may concurrently act to prevent pregnancy.

11.4.5 ROLE OF HORMONES AND HORMONE-RELATED PROTEINS IN INTRASPECIFIC CHEMICAL COMMUNICATION

Several chemosignals in mice influence the neuroendocrine circuits that mediate reproduction. In addition, chemosignals may modulate adrenocortical responses (Archer 1969); male chemosignals acutely increase corticosterone in estrus females (Marchlewska-Koj et al. 1990).

Since early reports (Bruce 1965), the production of male chemosignals has been linked to circulating androgens, which in turn depend on the social status of the male and on the possible presence of the female (Lombardi and Vandenbergh 1977). However, chemosignal production also depends on the functional presence of oxytocin and estrogen receptors alpha and beta (Kavaliers et al. 2004).

On the other hand, the processing of intraspecific chemical cues in main and accessory olfactory systems is modulated by hormones that act by modifying behavior, for example by increasing investigation or by enhancing the neural responses to chemosignals, to favor mating (Moffat 2003). The processing of chemical stimuli for social recognition appears dependent on central actions of oxytocin and vasopressin (Bielsky and Young 2004). Initial reports showed that male chemosignals facilitate secretion of the follicle-stimulating hormone (FSH) in females (Avery 1969) and favor the preovulatory luteinizing hormone (LH) surge, possibly after neuropeptide Y priming of the anterior pituitary (Xu et al. 2000); they may also facilitate maternal behavior acting via prolactin (Larsen et al. 2008). The sexually dimorphic effect of male chemosignals on LH release is not related to aromatase, and hence estrogen, action during brain development (Bakker et al. 2010). The interaction of chemosensory information with the hypothalamic gonadotropes is possibly modulated by the peptide kisspeptin (Jouhanneau et al. 2013).

Female urine induces LH release (Maruniak and Bronson 1976); it acts by stimulating the VNO to induce both LH and testosterone release in males (Coquelin et al. 1984; Wysocki et al. 1983), the magnitude of the effect being dependent on previous sexual experience (Clancy et al. 1988) but independent from activation of estrogen receptor alpha (Gore et al. 2000). Chemosignals from unknown females decrease prolactin in pregnant females, enhancing their anxiety after parturition (Larsen and Grattan 2010).

Males and females respond differently to the same stimulus both in the VNO and in the AOB (Halem et al. 2001). These responses are possibly modulated by centrifugal noradrenergic fibers from the superior cervical ganglion that modulate vasculature pumping in the VNO. However, their activity is not mandatory for the responses to chemosignals in the central accessory olfactory projection areas (Pankevich et al. 2003).

The complete circuit for neuroendocrine modulation by chemosignals has thus been unraveled: AOB projections activate receptors for excitatory amino acids in the amygdala that, through the stria terminalis, subsequently excite tuberoinfundibular dopaminergic neurons in the arcuate nucleus of the hypothalamus (Li et al. 1990). Both the main and accessory olfactory systems communicate with nearly 800 GnRH neurons that in turn may influence many different brain areas, in some cases with sex-specific connections (Boehm et al. 2005).

11.5 CONCLUDING REMARKS

This chapter referred to the mouse as a model organism that makes use of different chemoreceptors in various sensory organs, including classical olfactory receptors in the main olfactory epithelium, vomeronasal receptors in the VNO, chemoreceptors in the Grüneberg ganglion and in the septal organ of Masera, as well as chemoreceptors in the mouth and trigeminal nerve. The processing of intraspecific chemical signals does not require specialized pathways; instead, it requires a high-level processing in multiple areas of the brain, which decode intraspecific chemosensory inputs to extract complex features related to the signal emitter.

Several sources of data demonstrate that the VNO is sensitive to a great variety of chemical stimuli, some of which are involved in intraspecific communication. In addition to fast responses, stimulation of the VNO may result in long-lasting modifications of the firing rate in the AOB, and in slow-onset activity in the amygdala. The slow responses may be tentatively related to the modifications evoked in the hypothalamus-driven neuroendocrine pathway, which result in reproductive activity modulation.

However, the main and accessory olfactory systems interact in different central areas, where inputs from the two systems converge. In addition, centrifugal projections from caudal brain centers modulate the activity of both main and accessory olfactory systems. The functions of both systems partially overlap and are highly integrated; hence chemical communication is the result of merging various inputs already at early steps of central processing.

ACKNOWLEDGMENTS

This chapter was supported by the SNIFFER EU grant no. 285203. I am deeply indebted to A. Caretta, A. Cavaggioni, K.E. Kaissling, A. Lenz, M. Redaelli, M.J. Ricatti and C. Zancanaro for critical reading of the manuscript.

REFERENCES

Agustín-Pavón, C., Martínez-Ricós, J., Martínez-García, F. and E. Lanuza. 2009. Role of nitric oxide in pheromone-mediated intraspecific communication in mice. *Physiol Behav* 98:608–13.

Araneda, R.C. and S. Firestein. 2006. Adrenergic enhancement of inhibitory transmission in the accessory olfactory bulb. *J Neurosci* 26:3292–8.

Archer, J.E. 1969. Adrenocortical responses to olfactory social stimuli in male mice. *J Mammal* 50:839–41.

Avery, T.L. 1969. Pheromone-induced changes in the acidophil concentration of mouse pituitary glands. *Science* 164:423–4.

Bakker, J., Pierman, S. and D. González-Martínez. 2010. Effects of aromatase mutation (ArKO) on the sexual differentiation of kisspeptin neuronal numbers and their activation by same versus opposite sex urinary pheromones. *Horm Behav* 57:390–5.

Baum, M.J. 2009. Sexual differentiation of pheromone processing: Links to male-typical mating behavior and partner preference. *Horm Behav* 55:579–88.

Baum, M.J. 2012. Contribution of pheromones processed by the main olfactory system to mate recognition in female mammals. *Front Neuroanat* 6, 20.

Bellringer, J.F., Pratt, H.P. and E.B. Keverne. 1980. Involvement of the vomeronasal organ and prolactin in pheromonal induction of delayed implantation in mice. *J Reprod Fertil* 59:223–8.

Belluscio, L., Koentges, G., Axel, R. and C. Dulac. 1999. A map of pheromone receptor activation in the mammalian brain. *Cell* 97:209–20.

Ben-Shaul, Y., Katz, L.C., Mooney, R. and C. Dulac. 2010. In vivo vomeronasal stimulation reveals sensory encoding of conspecific and allospecific cues by the mouse accessory olfactory bulb. *Proc Natl Acad Sci U S A* 107:5172–7.

Bian, X., Yanagawa, Y., Chen, W.R. and M. Luo. 2008. Cortical-like functional organization of the pheromone-processing circuits in the medial amygdala. *J Neurophysiol* 99:77–86.

Bielsky, I.F. and L.J. Young. 2004. Oxytocin, vasopressin, and social recognition in mammals. *Peptides* 25:1565–74.

Binns, K.E. and P.A. Brennan. 2005. Changes in electrophysiological activity in the accessory olfactory bulb and medial amygdala associated with mate recognition in mice. *Eur J Neurosci* 21:2529–37.

Boehm, U., Zou, Z. and L.B. Buck. 2005. Feedback loops link odor and pheromone signaling with reproduction. *Cell* 123:683–95.

Brechbühl, J., Klaey, M. and M.C. Broillet. 2008. Grueneberg ganglion cells mediate alarm pheromone detection in mice. *Science* 321:1092–5.

Bredy, T.W. and M. Barad. 2008. Social modulation of associative fear learning by pheromone communication. *Learn Mem* 16:12–18.

Breer, H., Fleischer, J. and J. Strotmann. 2006. The sense of smell: Multiple olfactory subsystems. *Cell Mol Life Sci* 63:1465–75.

Brennan, P., Kaba, H. and E.B. Keverne. 1990. Olfactory recognition: A simple memory system. *Science* 250:1223–6.

Bronson, F.H. and W.K. Whitten. 1968. Oestrus-accelerating pheromone of mice: Assay, androgen-dependency and presence in bladder urine. *J Reprod Fertil* 15:131–4.

Brouette-Lahlou, I., Godinot, F. and E. Vernet-Maury. 1999. The mother rat's vomeronasal organ is involved in detection of dodecyl propionate, the pup's preputial gland phero-mone. *Physiol Behav* 66:427–36.

Bruce, H.M. 1960. A block to pregnancy in the mouse caused by proximity of strange males. *J Reprod Fertil* 1:96–103.

Bruce, H.M. 1965. Effect of castration on the reproductive pheromones of male mice. *J Reprod Fertil* 10:141–3.

Canavan, S.V, Mayes, L.C. and H.B. Treloar. 2011. Changes in maternal gene expression in olfactory circuits in the immediate postpartum period. *Front Psychiatry* 2:40.

Choi, G.B., Dong, H.W., Murphy, A.J., Valenzuela, D.M., Yancopoulos, G.D., Swanson, L.W. and D.J. Anderson. 2005. Lhx6 delineates a pathway mediating innate reproductive behaviors from the amygdala to the hypothalamus. *Neuron* 46:647–60.

Choi, G.B., Stettler, D.D., Kallman, B.R., Bhaskar, S.T., Fleischmann, A. and R. Axel. 2011. Driving opposing behaviors with ensembles of piriform neurons. *Cell* 146:1004–15.

Clancy, A.N., Singer, A.G., Macrides, F., Bronson, F.H. and W.C. Agosta. 1988. Experiential and endocrine dependence of gonadotropin responses in male mice to conspecific urine. *Biol Reprod* 38:183–91.

Coppola, D.M. and R.J. O'Connell. 1989. Stimulus access to olfactory and vomeronasal receptors in utero. *Neurosci Lett* 106:241–8.

Coquelin, A., Clancy, A.N., Macrides, F., Noble, E.P. and R.A. Gorski. 1984. Pheromonally induced release of luteinizing hormone in male mice: Involvement of the vomeronasal system. *J Neurosci* 4:2230–6.

Davison, I.G. and M.D. Ehlers. 2011. Neural circuit mechanisms for pattern detection and feature combination in olfactory cortex. *Neuron* 70:82–94.

de la Rosa-Prieto, C., Ubeda-Banon, I., Mohedano-Moriano, A., Pro-Sistiaga, P., Saiz-Sanchez, D., Insausti, R., and A. Martinez-Marcos. 2009. Subicular and CA1 hippo-campal projections to the accessory olfactory bulb. *Hippocampus* 19:124–9.

Del Punta, K., Puche, A., Adams, N.C., Rodriguez, I. and P. Mombaerts. 2002. A divergent pattern of sensory axonal projections is rendered convergent by second-order neurons in the accessory olfactory bulb. *Neuron* 35:1057–66.

De Olmos, J., Hardy, H. and L. Heimer. 1978. The afferent connections of the main and acces-sory olfactory bulb formations in the rat. An experimental HRP-study. *J Comp Neurol* 181:213–44.

Dluzen, D.E. and V.D. Ramirez.1983. Localized and discrete changes in neuropeptide (LHRH and TRH) and neurotransmitter (NE and DA) concentrations within the olfactory bulbs as a function of social interaction. *Horm Behav* 17:139–45.

Drickamer, L.C. 1999. Pregnancy in female house mice exposed to urinary chemosignals from other females. *J Reprod Fertil* 115:233–41.

Fan, S. and M. Luo. 2009. The organization of feedback projections in a pathway important for processing pheromonal signals. *Neuroscience* 161:489–500.

Ferrero D.M., Moeller, L.M., Osakada, T., Horio, N., Li, Q., Roy, D.S., Cichy, A., Spehr, M., Touhara, K. and S.D. Liberles. 2013. A juvenile mouse pheromone inhibits sexual behaviour through the vomeronasal system. *Nature* doi:10.1038/nature12579.

Fleischer, J., Schwarzenbacher, K., Besser, S., Hass, N. and H. Breer. 2006. Olfactory recep-tors and signalling elements in the Grueneberg ganglion. *J Neurochem* 98:543–54.

Gore, A.C., Wersinger, S.R. and E.F. Rissman. 2000. Effects of female pheromones on gonadotropin-releasing hormone gene expression and luteinizing hormone release in male wild-type and oestrogen receptor-alpha knockout mice. *J Neuroendocrinol* 12:1200–4.

Greer, C.A., Stewart, W.B., Teicher, M.H. and G.M. Shepherd. 1982. Functional development of the olfactory bulb and a unique glomerular complex in the neonatal rat. *J Neurosci* 2:1744–59.

Grosmaitre, X., Fuss, S.H., Lee, A.C., Adipietro, K.A., Matsunami, H., Mombaerts, P. and M. Ma. 2009. SR1, a mouse odorant receptor with an unusually broad response profile. *J Neurosci* 29:14545–52.

Guillamon, A. and S. Segovia. 1997. Sex differences in the vomeronasal system. *Brain Res Bull* 44:377–82.

Guo, J., Zhou, A. and R.L. Moss. 1997. Urine and urine-derive compounds induce c-fos mRNA expression in accessory olfactory bulb. *Neuroreport* 8:1679–83.

Guzzo, A.C., Jheon, J., Imtiaz, F. and D. deCatanzaro. 2012. Oestradiol transmission from males to females in the context of the Bruce and Vandenbergh effects in mice (*Mus musculus*). *Reproduction* 143:539–48.

Halem, H.A., Cherry, J.A. and M.J. Baum. 1999. Vomeronasal neuroepithelium and forebrain Fos responses to male pheromones in male and female mice. *J Neurobiol* 39:249–63.

Halem, H.A., Baum, M.J. and J.A. Cherry. 2001. Sex difference and steroid modulation of pheromone-induced immediate early genes in the two zones of the mouse accessory olfactory system. *J Neurosci* 21:2474–80.

Harvey, F.E. and J.J. Cowley. 1984. Effects of external chemical environment on the developing olfactory bulbs of the mouse (*Mus musculus*). *Brain Res Bull* 13:541–7.

Hasen, N.S. and S.C. Gammie. 2009. Trpc2 gene impacts on maternal aggression, accessory olfactory bulb anatomy and brain activity. *Genes Brain Behav* 8:639–49.

Hasen, N.S. and S.C. Gammie. 2011. Trpc2-deficient lactating mice exhibit altered brain and behavioral responses to bedding stimuli. *Behav Brain Res* 217:347–53.

Hendrickson, R.C., Krauthamer, S., Essenberg, J.M. and T.E. Holy. 2008. Inhibition shapes sex selectivity in the mouse accessory olfactory bulb. *J Neurosci* 28:12523–34.

Holy, T.E., Dulac, C. and M. Meister. 2000. Responses of vomeronasal neurons to natural stimuli. *Science* 289:1569–72.

Hoover, J.E. and L.C. Drickamer. 1979. Effects of urine from pregnant and lactating female house mice on oestrous cycles of adult females. *J Reprod Fertil* 55:297–301.

Hovis, K.R., Ramnath, R., Dahlen, J.E., Romanova, A.L., Larocca, G., Bier, M.E. and N.N. Urban. 2012. Activity regulates functional connectivity from the vomeronasal organ to the accessory olfactory bulb. *J Neurosci* 32:7907–16.

Illig, K.R. 2007. Developmental changes in odor-evoked activity in rat piriform cortex. *Neuroscience* 145:370–6.

Ishii, T. and P. Mombaerts. 2008. Expression of nonclassical class I major histocompatibility genes defines a tripartite organization of the mouse vomeronasal system. *J Neurosci* 28:2332–41.

Jacobson, L., Trotier, D. and K.B. Doving. 1998. 'Anatomical description of a new organ in the nose of the domesticated animals' by Ludvig Jacobson (1813). *Chem Senses* 23:743–54.

Jennings, J.H., Sparta, D.R., Stamatakis, A.M., Ung, R.L., Pleil, K.E., Kash, T.L. and G.D. Stuber. 2013. Distinct extended amygdala circuits for divergent motivational states. *Nature* 496:224–8.

Jouhanneau, M., Szymanski, L., Martini, M., Ella, A. and M. Keller. 2013. Kisspeptin: A new neuronal target of primer pheromones in the control of reproductive function in mammals. *Gen Comp Endocrinol* 188:3–8.

Kaba, H. and E.B. Keverne. 1988. The effect of microinfusions of drugs into the accessory olfactory bulb on the olfactory block to pregnancy. *Neuroscience* 25:1007–11.

Kaba, H., Rosser, A. and E.B. Keverne. 1989. Neural basis of olfactory memory in the context of pregnancy block. *Neuroscience* 32:657–62.

Kaba, H., Hayashi, Y., Higuchi, T. and S. Nakanishi. 1994. Induction of an olfactory memory by the activation of a metabotropic glutamate receptor. *Science* 265:262–4.

Kaluza, J.F., Gussing, F., Bohm, S., Breer, H. and J. Strotmann. 2004. Olfactory receptors in the mouse septal organ. *J Neurosci Res* 76:442–52.

Kang, N., Baum, M.J. and J.A. Cherry. 2009. A direct main olfactory bulb projection to the 'vomeronasal' amygdala in female mice selectively responds to volatile pheromones from males. *Eur J Neurosci* 29:624–34.

Kang, N., Baum, M.J. and J.A. Cherry. 2011. Different profiles of main and accessory olfactory bulb mitral/tufted cell projections revealed in mice using an anterograde tracer and a whole-mount, flattened cortex preparation. *Chem Senses* 36:251–60.

Kavaliers, M., Agmo, A., Choleris, E., Gustafsson, J.A., Korach, K.S., Muglia, L.J., Pfaff, D.W. and S. Ogawa. 2004. Oxytocin and estrogen receptor alpha and beta knockout mice provide discriminably different odor cues in behavioral assays. *Genes Brain Behav* 3:189–95.

Keller, M., Baum, M.J., Brock, O., Brennan, P.A. and J. Bakker. 2009. The main and the accessory olfactory systems interact in the control of mate recognition and sexual behavior. *Behav Brain Res* 200:268–76.

Kennedy, J.M. and K. Brown. 1970. Effects of male odor during infancy on the maturation, behavior, and reproduction of female mice. *Dev Psychobiol* 3:179–89.

Keverne, E.B. 2004. Importance of olfactory and vomeronasal systems for male sexual function. *Physiol Behav* 83:177–87.

Kim, S.Y., Adhikari, A., Lee, S.Y., Marshel, J.H., Kim, C.K., Mallory, C.S., Lo, M., Pak, S., Mattis, J., Lim, B.K., Malenka, R.C., Warden, M.R., Neve, R., Tye, K.M. and K. Deisseroth. 2013. Diverging neural pathways assemble a behavioural state from separable features in anxiety. *Nature* 496:219–23.

Kumar, A., Dudley, C.A. and R.L. Moss. 1999. Functional dichotomy within the vomeronasal system: Distinct zones of neuronal activity in the accessory olfactory bulb correlate with sex-specific behaviors. *J Neurosci* 19:RC32:1–6.

Lanuza, E., Novejarque, A., Martínez-Ricós, J., Martínez-Hernández, J., Agustín-Pavón, C. and F. Martínez-García. 2008. Sexual pheromones and the evolution of the reward system of the brain: The chemosensory function of the amygdala. *Brain Res Bull* 75:460–6.

Larriva-Sahd, J. 2008. The accessory olfactory bulb in the adult rat: A cytological study of its cell types, neuropil, neuronal modules, and interactions with the main olfactory system. *J Comp Neurol* 510:309–50.

Larsen, C.M. and D.R. Grattan. 2010. Exposure to female pheromones during pregnancy causes postpartum anxiety in mice. *Vitam Horm* 83:137–49.

Larsen, C.M., Kokaym, I.C. and D.R. Grattan. 2008. Male pheromones initiate prolactin-induced neurogenesis and advance maternal behavior in female mice. *Horm Behav* 53:509–17.

Lee, D.L. and J.L. Wilson. 2012. Urine from sexually mature intact male mice contributes to increased cardiovascular responses during free-roaming and restrained conditions. *ISRN Vet Sci* 2012:185461.

Lei, H., Mooney R. and L.C. Katz. 2006. Synaptic integration of olfactory information in mouse anterior olfactory nucleus. *J Neurosci* 26:12023–32.

Lèvai, O. and J. Strotmann. 2003. Projection pattern of nerve fibers from the septal organ: DiI-tracing studies with transgenic OMP mice. *Histochem Cell Biol* 120:483–92.

Lévai, O., Feistel, T., Breer, H. and J. Strotmann. 2006. Cells in the vomeronasal organ express odorant receptors but project to the accessory olfactory bulb. *J Comp Neurol* 498:476–90.

Lévy, F. and M. Keller. 2009. Olfactory mediation of maternal behavior in selected mammalian species. *Behav Brain Res* 200:336–45.

Li, C.S., Kaba, H., Saito, H. and K. Seto. 1990. Neural mechanisms underlying the action of primer pheromones in mice. *Neuroscience* 36:773–8.

Li, C.S., Kaba, H., Saito, H. and K. Seto. 1992. Cholecystokinin: Critical role in mediating olfactory influences on reproduction. *Neuroscience* 48:707–13.

Lin, D.Y., Zhang, S.Z., Block, E. and L.C. Katz. 2005. Encoding social signals in the mouse main olfactory bulb. *Nature* 434:470–7.

Lloyd-Thomas, A. and E.B. Keverne. 1982. Role of the brain and accessory olfactory system in the block to pregnancy in mice. *Neuroscience* 7:907–13.

Lombardi, J.R., and J.G. Vandenbergh. 1977. Pheromonally induced sexual maturation in females: Regulation by the social environment of the male. *Science* 196:545–6.

Luo, M. and L.C. Katz. 2004. Encoding pheromonal signals in the mammalian vomeronasal system. *Curr Opinion Neurobiol* 14:428–34.

Luo, M., Fee, M.S. and L.C. Katz. 2003. Encoding pheromonal signals in the accessory olfactory bulb of behaving mice. *Science* 299:1196–201.

Ma, M. 2007. Encoding olfactory signals via multiple chemosensory systems. *Crit Rev Biochem Mol Biol* 42:463–80.

Mak, G.K., Enwere, E.K., Gregg, C., Pakarainen, T., Poutanen, M., Huhtaniemi, I. and S. Weiss. 2007. Male pheromone-stimulated neurogenesis in the adult female brain: Possible role in mating behavior. *Nat Neurosci* 10:1003–11.

Marchlewska-Koj, A. and M. Jemioło. 1978. Evidence for the involvement of dopaminergic neurons in the pregnancy block effect. *Neuroendocrinology* 26:186–92.

Marchlewska-Koj, A. and M. Zacharczuk-Kakietek. 1990. Acute increase in plasma corticosterone level in female mice evoked by pheromones. *Physiol Behav* 48:577–80.

Martel, K.L. and M.J. Baum. 2007. Sexually dimorphic activation of the accessory, but not the main, olfactory bulb in mice by urinary volatiles. *Eur J Neurosci* 26:463–75.

Martel, K.L. and M.J. Baum. 2009. A centrifugal pathway to the mouse accessory olfactory bulb from the medial amygdala conveys gender-specific volatile pheromonal signals. *Eur J Neurosci* 29:368–76.

Martínez-García, F., Novejarque, A. and E. Lanuza. 2008. Two interconnected functional systems in the amygdala of amniote vertebrates. *Brain Res Bull* 75:206–13.

Martínez-Hernández, J., Lanuza, E. and F. Martínez-García. 2006. Selective dopaminergic lesions of the ventral tegmental area impair preference for sucrose but not for male sexual pheromones in female mice. *Eur J Neurosci* 24:885–93.

Martinez-Marcos, A. 2009. On the organization of olfactory and vomeronasal cortices. *Prog Neurobiol* 87:21–30.

Martínez-Ricós, J., Agustín-Pavón, C., Lanuza, E. and F. Martínez-García. 2008. Role of the vomeronasal system in intersexual attraction in female mice. *Neuroscience* 153:383–95.

Maruniak, J.A. and F.H. Bronson. 1976. Gonadotropic responses of male mice to female urine. *Endocrinology* 99:963–9.

McLean, J.H., Darby-King, A. and G.D. Paterno. 1995. Localization of 5-HT2A receptor mRNA by in situ hybridization in the olfactory bulb of the postnatal rat. *J Comp Neurol* 353:371–8.

Meeks, J.P., Arnson, H.A. and T.E. Holy. 2010. Representation and transformation of sensory information in the mouse accessory olfactory system. *Nat Neurosci* 13:723–30.

Miller, S.S. and N.E. Spear. 2008. Olfactory learning in the rat neonate soon after birth. *Dev Psychobiol* 50:554–65.

Miyamichi, K., Amat, F., Moussavi, F., Wang, C., Wickersham, I., Wall, N.R., Taniguchi, H., Tasic, B., Huang, Z.J., He, Z., Callaway, E.M., Horowitz, M.A. and L. Luo. 2011. Cortical representations of olfactory input by trans-synaptic tracing. *Nature* 472:191–6.

Moffatt, C.A. 2003. Steroid hormone modulation of olfactory processing in the context of socio-sexual behaviors in rodents and humans. *Brain Res Rev* 43:192–206.

Mohedano-Moriano, A., Pro-Sistiaga, P., Ubeda-Bañon, I., de la Rosa-Prieto, C., Saiz-Sanchez, D. and A. Martinez-Marcos. 2008. V1R and V2R segregated vomeronasal pathways to the hypothalamus. *Neuroreport* 19:1623–6.

Mohedano-Moriano, A., de la Rosa-Prieto, C., Saiz-Sanchez, D., Ubeda-Bañon, I., Pro-Sistiaga, P., de Moya-Pinilla, M. and A. Martinez-Marcos. 2012. Centrifugal telencephalic afferent connections to the main and accessory olfactory bulbs. *Front Neuroanat* 6:19.

Moncho-Bogani, J., Lanuza, E., Hernández, A., Novejarque, A. and F. Martínez-García. 2002. Attractive properties of sexual pheromones in mice: Innate or learned? *Physiol Behav* 77:167–76.

Moncho-Bogani, J., Martinez-Garcia, F., Novejarque, A. and E. Lanuza. 2005. Attraction to sexual pheromones and associated odorants in female mice involves activation of the reward system and basolateral amygdala. *Eur J Neurosci* 21:2186–98.

Moriceau, S., Wilson, D.A., Levine, S. and R.M. Sullivan. 2006. Dual circuitry for odor-shock conditioning during infancy: Corticosterone switches between fear and attraction via amygdala. *J Neurosci* 26:6737–48.

Moriya-Ito, K., Endoh, K., Fujiwara-Tsukamoto, Y. and M. Ichikawa. 2013. Three-dimensional reconstruction of electron micrographs reveals intrabulbar circuit differences between accessory and main olfactory bulbs. *Front Neuroanat* 7:5.

Mucignat-Caretta, C. 2010. The rodent accessory olfactory system. *J Comp Physiol A* 196:767–77.

Mucignat-Caretta, C., Caretta, A. and A. Cavaggioni. 1995. Pheromonally accelerated puberty is enhanced by previous experience of the same stimulus. *Physiol Behav* 57:901–3.

Mucignat-Caretta, C., Caretta, A. and E. Baldini. 1998. Protein-bound male urinary pheromones: Differential responses according to age and gender. *Chem Senses* 23:67–70.

Mucignat-Caretta, C., Cavaggioni, A. and A. Caretta. 2004. Male urinary chemosignals differentially affect aggressive behavior in male mice. *J Chem Ecol* 30:777–91.

Mucignat-Caretta, C., Colivicchi, M.A., Fattori, M., Ballini, C., Bianchi, L., Gabai, G., Cavaggioni, A. and L. Della Corte. 2006. Species-specific chemosignals evoke delayed excitation of the vomeronasal amygdala in freely-moving female rats. *J Neurochem* 99:881–91.

Mugford, R.A. and N.W. Nowell. 1970. Pheromones and their effect on aggression in mice. *Nature* 226:967–8.

Mugford, R.A. and N.W. Nowell. 1971. The preputial glands as a source of aggression-promoting odors in mice. *Physiol Behav* 6:247–9.

Niimi, K., Horie, S., Yokosuka, M., Kawakami-Mori, F., Tanaka, K., Fukayama, H. and Y. Sahara. 2012. Heterogeneous electrophysiological and morphological properties of neurons in the mouse medial amygdala in vitro. *Brain Res* 1480:41–52.

Noack, J., Richter, K., Laube, G., Haghgoo, H.A., Veh, R.W. and M. Engelmann. 2010. Different importance of the volatile and non-volatile fractions of an olfactory signature for individual social recognition in rats versus mice and short-term versus long-term memory. *Neurobiol Learn Mem* 94:568–75.

Noaín, D., Pérez-Millán, M.I., Bello, E.P., Luque, G.M., Casas Cordero, R., Gelman, D.M., Peper, M., Tornadu, I.G., Low, M.J., Becú-Villalobos, D. and M. Rubinstein. 2013. Central dopamine d2 receptors regulate growth-hormone-dependent body growth and pheromone signaling to conspecific males. *J Neurosci* 33:5834–42.

Novejarque, A., Gutiérrez-Castellanos, N., Lanuza, E. and F. Martínez-García. 2011. Amygdaloid projections to the ventral striatum in mice: Direct and indirect chemosensory inputs to the brain reward system. *Front Neuroanat* 5:54.

Okere, C.O. 2012. Differential plasticity of microglial cells in the rostrocaudal neuraxis of the accessory olfactory bulb of female mice following mating and stud male exposure. *Neurosci Lett* 514:116–21.

Pankevich, D., Baum, M.J. and J.A. Cherry. 2003. Removal of the superior cervical ganglia fails to block Fos induction in the accessory olfactory system of male mice after exposure to female odors. *Neurosci Lett* 345:13–16.

Pro-Sistiaga, P., Mohedano-Moriano, A., Ubeda-Bañon, I., Del Mar Arroyo-Jimenez, M., Marcos, P., Artacho-Pérula, E., Crespo, C., Insausti, R. and A. Martinez-Marcos. 2007. Convergence of olfactory and vomeronasal projections in the rat basal telencephalon. *J Comp Neurol* 504:346–62.

Raisman, G. 1972. An experimental study of the projection of the amygdala to the accessory olfactory bulb and its relationship to the concept of a dual olfactory system. *Exp Brain Res* 14:395–408.

Restrepo, D., Arellano, J., Oliva, A.M., Schaefer, M.L. and W. Lin. 2004. Emerging views on the distinct but related roles of the main and accessory olfactory systems in responsiveness to chemosensory signals in mice. *Horm Behav* 46:247–56.

Roberts, S.A., Davidson, A.J., McLean, L., Beynon, R.J. and J.L. Hurst. 2012. Pheromonal induction of spatial learning in mice. *Science* 338:1462–5.

Rodriguez, I., Feinstein, P. and P. Mombaerts. 1999. Variable patterns of axonal projections of sensory neurons in the mouse vomeronasal system. *Cell* 97:199–208.

Roppolo, D., Ribaud, V., Jungo, V.P., Lüscher, C. and I. Rodriguez. 2006. Projection of the Grüneberg ganglion to the mouse olfactory bulb. *Eur J Neurosci* 23:2887–94.

Rosser, A.E. and E.B. Keverne. 1985. The importance of central noradrenergic neurones in the formation of an olfactory memory in the prevention of pregnancy block. *Neuroscience* 15:1141–7.

Rottman, S.J. and C.T. Snowdon. 1972. Demonstration and analysis of an alarm pheromone in mice. *J Comp Physiol Psychol* 81:483–90.

Salazar, I., Sanchez-Quinteiro, P., Cifuentes, J.M. and P. Fernandez De Troconiz. 2006. General organization of the perinatal and adult accessory olfactory bulb in mice. *Anat Rec A Discov Mol Cell Evol Biol* 288:1009–25.

Samuelsen, C.L. and M. Meredith. 2009. Categorization of biologically relevant chemical signals in the medial amygdala. *Brain Res* 1263:33–42.

Scalia, F. and S.H. Winans. 1975. The differential projections of the olfactory bulb and accessory olfactory bulb in mammals. *J Comp Neurol* 161:31–56.

Scott J.W. and D.W. Pfaff. 1970. Behavioral and electrophysiological responses of female mice to male urine odors. *Physiol Behav* 5:407–11.

Serguera, C., Triaca, V., Kelly-Barrett, J., Banchaabouchi, M.A. and L. Minichiello. 2008. Increased dopamine after mating impairs olfaction and prevents odor interference with pregnancy. *Nat Neurosci* 11:949–56.

Shpak, G., Zylbertal, A., Yarom, Y. and S. Wagner. 2012. Calcium-activated sustained firing responses distinguish accessory from main olfactory bulb mitral cells. *J Neurosci* 32:6251–62.

Sosulski, D.L., Bloom, M.L., Cutforth, T., Axel, R. and S.R. Datta. 2011. Distinct representations of olfactory information in different cortical centres. *Nature* 472:213–16.

Suárez, R., García-González, D. and F. deCastro. 2012. Mutual influences between the main olfactory and vomeronasal systems in development and evolution. *Front Neuroanat* 6:50.

Sugai, T., Miyazawa, T., Yoshimura, H. and N. Onoda. 2005. Developmental changes in oscillatory and slow responses of the rat accessory olfactory bulb. *Neuroscience* 134:605–16.

Tachikawa, K.S., Yoshihara, Y. and K.O. Kuroda. 2013. Behavioral transition from attack to parenting in male mice: A crucial role of the vomeronasal system. *J Neurosci* 33:5120–6.

Takahashi, Y. and H. Kaba. 2010. Muscarinic receptor type 1 (M1) stimulation, probably through KCNQ/Kv7 channel closure, increases spontaneous GABA release at the dendrodendritic synapse in the mouse accessory olfactory bulb. *Brain Res* 1339:26–40.

Taniguchi, M., Yokoi, M., Shinohara, Y., Okutani, F., Murata, Y., Nakanishi, S. and H. Kaba. 2013. Regulation of synaptic currents by mGluR2 at reciprocal synapses in the mouse accessory olfactory bulb. *Eur J Neurosci* 37:351–8.

Teicher, M.H., Stewart, W.B., Kauer, J.S. and G.M. Shepherd. 1980. Suckling pheromone stimulation of a modified glomerular region in the developing rat olfactory bulb revealed by the 2-deoxyglucose method. *Brain Res* 194:530–5.

Thompson, R.N., Napier, A. and K.S. Wekesa. 2007. Chemosensory cues from the lacrimal and preputial glands stimulate production of IP3 in the vomeronasal organ and aggression in male mice. *Physiol Behav* 90:797–802.

Tian H. and M. Ma. 2008. Differential development of odorant receptor expression patterns in the olfactory epithelium: A quantitative analysis in the mouse septal organ. *Dev Neurobiol* 68:476–86.

Upadhya, S.C., Smith, T.K., Brennan, P.A., Mychaleckyj, J.C. and A.N. Hegde. 2011. Expression profiling reveals differential gene induction underlying specific and non-specific memory for pheromones in mice. *Neurochem Int* 59:787–803.

Vandenbergh, J.G. 1969. Male odor accelerates female sexual maturation in mice. *Endocrinology* 84:658–60.

Van der Lee, S. and L.M. Boot. 1955. Spontaneous pseudopregnancy in mice. *Acta Physiol Pharmacol Neerl* 4:442–4.

Varga, A.G. and D.W. Wesson. 2013. Distributed auditory sensory input within the mouse olfactory cortex. *Eur J Neurosci* 37:564–71.

Von Campenhausen, H. and K. Mori. 2000. Convergence of segregated pheromonal pathways from the accessory olfactory bulb to the cortex in the mouse. *Eur J Neurosci* 12:33–46.

Wacker, D.W. and M. Ludwig. 2012. Vasopressin, oxytocin, and social odor recognition. *Horm Behav* 61:259–65.

Wagner, S., Gresser, A.L., Torello, A.T. and C. Dulac. 2006. A multireceptor genetic approach uncovers an ordered integration of VNO sensory inputs in the accessory olfactory bulb. *Neuron* 50:697–709.

Wang, Z. and D.R. Storm. 2011. Maternal behavior is impaired in female mice lacking type 3 adenylyl cyclase. *Neuropsychopharmacology* 36:772–81.

Whitney, G., Alpern, M., Dizinno, G. and G. Horowitz. 1974. Female odors evoke ultrasounds from male mice. *Anim Learn Behav* 2:13–18.

Winans, S.S. and F. Scalia. 1970. Amygdaloid nucleus: New afferent input from the vomero-nasal organ. *Science* 170:330–2.

Wolff, J.O. 2007. Social biology of rodents. *Integrative Zool* 2:193–204.

Wysocki, C.J. and J.J. Lepri. Consequences of removing the vomeronasal organ. *J Steroid Biochem Mol Biol* 39:661–9.

Wysocki, C.J., Katz, Y. and R. Bernhard. 1983. Male vomeronasal organ mediates female-induced testosterone surges in mice. *Biol Reprod* 28:917–22.

Xu, F., Schaefer, M., Kida, I., Schafer, J., Liu, N., Rothman, D.L., Hyder, F., Restrepo, D. and G.M. Shepherd. 2005. Simultaneous activation of mouse main and accessory olfactory bulbs by odors or pheromones. *J Comp Neurol* 489:491–500.

Xu, M., Hill, J.W. and J.E. Levine. 2000. Attenuation of luteinizing hormone surges in neuro-peptide Y knockout mice. *Neuroendocrinology* 72:263–71.

Yokosuka, M., Matsuoka, M., Ohtani-Kaneko, R., Iigo, M., Hara, M., Hirata, K. and M. Ichikawa. 1999. Female-soiled bedding induced fos immunoreactivity in the ventral part of the premammillary nucleus (PMv) of the male mouse. *Physiol Behav* 68:257–61.

Zalaquett, C. and D. Thiessen. 1991. The effects of odors from stressed mice on conspecific behavior. *Physiol Behav* 50:221–7.

Zibman, S., Shpak, G. and S. Wagner. 2011. Distinct intrinsic membrane properties determine differential information processing between main and accessory olfactory bulb mitral cells. *Neuroscience* 189:51–67.

Zimnik, N.C., Treadway, T., Smith, R.S. and R.C. Araneda. 2013. α1A-Adrenergic regulation of inhibition in the olfactory bulb. *J Physiol* 591:1631–43.

12 Molecular and Neural Mechanisms of Pheromone Reception in the Rat Vomeronasal System and Changes in the Pheromonal Reception by the Maturation and Sexual Experiences

Makoto Kashiwayanagi

CONTENTS

12.1 Introduction 348
12.2 Neural Pathways of the Vomeronasal System 349
12.3 Minor Role of Cyclic Adenosine Monophosphate in Rat Pheromone
Reception 349
12.4 Pheromonal Transduction Mediated via Phospholipase C in the
Mammalian Vomeronasal System 350
12.5 Pheromonal Transduction Dependent and Independent on TRPC2 351
12.6 Selective Pheromone Reception in VSNs 353
12.7 Pheromone Receptors...355
12.8 Projection of Pheromonal Information to the AOB 355
12.9 Interaction between Sexual Behavior and Gonadal Hormones 357
12.10 Functional Changes in Neurons at the AOB Induced by Sexual
Experience..358
12.11 Gonadal Hormones Induces Changes in Brain Functions 358
12.12 Gonadal Hormones Modulate GABAergic Functions 359

12.13 Modulation of Reproductive Behaviors by Neurosteroid via GABA$_A$
         Receptors at the AOB 360
References 360

Pheromones affect gonadal functions and sexual behaviors, which are mainly received by the vomeronasal sensory neurons (VSNs) in the vomeronasal organ (VNO). The mechanism of discrimination and transduction in the pheromone reception is simple in contrast to the reception of general odorants in the main olfactory system. Single olfactory sensory neurons respond to various kinds of odorants that have various molecular structures and odor qualities. Most single VSNs, however, receive only one kind of pheromone. In rodents, pheromones are received with two types of pheromone receptors such as VR1s and VR2s, which exist at the upper and lower layers of the vomeronasal sensory epithelium. Binding of pheromones with VR1s and VR2s activates phospholipase C via Gi and Go, respectively, which in turn induces production diacylglycerol (DAG) and inositol-1,4,5-trisphosphate (IP$_3$). DAG activates TRPC2, a TRP channel, which induces depolarization of the VSN. In general, DAG has been considered as only a second messenger inducing depolarization by pheromones. However, in mice lacking TRPC2, 2-heptanone, a mouse pheromone, and urine containing various pheromones of high concentration induced pheromonal responses in the VSN. Dialysis of IP$_3$ into the VSNs of the rat and turtle induced inward currents with increases in membrane conductance. These results suggest that IP$_3$ also plays a role as second messengers in the pheromonal reception.

To find a good mating partner, it is necessary that animals have superior vomeronasal systems. Pheromonal information received by VSNs is transmitted to the accessory olfactory bulb (AOB). Sexually experienced male rats have been shown to prefer estrous to diestrous female urine, while sexually inexperienced males do not exhibit these preferences. In the localized region of the AOB, many more neurons were activated in the sexually experienced male rats than in the inexperienced rats after exposure to urinary pheromones excreted from estrous females, suggesting that sexual experience in males enhances the transmission of reproductively salient information concerning potential estrous status to a specific region of the AOB.

12.1 INTRODUCTION

Pheromones, which provide specific information concerning the identity, gender, and endocrine status in a variety of mammals have been found in saliva, skin gland secretions, and urine (Halpern and Martinez-Marcos 2003; Powers and Winans 1975; Wysocki and Meredith 1987). The modulation of gonadal function by the smell of urine has been well established in rodent VNOs (Halpern and Martinez-Marcos 2003; Keverne et al. 1986; Wysocki and Meredith 1987). In rats, pheromones in urine excreted from males and females induce various changes in gonadal functions such as reflex ovulation in the absence of coitus and mounting (Johns et al. 1978), a reduction in the estrous cycle of females from 5 to 4 days (Chateau et al. 1976), and estrous synchrony among females living together (McClintock 1978). In addition, exposure to male pheromones elicits the release of luteinizing hormone (Rajendren et al. 1990) and estrous synchrony among females that are living together (McClintock 1978).

The vomeronasal system is a second olfactory system organized in parallel with the main olfactory system. The VNO, which is the peripheral chemoreceptor organ of the vomeronasal system, forms a bilateral tubular structure in the ventral part of the nasal cavity of rats. The interaction of a pheromone with receptors of VNSs initiates a sequential molecular event leading to action potential initiation. In the present chapter, we first addressed the mechanism of pheromone reception by the VNO in rats.

Chemical signals excreted from animals affect the sexual behavior of conspecific male and female animals. Information regarding the females' endocrine state is transmitted to males by means of urinary pheromones. Sexually experienced male rats prefer estrous to diestrous urine odor (Lydell and Doty 1972; Pfaff and Pfaffmann 1969). The VSNs project information to the mitral cells (MTCs) in the AOB located on the dorsocaudal surface of the main olfactory bulb (MOB) (Rajendren et al. 1990). Pheromonal information transmitted via the MTC is modified by GABA-immunoreactive interneurons at the periglomerular cell (PGC) layer and by granule cells (GCs) (Keverne et al. 1986; Halpern and Martinez-Marcos 2003; Wysocki and Meredith 1987).

Comparison of c-Fos expression, which is correlated with the neural activity of sexually experienced and inexperienced males after exposure to estrous or diestrous urine shows changes in cellular responses to urinary pheromones as related to sexual experience (Honda et al. 2008). In the lateral and rostral sectors of the PGC, many more Fos-immunoreactive (Fos-ir) cells were observed in the sexually experienced rats than in the inexperienced rats. In the latter part of this chapter, I will describe possible mechanisms concerning changes in neural function with sexual experiences and aging.

12.2 NEURAL PATHWAYS OF THE VOMERONASAL SYSTEM

In 1974, Barber and Raisman confirmed the projection of the VNO to the AOB by an autoradiographic investigation (Barber and Raisman 1974). The amygdala receives pheromonal information from the AOB (Scalia and Winans 1975). Pheromonal information is relayed by the corticomedial amygdala to the medial zone of the hypothalamus, thalamic, cortical, and midbrain regions concerned with regulating eye and head movements, and, thus visuomotor and directed attentional mechanisms (Risold and Swanson 1995; Segovia and Guillamon 1993). The claustrum is interconnected with the various regions of the cortex including olfactory related regions such as the prepiriform olfactory cortex and entorhinal cortex and also projects to the hippocampus (Crick and Koch 2005) and the amygdala.

12.3 MINOR ROLE OF CYCLIC ADENOSINE MONOPHOSPHATE IN RAT PHEROMONE RECEPTION

The mechanisms of the generation of odor responses and of the discrimination of odorants in the main olfactory system have been well explored in the past three decades. It is widely considered that olfactory G-protein-coupled receptors (GPCRs) are receptors for odorants. Binding of an odorant to olfactory GPCRs induces stimulation of the cyclic adenosine monophosphate (cAMP)-mediated cascade via an olfactory-specific G protein (G_{olf}), G_s type of G proteins or the IP_3-mediated cascade (Buck and Axel 1991; Huque and Bruch 1986; Pace et al. 1985). An increase of intraciliary cAMP and IP_3 level activates cAMP- and IP_3-gated cation channels, respectively (Kashiwayanagi 1996; Nakamura and Gold 1987; Restrepo et al. 1992), causing cell depolarization and triggering the discharge of action potentials.

Application of forskolin, GTP, and GTPγS induced cAMP accumulation in the rat vomeronasal epithelium (VNE) (Sasaki et al. 1999). The dose-response relationships of cAMP accumulation in the rat VNE induced by forskolin were similar to those in the rat olfactory epithelium (Sasaki et al. 1999). These results indicate that there is a functional cAMP-synthetic pathway in the rat VNE. Stimulation with many odorants increased in cAMP levels in rat olfactory cilia preparations (Boekhoff et al. 1990). Urine preparations excreted from male and female Wistar rats and male Donryu rats, which contain various pheromones, increased impulse frequency in rat vomeronasal sensory neurons (Inamura et al. 1999b). However, none of these urine preparations changed the cAMP levels in the rat VNE (Sasaki et al. 1999).

There is a possibility that some untested pheromones increase cAMP in vomeronasal tissue preparations. However, dialysis of concentrations of cAMP as high as 0.5 mM or 1 mM did not induce any response in rat VSNs (Liman and Corey 1996; Sasaki et al. 1999). In olfactory neurons of various animals, application of 0.5 mM or 1 mM cAMP induced large responses, which are maximum levels of magnitudes. These results indicate that cAMP is not a primary second messenger in the excitatory pheromonal responses of rat VSNs.

12.4 PHEROMONAL TRANSDUCTION MEDIATED VIA PHOSPHOLIPASE C IN THE MAMMALIAN VOMERONASAL SYSTEM

Increases in the impulse frequency of the rat VSNs in response to urine were blocked by U-73122 or neomycin, phospholipase C (PLC) inhibitors (Figure 12.1) (Holy et al. 2000; Inamura et al. 1997b). These results suggest that rat pheromonal responses are generated via PLC. In early biochemical experiments, accumulation of IP_3 induced by pheromones was measured to explore the transduction mechanism because IP_3 plays a role as a second messenger in olfactory reception. Activation of PLC induced by pheromones metabolizes PIP_2 to IP_3 and DAG. Therefore, the DAG levels are elevated simultaneously with accumulation of IP_3 induced by pheromones. Application of GTPγS significantly increased the activity, while the application of 1% male Wistar urine increased the activity only slightly. The accumulation of IP_3 greatly increased in response to 1% urine in the presence of 100 mM GTPγS, suggesting that the urine induces IP_3 accumulation in a GTP-dependent manner (Figure 12.2a). In the presence of 10 mM GTP, the male Wistar urine induced IP_3 accumulation in a dose-dependent manner (Figure 12.2b). Male Wistar urine as well as female Wistar urine and male Donryu urine induced IP_3 accumulation in the female Wistar rat (Figure 12.2c) (Sasaki et al. 1999), which indicates that rat urine preparations also induced accumulation of DAG in rat VNE. IP_3 accumulation in response to pheromones in mammals has

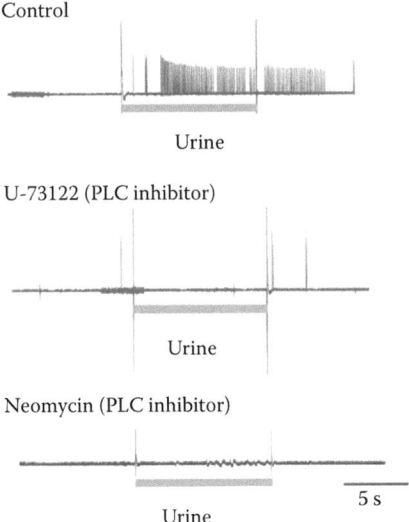

FIGURE 12.1   An increase in impulse frequency of a female Wistar vomeronasal sensory neuron recorded with the cell-attached mode in response to the male Wistar urine in normal tyrode solution without blockers, 10 mM U-73122 and 1 mM neomycin. (Modified from Inamura et al. (1997b) *Neurosci Lett* 233: 129–132.) Large electrical noise shown in the traces was induced by electrical actual valves at the "on" time of stimulation. PLC = phospholipase-C.

also been observed in the hamster and pig. That is, aphrodisin, a pheromone excreted from the female hamster, and semen fluid from the pig induce IP_3 accumulation in preparations of male hamster VNE (Kroner et al. 1996) and porcine VNE (Wekesa and Anholt 1997), respectively. These results suggest that various pheromones induce the generation of DAG and IP_3 in mammalian VSNs.

12.5   PHEROMONAL TRANSDUCTION DEPENDENT AND INDEPENDENT ON TRPC2

Various sensory stimuli such as temperature, touch, pain, osmolarity, pheromones, and taste are converted to electrical signals via TRP channels (Clapham 2003). TRPC2 localized at the sensory microvilli of rat VSNs (Liman et al. 1999). Application of low concentration of 2-heptanone or diluted female urine did not induce pheromonal responses and failed to initiate aggressive attacks to intruder males in mice lacking TRPC2 (Leypold et al. 2002), suggesting that TRPC2 is required to detect male-specific pheromones eliciting aggressive behaviors.

However, mice lacking TRPC2 responded to higher concentration of pheromones. The magnitudes of pheromonal responses to 0.1 mM 2-heptanone or female urine at dilution of $1/10^2$ in mice lacking TRPC2 were 25% and 37% of those of wild mice (Leypold et al. 2002). This suggests that responses to pheromones are also generated via

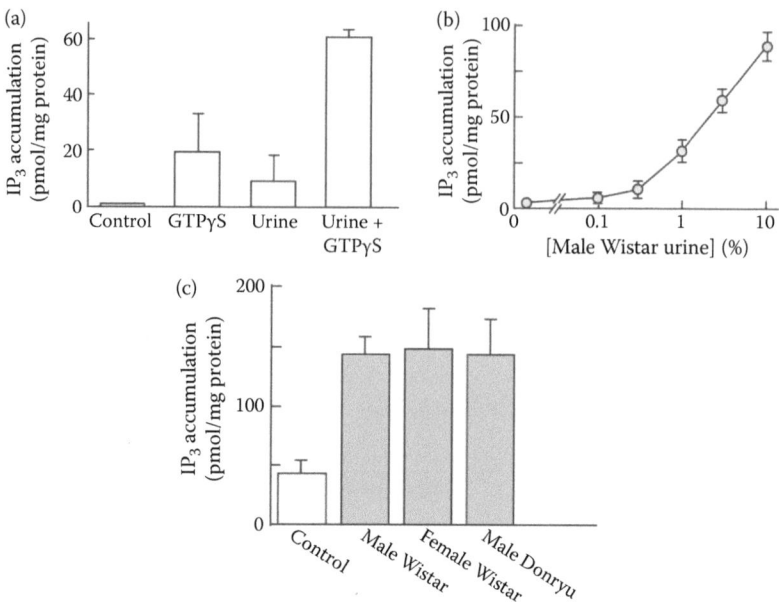

FIGURE 12.2 (a) Effects of 100 mM GTPγS, 1% male Wistar rat urine, and 1% male Wistar urine in the presence of 100 mM GTPγS on phospholipase-C activities. (b) The effects of various concentrations of male Wistar urine on phospholipase-C activity in the vomeronasal preparation in the presence of 10 mM GTP. (c) Effects of 5% male Wistar urine, 5% female Wistar urine, and 5% male Donryu urine on phospholipase-C activity. Urine preparations were applied in the presence of 10 mM GTP. (Modified from Sasaki et al. (1999) *Brain Res* 823: 161–168.)

the TRPC2-independent pathways in mice. In rats, dialysis of IP₃ into rat VSNs induced inward currents with increases in membrane conductance under the voltage-clamp condition (Figure 12.3a) (Inamura et al. 1997a). The average reversal potential of urine-induced current was similar to that of the IP₃-induced current in rat VSNs (Inamura and Kashiwayanagi 2000; Inamura et al. 1997a). These results indicate that the response of rat VSNs to urinary pheromones is generated not only via the DAG-dependent pathway but also via the IP₃-dependent pathway similar to mice (Figure 12.3b).

TRPC1 and TRPC3 have been demonstrated to interact with IP₃ receptors (IP₃-Rs) (Boulay et al. 1999; Rosado and Sage 2000). In rat VSNs, expression of IP₃R3 overlapped with that of TRPC2 (Brann et al. 2002). A peptide mimicking a domain on TRPC2 to which IP₃R3 binds reduced chemosignal-activated currents of musk turtle (Brann and Fadool 2006), suggesting that IP₃ generated by pheromones induce inward current via IP₃R3 and TRPC2 in rat VSNs. A recent study supports this idea. Homer 1b/c, an adaptor protein binding to proline-rich sequences, binds calcium signaling proteins (Brann and Fadool 2006). Homer 1b/c was immunoprecipitated with both IP₃R3 and TRPC2 (Mast et al. 2010). In the mouse VNE, expression of TRPC1 mRNA is confirmed by the RT-PCR (Matsui and Kashiwayanagi 2004). Therefore, it is possible that IP₃ itself, in addition to DAG, plays a role as a second messenger in pheromonal transduction.

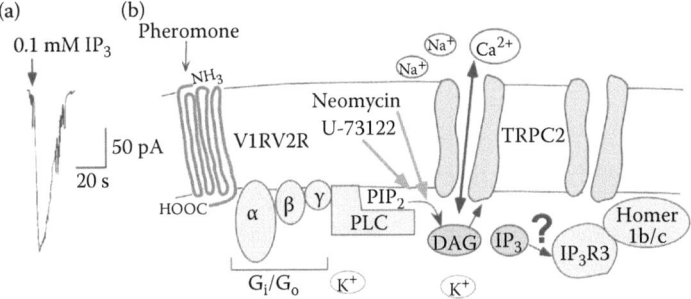

FIGURE 12.3 (a) IP_3-induced inward currents recorded from rat vomeronasal receptor neurons under the whole-cell voltage-damp condition. (Modified from Inamura et al. (1997a) *Chem Senses* 22: 93–103.) Membrane currents at the breakthrough measured with the pipette containing 0.1 mM IP_3 under voltage-damp with a holding potential at −70 mV. (b) Schematic representation of the transduction pathway in pheromonal reception.

12.6 SELECTIVE PHEROMONE RECEPTION IN VSNs

Selectivity of rat VSNs to pheromones is very high. Single VSNs of female Wistar rats responded selectively to urine from male and female Wistar rats and male Donryu and Sprague-Dawley rats (Figure 12.4) (Inamura et al. 1999b). Most each sensory neuron responded to only one class of urine. Thus, VSNs discriminate differences in

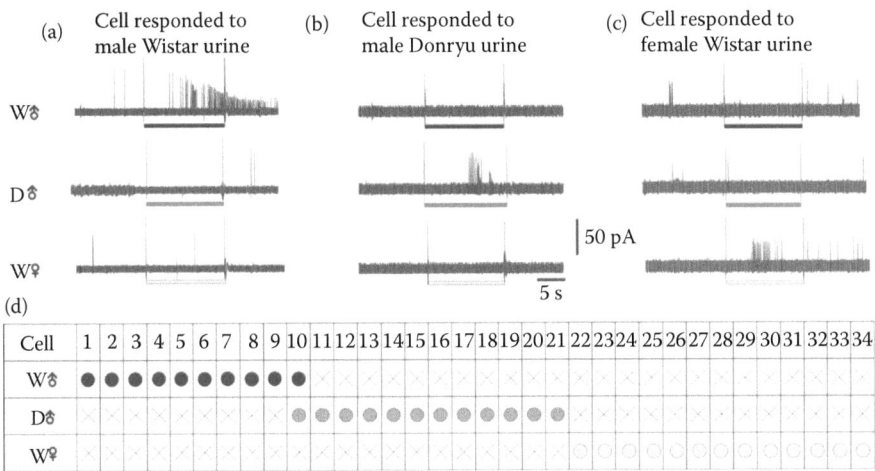

FIGURE 12.4 Responses of three typical neurons (a, b, and c) to the urine preparations from male Wistar, male Donryu, and female Wistar rats. (Modified from Inamura et al. (1999b) *J Physiol (Lond)* 517: 731–739.) Electrical noise shown in the traces was caused by electrically actuated valves at the "on" and "off" time of stimulation. (d) The response profiles of 34 single vomeronasal neurons of female Wistar rats to various urine preparations. Circle indicates that urine induced an increase in impulse frequency; X indicates no increase in impulse frequency on application of urine, and W and D indicate Wistar and Donryu rats, respectively.

sex and strains in urinary pheromones. High selectivity of VSNs to pheromones was also confirmed in mice (Leinders-Zufall et al. 2000). This is in contrast to olfactory neurons, which respond to many odorants with quite diverse molecular structures and odor qualities (Kang and Caprio 1995; Kashiwayanagi et al. 1996; Sicard and Holley 1984).

The accumulation of IP_3 induced by the male Wistar urine was inhibited by pertussis toxin (PTX) in the VNE of female Wistar rats (Figure 12.5a) (Sasaki et al. 1999). These results suggest that receptors to urinary pheromone(s) in the VNE are coupled with PTX-sensitive G proteins such as G_i and/or G_o. The male Wistar urine and male Donryu urine preferentially decreased ADP ribosylation of G_i and G_o with PTX in the vomeronasal sensory epithelium of female Wistar rats, respectively (Sasaki et al. 1999), suggesting that PLC are activated via G_i and G_o in response to the male Wistar urine and male Donryu urine, respectively.

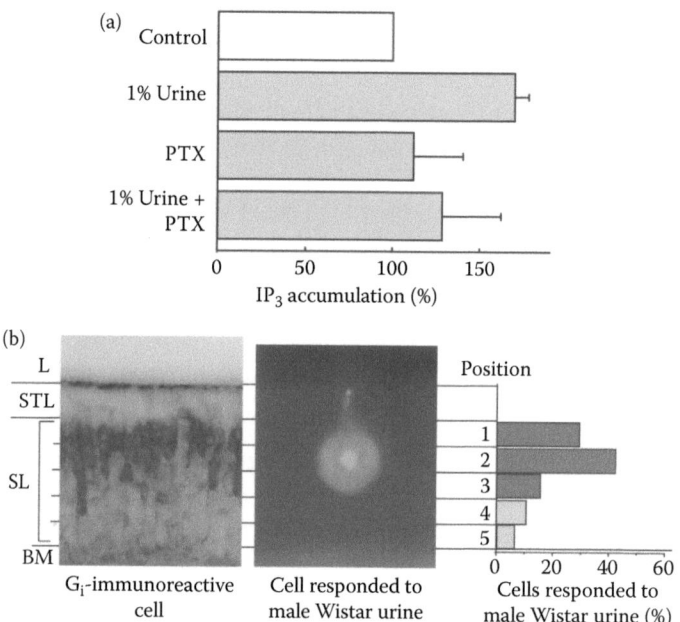

FIGURE 12.5 (**See color insert.**) (a) Inhibition of IP_3 accumulation in response to 1% male Wistar urine induced by the treatment with 50 mg/ml Pertussis toxin (PTX). (Modified from Sasaki et al. (1999) *Brain Res* 823: 161–168.) (b) Laminar distribution of neurons responding to various urine preparations. (Modified from Inamura et al. (1999b) *J Physiol (Lond)* 517: 731–739.) Left, sensory neurons immunoreactive to anti-$G_{i2\alpha}$ in the rat vomeronasal sensory epithelium. Center, vomeronasal sensory neuron dialyzed with 1% Lucifer yellow. Right, graphical representation of the lamina distribution of neurons responding to male Wistar rat urine. Abscissa indicates percentage of neurons responding to male Wistar urine. L = lumen of the vomeronasal organ, STL = layer of supporting cells, SL = layer of receptor neurons, BM = basal membrane.

Cell bodies of VSNs are located at various depths in the cellular layer of sensory epithelium. The VSNs at the apical and basal layers of the VNE of rats are immuno-reactive to anti-$G_{i2\alpha}$ and anti-$G_{o\alpha}$ proteins, respectively (Halpern et al. 1995; Jia and Halpern 1996). Localization of the cell bodies of VSNs of female Wistar rats, which responded to male Wistar and male Donryu urine, was examined using the slice patch-clamp technique (Figure 12.5b) (Inamura et al. 1999b). The VSNs of female rats, which respond to male Wistar urine, are localized in the apical layer of the epithelium where $G_{i2\alpha}$ is selectively expressed (Inamura et al. 1999b). The prepon-derance of neurons responding to conspecific female rat urine and male urine from a different strain (Donryu or Sprague-Dawley) were found in the G_o-positive laminae of the VSE. Thus, responses of VSNs in the apical portion of the female rat VSE to the male Wistar urine were mediated via G_i, while responses of neurons in the basal portion to the male Donryu urine were mediated via G_o.

12.7 PHEROMONE RECEPTORS

Two families (V1Rs and V2Rs) of vomeronasal GPCRs unrelated to olfactory GPCRs were cloned from the rat and mouse VNE (Dulac and Axel 1995; Herrada and Dulac 1997; Matsunami and Buck 1997; Ryba and Tirindelli 1997). Each family is composed of about 100 species of vomeronasal GPCR. V1Rs exist in the VSNs expressing $G_{i2\alpha}$ in the upper layer of the epithelium (Dulac and Axel 1995) and V2Rs exist in the neurons expressing $G_{o\alpha}$ (Herrada and Dulac 1997; Matsunami and Buck 1997; Ryba and Tirindelli 1997). There is good evidence that volatile mouse pheromones activate V1Rs and nonvolatiles activate V2Rs. Thus the VSN of the mutant mice lacking two of the 12 V1R gene families does not respond to some volatile pheromones (Del Punta et al. 2002), while application of a volatile phero-mone, 2-heptanone, induces an inward current in isolated V1Rb2 VSNs (Boschat et al. 2002). VSNs expressing a V2R respond to nine amino-acid peptide pheromones (Leinders-Zufall et al. 2004) and a male-specific 7 kDa peptide (ESP1) secreted from the extraorbital lacrimal gland stimulates V2R-expressing VSNs (Kimoto et al. 2005). These results suggest that volatile and nonvolatile pheromones are received by V1R- and V2R-expressing neurons of mice, respectively.

As described above, VSNs responding to male or female Wistar rat urine are localized in the apical or basal layer of the VNE of female Wistar rats, respectively (Inamura et al. 1999b), suggesting that the response to volatile compounds in male Wistar urine is induced via the former type of GPCR (i.e., V1R), whereas responses to nonvolatile compounds in female Wistar urine are induced via the latter type of GPCR (i.e., V2R). However, the pheromone reception in rats is more complicated, as described in the next section.

12.8 PROJECTION OF PHEROMONAL INFORMATION TO THE AOB

Mitral/tufted cells receive pheromonal information from VSNs at glomeruli in the AOB. The rostral region of the AOB is innervated by a population of V1R- and

$G_{i2\alpha}$-expressing VSNs whose cell bodies are located in the apical layer of the VSE (Halpern et al. 1995). The caudal region of the AOB glomerular layer is innervated by V2R- and $G_{o\alpha}$-expressing VSNs whose cell bodies are located in the basal layer of the VSE. The two regions of the AOB have a clear and sharp boundary midway along the rostrocaudal axis of the AOB. Electrical stimulation of the rostral vomeronasal nerve layer produces neural activity measured by the real-time optical imaging only within the rostral region of the glomerular layer, and electrical stimulation of the caudal vomeronasal nerve layer produces responses only within the caudal region (Sugai et al. 1997). This suggests that there is no anatomical connection between the rostral and caudal regions and that the AOB is functionally segregated into at least two subdivisions.

These two subregions receive different pheromonal information from VSNs (Brennan et al. 1999; Inamura et al. 1999a; Matsuoka et al. 1999). As described above, the VSNs that respond to the male Wistar urine are localized in the apical layer of the sensory epithelium, where V1Rs and $G_{i2\alpha}$ are selectively expressed, in female Wistar rats (Inamura et al. 1999b). Exposure of the VNO of female Wistar rat to male Wistar urine induced the appearance of many more Fos-ir cells in the rostral portion of the AOB than in the caudal portion (Inamura et al. 1999a). These results suggest that the response to pheromones in the male Wistar rat urine is preferentially transmitted to the rostral part of the AOB of female Wistar rats (Figure 12.6).

In rats, the activity of the component in male urine that induces expression of Fos-immunoreactivity in the rostral region as well as in the caudal region of the AOB of females is abolished by pronase treatment (Tsujikawa and Kashiwayanagi 1999).

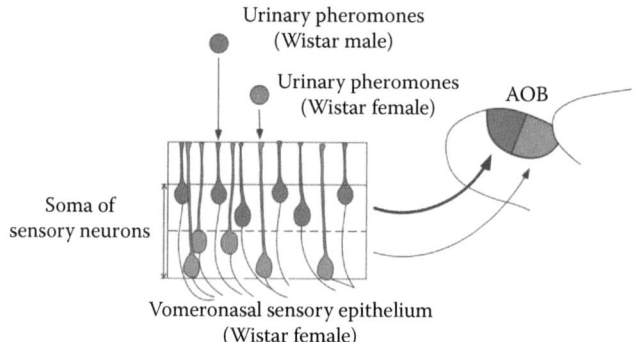

FIGURE 12.6 **(See color insert.)** Schematic representation of the transmission pathway of male Wistar urinary pheromone from the vomeronasal sensory neurons at the apical layer of the epithelium to the rostral region of the accessory olfactory bulb. The vomeronasal pathway in rodents. Pheromonal information of male and female Wistar rats are received by vomeronasal sensory neurons at upper and under layers of female Wistar rats, respectively. (From Inamura et al. (1999b) *J Physiol (Lond)* 517: 731–739.) Pheromonal information received by sensory neurons having Gi and G1Rs are transmitted to the rostral region of the AOB. (From Inamura et al. (1999a) *Eur J Neurosci* 11: 2254–2260.)

Exposure to crude urine excreted from male rats induces an increase in the plasma progesterone concentration in female rats. However, exposure of females in the estrous state to urine preparations treated with protease does not induce increases in plasma progesterone, suggesting that the presence of a protease-sensitive component in male urine exerts an influence on the endocrine state of estrous females of rats (Tomioka et al. 2005).

Exposure of the female rat VNO to either a dialyzed urine preparation (<500 Da) or the remaining constituents (>500 Da) of male rat urine does not induce expression of Fos-ir cells in the rostral and caudal regions of the AOB, whereas exposure to a mixture of these preparations induces expression of Fos-ir cells in the rostral and caudal regions (Yamaguchi et al. 2000). These findings suggest that a combination of low and high molecular weight substances is necessary for the activation of V1R- and V2R-expressing neurons and/or the increases in Fos-immunoreactivity in the rostral and caudal regions of the AOB of rats.

12.9 INTERACTION BETWEEN SEXUAL BEHAVIOR AND GONADAL HORMONES

Sexually experienced male rats prefer estrous to diestrous urine while sexually inexperienced males do not exhibit these preferences (Lydell and Doty 1972; Pfaff and Pfaffmann 1969). Mating itself and/or odors (pheromones) of mating partners induce(s) changes in levels of gonadal steroid hormones in male rats (Fowler et al. 2003; Frankel 1984; Smith et al. 1992). Plasma testosterone levels are higher in old mating rats than in old nonmating rats (Frankel 1984; Smith et al. 1992). Sexually experienced male rats show increases in levels of luteinizing hormone (LH), prolactin, and testosterone after mating with estrous females (Kamel et al. 1977). In contrast, LH levels of male rats are significantly elevated by exposure to the odor of estrous females while testosterone levels are unaffected (Kamel et al. 1977). These observations suggest that mating itself and/or contact with nonvolatile compounds induces an elevation of testosterone levels in male rats.

Olfactory function is markedly altered in old age and in a number of age-related diseases. Gonadal conditions are also altered with aging. Plasma concentrations of estradiol and allopregnanorone in young male rats are lower than those in old males (Bernardi et al. 1998; Luine et al. 2007). Levels of testosterone in young male rats (2.5–6 months old) and estradiol in young female rats are higher than those in old rats (18–20 months old) (Bernardi et al. 1998; Gruenewald et al. 2000; Luine et al. 2007). Endogenous changes in endocrine state, as well as the exogenous application of hormones, affect various sexual behaviors. Testosterone regulates the mating behavior of male hamsters by maintaining the responsiveness of magnocellular division of the medial preoptic nucleus to vaginal secretions from female hamsters (Swann and Fiber 1997). Pregnenolone sulfate exerts inhibitory effects on olfactory-mediated male mouse sexual interest, preference, and motivation (Kavaliers and Kinsella 1995). Therefore, it is possible that changes in gonadal function induced by sexual experience and aging affect behaviors that relate to olfaction.

12.10   FUNCTIONAL CHANGES IN NEURONS AT THE AOB INDUCED BY SEXUAL EXPERIENCE

Sexual experience induces an augmentation of Fos-immunoreactivity to estrous urine in the PGC layer in a localized region of the male AOB (Honda et al. 2008). In sexually experienced male rats, the Fos-ir cell density in the lateral-rostral sectors of the PGL of the AOB after exposure to estrous urine is much higher than that after exposure to diestrous urine (Figure 12.7a). In sexually inexperienced males, exposure to estrous urine is not associated with an increase in Fos-ir cell density in these sectors (Figure 12.7b). Thus, the preference of sexually experienced males for estrous urine may be brought about by an increase in the efficiency of synaptic transmission, which itself is a kind of memory, at the PGC in the localized region of the AOB.

12.11   GONADAL HORMONES INDUCES CHANGES IN BRAIN FUNCTIONS

The nervous system is widely influenced by gonadal hormones and the brain is continuously shaped by the changing hormone. For example, gonadal steroid hormones modulate the dendrites of neurons in the adult rat central nervous system. Testosterone induces an elongation of dendrites of the medial nucleus of the amygdala (Cooke and Woolley 2005). The pyramidal cells of the CA1 region in the dentate gyrus in the hippocampus also undergo phasic changes in dendritic spine and synapse density across the estrous cycle (Cooke and Woolley 2005), indicating that

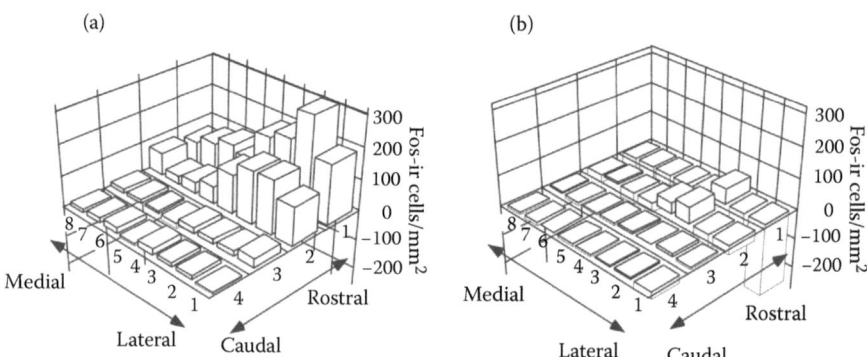

FIGURE 12.7   Localized difference in the expression of Fos-ir cells in the PGC layer when sexually experienced males (a) and inexperienced males (b) were stimulated with diestrous or estrous urine. (Modified from Honda et al. (2008) *Eur J Neurosci* 27: 1980–1988.) The vertical axis indicates a subtraction of the mean number of Fos-ir cells/mm^2 obtained from three animals after exposure to diestrous urine from that after exposure of to estrous urine. The rostral and caudal halves of the G_{i2}-positive region were defined as regions 1 and 2, respectively. The rostral and caudal halves of the G_{i2}-negative region were defined as regions 3 and 4, respectively.

gonadal steroids modulate the morphology of dendrites and patterns of synaptic connectivity.

In the mammalian brain, adult neurogenesis has been found to occur primarily in the forebrain subventricular zone (SVZ) and dentate gyrus of the hippocampus throughout the entire lifespan (Fowler et al. 2002). The SVZ generates an immense number of neurons (Lois and Alvarez-Buylla 1994), which migrate via the rostral migratory stream to the MOB. Morphological and electrophysiological studies have shown that these cells differentiate into GCs and PGCs in the MOB and AOB, even during adulthood (Bonfanti et al. 1997; Carleton et al. 2003; Graziadei and Graziadei 1979; Lois and Alvarez-Buylla 1994; Luskin 1993). Newly generated neurons play important roles in odor discrimination (Gheusi et al. 2000) and odor memory (Rochefort et al. 2002).

The 23-month-old male rats revealed deficits in sexual responses such as mounting and penile reflex behavior (Smith et al. 1992). In the adult hippocampus of rats, neurogenesis occurs at least up to 11 months of age (Kaplan and Bell 1984). The density of newly generated cells in the MOB of old rats is remarkably lower than that in young rats. The density of newly generated cells near the region of the AOB in the MOB of young rats is higher than that distant from the AOB. It is interesting that the density of newly generated cells near the AOB is lower than that distant from the AOB (Honda et al. 2009). These results suggest that the density of newly generated cells at the region near the AOB in the MOB may be concerned with age-dependent changes in ability of discrimination and memory of general odors and odors related to the reproduction. Gonadal steroid hormones play an important role in the proliferation, survival, and activation of neurons. Pregnenolone increases cell proliferation in the rat SVZ (Mayo et al. 2005).

12.12   GONADAL HORMONES MODULATE GABAergic FUNCTIONS

In the brain, gonadal hormones modulate expression and functions of GABA receptors. $GABA_A$ receptor subunit mRNA levels in the medial amygdala, medial preoptic area, and the ventromedial nucleus of the mouse hypothalamus undergo significant changes from puberty to adult (Bernardi et al. 1998). There are significant changes in expression of $GABA_A$ receptors and plasma concentration of gonadal hormones during aging (Henderson et al. 2006; Ramsey et al. 2004).

Expression of $GABA_A$ receptors are changed with levels of plasma gonadal hormone concentration in rat hippocampus during pregnancy and after delivery (Sanna et al. 2009). The stimulatory effect of muscimol, a $GABA_A$ receptor agonist, on $^{36}Cl^-$ uptake by cerebrocortical membrane vesicles decreased on pregnancy days 15 and 19 and increased two days after delivery (Concas et al. 1998). Treatment with 17β-estradiol induces an increase of inhibitory axosomatic synapses in the anteroventral periventricular nucleus (Kurunczi et al. 2009). Testosterone treatment enhances GABA-induced $^{36}Cl^-$ uptake into corticohippocampal synaptoneurosomes of male rats (Svensson et al. 2003).

12.13 MODULATION OF REPRODUCTIVE BEHAVIORS BY NEUROSTEROID VIA GABA$_A$ RECEPTORS AT THE AOB

The GABA$_A$ receptor is composed of five transmembrane subunits, such as $\alpha 1-6$, $\beta 1-3$, $\gamma 1-3$, δ, ϵ, θ, π, and $\rho 1-3$, arranged to form an intrinsic anion-conducting channel (Mitchell et al. 2008). In the GC layer of the rat AOB, mRNA for $\alpha 2$ and $\beta 3$ is highly expressed, while $\alpha 5$ mRNA is present to a much smaller degree (Laurie et al. 1992). Low to moderate levels of $\alpha 4$, $\gamma 1$, $\gamma 2$, $\gamma 3$, and δ mRNAs are found, while $\alpha 1$, $\alpha 3$, $\beta 1$, and $\beta 2$ mRNAs are absent from GCs in the AOB. A high degree of expression is observed in cells of the MTC layer complex of transcripts for $\alpha 1$, $\beta 1$, $\beta 2$, $\beta 3$, and $\gamma 2$, with mild expression of $\alpha 2$ and $\alpha 3$ mRNAs, with other subunit mRNAs being undetectable. A few cells of the glomerular layer contain moderate amounts of $\beta 2$ mRNA and low amounts of mRNAs for $\alpha 1$, $\alpha 3$, $\beta 3$, and $\gamma 1$. In the MTC of the mouse AOB, GABA$_A$-mediated inhibitory postsynaptic currents are observed (Taniguchi and Kaba 2001).

Intracerebroventricular administrations of the gonadal and neurosteroid, 3α-hydroxy-4-pregnen-20-one (3αHP), enhances male mice preference for the odors of estrous females while having no significant effect on the responses to the nonestrous female odors (Kavaliers et al. 1994). The effects of 3αHP are significantly reduced by peripheral administrations of the GABA$_A$ antagonists, bicuculline and picrotoxin. Androgen and estrogen receptor immunoreactive cells exist in the AOB, the medial amygdala, and the bed nucleus of the stria terminalis of rats (Portillo et al. 2006). The feedback projections from the bed nucleus of stria terminalis and the vomeronasal amygdala to the AOB are topographically organized in mice (Fan and Luo 2009). Thus, the gonadal and neurosteroids, such as 3αHP, have facilitatory effects on olfactory-mediated male sexual interest or motivation that involve interactions with the GABA$_A$ receptor at the AOB.

REFERENCES

Barber PC and Raisman G (1974) An autoradiographic investigation of the projection of the vomeronasal organ to the accessory olfactory bulb in the mouse. *Brain Res* 81: 21–30.

Bernardi F, Salvestroni C, Casarosa E, Nappi RE, Lanzone A, Luisi S, Purdy RH, Petraglia F and Genazzani AR (1998) Aging is associated with changes in allopregnanolone concentrations in brain, endocrine glands and serum in male rats. *Eur J Endocrinol* 138: 316–321.

Boekhoff I, Tareilus E, Strotmann J and Breer H (1990) Rapid activation of alternative second messenger pathways in olfactory cilia from rats by different odorants. *EMBO J* 9: 2453–2458.

Bonfanti L, Peretto P, Merighi A and Fasolo A (1997) Newly-generated cells from the rostral migratory stream in the accessory olfactory bulb of the adult rat. *Neuroscience* 81: 489–502.

Boschat C, Pelofi C, Randin O, Roppolo D, Luscher C, Broillet MC and Rodriguez I (2002) Pheromone detection mediated by a V1r vomeronasal receptor. *Nat Neurosci* 5: 1261–1262.

Boulay G, Brown DM, Qin N, Jiang M, Dietrich A, Zhu MX, Chen Z, Birnbaumer M, Mikoshiba K and Birnbaumer L (1999) Modulation of Ca(2+) entry by polypeptides of the inositol 1,4,5-trisphosphate receptor (IP3R) that bind transient receptor potential

(TRP): Evidence for roles of TRP and IP3R in store depletion-activated Ca(2+) entry. *Proc Natl Acad Sci U S A* 96: 14955–14960.

Brann JH and Fadool DA (2006) Vomeronasal sensory neurons from *Sternotherus odoratus* (stinkpot/musk turtle) respond to chemosignals via the phospholipase C system. *J Exp Biol* 209: 1914–1927.

Brann JH, Dennis JC, Morrison EE and Fadool DA (2002) Type-specific inositol 1,4,5-tris-phosphate receptor localization in the vomeronasal organ and its interaction with a transient receptor potential channel, TRPC2. *J Neurochem* 83: 1452–1460.

Brennan PA, Schellinck HM and Keverne EB (1999) Patterns of expression of the immediate-early gene egr-1 in the accessory olfactory bulb of female mice exposed to pheromonal constituents of male urine. *Neuroscience* 90: 1463–1470.

Buck L and Axel R (1991) A novel multigene family may encode odorant receptors: A molecular basis for odor recognition. *Cell* 65: 175–187.

Carleton A, Petreanu LT, Lansford R, Alvarez-Buylla A and Lledo PM (2003) Becoming a new neuron in the adult olfactory bulb. *Nat Neurosci* 6: 507–518.

Chateau D, Roos J, Plas-Roser S, Roos M and Aron C (1976) Hormonal mechanisms involved in the control of oestrous cycle duration by the odour of urine in the rat. *Acta Endcrinol* 82: 426–435.

Clapham DE (2003) TRP channels as cellular sensors. *Nature* 426: 517–524.

Concas A, Mostallino MC, Porcu P, Follesa P, Barbaccia ML, Trabucchi M, Purdy RH, Grisenti P and Biggio G (1998) Role of brain allopregnanolone in the plasticity of gamma-aminobutyric acid type A receptor in rat brain during pregnancy and after delivery. *Proc Natl Acad Sci U S A* 95: 13284–13289.

Cooke BM and Woolley CS (2005) Gonadal hormone modulation of dendrites in the mammalian CNS. *J Neurobiol* 64: 34–46.

Crick FC and Koch C (2005) What is the function of the claustrum? *Philos Trans R Soc Lond B Biol Sci* 360: 1271–1279.

Del Punta K, Leinders-Zufall T, Rodriguez I, Jukam D, Wysocki CJ, Ogawa S, Zufall F and Mombaerts P (2002) Deficient pheromone responses in mice lacking a cluster of vomeronasal receptor genes. *Nature* 419: 70–74.

Dulac C and Axel R (1995) A novel family of genes encoding putative pheromone receptors in mammals. *Cell* 83: 195–206.

Fan S and Luo M (2009) The organization of feedback projections in a pathway important for processing pheromonal signals. *Neuroscience* 161: 489–500.

Fowler CD, Freeman ME and Wang Z (2003) Newly proliferated cells in the adult male amygdala are affected by gonadal steroid hormones. *J Neurobiol* 57: 257–269.

Fowler CD, Liu Y, Ouimet C and Wang Z (2002) The effects of social environment on adult neurogenesis in the female prairie vole. *J Neurobiol* 51: 115–128.

Frankel AI (1984) Plasma testosterone levels are higher in old mating rats than in old non-mating rats. *Exp Gerontol* 19: 345–348.

Gheusi G, Cremer H, McLean H, Chazal G, Vincent JD and Lledo PM (2000) Importance of newly generated neurons in the adult olfactory bulb for odor discrimination. *Proc Natl Acad Sci U S A* 97: 1823–1828.

Graziadei PP and Graziadei GA (1979) Neurogenesis and neuron regeneration in the olfactory system of mammals. I. Morphological aspects of differentiation and structural organization of the olfactory sensory neurons. *J Neurocytol* 8: 1–18.

Gruenewald DA, Naai MA, Marck BT and Matsumoto AM (2000) Age-related decrease in hypothalamic gonadotropin-releasing hormone (GnRH) gene expression, but not pituitary responsiveness to GnRH, in the male Brown Norway rat. *J Androl* 21: 72–84.

Halpern M and Martinez-Marcos A (2003) Structure and function of the vomeronasal system: An update. *Prog Neurobiol* 70: 245–318.

Halpern M, Shapiro LS and Jia C (1995) Differential localization of G proteins in the opossum vomeronasal system. *Brain Res* 677: 157–161.

Henderson LP, Penatti CA, Jones BL, Yang P and Clark AS (2006) Anabolic androgenic steroids and forebrain GABAergic transmission. *Neuroscience* 138: 793–799.

Herrada G and Dulac C (1997) A novel family of putative pheromone receptors in mammals with a topographically organized and sexually dimorphic distribution. *Cell* 90: 763–773.

Holy TE, Dulac C and Meister M (2000) Responses of vomeronasal neurons to natural stimuli. *Science* 289: 1569–1572.

Honda N, Sakamoto H, Inamura K and Kashiwayanagi M (2008) Changes in Fos expression in the accessory olfactory bulb of sexually experienced male rats after exposure to female urinary pheromones. *Eur J Neurosci* 27: 1980–1988.

Honda N, Sakamoto H, Inamura K and Kashiwayanagi M (2009) Age-dependent spatial distribution of bromodeoxyuridine-immunoreactive cells in the main olfactory bulb. *Biol Pharm Bull* 32: 627–630.

Huque T and Bruch RC (1986) Odorant- and guanine nucleotide-stimulated phophoinositide turnover in olfactory cilia. *Biochem Biophys Res Commun* 137: 36–42.

Inamura K and Kashiwayanagi M (2000) Inward current responses to urinary substances in rat vomeronasal sensory neurons. *Eur J Neurosci* 12: 3529–3536.

Inamura K, Kashiwayanagi M and Kurihara K (1997a) Inositol-1,4,5-trisphosphate induces responses in receptor neurons in rat vomeronasal sensory slices. *Chem Senses* 22: 93–103.

Inamura K, Kashiwayanagi M and Kurihara K (1997b) Blockage of urinary responses by inhibitors for IP_3-mediated pathway in rat vomeronasal sensory neurons. *Neurosci Lett* 233: 129–132.

Inamura K, Kashiwayanagi M and Kurihara K (1999a) Regionalization of Fos immunostaining in rat accessory olfactory bulb when the vomeronasal organ was exposed to urine. *Eur J Neurosci* 11: 2254–2260.

Inamura K, Matsumoto Y, Kashiwayanagi M and Kurihara K (1999b) Laminar distribution of pheromone-receptive neurons in rat vomeronasal epithelium. *J Physiol (Lond)* 517: 731–739.

Jia CP and Halpern M (1996) Subclasses of vomeronasal receptor neurons: Differential expression of G proteins ($G_{i\alpha2}$ and $G_{o\alpha}$) and segregated projections to the accessory olfactory bulb. *Brain Res* 719: 117–128.

Johns MA, Feder HH, Komisaruk BR and Mayer AD (1978) Urine-induced reflex ovulation in anovulatory rats may be a vomeronasal effect. *Nature* 272: 446–448.

Kamel F, Wright WW, Mock EJ and Frankel AI (1977) The influence of mating and related stimuli on plasma levels of luteinizing hormone, follicle stimulating hormone, prolactin, and testosterone in the male rat. *Endocrinology* 101: 421–429.

Kang J and Caprio J (1995) In vivo responses of single olfactory receptor neurons in the channel catfish, *Ictalurus punctatus*. *J Neurophysiol* 73: 172–177.

Kaplan MS and Bell DH (1984) Mitotic neuroblasts in the 9-day-old and 11-month-old rodent hippocampus. *J Neurosci* 4: 1429–1441.

Kashiwayanagi M (1996) Dialysis of inositol 1,4,5-trisphosphate induces inward currents and Ca^{2+} uptake in frog olfactory receptor cells. *Biochem Biophys Res Commun* 225: 666–671.

Kashiwayanagi M, Shimano K and Kurihara K (1996) Existence of multiple receptors in single neurons: Responses of single bullfrog olfactory neurons to many cAMP-dependent and independent odorants. *Brain Res* 738: 222–228.

Kavaliers M and Kinsella DM (1995) Male preference for the odors of estrous female mice is reduced by the neurosteroid pregnenolone sulfate. *Brain Res* 682: 222–226.

Kavaliers M, Wiebe JP and Galea LAM (1994) Male preference for the odors of estrous female mice is enhanced by the neurosteroid 3α-hydroxy-4-pregnen-20-one (3αHP). *Brain Res* 646: 140–144.

Keverne EB, Murphy CL, Silver WL, Wysocki CJ and Meredith M (1986) Non-olfactory chemoreceptors of the nose: Recent advances in understanding the vomeronasal and trigeminal systems. *Chem Senses* 11: 119–133.

Kimoto H, Haga S, Sato K and Touhara K (2005) Sex-specific peptides from exocrine glands stimulate mouse vomeronasal sensory neurons. *Nature* 437: 898–901.

Kroner C, Breer H, Singer AG and O'Connell RJ (1996) Pheromone-induced second messenger signaling in the hamster vomeronasal organ. *Neuroreport* 7: 2989–2992.

Kurunczi A, Hoyk Z, Csakvari E, Gyenes A and Parducz A (2009) 17beta-Estradiol-induced remodeling of GABAergic axo-somatic synapses on estrogen receptor expressing neurons in the anteroventral periventricular nucleus of adult female rats. *Neuroscience* 158: 553–557.

Laurie DJ, Seeburg PH and Wisden W (1992) The distribution of 13 GABA$_A$ receptor subunit mRNAs in the rat brain. II. Olfactory bulb and cerebellum. *J Neurosci* 12: 1063–1076.

Leinders-Zufall T, Brennan P, Widmayer P, PC S, Maul-Pavicic A, Jager M, Li XH, Breer H, Zufall F and Boehm T (2004) MHC class I peptides as chemosensory signals in the vomeronasal organ. *Science* 306: 1033–1037.

Leinders-Zufall T, Lane AP, Puche AC, Ma W, Novotny MV, Shipley MT and Zufall F (2000) Ultrasensitive pheromone detection by mammalian vomeronasal neurons. *Nature* 405: 792–796.

Leypold BG, Yu CR, Leinders-Zufall T, Kim MM, Zufall F and Axel R (2002) Altered sexual and social behaviors in trp2 mutant mice. *Proc Natl Acad Sci U S A* 99: 6376–6381.

Liman ER and Corey DP (1996) Electrophysiological characterization of chemosensory neurons from the mouse vomeronasal organ. *J Neurosci* 16: 4625–4637.

Liman ER, Corey DP and Dulac C (1999) TRP2: A candidate transduction channel for mammalian pheromone sensory signaling. *Proc Natl Acad Sci U S A* 96: 5791–5796.

Lois C and Alvarez-Buylla A (1994) Long-distance neuronal migration in the adult mammalian brain. *Science* 264: 1145–1148.

Luine VN, Beck KD, Bowman RE, Frankfurt M and Maclusky NJ (2007) Chronic stress and neural function: accounting for sex and age. *J Neuroendocrinol* 19: 743–751.

Luskin MB (1993) Restricted proliferation and migration of postnatally generated neurons derived from the forebrain subventricular zone. *Neuron* 11: 173–189.

Lydell K and Doty RL (1972) Male rat odor preferences for female urine as a function of sexual experience, urine age, and urine source. *Horm Behav* 3: 205–212.

Mast TG, Brann JH and Fadool DA (2010) The TRPC2 channel forms protein-protein interactions with Homer and RTP in the rat vomeronasal organ. *BMC Neurosci* 11: 61.

Matsunami H and Buck LB (1997) A multigene family encoding a diverse array of putative pheromone receptors in mammals. *Cell* 90: 775–784.

Matsuoka M, Yokosuka M, Mori Y and Ichikawa M (1999) Specific expression pattern of Fos in the accessory olfactory bulb of male mice after exposure to soiled bedding of females. *Neurosci Res* 35: 189–195.

Mayo W, Lemaire V, Malaterre J, Rodriguez JJ, Cayre M, Stewart MG, Kharouby M, Rougon G, Le Moal M, Piazza PV and Abrous DN (2005) Pregnenolone sulfate enhances neurogenesis and PSA-NCAM in young and aged hippocampus. *Neurobiol Aging* 26: 103–114.

McClintock MK (1978) Estrous synchrony and its mediation by airborne chemical communication (*Rattus norvegicus*). *Horm Behav* 10: 264–276.

Mitchell EA, Herd MB, Gunn BG, Lambert JJ and Belelli D (2008) Neurosteroid modulation of GABAA receptors: Molecular determinants and significance in health and disease. *Neurochem Int* 52: 588–595.

Nakamura T and Gold GH (1987) A cyclic nucleotide-gated conductance in olfactory receptor cilia. *Nature* 325: 442–444.

Pace U, Hanski E, Salomon Y and Lancet D (1985) Odorant-sensitive adenylate cyclase may mediate olfactory reception. *Nature* 316: 255–258.

Pfaff D and Pfaffmann C (1969) Behavioral and electrophysiological responses of male rats to female rat utine odors. In: *Olfaction and Taste*, Vol. III (Pfaffmann C, ed.), pp. 258–267. New York: Rockefeller University Press.

Portillo W, Diaz NF, Cabrera EA, Fernandez-Guasti A and Paredes RG (2006) Comparative analysis of immunoreactive cells for androgen receptors and oestrogen receptor alpha in copulating and non-copulating male rats. *J Neuroendocrinol* 18: 168–176.

Powers JB and Winans SS (1975) Vomeronasal organ: Critical role in mediating sexual behavior of the male hamster. *Science* 187: 961–963.

Rajendren G, Dudley CA and Moss RL (1990) Role of the vomeronasal organ in the male-induced enhancement of sexual receptivity in female rats. *Neuroendocrinology* 52: 368–372.

Ramsey MM, Weiner JL, Moore TP, Carter CS and Sonntag WE (2004) Growth hormone treatment attenuates age-related changes in hippocampal short-term plasticity and spatial learning. *Neuroscience* 129: 119–127.

Restrepo D, Boekhoff I, Miyamoto T, Teeter JH and Breer H (1992) Rapid kinetic measurements of second messenger formation in isolated olfactory cilia from the channel catfish (*I. punctatus*). *Chem Senses* 17: 690.

Risold PY and Swanson LW (1995) Evidence for a hypothalamothalamocortical circuit mediating pheromonal influences on eye and head movements. *Proc Natl Acad Sci U S A* 92: 3898–3902.

Rochefort C, Gheusi G, Vincent JD and Lledo PM (2002) Enriched odor exposure increases the number of newborn neurons in the adult olfactory bulb and improves odor memory. *J Neurosci* 22: 2679–2689.

Rosado JA and Sage SO (2000) Coupling between inositol 1,4,5-trisphosphate receptors and human transient receptor potential channel 1 when intracellular Ca^{2+} stores are depleted. *Biochem J* 350 Pt 3: 631–635.

Ryba NJP and Tirindelli R (1997) A new multigene family of putative pheromone receptors. *Neuron* 19: 371–379.

Sanna E, Mostallino MC, Murru L, Carta M, Talani G, Zucca S, Mura ML, Maciocco E and Biggio G (2009) Changes in expression and function of extrasynaptic $GABA_A$ receptors in the rat hippocampus during pregnancy and after delivery. *J Neurosci* 29: 1755–1765.

Sasaki K, Okamoto K, Inamura K, Tokumitsu Y and Kashiwayanagi M (1999) Inositol-1,4,5-trisphosphate accumulation induced by urinary pheromones in female rat vomeronasal epithelium. *Brain Res* 823: 161–168.

Scalia F and Winans SS (1975) The differential projections of the olfactory bulb and accessory olfactory bulb in mammals. *J Comp Neurol* 161: 31–55.

Segovia S and Guillamon A (1993) Sexual dimorphism in the vomeronasal pathway and sex differences in reproductive behaviors. *Brain Res Brain Res Rev* 18: 51–74.

Sicard G and Holley A (1984) Receptor cell responses to odorants: Similarities and differences among odorants. *Brain Res* 292: 283–296.

Smith ER, Stefanick ML, Clark JT and Davidson JM (1992) Hormones and sexual behavior in relationship to aging in male rats. *Horm Behav* 26: 110–135.

Sugai T, Sugitani M and Onoda N (1997) Subdivisions of the guinea-pig accessory olfactory bulb revealed by the combined method with immunohistochemistry, electrophysiological, and optical recordings. *Neuroscience* 79: 871–885.

Svensson AI, Akesson P, Engel JA and Soderpalm B (2003) Testosterone treatment induces behavioral disinhibition in adult male rats. *Pharmacol Biochem Behav* 75: 481–490.

Swann J and Fiber JM (1997) Sex differences in function of a pheromonally stimulated pathway: Role of steroids and the main olfactory system. *Brain Res Bull* 44: 409–413.

Taniguchi M and Kaba H (2001) Properties of reciprocal synapses in the mouse accessory olfactory bulb. *Neuroscience* 108: 365–370.

Tomioka M, Murayama T and Kashiwayanagi M (2005) Increases in plasma concentration of progesterone by protease-sensitive urinary pheromones in female rats. *Biol Pharm Bull* 28: 1770–1772.

Tsujikawa K and Kashiwayanagi M (1999) Protease-sensitive urinary pheromones induce region-specific Fos-expression in rat accessory olfactory bulb. *Biochem Biophys Res Commun* 260: 222–224.

Wekesa KS and Anholt RRH (1997) Pheromone regulated production of inositol-(1, 4, 5) trisphosphate in the mammalian vomeronasal organ. *Endocrinology* 138: 3497–3504.

Wysocki CJ and Meredith M (1987) The vomeronasal system. In: *Neurobiology of Taste and Smell* (Finger TE, Silver WL, eds.), pp. 125–150. New York: John Wiley.

Yamaguchi T, Inamura K and Kashiwayanagi M (2000) Increases in Fos-immunoreactivity after exposure to a combination of two male urinary components in the accessory olfactory bulb of the female rat. *Brain Res* 876: 211–214.

13 Social Cues, Adult Neurogenesis, and Reproductive Behavior

Paolo Peretto and Raúl G. Paredes

CONTENTS

13.1 Introduction 367
13.2 Reproductive and Social Behavior Modulates Adult Neurogenesis 369
13.3 Role of AOB Newborn Neurons in the Pheromonal Mating-Induced
Imprinting 371
    13.3.1 Olfactory Pregnancy Block in Mice 371
    13.3.2 Male Pheromones Affect Integration of Newborn Neurons in
    the AOB of Adult Female Mice 372
    13.3.3 AOB Newborn Neurons Are Functionally Recruited by
    Individual Male Odors 372
    13.3.4 Molecules and Pathways Affecting Sensory-Driven Survival of
    AOB Newborn Neurons 375
    13.3.5 Functional Role of AOB Newborn Neurons 376
13.4 Paced Mating and Neurogenesis in the AOB of Adult Rats 377
    13.4.1 Mating Behavior in Rats 377
    13.4.2 Female Sexual Behavior and Neurogenesis 378
    13.4.3 Male Sexual Behavior and Neurogenesis 381
    13.4.4 Opioids and Neurogenesis 382
13.5 Concluding Remarks 382
Acknowledgments 383
References 383

13.1 INTRODUCTION

Adult neurogenesis is a striking form of neural plasticity occurring in restricted regions of the mammalian brain. The past decades have witnessed tremendous research efforts in this field providing significant information regarding the anatomical, molecular, and functional mechanisms underlying neurogenesis in the adult brain. New neuron production regulates integrated brain functions, learning and memory, and adapts the brain to the changing world. Recent data in rodents indicates a link between adult neurogenesis and reproductive and social behavior. This provides the opportunity to unravel the function of this form of neural plasticity in ethologically relevant contexts and opens new perspectives to explore how the brain processes social stimuli. In this chapter we will summarize some of the major key points regarding the cues and mechanisms modulating adult neurogenesis during social interaction and possible role/s played by newborn neurons in this context. To achieve this goal we will give an overview of past and ongoing literature showing this link, with particular emphasis on our recent studies on two examples of sexual behavior: mate pheromonal imprinting in female mice, and paced mating in rats.

The early conception of the function of the brain postulated that once the brain developed it became stable and no new neurons were added in adulthood. This dogma has gradually been dropped over the past 40 years with the clear demonstration that adult neurogenesis is a striking form of structural remodeling characterizing the brain of vertebrates, though with significant differences between groups (Lindsey and Tropepe 2006; Bonfanti and Peretto 2011). In the 1970s, the issue of adult neurogenesis was regarded with skepticism although during the previous decade some proliferative activity was reported in the brain by Altman and colleagues (Altman 1963; Altman and Das 1965). Only years later, with the progress of neuroanatomical techniques and the demonstration of genesis and integration of new neurons in the adult brain of canaries (Nottebohm 1985), adult neurogenesis regained attention.

A new era in this field occurred starting from two simultaneous findings: the occurrence of a massive cell migration toward the rodents' olfactory bulb (Luskin 1993; Lois and Alvarez-Buylla 1994) and the first isolation of adult neural stem cells (Reynolds and Weiss 1992). These studies strengthened the idea that neural plasticity in adult mammals not only occurs through synaptic remodeling but also through the addition of new neurons in the mature preexisting circuits. During the last two decades, this intriguingly persisting process in the mammalian brain was intensely investigated, and several review articles progressively made the point on its extension, features, and significance under physiological and pathological conditions (Emsley et al. 2005; Sohur et al. 2006; Gould 2007; Migaud et al. 2010; Bonfanti and Peretto 2011; Curtis et al. 2011; Fuentealba et al. 2012).

It is now clear that a constitutive/physiologic neurogenesis in adult mammals mostly occurs within two telencephalic regions, the subventricular zone–olfactory bulb system (SVZ-OB) (Lois and Alvarez-Buylla 1994) and the subgranular zone (SGZ) of dentate gyrus (DG) of the hippocampus (Kempermann et al. 2004). The neurogenic process in these regions is orchestrated by a complex interplay between intrinsic and extrinsic environmental cues. Several developmental signals, morphogens, growth factors, neurotransmitters, hormones, transcription factors, and epigenetic regulators have been described to tightly regulate specification and activity of proliferating progenitors, as well as the migration and integration of neuronal precursors within functional circuits (Hagg 2005; Faigle and Song 2013). The signaling mechanisms supporting adult neurogenesis are dynamically regulated by many environmental cues that can either positively or negatively influence the neurogenic process at the level of progenitor cells and during the integration of newborn neurons within circuits (Ma et al. 2009). This activity-dependent regulation is only beginning to be unraveled. Importantly, adult-born neurons in neurogenic regions exhibit critical periods of plasticity during a specific time window of their maturation (Nissant et al. 2009; Ming and Song 2011), and high responsiveness toward the same stimuli driving their integration/selection into circuits (Magavi et al. 2005; Kee et al. 2007). This supports a role of newborn neurons in sensory processing in DG and OB. Accordingly, several sources of data indicate that adult neurogenesis contributes to mechanisms of learning and memory (Lazarini et al. 2009; Moreno et al. 2009), and more recent hypotheses suggest that it also contributes to enhancing pattern separation (Aimone et al. 2011; Sahay et al. 2011). In this view, the continuous addition of new neurons in the olfactory bulb and hippocampus rather than being a simple mechanism of renewal of preexisting cells expands the capacity for plasticity in these regions.

In this chapter we will describe recent findings (see Feierstein et al. 2012 for review) that link reproductive and social stimuli to adult neurogenesis, particularly in the olfactory system. We will address first a brief description of the main results supporting this connection, and then, taking into account our recent work (Oboti et al. 2009, 2011; Corona et al. 2011; Portillo et al. 2012), we will focus on two striking examples of sexual behavior: (1) mate pheromonal imprinting to avoid pregnancy block in female mice and (2) paced mating in rats, namely the ability to control the sexual interaction. In both cases we demonstrate a link between adult neurogenesis and social activities underlying the reproductive function.

13.2 REPRODUCTIVE AND SOCIAL BEHAVIOR MODULATES ADULT NEUROGENESIS

A growing number of studies indicates that reproductive and social behavior modulates adult neurogenesis (Smith et al. 2001; Huang and Bittman 2002; Shingo et al. 2003; Mak et al. 2007; Larsen et al. 2008; Ruscio et al. 2008; Oboti et al. 2009; Furuta and Bridges 2009; Feierstein et al. 2010; Mak and Weiss 2010; Corona et al. 2011; Oboti et al. 2011; Sakamoto et al. 2011; Portillo et al. 2012; Brus et al. 2013). This provides the opportunity to unravel the function of this form of adult neural plasticity in ethologically relevant contexts, and in parallel, the chance to investigate how sensory stimuli underlying reproductive and social behavior are processed in the adult brain. Although several unanswered questions on the link between reproductive/social stimuli and adult neurogenesis remain to be clarified, some major/common key points can be extrapolated from the current data.

Pheromonal cues in rodents convey information about species-specificity, gender, social status, health, genetic advantage and individual recognition (see Tirindelli et al. 2009 for review). Accordingly, they have been demonstrated as major stimuli to enhance neurogenesis in both the olfactory bulb region and hippocampus (Mak et al. 2007; Larsen et al. 2008; Oboti et al. 2009, 2011; Mak and Weiss 2010). Nevertheless, other sensory stimuli/pathways are potentially involved in the modulation of adult neurogenesis occurring during reproductive and social experiences. Indeed, neurogenesis is also enhanced in pregnancy or pseudopregnancy and lactation (Shingo 2003; Larsen 2010) in male mice upon interaction with their offspring (Mak and Weiss 2010) and during pacing behavior (Corona et al. 2011; Portillo et al. 2012). These conditions and activities are driven by multiple sensory pathways and involve complex levels of brain integration and elaboration.

A direct contact (fully exploration of pheromonal cues and/or pairing) between subjects or with the stimulus (bedding or urine) is necessary to affect adult neurogenesis (Huang and Bittman 2001; Smith et al. 2001; Shingo et al. 2003; Mak et al. 2007; Larsen et al. 2008; Ruscio et al. 2008; Furuta and Bridges 2009; Oboti et al. 2009; Brus et al. 2010; Feierstein et al. 2010; Mak and Weiss 2010; Corona et al. 2011; Oboti et al. 2011; Sakamoto et al. 2011; Portillo et al. 2012). This implies activation of both the main olfactory and vomeronasal systems, which cooperate in the control of social/reproductive behavior (Keller et al. 2006). However, their relative contribution to adult neurogenesis in different social contexts still remains largely unexplored.

Interestingly, reproductive/social stimuli affect neurogenesis at two different levels of the neurogenic process, increasing proliferation of progenitor cells in the neurogenic niches and survival/integration of newborn neurons within the functional circuits. The anterior pituitary hormone prolactin (PRL) appears as a key factor promoting proliferation of SVZ progenitors during social interaction (Shingo et al. 2003; Mak et al. 2007; Larsen et al. 2008, 2010; Mak and Weiss 2010). Such enhanced proliferation per se results in increased incorporation of newborn neurons into the olfactory bulb circuits 15–20 days after the social experience (Shingo et al. 2003; Mak et al. 2007; Larsen et al. 2008). PRL in female rodents rises during mating (Erskine 1995), pregnancy and lactation (Grattan and Kokay 2008) and after prolonged exposure to male pheromones (Larsen 2008). This is crucial for adaptation and survival of the mother and fetus during pregnancy and postpartum period (Grattan and Kokay 2008). Thus, based on these functions it has been proposed that PRL-induced neurogenesis in the maternal brain early in gestation favors the parental care (Larsen et al. 2010). Similarly, PRL also increases neurogenesis in the paternal brain, possibly influencing paternal offspring recognition (Mak and Weiss 2010). Accordingly, inactivation of the PRL-enhanced neurogenesis seems to negatively affect some aspects of the parental behavior (Larsen et al. 2010; Mak and Weiss 2010), although conflicting results have been obtained after disruption of olfactory bulb neurogenesis via x-ray irradiation (Feierstein et al. 2010) or through genetically targeted ablation of newborn neurons (Sakamoto et al. 2011). Further work needs to clarify how PRL acts to stimulate proliferation in adult SVZ neurogenic niche (see for example Mak et al. 2007; Larsen et al. 2008), and to explore the involvement of other predictable factors influencing/mediating such activity during social interaction (see the following paragraphs describing the putative role of opioids released in pacing behavior).

As mentioned above, social stimuli and in particular pheromonal cues can also promote adult neurogenesis favoring the survival/integration of newborn neurons during a critical time window of their maturation. This sensory-driven activity, as detailed in the next paragraphs, appears independent from the proliferative effect exerted on progenitor cells and it is prominent in the accessory olfactory bulb (AOB) of female mice. Importantly, in this region newborn neurons are preferentially activated (show higher level of c-Fos expression) shortly after their integration by the same pheromonal cues that enhance their survival (Oboti et al. 2011). This supports a rapid functional recruitment of these cells in circuits activated by the social experience. Accordingly, depletion of these "young and excitable" neurons leads to abnormal social interaction between sexes (Feierstein et al. 2010; Sakamoto et al. 2011) and inability to recognize the mating partner (Oboti et al. 2011). Thus, during reproductive and social experiences neurogenesis increases through a double mechanism: (1) enhancing proliferation of progenitor cells, which later on provide a pool of newborn neurons potentially involved in the parental behavior, and (2) favoring survival of integrating neurons, which are rapidly involved in mechanisms of individual/partner recognition. Although further analyses are needed to clarify this mechanism, it appears committed to optimize reproductive success.

In the following paragraphs we will give two striking examples showing (1) how social stimuli/interaction can influence neurogenesis affecting SVZ progenitors proliferation and survival of newborn neurons, (2) how this process significantly increases incorporation of new cells in the AOB of rodents, and (3) how newborn neurons in the AOB of female mice are involved in processing male pheromonal cues.

13.3 ROLE OF AOB NEWBORN NEURONS IN THE PHEROMONAL MATING-INDUCED IMPRINTING

13.3.1 OLFACTORY PREGNANCY BLOCK IN MICE

In mice, when a recently mated female is exposed to chemosignals from an unfamiliar male, a neuroendocrine reflex leads the female to pregnancy block and in turn a return to estrous.

This reflex known as the Bruce effect (Bruce 1959) is one of the best-known examples of behavior driven by pheromones that involves the vomeronasal system (Brennan and Zufall 2006). Indeed, the Bruce effect is mediated by vomeronasal (VN) excitatory projections from the accessory olfactory bulb to the medial amygdala (MeA), the bed nucleus of the stria terminalis, the medial hypothalamus, and ultimately the dopaminergic neurons of the arcuate nucleus that control prolactin release by the anterior pituitary (Li et al. 1989). Prolactin in mice is luteotrophic and its dopaminergic inhibition, mediated by the arcuate nucleus during critical post-mating stages, prevents blastocyst implantation (Brennan 2009). Pregnancy block depends on male chemosensory cues contained in urine since both direct exposure to male soiled bedding or application of male urine to the nose of a recently mated female induces pregnancy failure (Rosser et al. 1989; Leinders-Zufall et al. 2004). Notably, exteroceptive estrous induction is lost by stud-male odors through enhancement of granule-to-mitral synaptic inhibition occurring in the AOB (see also Chapter 11) during a sensitive period around mating (Brennan et al. 1990; Matsuoka et al. 1997). This male-specific pheromone recognition/imprinting process involves a restricted pool of granule cells in the AOB, which actually inhibits mitral cell signal transmission to the forebrain areas involved in estrous induction for several weeks (50–60 days) (Brennan et al. 1990; Matsuoka et al. 1997). Long-term maintenance of this inhibition implies that, since pairings may occur at shorter intervals, different or partially overlapping cohorts of cells may be necessary to each imprinting process. These features drove us to hypothesize a role of adult olfactory bulb neurogenesis in the Bruce effect since our early studies showing that SVZ-derived neuroblasts also reach the AOB (Bonfanti et al. 1997; Peretto et al. 1999). Indeed, adult neurogenesis continuously refills this region with pools of young newborn inhibitory interneurons that show unique functional properties such as enhanced synaptic plasticity and increased responsiveness to recently experienced odors (Magavi et al. 2005; Nissant et al. 2009).

13.3.2 MALE PHEROMONES AFFECT INTEGRATION OF NEWBORN NEURONS IN THE AOB OF ADULT FEMALE MICE

Preliminary investigation of olfactory bulb neurogenesis in adult mice confirmed that a subpopulation of SVZ-derived neuroblasts acquires proper neurochemical and morphological profiles of mature inhibitory GABAergic interneurons in the AOB of both sexes (Figure 13.1) (Oboti et al. 2009). This data definitely demonstrated the AOB, just as the MOB and the hippocampus, represents a site of constitutive adult neurogenesis. Then, we showed that chronic exposure (28 days) to male soiled bedding, which contains semiochemicals present in urine and exocrine glands secretion (Brennan and Keverne 2004), significantly increases the number of new neurons in the AOB of adult females (Oboti et al. 2009). This effect was elicited only by direct contact with male bedding and not by its volatile compounds, thus supporting such experience-dependent regulation of neurogenesis requires vomeronasal organ activity.

One major point was to clarify whether AOB-enhanced neurogenesis induced by long-term exposure to male pheromones depends on early proliferative effects on SVZ progenitors, as shown in other studies (Shingo et al. 2003; Mak et al. 2007; Larsen et al. 2008), or by increased survival of newborn neurons. According to the occurrence of critical periods for sensory experience-dependent survival of newly generated granule cells in MOB (Petreanu and Alvarez-Buylla 2002; Yamaguchi and Mori 2005), we found that 1-week-long familiarization/exposure to male bedding/urine enhances survival of newborn neurons during their selection/integration within circuits (affecting cells aged between 7 and 14 days) (Figure 13.2a). This effect was most prominent in the granule cell layer of the AOB and much weaker in the MOB. In addition, it was restricted to postpubertal females (absent in prepubertal females or in adult males) (Figure 13.2b) (Oboti et al. 2011). This indicated that neuronal integration in the AOB of adult female mice tightly correlates with the activity elicited in this region by exposure to male odors.

13.3.3 AOB NEWBORN NEURONS ARE FUNCTIONALLY RECRUITED BY INDIVIDUAL MALE ODORS

In order to establish whether newborn neurons in the AOB could represent a competent cellular substrate responding specifically to male individual odors, we combined immunostaining for c-Fos as a marker for neuronal excitation (Morgan et al. 1987) and BrdU labeling. Such an approach was previously used to visualize the functional recruitment of newborn cells in defined sensory tasks in the main olfactory bulb and hippocampus (Magavi et al. 2005; Kee et al. 2007). We found that the percentage

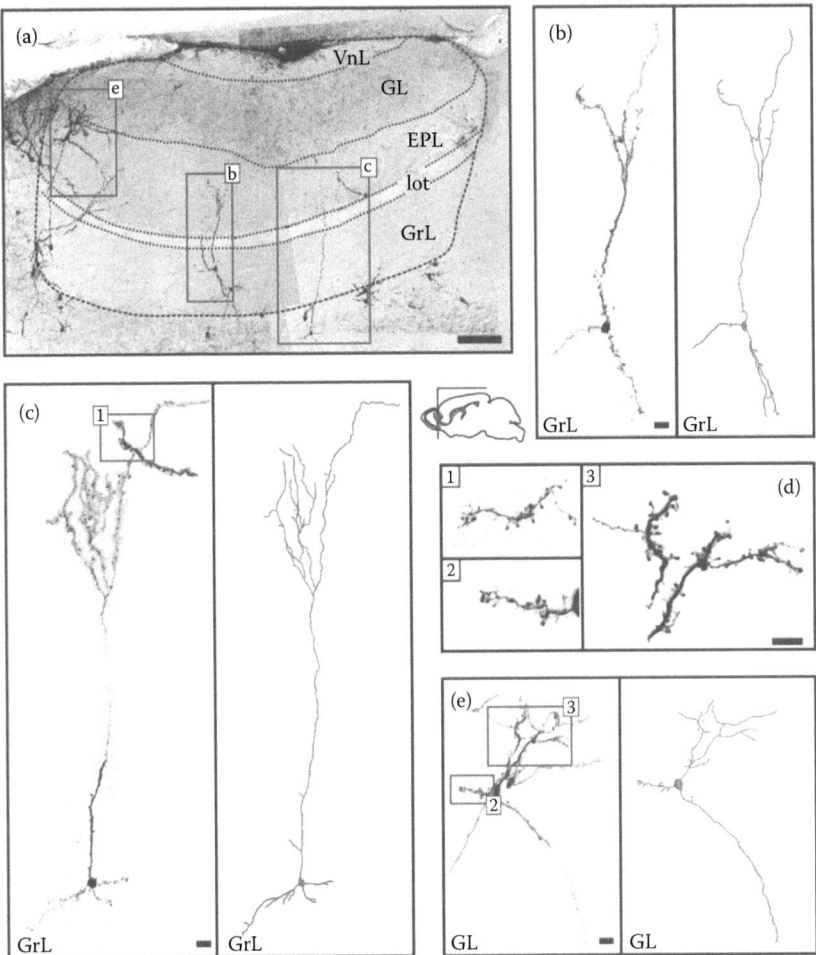

FIGURE 13.1 Adult-generated SVZ-derived neurons in the AOB of mice: morphological analysis. 3-D reconstruction of EGFP-positive SVZ-derived precursors at 60 days after homotopic engraftment. EGFP-positive cells can be observed in the AOB layers and in the MOB GrL. Some of the EGFP-positive cells with cell soma localized in the MOB GrL send dendritic processes into the AOB layers. All cells show features of mature interneurons with well-developed dendritic arborizations and spines. (a) Projection on the Z-plane of the section series analyzed. (b–e) Reconstruction at higher magnification of single cells entangled in squares in (a). In (d) are details of spiny cell processes. Scale bar 100 μm in (a), 10 μm in (b–e). EPL = external plexiform layer, GL/VnL = glomerular and vomeronasal nerve layer, GrL = granular layer, lot = lateral olfactory tract. (Modified from Oboti, L. et al. 2009. *Eur. J. Neurosci.* 29:679–92.)

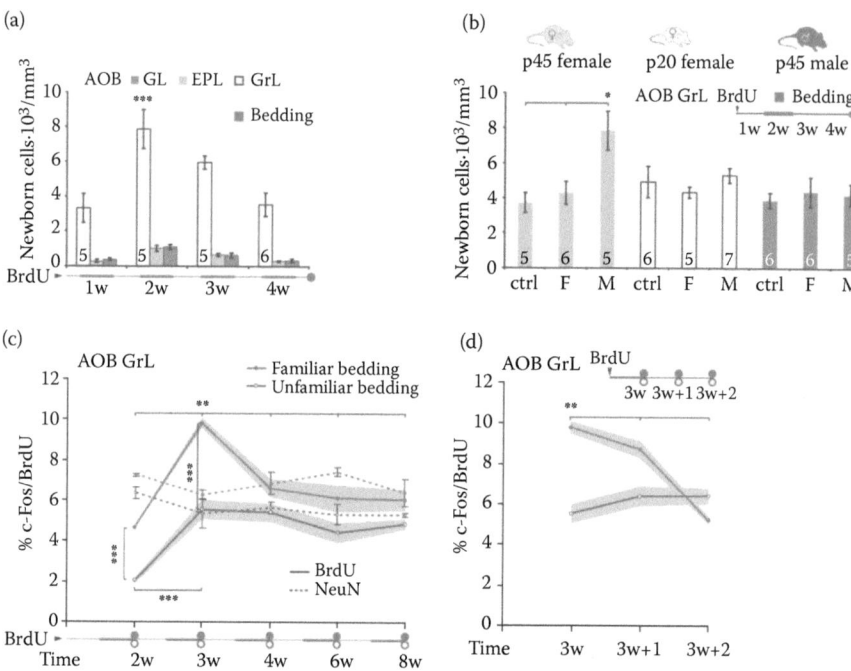

FIGURE 13.2 **(See color insert.)** Sensory-driven integration and function of newborn neurons in the AOB of adult female mice. (a) Quantification of BrdU-labeled cells at 28 dpi in multiple AOB layers of p45 female mice after familiarization with male bedding performed in four different weeks. Male bedding exposure enhances integration of newborn neurons aged between 7 and 14 days. (b) Density of newborn granule cells in adult females, prepubertal females, and adult males evaluated at 28 dpi of BrdU and after 1-week exposure to female (F) or male (M) bedding (from 7th to 14th dpi of BrdU). (c, d) Percentage of c-Fos/BrdU (continuous lines) and c-Fos/NeuN-positive cells (dotted lines) induced by familiar (red) and unfamiliar (blue) bedding/pheromones in AOB GrL. Notably, AOB newborn granule neurons, but not preexisting ones (NeuN-positive, mature neurons), are preferentially activated (c-Fos-positive) by experienced pheromones. This activity is transient and picks around the third week of age of newborn cells. Data is means ± sem (represented by gray areas in c, d) and the numbers in the graph bars indicate the amount of animals used. *$P < 0.05$, **$P < 0.01$, ***$P < 0.001$. (Modified from Oboti, L. et al. 2011. *Front. Neurosci.* 5:113.)

of c-Fos/BrdU coexpression induced by familiar (experienced for 1 week) versus unfamiliar (never experienced before) male soiled bedding was significantly higher in AOB newborn cells aged 3 weeks (Figure 13.2c). Moreover, this activity was transient since it started to vanish after 7 days (Figure 13.2d) (Oboti et al. 2011). These results indicate that, in female mice, AOB newborn granule cells are functionally recruited by male pheromones soon after their integration and preferentially by experienced male cues. The transient nature of this activation is consistent with a privileged involvement of young newborn neurons in the elaboration of specific chemosensory signals. Interestingly, exposure to experienced male cues, but not unfamiliar ones, elicited attenuated responses in nuclei of the VNS involved in estrous

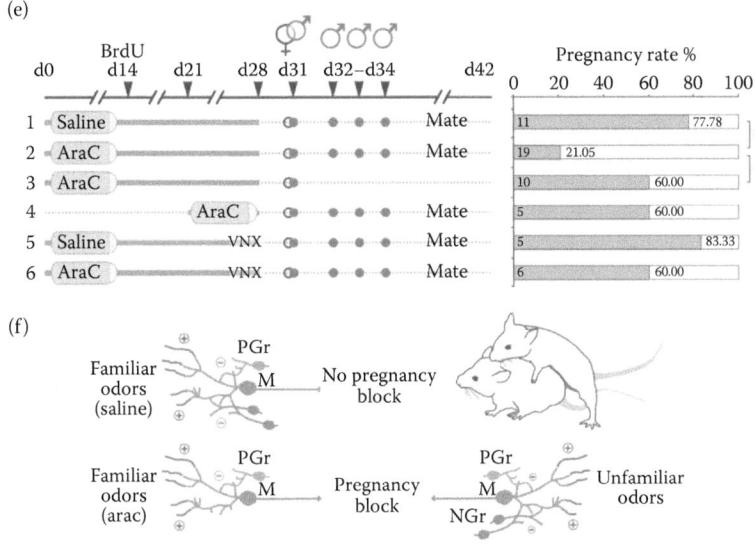

FIGURE 13.2 (**See color insert.**) (*Continued*) Sensory-driven integration and function of newborn neurons in the AOB of adult female mice. (e) Matings and male stimulations on female mice after different protocols of Ara-C/saline treatment in normal and vomeronasal nerve-lesioned mice (VNX). Shown in the graph are the pregnancy rates (in percentage) as a function of the different treatment conditions evaluated 11 days after mating. (f) Schematic diagram illustrating the role of AOB newborn granule cells (NGr) in the modulation of a mate's familiar signals (left side) and unfamiliar ones (right side): granule cells are preferentially involved in the detection of male individual odors once integrated into preexisting circuits. When highly responsive newborn granule cells (NGr) are eliminated after Ara-C treatment (left side, bottom), preexisting granule cells (PGr) are not sufficient to prevent pregnancy block by a mate's familiar odors (red arrows). *$P < 0.05$. (Modified from Oboti, L. et al. 2011. *Front. Neurosci.* 5:113.)

induction such as the medial hypothalamus, medial amygdala, and arcuate nucleus (Oboti et al. 2011). This supported the theory that the enhanced response shown by AOB new neurons to male individual odors could impact on the forebrain activity induced by these cues.

13.3.4 MOLECULES AND PATHWAYS AFFECTING SENSORY-DRIVEN SURVIVAL OF AOB NEWBORN NEURONS

In order to investigate the nature of the cues affecting enhanced integration of newborn neurons in the AOB, we employed molecules present in the low molecular weight (LMW) fraction of urine (containing small organic molecules and small peptides) as well as high molecular weight (HMW) major urinary proteins (MUPs) contained in the HMW fraction) (Brennan and Zufall 2006). We found that urine deprived of MUPs by protease treatment was still capable of inducing newborn cell survival in the AOB. Accordingly, the HMW fraction of male urine treated with menadion to deprive it from all volatile ligands (Chamero et al. 2007) and loaded onto female LMW urine fraction to stimulate investigation was uneffective. Importantly, these effects were specific for the AOB and not MOB (Oboti et al. 2011). These results indicated the cues affecting adult neurogenesis in the AOB of females mice are comprised in the LMW fraction of male urine, which is primarily important for mate recognition/Bruce effect (Peele et al. 2003; Leinders-Zufall et al. 2004). Pheromones contained in the LMW fraction of male urine are sensed by both the main and accessory olfactory systems (Brennan and Zufall 2006). Therefore, to investigate which way male olfactory cues influence AOB neurogenesis in female mice, we compared the effects of genetic deletion of the Trpc2 cation channel (which leads to impaired VN function (Leypold et al. 2002) with lesions of the MOE caused by intranasal irrigation of zinc sulfate ($ZnSO_4$) (McBride et al. 2003). We found that enhanced AOB granule cell survival was absent in $trpc2^{-/-}$ mice. By contrast, $ZnSO_4$ lesions of the MOE did not abolish enhanced neuronal survival in the AOB, confirming that vomeronasal contact is necessary and sufficient to increase new AOB granule cells. Since AOB activity can also be induced by centrifugal inputs from the medial amygdala, even in absence of VNO stimulation (Pankevich et al. 2006; Martel and Baum 2009), we delivered excitotoxic lesions to the MeA by injections of ibotenic acid (Chauveau et al. 2008), a glutamatergic agonist, and stimulated both lesioned and sham-lesioned mice with male bedding. Exposures increased cell survival in the AOB of sham-lesioned mice but not in lesioned ones. Notably, granule cell survival in the MOB was unaffected by either bedding exposure or lesioning with ibotenic acid (Oboti et al. 2011). Together these results indicated that male pheromones trigger integration of new AOB granule cells through VN centripetal and centrifugal sensory activity.

13.3.5 Functional Role of AOB Newborn Neurons

Sensory mechanisms and molecular cues driving survival and functional recruitment of newborn granule cells in the AOB of female mice support that adult neurogenesis in this region could be involved in the mate pheromonal imprinting. To test this hypothesis, we blocked the renewal of adult-born interneurons by administrating the antimitotic drug Ara-C using osmotic minipumps, as previously done in other studies (Breton-Provencher et al. 2009). After a 4-week-long delivery of the antimitotic drugs (a period of time covering the pick responsiveness to familiar stimuli of AOB newborn neurons), which completely eliminated newborn cells in both the main and accessory olfactory bulbs, we tested the ability of adult female mice to recognize their mating partners in order to avoid exteroceptive implantation failure (Figure 13.2e). In other words, we tested whether Ara-C-treated females exposed during the postmating critical period (3 days after the beginning of mating in coincidence with the prolactin peaks) (Peele et al. 2003; Leinders-Zufall et al. 2004) to the same partners undergo pregnancy block.

In contrast to saline-treated females, we found a high rate of pregnancy failure in Ara-C-treated mice, meaning the treatment switched the effect of familiar odor to that of an unfamiliar one. To rule out that this effect of Ara-C was caused by induced infertility, we analyzed the pregnancy failure rate after Ara-C treatment without subsequent exposure to familiar odor. In this case, the pregnancy failure rate was low, indicating that the high pregnancy failure was due to odor exposure after mating and not to Ara-C-induced infertility. These experiments demonstrated that ablation of bulbar neurogenesis compromises the formation of the stud male olfactory memory in female mice. To rule out a potential involvement of MOB newborn neurons in this memory, we tested Ara-C-treated females after surgical lesion of the vomeronasal nerves, a condition known to eliminate exteroceptive pregnancy block alone (Bellringer et al. 1980; Matsuoka et al. 2005). This procedure was sufficient to prevent the high rate of pregnancy block by stud male exposure (Oboti et al. 2011), further supporting the key role of AOB newborn interneurons in this process (Figure 13.2f).

13.4 PACED MATING AND NEUROGENESIS IN THE AOB OF ADULT RATS

13.4.1 Mating Behavior in Rats

As should be evident by now, it is clear that pheromones induce neurogenesis in the adult brain of mammals and that these neurons can be involved in the modulation of reproductive and social behavior. In the following section we will describe another type of stimuli that also induces neurogenesis in the olfactory bulb and which is also crucial for reproduction and paced mating. There are only a few studies that have evaluated the effects of sexual behavior on neurogenesis and most of them have been done in rats. A detailed description of mating behavior is beyond the scope of this chapter but some important aspects will be briefly described. In rats, mating consists of a series of stereotyped behavioral patterns that are easily distinguishable. In the case of the male, they display several mounts and intromissions that would eventually lead to ejaculation. If the female is receptive she would display proceptive behaviors characterized by ear wiggling, hoping, and darting. Another important component of the behavior displayed by the female is the receptive posture in response to a mount from the male (a detailed description of the rat male and female sexual behavior can be found in Blaustein and Erskine 2002 and Meisel and Sachs 1994). A key factor in the sexual interaction is the ability that the females have to space the sexual interaction depending on the stimulation they receive. The vaginal stimulation that the female receives during mating is more intense after ejaculation than after intromission and mounts (Erskine 1989). Classic studies have shown that if given the opportunity, females will escape from the male side with a higher frequency after ejaculation than intromission and mounts. The possibility to space or control the sexual interaction is known as paced mating (Erskine 1989) and this is what usually occurs when rats mate in seminatural and natural conditions (Robitaille and Bouvet 1976; McClintock and Adler 1978; McClintock and Anisko 1982; McClintock et al. 1982).

Mating in rats is highly promiscuous and it occurs in groups where several males and females mate at the same time. In this way a female may receive several mounts, intromissions, and ejaculations from different males. As well, males could display mounts, intromissions, and ejaculations with several females (McClintock and Anisko 1982) and hence both males and females pace the sexual interaction (see Martinez and Paredes 2001 for a discussion). The possibility to pace the sexual interaction has several physiological and behavioral advantages over nonpaced mating (Erskine and Baum 1982; Erskine 1989; Erskine et al. 1989; Paredes and Alonso 1997; Paredes and Vazquez 1999). For example, only when males and females pace the sexual interaction does sexual behavior induce a reward state as evaluated by conditioned place preference (Paredes and Vazquez 1999; Martinez and Paredes 2001; Paredes 2009) that assures that the behavior will be repeated in the future. Under laboratory conditions paced mating can be easily observed when the mating cage is divided by a partition, with a hole in the bottom large enough to allow the female, but not the male, to move back and forth to the compartment in which the male is confined (Erskine 1989). In this way the female controls or paces the sexual interactions; when the animals mate without the partition the male controls the sexual interaction. In the following section we will describe how our studies suggest that paced mating, but not nonpaced mating, induces an increase in the number of new neurons in the adult brain by affecting proliferation of SVZ progenitors.

13.4.2   FEMALE SEXUAL BEHAVIOR AND NEUROGENESIS

Few studies have evaluated the effects of mating on neurogenesis (see Table 13.1). One of the first studies that evaluated if the new cells generated in the adult are activated during sexual behavior was done in hamsters. For that purpose male hamsters were injected with BrdU either 10 days, 3 weeks, or 7 weeks before they were allowed to mate with a receptive female. After 90 minutes of mating subjects were sacrificed and c-Fos expression and BrdU labeling analyzed. The new neurons in the olfactory bulbs expressed Fos in those males injected with the synthetic marker 3 and 7 weeks before mating. No significant increase in the number of double-labeled neurons was observed when male hamsters were exposed to female hamster vaginal secretions, to an aggressive male, or to peppermint odor, suggesting that neurons born in adulthood are incorporated in the functional circuits that control mating behavior in the hamster (Huang and Bittman 2002).

Since the olfactory bulbs and the processing of chemosensory cues are crucial for a successful reproduction we decided to investigate if mating itself can induce the formation of new neurons in the adult olfactory bulb region. In the first study ovariectomized females, hormonally primed to induce sexual behavior, were tested under different behavioral conditions. One group of females was exposed to a sexually active male, another group of females mated in a condition where they were not able to pace the sexual interaction, and one group of females that paced the sexual interaction. All females received BrdU injections 1 hour before, immediately after, and 1 hour after the behavioral tests and sacrificed 15 days later. This would let us determine the survival of neurons produced around the time of mating. We observed a significant increase in the density of BrdU-positive cells in the granular layer of AOB when females were allowed to pace the sexual interaction in comparison to the other groups. No differences in cell density in the main

TABLE 13.1

Studies That Have Evaluated Cell Proliferation and Neurogenesis with Different Aspects of Mating

Species	Treatment	Region Analyzed	Effect	Reference
		Females:		
Prairie voles	Exposed to sexually mature males	SVZ	Increased number of new cells	Smith et al. 2001
Sheep	Exposed to a novel male	DG hippocampus	Increased number of new cells	Hawken et al. 2009
Rats	BrdU before and after mating, sacrificed at 15 days	OB	Increased number of neurons in the AOB in females that paced mating	Corona et al. 2011
Rats	BrdU before and after mating, sacrificed at 15 days, mated three more times	OB	Increased number of neurons in the AOB and MOB in females that paced mating	Arzate et al. 2013
		Males:		
Hamster	Fos evaluated 3 or 7 weeks after BrdU injection	Olfactory bulbs	New neurons express Fos after mating	Huang and Bittman 2002
Hamster	BrdU and mated weekly for 7 days	Medial preoptic area and amygdala	No effect	Antzoulatos et al. 2008
Rats	BrdU after one intromission, sacrificed at 2 hours	Hippocampus	Increased number of new neurons	Leuner et al. 2010
Rats	BrdU after one intromission, sacrificed at 2 weeks	Hippocampus	Increased number of new neurons	Leuner et al. 2010
Rats	Brdu before and after mating, sacrificed at 15 days	OB	Increased number of neurons in the AOB in males that paced mating	Portillo et al. 2012

Note: AOB = accessory olfactory bulb, DG = dentate gyrus, MOB = main olfactory bulb, OB = olfactory bulb, SVZ = subventricular zone.

olfactory bulb were found. These results indicate that pacing behavior promotes an increase in SVZ proliferation that in turn leads to a higher density of the new cells in the accessory olfactory bulb (Corona et al. 2011). This effect is specifically associated with the ability of controlling and pacing the sexual interaction since nonpaced mating did not induce changes in cell density. As well, the increase in cell density is not associated with different levels of estradiol and progesterone or behavioral differences because all groups had the same hormone (Arzate et al. 2013) and behavioral levels.

In a followup study we tested if the repetition of the stimulus could increase the number of new neurons in the olfactory bulbs after the first sexual encounter. For that purpose females were randomly assigned to one of the following groups: (1) females without sexual contact, (2) females that were given one session of paced mating, (3) females given four sessions of nonpaced mating, and (4) females given four sessions of paced mating. Some sections were analyzed for BrdU immunohistochemistry and others were double-labeled with immunofluorescence for BrdU and NeuN to label mature neurons. As in our previous experiment all groups were injected with BrdU 1 hour before, immediately after, and 1 hour after the behavioral tests. Females were sacrificed 15 days later and the density of new neurons was analyzed in the olfactory bulbs. The results of this experiment further confirmed our previous result; that is, the females that paced the sexual interaction in one session showed a higher number of mature (NeuN-positive) neurons in the granular cell layer of the AOB. The group that mated four times also had a higher number of neurons in the granular cell layer. Moreover, the group that mated four times pacing the sexual interaction also showed a significant increase in the number of neurons in the granular layer of the MOB (Arzate et al. 2013). It is clear then that paced mating induces plastic changes that eventually result in an increase in the number of new neurons in the OB 15 days after the first mating encounter. If the female mates once the increase is observed in the granular cell layer of the AOB, and if the stimulation is repeated an increase in the number of neurons is observed in the AOB and the MOB.

In our design females were sacrificed 15 days after copulation and BrdU injection, suggesting that a higher number of neurons survive the 2-week period. As already described, there is clear evidence indicating that pheromones induce cell proliferation in the adult SVZ. For example, female prairie voles exposed to a sexually mature male in tests where they could smell, see, and hear but had limited physical contact with the male for 48 hours showed an increase in the proliferation of new cells with a neuronal phenotype in the SVZ. Ovariectomized females exposed to males or intact females exposed to females showed no increase in BrdU labeling (Smith et al. 2001). It has also been shown that pheromones of the preferred dominant male stimulate cellular proliferation in the SVZ and neuronal production in the MOB (Mak et al. 2007). Since we sacrificed the females 15 days after copulation, it is possible that the time frame to observe differences in the number of cells in the SVZ had passed. Therefore, studies are now underway to evaluate cell proliferation in the SVZ and the rostral migratory stream (RMS) after exposure to sexually relevant olfactory cues and after mating. Preliminary data indicates that 2 days after females are exposed to a sexually experienced male or to amyl acetate there is a significant increase in the number of BrdU labeled cells in the SVZ (Paredes et al. 2013, manuscript in preparation). These results could indicate that in our experimental conditions, 1 hour of exposure to male pheromones or to amyl acetate is sufficient to induce cell proliferation in the SVZ. These results are in agreement with observations indicating that after the new neurons area born, their survival is activity-dependent (Alonso et al. 2006; Mandairon et al. 2006; Lazarini et al. 2009; Moreno et al. 2009). For example, it has been shown that a spaced learning paradigm as opposed to a massed paradigm increased the survival of adult-born neurons in the olfactory bulb, allowing long-term consolidation of the olfactory task (Kermen et al. 2010). As well, discrimination learning increases the number of newborn neurons in the adult OB, prolonging their survival (Alonso et al. 2006). In male mice the prolonged exposure (40 days) to an odor-enriched environment increases the number of new cells in the glomerular cell layer of the MOB, facilitating odor discrimination (Rochefort et al. 2002; Rochefort and Lledo 2005). It is also documented that the survival of newborn cells is significantly reduced in mice unilaterally deprived from sensory input by naris occlusion, suggesting that the survival of the new cells in the OB depends on sensory input (Mandairon et al. 2006). Future studies will need to determine if repeated exposures to male pheromones or to amyl acetate increases the survival of the cells observed in the SVZ. We also need to determine the functional integration of the BrdU-labeled neurons 45 days after paced mating once the new neurons integrate into functional circuits.

13.4.3 Male Sexual Behavior and Neurogenesis

The first studies that evaluated if sexual behavior in males could induce cell proliferation in the adult brain were done in the hamster (Antzoulatos et al. 2008). Sexually experienced males were injected with BrdU and mated weekly for 7 weeks. No enhancement of cell proliferation was found in the medial preoptic area or the medial amygdala, two structures crucial for the expression of male sexual behavior. In another study male rats were exposed to sexually receptive females with whom they could copulate for 1 day or for 14 consecutive days. Male rats that mated once (acute) or for 14 consecutive days (chronic) with receptive females showed an increase in cell proliferation and neurogenesis in the dentate gyrus of the hippocampus. No changes were observed in males exposed to receptive females, suggesting that a rewarding experience, sex, promotes neurogenesis (Leuner et al. 2010). In both of these studies, the authors did not evaluate cell proliferation in the SVZ, the RMS, or the olfactory bulbs. Therefore, we decided to investigate if sexual behavior can induce SVZ neurogenesis in males. Our design was similar to what we have done in females. Basically, BrdU was injected 1 hour before, at the end, and 1 hour after the behavioral tests, sacrificing the subjects 15 days later. The groups included (1) males without sexual stimulation, (2) males exposed to female odors, and (3) males that mated for 1 hour without pacing the sexual interaction and males that paced the sexual interaction until achieving 1 or 3 ejaculations. As in the case of the female, we observed a significant increase in the number of newborn neurons in the granular cell layer of the AOB in the groups of males that ejaculated once or three times pacing (controlling) the sexual interaction. No differences between groups were found in the other layers of the AOB or in the MOB. We also showed that around 40% of the new cells differentiated into neurons. The group of males not allowed to pace the sexual interaction ejaculated a mean of 3.4 times during the 1-hour test. Despite the fact that the nonpaced group had more ejaculations than the groups that ejaculated 1 or 3 times no increase in the number of new neurons was observed. These results clearly indicate the quality of the stimulation received during paced mating but not the intensity of the stimulation (number of intromissions) is a crucial factor to induce neurogenesis in the AOB (Portillo et al. 2012).

13.4.4 OPIOIDS AND NEUROGENESIS

Another important difference between paced and nonpaced mating is the reward-ing value of the sexual interaction. We have repeatedly shown that only if subjects, both males and females, pace the sexual interaction, a conditioned place preference indicative of a reward state is developed (Martinez and Paredes 2001; Camacho et al. 2009). This reward state is mediated by a common opioid system because administration of the opioid antagonist naloxone blocks the rewarding effects induced by sexual behavior in both males (Agmo and Gomez 1993) and females (Paredes and Martinez 2001). In fact several lines of evidence suggest that opioids are released during sexual behavior thereby reducing the aversive consequences of repeated sexual stimulation and enabling the eventual development of a reward or positive affective estate (Agmo 2007; Paredes and Fernández-Guasti 2008; Paredes 2009). Other appetitive behaviors also induce cell proliferation in the hippocam-pus. Rats emit 50-KHz ultrasonic vocalizations in appetitive situations, like tickle, while 22-KHz calls are associated with aversive situations (see Wohr et al. 2009 and references therein). The rate of hippocampal cell proliferation was analyzed in rats that perceived tickling as appetitive or aversive and in nontickled rats. Repeated tickling increased cell proliferation in the hippocampus in the rats that experienced tickling as appetitive (Wohr et al. 2009). Another activity that could be considered appetitive and is mediated by opioids is exercise. It has also been shown that exer-cise, either treadmill or swimming, stimulates neurogenesis in the hippocampus of adult rats (Chae et al. 2012). Further studies need to address whether cell prolifera-tion and neurogenesis in appetitive behaviors are mediated by opioids. This indeed is a possibility considering that opioids are also involved in the proliferation and survival of the new cells in the SVZ (Sargeant et al. 2008) and that morphine treat-ment increases the number of new cells in the SVZ of adult male rats (Messing et al. 1979).

13.5   CONCLUDING REMARKS

Here we have shown that social stimuli underlying reproductive behavior in rodents enhance incorporation of newborn neurons in adult neurogenic regions, particularly in the olfactory bulb region. This occurs through a double action that influences prolif-eration of progenitors/precursors and incorporation of newborn neurons within func-tional circuits. Pheromonal cues contained in urine and hormones such as PRL are important mediators of this mechanism. Nevertheless, additional studies are needed to better characterize the nature and source of the sensory cues driving increased incorporation of newborn neurons as well as the molecular players and brain cir-cuits/systems mediating this mechanism during social interaction. For example, our experiments of urine fractionation suggest the survival of new neurons in the AOB is regulated by molecules included in the LMW urine fraction, but whether MHC peptide ligands (already known to convey individuality in the Bruce effect) or MUPs testosterone-dependent volatile ligands are involved in such activity remains to be determined. Similarly, the possible role of opioids and the involvement of the reward state in controlling proliferation of SVZ progenitors deserve further investigation.

Concerning the role played by OB newborn neurons, the data reviewed in this chapter supports that they are involved in recognizing the mating partner, which is critical to avoiding pregnancy block in mice, and learning the odor of the offspring, to favor selective care and preventing inbreeding. Thus, waiting for further feasible confirmations, the occurrence of adult neurogenesis in key regions controlling social stimuli appears of extraordinary functional importance in the context of the reproductive behavior.

ACKNOWLEDGMENTS

This chapter was supported by PRIN 2010-2011 to Paolo Peretto and by CONACyT 167101 and DGAPA IN200512 to Raúl G. Paredes.

REFERENCES

Agmo, A. 2007. *Learning and Sex: Sexual Activity as Reinforcement and Reward. Functional and Dysfunctional Sexual Behavior.* London: Academic Press.

Agmo, A. and M. Gomez. 1993. Sexual reinforcement is blocked by infusion of naloxone into the medial preoptic area. *Behav. Neurosci.* 107:812–18.

Aimone, J.B., W. Deng and F.H. Gage. 2011. Resolving new memories: A critical look at the dentate gyrus, adult neurogenesis, and pattern separation. *Neuron* 70(4):589–96.

Alonso, M., C. Viollet, M.M. Gabellec, V. Meas-Yedid, J.C. Olivo-Marin and P.M. Lledo. 2006. Olfactory discrimination learning increases the survival of adult-born neurons in the olfactory bulb. *J. Neurosci.* 26:10508–13.

Altman, J. 1963. Autoradiographic investigation of cell proliferation in the brains of rats and cats. *Anat. Rec.* 145:573–91.

Altman, J. and G.D. Das. 1965. Post-natal origin of microneurones in the rat brain. *Nature* 207:953–56.

Antzoulatos, E., J.E. Magorien and R.I. Wood. 2008. Cell proliferation and survival in the mating circuit of adult male hamsters: Effects of testosterone and sexual behavior. *Horm. Behav.* 54:735–40.

Arzate, D.M., W. Portillo, R. Corona and R.G. Paredes. 2013. Repeated paced mating promotes the arrival of more newborn neurons in the main and accessory olfactory bulbs of adult female rats. *Neuroscience* 232:151–60.

Bellringer, J.F., H.P. Pratt and E.B. Keverne. 1980. Involvement of the vomeronasal organ and prolactin in pheromonal induction of delayed implantation in mice. *J. Reprod. Fertil.* 59:223–28.

Blaustein, J.D. and M.S. Erskine. 2002. *Feminine Sexual Behavior: Cellular Integration of Hormonal and Afferent Information in the Rodent Brain.* New York: Academic Press.

Bonfanti, L. and P. Peretto. 2011. Adult neurogenesis in mammals: A theme with many variations. *Eur. J. Neurosci.* 34:930–50.

Bonfanti, L., P. Peretto, A. Merighi and A. Fasolo. 1997. Newly-generated cells from the rostral migratory stream in the accessory olfactory bulb of the adult rat. *Neuroscience* 81:489–502.

Brennan, P.A. 2009. Outstanding issues surrounding vomeronasal mechanisms of pregnancy block and individual recognition in mice. *Behav. Brain Res.* 200:287–94.

Brennan, P.A. and E.B. Keverne. 2004. Something in the air? New insights into mammalian pheromones. *Curr. Biol.* 14:81–89.

Brennan, P.A. and F. Zufall. 2006. Pheromonal communication in vertebrates. *Nature* 444:308–15.

Brennan, P., H. Kaba and E.B. Keverne. 1990. Olfactory recognition: A simple memory system. *Science* 250:1223–26.

Breton-Provencher, V., M. Lemasson, M.R. Peralta and A. Saghatelyan. 2009. Interneurons produced in adulthood are required for the normal functioning of the olfactory bulb network and for the execution of selected olfactory behaviors. *J. Neurosci.* 29:15245–57.

Bruce, H. 1959. An exteroceptive block to pregnancy in the mouse. *Nature* 184:105.

Brus, M., M. Meurisse, G. Gheusi, M. Keller, P.M. Lledo and F. Levy. 2013. Dynamics of olfactory and hippocampal neurogenesis in adult sheep. *J. Comp. Neur.* 521:169–88.

Camacho, F.J., W. Portillo, O. Quintero-Enriquez and R.G. Paredes. 2009. Reward value of intromissions and morphine in male rats evaluated by conditioned place preference. *Physiol. Behav.* 98:602–27.

Chae, C.H., H.C. Lee, S.L. Jung et al. 2012. Swimming exercise increases the level of nerve growth factor and stimulates neurogenesis in adult rat hippocampus. *Neuroscience.* 212:30–37.

Chamero, P., T.F. Marton, D.W. Logan et al. 2007. Identification of protein pheromones that promote aggressive behaviour. *Nature.* 450:899–902.

Chauveau, F., C. Pierard, M. Coutan et al. 2008. Prefrontal cortex or basolateral amygdala lesions blocked the stress-induced inversion of serial memory retrieval pattern in mice. *Neurobiol. Learn. Mem.* 90:395–403.

Corona, R., J. Larriva-Sahd and R.G. Paredes. 2011. Paced-mating increases the number of adult new born cells in the internal cellular (granular) layer of the accessory olfactory bulb. *PloS One* 6:e19380.

Curtis, M.A., M. Kam and R.L. Faull. 2011. Neurogenesis in humans. *Eur. J. Neurosci.* 33: 1170–74.

Emsley, J.G., B.D. Mitchell, G. Kempermann and J.D. Macklis. 2005. Adult neurogenesis and repair of the adult CNS with neural progenitors, precursors, and stem cells. *Prog. Neurobiol.* 75:321–41.

Erskine, M.S. 1989. Solicitation behavior in the estrous female rat: A review. *Horm. Behav.* 23:473–502.

Erskine, M.S. 1995. Prolactin release after mating and genitosensory stimulation in females. *Endocr. Rev.* 16(4):508–28.

Erskine, M.S. and M.J. Baum. 1982. Effects of paced coital stimulation on termination of estrus and brain indoleamine levels in female rats. *Pharmacol. Biochem. Behav.* 17:857–61.

Erskine, M.S., E. Kornberg and J.A. Cherry. 1989. Paced copulation in rats: Effects of intromission frequency and duration on luteal activation and estrus length. *Physiol. Behav.* 45:33–39.

Faigle, R. and H. Song. 2013. Signaling mechanisms regulating adult neural stem cells and neurogenesis. *Biochim. Biophys. Acta* 1830(2):2435–48.

Feierstein, C.E. 2012. Linking adult olfactory neurogenesis to social behavior. *Front. Neurosci.* 6:173.

Feierstein, C.E., F. Lazarini, S. Wagner et al. 2010. Disruption of adult neurogenesis in the olfactory bulb affects social interaction but not maternal behavior. *Front. Behav. Neurosci.* 4:176.

Fuentealba, L.C., K. Obernier and A. Alvarez-Buylla. 2012. Adult neural stem cells bridge their niche. *Cell Stem Cell* 10(6):698–708.

Furuta, M. and R.S. Bridges. 2009. Effects of maternal behavior induction and pup exposure on neurogenesis in adult, virgin female rats. *Brain Res. Bull.* 80(6):408–13.

Gould, E. 2007. How widespread is adult neurogenesis in mammals? *Nat. Rev. Neurosci.* 8: 481–88.

Grattan, D.R. and I.C. Kokay. 2008. Prolactin: A pleiotropic neuroendocrine hormone. *J. Neuroendocrinol.* 20(6):752–63.

Hagg, T. 2005. Molecular regulation of adult CNS neurogenesis: An integrated view. *Trends Neurosci.* 28(11):589–95.

Huang, L. and E.L. Bittman. 2002. Olfactory bulb cells generated in adult male golden hamsters are specifically activated by exposure to estrous females. *Horm. Behav.* 41:343–50.

Kee, N., C.M. Teixeira, A.H. Wang and P.W. Frankland. 2007. Preferential incorporation of adult-generated granule cells into spatial memory networks in the dentate gyrus. *Nat. Neurosci.* 10:355–62.

Keller, M., Q. Douhard, M.J. Baum and J. Bakker. 2006. Destruction of the main olfactory epithelium reduces female sexual behavior and olfactory investigation in female mice. *Chem. Senses* 31(4):315–23.

Kempermann, G., L. Wiskott and F.H. Gage. 2004. Functional significance of adult neurogenesis. *Curr. Opin. Neurobiol.* 14:186–91.

Kermen, F., S. Sultan, J. Sacquet, N. Mandairon and A. Didier. 2010. Consolidation of an olfactory memory trace in the olfactory bulb is required for learning-induced survival of adult-born neurons and long-term memory. *PloS One* 5:e12118.

Larsen, C.M. and D.R. Grattan. 2010. Prolactin-induced mitogenesis in the subventricular zone of the maternal brain during early pregnancy is essential for normal postpartum behavioral responses in the mother. *Endocrinology* 151(8):3805–14.

Larsen, C.M., I.C. Kokay and D.R. Grattan. 2008. Male pheromones initiate prolactin-induced neurogenesis and advance maternal behavior in female mice. *Horm. Behav.* 53:509–17.

Lazarini, F., M.A. Mouthon, G. Gheusi et al. 2009. Cellular and behavioral effects of cranial irradiation of the subventricular zone in adult mice. *PloS One* 4:e7017.

Leinders-Zufall, T., P. Brennan, P. Widmayer et al. 2004. MHC class I peptides as chemosensory signals in the vomeronasal organ. *Science* 306:1033–37.

Leuner, B., E.R. Glasper and E. Gould. 2010. Sexual experience promotes adult neurogenesis in the hippocampus despite an initial elevation in stress hormones. *PloS One* 5:e11597.

Leypold, B.G., C.R. Yu, T. Leinders-Zufall et al. 2002. Altered sexual and social behaviors in trp2 mutant mice. *Proc. Natl. Acad. Sci. U S A* 99:6376–81.

Li, C.S., H. Kaba, H. Saito and K. Seto. 1989. Excitatory influence of the accessory olfactory bulb on tuberoinfundibular arcuate neurons of female mice and its modulation by oestrogen. *Neuroscience* 29:201–8.

Lindsey, B.W. and V. Tropepe. 2006. A comparative framework for understanding the biological principles of adult neurogenesis. *Prog. Neurobiol.* 80:281–307.

Lois, C. and A. Alvarez-Buylla. 1994. Long-distance neuronal migration in the adult mammalian brain. *Science* 264:1145–48.

Luskin, M.B. 1993. Restricted proliferation and migration of postnatally generated neurons derived from the forebrain subventricular zone. *Neuron* 11(1):173–89.

Ma, D.K., W.R. Kim, G.L. Ming and H. Song. 2009. Activity-dependent extrinsic regulation of adult olfactory bulb and hippocampal neurogenesis. *Ann. N. Y. Acad. Sci.* 1170:664–73.

Magavi, S.S., B.D. Mitchell, O. Szentirmai, B.S. Carter and J.D. Macklis. 2005. Adult-born and preexisting olfactory granule neurons undergo distinct experience-dependent modifications of their olfactory responses in vivo. *J. Neurosci.* 25:10729–39.

Mak, G.K. and S. Weiss. 2010. Paternal recognition of adult offspring mediated by newly generated CNS neurons. *Nat. Neurosci.* 13:753–58.

Mak, G.K., E.K. Enwere, C. Gregg et al. 2007. Male pheromone-stimulated neurogenesis in the adult female brain: Possible role in mating behavior. *Nat. Neurosci.* 10:1003–11.

Mandairon, N., J. Sacquet, F. Jourdan and A. Didier. 2006. Long-term fate and distribution of newborn cells in the adult mouse olfactory bulb: Influences of olfactory deprivation. *Neuroscience* 141:443–51.

Martel, K.L. and M.J. Baum, 2009. A centrifugal pathway to the mouse accessory olfactory bulb from the medial amygdala conveys gender-specific volatile pheromonal signals. *Eur. J. Neurosci.* 29:368–76.

Martinez, I. and R.G. Paredes. 2001. Only self-paced mating is rewarding in rats of both sexes. *Horm. Behav.* 40:510–17.

Matsuoka, M., H. Kaba, Y. Mori and M. Ichikawa. 1997. Synaptic plasticity in olfactory memory formation in female mice. *Neuroreport* 8:2501–4.

Matsuoka, M., M. Norita and R.M. Costanzo. 2005. A new surgical approach to the study of vomeronasal system regeneration. *Chem. Senses* 30Suppl:129–30.

McBride, K., B. Slotnick and F.L. Margolis. 2003. Does intranasal application of zinc sulfate produce anosmia in the mouse? An olfactometric and anatomical study. *Chem. Senses* 28:659–70.

McClintock, M.K. and N.T. Adler. 1978. The role of the female during copulation in wild and domestic Norway rats (*Rattus norvegicus*). *Behaviour* 67:67–96.

McClintock, M.K. and J.J. Anisko. 1982. Group mating among Norway rats. I. Sex differences in the pattern and neuroendocrine consequences of copulation. *Anim. Behav.* 30:398–409.

McClintock, M.K., J.J. Anisko and N.T. Adler. 1982. Group mating among Norway rats. II. The social dynamics of copulation: Competition, cooperation, and mate choice. *Anim. Behav.* 30:410–25.

Meisel, R.L. and B.D. Sachs. 1994. The physiology of male sexual behavior. In *The Physiology of Reproduction*, Third Edition, E. Knobil and J.D. Neill, eds., 3–105. New York: Raven Press.

Messing, R.B., C. Dodge, J.C. Waymire, G.S. Lynch and S.A. Deadwyler. 1979. Morphine induced increases in the incorporation of 3H-thymidine into brain striatal DNA. *Brain Res. Bulletin.* 4:615–19.

Migaud, M., M. Batailler, S. Segura, A. Duittoz, I. Franceschini and D. Pillon. 2010. Emerging new sites for adult neurogenesis in the mammalian brain: A comparative study between the hypothalamus and the classical neurogenic zones. *Eur. J. Neurosci.* 32:2042–52.

Ming, G.L. and H. Song. 2011. Adult neurogenesis in the mammalian brain: Significant answers and significant questions. *Neuron* 70(4):687–702.

Moreno, M.M., C. Linster, O. Escanilla, J. Sacquet, A. Didier and N. Mandairon. 2009. Olfactory perceptual learning requires adult neurogenesis. *Proc. Natl. Acad. Sci. U S A* 106:17980–5.

Morgan, J.I., D.R. Cohen, J.L. Hempstead and T. Curran. 1987. Mapping patterns of c-Fos expression in the central nervous system after seizure. *Science* 237:192–97.

Nissant, A., C. Bardy, H. Katagiri, K. Murray and P.M. Lledo. 2009. Adult neurogenesis promotes synaptic plasticity in the olfactory bulb. *Nat. Neurosci.* 12:728–30.

Nottebohm, F. 1985. Neuronal replacement in adulthood. *Ann. N. Y. Acad. Sci.* 457:143–61.

Oboti, L., G. Savalli, C. Giachino et al. 2009. Integration and sensory experience-dependent survival of newly-generated neurons in the accessory olfactory bulb of female mice. *Eur. J. Neurosci.* 29:679–92.

Oboti, L., R. Schellino, C. Giachino et al. 2011. Newborn interneurons in the accessory olfactory bulb promote mate recognition in female mice. *Front. Neurosci.* 5:113.

Pankevich, D.E., J.A. Cherry and M.J. Baum 2006. Accessory olfactory neural Fos responses to a conditioned environment are blocked in male mice by vomeronasal organ removal. *Physiol. Behav.* 87:781–88.

Paredes, R.G. 2009. Evaluating the neurobiology of sexual reward. *ILAR J.* 50:15–27.

Paredes, R.G. and A. Alonso. 1997. Sexual behavior regulated (paced) by the female induces conditioned place preference. *Behav. Neurosci.* 111:123–28.

Paredes, R.G. and A. Fernández-Guasti. 2008. Rewarding properties of mating. In *Neural Mechanisms of Drugs of Abuse and natural Reinforcers*, M. Méndez and R. Mondragón-Ceballos, eds., 159–170. Trivandrum Kerala, India: Research Signpost.

Paredes, R.G. and B. Vazquez. 1999. What do female rats like about sex? Paced mating. *Behav. Brain Res.* 105:117–27.

Peele, P., I. Salazar, M. Mimmack, E.B. Keverne and P.A. Brennan. 2003. Low molecular weight constituents of male mouse urine mediate the pregnancy block effect and convey information about the identity of the mating male. *Eur. J. Neurosci.* 18:622–28.

Peretto, P., A. Merighi, A. Fasolo and L. Bonfanti. 1999. The subependymal layer in rodents: A site of structural plasticity and cell migration in the adult mammalian brain. *Brain Res. Bull.* 49:221–43.

Petreanu, L. and A. Alvarez-Buylla. 2002. Maturation and death of adult-born olfactory bulb granule neurons: Role of olfaction. *J. Neurosci.* 22:6106–13.

Portillo, W., N. Unda, F.J. Camacho et al. 2012. Sexual activity increases the number of new-born neurons in the accessory olfactory bulb of male rats. *Front. Neuroanat.* 6: Art 25.

Reynolds, B.A. and S. Weiss. 1992. Generation of neurons and astrocytes from isolated cells of the adult mammalian central nervous system. *Science* 225:1707–10.

Robitaille, J.A. and J. Bouvet. 1976. Field observations on the social behavior of the Norway rat, *Rattus norvegicus* (Berkenhout). *Biol Behav.* 1:289–308.

Rochefort, C. and P.M. Lledo. 2005. Short-term survival of newborn neurons in the adult olfactory bulb after exposure to a complex odor environment. *Eur. J. Neurosci.* 22:2863–70.

Rochefort, C., G. Gheusi, J.D. Vincent and P.M. Lledo. 2002. Enriched odor exposure increases the number of newborn neurons in the adult olfactory bulb and improves odor memory. *J. Neurosci.* 22:2679–89.

Rosser, A.E., C.J. Remfry and E.B. Keverne. 1989. Restricted exposure of mice to primer pheromones coincident with prolactin surges blocks pregnancy by changing hypotha-lamic dopamine release. *J. Reprod. Fertil.* 87:553–59.

Ruscio, M.G., T.D. Sweeny, J.L. Hazelton, P. Suppatkul, E. Boothe and C.S. Carter. 2008. Pup exposure elicits hippocampal cell proliferation in the prairie vole. *Behav. Brain Res.* 187(1):9–16.

Sahay, A., D.A. Wilson and R. Hen. 2011. Pattern separation: A common function for new neurons in hippocampus and olfactory bulb. *Neuron* 70(4):582–8.

Sakamoto, M., I. Imayoshi, T. Ohtsuka, M. Yamaguchi, K. Mori and R. Kageyama. 2011. Continuous neurogenesis in the adult forebrain is required for innate olfactory responses. *Proc. Natl. Acad. Sci. U S A* 108:8479–84.

Sargeant, T.J., J.H. Miller and D.J. Day. 2008. Opioidergic regulation of astroglial/neuronal proliferation: Where are we now? *J. Neurochem.* 107:883–97.

Shingo, T., C. Gregg, E. Enwere et al. 2003. Pregnancy-stimulated neurogenesis in the adult female forebrain mediated by prolactin. *Science* 299:117–20.

Smith, M.T., V. Pencea, Z. Wang, M.B. Luskin and T.R. Insel. 2001. Increased number of BrdU-labeled neurons in the rostral migratory stream of the estrous prairie vole. *Horm. Behav.* 39:11–21.

Sohur, U.S., J.G. Emsley, B.D. Mitchell and J.D. Macklis. 2006. Adult neurogenesis and cel-lular brain repair with neural progenitors, precursors and stem cells. *Philos. Trans. R. Soc. Lond. B. Biol. Sci.* 361:1477–97.

Tirindelli, R., M. Dibattista, S. Pifferi and A. Menini. 2009. From pheromones to behavior. *Physiol Rev.* 89(3):921–56.

Wohr, M., M. Kehl, A. Borta, A. Schanzer, R.K. Schwarting and G.U. Hoglinger. 2009. New insights into the relationship of neurogenesis and affect: Tickling induces hip-pocampal cell proliferation in rats emitting appetitive 50-kHz ultrasonic vocalizations. *Neuroscience* 163:1024–30.

Yamaguchi, M. and K. Mori. 2005. Critical period for sensory experience-dependent survival of newly generated granule cells in the adult mouse olfactory bulb. *Proc. Natl. Acad. Sci. U S A* 102:9697–702.

14 Influence of Cat Odor on Reproductive Behavior and Physiology in the House Mouse (*Mus Musculus*)

Vera V. Voznessenskaya

CONTENTS

14.1 Introduction 389
14.2 Methods and Materials 392
    14.2.1 Test Subjects 392
    14.2.2 Assay for Fecal Corticosterone Metabolites...393
    14.2.3 Immunohistochemistry Assay 394
    14.2.4 Vomeronasal Surgery...394
14.3 Results and Discussion 395
14.4 Conclusions..402
Acknowledgments...402
References..402

14.1 INTRODUCTION

Closely related Mus species *Mus musculus* and *Mus domesticus* are the most popular objects in the study of mammalian chemical communication. The understanding of pheromone influences on mammalian behavior has advanced dramatically since the term "pheromone" was introduced. The major advances in recent years have been based mainly on a single species—the mouse (laboratory form of *Mus musculus domesticus*). Genetic technologies have revealed a surprisingly large repertoire of chemosensory receptors in mice that potentially detect pheromones (Brennan 2010). However, interspecies chemical communication in the house mouse remains the least investigated area. Use of the laboratory inbred strains of mice makes understanding of the behavioral effects elicited by chemical signals from other species even more complicated.

Predator-prey relationships provide an excellent model for the study of interspecies chemical communication. Small mammals in general are frequently at risk to be caught by mammalian, avian, or reptilian predators. In turn their prey species developed a variety of specific adaptations to facilitate recognition, avoidance, and defense against predators. Such antipredator behavioral systems are critical for survival (see review Apfelbach et al. 2005). Chemosensory detection is a very important aspect for predator avoidance strategy for many mammals including the house mouse. Odors from carnivores may elicit fear-induced stereotypic behaviors, change activity patterns and feeding rate, and affect the neuroendocrine system, reproductive behavior, and reproductive output in potential prey (Apfelbach et al. 2005; Dielenberg and McGregor 2001; Harvell 1990; Hayes 2008; Hayes et al. 2006; Kats and Dill 1998; Müller-Schwarze 2006). A number of studies (see Table 14.1) showed effects of odors derived from different predators on behavior and physiology of the

TABLE 14.1
Examples of Studies on Responses to Predator Odors in the House Mouse (*Mus domesticus, Mus musculus*)

House Mouse Species	Predator Species/ Odor Source	Effect Described/ No Effect	Reference
Mus musculus domesticus/lab	*Elaphe obsoleta* (rat snake)/shed skin extract	Feeding rate, defecation rate	Weldon et al. (1987)
Mus musculus domesticus	*Virginia striatula* (rough earth snake)/body scent	No effect on feeding rate, ambulation, defecation rate	Weldon et al. (1987)
Mus domesticus	*Felis catus* (cat)/feces	Space use, trapping rate	Dickman (1992)
Mus domesticus	*Felis catus* (cat)/feces	Trapping rate	Drickamer et al. (1992)
Mus domesticus	*Canis familiarus*(dog)/ feces	No effect on trapping rate	Drickamer et al. (1992)
Mus domesticus	*Vulpes vulpes* (red fox)/ feces	Feeding rate	Coulston et al. (1993)
Mus domesticus	*Mustela erminea* (stoat)/ anal gland secretion	Feeding rate	Coulston et al. (1993)
Mus domesticus	*Canis latrans* (coyote)/ urine	Feeding rate	Nolte et al. (1994)
Mus domesticus	*Felis catus* (cat)/feces	Behavior	Berton et al. (1998)
Mus musculus	*Mustela putorius furo* (ferret)/urine	Behavior	Roberts et al. (2001)
Mus musculus	*Felis catus* (cat)/urine	Litter size, postnatal development	Vasilieva et al. (2001)
Mus musculus	*Felis catus* (cat)/urine	Altered estrous cycle	Feoktistova et al. (2002)
Mus domesticus	*Vulpes vulpes* (red fox)/ feces	No effect on feeding rate, population dynamics	Banks and Powell (2004)
Mus musculus	*Lampropeltis getula* (kingsnake)/fecal material, shed skins	No effect on total number of offspring, litter size, weight of pups	Starke and Ferkin (2013)
Mus musculus domesticus/lab	*Canis lupus* (wolf)/urine volatiles	Aversive and fear-related responses	Osada et al. (2013)

house mouse. It implies the existence of shared signal properties through a number of predator species. This idea about generalized "leitmotif" of predator odors was suggested even much earlier (Stoddart 1980).

The idea about the existence of a common carnivore signal was experimentally tested for the first time by Nolte et al. (1994). Manipulations with predator diet as well as chemical removal from carnivore urine of the sulfurous compounds and amines revealed their key role in the effects of coyote urine (*Canis latrans*) on feeding rates in wild living *Mus domesticus* (Nolte et al. 1994). Berton et al. (1998) also demonstrated that the diet of a cat strongly affects the behavior of mice towards its feces. Using similar chemical manipulations with cat urine (*Felis catus*) and manipulating with the diet of urine donors, it has been shown that sulfurous compounds and amines are critical for reproductive inhibitory effects of the cat urine in rodents (Voznessenskaya et al. 2002). Another study (Fendt 2006) indicates that only exposure to urine of canids and felids but not of herbivores induces defensive behavior in laboratory rats (Fendt 2006). The term "kairomone" was widely adopted to name predator chemical signals: "kairomones, such as those that elicit fear behavior, are cues transmitted between species that selectively disadvantage the signaler and advantage the receiver" (Wyatt 2003). In search of the molecular nature of kairomones, Papes et al. (2010) isolated the salient molecules from two species (rat and cat) using a combination of behavioral assays in naïve laboratory mice, calcium imaging and c-Fos induction. The defensive behavior-promoting activity released by other animals is encoded by species-specific proteins belonging to the major urinary protein (MUP) family, homologs of aggression-promoting mouse pheromones and mediated through the vomeronasal organ (VNO) (Papes et al. 2010).

The trace-amine-associated receptors (TAARs) form a specific family of G protein-coupled receptors in vertebrates that was initially considered to be neurotransmitter receptors before it was discovered that mouse TAARs function as chemosensory receptors in the olfactory epithelium (Liberles and Buck 2006). Discovery of a new function of TAARs stimulated the search for the potential ligands. More recent studies (Liberles 2009) showed that ligands for mouse TAARs include a number of volatile amines, several of which are natural constituents of mouse urine. One chemical, 2-phenylethylamine, is reported to be enriched in the urine of stressed animals, and two others, trimethylamine and isoamylamine, are enriched in male versus female urine. These findings raised the possibility that some TAARs are pheromone receptors (Liberles 2009). Further studies (Ferrero et al. 2011) revealed that 2-phenylethylamine is a key component of a predator odor blend that triggers hardwired aversion circuits in the rodent brain. Neurons expressing TAARs project to discrete glomeruli predominantly localized to a confined bulb region (Johnson et al. 2012). TAARs expression involves different regulatory logic than OR expression. Moreover, the epigenetic signature of OR gene choice is absent from TAAR genes. The unique molecular and anatomical features of the TAAR neurons suggest that they constitute a distinct olfactory subsystem (Johnson et al. 2012). Initially 2-phenylethylamine was purified from bobcat urine; quantitative HPLC analysis across 38 mammalian species indicated enriched 2-phenylethylamine production by numerous carnivores. Rats and mice avoid a 2-phenylethylamine odor source; enzymatic depletion of 2-phenylethylamine from a carnivore odor showed that it is required for full avoidance behavior (Ferrero et al. 2011). This study clearly demonstrated that rodent olfactory sensory neurons have the capacity for recognizing interspecies odors.

Findings of universal carnivore signals may explain why potential prey respond to odors from allopatric predators with which they do not have evolutionary links and never encountered in their lives, on one hand. On the other hand, the ability of predator odors to produce profound effects on the behavior of prey in general and especially on the reproductive behavior and neuroendocrine system is associated with natural predators only, which suggests an essential role of the evolutionary link between signaling predator and potential prey. First of all, it means that potential prey (in our case, mice) are able to distinguish predator species on a chemosensory basis. Numerous studies (Table 14.1) support this observation (also see review in Apfelbach et al. 2005). It raises a question about the multicompound nature of the kairomones as well as about the existence of species-specific predator chemical cues. One of the most specialized predators toward the house mouse is the domestic cat *Felis catus*. A long history of coexistence in the same environments led to the development of mutual adaptations at the genetic level. These two species provide a perfect model for the study of innate responses to predator odors.

Felinine is a unique sulfur-containing amino acid found in the urine of domestic cats (Rutherfurd et al. 2002). Sulfur-containing volatile compounds 3-mercapto-3-methyl-1-butanol, 3-mercapto-3-methylbutyl formate, 3-methyl-3-methylthio-1-butanol, and 3-methyl-3-(2-methyl-disulfanyl)-1-butanol are identified as species-specific odorants and candidates of felinine derivatives from the cat urine. The levels of these compounds were found to be sex- and age-dependent (Miyazaki et al. 2006a, b). These cat-specific volatile compounds may represent pheromones used as territorial markers for conspecific recognition or reproductive purposes by mature cats (Miyazaki et al. 2008). Species-specific compounds may be used also by other species to recognize potential predators and their physiological status. We now present evidence to support bioactivity of L-felinine and its derivates with the house mouse (*Mus musculus*).

14.2 METHODS AND MATERIALS

14.2.1 TEST SUBJECTS

Test subjects were 4–6-month-old mice (*Mus musculus*) from an outbred laboratory population as well as 2–3-day-old laboratory generation of mice trapped in natural biotopes in the Moscow region (*Mus musculus musculus*). We didn't use mice trapped directly from the wild since it limited our knowledge about their experience with predator odors. Before the start of the experiments, females and males were housed singly. Experimental rooms were illuminated on a 14:10-hour light:dark schedule, and maintained at 20°C–22°C. Food and tap water were provided *at libitum*. Virgin females in proestrus/estrus, as determined by vaginal cytology, were chosen for the mating experiments. Sexually experienced males that were not mated in the 14 days before the test were used as sires. The morning after pairing, the females were checked for successful mating, as indicated by the presence of a vaginal plug. Successfully mated females were then housed singly or placed in enclosures of 12 females each.

The experimental method consisted of applying 0.1 ml of a test solution (cat urine or 0.05% L-felinine, US Biologicals) to the bedding of pregnant mice every other day for different time durations. This application maximized the likelihood of physical and odor exposure of the test stimulus to the female. In experiments, four treatments were used: tap water (WAT), as a negative control; urine from guinea pigs maintained on a vegetarian diet (vegetables, grains, and water *ad libitum*), as a urine control (GPU); urine from domestic cats maintained on a meat diet and that normally hunt for mice (CU), as a model stimulus representing unadultered predator urine; L-felinine (US Biologicals) in concentration 0.05%, comparable with intact cat urine, as a potential active ingredient. Cats were maintained on a meat diet for 14 days before urine collection. After mating, females were randomly assigned to treatment groups. Mean differences among treatment groups were determined in separate analyses for the number of pups and sex ratios using the software STATISTICA8.

For each experimental group, the total number of offspring was counted as well as number of pups per female; sex ratio was determined.

Urine from domestic cats (*Felis catus*) was used as a source of sympatric predator chemical cues. These cats normally hunt for mice and have mice as part of their diet. If needed, additional meat was added to their diet. Freshly voided urine was frozen (–22°C). Once defrosted, urine was used only once. Nonpredator urine was obtained from guinea pigs. Individuals of these species were placed into metabolic stainless steel cages overnight, and urine was collected and stored using the method described above. Urine was stored at –22°C.

An open arena (D = 0.7 m) with bright lights was used as a positive control for corticosterone assay. Pregnant females were placed for 15 minutes in the center of the arena on the first, fourth, and seventh day of gestation. During the test, we also used a buzzer, which made a loud noise, every 5 minutes. In addition, handling of mice physically also induced additional stress. Blood samples from orbital area were drawn after each test for corticosterone assay.

Animals within each treatment were randomly assigned to one of four cohorts. Blood samples (50 µl) were obtained once in 3 days for each cohort for each of the treatment for the first 7 days of gestation. This minimized the handling and sampling of individual mice while allowing a detailed study of changes in hormonal pattern as a function of time and treatment. Our experience shows that this method of repeated blood sampling has no long-term effect on visible scarring associated with traditional tail sampling technologies. Samples were centrifuged and the plasma frozen at –20°C until subsequent analysis. Plasma corticosterone was assayed (in duplicate) by enzyme immunoassay (EIA) method (DRG, Springfield, NJ, USA).

14.2.2 Assay for Fecal Corticosterone Metabolites

In small animals like mice, the monitoring of endocrine functions over time is constrained seriously by the adverse effects of blood sampling. Therefore, we used noninvasive technique to monitor glucocorticoids with recently established

5a-pregnane-3ß,11-ethol,21-triol-20-one enzyme immunoassay (Touma et al. 2004) to assess adrenal activity in mice under conditions of long-lasting exposures to predator odors. Mice were exposed to cat urine or L-felinine (0.05%) on an everyday basis for a period of 2 weeks. On completion of exposures fecal material was collected from each animal over 24 hours. Extraction procedure was performed with 80% methanol. Concentration of corticosterone metabolites was measured with spectrophotometer (Spectramax340, Molecular Devices, LLC, Sunnyvale, USA) at 450 and 670 nm. Specific antibodies were received from Prof. E. Möstl's laboratory (University of Veterinary Medicine, Vienna).

14.2.3 IMMUNOHISTOCHEMISTRY ASSAY

To visualize activated neurons on olfactory bulbs sections in response to stimulation, Fos protein immunohistochemistry was used (Flavell and Greenberg 2008). Fos protein is a product of *c-fos* known as an immediate early gene that is induced quickly by different stimuli including cell depolarization (Sheng and Greenberg 1990). Labeling Fos provides a physiological marker of neurons activated in response to specific stimuli. The half-life span of protein Fos is 2 hours: depending on specific characteristics and neural cell localization, optimal exposure time for maximal Fos detection may range from 45 to 90 minutes. In our experiment for vomeronasal (VNO) receptor epithelium optimal time exposure was determined as 90 minutes (Voznessenskaya et al. 2010). To stimulate the main and accessory olfactory systems mice were exposed to L-felinine (0.05% in water) for 40 minutes using half-duty cycle (one minute—specific odor, one minute—clean air). Immediately after exposure mice were perfused with 3% paraformaldehyde in phosphate buffer. Olfactory bulbs were removed and postfixed in paraformaldehyde for 16 hours. We used standard procedure for fixation of olfactory bulbs, cryoprotection, and immunohistochemical staining of olfactory bulbs sections (DellaCorte 1995). We used the indirect avidin/biotin method; horseradish peroxidase was used as an enzymatic label and diaminobenzidine (DAB) was used as a chromogen. Sections were made at 20 μm using cryostat (Triangle Biomedical, Durtham, USA). Immunostaining was made according to standard 3-day protocol using primary antibodies (Santa Cruz Biotechnology, Dallas, USA): c-fos (4) sc-52, dilution 1: 500. For visualization and counting of Fos-positive cells we used a Nikon©Eclipse E400 microscope with a Nikon©Coolpix 990 camera. For picture analyses we used ImageJ (http://rsbweb.nih.gov/ij/index.html).

14.2.4 VOMERONASAL SURGERY

Vomeronasal surgery (VNX) via an expanded nasopalatine foramen was performed as previously described (Wysocki and Wysocki 1995). Control animals underwent sham operations. We used soybean agglutinin-horseradish peroxidase (SBA-HRP) immunohistochemistry of the accessory olfactory bulb to verify VNX (Wysocki and Wysocki 1995). In mice, SBA-HRP stains the glomeruli in the accessory, but not main, olfactory bulb. Stain is absent after a successful VNX.

All experimental procedures were approved by vivarium ethical committee by Institute of Ecology & Evolution, Russian Academy of Sciences.

14.3   RESULTS AND DISCUSSION

We observed a block of pregnancy in experimental groups of mice exposed after mating to intact cat urine. The percentage of females with pregnancy block ranged from 31.25% to 68.75% depending on the season (Figure 14.1) with the largest effect in the autumn-winter period. This reflects the seasonal dynamics of reproduction in the house mouse. Even in the laboratory, numbers of cycling and accordingly breeding females are lower in the autumn-winter period. Hormonal status also experiences seasonal changes, making females more vulnerable to any stress event during the autumn-winter season. In control groups of mice we also observed some seasonal variation in the percentage of pregnancy block: from 12.5% in spring-summer to 18.75% in autumn-winter, but differences were not significant. We did not observe such robust seasonal differences in sensitivity to potential active ingredient from cat urine the unique Felidae family amino acid L-felinine. Exposures of mated females to L-felinine (0.05% in water) provoked pregnancy block in 67.85% female mice while in the control group we observed only 17.86% ($n = 28$, $p < 0.01$, Fisher test). For the remaining females we counted the total number of pups to compare litter sizes for experimental and control animals. Differences in litter size in the control and experimental groups were not significant; we only observed a tendency for lower litter size in the felinine treatment group. This may be explained by a considerable percentage (67.8%) of females with pregnancy block in the felinine treatment group. We used the total number of pups per fertile female as a cumulative indicator that takes into consideration both effects: block of pregnancy and litter size reductions. In the felinine treatment group the total number of pups per fertile female was 2.5 ± 1 while in the control group it was 5.70 ± 1.00 ($n = 28$, $p = 0.046$ Mann-Whitney U test). This clearly indicates significant suppression of reproduction. Another significant effect was observed under cat urine exposures: skewed sex ratios in favor of males ($p < 0.001$) (Figure 14.2a). A similar effect was observed for the L-felinine exposure

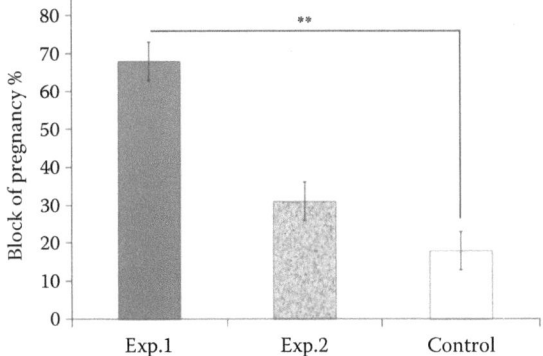

FIGURE 14.1   The influence of exposures of cat (*Felis catus*) urine during first week of gestation on the percent of pregnancy block in the house mouse *Mus musculus*; exp.1 = autumn-winter season; exp.2 = spring-summer season; control = autumn-winter season. $p < 0.01$, $n = 16$.

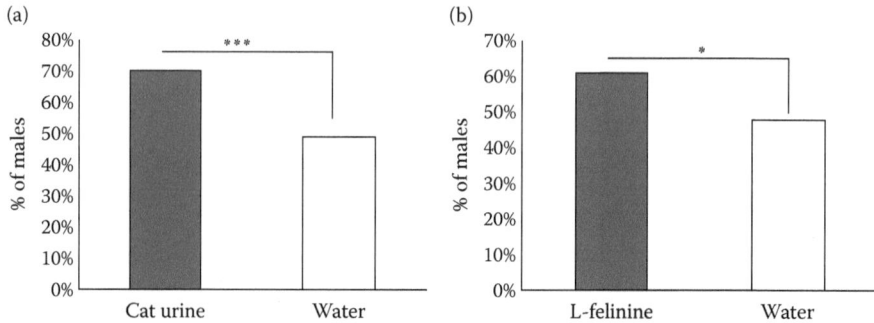

FIGURE 14.2 (a) The influence of cat (*Felis catus*) urine exposures during gestation on sex ratio in house mouse *Mus musculus* (***$p \leq 0.001$, n (cat urine) = 52, n (water) = 118, Fisher test). (b) The influence of the L-felinine (0.05%) exposures during gestation on sex ratio in house mouse *Mus musculus* (*$p \leq 0.05$, n (L-felinine) = 72; n (water) = 160, Fisher test).

group ($p < 0.05$) (Figure 14.2b). A skewed sex ratio in favor of males was interpreted as an adaptive response. A high concentration of felinine provides information about the high population density of a very specialized predator, *Felis catus*. Cushing (1985) was the first who proposed that females in a reproductive condition are more vulnerable to predation than nonreproductive females, and thus he suggested that it would be adaptive for them to suppress reproduction in case of high predation risk. The generation time of rodents is relatively short; complete reproductive inhibition may not be adaptive. However, reduced reproduction may be beneficial. Reduced reproduction would relieve energetic constraints on lactating females that might otherwise jeopardize survival if a full litter size was attempted. In accordance with theory on reproductive value, the probability to succeed reproductively under deteriorating environmental conditions is higher for males. First, males are more mobile and cover longer distances escaping from unfavorable territory. Even with reduced litter size, females may still experience lower survival probabilities during reproduction and lactation in food-limiting or predator-overpopulated environments because of energetic constraints. However, males would be less constrained by such energetic considerations. Thus, their survivorship probabilities may be higher than females, and by implication their value in contributing to fitness would also be higher.

Such indicators like litter size or number of pups per fertile female prior to weaning characterizes fecundity rates rather than reproductive success. To assess reproductive success we used such indicators as survivorship of pups after weaning. In another set of experiments we compared the reproductive output of *Mus musculus musculus* during two spring-summer months (mid-May–mid-July) in four fully covered small enclosures (1.5 m × 2.0 m). We placed nest building material and wood shavings in each module. In two of them (exp. 1, exp. 2) we placed plastic containers with a cotton swab soaked with cat urine. Additionally we placed a wooden box (0.3 × 0.3 m) that served as a shelter in one control (contr. 2) and one experimental (exp. 2) module. Taking into consideration frequency of scent marking in *Felis catus* under natural conditions, we renewed cotton swabs once a week. We placed equal numbers of cycling females of the same age and the same weight in each module. We also placed two males of the same age and weight in each enclosure. By the end of experiment we counted the number of live pups and the number of adult live females for each enclosure (Figure 14.3a, b). Cat urine significantly affected reproductive output in the house mouse. In experimental module 2 (no shelter inside) we did not find any live pups, while in control 1 the number of pups per fertile female was 2.9 ($p < 0.001$, $n = 10$). In experimental module 2 (shelter inside), the number of pups per fertile female was 1.64 while in the control group it was 2–3.36. By the end of the experiment, the age of pups ranged from 3 to 5 weeks. This data indicated that responses to predator odors even under seminatural conditions may be significantly modified by the availability of shelter. Early field studies on the effectiveness of predator odors as natural repellents revealed the importance of characteristics of habitat such as availability of cover (Epple et al. 1993).

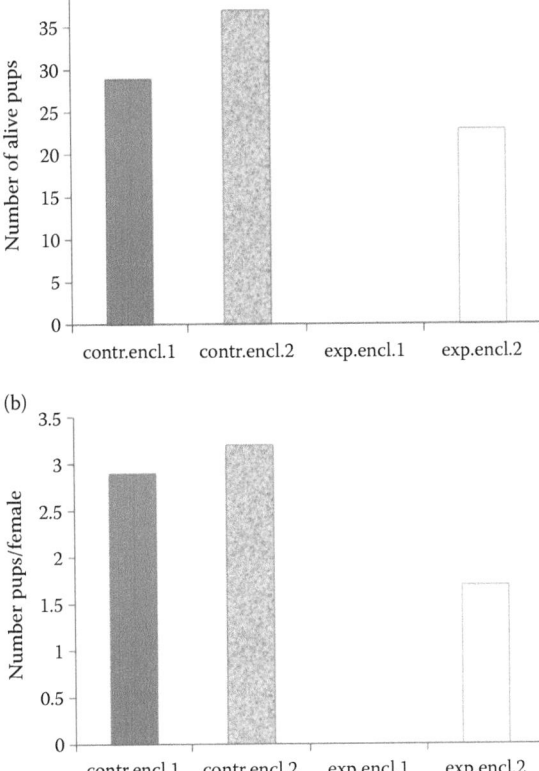

FIGURE 14.3 (a) The influence of cat (*Felis catus*) urine on reproductive output of the house mouse (*Mus musculus musculus*) under seminatural conditions. Y-axis indicates number of live pups by the end of the experiment (7 weeks). (b) The influence of cat (*Felis catus*) urine on reproductive output of the house mouse (*Mus musculus musculus*) under seminatural conditions. Y-axis indicates number of live pups per female by the end of the experiment (7 weeks).

In natural environments mice frequently undergo long-lasting and chronic exposures to predator odors. Taking these circumstances into account we evaluated chemosensory behavior of mice exposed to cat odor for extended periods. Male mice were
exposed to cat urine for 10 days; 24 hours after completion of the exposures they
were tested in a number of tests. Those animals discriminated female odor from
water but did not show preference for receptive female odor versus diestrus female
odor in a standard odor preference test (Figure 14.4a, b). Striking similarity was
observed in the same test when mice were exposed for 10 days to L-felinine (0.05%).
Male mice discriminated female urine versus male urine but not estrus female urine
versus diestrus female urine. We compared the sexual behavior of males before long-
lasting exposure to cat urine and after exposure in a standard pairing test with a
receptive female. Exposure to cat urine significantly ($p < 0.05$) diminished the number of nasal-nasal contacts, attempted mountings, and mountings with intromissions
(Figure 14.5). Our results indicate that extended exposures to cat odor suppresses
sexual behavior in male mice.

To study possible mechanisms underlying suppressed sexual behavior and
reduced reproduction effort we monitored plasma corticosterone, the major glucocorticoid in mice. We observed clear elevation of plasma corticosterone ($p < 0.001$,
$n = 8$, Tukey test) in response to cat urine in female mice (Figure 14.6a, d). As a
positive control we used an open arena test with added stress (Figure 14.6c). Mice
responded to this kind of treatment with elevated corticosterone but we observed
habituation during the course of consecutive placements (days 1–7). At the same
time mice did not habituate to consecutive exposures to cat urine (Figure 14.6a).
We also observed such a habituation in mice introduced to another novel stimulus:
guinea pig urine (Figure 14.6b). To explore for how long predator chemical cues may
provoke elevated corticosterone we exposed male mice to L-felinine (0.05% in water)
for 2 weeks. On completion of the exposures, feces from each animal were collected
over 24 hours and glucocorticoid metabolites were measured for each animal. In
the control group of male mice, the concentration of corticosterone metabolites was

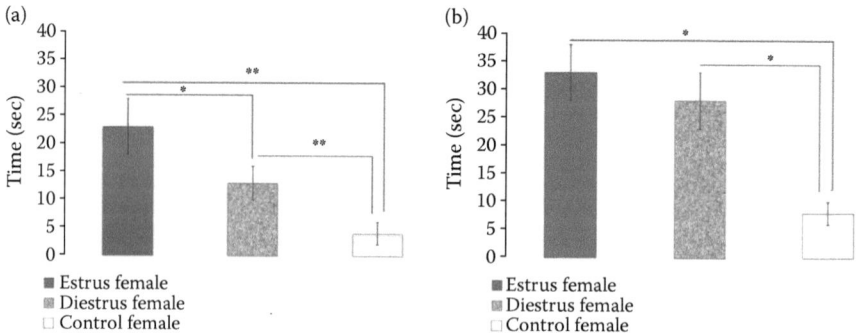

FIGURE 14.4   (a) Performance of males in standard odor preference test in the house mouse
(*Mus musculus*). (b) Performance of males in standard odor preference test in the house mouse
(*Mus musculus*) after 10 days of exposure to cat urine (*Felis catus*). *$p \le 0.05$, **$p \le 0.01$
Wilcoxon matched pairs test, T-SEM; $n = 8$ each group.

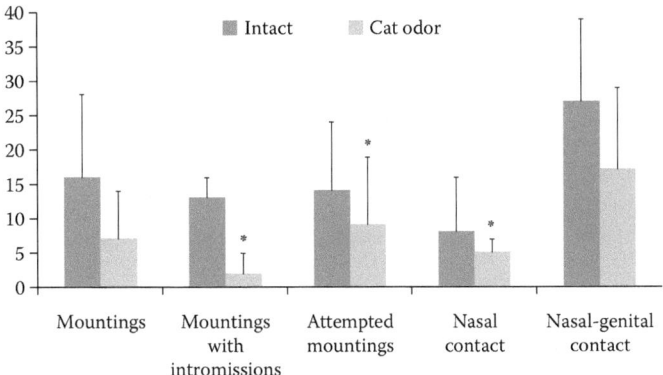

FIGURE 14.5   The influence of long-lasting (10 days) exposure of cat urine on sexual behavior of males *Mus musculus* in standard pairing test with receptive female. *$p \le 0.05$ Wilcoxon matched pairs test, T-SD; $n = 10$; $t = 60$ min.

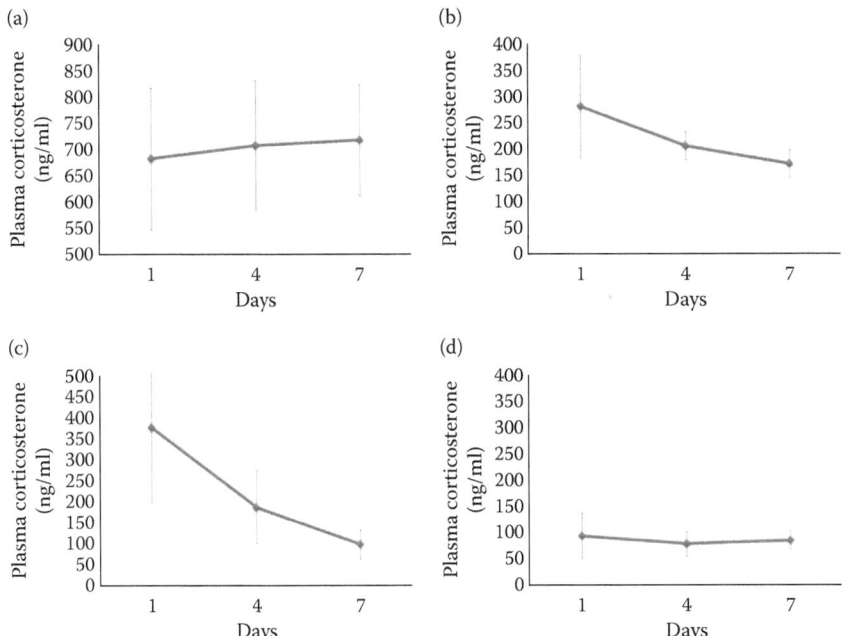

FIGURE 14.6   (a) The influence of short-term exposures (15 min) to cat urine on plasma corticosterone in the house mouse (*Mus musculus*). X-axis indicates days of treatment during one week. Mean ± SD; $n = 8$. (b) The influence of short-term exposures (15 min) to nonpredator urine (guinea pig) on plasma corticosterone in the house mouse (*Mus musculus*). X-axis indicates days of treatment during one week. Mean ± SD; $n = 8$. (c) The influence of introduction to "open field" (15 min) on plasma corticosterone in the house mouse (*Mus musculus*). X-axis indicates days of experiment during one week. Mean ± SD; $n = 8$. (d) The influence of short-term exposures (15 min) to tap water (control) on plasma corticosterone in the house mouse (*Mus musculus*). X-axis indicates days of treatment during one week. Mean ± SD; $n = 8$.

203.85 ± 47.74 ng/0.2g feces; in the felinine treatment group, 702.15 ± 122.24 ng/0.2 g feces ($n = 13$, $p < 0.001$, t-test). The response of laboratory naive animals to predator scents and failure to habituate to the stimulus indicate the innate nature of the response. Chronically elevated plasma corticosterone in response to cat odor exposure, especially at early stages of pregnancy, may be a reason for the induction of pregnancy block. Early studies by McNiven and de Catanzaro (1990) showed that diverse psychological stressors, including exposures to predator odor (rat), correlated with elevated plasma corticosterone, provoke pregnancy block in the house mouse.

There have been a number of studies investigating the effects of cat odor exposure on the glucocorticoid and ACTH production of the potential prey (Blanchard et al. 1998; Cohen et al. 2000; Figueiredo et al. 2003; Papes et al. 2010; Sullivan and Gratton 1998; Voznessenskaya et al. 2002). The first study, performed in rats (File et al. 1993), showed that a cloth that had been rubbed on a cat caused an increase in circulating corticosterone. However, with repeated exposure to the cat odor stimulus this endocrine response habituated, indicating the role of learning. In our studies using cat urine as a source of odor, we also observed habituation to cat odor at the level of plasma coticosterone response in Wistar rats (Voznessenskaya et al. 2002). In contrast, mice did not habituate to repeated exposures of the cat odor. It may imply a hardwired processing of a cat odor as a pheromone in the house mouse. It seems likely that a profound and aversive effect of predator odors and lack of habituation under repeated exposures might only exist "if predator and prey have a long evolutionary history in parallel so that a prey species becomes genetically pre-disposed to avoid the odors of sympatric predators" (Stoddart 1980b); another important condition of such a response: an extremely high risk of fatal outcome for the direct interactions of predator and prey. Obviously, rats are less vulnerable to such predators as cats (*Felis catus*) compared with mice (*Mus musculus, Mus domesticus*).

The well-known Bruce effect (Bruce 1959)—a pregnancy block in rodents caused by exposure to a strange male after mating—requires an intact VNO. This phenomenon relies on an olfactory memory formed in the accessory olfactory bulb (Bellringer et al. 1980; Brennan 2004). We performed a VNX to test a hypothesis about the involvement of VNO in suppression of reproduction in mice under cat odor exposures. The results of the experiment are presented in Figure 14.7. Female mice with VNO removal did not show significant reductions in litter size under cat odor exposures while in sham-operated animals we observed significant reductions in litter size ($p < 0.05$, $n = 12$, Mann-Whitney test). Our findings are in good agreement with other studies on involvement of VNO in reception and analysis of predator odors. Papes et al. (2010) showed that VNO function is necessary for the display of innate behavior induced by odors from the cat, rat, and snake. Our earlier research also showed the involvement of the VNO in reproductive inhibitory effects elicited by cat odor in rats and mice (Kassesinova and Voznessenskaya 2009; Voznessenskaya et al. 1992, 2006). To explore further whether potential active ingredient from the cat urine L-felinine could be analyzed by VNO, we performed Fos immunohistochemistry. Exposure of L-felinine (0.05% in water) intermittently (1 minute on, 1 minute off) for 40 minutes to male mice as well as to female mice ($n = 4$, two independent experiments) resulted in elevated Fos immunoreactivity in the

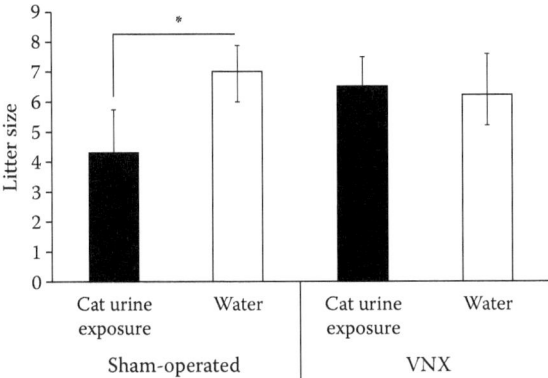

FIGURE 14.7   The influence of vomeronasal organ removal (VNX) on litter size in the house mouse under cat (*Felis catus*) urine exposures. SHAM = sham-operated animals; *p ≤ 0.05, n = 12.

caudal part of the accessory olfactory bulb (AOB). We also observed a variable pattern of activation in the main olfactory bulb (MOB). More prolonged exposures of L-felinine (90 minutes) stimulated Fos immunoreactivity mainly in the basal part of VNO epithelial tissue. VNO neuroepithelial tissue is subdivided into two anatomically and functionally distinct subpopulations of neurons (Rodriguez et al. 2002). Apical sensory neurons are located closer to the lumen and express V1Rs. Basal neurons are located more peripherally and express V2Rs. V1Rs are mainly activated by volatile compounds and V2Rs by substances of higher molecular weight and peptides (Halpern and Martinez-Marcos 2003). Neuroanatomical projections of V1R and V2R neurons also differ. V2R neurons project to the caudal part of the AOB, which sends its projections to the vomeronasal amygdala where myriad steroid receptors are located (Halem et al. 2001). V1R neurons send their projections to the rostral part of the AOB, which in turn projects to the rostral amygdala (Rodriguez et al. 1999). Activation in the caudal part of the AOB may indicate that L-felinine is analyzed rather as a pheromone. Felinine in the presence of water is quite unstable and exists in the form of a mixture of felinine and sulfur-containing volatile compounds; 3-mercapto-3-methyl-1-butanol is one of the candidates for the pheromone role in cats for which behavioral response was observed, although the significance of the response is still unclear (Miyazaki et al. 2008). As far as volatile compounds are released eventually, this may explain the variable pattern of activation in MOB.

Our data supports the hypothesis that species-specific molecules in cats involved in communication between conspecifics may play the role of kairomones for potential prey—the house mouse. On the basis of the cat scent marks, gender, age, physiological status, location, and population density of the predator could be detected which provides essential information for the adaptive response in mice. In conclusion it should be noted that we are at the very beginning of identifying species-specific molecules from different predators. A paper that was just recently published presents evidence that wolf (*Canis lupus*) urinary volatiles can engender aversive and fear-related responses in mice and pyrazine analogues were identified as the predominant active components among these volatiles to induce avoidance and freezing behaviors via stimulation of the murine AOB (Osada et al. 2013). More discoveries on the signals of particular predators are expected in the near future.

14.4 CONCLUSIONS

Accumulated up-to-date research shows a complex nature of predator odors that elicit adaptive responses in prey species. An evolutionary link between predator and prey is essential for full defensive response in the potential prey. Only sympatric species develop a full set of adaptations against a predator. Another important aspect is the ecology of the predator and the potential prey. Even if sympatric, predator and prey may have different ecological niches. Though predator odors induce innate responses, learning is still important. Rodents, though slowly, still habituate to predator odors at the level of the behavior, which is the major limitation of using predator odors as natural repellents. Recent studies revealed a multicompound nature of predator odors that characterize them rather than as a pheromonal blend. Defensive responses in prey also depend on presentation context, gender, age, social status, and the physiological state of the signal recipient. None of the known molecules may produce a full set of defensive behaviors. The hardwired nature of the predator-prey responses is a laboratory phenomenon rather than what is observed under natural conditions. Nevertheless, laboratory research is a very important stage in identifying active ingredients to be tested in the field.

ACKNOWLEDGMENTS

Supported by the Russian Foundation for basic research 10-04-01599a and by a project from the RAS Program "Zhivaya priroda" (Live Nature).

REFERENCES

Apfelbach, R., Blanchard, C.D., Blanchard, R.J., Hayes, R.A. and I.S. McGregor. 2005. The effects of predator odors in mammalian prey species: A review of field and laboratory studies. *Neurosci Biobehav Rev* 29: 1123–44.

Bellringer, J.F., Pratt, H.P. and E.B. Keverne. 1980. Involvement of the vomeronasal organ and prolactin in pheromonal induction of delayed implantation in mice. *J Reprod Fertil* 59: 223–28.

Berton, F., Vogel, E. and C. Belzung. 1998. Modulation of mice anxiety in response to cat odor as a consequence of predators diet. *Physiol Beha* 65: 247–54.

Blanchard, R.J., Nikulina, J.N., Sakai, R.R., McKittrick, C., McEwen, B. and D.C. Blanchard. 1998. Behavioral and endocrine change following chronic predatory stress. *Physiol Behav* 63: 561–69.

Brennan, P.A. 2004. The nose knows who's who: Chemosensory individuality and mate recognition in mice. *Horm Behav* 46: 231–40.

Brennan, P.A. 2010. Pheromones and mammalian behavior. In: A. Menini (ed.), *The Neurobiology of Olfaction*. CRC Press: Boca Raton, FL.

Bruce, H.M. 1959. An exteroceptive block to pregnancy in the mouse. *Nature* 184: 105.

Buron, G., Hacquemand, R., Pourié, G., Lucarz, A., Jacquot, L. and G. Brand. 2007. Comparative behavioral effects between synthetic 2,4,5-trimethylthiazoline (TMT) and the odor of natural fox feces in mice. *Behav Neurosci* 121: 1063–72.

Burwash, M., Tobin, M., Woolhouse, A. and T. Sullivan. 1998. Field testing synthetic predator odors for roof rats (Rattus rattus) in Hawaiian macadamia nut orchards. *J Chem Ecol* 24: 603–30.

Cohen, H., Benjamin, J., Kaplan, Z. and M. Kotler. 2000. Administration of high-dose ketoconazole, an inhibitor of steroid synthesis, prevents posttraumatic anxiety in an animal model. *Eur Neuropsychopharmacol* 10: 429–35.

Coulston, S., Stoddart, D. and D. Cump. 1993. Use of predator odors to protect chick-peas from predation by laboratory and wild mice. *J Chem Ecol* 19: 607–12.

Cushing, B.S. 1985. Estrous mice and vulnerability to weasel predation. *Ecology* 66: 1976–78.

DellaCorte, C. 1995. Immunohistochemistry. In: A.I. Spielman and J.G. Brand (eds.), *Experimental Cell Biology of Taste and Olfaction: Current Techniques and Protocols*, Boca Raton, FL: CRC, p. 145.

Dickman, C. 1992. Predation and habitat shift in the house mouse, *Mus domesticus. Ecology* 73: 313–22.

Dielenberg, R.A. and I.S. McGregor. 2001. Defensive behavior in rats towards predatory odors: A review. *Neurosci Biobehav Rev* 25: 597–609.

Drickamer, L., Mikesic, D. and K. Shaffer. 1992. Use of odor baits in traps to test reactions to intra- and inter-specific cues in house mice living in outdoor enclosures. *J Chem Ecol* 18: 2223–50.

Epple, G., Mason, J., Nolte, D. and D. Campbell. 1993. Effects of predator odors on feeding in the mountain beaver *Aplodontia rufa. J Mammalogy* 74: 715–22.

Fendt, M. 2006. Exposure to urine of canids and felids, but not of herbivores, induces defensive behavior in laboratory rats. *J Chem Ecol* 32: 2617–27.

Feoktistova, N.Yu., Naidenko, Sv.V., Voznesenskaya, A.E., Krivomazov, G.J., Clark, L. and V.V. Voznessenskaya. 2002. The influence of predator odors and overcrowded mouse odors on regulation of estrus cycles in house mice (*Mus musculus*). In: G.R. Singleton, L.A. Hinds, C.J. Krebs and D.M. Spratt (eds.), *Rats, Mice and People: Rodent Biology and Management*, ACIAR: Canberra, Australia, pp. 173–75.

Ferrero, D.M., Lemon, J.K., Fluegge, D., Pashkovski, S.L., Korzan, W.J., Datta, S.R., Spehr, M., Fendt, M. and S.D. Liberles. 2011. Detection and avoidance of a carnivore odor by prey. *Proc Natl Acad Sci U S A* 108: 11235–40.

Figueiredo, H.F., Bodie, B.L., Tauchi, M., Dolgas, C.M. and J.P. Herman. 2003. Stress integration after acute and chronic predator stress: Differential activation of central stress circuitry and sensitization of the hypothalamo–pituitary–adrenocortical axis. *Endocrinology* 144: 5249–58.

File, S.E., Zangrossi, Jr. H., Sanders, F.L. and P.S. Mabbutt. 1993. Dissociation between behavioral and corticosterone responses on repeated exposures to cat odor. *Physiol Behav* 54: 1109–11.

Flawell, S.W. and M.E. Greenberg. 2008. Signaling mechanisms linking neuronal activity to gene expression and plasticity of the nervous system. *Annu Rev Neurosci* 31: 563–90.

Halem, H.A, Baum, M.J. and J.A. Cherry. 2001. Sex difference and steroid modulation of pheromone-induced immediate early genes in the two zones of the mouse accessory olfactory system. *J Neurosci* 21: 2474–80.

Halpern, M. and A. Martinez-Marcos. 2003. Structure and function of the vomeronasal system: An update. *Prog Neurobiol* 70: 245–318.

Harvel, C.D. 1990. The ecology and evolution of inducible defenses. *Quart Rev Biol* 65: 323–40.

Hayes, R.A. 2008. Seasonal responses to predator fecal odors in Australian native rodents vary between species. In: J.L. Hurst, R.J. Beynon, S.C. Roberts, T.D. Wyatt (eds.), *Chemical Signals in Vertebrates 11*, Springer: New York, pp. 379–87.

Hayes, R.A., Nahrung, H.F. and J.C. Wilson. 2006. The response of native Australian rodents to predator odours varies seasonally: A by-product of life-history variations? *Anim Beha* 71: 1307–14.

Johnson, M.A., Tsai, L., Roy, D.S., Valenzuela, D.H., Mosley, C., Magklara, A., Lomvardas, S., Liberles, S.D. and G. Barnea. 2012. Neurons expressing trace amino-associated receptors project to discrete glomeruli and constitute an olfactory subsystem. *Proc Natl Acad Sci U S A* 109: 13410–15.

Kassesinova, E. and V. Voznessenskaya. 2009. The role of predator odors in regulation of oestrous cycles in house mouse. *Chem Senses* 34(3): E.35.

Kats, L.B. and L.M. Dill. 1998. The scent of death: Chemosensory assessment of predation risk by prey animals. *Ecosci* 5: 361–94.

Liberles, S.D. 2009. Trace amine-associated receptors are olfactory receptors in vertebrates. *Ann N Y Acad Sci* 1170: 168–72.

Liberles, S.D. and L.B. Buck. 2006. A second class of chemosensory receptors in the olfactory epithelium. *Nature* 442: 645–50.

MacNiven, E. and D. de Catanzaro. 1990. Reversal of stress-induced pregnancy blocks in mice by progesterone and metyrapone. *Physiol Behav* 47: 443–48.

Miyazaki, M., Yamashita, T., Hosokawa, M., Taira, H. and A. Suzuki. 2006a. Species-, sex-, and age-dependent urinary excretion of cauxin, a mammalian carboxylesterase family. *Comp Biochem Physiol B Biochem Mol Biol* 145: 270–77.

Miyazaki, M., Yamashita, T., Suzuki, Y., Soeta, S., Taira, H. and A. Suzuki. 2006b. A major urinary protein of the domestic cat regulates the production of felinine, a putative phero-mone precursor. *Chem Biol* 13: 1071–79.

Miyazaki, M., Yamashita, T., Taira, H. and A. Suzuki. 2008. The biological function of cauxin, a major urinary protein of the domestic cat (*Felis catus*). In: J.L. Hurst, R.J. Beynon, S.C. Roberts and T.D. Wyatt (eds.), *Chemical Signals in Vertebrates 11*, Springer: New York, pp. 51–60.

Müller-Schwarze, D. 2006. *Chemical Ecology of Vertebrates*. Cambridge University Press: New York.

Nolte, D., Mason, J., Epple, G., Aronov, E. and D. Campbell. 1994. Why are predator urines aversive to prey? *J Chem Ecol* 20: 1505–16.

Osada, K., Kurihara, K., Izumi, H. and M. Kashiwayanagi. 2013. Pyrazine analogues are active components of wolf urine that induce avoidance and freezing behaviours in mice. *PLoS ONE* 8(4): e61753.

Papes, F., Logan, D.W. and L. Stowers. 2010. The vomeronasal organ mediates interspe-cies defensive behaviors through detection of protein pheromone homologs. *Cell* 141: 692–703.

Roberts, S., Gosling, L., Thornton, E. and J. McChung. 2001. Scentmarking by male mice under the risk of predation. *Behav Ecol* 12: 698–705.

Rodriguez, I., Del Punta, K., Rothman, A., Ishii, T. and P. Mombaerts. 2002. Multiple new and isolated families within the mouse superfamily of V1r vomeronasal receptors. *Nat Neurosci* 5: 134–44.

Rodriguez, I., Feinstein, P. and P. Mombaerts. 1999. Variable patterns of axonal projections of sensory neurons in the mouse vomeronasal system. *Cell* 97: 199–208.

Rutherfurd, K.J., Rutherfurd, S.M., Moughan, P.J. and W.H. Hendriks. 2002. Isolation and characterization of a felinine-containing peptide from the blood of the domestic cat (*Felis catus*). *J Biol Chem* 277: 114–19.

Sheng, M. and M.E. Greenberg. 1990. The regulation and function of c-fos and other immedi-ate early genes in the nervous system. *Neuron* 4: 477–85.

Starke, W.W., III and M.H. Ferkin. 2013. The effects of cues from kingsnakes on the reproduc-tive effort of house mice. *Current Zool* 59: 135–41.

Stoddart, D. 1980a. Some responses of a free-living community of rodents to the odors of predators. In: D. Muller-Schwarze and R. Silverstein (eds.), *Chemical Signals: Vertebrates and Aquatic Invertebrates*, Plenum Publishing: New York.

Stoddart, M. 1980b. *The Ecology of Vertebrate Olfaction*. Chapman & Hall: London.

Sullivan, R.M. and A. Gratton. 1998. Relationships between stress-induced increases in medial prefrontal cortical dopamine and plasma corticosterone levels in rats: Role of cerebral laterality. *Neuroscience* 83: 81–91.

Touma, C., Palme, R. and N. Sachser. 2004. Analyzing corticosterone metabolites in fecal samples of mice: A noninvasive technique to monitor stress hormones. *Horm Behav* 45: 10–22.

Vasilieva, N.Y., Parfenova, V. and R. Apfelbach. 2001. The effect of predator odour on reproduction success in three rodent species. In: H.-J. Pelz, D.P. Cowan and C.J. Feare (eds.), *Advances in Vertebrate Pest Managmenent II*, Filander Verlag: Furth.

Voznessenskaya, V.V., Klyuchnikova, M.A. and C.J. Wysocki. 2010. Roles of the main olfactory and vomeronasal systems in the detection of androstenone in inbred strains of mice. *Current Zool* 56: 813–18.

Voznessenskaya, V.V., Naidenko, S.V., Feoktistova, N.Yu., Krivomazov, G.J., Miller, L.A. and L. Clark. 2002. Predator odours as reproductive inhibitors for Norway rats. In: G.R. Singleton, L.A. Hinds, C.J. Krebs and D.M. Spratt (eds.), *Rats, Mice and People: Rodent Biology and Management*, ACIAR Monograph, Canberra, Australia, Vol. 96, pp. 131–36.

Voznessenskaya, V.V., Voznesenskaia, A.E. and M.A. Klyuchnikova. 2006. The role of vomeronasal organ in reception of predator scents. *Chem Senses* 31(8): E43–E44.

Voznessenskaya, V.V., Wysocki C.J. and E.P. Zinkevich. 1992. Regulation of the estrous cycle by predator odors: Role of the vomeronasal organ. In: R.L. Doty and D. Muller-Schwarze (eds.), *Chemical Signals in Vertebrates 6*, Plenum Press: New York, pp. 281–84.

Weldon, J.D., Divita, F.M. and G.A. Middendorf. 1987. Responses to snake odours by laboratory mice. *Behav Processes* 14: 137–46.

Wyatt, T.D. 2003. *Pheromones and Animal Behaviour: Communication by Smell and Taste*. Cambridge University Press: Cambridge.

Wysocki, C.J. and L.M. Wysocki. 1995. Surgical removal of the vomeronasal organ and its verificaiton. In: A. Spielman and J.G. Brand (eds.), *Experimental Cell Biology of Taste and Olfaction*. CRC Press: Boca Raton, FL, pp. 49–57.

15 Pheromones of Tiger and Other Big Cats

Mousumi Poddar-Sarkar and Ratan Lal Brahmachary

CONTENTS

15.1 Introduction 408
15.2 Historical Aspects 409
15.3 Phenomenon of Chemical Signaling: Physiological Implications 410
15.4 Semiochemicals as Stimuli 411
15.5 Home Range and Territories 412
15.6 Scent Marking in Big Cats 413
    15.6.1 Unknown Spray 413
    15.6.2 Primary Source of Feline Pheromone 415
        15.6.2.1 MF: From Anal Gland or from the Urinary Tract?
            A Confusion Lasting Over Decades 415
    15.6.3 Other Sources of Pheromones in Big Cats: Hair, Mane,
        Saliva, Interdigital Gland and Anal Gland Secretion: A
        Comparatively Unexplored Area in Big Cat Research 416
    15.6.4 Significance of MF in the Social Life of the Tiger 417
    15.6.5 Olfactory Sense of Big Cats: An Ability Denied Earlier 418
    15.6.6 Feline Attractants: Traditional Knowledge and Modern Science... 419
    15.6.7 Flehmen, the Characteristic Grimace 419
    15.6.8 Mechanism for the Long Persistence of Pheromones of Big
        Cats in Nature 420
15.7 Major Histocompatibility Complex of Genes and Individuality in
    Pheromonal Signals of Big Cats 421
15.8 Genomics, Proteomics, and Metabolomics in Pheromone Research
    of Big Cats: The Search for the Evolutionary Lineage and Linkage of
    Big-Cat Population 422
15.9 Quantitative Approach for Understanding the Territory and Home-
    Range in the Tiger Community 424
    15.9.1 Territoriality in the Tiger 424
        15.9.1.1 Study Design and Strategy 426
        15.9.1.2 Overview on Marking Patterns 429
    15.9.2 MF Spraying versus Ordinary Urination 434
    15.9.3 MF Spraying in Proestrous, Estrous, and Postestrous Periods 435
    15.9.4 MF Spray and Flehmen 436

15.9.5   Territoriality in the Lion..436
15.9.6   Territoriality and the Cheetah 437
15.9.7   Territoriality and the Leopard 437
15.9.8   Territoriality and the Puma (Cougar) 438
15.9.9   Territoriality and the Jaguar...438
15.9.10 Ontogeny of Different Physiological Phenomena in Cubs of
          Big Cats 439
15.10 Chemistry Related to MF of the Tiger and Other Big Cats 439
        15.10.1 Collection of MF 439
        15.10.2 Chemical Analysis of MF...440
                 15.10.2.1 Volatile and Nonvolatile Compounds Identified in
                            MF...440
        15.10.3 Natural Fixative of MF...447
15.11 Review and Conclusions: Fifty Years of Pheromone Research of
      Big Cats 448
15.12 Many Unsolved Problems 449
Acknowledgments..451
References...452

This chapter is dedicated to the memory of the late Prof. J. Dutta of Bose Institute, Calcutta—our longstanding collaborator.

15.1   INTRODUCTION

This chapter is based on a long-standing quest initiated by one of us (RLB) in 1964 when George Schaller undertook the first detailed scientific study of the tiger (Schaller 1967). It has been rather like chasing a crooked shadow through a maze, for the concept of pheromones in mammals was not well understood at that time and many misconceptions on the social life of the tiger, and particularly regarding the question of olfactory ability of the tiger, obscured the views of latter-day researchers. We have attempted to take into account implications of evolution, ethology, etho-chemistry, and ethogenomics while studying the strategies for documenting the different forms of chemical communication in the world of big cats, especially the tiger. Our knowledge in this context was enriched by experiences during several periods of fieldwork in different ecological terrain, in India and Africa, in most of the cases by one of us (RLB). The subject of chemical signaling, only one aspect of which is pheromone, is very wide-ranging. For the sake of clarity and the self-sufficiency of this chapter, we project our views in two ways: in the first part we present a general treatment of chemical signals concentrating mainly on pheromonal signals (communication) in the tiger and other big cats, the lion (Asiatic and African), the Indian leopard, and the African cheetah, and in the second part we will try to substantiate our views with quantitative data records that we have gathered over decades in different phases. Our study on the latter three big cats is less exhaustive than that on the tiger and we have had no opportunity of investigating the jaguar and the cougar (puma).

A major breakthrough in the subject of animal psychology occurred when in 1973, Konrad Lorenz, Niko Tinbergen, and Karl von Frisch won the Nobel Prize for Physiology/Medicine for their pioneering work on animal behavior. A new terminology—ethology—came of age; the terms "ethology" and "animal behavior" became synonymous and now also include sociobiology and related analysis and modeling on behavioral studies from the evolutionary perspective, comparative psychology, and very recently, signal engineering and ethochemistry (as developed in the context of our work on big cats), and so forth.

15.2 HISTORICAL ASPECTS

People living in and around jungles or even in the country (rather than the urban population) had gained inklings of what we today call pheromones and their physiological/behavioral aspects. The strong taint of vixen in reproductive stage (which the dog-fox responds to) well known to people in the British countryside and has been clearly described in a poem by John Maesfield. Likewise, South African farmers were familiar with jackals' smell in the right season (Van Der Merwe 1953). Such knowledge has been part and parcel of the poacher's professional skill.

At a more scientific level we can refer to Darwin (1871). In *The Descent of Man in Selection in Relation to Sex*, he described the strong smell of breeding crocodiles, many mammals, and hinted at a possible sexual selection of such smells. To quote from Darwin, "...if the most odoriferous males are the most successful in winning over females and leaving offspring to inherit the gradually perfected glands and odour..." He described how competition for mates (i.e., sexual selection) would lead to the evolution of traits that either helped a male to fight off other males or to make it particularly attractive to females, or both. He believed that it could be such a powerful force that such a trait could evolve with time. He saw sexual selection as a special case of natural selection with emphasis on mating success. This can be reinterpreted today as the concept of sex-attractant pheromone evolving through sexual selection.

In his book, Darwin also argued that there is an evolutionary continuity and that the root of human behavior patterns lies in many nonhuman animals. The study of pheromones reveals the truth of Darwin's reasoning. It is now known that pheromones are used by various animals and vestiges of human pheromone have also been traced.

Tinbergen (1951) raised four questions of ethology in his book *The Study of Instinct*: (1) What is its function and survival value? (2) How has it evolved over time? (3) How has it developed in the individual? (4) What is the physiological causation?

In the following sections we will attempt to address some of the questions raised by him with respect to our findings on big cats.

Animal behavior is not explainable by fossil records. Nonetheless, Rasmussen (1999) indulges in interesting speculations. On the basis of an old report on rock engravings by Pocock (Rasmussen 1999), she suggests that the temporal gland in mammoths was larger in size that that of the Asian elephant. It is possible that in mammoths both males and females utilized this as a source of pheromones. But today it is a major pheromonal source in the male Asian elephant only.*

We must treat the concepts of instinct, ontogeny of specific behavior related to pheromonal communication, strategies of signaling and their impact on reproduction, phylogenetic relationship of specific behavioral patterns in big cats, in order to understand the facts and hypotheses in the light of evolution. Here we face many problems, some of which are hard nuts to crack.

15.3 PHENOMENON OF CHEMICAL SIGNALING: PHYSIOLOGICAL IMPLICATIONS

The phenomenon of sexual selection is correlated with signaling in most animals. Social signals are diverse; besides visible and auditory signaling, chemical signals (olfactory signals) play an important role in the world of many animals including mammals. These could be urine, feces, glandular secretions, and so forth. Whereas a visual or auditory signal is functional only during the physical presence of the animal, a chemical (olfactory) signal persists even in its absence. Brahmachary (1986) pointed out how such signals might convey information on species specificity, sexual status, and so forth, and Wyatt (2003) lists more information such as on age, health, and fear. Penn (2006) calls such signals an extended phenotype, being inspired by a famous book (Dawkins 1983). Dawkins used this expression to mean "action at a distance" as in the case of the Bruce effect; here the pheromone helps the second male (physically absent) to block pregnancy of the female due to the sperm of the first male, thus acting to the advantage of the second. We feel, however, that in a more general sense pheromones are also extended phenotypes leaving their "individuality imprint" at a distance, such as in the scat, dung, urine, and glandular secretion. We will return to this theme in the context of big cats.

Maynard and Harper (1988) divided signals into two categories, assessment and conventional. They explained that assessment signals are "honest" in function and involve a large expenditure of energy (e.g., the loud call of a barking deer). Conventional signals are generally prominent display badges such as the tusks of the male Asian elephant, large shaggy mane of the African lion, brightly colored feathers/wings of birds like that of widow bird or peacock; these are signals or certificates attesting good health, virility, high nutritional status, and so forth in the organisms (Brahmachary 2011). Sometimes these signals are misleading (e.g., the black bib of the house sparrow is not always correlated with male dominance) (Maynard and Harper 1988). As we will see, in the big cats the major pheromonal signal is based on metabolic expenditure, namely the loss of a large amount of lipids, presence of hormones, or derivatives thereof in urine and marking fluid (MF) (see Section 15.6.1). This ability of metabolizing energy as well as the traces of hormones in urine, MF, and so forth can claim to be honest signals, as we will describe later.

---

* The male African elephant also utilizes the temporal pheromone as studied by Rasmussen and others. The African females and, more rarely, the Indian she elephants are known to secrete a watery (less viscous) fluid from the temporal gland, which is a sign of agitation rather than of pheromonal function.

The difference between hormone and pheromone is generally clear. Hormone, which means "I excite," is restricted to an individual body, while pheromone, which connotes a sense of "carrying" (pherein) excitement, is concerned with transmission of "excitement" (here, a chemical signal) to another individual. However, a comparative study on heterosexual and homosexual men and heterosexual and lesbian women by Berglund et al. (2006) suggests that certain sex hormones and their breakdown products also act as pheromones. He reported that two compounds, progesterone derivative 4,16-androstadien-3-one (AND) and estrogen-like steroid estra-1,3,5(10),16-tetraen-3-ol (EST) induce sex-specific effects on the autonomic nervous system, mood, and context-dependent sexual arousal. Thus AND and EST are supposed to be candidate compounds for human pheromones. Therefore, we feel it is unnecessary to take recourse to hair-splitting while defining terminologies.

15.4 SEMIOCHEMICALS AS STIMULI

"Semeion" means signal (as in semaphore signal) in Greek. So, semiochemicals include all signals in our biological context; it may be the mRNA, a signal from DNA, an osmic (smelly) chemical signal emitted by an animal that excites another animal of the same species (pheromone) or alerts the prey, or the smell of the prey can betray it to the advantage of the predator. The last two cases have been described as allelochemicals by Nordlund and Lewis (1976) and are subdivided into two other classes—kairomones and allomones. However, these two terms were well defined by Claesson and Silverstein (1977) and in *Chemical Signals in Vertebrates I* (1977, Plenum Press, London) and we feel the term allelochemical might lead to confusion for this is well known in modern Botanical literature as the classes of substances emitted through root exudates or material leaching out of leaves shed on ground and affecting the growth of plants of the same or other species. We, therefore, are of the opinion that the terms pheromone, allomone, and kairomone are sufficient to describe the relevant issues.

The concept of the pheromone has further been categorized into "releaser" and "primer" depending on the time gap between exposure to the pheromone and the response in the receiver, either immediately or after a longer duration (Albone 1984). The interim period between stimulus and the response may vary; action can be instantaneous or delayed depending on orientation of stimuli, gradual accumulation of information, or controlled response with the influence of environment. Stimulus of smelling the young triggers the final letdown of milk from a lactating mother, although prior growth of the mammary gland is accelerated by the cumulative arousal action of the pregnant female by grooming or licking the nipples in rodents (Manning and Dawkins 1992). Ewert and Trand (1979) correlated the selective responsiveness to key stimuli at a behavioral level with the underlying neurobehavioral mechanism. In other words, the strength of responsiveness or motivation to the stimuli depends on various internal and external factors of the receiving animal.

"Communication" has generally but not always evolved for the sake of mutual benefit of sender and receiver (Manning and Dawkins 1992). In our view, when interindividual communication takes place within the same sex to reduce direct confrontation with rivals, it is beneficial to the sender but not to the receiver. On the other hand, communication between the opposite sex is beneficial for both individuals.

Communication (i.e., the passage of 'information) may also be disadvantageous in the context of the predator-prey relation. However, the latter aspect will not be treated because our focus is on intraspecific communication through pheromones. Chemical communication as a sex attractant or warning signal in the context of territorial connotations necessarily includes a self–nonself distinction (i.e., those signal(s) as those that bear the characteristic individual signature). Wyatt (2003) emphasizes this aspect as well as that of kin or clan recognition as distinct from pheromones, but it would be less complicated to retain the term pheromone and interpret it in a wider context.

15.5   HOME RANGE AND TERRITORIES

The aspects of home range and territories include a study of sex allocation and amenities such as availability of food, water, and abode, and are related to social conduct. Territory may be permanent or temporary and a territory holder is perhaps dominant over an intruder. Most territories and home ranges can be more clearly understood in the description by earlier authors as mentioned below. More data has been furnished later.

Many hunters in the past talked of tiger beats rather like the beat areas of police officers. When a tiger was killed, very soon its particular niche was filled up by another. "A good tiger beat is sure to continue to be so year after year" even when the "predecessor may have been killed but recently" (Fayrer 1875). Likewise, Lyddeker (1893–94) mentions observations docketing the fact that when "a tiger with a restricted beat is killed…another will occupy its place frequenting the same lairs and drinking at the same pools." This view, based on a good deal of experience and also verbally communicated (to RLB) by Wakefield, an old-time hunter naturalist of vast experience of Indian wildlife, implies territoriality or home range but in the scientific world the concept was formulated by Howard (1922) while describing British song-birds (their song is a vocal communication). Uxkuell (1934) used two words, Heim and Heimat, both of which in German have the connotation of home. Taking this cue, ecologists and ethologists later on introduced two terms, territory and home range. *A territory may be defined as that part of home range which is actively defended* (Burt 1943), if need be, by fighting. In the 1940s and 1950s Hediger clearly propounded the idea that animals reserve separate sections of their home range for specific, different purposes, such as sleeping, resting, defecating, and wallowing (Hediger 1977). Only some part(s) of the home range are actively defended but signals with agonistic con-notation such as roaring or the olfactory cues based on fresh pheromones minimize the chances of an all-out fight. However, in other parts of the home range the presence of others is tolerated or even encouraged. Some authors have used the expression "land tenure" rather than territory/home range in the context of big cats.

Barnett (1981) mentioned four kinds of factors for confinement of an animal in its territory: (1) existence of structural barriers for movement, (2) the attraction of the place due to various reasons, (3) due to specific apotreptic behavior that keeps animals apart, and (4) the animal may withdraw itself on detecting the presence of conspecifics. We feel the first factor (structural barrier) is an exceptional case. Also, the difference between (3) and (4) is not clear-cut. According to Powell (2000) home range represents an interplay between the environment and an animal's understanding of that environment (i.e., its cognitive map). Home-range behavior is the product of a decision-making process shaped by natural selection to increase the contributions of spatially distributed resources to fitness (Börger et al. 2008; Mitchell and Powell 2004; Spencer et al. 1990). For demarcation of territory, animals use optical, acoustic, olfactory, and chemical cues or a combination of these so that other conspecific members can recognize it in several ways. Anderson (1994) classified territories under different categories: (a) a region in which most or all activities of the animals are carried out, (b) a region smaller than the total area of movement of the animal (i.e., territory is a subset within a big set of home range), and (c) a region in which a female raises her offspring.

In a natural or stable environment, territorial behavior, though diverse, entails an orderly and often peaceful spacing out of the population so that most individuals can carry out essential functions like searching for food, prospective mates, raising young, and so forth without much overt aggression. Thus in the tiger and lion territorial behavior is linked with breeding and raising offspring (Locke 1954; Schaller 1967; Schaller 1972).

15.6 SCENT MARKING IN BIG CATS

15.6.1 UNKNOWN SPRAY

In about a century and a half of numerous blood sport literature, the behavior pattern of the tiger, originally called the unknown spray and now known as scent marking, was conspicuous by its absence. Even stalwart oldtimers like Corbett (1944), James Inglis (1892), and Dunbar Brander (1923) apparently never noticed this now so familiar phenomenon (today it is well known that tigers, lions, leopards, etc., frequently eject a fluid through the urinary channel like a spray directed upward). Schaller (1967) unearthed a single reference, that of Locke (1954). This discerning military officer wrote tersely but to the point, "From time to time, presumably when in quest of a mate or when wishing to indicate that he regards this area as his own particular hunting ground, the adult male is capable of ejecting a strong smelling secretion from beneath the tail, which is raised vertically during this process. The fluid is expelled upwards and backwards with surprising force. The spot which the tiger has chosen for this purpose can easily be recognized by the odour. Traces of the fluid may also sometimes be found on surrounding vegetation including the undersides of the leaves on low hanging boughs." His views essentially are valid even today. But he saw only the male tiger spraying. In 1964, Schaller in Kanha, India, noticed a tigress raising her tail and spraying a fluid through the urinary channel upward and backward and later he observed this behavior a few more times. Schaller was toying with a theory that this spray might encode certain information to be decoded by other tigers (Schaller 1964, personal communication). One day in 1964, Schaller and one of us (RLB) noticed a long file of spotted deer pausing and smelling the leaves of a stunted *Butea monosperma* tree that was apparently sprayed by the resident tigress (Brahmachary 1964, personal field diary). In 1967, Schaller first brought this spray to the notice of scientists through his book (Schaller 1967). At a later date, while watching a tame tigress in a jungle in Orissa, India, 52 sprays were noticed in a single day (Brahmachary 1976, unpublished personal field diary) and hundreds of sprayings were subsequently observed in a very large compound where the tigress had a free run. Much of this data was documented by S.R. Choudhury, who had been studying this tiger from cubhood (Figure 15.1) since 1974. (Some of this data was published posthumously [Choudhury 1999].) Around the 1980s many naturalists and even casual tourists in Indian National Parks became familiar with the spray of tigers.

The tiger indeed employs this fluid for attracting mates and staking a claim on its hunting territory. Actually, unlike the tomcat, the female (known as the queen), sprays very infrequently. The same is valid for lionesses (Brahmachary and Singh 2000; McBride 1977; Schaller 1972), leopards (Poddar-Sarkar and Brahmachary 2004), and cheetahs (Caro 1994; Eaton 1974; Poddar-Sarkar and Brahmachary 1997). In these species the female rarely marks, generally only during the onset of the estrous stage, but the tigress marks very frequently, almost rivaling the male as we will see later. Of the other members of the cat family the female serval sprays frequently (personal communication from staff at Moholoholo, South Africa, in 2005).

Comparative ethology of spraying MF has been studied in the tiger and lion. In the African lion Schaller (1972) noticed that the direction of the spray varies rather widely, upward and backward, horizontally backward, and simply downward

FIGURE 15.1 "Khairi" (born August 1972; died September 1977): the pet tigress of Simlipal forest, which initiated scientific research in tiger pheromones. (Photo courtesy of Mr. Dilip Bhattacherya.)

(without assuming the squatting posture adopted during urination). In the Asiatic lion Brahmachary et al. (1999) also detected such modalities. In the tiger the direction is more fixed—almost always upward and backward (Brahmachary et al. 1999; Poddar-Sarkar et al. 2013). Barja and Miguel (2010) observed Siberian tigers (*P.t. altaica*) and Barbary lions (*P. leo leo*) in a Madrid zoo. They reported what they consider to be significant differences in the marking behavior of the tiger and lion, namely that the frequency of marking occurs more in the tiger while the marking duration act occurs more in the lion. They have attempted to correlate these (and other) differences with the social/asocial nature of the two species as well as with their habitat differences (forest versus open area). The tiger shows seasonal variation in marking patterns whereas the lion does not. They also correlated many environmental factors in addition to their reproductive physiology (seasonal polyestrous versus annual polyestrous). Such extrapolations from a European zoo to the natural conditions of Siberia and Barbary of Africa are difficult. Moreover, the now-extinct Barbary lion was an unusual race of the African lion.

15.6.2 Primary Source of Feline Pheromone

15.6.2.1 MF: From Anal Gland or from the Urinary Tract? A Confusion Lasting Over Decades

Descriptions of the scent-marking behavioral trait have been confusing. In the relevant literature we note a number of different terms like marking, urine marking, urine spraying, scent marking, spraying marking fluid, or simply spraying. Schaller (1967) and then again while studying the MF of the African lion (Schaller 1972) wrongly asserted that the spray is a mixture of urine and anal gland secretion and this error was repeated *ad nauseum*. It was repeated by McDougal (1977) and in many other publications (Albone 1984). Ewer (1968, 1973) cast doubt on this but even as late as in 2006 the mistake persisted (Thapar 2006). There is, anatomically, no connecting link between the urinary tract and the anal gland of the tiger (Hashimoto et al. 1963). The same must be valid for other big cats. The differences between anal gland secretion and spray fluid were evident on comparing the activities of a striped Indian hyena and a tigress (Choudhury and Brahmachary 1977, unpublished). That hyena ejects anal gland secretion is very well known in the African spotted hyena (Mills et al. 1990) and the tigress sprays MF through the urinary channel. The male tiger can reverse its penis and even the female can spray upward, though at a lower angle. Asa (1993) marked anal sac secretions of various felids with an inert dye and found no mixing of anal secretion and urine. But the confusion continued until Andersen and Vulpius (1999) pointed out that it had finally been laid to rest by Brahmachary and Dutta (1987). Up to 1981 they too had been confused (Brahmachary and Dutta 1981). The chemical contents of the anal gland of the Asiatic lion turned out to be different from the spray fluid (Brahmachary and Singh 2000), thus suggesting different sources. Therefore, the spray of big cats ejected upward has been referred to as MF by our group. MF of the lion, leopard, and cheetah is yellowish, being contaminated with urinary urochrome but that of the tiger is generally white. Van Hurk (2007) also coined the term MF as we have. Burger et al. (2008) also mentioned the tiger and cheetah territorial MF as a mixture of urine and lipidic materials.

In addition to MF, feces, scats, dung, droppings, and so forth are prized by captive animals as has been noticed by zookeepers and the attention attracted by urine has also been discussed earlier. Apparently the economy of nature follows a "waste not, want not" strategy; excretory products serve the animals as valuable chemical signals involved in their physiological and ethological function(s). Brahmachary (1986) points out that urine, full of many metabolic products including hormones and their breakdown products, may well bear the print of individuality, the health status, estrous, preestrous stage, and so forth. Feces, too, bear similar imprints and characteristic bile salts. Mooney (1984) emphasized the bile salts, which are species-specific and might even be individual-specific. All these justify the concept of the pheromone as an extended phenotype.

15.6.3 OTHER SOURCES OF PHEROMONES IN BIG CATS: HAIR, MANE, SALIVA, INTERDIGITAL GLAND AND ANAL GLAND SECRETION: A COMPARATIVELY UNEXPLORED AREA IN BIG CAT RESEARCH

The lion, unlike the tiger, head-rubs on tree trunks just before spraying (Schaller 1972). Poddar-Sarkar et al. (2008) pointed out that since head-rubbing on tree trunks is as frequent as spraying MF, both MF and the mane are likely to leave two osmic signals simultaneously at two neighboring sites. After observing the normal behavior of lions in a breeding center of South Africa (head-rubbing followed by MF spraying) a chemical analysis of the mane was undertaken. C9-C24 fatty acids were detected and these molecules with high and low volatility might play the role of pheromones. The lion is a social cat and as Schaller (1972) pointed out, head-rubbing between the female and the lion could transfer the imprint of the male to cubs because head-rubbing of the cubs with the lioness is very frequent. For brief periods tigers, tigresses, or leopards/leopardesses also rub heads and lick. In this male-female and mother-cub interaction pheromonal roles of hair/skin secretions and saliva are a distinct possibility.

According to the experience of Schaller (1967) and Brahmachary (1964, unpublished), deer are not particularly scared of the MF smell, they only show curiosity, but on detecting the scent of the body smell of the tiger or lion, deer, antelopes, and even a mother elephant with calf become alert or they panic. Body smell may be a pheromone for a mating pair and a kairomone for the prey species. A mating tigress may rub her body on trees (Thapar 2006). Sankhala (1993) stated that he detected no odor of tigers most of the time but sometimes a strong smell was perceptible. An old-time hunter (Dunbar Brander 1923) mentions the strong body odor of freshly killed tigers and three wildlife officers detected a strong smell in the tranquilized marsh tigers of Sunderbans (G. Tanti and S. PalChoudhury 2008, personal communication). We investigated the skin secretions of three such tranquilized tigers and detected saturated, unbranched fatty acid methyl esters and a few benzenoid compounds (Poddar-Sarkar et al. 2013).

Lipocalin is a family of proteins including aphrodisin, known to be a nonvolatile pheromone in rodents (Singer et al. 1986; Vincent et al. 2001). De and his group detected 20-KDa lipocalins in the saliva (and even in the tear gland) of the hamster (De 1996; Thavathiru et al. 1999). The fatty acid of hair/skin may well serve as ligands of lipocalins, and while licking each other big cats also might use this mechanism (Poddar-Sarkar et al. 2013).

Tigers and other big cats rear up leaning against trees and scratch the bark. Inglis wrote that big tigers leave such scratch marks on tree trunks at 11 feet (and more) above the ground level (Inglis 1892). Mountfort (1981) published a photograph of such scratching on a tree trunk at a height of 10 ft. A soft-barked tree was very frequently scratched so severely by a tiger in Nandankanan Biological Park that it wilted and died. Formerly most hunters and trekkers talked of big cats sharpening their claws in this fashion. Choudhury and Brahmachary, while watching the pet tigress in a large compound, noticed a sort of outer scale of the claw left in the scratch mark (Brahmachary, personal field diary, 1977). These scratch marks may also serve as communicatory signals. This theory is strengthened if we take into account the secretion of interdigital glands. Nothing is known at present regarding the olfactory signals of interdigital glands.

A brief note on lipidic composition of MF and anal gland secretion of the Asiatic lion was reported by Brahmachary and Singh (2000). As late as 2001 Bininda-Emonds et al. (2001) studied the lipid composition of the anal sac of different felidae (other carnivores); we will return to this theme in Section 15.8.

15.6.4 Significance of MF in the Social Life of the Tiger

A mass of data collected by us while studying 12 tigers in an open air zoo (Brahmachary et al. 1992; Poddar-Sarkar et al. 1995) and some of the results gathered by the Smithsonian group on wild tigers over decades in Chitawan (Smith et al. 1987, 1989) strongly suggest that pheromones sprayed through the urinary channel are used in communication among tigers including decoding information pertaining to territory and mating strategy. Sprayings by wild tigers have frequently been observed (McDougal 1977; Thapar 1986, 2006) and these support the view developed above. A tigress ready to drop a litter and raise cubs in privacy sprayed on six trees as if marking a ring of posts encircling this private area with a diameter of 90 m (Singh 1981). Ewer (1968, 1973) pointed out that without exception female mammals markedly increase urination at or just before the estrous stage, evidently for advertising the sexual status. This has been recorded by us in the case of two tigresses (Poddar-Sarkar et al. 1995; vide Section 15.9.3). After this, marking abruptly stopped altogether at mid or full estrous and then again rose to the normal level. That pheromonal secretion plays a role in reproductive/mating strategy is implied in our findings on the Indian (Asiatic) lion (Brahmachary and Singh 2000). The findings of Joy Adamson and George Adamson (Adamson 1960, 1986) and of more rigorous observers (McBride 1977; Schaller 1972) indicated the same fact.

Eaton (1974) described the pheromonal role of urinary sprays in the cheetah. The observations strongly suggest that informational content pertaining to territorial and mating strategies is encoded in the pheromonal secretion through MF that persists, however, for only about 24 hours.

Caro (1994) states that nonresident cheetahs moving as transients in others' established territories suffer from stress, apparently after perceiving fresh MF, urine, and so forth of the latter. This is an appropriate response supporting the concept of communication through the chemicals of MF, urine, and so forth.

Zoo tigers strongly react to the smell of the opposite sex introduced in their quarters during their temporary absence (Brahmachary 1988, unpublished observation). In nature a male reacted strongly on sniffing fresh MF of a tigress as reported by Panwar (McDougal 1977). In snow leopards in captivity (in Darjeeling Himalayan Zoo, India) it was noted that urine and scent-marking-borne pheromones of both sexes mutually induced the animals to come into the reproductive stage (Rishi 2013).

In the lion, the most social cat, synchronous estrous of females in a pride is another social ethos in which we note the implication of pheromones. Bertram (1975) and Pusey and Packer (1983) point out that most lionesses of the pride belonging to the right age group come to the estrous stage simultaneously. Lions may breed throughout the year but although different prides may breed at different times, within a pride all the females tend to come into estrous at about the same time. This fact reminds one of the synchronous period of female students living in proximity in hostels/dormitories (McClintock 1971). The report was challenged but later investigations apparently supported the claim but controversies continue. Also relevant in this context is a study on volatile fatty acids of the human vagina (Michael et al. 1974). Graham (1991) concedes that "there is a broad consensus" that women living together show such a tendency and that olfactory cues may be involved. Observation suggests that females of a family of humans have synchronous menstruation cycles. Stern and McClintock (1998) adduce further proof that armpit pheromones play a role but doubts remain. In the lioness the role of pheromones may well be stronger than in the human female. Wyatt (2003) furnishes some more recent data.

In the domestic cat body smell, presumably through dermal secretion, plays an important role. On the basis of a 25-year study of feral cats in Calcutta, Jayashri Datta (2010, personal communication) states that a tomcat refrains from killing its own young by smelling the body; it sometimes kills nonkin young apparently only when these kittens seem to be a hindrance to copulating with the female.

However, a coordinated approach involving the study of understanding pheromone receptor families, the network of signal transduction and neural circuitry, approaches from functional genomics, and aspects of electrophysiology and imaging techniques might be helpful for unraveling more facts regarding the role of pheromones in big cats' social lives. This is a challenge for future workers in this field.

15.6.5 Olfactory Sense of Big Cats: An Ability Denied Earlier

We point out that even in the 1970s and early 1980s the idea of pheromonal communication in the world of tigers was met with stiff resistance in the wildlife circle, largely due to the dead weight of Jim Corbett, the celebrated hunter naturalist who strongly maintained the opinion that tigers have no sense of smell (Corbett 1944, 1954). We faced fierce antagonism to Schaller's concept and ours that tigers communicate through the smell of MF. However, even in the very early volumes of the *Bombay Natural History Society Journal* (the most important repository of knowledge on wildlife in India in that period), we detect many interesting descriptions for and against the olfactory ability of the tiger (Brahmachari and Brahmachary

1980). It is no longer necessary to review this old data for the very close observation of pet tigresses in nature or near-natural conditions such as of Lindbland (1984) or S.R. Choudhury in the 1970s (part of which was posthumously published much later [Choudhury 1999]) and of Singh (1986) prove beyond doubt the olfactory ability of the tiger. The tiger, according to these findings, which cannot be duplicated with wild tigers, might well utilize olfactory power for spooring prey under certain circumstances apart from responding to pheromonal cues. Later, in 1986–1989, we had the opportunity of observing a subadult male tiger at our disposal and could confirm its olfactory faculty. Even a 5-month-old pet tiger could perceive groundborne smell (Brahmachary and Poddar-Sarkar 1988, unpublished). More recently, Thapar (2006) furnishes modern physiological data on the tiger's olfactory sense. In three African lion cubs of George Adamson observed in 1988–89, the olfactory sense, together with the flehmen gesture, seemed to suddenly emerge in the sixth month (Brahmachary 1980, personal diary). In Kora, Kenya, Adamson and Brahmachary observed that a lioness could detect the smell of stale meat from a 25-ft distance and could also smell the MF of a male lion from a 30-ft distance (accurately measured by tape) 3 hours after ejection (Brahmachary personal diary).

15.6.6 FELINE ATTRACTANTS: TRADITIONAL KNOWLEDGE AND MODERN SCIENCE

In India valerian oil was formerly used to trap wild cats. Interestingly, valeric acid is a significant compound in the vaginal secretion of cats (Bland 1979) and cats are attracted by the plant *Valerina officinalis* (Bland 1979). It occurs also in the MF of tigers and cheetah (Poddar-Sarkar and Brahmachary 1997; Poddar-Sarkar et al. 1991). Albone (1984) described the putative components of catnip (*Nepeta cataria*) that attract or stimulate cats. Certain persons brought catnip from England to George Adamson's lion camp in Kora, Kenya, and tried to study the effect of this plant on the African lion but no conclusive results were obtained (Adamson 1988, personal communication). In India the Indian spikenard or Jatamanshi (*Nardostachys jatamanshi*) and *Acalypha indica* (local name Muktajhuri) are well-known plants that attract/stimulate cats. There is no report on their effect on tigers. Olden-day hunters in India mentioned a certain grass they called "Balachar" that apparently attracted tigers. One such hunter said that this grass, moistened with water, strongly attracted male tigers, A sample was procured and it turned out to be not grass but a dicot plant *Cassia sp*. On moistening freshly collected leaves with water they emitted an unusual sweet fruity smell (Brahmachary 1987, unpublished) but at that time there was no opportunity of investigating the effect of this smelly water on a tiger. This raises a possible line of research and the volatile flavor molecules could be studied with headspace gas chromatography mass spectrometry (GCMS).

15.6.7 FLEHMEN, THE CHARACTERISTIC GRIMACE

Many animals carry out a grimacing gesture with protruded tongue while encountering certain types of smell. Schneider studied this aspect as early as in 1932 (see Verberne 1970) and Verberne (1970) studied flehmen in the cat family in detail. It seems to be most prominent in the tiger (and tigress) and we have personally confirmed that in the tiger this grimace is far more pronounced than in the leopard and lion. Schaller wrote (1967), "Tigress ejects a rather wide spray. On two occasions a tiger stopped and sniffed the scent, grimacing afterwards with nose wrinkled, and tongue hanging out, a gesture described as 'flehmen' by Leyhausen." Flehmen is very well known in many other animals but here we focus on the cat family. The works of Schilling (1970), Estes (1972), and others now clearly indicate that the vomeronasal organ (VNO) (Jacobson's organ) discovered as early as 1703 began to reveal its functional importance from 1972 onward. It is now generally accepted that nonvolatile, heavy molecules pass through the tip of the tongue into the orifices of the VNO and transmit signals to such regions on the brain where nerves from the nasal epithelium do not reach. Airborne odorant molecules may be transferred to the VNO during the flehmen gesture (Verberne 1970). Of all the members of the cat family it is most prominent in the tiger and the tip of the tongue may transport relatively large amounts of nonvolatile substances into the VNO. Molecules like aphrodisin might play a role but this aspect has not been studied in the big cats.

15.6.8 MECHANISM FOR THE LONG PERSISTENCE OF PHEROMONES OF BIG CATS IN NATURE

A tiger would be very hard-pressed indeed to repeatedly renew territorial marks and/ or sexual attractants in a large territory. Nature has therefore devised a mechanism for allowing the volatiles to linger with the help of lipids and proteins. Even in the classical perfume industry, as in processing rose essence in India centuries ago, oil (i.e., lipid of vegetal origin) was used as a fixative (in this case, sesame oil). Schaller (1967) wrote "Several clumps of a granular whitish precipitate were in it… (apparently urine)… tiger urine by itself however did not have particularly strong odor whereas this fluid smell, very musky, readily discernible to human nose at a distance of 10–15 feet. The scent adhered to the vegetation for a long time." Schaller (1967) noticed the smell of MF on a tree trunk after a few weeks even in the rainy season. Sankhala (1978, 1993) denied any importance of the smell of MF because he thought that within a few hours of strong sunshine or even after a very light rain the smell would vanish. Apparently, the lipid fixatives hold the volatile molecules for a much longer time but changes in the aroma quality with time are perceptible. The proportions of more or less volatile molecules alter with time; more volatiles escape earlier. This is a possible mechanism for distinguishing fresh from old markings; animals respond to them differentially. But how much of the "individuality imprint" can last in old marks is a difficult question.

For a preliminary observation certain leaves in a clump of mangrove leaves were smeared with tiger MF. After drying the clump was immersed in estuarine tidal water in Bhitar Kanika, Orissa, India. After 22 hours the leaves were raised and two unbiased persons could instantaneously identify by smell those particular leaves that had been smeared with MF. The marking is subject to twice-daily tidal inundation, and is thus likely to be washed away. Perhaps together with the fixative lipids of MF, the waxy surface of certain mangroves further help to slow down the release of the odor molecules. Of three mangroves *Excoecaria sp.*, *Sonneratia sp.*, and *Heritiera sp.* studied by us, *Heritiera* sp. best retained the smell. The marsh tigers of Sundarban spray MF in such a tidal estuarine habitat. Therefore, the lipidic composition of the MF might be correlated with the wax coating of the leaf for sustenance of the aroma. This is an object of our future plan for a detailed study of the lipidic part of MF and of mangrove leaves in detail. Rutherfurd (2004) demonstrated that felinine, one of the key components of major urinary protein (MUP) involved in scent marking in the small cat family and regulated by sex hormones, persists in nature at normal temperature for 30 days. The persistence of pheromonally important components in nature is the key factor for sustenance and has a survival value.

15.7 MAJOR HISTOCOMPATIBILITY COMPLEX OF GENES AND INDIVIDUALITY IN PHEROMONAL SIGNALS OF BIG CATS

A very relevant question is whether and how the mark of individuality is imprinted in the chemicals purported to be pheromones, for essentially all territorial connotations of pheromones are based on the distinction between self and nonself. Brennan (2008) reported that, "...recognition of individual identity and relatedness of individuals play vital roles in mammalian social behaviour (for most vertebrates). Individual and kin recognition depend on being able to detect and discriminate differences in genetically determined chemosensory signals. In identifying these chemosensory signals of individual identity...the best known are the genes of MHC." Compounds produced through the expression of polymorphic major histocompatibility complex (MHC) genes are thought to influence individual odor and pheromone production, thereby facilitating an olfactory cue for kin recognition and mate choice (Chong 2009). Jeffery and Bangham (2000) and Kelley et al. (2005) proposed that MHC diversity is maintained through two forms of a balancing equation: heterozygote advantage and natural-selection-dependent frequency over long periods of evolutionary time. These facts imply distinguishing self from nonself (Brahmachary et al. 1993; Brennan 2008; Hurst et al. 2001; Yamazaki et al. 2000). Although "self/nonself" recognition is controlled by diversified MHC class I and class II MHC genes (the most polymorphic loci known in vertebrates), many species show limited or no MHC diversity; for example, experiments for skin grafting between individuals of unrelated cheetahs (*Acinonyx jubatus*) are successful, indicating that these African cats have little MHC diversity (O'Brien et al. 1986). Similarly, Asiatic lions (*Panthera leo persica*) show very low MHC diversity, indicating genetic bottleneck (Penn 2002).

However, genetically identical inbred mice have a significant variability in the proportion of volatile urinary components, suggesting that nongenetic factors, such as nutrition and environmental condition, also have significant effects on individual urine odor (Rock et al. 2007). MHC-related genome study of big-cat lineage (in which species divergence began 5–8 million years ago) revealed that there is 93% nucleotoide sequence similarity between class I transcripts in the cheetah, 93%–99% similarity in the ocelot, and 92%–100% similarity in the domestic cat when compared between species and only a 13% variation exists in all felidae for species-specificity (O'Brien and Yuhki 1999). According to Brennan (2010), no specific explanation is now valid for establishing the mechanism by which MHC genotype could affect metabolic pathways to account for the reported quantitative differences in urinary volatiles of individuals. Urine samples from MHC-congenic mice have consistently different proportions of volatile carboxylic acids (Singer et al. 1997).

The use of police dogs in hounding out criminals on the basis of their distinctive body odor is well known. We may here consider the old data on establishing 41 different parameters as star-shaped lines representing such ensembles of compound on human beings that apparently are characteristic for each individual (Strauss 1960). The police dog might be working on some such basis. Distinguishing an individual mongoose among 24 on the basis of ratios of 12 carboxylic acids characteristic for every individual mongoose was reported (Gorman 1976; Gorman et al. 1974). In a preliminary attempt we tried to investigate this aspect in the tiger (Poddar-Sarkar and Brahmachary 1999). It revealed a more complex situation in the tiger. Seasonal variations in the proportions of 10 carboxylic acids were noted but the patterns of a mother and son were much closer than that of a distantly related tigress. More recently, a more detailed study of such proportions has been carried out on *Lemur catta* (Palagi and Dapporto 2006). In view of recent findings on dogs smelling out and distinguishing different scats, trained dogs might be used for matching the smell of individual tigers by sampling fresh scratch marks, MF, and scats in a tiger census (Wasser 2009).

15.8   GENOMICS, PROTEOMICS, AND METABOLOMICS IN PHEROMONE RESEARCH OF BIG CATS: THE SEARCH FOR THE EVOLUTIONARY LINEAGE AND LINKAGE OF BIG-CAT POPULATION

A number of attempts have been made to quantitate the relatedness of different big cats with the help of recent molecular techniques but unequivocal results have not been obtained. For example, Davis et al. (2010) pointed out that "despite multiple publications on the subject no two molecular studies have reconstructed *Panthera* with the same topology." To trace evolutionary history among the eight subspecies of tiger (*Panthera tigris*), three of which are extinct, a number of genetic markers like 4-kb mtDNA sequence, allele variation in the nuclear MHC class II DRB gene, and composite nuclear microsatellite genotypes based on 30 loci have recently been carried out (Davis et al. 2010). Amplification, cloning, and sequencing of alpha-1 and alpha-2 domain of MHC class I and beta-1 domain of MHC class II DRB genes from scat of 16 wild and captive Bengal tigers (*Panthera tigris tigris*) of different geographic origins of India reveal a low number of MHC DRB alleles but high variability in peptide-binding sites (Pokorny 2011).

In 13 different attempts, the tiger, lion, snow leopard, clouded leopard, jaguar, cheetah, and cougar (puma) have been considered and the results are far from unequivocal (only one attempt included the cheetah and cougar). However, Davis et al. (2010) considered all old data as well as new sequences "generated for 3 single copy regions of the Felid Y chromosome as well as 4 mitochondrial and 4 autosomal gene segments" and Bayesian statistics to establish phylogenetic trees (Bayesian estimation of species trees). On the basis of this attempt they claim "monophyletic origin of lion and leopard, with jaguar sister to these species as well as a sister species relationship of tiger and snow leopard." Likewise, the subspecies/races of the tiger have also been studied from the genomic perspective, and with certain reservations, it has been accepted that the Amur (Siberian) race and the now-extinct Caspian race (samples of which were obtained from a museum specimen) are phylogenetically close as determined with the help of mitochondrial DNA sequences (Driscoll et al. 2009). The Bengal tiger (*P. tigris tigris*), the Sumatran tiger (*P.t. sumatrae*), and the Chinese tiger (*P.t. amoyensis*) are distant branches. Cracraft et al. (1998) had earlier suggested that the Sumatran tiger is distinct from the Asian mainland races. There is an indirect implication of this race genomics on pheromones.

From the biochemical perspective of metabolic products acting as pheromones, cauxin, a 70-kD MUP protein that belongs to the carboxylesterase (CES) superfamily attracts our attention. Cauxin answers for about 90% of total MUP in the smaller felids like the domestic cat. but it has not been detected in nonfelids. It is fivefold less in the Pantherine line (e.g., lion, jaguar, tiger, and leopard) (Li et al. 2011; McLean 2007; Miyazaki et al. 2006). Datta and Harris (1951) and Westall (1953) reported felinine, a characteristic molecule in urine of both the sexes of domestic cat which, as it now turns out, is a hydrolyzed product of cauxin. Interestingly, felinine has not been detected in *Panthera* (Hendrik et al. 1995). The exact role of cauxin is uncertain; we do not know whether it plays a role as a nonvolatile pheromone such as aphrodisin or a pheromone-binding protein. Nonetheless, genomics of and possible selection pressures, both positive and negative, on this protein are interesting from the perspective of evolution of the cat CES protein family. This might be worth pursuing in the future. These facts remind us of the Darwinian concept of natural selection between and within cat families. However, as already mentioned, the exact role of cauxin is uncertain.

Metabolomics in the context of 2 acetyl-1-pyrroline (2AP), an interesting component of MF and urine of both sexes of the tiger and leopard but not of the lion and cheetah, has been treated in Section 15.10.2.1.1. The two main large cat clades, one with lion-leopard-jaguar and the second with tiger-snow leopard as proposed by Hemmer (1974), Johnson et al. (1996), Bininda-Emonds et al. (2001), and Nowak (2010) on the basis of mitochondrial RFLP analysis and on excretory chemical signals may open up a new vista for establishing evolutionary lineage of felids. Bininda-Edmond's approach by studying the lipid chemistry of anal gland secretion of 32 lions and 22 tigers (though the pedigree or purebreed was not mentioned and the sample sizes from other felids were relatively small) established the nearest relationship between the lion and leopard. Thus, in the future, after analyzing more metabolites one might find a different metabolic picture.

Therefore, the evolutionary history of the emergence of the cat lineage still remains incomplete and we are yet to write it in a more rational and logical way. Many authors proposed that the pantherine line has a comparatively recent origin of about 3.8–4.6 million years but the individual speciation event occurred only 1 million years ago by gradual amalgamation of introgressive characters and favorable divergence.

15.9 QUANTITATIVE APPROACH FOR UNDERSTANDING THE TERRITORY AND HOME-RANGE IN THE TIGER COMMUNITY

15.9.1 TERRITORIALITY IN THE TIGER

Baker wrote that the tiger is not the unsociable creature it is commonly understood to be; on the contrary it is fond of consorting with others (Baker 1887). Forsyth reported seven adult tigers in a patch of cover (Forsyth 1872). Schaller also saw seven adult tigers around a kill (Schaller 1967). McDougal (1977) refers to a larger association, nine tigers, but an impressive example that has apparently escaped the notice of wildlifers and tiger specialists was reported by Eden, namely as many as 15 tigers in a patch of cover (Eden 1837–38). In more recent times an association of nine tigers (including both males and females) amicably feeding has been reported in Ranthambhore, India (Thapar 1986). McDougal (1977) noted while offering a ready-made supply of food (bait) that more than one tiger would feed from this source, though generally not simultaneously. Schaller (1967) and McDougal (1977) described the home range/territorial system of tigers; a tiger/tigress has a center (or more than one center) of activities within its range where it spends most of the time. The extensive data on tiger territory gained by the Smithsonian group in Nepal revealed that males establish territories ranging from 19 to 151 sq km while female territories range from 10 to 50 sq km. In seven adjacent areas of seven tigresses in Nepal, the Smithsonian scientists noticed overlaps between the areas of tigers 2 and 3, tigers 5 and 7 and tigers 6 and 7. They studied the details of gradual separation of mother and daughter into different territories after 2 years of total overlap.

Ranges/territories may extend from a few square miles (Schaller 1967) to 1200 sq km for the Siberian tiger (Thapar 2006). Home range/territory is determined by the number of animals, assurance of food such as prey density, and for the tigress, certain amenities (including food) and for properly raising cubs. As the tiger generally stakes out his claims over an area in which more than one tigress have their smaller areas, the tiger automatically ensures an area with prey. This trend would basically be applicable to the other big cats. In India, Panwar (1987) recorded that a male territory may enclose more than one female territory. In Chitawan, Nepal, a male tiger's territory enclosed those of seven tigresses. We have seen that tiger territories/home ranges vary widely. Very recently much detailed data has been gathered in the relatively small Indian National Parks. One example is that of a radio-collared tiger translocated into Sariska National Park which for about 1 year held an average home range of 140 sq km, including a largely overlapping smaller range of a female (Bhattacharjee et al. 2012; Sankar et al. 2012).

More pertinent to our theme is the role of pheromones in territorial and mating strategies. As McDougal (1977) stated, "Defence—rarely means fighting tooth and claw with intruders. The occupant advertises its presence through various means of marking its environment." Theoretically, the tiger, or for that matter, a lion or any other big cat, can ill afford to indulge in an all-out fight because even the winner will sustain injuries resulting in a handicap so far as catching prey is concerned. This may thus spell the doom for the winner, too, though in case of the group or pride of lions it will be less disastrous. Overt proclamations such as roaring or pheromonal messages are warnings likely to minimize the chances of such fighting.

It might be relevant to mention here that in Panna and Ranthambhore, India, about 25% and 35%, respectively, of male tigers are killed by fighting among themselves (Chandawat 2002, 2006; Thapar 2006). In view of the vast potential for inflicting injuries the toll of tigers could have been much higher unless a certain mechanism of restraint had been operant. The same must be valid for other big cats but in the lion one has to consider collective defense of the territory and pride and that collective attempt of a "bachelor gang" win over a pride/territory (McBride 1977; Packer et al. 1991; Packer and Pusey 1997; Schaller 1972). In a more popular vein, Jackman recorded such details (Jackman et al. 1982).

Carrington-Turner stated that tigers in their postprime period of life tend to be displaced toward the periphery; they have to abdicate and opt for suboptimal areas (Carrington-Turner 1959). More precise and modern reports are those by McDougal (1977), Smith et al. (1987) (Nepal), and Panwar (1987) and Thapar (2006) (India). It seems tigers in pre- and postprime conditions are unable to establish territories in the most favorable areas; they are forced to occupy peripheral zones. Thapar (2006) reported that in Ranthambhore, India, the young male tigers are forced to move to the peripheral areas; later they return and try to occupy territories in optimal areas. An example of forming a boundary by marking six trees with a diameter of 90 m by a female while litter-dropping and raising cubs was reported by Singh (Singh 1981).

To understand the territorial behavior of the coinhabitant and neighboring tigers by marking MF under a common open-air condition of a tropical climate we formulated several models within many constraints, problems of logistics, and limitations beyond our control.

We have selected an open-air enclosure for showcasing a nearly natural wildlife situation under captivity where animals have developed zoo-specific cognitive behavior in addition to their instinctive behavior. We have recorded data on MF spray, urination, scat deposition, and flehmen over about 5 years (with some gaps) at Nandankanan Zoological Garden (20°23′45.59″ N, 85°49′21.59″ E), Orissa, India. Detailed answers to the following questions will be attempted:

1. Whether marking is preferential or random
2. Whether there is any differential approach for marking by females and males due to the presence or absence of same or opposite sex in the neighborhood
3. Whether the territorial demarcation of one individual overlaps with another when they inhabit the same territorial zone
4. Whether the total area covered is a determining factor for frequency of marking
5. What the relative difference between the frequency of ordinary urination and MF spray and the respective difference between male and female is
6. Whether frequency of MF spraying has any correlation with the reproductive status of the female and male
7. Whether there is any correlation between MF spray and flehmen
8. Whether the incidence of MF spray, flehmen, etc. occurs at an early age in big cats

To test the concept of putting differential emphasis (actively defended territory and less actively defended territory) over different parts of home range, two experimental approaches have been framed:

1. Spatial distribution of marking at different locations of the enclosures by considering (a) the frequency of marking on each specific location and (b) MF spray per unit length of the boundary
2. By considering location-wise tree-marking to conceptualize a miniature home range under this captive condition and thus to extrapolate the basic strategy of MF spray for maintaining the territory and home range in the Bengal tiger

15.9.1.1   Study Design and Strategy

Data on the MF spraying of 12 tigers confined in four different groups (Gr I, II, III, and IV) in four enclosures (En. 1, 2, 3, and 4) were considered for the study (Figure 15.2; Tables 15.1 and 15.2). Each enclosure has a brick-built structure for dropping food and as a shelter (S) with two-way doors through which tigers can go in and out, a water-filled moat for swimming, a grassy sward, and many trees. A continuous high iron chain-link mesh (demarcated as "C" in Figure 15.2) divided En. 1 and En. 2 and similarly En. 3 and En. 3A. Each of this mesh runs from one corner of the enclosure, passes through the water moat, and ends on the other side of the enclosure (the "B" side of Figure 15.2) from where observation was recorded. Observation was carried out on Gr I comprising male M1 (stud book* No. W363), and females F1 (No. 356), F2 (No. 358), and F3 (No. 357) for 13 months (duration of study: June 1987–June 1988); Gr II comprising F4 (No. W360), F5 (No. W362), and F6 (No. W359) was observed for 7 months (study duration: December 1987–June 1988) following the previous schedule; Gr III comprising M2 (No. W325), F7 (No. W296),

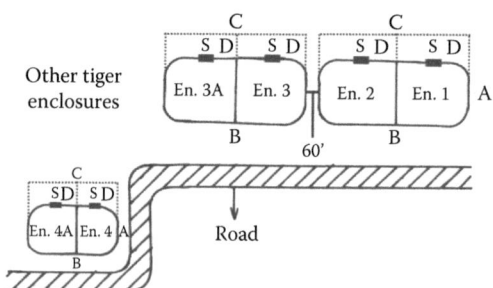

FIGURE 15.2   Site plan of enclosures (not to scale) for observation at Nandankanan (20°23′45.59″ N, 5°49′21.59″ E) Zoological Garden, Orissa, India. (A detailed design is available at www.panoramio.com/photo/23742383.) A = Vacant land (with no member boundary), B = side from where observation was taken, C = common boundary (with neighbor on opposite side), D = annex part of enclosures; Boundaries made up of iron chain-link mesh.

---

* International Tiger Stud Book published by Zoologischen Garten Leipzing, December 15, 1990 (Prof. Dr. rer.nat Siegfried Seifert, stud book keeper).

TABLE 15.1
Spatial Distribution of MF Spraying by Different Members of Gr I, II and III

Studbook No.[a] (Name)	Date of Birth	Sex and No. Used in the Text	C (Common Boundary)		A (No Member Boundary)		S (Shelter)			Total No. of Observations
			Percentage of Marking	Marking per Unit Length	Percentage of Marking	Marking per Unit Length	Percentage of Marking	Marking per Unit Length	Tree	
W363 (Sangram)	29.12.1983	M1	28.64	8.63	7.83	2.46	32.69	44.77	30.83	3013
356 (Janhavi)	11.12.1983	F1	16.37	4.37	12.66	3.52	57.23	69.45	13.74	2670
358 (Rohini)	11.12.1983	F2	26.86	4.61	7.34	1.31	48.13	37.54	17.66	1716
357 (Kaberi)	11.12.1983	F3	5.62	0.45	8.12	0.68	68.87	25.04	17.37	800
W360 (Sipra)	29.12.1983	F4	15.36	0.47	32.02	1.02	30.72	4.27	21.89	306
W362 (Sweta)	29.12.1983	F5	14.03	0.24	25.14	0.45	32.75	2.54	28.07	171
W359 (Swapna)	29.12.1983	F6	57.75	1.49	9.3	0.25	17.05	2	15.89	228
There is 60-ft Gap Between En. 2 and En. 3										
W325 (Aswini)	20.10.1981	M2	53.06	2.68	12.42	0.47	12.47	2.86	20.16	491
W296 (Alaka)	08.01.1980	F7	5.9	0.17	76.74	1.81	11.8	1.54	5.56	288
W331 (Jamuna)	08.12.1981	F8	18.89	0.51	57.78	1.27	12.59	1.54	10.74	270

[a] International Tiger Stud Book published by Zoologischen Garten Leipzing, December 15,1990 (Prof. dr. rer.nat Siegfried Seifert, stud book keeper).

TABLE 15.2
Spatial Distribution of MF Spraying by Members of Gr IV

Studbook No.[a] (Name)	Date of Birth	Sex and No. Used in the Text	C (Common Boundary)		A (No member Boundary)		D (Annex En. 4)		S (Shelter)			
			Percentage of Marking	Marking per Unit Length	Percentage of Marking	Marking per Unit Length	Percentage of Marking	Marking per Unit Length	Percentage of Marking	Marking per Unit Length	Tree	Total No. of Observations
(Bharat)	06.06.1977	M3	17.63	2.2	14.03	2.33	8.62	2.87	54.91	18.26	4.81	499
(Bharati)	08.01.1980	F9	20.59	3.65	25.1	5.94	3.95	1.87	48.09	22.73	2.26	709

Note: Total duration of observation 1547 hours. Total number of MF spraying and Urination recorded from 12 members of Gr I, Gr II, Gr III and Gr IV are 11161 and 230 respectively.

[a] Not registered in stud book.

and F8 (No. W331) for 4 months (duration of study: August 1988–November 1988); and Gr IV comprising M3* and F9† for 5 months (duration of study: October and November 1988–February 1989) on a 3-hour daily basis from 8:30–11:30 AM after ascertaining the period of brisk activity. The members of Gr I and Gr II resided side by side in these two adjacent enclosures (En. 1 and En. 2) and could communicate with each other only through visible and audible cues. Similarly, Gr III resided in En. 3 who had as neighbors one male and one female in En. 3A. Each enclosure had an annexure part (Annex En. 1, Annex En. 3, etc.; "D" in Figure 15.2) separated from the main enclosure with the same type of chain-link mesh boundary. Members of main enclosure and the Annex enclosure could also communicate by visual and tactile cues through the chain-link mesh. During collection of MF for chemical analysis the animals usually had been driven to that annex part (Figure 15.2). There was a huge vacant piece of land beyond a similar type of iron chain-link boundary (the A side of Figure 15.2) of En 1 and therefore shared with no other members. There was a 60-ft gap between En 2 and En 3. The approximate area covered by En. 1, En. 2, and En. 3 was about 10^4 square feet. The situation of En. 4 was quite different and smaller, about 1200 square feet with a tiny triangular water body inside, surrounded by a similar type of iron chain-link mesh traversing in between En. 4 and En. 4A and having an annex part in one side as in Annex En. 4 ("D" in Figure 15.2). There was a lone male on that side. En. 4A was occupied by a male and a female.

15.9.1.2   Overview on Marking Patterns

15.9.1.2.1   Test Hypothesis I

To address questions 1, 2, 3, and 4 above, we considered five locations (shelter (S), common boundary (C), no-member boundary (A), boundary in Annex part (D), and trees [T]) for recording data. Observation from the B side of specific enclosures (Figure 15.2) was considered only when certain tigers roamed at ease and least disturbed by visitors. It is to be noted that tigers in different assortments were arranged by zoo authorities and could not be altered at our will, so the tigers had to be studied in accordance with the regime of this public zoo.

Under these circumstances (vide Case studies 1–4), certain relevant data in relation to questions 1–4 have been furnished in Tables 15.1 and 15.2.

The answers to the above questions are as follows:

1. The frequency of MF spray is differential (i.e., higher in males than in female) (see also Figure 15.3).
2. The very first conclusion that emerges is that tigers prefer to mark maximally at the shelter where they get food and retire during inclement weather in all cases.
3. The length of the boundary is not the prime factor for marking by the tigers; however, the presence of neighbors (and their sex) beyond the common boundary (which can be conceptualized as possible "intruders") is the most

---

* Not registered in the stud book.
† Not registered in the stud book.

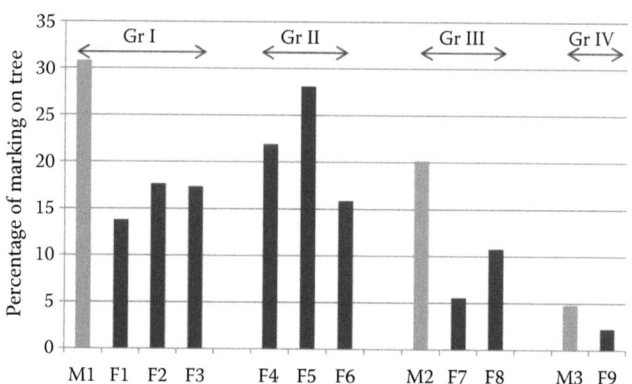

FIGURE 15.3 Tree-marking by different members of Groups I, II, III and IV.

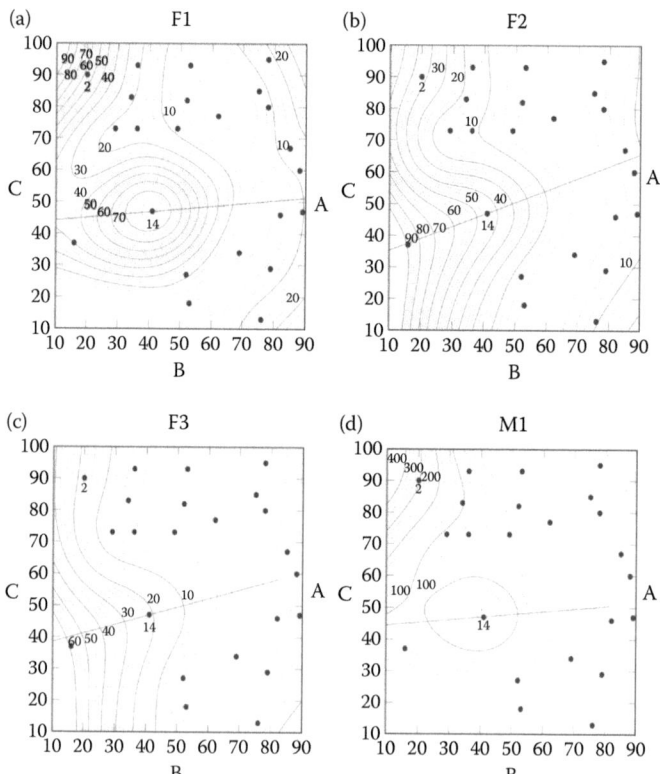

FIGURE 15.4 a–d: Contour map drawn on the basis of frequency of MF spray on 23 trees by Gr I members (a = F1, b = F2, c = F3 and d = M1) of En. 1 during June 1987–June 1988. (●) = location of trees, tree no. (●14), tree no. (●2); Y-axis = side of common boundary (C), Side 'A' and 'B' as per Figure 15.2.

important factor for frequency of marking. The boundary beyond which there is no neighbor was less important for the resident members.

4. Under restricted captive condition the combination of sexes of coinhabitant members and neighboring members exert a direct influence on frequency of marking. When the combination of sexes in a neighborhood has been altered the spatial distribution pattern of MF spray changes. The above behavioral pattern may project an insight in the breeding strategy of zoo tigers.

5. From Tables 15.1 and 15.2 it is also revealed that under the limitations of captive situation, within a common home range tigers maintain specific location-wise nonoverlapping territories. When they are in combination they mutually select the location of preference for spraying MF depending on the combination of resident and nonresident members. It is also explained through a contour map drawn on the basis of frequency of marking on trees (Figure 15.4a to d).

6. The synchronization in frequency of marking among the coinhabitant members during proestrous, estrous, and postestrous have been observed (Figure 15.5a,b). It was also observed that the rate of marking by a tigress increases during proestrous and suddenly falls during estrous and again rises to normal at postestrous. The coinhabiting male also behaves in a similar way.

7. Synchronization in MF spray and flehmen incidence was also observed in some cases (Figure 15.6). Therefore, all the observations support the hypothesis that MF does act as a means of communication among tigers.

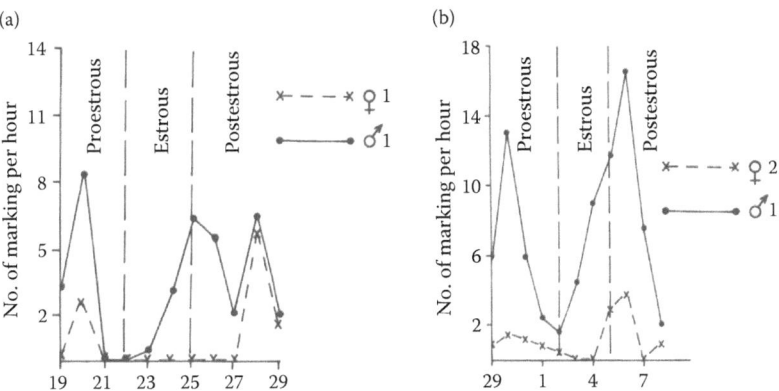

FIGURE 15.5 (a) Differential marking pattern of M1 and F1 during different reproductive phases. X-axis indicates specific date from 04/19/88 to 04/29/88. Mating event occurred during April 22 to 25, 1988. (b) Differential marking pattern of M1 and F2 during different reproductive stages. X-axis indicates specific date from 01/29/88 to 02/08/88. Mating event occured during February 2 to 5, 1988.

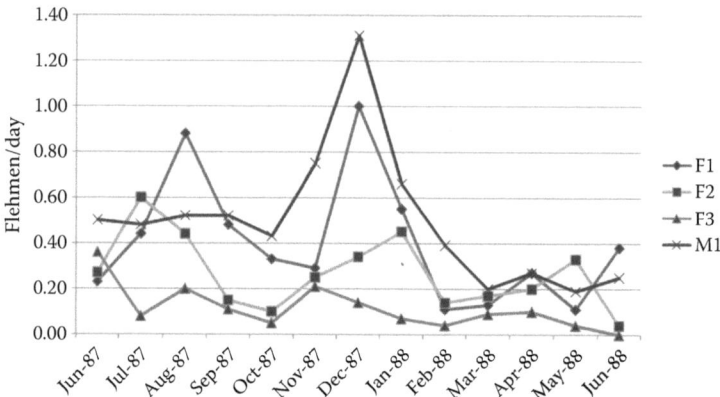

FIGURE 15.6 Correlation of MF spray and Flehmen incidence of Gr I members during every month of observation. Observation was taken maintaining the schedule mentioned in Section 15.9.1.1.

Case Study 1: En. I with Gr 1

The introduction of a member beyond one boundary changed the spatial distributional patterns of marking by the resident members and thereby strengthened the above view. For a short while F1 was transferred for a few months to Annex En. I. M1 of Gr 1 could interact with her through the chain-link mesh (i.e., the boundary between En. I and Annex En. I; vide "D" in Figure 15.2). Now 448 out of 1442 total markings of M1 was noted to be on that side, which was never marked before (duration of observation: 137 hours 20 minutes).

Among the three tigresses of En. I who were litter mates and so of the same age, F3 marked significantly less (Table 15.1), which might be correlated with a low dominance rank.

Case Study 2: En. II with Gr 2 and En. III with Gr 3

Another experimental model revealed a differential pattern of MF spray among the resident members of a single group. This depended on the presence and particular sex combinations of a neighboring group in an enclosure 60 feet apart (vide Figure 15.2 and Table 15.1 for the combination of resident and neighboring members in these two enclosures; members could communicate by vocalization but not by visible cues). The preference for MF spray depended on the strategy of "defense against the same sex."

Case Study 3: En. IV with Gr 4

The composition of this enclosure was quite different and the members were very attentive to spray at both the common boundaries ("C" and "D" of En. 4 of Figure 15.2; Table 15.2) because there were neighbors on both sides.

Case Study 4: En. I with Gr 1 and En. IV with Gr V

In order to test the hypothesis that the enclosure area (represented here as length X breadth of boundary) has no bearing on the frequency of marking, we considered En. 1 (~10,000 sq ft) and En. 4 (~1200 sq ft). Data on frequency of MF spray per unit length on the common boundary and on the boundary without neighbors revealed an interesting feature, namely, marking frequency does not depend on the length (vide Tables 5.1 and 5.2). The equivalent statement is that MF frequency is not higher on longer boundaries or less on shorter boundaries. So, once again we see that the assortment of neighbors/combination of sex, rather than the area, is the factor that determines the frequency of MF spraying.

15.9.1.2.2   Test Hypothesis II

After getting a preferential overview on marking pattern we have undertaken the second experimental approach to map the location-wise distributional trend of marking on trees by each individual of Gr I tigers throughout the year in different seasons. It was possible to consider the MF spray on trees only within En. I (between June 1987–June 1988) containing 23 trees of different families, shape, height, and width of the tree trunk and canopy. Three females of Gr II resided in adjacent En. 2 as mentioned previously. The spatial distribution of markings over 23 trees by three tigresses (F1, F2, F3) and one tiger (M1) and the density of spraying of MF on each tree were calculated. Two-dimensional contour maps for each tiger was developed by moving average isopleths (Davis 1973) using SYSTAT 7.0 (Figure 15.4a to d) by placing 23 trees on an x and y locational grid and z as marking per tree. The area of the enclosure comprising the trees was divided into square grids of equal size in such a manner that each grid would contain at least one tree. All the data within this grid were averaged and the mean value assigned to the central point of the grid and subsequently the mapped area was contoured at suitable contour intervals (here, 5-15-25-35— — — — n) for each member (Figure 15.4a to d).

Of these 23 trees most markings were on tree numbers 1, 2, 4, and 14. The data indicated that tree T 2, which had a thin trunk and small canopy, received the highest degree of markings (446 by all the members) possibly because of its location, namely proximity to the common chain-link mesh as well as shelter. T 14 (vide Figure 15.3) received the second highest number of markings (344 by four members) although it had the thickest trunk and widest canopy. No preference for a particular tree species was evident from our observation. That both the trees were frequently marked is an indicator that the width of the trunk is not an innate releasing mechanism (IRM), it does not elicit more spraying. (This is also relevant to the wall of the shelter, which resembles a very wide tree trunk.) Although Smith et al. (1989) suggested that the rate of marking on trees may depend on the diameter of tree trunk and angle of lean, our results indicated that the location of the trees rather than their width and canopy spread is important with respect to spraying and that within this common area each tiger had staked out a preferential zone.

The contour lines further suggest that markings are preferential rather than random because all the members of this group preferred to mark on trees located near the common iron chain-link mesh where there were three female tigers beyond the C side of the *Y*-axis (Figure 15.4a to d). Tigers were not very attentive to sprays on the trees beyond the opposite side (i.e., the A side of the *Y*-axis) which was lying vacant. Therefore, trees located near that side received less density of marking.

It is evident from Figure 15.4 that every tiger had a nearly nonoverlapping contour and the contour lines with highest value were closer to the shelter and common boundary. The highest contour value of the male along the C side of the *Y*-axis covering a wider zone (almost the entire length of the common boundary) also signified that he not only advertised himself to the females of his own enclosure but also to the neighboring females of the adjacent enclosure. Theoretically, there are two possibilities:

1. Straight lines running through the locations of maximum value along contour lines of different tigers lie at different points along the common boundary in every case. That would indicate the view of a group behavior. In the case of group behavior, we would expect that every tiger would prefer to mark at the common boundary (C, i.e., the *Y*-axis).
2. Contour maps having unique features signify uniqueness for individuality. Figure 15.4a to d indeed suggests both these phenomena.

In a common miniature home range, territoriality is always maintained for every individual member of a group. In nature tigers holding individual territories as well as tolerating certain overlaps are well known (McDougal 1977; Singh 1981; Thapar 1986). They describe amicable feeding in free-ranging tigers and Smith et al. (1987) show overlapping and shifting of home range/territory. It is worth noting that even within the outdoor enclosure shared by four captive tigers, there is a tendency of tree marking that reflects a sort of partitioning of this restricted home range for each individual as evident from the contour analysis data.

All these facts strengthen the hypothesis that scent-marking has a territorial and sexual connotation and that in the enclosures both group behavior and individual uniqueness are evident in the territory marking context. Furthermore, the width of a tree trunk is not an IRM-eliciting marking; rather, the location of the tree is important (i.e., in the context of territorial boundaries).

15.9.2   MF Spraying versus Ordinary Urination

According to our observations, the average ratio of MF:urination was ~50:1 in 12 tigers and tigresses, with individual variation ranging from 10:1 in one male to 600:1 in another. In two Asiatic lions the ratio was ~6:1 and ~25:1 and in a cheetah in Namibia it was ~89:9. In a leopard in India it was ~28:9. This high frequency of MF ejection makes sense to the evolutionary biologist: MF must have evolved to serve a purpose, or else that much wastage of energy would have been selected against. Of all the overt possible sources of pheromone such as scats, urine, and MF the latter seems to be the most important considering frequency.

It was also observed that males always mark more on trees than the females (Figure 15.3). No behavioral implications can be made from this observation at this stage, except that the frequency of spraying is less in tigresses.

A mature Asiatic lion (Gir, India, 1.7 years of age) sprayed MF 253 times while he urinated only 11 times during a certain period. The distribution pattern of MF was random. During the same period a mature lioness in the same enclosure was never observed to spray but she urinated 176 times in a nonrandom manner. The ratio of MF:urine in the male partner was 23:1.

Observations on a leopard during September to December 1996 on a daily hourly basis in a rescue camp of Rajabhatkhawa, West Bengal, India, revealed that in this male leopard the ratio of frequency of MF spray:urination:scat deposition is about 18:6:1.

15.9.3 MF Spraying in Proestrous, Estrous, and Postestrous Periods

Datewise observations in several occasions reveal that at different reproductive phases of females (during proestrous, estrous, and postestrous) the rate of marking (number/hour) sharply varies. In general, there is a sharp rise in rate of MF spray during proestrous that sharply falls to zero and again sharply rises during post estrous (Figure 15.5a, b). A similar feature in the rate of MF spray has been observed in the mating partner. This trend makes sense; if the female finds a partner and copulation takes place, she no longer needs to attract a new mate but if copulation does not occur, she will continue to mark the old territory and enhance advertisement for a mate. These facts also indicate mutual stimulation by the two sexes for entering into the reproductive stage (see the case of snow leopards in Section 15.9.7).

Brahmachary and Singh (2000) observed Asiatic lions in enclosures within the natural environment of Gir National Park, India (each enclosure is of 81 sq m and occasionally the lions are released in a 420-ha fenced forest for roaming and natural hunting of prey). When a lioness of this group exhibited overt estrous or proestrous behavior such as rolling on her back, she sprayed four times in 4 days (2 hours observation time per day). After 4 days the behavior mentioned above as well as spraying was no longer observed. During the same period the male in the same enclosure sprayed MF 60 times. It is worth noting that although the lioness only rarely spray-marked, she urinated markedly more frequently as already mentioned in Section 15.9.2. Ewer (1968, 1973) stated that the surge of reproductive hormones during the proestrous stage stimulates both tiger and lion (of both sexes) to urinate and/or mark at an increased rate.

It is very probable that the pattern of spraying in cohabiting partners (in the case of tigers and lions) is associated with the reproductive cycle and so, directly or indirectly, might be influenced by reproductive hormones. Therefore, the marking pattern of a tigress also influences the marking behavior of a cohabiting male. This ethophysiological implication may play an important role in MF spray of big cats.

We also noted that a male leopard, as it came to the right reproductive stage (confirmed by the peak period of sawing calls, the equivalent of roaring in leopards) revealed a beautifully correlated peak with the increase in frequency of MF by four times and gradual decline in both sawing calls and MF spraying (Brahmachary 2004, unpublished).

15.9.4 MF Spray and Flehmen

We have recorded the data on incidence of flehmen by Gr I tigers throughout the year. On comparing with the data of MF spraying we note an interesting feature, namely the tiger who sprays MF more frequently also shows more flehmen incidence (Figure 15.6). Synchronization of flehmen of coinhabitant members is also revealed. F3 is an exception as already mentioned.

15.9.5 Territoriality in the Lion

The first exhaustive study of the East African lion (more specifically, the Serengeti lion) was carried out by Schaller (1972). Despite later investigations such as those of Bertram (1975), who continued the study initiated by Schaller (1972), and the numerous papers of Craig Packer and his school (some of which have been mentioned below), Schaller (1972) sums up the essence of land tenure (a term that is more general than that of territory or home range) of lions in East Africa. Among the Felidae, "the lion is unique in the extent of its social life" and this pivots around the pride, which is very important in the evolution of lion sociality. In the lion, prides rather than merely pairs or loners are the pivot of the social life (Schaller 1972). Males primarily take part in group-territorial competition, and females' reproductive success, mortality, and so forth are significantly associated with male neighbors (Heinsohn 1997; Packer et al. 1990, 2009). For the lionesses the area probably connotes a hunting ground and a place for litter dropping; for the lion the territory is a place containing lionesses available for mating. In Serengeti pride ranges of 120–275 sq km were noted (Schaller 1972). According to Mosser and Packer (2008), Serengeti lions defend territories of a mean size of 56 sq km with a range of 15–219 sq km. Core areas are generally exclusive with different degree of overlap, pride fission and male coalition from the neighborhood (VanderWaal et al. 2009).

As opposed to regular pride members there are also nomads, mostly coalitions of bachelor males but there are nomadic lionesses and nomadic pairs (i.e., a twosome pride consisting of a lion and lioness) as well (Schaller 1972). A nomadic bachelor group may later oust the male(s) of an established pride and win over the females and the pride area. The land tenure system of the pride was termed by Schaller as the pride area and of the nomads as a range.

Nomads wander widely but the resident prides remain within their limited areas. Of the 14 prides around Seronera in Serengeti, prides 1 and 3 ranged over 400 sq km each while pride 2 moved over at least 210 sq km. Pride areas overlapped extensively, despite occasional aggressive encounters when the residents generally won. In two small pride areas of about 39 sq km each in Lake Manyara, also studied by Schaller (1972), about 20 sq km (>50%) was the overlap area. Nomads follow large migratory herds of herbivores and move over vast ranges. One male nomad was estimated to have wandered over 4700 sq km.

In general, as Schaller (1972) concludes, despite large overlaps in pride areas, "direct confrontations are remarkably infrequent." Generally lion prides avoid each other as they become aware of the presence of the other group. Agonistic interactions such as on perceiving the presence of the neighbors or strangers close by frequently elicit scent-marking. This is relevant to the function of pheromones in the context of territory. McBride (1977) reported from his white lion camp of Timbavati that lionesses often remain with their pride of birth and if they were rejected for any reason they would be at risk (McBride 1977, 1981).

Thus we note that in the lion as in the tiger, the concept of territory and home range (or pride area or range) is rather elastic. The land tenure system is influenced by a number of environmental factors, and the territory or pride area may quantitatively vary to a large extent.

Chellam (1993) reported that in the Asiatic lion (Gir, India) annual home ranges of lions are ~200 and ~120 sq km, respectively, for the male and female. Brahmachary and Singh (1998, unpublished) mapped the land tenure system in lions based on data made available by Gir authorities. Six lion prides were frequently seen around a nes (i.e., settlement of local herdsmen). Sometimes more than one pride shared the nes. About 5% of their cattle are annually culled by the lions and none of these lions was radio-collared but they were visible. Brahmachary (2011) also furnished a map of three lion prides of different composition that reveals partially overlapping home ranges/territories with approximate areas of ~30 sq km or less. None of these lions was radio-collared but they were frequently visible and hence the minimum polygonal areas could be approximately estimated by sightings.

15.9.6 TERRITORIALITY AND THE CHEETAH

Caro and Collins (1987) pointed out that the male cheetah generally establishes territory by spraying very frequently depending on prey density and to attract females. Males may form coalitions between two to four members but fierce fighting between rival coalitions for mating with partner and territorial disputes have been noted. Cheetah females, when adult, range alone. Caro recorded territories of cheetahs in Serengeti. The male territories were ~37–48 sq km, the largest being ~75 sq km. Female territories were much more extensive; the smallest was ~400 sq km—an unusual case in big cats (Caro 1994). Caro (1994) states that nonresident cheetahs moving as transients in others' established territories suffer from stress, apparently after perceiving fresh MF, urine, and so forth of the latter. This is an appropriate response supporting the concept of communication through the chemicals of MF, urine, and so forth. The sizes of home-ranges vary immensely between the studies that have been carried out in different areas. In Kruger National Park and Matusadona National Park home range was recorded as >200 sq km for both seminomadic males and females (Broekhuis 2007); in Namibia it is on average 1647 sq km (Muntifering et al. 2006), whereas in Botswana the home range size of a single male is 494–663 sq km and for coalition of two females it is 241–361 sq km. By fitting cell/GPSs and VHF collars it was revealed that females traverse on an average a distance of 0–20 km/day, which increases when the cubs leave the den, and males traverses 0–39 km/day (Houser et al. 2009).

15.9.7 TERRITORIALITY AND THE LEOPARD

The male leopard generally occupies a large area that usually overlaps with one or more females' area. In an arid desert area where the prey is limited, home range is larger and sometimes overlaps with that of the same sex (Jenny 1976). Leopard ranges of 6–18–26 sq km by females and 17–76–137–260 sq km by males have been reported from different forests of Chitawan, Serengeti, Kruger, Thailand, and the Israeli desert (Bailey 1993; Ilany 1981; Strander et al. 1997; Turnbull-Kemp 1967). Overlapping home ranges of five male and two female leopards were found to be restricted within ~0.17° × 0.15° longitude/latitude in Namibia (Africat News 2000). In the Russian Far East Amur leopards stake home ranges of about 33–63 sq km (females) and up to 280 sq km (males) as reported by Miquelle et al. (1996). In Rhodesia leopards moved within an area of only 10–19 sq km (Smith 1978). According to the observation of Jackson and Ahlborn (1989) on the snow leopard of Nepal, home range size varied widely among individuals, from ~12–39 sq km; individuals may share a common core area. However, due to lack of sufficient data they could not exactly determine exclusive home ranges of individual males or females.

15.9.8 TERRITORIALITY AND THE PUMA (COUGAR)

Mountain lions, *Puma concolor,* have individual and apparently undefended territory because they mutually avoid each other (Maser et al. 1976). On the basis of his observation in Idaho, Horonecker (1970) recorded that the mountain lion often uses claw marks on a trail and on high ridges and then urinates and defecates over them, and that male home range varies from 65–250 sq km and female home range from 13–52 sq km. However, they often cross their own territories. Territories of males may overlap with the territories of many females so that the male has access to them in breeding season. As we have mentioned earlier this is also valid for tigers. According to Russell (1978), male home ranges usually are a minimum of 40 sq km and female ranges are 8–32 sq km. It was recorded during the study in southern Utah by Hemker and his colleagues (1984) that males occupied areas of up to 513 square miles and females up to 426 square miles. Sitton and Wallen (1976) studied cougars in Big Sur, California, and documented the average home ranges, which varied from 25 to 35 square miles for males and from 18 to 25 square miles for females. Actually, the density of the socially tolerant cougar depends on home range size and degree of overlap.

15.9.9 TERRITORIALITY AND THE JAGUAR

Rabinowitz and Nottingham (1986) proposed that there is a dynamic equilibrium in the relatively dense population of jaguars when they observed this elusive big cat in Cockscomb Basin, Belize. The land tenure system/home range of one adult male overlaps with that of the adjacent male whereas home ranges of female do not overlap and their movements were restricted within the ranges of individual adult males. Spacing patterns were based on regions of exclusive use "core area" within a home range The average home range was reported from the radio-collared males and females as 33.4 sq km and 10 sq km, respectively (Rabinowitz and Nottingham 1986). But territory and home range size depends on habitat and density of prey. Territory size is reported as 2 to 5 sq km in Mexico, 390 sq km in Brazil, 65 sq km for males, and up to 29 sq km for females in Kaa-Iya del Gran Chaco's reserve of Bolivia and Paraguay and over 1359 sq km for one adult male in Arizona. Besides vocalization, backward urine spraying on prominent locations, claw scratching and cheek-rubbing are also very common in the jaguar (Baker 2002).

So, from tiger to jaguar we find that territory/home range varies widely.

15.9.10 ONTOGENY OF DIFFERENT PHYSIOLOGICAL PHENOMENA IN CUBS OF BIG CATS

We recorded the data on the ontogeny of different physiological phenomena by rearing a tiger cub (Stud book no. 446, Dora III) born on July 25, 1987 at Nandankanan zoo. The cub was closely observed daily for 5 hours in the morning and 2 hours in the afternoon. An indication of the flehmen gesture was first noticed on November 15, 1987 at about the age of 3-1/2 months as recorded in our field diary of 1987. The cub was observed to sniff various objects lying here and there within the enclosure and sometimes showed a flehmen gesture, mostly on sniffing his own urine. In three female cubs in another Indian zoo the appearance of the first flehmen was observed at the beginning of the fourth month (Brahmachary, Walker and Mallya 1985, unpublished). The initiation of squirting by raising the tail was noticed in Dora III at the age of 7 months. He tried to eject MF in small burst with very small quantity. This happened very infrequently. The animal made its first regular squirting when it reached the age of 1 year. The pet tigress Khairi did so at the age of 1 year, too (Choudhury 1999). In two leopard cubs the flehmen gesture appeared before the age of 3-1/2 months (Brahmachary 1980, unpublished).

Chemical analysis of the urine of a tiger cub revealed traces of free fatty acids in the urine at 3 months of age. The presence of monoamine, diamine, and polyamines were detected in the urine of 5-month-old cub. The characteristic aroma was faintly perceptible in the urine sample collected at 7 months while a good musky aroma was detectable at the age of 1 year (Brahmachary 1990; Poddar-Sarkar 1995). It is worth noting that the amount of lipids in the urine increases with age, though even at 1 year of age, it is much less than that of the adult MF.

In the leopard cubs the aroma appears only at about the age of 3 months (Brahmachary 1980, 1988, unpublished). As mentioned earlier, in three lion cubs of George Adamson, Brahmachary (1988, unpublished) observed the first appearance of the flehmen gesture in the sixth month correlated with the sudden incidence of sniffing and flehmening.

15.10 CHEMISTRY RELATED TO MF OF THE TIGER AND OTHER BIG CATS

15.10.1 COLLECTION OF MF

The collection of MF was very easy in the case of a pet tigress (Khairi) and a pet cheetah, but generally, for chemical analysis of MF of the tiger, leopard, lion, and cheetah we adopted a devise for collection by walking with a clean tray behind the chain-link mesh and waited patiently for ejection by the animal and ultimately collected a part of it while in air. In general it was noted that when the animal is introduced into a new area it has a tendency to spray several times for establishing the territory, and if MF/urine of one animal is allowed to be smelled by another, he or she sprays instantly. We have utilized these two innate behavioral aspects for collection. After collection we added hexane in the field to prevent bacterial infection and the sample was kept under refrigeration for future analysis.

In a similar manner leopard MF (Poddar-Sarkar and Brahmachary 2004) and lion MF (Brahmachary and Singh 2000) were collected by placing cotton wads impregnated with MF or urine of another leopard or lion in the interstices of the chain-link mesh. (The Asiatic lion MF, unlike that of the tiger, is not always ejected upward and backward; sometimes the jet is aimed horizontally backward and sometimes even downward, in which case the MF cannot be collected [Brahmachary and Singh, 2000].) In the case of the cheetah, as the animal was tame and had a free run in a very large enclosure in Namibia, MF could be collected in a wide-mouthed beaker as the spray was aimed against a tree (Poddar-Sarkar and Brahmachary 1997). All the samples were collected during daytime hours. Of all these big cats, the cheetah is most diurnal.

15.10.2 Chemical Analysis of MF

15.10.2.1 Volatile and Nonvolatile Compounds Identified in MF

Van den Hurk (2007) sums up in tabular form much of the findings on pheromones of the small cats and big cats including our results. During the late 1970s, 1980s, and 1990s we primarily isolated different chemical groups on the basis of pH difference during steam distillation from MF, such as free fatty acids (FFAs) in the acidic fraction, 2AP, aldehydes and ketones in neutral fraction, and amines in basic fractions. The fractions were rendered into salt or derivatized according to their functional groups and then subjected to different chromatographic techniques. Most of the compounds were identified by following classical methods and using modern-day instruments (Table 15.3). Burger et al. (2008) identified more compounds from MF of tiger by headspace solid phase microextraction (SPME) GCMS, which has the advantage of direct application to the instrument. The important point to note is that in the tiger volatile free fatty acids (FFAs); (C2-C10), which are known to be pheromones in many mammals, comprise only branched and unbranched FFAs; no unsaturated and antiso FFAs have been identified in tiger MF (Figures 15.7 and 15.8a, b; Table 15.3). Primary, secondary, and tertiary amines as well as carbonyl compounds were identified in MF of tigers. β-phenylethylamine was detected in MF of four tigers and in MF of leopard and cheetah (Figure 15.9). β-phenylethylamine may be a common urinary excretory product but it was interesting to note its role in depressed persons and fierce criminals (Brahmachary and Dutta 1981). Van den Hurk (2007) reevaluated the importance of β-phenylethylamine in this context.

Aldehydes and ketones have not been detected in cheetah MF although these have been identified in the tiger and leopard (Figure 15.10). In the context of genomics and metabolomics, we also mention the findings on cheetah MF. Unfortunately we could work only on a single cheetah, a subadult almost attaining the adult stage, but if this is indeed a characteristic chemical feature of the cheetah (as opposed to that of the tiger, lion, and leopard), then we face a question that is intriguing *per se*. Moreover,

TABLE 15.3

Volatile Compounds Identified from MF of Tiger, Leopard, Cheetah, and Lion

Volatiles	Leopard	Cheetah	Tiger	Lion
Free fatty acids	Acetic, propionic, isobutyric, butyric, isovaleric, isohexanoic, hexanoic, isoheptanoic, hepta, isoocta, octa and nonanoic acid (Poddar-Sarkar and Brahmachary 2004)	Acetic, isopropionic, propionic, isobutyric, butyric, isovaleric, valeric, hexanoic, isooctanoic, octanoic (isohexanoic and heptanoic acid absent) (Poddar-Sarkar and Brahmachary 1997)	Acetic, propionic, isobutyric, butyric, isovaleric, valeric, isohexanoic, hexanoic, isoheptanoic, iso-octanoic, octanoic, isononanoic and nonanoic (Poddar-Sarkar, Brahmachary and Dutta 1991)	No free fatty acid has been identified (Andersen and Vulpius 1999) (from urine sample)
Aldehyde, ketones, and alcohol	Acetaldehyde, propional-dehyde, isohexanal-dehyde, hexanaldehyde, and heptanaldehyde (Brahmachary and Dutta 1987) Acetone	Aldehyde and ketone not detected (Poddar-Sarkar and Brahmachary 1997)	Acetaldehyde (Poddar-Sarkar 1995), Hexanal, heptanal, octanal, 2-octenal, nonanal, benzaldehyde, decanal, undecanal (Burger et al. 2008) Acetone (Poddar-Sarkar 1995); 3-heptanone, 2-heptanone, 2-octanone, 3-methyl-2-octanone, 2-nonanone, 3-methyl-2-nonanone, 3-methyl-2-decanone, 2-hepta decanone (Burger et al. 2008)[a] Hexadecanol, 2 hexanol	3-methylbutanal, hexanal, heptanal, octanal, nonanal, benzaldehyde (Andersen and Vulpius 1999) (from urine sample) Acetone, 2-butanone, 2-pentanone, 3-hexanone, 4-heptanone, 2-heptanone
Amines	Dimethylamine, trimethyl-amine, ethylenediamine, putrescein, Cadaverine 2-phenylethyl amine (Poddar-Sarkar and Brahmachary 2004)	No data	Ammonia, 2-phenylethyl-amine (Brahmachary and Dutta 1979) Ammonia, methylamine, dimethylamines, trimethylamines, triethylamine, propylamine butane-1,4-diamine (Bank et al. 1992)	Trimethylamine,3-ethyl-1-butylamine (Andersen and Vulpius 1999) (from urine sample)

(continued)

TABLE 15.3 (Continued)
Volatile Compounds Identified from MF of Tiger, Leopard, Cheetah, and Lion

Volatiles	Leopard	Cheetah	Tiger	Lion
Aroma molecules	2-acetyl-1-pyrroline (Brahmachary and Dutta [1978, unpublished])	Absent	2-acetyl-1-pyrroline (Brahmachary and Dutta 1987, Brahmachary, Poddar-Sarkar and Dutta 1990, Brahmachary, Poddar-Sarkar and Dutta 1992, Brahmachary 1996)	Absent
Total lipid	1.15 mg/ml	3.87 ± 0.58 mg/ml	1.88 ± 0.75 mg/ml	
Neutral lipid		Diglyceride, triglyceride, sterols (Poddar-Sarkar and Brahmachary 1997)	Cholesterol ester, wax ester, monoglyceride, diglyceride, triglyceride, sterols, phospholipids (Poddar-Sarkar 1996)	Wax ester, sterol esters (but relative paucity of free sterol), monoglyceride, diglyceride (Brahmachary and Singh 2000)

Note: Besides these chemical groups, several volatile compounds with different structure and functional groups were also identified from the lion and tiger by the authors of 5 and 7.

[a] It is not clear from the explanation whether that the analysis was undertaken with the urine, MF, or both. However, some of the molecules likely may be common to both because MF is sprayed through the urinary channel.

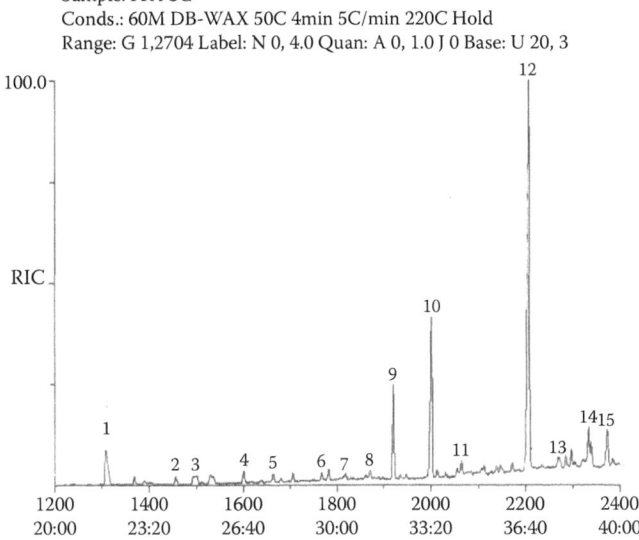

FIGURE 15.7 Gas chromatogram of free fatty acids identified from MF of tiger (M1). Peak No. 1 = acetic acid, 2 = propionic acid, 3 = isobutyric, 4 = butyric, 5 = isovaleric acid, 6 = valeric, 7 = isohexanoic, 8 = hexanoic, 9 = isoheptanoic, 10 = heptanoic, 11 = isooctanoic, 12 = octanoic, 13 = isononanoic, 14 = nonanoic, 15 = decanoic acid.

it has a bearing on the problem of metabolomics. This fact might be exploited while studying the genomics of the different cat species—and the cheetah is an unusual member of the cat family—and furthermore, while considering the putative differences in MF at the species level (see Section 15.8 for details).

15.10.2.1.1   2AP: The Elusive Aroma Molecule of Tiger MF, Basmati Rice, Bassia Flower, and a Certain Pulse (Mung Bean)

In 1982 this aroma molecule 2AP was identified in boiled Basmati rice (Buttery et al. 1982) and the next year it was detected in the leaves of *Pandanus amaryllifolius* (Roxb. = *P. faetoedus*; family: Pandanaceae) by Buttery's group (Buttery et al. 1983). This plant was long known as producing a smell equivalent to that of fragrant rice and in Bengal it was locally known as Payes leaf. In 1977 Brahmachary and Dutta detected this aroma from MF of the pet tigress Khairi and noticed that on acidification of MF, fragrant rice water, and *P. amaryllifolius* Roxb. leaf extract the aroma disappears but it reappears as the fluid is rendered alkaline, and they had an inkling that the three fragrance molecules have some chemical similarity and that civetone, which also has a ricelike fragrance, must be different from the rice/tiger aroma (Brahmachary and Dutta 1979, unpublished note). In the 1980s and 1990s a large number of experiments with paper chromatography (PC) and gas chromatography (GC), using two solvents and two GC columns and cochromatography of GC, indicated the closely similar nature of rice aroma and tiger aroma (Brahmachary et al. 1990). Later comparisons with a synthetic sample gifted by Schieberle revealed that

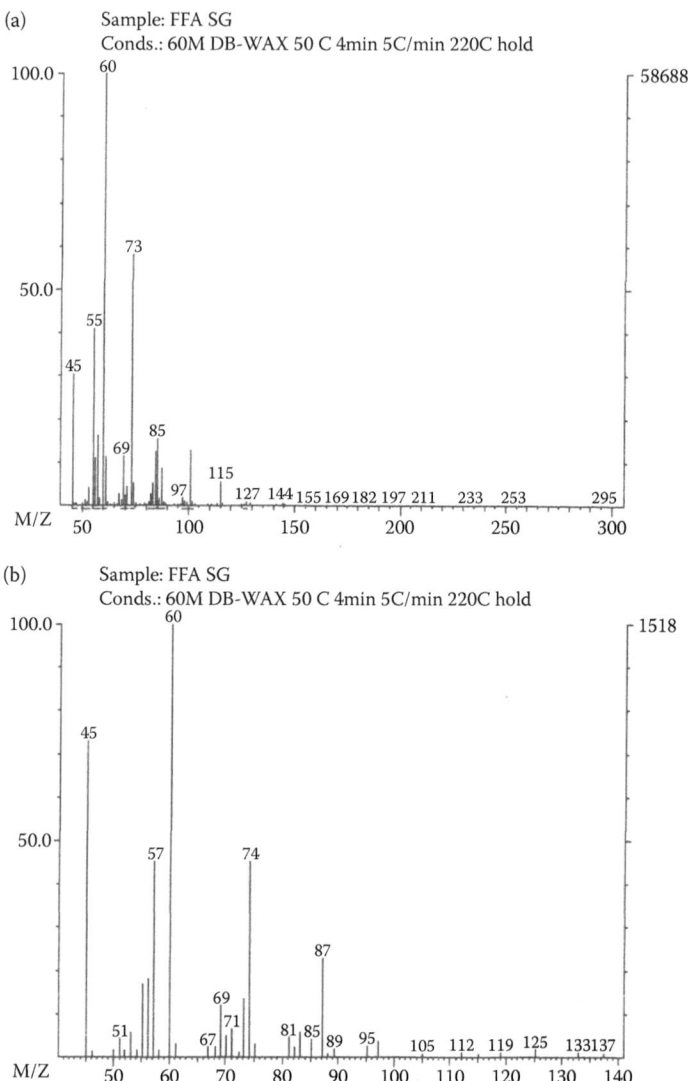

FIGURE 15.8 (a) Mass fragments of octanoic acid identified from the acidic fraction of steam distillate of MF of the tiger. (b) Mass fragments of isovaleric acid identified from acidic fraction of steam distillate of MF of the tiger.

the tiger aroma is in fact 2AP (Brahmachary 1996). Schieberle (1985) synthesized 2AP following an easy technique based on the Maillard reaction (MR) at a temperature of 170°C and our group could bring it down to 128°C–135°C (Poddar-Sarkar et al. 1992) and ultimately to 105°C but no less (Poddar-Sarkar et al. 1993, unpublished). The biosynthesis of 2AP in rice may be enzymatic by a metabolic process within the system and not by MR. Stable isotope labeling (C^{13} and C^{15}) shows that the nitrogen of 2AP is derived from proline but the carbon source of the acetyl group

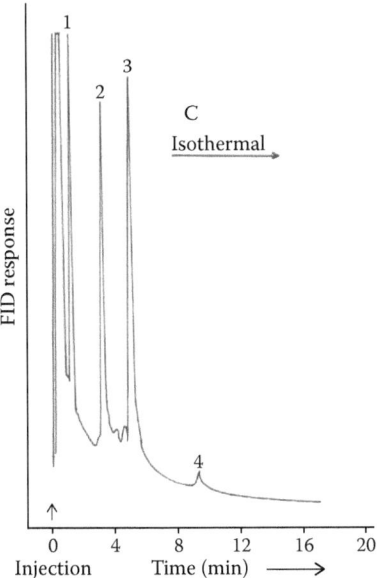

FIGURE 15.9 Gas chromatogram of amines present in MF of the cheetah. Peak numbers: 1 = ethylenediamine, 2 = putresceine, 3 = cadaverine, 4 = β-phenylethylamine. Column: 10% Carbowax 20M + 5% KOH (3m × 3mm) packed stainless steel metal column, program: column temp. 150°C (isothermal), C = Cheetah.

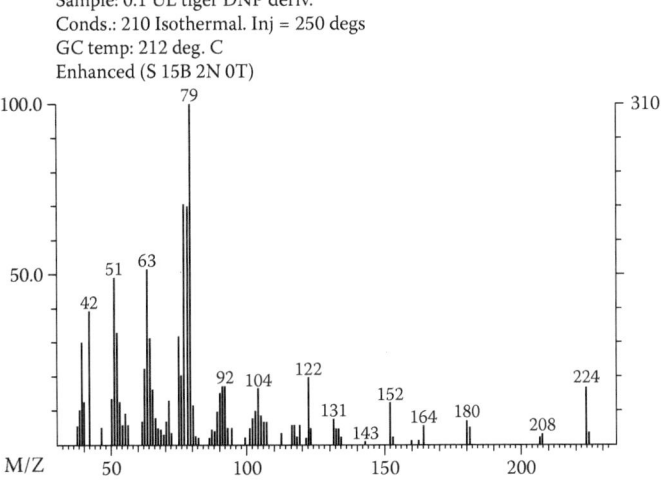

FIGURE 15.10 Mass fragments of acetaldehyde-2,4 dinitrophenyl-hydrazone derivatized from the neutral fraction of steam distillate MF of the tiger. After derivatization the sample was purified through thin-layer chromatography before subjected to gas chromatography.

is some other molecule and that the reaction occurs at a lower temperature than required for MR (Yoshihashi et al. 2002). 2AP was also detected in flowers of *Bassia latifolia* Roxb. = *Maduca indica* (local name Mahua) (Midya and Brahmachary 1997), in flowers of *Vallaris solanacea* (Basu et al. 2007), and is also a component of a fragrant pulse of Bengal (local name Sona Mung) (Brahmachary and Ghosh 2002). However, Burger et al. could not confirm the presence of 2AP in Bengal tiger MF. Apps reports no 2AP smell in African leopard (Brahmachary 2013, personal communication). We have detected 2AP in the MF of the Siberian tiger when the sample was available to us under an Indo-U.S. exchange program (Soso et al. 2012). The characteristic smell is absent in adult African lions and cubs and adult Asiatic lions and an almost adult cheetah (free-ranging adult African lions and cubs introduced into nature for a few hours every day at Kampi ya Simba, George Adamson's camp at Kora, Kenya, and a big cub in captivity at Lulimbi, Zaire, were the source of MF and urine in this context. The Asiatic adult lions in captivity in the heart of the forest in Gir, India, and an almost adult cheetah free-ranging in about 10,000 square meter in Namibia were observed).

Our findings on the presence of 2AP, the most uncommon compound among many candidates for pheromones present in both MF and urine of both the sexes of tiger and Indian leopard (but not in lion and cheetah), might project a new line of thought for understanding the two distinct phylogenetic clades of the cat family. The tiger and the leopard, rather than the lion, might be near neighbors because of

FIGURE 15.11   Tranquillized mangrove–marsh Bengal tiger of Sundarban being released back to nature after collection of sample for GCMS for the analysis of chemical compounds of body odor. (Photo courtesy of Subroto Pal Chowdhury.)

the metabolite 2AP. This aspect of metabolomics can indirectly shed light on the relevance of genomics to tiger pheromone. As already mentioned we have recently traced 2AP in the Siberian tiger (Soso et al. 2012) and also in the so-called marsh mangrove tiger of the Sunderbans, a race of the Bengal tiger (Poddar-Sarkar and Brahmachary 2012, unpublished). Since, as mentioned above, the Siberian/Amur race is significantly different from the Bengal tiger, 2AP in the tiger is probably of ancient origin. In that case the presence of 2AP in the marsh tiger, a recent adaptation in the Sundarban mangrove swamp is not surprising (Figure 15.11).

It would be of interest to investigate the presence of 2AP in the Chinese and Sumatran races. In our views, the origin of this significant aroma molecule may be helpful for tracing the tiger lineage.

15.10.3 Natural Fixative of MF

Ii is interesting to note that during or just after the Second World War, when the distinction between MF and urine was not clear, it was calculated that a tiger threw away (through urine/MF) daily an amount of lipid equivalent to one week's butter ration (20g) allowed to a British soldier in wartime (Hewer et al. 1948; Mathews 1969). According to our findings the lipid content of a tiger MF is about 2g/L of MF. For the leopard and cheetah the corresponding values are 1.15g/L and >3g/L, respectively. As was found long ago (Brahmachary and Choudhuri 1979, unpublished), on steam distillation the volatile molecules separated from the larger lipids disperse very rapidly (in minutes) whereas MF dried on a leaf bears a perceptible smell even after 10 days (Brahmachary 1964, unpublished). The lipid fraction of MF of the tiger is comprised of cholesterol ester, wax ester, triglyceride, FFAs, diglycerides, monoglycerides, free sterol, and phospholipids (Table 15.3). Nonvolatile lipid fraction contains saturated, monounsaturated, and polyunsaturated fatty acids of mostly 14, 16, 18, 20, and 22 carbon number. The percentage of saturated fatty acids in the wax lipid of MF is about 55%, which is more than the triglyceride and sterol ester. Such a composition might be the basis of fixative action (Poddar-Sarkar 1996) and can further be correlated with the wax coating of a leaf for sustenance of aroma.

The wildlifers in the Sunderbans are divided on the issue of MF smell surviving twice-daily tidal inundations; some hold the view that marking does not last at all, while others are of the opinion that tigers spray MF only on trees above the high-water mark (Montgomery 1995). The findings on the mangrove leaves immersed in estuarine water mentioned previously, are however, relevant in this context.

All of the above suggest the role of fixative lipids in the tiger. The lipid content of the cheetah is higher than that of the tiger (Table 15.3) but cheetah MF loses its attraction to other cheetahs after only 24 hours (Eaton 1974).

Proteins as fixatives of pheromones in rodents (Pelosi 1998), as enzymes for synthesis of pheromone precursor (Miyazaki et al. 2006), or directly as a means for chemical communication (Novotony 2003) are known. In the tiger and other big cats the protein content of MF is slight but of course various proteins are likely to occur in small concentrations. In the cat urine cauxin may be present in amounts less than 1g/L (Miyazaki et al. 2008) and it was detected in the urine of the Asiatic lion, Amur tiger, Persian leopard, clouded leopard, and jaguar (McLean et al. 2007). Burger et al. (2008) also identified cauxin from a Bengal tiger. In the big cats, lipids as fixative agents are probably more important than proteins.

15.11 REVIEW AND CONCLUSIONS: FIFTY YEARS OF PHEROMONE RESEARCH OF BIG CATS

The study of pheromones of big cats is undoubtedly one of the more difficult or less tractable propositions but as we have endeavored to show, since 1964, we have made considerable headway. The lacunae in the field have also been discussed. The question of pheromone-based individual recognition is a formidable one in any mammal and it is compounded by the practical difficulties implicit in investigations on big cats. But nonetheless, we feel we can answer some of the basic questions. It transpires from our attempts that the major source of pheromones in big cats is the MF and this is an ensemble of various chemical compounds.

As concluding remarks we may sum up the process of understanding MF chronologically. Around 1960 the concept of pheromones in big cats emerged apparently in the widely read popular accounts of Joy Adamson on Elsa the lioness and later on Pippa the cheetah. During the late 1960s Schaller (1967) brought it to the attention of scientists and hinted at the possibility of decoding the information encoded in this message. For a long time after Schaller described MF spray as a mixture of urine and anal gland secretion, this misconception continued to persist in the scientific literature, even as late as 2006 (Thapar 2006). We have described a mass of evidence (from the early 1980s to 2013) suggesting that MF has evolved to meet a purpose—to communicate with conspecific neighbors—for otherwise, this loss of energy (including, in particular, throwing away a large amount of lipids that are metabolically costly) would have been selected against through the evolutionary time scale. However, compared to the tigress and female serval all other female cats spray rarely, only during pro- or early estrous but all urinate significantly more during or before estrous. In addition, the related facts of different modalities of MF spray and urination in different big cat families like the tiger, lion, cheetah, and leopard were reported during the 1980s in several papers (Brahmachary and Dutta 1981, 1986, 1987). An analogy in the plant world can be drawn if we consider root exudates, the plant equivalent of excretory products like urine, MF, and feces of the big cats or other animals. Likewise certain leaf volatiles are the equivalent of animal body odor.

Around the 1980s and 1990s, in a number of papers Brahmachary, Dutta, and Poddar-Sarkar detected 30–40 compounds in the tiger, and similar types of compounds in the cheetah, leopard, and lion MF and an unusual molecule 2AP in the tiger and leopard only (Brahmachary et al. 1990; Poddar-Sarkar and Brahmachary 2004; Poddar-Sarkar et al. 1995; Brahmachary, 1996). Andersen and Vulpius (1999) identified several volatile compounds from zoo lions. Burger et al. (2008) identified more than 100 compounds in a zoo tiger MF, with the help of headspace SPME-GCMS, but they failed to find 2AP. We have meanwhile detected 2AP in the marsh-mangrove tiger of the Sunderbans, India (Poddar-Sarkar and Brahmachary 2011, unpublished) and also in the Siberian tiger in 2012 (Soso et al. 2012). Chemical analysis of MF of the cheetah and leopard have been reported as mentioned earlier. McLean's group highlighted the presence of cauxin protein in urine (McLean et al.

2007). Proteomics- and genomics-related work for tracing the phylogeny of big cats was carried out by several schools in recent years. (In Section 15.8 we discussed the genomics-related work as well as the metabolomics perspective.) Preliminary findings on the body odor of three tranquilized mangrove-marsh Bengal tigers of Sundarban were reported in 2013 (Poddar-Sarkar et al. 2013).

Most or all mammals rely on olfactory signals and this ability must have evolved early in the evolutionary history. Even an *E. coli* bacterium is attracted to certain molecules like amino acids in the medium that are associated with decaying matter indicating a food source.

More specifically, we may try to answer some of the questions raised by Tinbergen's remarks in Section 15.2. What exactly is the survival value of the instinct of spraying MF by big cats? This was the first query of Tinbergen. The answer in popular terms would be "waste not, want not." Such is nature's economy that nothing is wasted. Materials like scats, urine, or MF rejected by the body are not necessarily wasted (true, some of these substances may be utilized as food by other organisms ranging from bacteria but here we are concerned only with the benefit to the big cats themselves). MF, as we have seen, plays a vital role in attracting the opposite sex, thereby ensuring reproduction, a vital issue, and proclaiming a territory. In a sense this material ejected from the body may be considered as an extended phenotype and have a great survival value.

Likewise, the second and third questions of Tinbergen (namely, how has MF developed over time and how has it developed in the individual?) are also amenable to the evolutionary perspective. As we have mentioned earlier, Darwin suggested that "...if the more odoriferous males are the most successful in winning over females...." then natural (in this case sexual) selection would lead to the evolution of the trait (sending this odor signal). This would be equally valid for the females and their olfactory abilities. This might answer both the second and third questions.

The fourth question (What is the physiological causation?) is more difficult to address at present. We do not know exactly how the lipid is generated in the urine of MF and by which metabolic pathway, or how an unusual molecule like 2AP arises. But production of these metabolites and their excretion is not against the laws of physiology.

15.12 MANY UNSOLVED PROBLEMS

1. We have as yet no inkling of the natural chemicals that undoubtedly send signals to the VNO of the tiger. These must be relatively heavy molecules with little volatility. These might be proteins like aphrodisin, cauxin, or other, totally different molecules. That the VNO is very active in the big cats is well evident through frequent flehmen gestures and touching the nostril with the tongue, though in this respect the big cat tongue is far more ill-adapted than the bifurcated narrow tongue of snakes or *Varanus* (monitor lizard).

2. We have to cross another hurdle regarding the recognition of an individual-specific set of pheromones(s) or osmic signals. Despite the possibility of individual recognition based on an ensemble of a number of chemical

compounds that vary quantitatively in the different individuals, as explained earlier, a tricky question arises. The fading of the more volatile compounds convey information to the big cat, namely that a rival/intruder left the site some time ago and so there is no immediate concern/fear/stress; but can the animal perceive whether it is its own old mark or that of another? This question arises because (a) the degree of fading, or in other words, the intensity of disappearance of many molecules depends on several environmental and climatic factors, and (b) the lesser the number of different molecules in the pheromonal potpourri, the more difficult it is to distinguish "individuality."

3. Again, even though the tiger might distinguish a lion's smell (formerly the lion and the tiger coexisted in many parts of India) because of the absence of 2AP, can it possibly have any inkling that a leopard is different from a tiger? All the currently known pheromones are chemically the same in both species and so despite quantitative differences a leopard may be perceived as another tiger. Observational results cannot answer the question because both intruders will be repelled by the resident tiger. There should be species-specific ratios of chemical compounds or some other mechanism of which we have no insight at present. One may legitimately ask which of the many putative candidates for pheromones are actually functional as pheromones. Apps et al. (2013) have accepted this challenging task with the hundred-odd urinary compounds of the African wild dog. They point out that many compounds such as carboxylic acids are common in sympatric species and so unlikely to be species-specific pheromones and therefore unique compounds should claim priority in this context. They have gone far along this line but we feel that because of essential genetic/physiological/biochemical reasons every species may not have specific, unique compounds in their urine, MF, glandular secretions, and so forth (we have, however, previously discussed the possibility of bile salts in scat characteristic for each species). On the other hand, the quantitative proportions of different carboxylic acids distinguish the individuality of the animal, as we have already noted. Might not such ratios and proportions characterize species as well? In a later personal communication by Apps, certain difficulties are evident. The data on the lion, leopard, and cheetah are still scant and his group detected no aldehyde or ketone even in the leopard or lion. Furthermore, they also did not detect 2AP in the African leopard. We will discuss elsewhere the possible reasons for these results but at the moment, neither we nor they can state how the sympatric big cats might be distinguished at the pheromonal level.

Voznessenskaya et al. (1992) point out that rats find it increasingly difficult to distinguish among more than 5–6 individual odors in a mixup. Even if they can distinguish only 5–6 individual odors, the same faculty might help the animals to distinguish 5–6 sympatric species and that would suffice in most cases.

4. A tiger overmarks the marking of an intruder and if the latter does not decamp immediately, the first tiger may sooner or later overmark the overmarking. In the case of certain rodents there is some evidence that they can detect the difference between the top and bottom overmarking. There is

evidence that the animal marking on top enjoys an advantage. It has been shown that a third animal can distinguish the two (Ferkin and Pierce 2007, 2010). Even in a mixture of urine of 6–7 mongoose individuality was distinguishable (Jordan et al. 2011). Therefore, it is in the interest of the resident to overmark once again. How the rodent or mongoose or the tiger can perceive this is a well-nigh incredible phenomenon.

ACKNOWLEDGMENTS

We are indebted to George Schaller, George Adamson, P. Amte and their colleagues, and many zoo authorities, private reserves, and keepers who have been associated with the big cats for many years in India and Africa. We are grateful to the-then Prime Minister of India Mrs. Indira Gandhi for her kind help for sanctioning a special grant for tiger research from the Department of Environment, the Government of India at Indian Statistical Institute, Calcutta, and also to other funding agencies of the Government of India like the Council of Scientific and Industrial Research and University Grant Commission. We are grateful to Prof. D. Muller Schwarze and Prof. Francis Webster of State University of New York, Syracuse, who extended their kind help for running our sample in GCMS in 1991 for free fatty acids. We extend our sincere thanks to Prof. Barbara Sommerville of Cambridge University and Prof. J. Waterhouse of Anglia Polytechnic, Cambridge, for their help with sniff GC in 1992 and valuable suggestions and critical review from time to time. RLB thanks Sally Walker, Sarada Mallya, Sabita Rai, and Katyaani of Mysore. MPS is thankful to her cofield worker Prof. S. S. Sarkar (her geologist husband) and her photographer friend Dr. N. Das during many field sessions. We acknowledge the kind help from our many collaborators and many reviewers for giving a definite shape to our findings. Surendranath College, Calcutta, University of Calcutta are gratefully acknowledged for extending infrastructural facilities. We appreciate the continuous inspiration and criticism coming from our research scholars including our American guest student Simone Soso, who participated in many constructive and stimulating discussions. We thank Dilip Bhattacherya from whose collection we obtained the photograph of Khairi, the pet tigress of Simlipal, who initiated scientific research on tiger pheromones and other behavioral aspects (Figure 15.12). We express our sincere gratitude to the late Mr. Saroj Raj Choudhuri and Ms. Nihar Nalini-who tamed Khairi- S. K. Patnaik, Director at Nandan Kanan Biological Park, Dr. L. N. Acharjo, veterinarian, Nandan Kanan and Shib Mohanti, Jibonaloke (who raised Dora III), office staff of Nandan Kanan; our many animal lover friends, and S. B. Mondal, Principal Chief Conservator of Forest, Government of West for kindly giving us permission. We thank Dr. E. Ali, Dr. Asish Sen, and the late Dr. P. Bhattyacharya of the Indian Institute of Chemical Biology, Kolkata, and Prof. Sibdas Ray, Department of Chemistry, University of Calcutta (CU). MPS is grateful to the Hon'ble Vice Chancellor (CU), Pro-VC and Registrar, CU for giving her permission to participate in Indo-US, 2011 and Indian National Science Academy-Hungary Academy of Science Bilateral Exchange Programme, 2012. We are indebted to many contributors during our four-decades-long study—our two pet tiger cubs Dora II and Dora III and many tigers and tigresses of Nandan Kanan, and Bipasa, the tigress of South Khairabari, North Bengal—who are no longer with us.

FIGURE 15.12 'Khairi' with her human mother (Ms. Nihar Nalini) and Mr. Dilip Bhattacherya. (From the collection of Mrs. Susmita Ghosh, Salt Lake, Calcutta.)

REFERENCES

Adamson, G. 1986. *My Pride and Joy*. Collins, London.

Adamson, J. 1960. *Born Free*. Collins, London.

Albone, E. 1984. *Mammalian Semiochemicals*. John Wiley, Chichester, United Kingdom.

Andersen, K.F. and Vulpius, T. 1999. Urinary volatile constituents of the lion. *Chem. Senses* 24: 179–189.

Anderson, M.B. 1994. *Sexual Selection*. Princeton University Press, Princeton, NJ.

Apps, P., Mmualefe, L. and McNutt, J.W. 2013. A reverse engineering approach to identifying which compounds to bioassay for signaling activity in the scent marks of African wild dogs (*Lycanon pictus*). In *Chemical Signals in Vertebrates 12*, M.L. East and M. Dehnhard (eds.), Springer, New York.

Asa, C.S. 1993. Relative contributions of urine and anal-sac secretions in scent marks of large felids. *Amer. Zool.* 33: 167–172.

Bailey, T.N. 1993. *The African Leopard: A Study of the Ecology and Behavior of a Solitary Felid*. Columbia University, New York.

Baker, E.B. 1887. *Sport in Bengal*. Ledger, Smith & Co., London.

Baker, W., Deem, S., Hunt, A., Munson, L. and Johnson, S. 2002. Jaguar species survival plan. In *Guidelines for Captive Management of Jaguars*, Vol. 1/1, C. Law (ed.). Jaguar Species Survival Plan Management Group, Forth Worth, TX, pp. 9–13.

Bank, G.R., Buglass, A.J. and Waterhouse, J.S. 1992. Amines in the marking fluid and anal sac secretion of the tiger, *Panthera tigris*. *Z. Naturforsch C.* 47(7–8): 618–620.

Barja, I. and Miguel, F.J. 2010. Chemical communication in large carnivores: Urine-marking frequencies in captive tigers and lions. *Pol. J. Ecol.* 58(2): 397–400.

Barnett, S.A. 1981. *Modern Ethology: The Science of Animal Behavior*. Oxford University Press, New York.

Basu, S.K. Mitra, A. Pal, D.C. and Datta, J. (eds.). 2007. *Encyclopedia of Himalayan Medicinal Flora*, Vol. 3. Horticulture Development Foundation, Kolkata.

Berglund, H., Lindström, P. and Savic, I. 2006. Brain response to putative pheromones in lesbian women. *Proc. Natl. Acad. Sci. U. S. A.* 103(21): 8269–8274.

Bertram, B.C.R. 1975. Social system of lions. *Sci. Am.* 232: 54–65.

Bhattacharjee, S., Mallik, P.K., Shekhawat, R.S., Anoop, K.R. and Sankar, K. 2012. Tale of a travelling tiger. *Sanctuary Asia* XXXIII(3), June.

Bininda-Emonds, O.R.P. and Decker-Flum, D.M. 2001. The utility of chemical signals as phylogenetic characters; an example from the felidae. *Biol. J. Linn. Soc. Lond.* 72: 1–15.

Bland, K.P. 1979. Tomcat odour and other pheromones. *Vet. Sci. Commun.* 3: 125–136.

Börger, L. Dalziel, B.D. and Fryxell. J.M. 2008. Are there general mechanisms of animal home range behaviour? A review and prospects for future research. *Ecol. Lett.* 11: 637–650.

Brahmachari, G.K. and Brahmachary, R.L. 1980. Mimeo note on olfaction in the tiger, Calcutta, India.

Brahmachary, R.L. and Dutta, J. 1979. Phenylethylamine as a biochemical marker of tiger. *Z. Naturforsch. C.* 34: 632–633.

Brahmachary, R.L. and Dutta, J. 1981. On the pheromones of tigers: Experiments and theory. *Am. Nat.* 118: 561–567.

Brahmachary, R.L. and Dutta, J. 1984. Pheromones of leopards: Facts and theory. *Tiger Paper* II(3): 18–23.

Brahmachary, R.L. 1986. Ecology and chemistry of mammalian pheromones. *Endeavour* 10(2): 65–68.

Brahmachary, R.L. and Dutta, J. 1987. Chemical communication in the tiger and leopard. In *Tigers of the World: The Biology, Biopolitics, Management, and Conservation of an Endangered Species*, R.L. Tilson and U.S. Seal (eds.). Noyes Publications, Park Ridge, NJ.

Brahmachary, R.L. 1990. Final report (Mimeo) on "Chemical and ecological aspects of pheromones of the tiger," submitted to Department of Environment, Govt. of India.

Brahmachary, R.L., Poddar-Sarkar M. and Dutta, J. 1990. The aroma of tiger...and rice. *Nature* 344: 26.

Brahmachary, R.L., Dutta, J. and Poddar-Sarkar, M. 1991. The marking fluid of the tiger. *Mammalia* 55(1): 150–152.

Brahmachary, R.L., Poddar-Sarkar, M. and Dutta, J. 1992. Chemical signal in the tiger. In *Chemical Signals in Vertebrates 6*, D. Muller-Schware and R.L. Doty (eds.). Plenum Press, New York, pp. 477–479.

Brahmachary, R.L., Poddar-Sarkar, M. and Dutta, J. 1993. Evolution of chemical signals. In *Human Population Genetics*, P.P. Majumdar (ed.). Plenum, New York.

Brahmachary, R.L. 1996. The expanding world of 2-acetyl-1-pyrroline. *Curr. Sci.* 71: 257–258.

Brahmachary, R.L., Singh, M. and Rajput, M. 1999. Scent marking in the Asiatic Lion. *Curr. Sci.* 78(2): 480–481.

Brahmachary, R.L. and Singh, M. 2000. Behavioral and chemical aspects of scent marking in the Asiatic lion. *Curr. Sci.* 78(6): 680–682.

Brahmachary, R.L. and Ghosh, M. 2002. Vaginal pheromone and other compounds in mung bean aroma. *J. Sci. Industr. Res.* 61: 625–629.

Brahmachary, R.L. 2011. *Animal Behaviour.* Naturism, Kolkata.

Brennan, P.A. 2008. MHC associated chemosignals and individual identity. In *Chemical Signals in Vertebrates 11*, J.L. Hurst, R.J. Beynon, S.C. Roberts and T.D. Wyatt (eds.). Springer, New York, pp. 131–140.

Broekhuis, F. 2007. Habitat selection patterns of cheetahs *Acinonyx jubatus* in the Serengeti, Tanzania. M.Sc. research project, The Royal Veterinary College, University of London.

Burger, B.V., Viviers, M.Z., Bekker, P.I., le Roux, M., Fish, N., Fourie, W.B and Weibchen, G. 2008. Chemical characterization of territorial marking fluid of male Bengal tiger, *Panthera tigris. J. Chem. Ecol.* 34: 659–671.

Burt, W.H. 1943. Territoriality and home range concepts as applied to mammals. *J. Mammal.* 24: 346–352.

Buttery, R.G., Ling, L.C. and Juliano, B.O. 1982. 2-acetyl-1-pyrroline: An important aroma component of cooking rice. *Chem. Indus.* 4: 958–959.

Buttery, R.G., Ling, L.C., Juliano, B.O. and Turnbaugh, J.C. 1983. Cooked rice aroma and 2-acetyl-1-pyrroline. *J. Agric. Food Chem.* 31: 823–826.

Caro, T.M. and Collins, D.A. 1987. Ecological characteristics of territories of male cheetahs (*Acinonyx jubatus*). *J. Zool. Lond.* 211: 89–105.

Caro, T.M. 1994. Cheetahs of the Serengeti plains: Group living in an asocial species. Ph.D. thesis, University of Chicago.

Carrington-Turner, J.E. 1959. *Man-Eaters and Memories*. Jaico, Bombay, India.

Chandawat, R.S. 2002. Great cats of India, TV episode (as consultant), Animal Planet.

Chandawat, R.S. 2006. Tigers of Dry Forest: Do they have enough space to survive? In *Tiger: The Ultimate Guide*, V. Thaper (ed). Oxford University Press, New Delhi.

Chellam, R. 1993. Ecology of Asiatic lion. Ph.D thesis, Saurashtra University, Rajkot.

Chong, A.Y.Y. 2009. Genetic variation in the MHC of the Collared peccary: A potential model for the effects of captive breeding on the MHC. *Orbit: The University of Sydney Undergraduate Research Journal* 1(1): 98–116.

Choudhury, S.R. 1999. *Khairi, the Beloved Tigress* (posthumously published). Natarajan Publisher, Dehra Dun, India.

Claesson, A. and Silverstein, R.M. 1977. Chemical methodology in the study of mammalian communications. In *Chemical Signals in Vertebrates*, D. Müller-Schware and M.M. Mozell (eds,). Plenum Press, New York, pp. 71–93.

Cracraft, J., Feinstein, J., Vaughan, J. and Helm-Bychowski, K. 1998. Sorting out tigers (*Panthera tigris*): Mitochondrial sequences, nuclear inserts, systematics and conservation genetics. *Anim. Conserv.* 1: 139–150.

Corbett, J. 1944. *Man-Eaters of Kumaon*. Oxford University Press, Bombay.

Corbett, J. 1954. *The Temple Tiger.* (Reprinted by Oxford University Press, India, 1988.)

Darwin, C. 1871. *The Descent of Man.* D. Appleton & Co., New York.

Datta, S.P. and Harris, H. 1951. A convenient apparatus for paper chromatography; results of a survey of the urinary amino-acid patterns of some animals. *J. Physiol. (Lond.).* 114: 39–41.

Davis, B.W., Li., G. and Murphy, W.J. 2010. Supermatrix and species tree methods resolve phylogenetic relationships within the big cats, *Panthera* (Carnivora: Felidae). *Mol. Phylogenet. Evol.* 56: 64–76.

Dawkins, R. 1983. *The Extended Phenotype.* Oxford University Press, Oxford.

De, P.K. 1996. Sex hormonal regulation of 20.5 and 24 kDa major male-specific proteins in Syrian hamster submandibular gland. *J. Steroid. Biochem. Mol. Biol.* 58(2): 183–187.

Driscoll, C.A., McDonald, D.W. and O'Brien, S.J. 2009. From wild animals to domestic pets, an evolutionary view of domestication. *PNAS* 106: 9971–9978.

Dunbar Brander, A.A. 1923. *Wild Animals in Central India.* E. Arnold, London.

Dunbar, J. 1989. *Tigers, Durbars and Kings: Fanny Eden's Indian Journals, 1837–1838.* John Murray Publishers, London.

Eaton, R. 1974. *The Cheetah: The Biology, Ecology and Behavior of Endangered Species.* Van Nostrand and Reinhold, New York.

Eden, F., 1837–1838. Tigers, Durbars and Kings: Fanny Eden's Indian Journals.

Estes, R.D. 1972. The role of vomeronasal organ in mammalian reproduction. *Mammalia* 36: 315–341.

Ewer, R.F. 1968. *Ethology of Mammals.* Elek Science, London.

Ewer, R.F. 1973. *Carnivores.* Weidenfeld and Nicholson, London.

Ewert, J.P. and Traud, R. 1979. Releasing stimuli for antipredator behavior in common toad, *Bufo bufo. Behaviour* 68: 170–180.

Fayer, J. 1875. *The Royal Tiger of Bengal: His Life and Death.* J & A Churchill, London.

Ferkin, M.H. and Pierce, A.A. 2007. Perspectives on over-marking: Is it good to be on top? *J. Ethol.* 25: 107–116.

Ferkin, M.H., Ferkin, A., Ferkin, B.D. and Vlautin, C.T. 2010. Olfactory experience affects response of meadow voles to the opposite-sex scent donor of mixed-sex overmarks. *Ethology* 116: 821–831.

Forsyth, J. 1872. *The Highlands of Central India.* Chapman and Hall, London.

Gorman, M.L., Nedwell, D.B. and Smith, R.M. 1974. An analysis of contents of anal scent pockets of *Herpestes. J. Zool. Lond.* 172: 384.

Gorman, M.L. 1976. A mechanism for individual recognition by odour in *Herpestes. Anim. Behav.* 24: 141.

Graham, C. 1991. Menstrual synchrony: An update and review. *Hum. Nat.* 2(4): 293–311.

Hashimoto, Y., Eguchi, Y. and Arakawa, J.A. 1963. Historical observation of the anal sac and its glands in a tiger. *Jap. J. Vet. Sci.* 25: 29–32.

Hediger, H. 1977. Ethology and study of animals in captivity. In *Grzimek's Encyclopedia of Ethology.* (ed). B. Grzimek, Van Nostrand and Reinhold, London.

Heinsohn, R. 1997. Group territoriality in two populations of African lions. *Anim. Behav.* 53: 1143–1147.

Hemker, T.P., Lindzey, F.G. and Ackerman, B.B. 1984. Population characteristics and movement patterns of cougars in Southern Utah. *J. Wildl. Mgmt.* 48: 1275–84.

Hemmer, H. 1974. Untersuchungen zur Stammesgeschichte der Pantherkatzen 839 (Pantherinae). Teil III. Zur Artgeschichte des Löwen, Panthera leo (Linnaeus 1758). *Veröffentlichungen der Zoologischen Staatssammlung* 17: 167–280.

Hendriks, W.H., Moughan, P.J., Tarttelin, M.F., and Woolhouse, A.D. 1995. Felinine: A urinary amino acid of Felidae. *Comp. Biochem. Physiol. B.* 112: 581–588.

Hewer, T.F., Matthews, L.H. and Malkin, T. 1948. Lipuria in tigers. *Proc. Zool. Soc. Lond.* 118: 924–928.

Horonecker, M. 1970. An analysis of mountain lion predation upon mule deer and elk in the Idaho primitive area. *Wildl. Monogr.* 21: 1–39.

Houser, A., Somers, M.J. and Boast, L.K. 2009. Home range use of free-ranging cheetah on farm and conservation land in Botswana. *S. Afr. J. Wildl. Res.* 39(1): 11–22.

Howard, H.E. 1992. *Territory in Bird Life.* John Murray Publishers, London.

Hurst, J.L., Payne, C.E., Nevison, C.M., Marie, A.D., Humphire, R.E., Robertson, D.H., Cavaggioni, A. and Beynon, R.J. 2001. Individual recognition in mice mediated by major urinary proteins. *Nature* 414: 631–634.

Ilany, G. 1981. The leopard of the Judean desert. *Israel Land. Nat.* 6: 59–71.

Inglis, J. 1892. *Tent Life in Tigerland, with which Is Incorporated Sport and Work in the Nepaul Frontier.* Sampson Low, Marston, and Co., London.

Jackman, B., Scott, J. and Scott, A. 1982. *The Marsh Lions: The Story of an African Pride.* Elm Tree Books, London.

Jackson, R. and Ahlborn, G. 1989. Snow leopards (*Panthera uncia*), in Nepal-home-range and movements. *Nat. Geogr. Res.* 5(2): 161–175.

Jeffery, K.J.M. and Bangham, C.R.M. 2000. Do infectious diseases drive MHC diversity? *Microbes Infect.* 2: 1335–1341.

Johnson, W.E., Dratch, P.A., Martenson, J.S. and O'Brien, S.J. 1996. Resolution of recent radiations within three evolutionary lineages of Felidae using mitochondrial restriction fragment length polymorphism variation. *J. Mammal. Evol.* 3: 97–120.

Jordan, N.R., Mwanguhya, F., Kyabulima, S., Rüedi, P., Hodge, S.J. and Cant, M.A. 2011. Scent marking in wild banded mongooses: Intrasexual overmarking in females. *Anim. Behav.* 81: 51–60.

Kelley, J., Walter, R. and Trowsdale, J. 2005. Comparative genomics of major histocompatibility complexes. *Immunogenetics* 56: 683–695.

Locke, A. 1954. *Tigers of Trengganu.* Museum Press, London.

Li, G., Janecka, J.E. and Murphy. W.J. 2011. Accelerated evolution of CES7, a gene encoding a novel major urinary protein in the cat family. *Mol. Biol. Evol.* 28: 911–920.

Lyddeker, R. (ed.). 1894. *The Royal Natural History,* Frederick Warne, London.

Manning, A. and Stamp Dawkins, M. 1992. *An Introduction to Animal Behaviour.* Cambridge University Press, Cambridge.

Maser, C., Mate, B.R., Franklin, J.F. and Dyrness, C.T. 1981. Natural history of Oregon coast mammals. USDA Forest Service, Forest Service, Pacific Northwest Forest and Range Experiment Station, General Technical Report PNW 133.

Mathews, L.H. 1969. *The Life of Mammals.* Weidenfeld and Nicholson, London.

Maynard Smith, J. and Harper, D.G.C. 1988. The evolution of aggression: Can selection generate variability? *Phil. Trans. R. Soc. Lond. B.* 319: 557–570.

McBride, C. 1977. *The White Lions of Timbavati.* Paddington Press, New York.

McBride, C. 1981. *Operation White Lion.* St Martin's Press, New York.

McClintock, M. 1971. Menstrual synchrony and suppression. *Nature* 229: 244–245.

McClintock, M.K. and Stern, K. 1998. Regulation of ovulation by human pheromones. *Nature* 392(6672): 177–179.

McDougal, C.W. 1977. *The Face of the Tiger.* Rivington and Krutsch, London.

McLean, L., Hurst, J.L., Gaskell, J.C., Lewis, J.C.M. and Beynon, R.J. 2007. Characterization of cauxin in the urine of domestic and big cats. *J. Chem. Ecol.* 33(10): 1997–2009.

Michael, R.P., Bonsall, R.W. and Warner, P. 1974. Human vaginal secretions volatile fatty acid content. *Science* 186: 1217–1219.

Midya, S. and Brahmachary, R.L. 1996. The aroma of Bassia flower. *Curr. Sci.* 71(6): 430.

Midya, S. and Brahmachary, R.L. 1997. A method of producing basmati rice aroma from *Bassia latifolia* flowers. *IRRN* 22: 21.

Mills, M.G.L. 1989. Comparative behavior and ecology of hyenas. In *Carnivore Behavior, Ecology and Evolution,* J.L. Gittelman (ed.). Cornell University Press, Ithaca, NY.

Mills, M.G.L. 1990. The lion (*Panthera leo*) and cheetah (*Acinonyx jubatus*) in Kruger National Park, South Africa. *Felid* 4(1): 13.

Miquelle, D., Arzhanova, T., Solkin, V. (eds.). 1996. A recovery plan for conservation of the Far Eastern leopard: Results of an international conference held in Vladivostok, Russia. Vladivostok, Russia. USAID Russian Far Eastern EPT Project.

Mitchell, M.S. and Powell, R.A. 2004. A mechanistic home range model for optimal use of spatially distributed resources. *Ecol. Model.* 177: 209–232.

Miyazaki, M., Yamashita, T., Suzuki, Y., Saito,Y., Soeta, S., Tiara, H. and Suzuki, A. 2006. Major urinary protein of the domestic cat regulates the production of felinine, a putative pheromone precursor. *Chem. Biol.* 13: 1071–1079.

Miyazaki, M., Yamashita, T., Taira, H. and Suzuki, A. 2008. The biological function of cauxin, a major urinary protein of the domestic cat. In *Chemical Signals in Vertebrates 11,* J.L. Hurst, R.J. Beynon, S.C. Roberts and C. Wyatt (eds.). Springer, New York.

Montgomery, S. 1995. *Spell of the Tiger.* Houghton Mifflin, Boston.

Mooney, N. 1984. Tasmanian tiger sighting. *Aust. J. Nat. Hist.* 21: 177–180.

Mosser, A. and Packer, C. 2009. Group territoriality and the benefits of sociality in the African lion, *Panthera leo. Anim. Behav.* 78: 359–370.

Mountfort, G. 1981. *Saving the Tiger.* Viking Press, New York.

Muntifering, J.R., Dickman, A.J., Perlow, M.L., Hruska, T., Ryan, P.G., Marker, L.L. and Jeo, R.N. 2006. Managing the matrix for large carnivores: A novel approach and perspective from cheetah (*Acinonyx jubatus*) habitat suitability modelling. *Anim. Conserv.* 9: 103–112.

Nordlund, D.A. and Lewis, W.J. 1976. Terminology of chemical releasing stimuli in intraspecific and interspecific interactions. *J. Chem. Ecol.* 2: 211–220.

Novotny, M.V. 2003. Pheromones, binding proteins and receptor responses in rodents. *Biochem. Soc. Trans.* 31: 117–122.

Nowak, R.M. 2010. *Walker's Mammals of the World,* Volume II. Johns Hopkins University Press, Baltimore, MD.

O'Brien, S.J., Wildt, D.F. and Bush, M. 1986. The cheetah in genetic peril. *Sci. Am.* 254: 84–92.

O'Brien, S.J. and Yuhki, N. 1999. Comparative genome organization of the major histocompatibility complex: Lessons from the Felidae. *Immunol. Rev.* 167: 133–144.

Packer, C., Scheel, D. and Pusey, A.E. 1990. Why lions form groups: Food is not enough. *Am. Nat.* 136: 1–79.

Packer, C., Kosmala, M., Cooley, H.S., Brink, H., Pintea, L., Garshelis, D., Purchase, G., Strauss, M., Swanson, A., Balme, G., Hunter, L. and Nowell, K. 2009. Sport hunting, predator control and conservation of large carnivores. *PLoS ONE* 4(6): e5941.

Palagi, E. and Dapporto, L. 2006. Beyond odor discrimination: Demonstrating individual recognition by scent in *Lemur catta. Chem. Senses* 31: 437–443.

Panwar, H.S. 1987. Project Tiger: The reserves, the tigers and their future. In *Tigers of the World: The Biology, Biopolitics, Management, and Conservation of an Endangered Species*, R.L. Tilson and U.S. Seal (eds.). Noyes Publications, Park Ridge, NJ.

Pelosi, P. 1998. Odorant-binding proteins: Structural aspects. *Ann. N. Y. Acad. Sci.* 855: 281–293.

Penn, D.J. 2002. The scent of genetic compatibility: Sexual selection and the major histocompatibility complex. *Ethology* 108: 1–21.

Penn, D.J. 2006. Chemical communication. In *Chemical Ecology: from Gene to Ecosystem*, M. Dicke and W. Takken (eds.). Springer, Dordrecht.

Poddar-Sarkar, M., Brahmachary, R.L. and Dutta, J. 1991. Short chain free fatty acid as a putative pheromone in the marking fluid of tiger. *J. Indian Chem. Soc.* 68: 255–256.

Poddar-Sarkar, M., Sarkar, N.D., Dutta, J. and Brahmachary, R.L. 1992. A method to synthesize the aroma of certain varieties of fragrant Indian rice. *IRRN* 17(5): 9.

Poddar-Sarkar, M. 1995. Mammalian semiochemicals: Chemical and behavioural aspects with special reference to tiger. PhD thesis, University of Calcutta.

Poddar-Sarkar, M. 1996. The fixative lipid of tiger pheromone. *J. Lipid Mediator Cell Signal.* 15: 89–101.

Poddar-Sarkar, M. and Brahmachary, R.L. 1997. Putative semiochemicals in the African cheetah. *J. Lipid Mediat. Cell Signal.* 15: 285–287.

Poddar-Sarkar, M. and Brahmachary, R.L. 1999. Can free fatty acids in the tiger pheromone act as an individual finger print? *Curr. Sci.* 76(2): 141–142.

Poddar-Sarkar, M. and Brahmachary, R.L. 2004. Putative chemical signals of leopard. *Anim. Biol.* 54(3): 255–259.

Poddar-Sarkar, M., Chakroborty, A., Bhar, R. and Brahmachary, R.L. 2008. Putative pheromones of lion mane and its ultrastructure. In *Chemical Signals in Vertebrates 11*, J.L. Hurst, R.J. Beynon, S.C. Roberts and T.D. Wyatt (eds.). Springer, New York, pp. 61–67.

Poddar-Sarkar, M., Ray, S., Chowdhury, P., Samanta, G. and Brahmachary, R.L. 2013. On the body odour of wild-caught mangrove marsh Bengal tiger of Sundarban. In *Chemical Signals in Vertebrates 12*, M.L. East and M. Dehnhard (eds.). Springer, New York.

Pokorny, I., Sharma, R., Goyal, S.P., Mishra, S. and Tiedemann, R. 2011. MHC class I and MHC class II DRB gene variability in wild and captive Bengal tigers (*Panthera tigris tigris*). *Immunogenetics* 63(2): 121.

Powell, R.A. 2000. Animal home ranges and territories and home range estimators. In *Research Technologies in Animal Ecology: Controversies and Consequences*, L. Boitani and T.K. Fuller (eds.). Columbia University Press, New York, pp. 65–110.

Pusey, A. and Packer, C. 1983. Once and future kings. *Nat. Hist.* August: 55–62.

Rabinowitz, A.R. and Nottingham, B.G. 1986. Ecology and behavior of the jaguar (*Panthera onca*) in Belize, Central America. *J. Zool. Lond.* 210: 149–159.

Rasmussen, L.E.L. 1999. Evolution of chemical signals in the Asian elephant, *Elephas maximus*: Behavioural and ecological influences. *J. Biosci.* 24(2): 241–251.

Rishi, V. 2012. The role of scent marking in the breeding behavior of tiger and other big cats. *The Indian Forester* 138(10): 910–914.

Röck, F., Hadeler, K.-P., Rammensee, H.-G. and Overath, P. 2007. Quantitative analysis of mouse urine volatiles: In search of MHC-dependent differences. *PLoS ONE* 2(5): e429.

Russell, K.P. 1978. Mountain lion. In *Big Game of North America*, T.L. Schmidt and D.L. Gilbert (eds.). Stackpole Books, Harrisburg, PA, p. 494.

Rutherfurd, S.M., Zhang, F., Harding, D.R., Woolhouse, A.D. and Hendriks, W.H. 2004. Use of capillary (zone) electrophoresis for determining felinine and its application to investigate the stability of felinine. *Amino Acids* 27(1): 49–55.

Sankar, K., Qamar, Q., Nigam, P., Malik, P.K., Sinha, P.R., Mehrotra, R.N., Gopal, R., Bhattacharjee, S., Mondal, K. and Gupta, S. 2010. Monitoring of reintroduced tigers in Sariska Tiger Reserve, Western India: Preliminary findings on home range, prey selection and food habits. *Trop. Conserv. Sci.* 3: 301–318.

Sankhala, K.S. 1978. *Tiger.* Collins, London.

Sankhala, K.S. 1993. *Tiger.* Lustre, New Delhi.

Schaller, G. 1967. *The Deer and the Tiger.* University of Chicago Press, Chicago, IL.

Schaller, G. 1972. *The Serengeti Lion.* University of Chicago Press, Chicago, IL.

Schieberle, P. 1995. Quantitation of important roast smelling odourants in popcorn by stable-isotope dilution assays and model studies on flavor formation during popping. *J. Agric. Food Chem.* 50: 2442–2448.

Schilling, A. 1970. Organe de Jacobson du Lemurien Malagache. Memoire du Museum National d'Histoire Naturelle, Serie A. *Zoologie.* 61: 206–280.

Singer, A.G., Macrides, F., Clancy, A.N. and Agosta, W.C. 1986. Purification and analysis of a proteinaceous aphrodisiac pheromone from hamster vaginal discharge. *J. Biol. Chem.* 261: 13323–13326.

Singer, A.G., Beauchamp, G.K. and Yamazaki, K. 1997. Volatile signals of the major histocompatibility complex in male mouse urine. *Proc. Natl. Acad. Sci. U. S. A.* 94: 2210–2214.

Singh, A. 1981. *Tara, a Tigress.* Quartet, London.

Sitton, L.W. and Wallen, S. 1976. California mountain lion study. Calif. Department of Fish and Game, Sacramento, CA.

Smith, J.L.D., McDougal, C.W. and Sunquist, M.R. 1987. Female land tenure system in tigers. In *Tigers of the World: The Biology, Biopolitics, Management, and Conservation of an Endangered Species*, R.L. Tilson and U.S. Seal (eds.). Noyes Publications, Park Ridge, NJ.

Smith, J.L.D., McDougal, C. and Miquelle, D. 1989. Scent marking in free-ranging tigers, *Panthera tigris. Anim. Behav.* 37(1): 1–10.

Smith, R.M. 1978. Movement patterns and feeding behavior of the leopard in the Rhodes Matopos National Park, Rhodesia. *Carnivore* 1(3): 58–69.

Soso, S.B., Poddar-Sarkar M., Koziel, J. and Brahmachary, R.L. 2012. More on "Aroma of tiger and…rice," (abstract), 36th Annual Conference of Ethological Society of India and National Symposium on "Live organisms and their expression in the environment," November 26–28, 2012, Kolkata, India, p. 54.

Spencer, S.R., Cameron, G.N. and Swihart, R.K. 1990. Operationally defining home range: Temporal dependence exhibited by Hispid cotton rats. *Ecology* 71(5): 1817–1822.

Stander, P.E., Haden, P.J., Kaqece, I.I. and Ghau, I.I. 1997. The ecology of asociality in Namibian leopards. *J. Zool. Lond.* 242: 343–364.

Strauss, B.S. 1960. *An Outline of Chemical Genetics.* W.B. Saunders, Philadelphia, PA.

Thapar, V. 1986. *Tiger, the Portrait of a Predator.* Collins, London.

Thapar, V. 2006. *Tiger: The Ultimate Guide.* Oxford University Press, New Delhi.

Thavathiru, E., Jana, N.R. and De, P.K. 1999. Abundant secretory lipocalins displaying male and lactation-specific expression in adult hamster submandibular gland, cDNA cloning and sex hormone-regulated repression. *Eur. J. Biochem.* 266: 467–76.

Tinbergen, N. 1951. *The Study of Instinct*. Oxford University Press, New York.

Turnbull-Kemp, P. 1967. *The Leopard*. Howard Timmins, London.

Van den Hurk, R. 2007. Intraspecific chemical communication in vertebrates. Pheromone Information Centre, Brugakker, Zeist, Netherlands.

Van Der Merwe, N.J. 1953. *The Jackal. Fauna and Flora of Transvaal*, Prov. Admin., Publication No. 4. South Africa.

VanderWaal, K.L., Mosser, A. and Packer, C. 2009. Optimal group size, dispersal decisions and postdispersal relationships in female African lions. *Anim. Behav.* 77: 949–954.

Verberne, G. 1970. Beobachtungen und Versuche über das flehmen Katzenartiger Raubtiere. *Z. f. Tierpsychol.* 27: 807–827.

Vincent, F., Lobel, D., Brown, K., Spinelli, S. Grote, P., Breer, H., Cambillau, C. and Tegoni, M. 2001. Crystal structure of aphrodiscin, a sex pheromone from female hamster. *J. Mol. Biol.* 305: 459–469.

von Uexküll, J. 1934. A stroll through the world of animals and men. In *Instinctive Behavior: The Development of a Modern Concept*, C.H. Schiller (ed.). International Universities Press, New York.

Voznessenskaya, V.V., Parfyonova, V.M. and Zinkevich, E.P. 1992. Individual odor types. In *Chemical Signals in Vertebrates 6*, R.L. Doty and D. Müller-Schwarze (eds.). Plenum Press, New York, pp. 503–508.

Wasser, S.K., Smith, H., Madden, L., Marks, N. and Vynne, C. 2009. Scent-matching dogs determine number of unique individuals from scat. *J. Wildl. Manage.* 73: 1233–1240.

Westall, R.G. 1953. The amino acids and other ampholytes of urine. 2. The isolation of a new sulphur-containing amino acid from cat urine. *Biochem. J.* 55: 244–248.

Wyatt, T.D. 2003. *Pheromones and Animal Behaviour: Communication by Smell and Taste*. Cambridge University Press, Cambridge.

Yamazaki, K., Beauchamp, G.K., Curran, M., Bard, J. and Boyse, E.A. 2000 Parent-progeny recognition as a function of MHC odor type identity. *Proc. Natl. Acad. Sci. U. S.A.* 97: 10500–10502.

Yoshihashi, T., Hung, N.T.T. and Inatomi, H. 2002. Precursors of 2AP, a potent flavor compound of an aromatic rice variety. *J. Agri. Food Chem.* 50: 2001–2004.

16 Cattle Pheromones

Govindaraju Archunan,
Swamynathan Rajanarayanan, and
Kandasamy Karthikeyan

CONTENTS

16.1 Abstract 461
16.2 Introduction 462
    16.2.1 The Reproductive Cycle and Behavior in Cattle 463
    16.2.2 Male Flehmen Behavior and Its Role in Cattle Estrus Detection ... 463
        16.2.2.1 Cow Flehmen Behavior 465
        16.2.2.2 Buffalo Flehmen Behavior 466
        16.2.2.3 Flehmen Behavior in Cattle Other than Cow
               and Buffalo 467
        16.2.2.4 Chemoreceptor System in Cattle 467
16.3 Biostimulation 469
16.4 Cow Pheromones: Sources and Identification of Chemosignals 470
16.5 Buffalo Pheromones 472
16.6 Boar Pheromones 473
16.7 Small Ruminant Pheromones: Goat and Sheep 474
16.8 Horse Pheromones 475
16.9 Odorant Binding Proteins in Cattle 476
16.10 Application of Pheromones in Cattle 477
16.11 Future Perspectives 479
Acknowledgments 479
References 479

16.1 ABSTRACT

Chemical signals play a major role in mammalian reproduction and behavior. While the existence of sexual pheromones and their significance in reproductive behavior are very well known in laboratory mammals, there is little information about the species-specific chemosignal in cattle. However, in the last decade there have been some important and relevant reports for cattle pheromones. This chapter provides a comprehensive account on cattle pheromones and their role in reproduction and social behaviors. Sexual attraction, mother-young interactions, estrus indication, estrus induction, puberty acceleration, reducing the postpartum anestrus, hormonal stimulation, and enhancing the penis erection are some of the classical events influenced by pheromones. During the reproductive cycle of cattle, the male exhibits a specific behavior of flehmen that is used as an indicator to detect estrus. Furthermore, the sensory system involved in the perception of pheromones is mediated through the vomeronasal organ (VNO) and main olfactory system (MOS). The sex attractant volatile compounds have been identified in the urine of the cow, buffalo, and horse. The urinary volatile compounds identified in female buffalo are capable of enhancing the sperm quantity, which can be considered as a significant finding for cattle pheromones. The pig appeasing pheromone (PAP) is characterized from the mother skin used to reduce the fighting behavior among the piglets whereas the boar salivary pheromone is found to induce puberty attainment and boar mate. Similarly, PAP is also present in the horse, cat, dog, and human and plays a significant role. In the goat and sheep, the volatiles identified in the ram and buck wool fleece to stimulate LH and ovulation induction. There is plenty of scope for the application of cattle pheromones in the animal reproduction and management.

16.2 INTRODUCTION

The pheromone signals are known to have a potential role in animal reproduction and management (Archunan 2009; Buck 2000; Dominic 1991; Rekwot et al. 2001; Tirindelli et al. 2009). The sources of chemosignals are urine, feces, vaginal secretions, saliva, and specialized scent glands including the odor produced from hair and wool (Albone 1984; Aron 1979; Patra et al. 2012; Van den Hurk 2007). Communication of the timing of the physiological event of ovulation and coordination of sexual behavior are important for successful fertilization (Schams et al. 1977; Ziegler et al. 1993); the success rate of artificial insemination in cattle mainly depends on the time of estrus it is inseminated. If the female fails to conceive when it is inseminated in the nonestrus period, it is a great economic loss estimated around $300 million to the dairy industry in the United States and in India the approximate loss would be several crores. The detection of estrus and diagnosing early pregnancy are the major problems in farm animals, particularly in buffalo. The other problems like irregular or prolonging of the estrous cycle, anestrus, fighting behavior among young ones, mother-young bond, unmotivated males, and poor farm animal management including stress are considered as major issues in farm animals (Rekwot et al. 2001) that need to be addressed. Moreover, pheromones have not been exploited for the purpose of enhancing livestock production.

The animal releases volatile odors into the surrounding atmosphere, most of which are waste products of metabolism in which emission of some compounds closely related to reproductive activities are termed as chemical signals (Hradecky 1975). The estrus-related odors are present only during the preestrus and estrus stages. These chemical signals have been reported to be volatile and nonvolatile molecules that are perceived through the main or accessory olfactory system (Brennan and Keverne 1997; Tirindelli et al. 1998). In mammals, structurally and anatomically, the olfactory systems are classified into two types: the MOS and the accessory olfactory system (AOS), specialized for the detection and transmission of pheromonal information (Halpern and Marinez-Marcos 2003; Mucignat-Caretta et al. 2012).

In the last two decades, there has been a considerable increase in the knowledge of the chemistry of pheromones in cattle (Rajanarayanan and Archunan 2011; Rameshkumar et al. 2000; Sankar and Archunan 2004; Sankar et al. 2007), pig

(Signoret 1970), horse (Kimura 2001), goat (Delgadillo et al. 2006), and sheep (Gelez and Fabre-Nys 2006). These findings suggest that cattle pheromones may act together for influencing precopulatory behavior and successful reproduction. It further indicates that cattle pheromones may be a single compound or a mixture of compounds and that each of the major fractions was faithfully involved in conveying specific signals related to reproductive and social behaviors. The pheromones have not been exploited as much as they can for farmer utility although they have a tremendous role in reproduction. The present review provides insights about the nature of the compounds, behavioral characterization, and practical utility of cattle pheromones that are available (Table 16.1).

16.2.1 THE REPRODUCTIVE CYCLE AND BEHAVIOR IN CATTLE

Cattle reproduction is vital for successful management of farm animals. In the reproductive cycle estrus has a central role in cattle reproduction. The period of the estrous cycle of cattle is calculated from one estrus (heat or phase of sexual receptivity) to the next estrus. The average estrous cycle periods are 21 days in bovine and buffalo and 17 days in ewes (Hall et al. 1959; Tilbrook and Cameron 1989). The estrous cycle is subdivided into four phases based on the dominant hormone or ovarian structure and reproductive behavior during each phase. The stages of the bovine cycle are preestrus, estrous, metestrus, and diestrus. The days are calculated as follows: day 0 is considered to be estrus, days 1–5 are metestrus, days 6–17 are diestrus, and days 18–20 are preestrus (James and Ireland 1980).

Generally, the estrous cycle of cattle is divided into two phases: (1) the folicular phase and (2) the luteal phase. During the follicular phase, under the influence of FSH the follicle starts to develop and undergoes ovulation; when the LH surge occurs the estrogen level is also high. The luteal phase begins when the corpus luteum (CL) is formed about 5 to 6 days after the cow is in heat and ends when the CL regresses from about 17 to 19 days of the cycle. Progesterone levels are high during this phase of the cycle and estrogen levels are low (De Jarnette et al. 2009). The season is also reported to influence the length of the estrous cycle in bovine (Sankar and Archunan 2012).

The female expressed estrus symptoms include vaginal swelling, mucus discharge, bellowing, frequent urination, and mounting on other animals during heat (estrus) period (Karthikeyan 2011; Rajanarayanan 2005; Rajanarayanan and Archunan 2004; Rameshkumar 2000). In order to attain optimum reproductive performance in dairy farms it is essential that herdsmen be aware of signs of estrus and their reliability, and then only they can accurately select females for artificial insemination, the most crucial aspect of successful conception being the timing of the insemination. Maximum conception rates are obtained when the insemination is done from 8 to 12 hours after the onset of estrus (also known as standing heat). Hence, detection of estrus is very much essential for achieving success in artificial insemination service.

16.2.2 MALE FLEHMEN BEHAVIOR AND ITS ROLE IN CATTLE ESTRUS DETECTION

The role and importance of chemical signals in reproductive behavior has been well studied in several species of mammals (Archunan 2009; Balakrishnan and

TABLE 16.1

Cattle Pheromones: Identification, Source, and Functions

Animal	Compounds	Sources	Functions	Reference
Cow	6-methyl-1-heptanol, 2-methyl-7-hydroxy-3-4, and heptene	Vaginal secretions	Indicating estrus	Preti 1984
	Acetaldehyde	Milk and blood	Attraction	Klemm et al. 1987
	Trimethylamine, acetic acid, phenol 4-propyl, pentanoic acid, and propionic acid	Saliva	Attraction	Sankar et al. 2007
	Acetic acid, 1-iodo undecane, and propionic acid	Feces	Attraction	Sankar and Archunan 2008
	Trimethylamine, acetic acid, phenol, propionic acid, and 3-hexanol	Vaginal fluid	Attraction and mounting	Sankar and Archunan 2011
	1-iodo undecane	Urine	Attraction	Archunan and Rameshkumar 2012
Buffalo	4-methyl phenol, 9-octa decenoic acid, and 1-chlorooctane	Urine	Attraction and mounting	Rajanarayanan and Archunan 2011
	9-octadecenoic acid	Vaginal fluid	Mounting	Karthikeyan and Archunan 2013
Sheep (ram)	Fatty acid I and II	Wool or hairs	LH stimulation	Over et al. 1990
	C16 and diols	Fleece	LH stimulation and ovulation	Signoret 1991
	Amniotic fluid	Placenta	Accelerate the maternal response	Poindron et al. 2010
Goat	4-ethyl octanoic acid, octanoic acid, and 2,6-di-t-buthyl-4-methyl phenol	Fleece	LH stimulation	Sugiyama et al. 1981, Iwata et al. 2003
Pig	16-androstene steroid	Male Saliva	Puberty attainment in female pig	Booth 1984
	Fatty acids (hexadecanoic acid, cis-9-octa decenic acid, 9,12-ctyladecanoic acid, dodecanoic acid, tetradecanoic acid, and decanoic acid)	Mother skin	Reducing stress in piglets (PAP)	Pageat 2001, McGlone and Anderson 2001
	5 α androsterone, 3 α androsterone	Saliva	Mating stance	Booth 1984
Horse	p-Cresol and m-Cresol	Urine	Ovulation marker	Mozuraitis et al. 2012
	p-Cresol	Urine	Enhance the erection	Buda et al. 2012

Alexander 1985; Dominic 1991; Rasmussen et al. 1996; Rekwot et al. 2001). One of the striking examples is that the male identifies the estrus through pheromonal signals released from the female. The most likely sources of such signals are urine and cervical mucus secretion (Archunan 2009). For example, in cattle, bulls can detect pheromone odors and differentiate between estrus and nonestrus urine (Sambraus and Waring 1975).

The ultimate criterion for effective pheromonal communication of the sexual status of a female is a clear message emitted from her body fluids (Rivard and Klemm 1989) during the estrus period by which the male is being attracted, followed by mating (Hradecky et al. 1983; Klemm et al. 1987). When a male is brought to a female in estrus the male generally sniffs the genital region of the female and exhibits several behaviors. One such significant behavior is the retraction of the upper lip, wrinkling of the nose, and baring gums. This is always accompanied by deep breathing and may also include the elevation and extension of the head. This specific behavior was observed by German scientist and postulated the term "flehmen" (Hradecky et al. 1983). The main function of flehmen has been hypothesized to be the discrimination of chemical signals in females (Bland and Jubilan 1987; Rajanarayanan and Archunan 2004; Sankar and Archunan 2004). Pheromonal compounds present in the body fluids during estrus periods provoke the manifestation of premating behaviors in the males including flehmen. During flehmen, the attractants from the female enter the olfactory organ of the bull, which in turn relays the signals to the hypothalamic areas of the brain (Vandenbergh 1999). The repeated flehmen behavior show is the most definite and integral part of the premating scenario. This unique behavior response has been reported in the bovine (Dehnhard et al. 1991; Hradecky et al. 1983; Rajanarayanan and Archunan 2004; Sankar and Archunan 2004), sheep (Blissitt et al. 1994), and horse (Ma and Klemm 1997). Hence, based on the available reports flehmen behavior can be considered as one of the best indicators for the female in estrus.

16.2.2.1 Cow Flehmen Behavior

Bulls exhibited flehmen behavior following the perception of a cow's body fluids like urine (Archunan and Rameshkumar 2012; Rameshkumar 2000), saliva (Sankar et al. 2007), feces (Sankar and Archunan 2008), and milk (Sankar and Archunan 2004) of the estrus stage. Our study demonstrated that the bull exhibited flehmen behavior when exposed to dummy cows sprayed with estrus urine ($p < 0.001$) as compared to that of nonestrus urine, and suggested that estrus urine was capable of inducing the flehmen behavior in the bull (Archunan and Rameshkumar 2012). The chemosensory responses and premating behavior of adult males to the female vaginal fluid have been well demonstrated in *B. taurus* (Klemm et al. 1987; Paleologou 1979; Sankar and Archunan 2011).

Under field conditions, visual stimuli probably play a role in stimulating male sexual behavior, but *in vitro* tests with cervicovaginal mucus clearly show that there is an olfactory component, independent of vision, in the stimulation of male reproductive behaviors, such as sniffing, licking, and repeated flehmen in response to vaginal chemical cues (Hradecky et al. 1983; Sankar and Archunan 2004, 2011). The flehmen behavior is the prime response against the olfactory signals of an estrus female leading to other behavioral responses. Based on this observation, it is strongly believed that there is a correlation between flehmen and other behavioral responses to olfactory signals of estrus females.

16.2.2.2 Buffalo Flehmen Behavior

The female reproductive system is generally complicated due to variation in the endocrine response, and this is more so in buffalo reproduction. The main problem in buffalo reproduction is the occurrence of silent ovulation without any behavioral signs of estrus (Dobson and Kamonpatana 1986), thereby making the detection of heat in buffaloes more difficult than in cows (Danell et al. 1984).

The problem of estrus detection in the buffalo is overcome by using the herd bulls that monitor olfactory and gustatory cues routinely (Reinhardt 1983; Williamson et al. 1972). Bulls receive the estrus-specific chemical signal from females and then exhibit flehmen behavior. This is preceded by fast tongue strokes over the rostral and medial part of the palate that have separate innervations and yield strokes that actually trigger the flehmen behavior (Jacobs et al. 1980). During flehmen, the attractants from the female enter the olfactory organ of the bull which, in turn, relay the signals to the hypothalamic areas of the brain (Vandenbergh 1999).

Rajanarayanan and Archunan (2004) reported that flehmen behavior increased from day 4 (preestrus period), peaked at day 0 (estrus), and decreased in late estrus. Further, the bulls exhibit a significant repeated flehmen behavior when exposed to estrus females. It is also interesting to note that even though the diestrus phase elicited the flehmen behavior, it is significantly lower ($p < 0.001$) when compared to estrus. From this report it was concluded that flehmen behavior is controlled by certain sex pheromones. The presence of such pheromones, however, may be negligible and may have a connection with some other events of the estrous cycle, such as interovulatory follicle growth (Hradecky et al. 1983). The flehmen behavior during the diestrus stage, nevertheless, had either singled or was nonresponsive. It is well known that when bulls identify the specific pheromonal compound(s) they continue to exhibit repeated flehmen and whenever such specific compound(s) are not

FIGURE 16.1 Group of male buffaloes showing the flehmen response toward estrus urine.

identifiable, the bulls will stop with a single flehmen (Estes 1972; Hradecky et al. 1983). We carried out a detailed study on flehmen behavior with regard to urinary compounds and obtained an interesting result. The flehmen behavior of the bull was greatly influenced by the estrus-specific compound as compared to that of diestrus compound (Karthikeyan and Archunan 2013; Rajanarayanan and Archunan 2011). The finding clearly indicates that estrus-specific compound is closely associated with the expression of flehmen behavior. Figure 16.1 shows the exhibition of flehmen by male buffaloes as soon as inhaling the estrus urine.

16.2.2.3 Flehmen Behavior in Cattle Other than Cow and Buffalo

Flehmen behavior is also reported in several ungulates other than the cow and buffalo. Bland and Jubilan (1987) reported ram flehmen after a ram sniffed urine voided by the female. Flehmen in response to urine was exhibited least often in the presence of the estrus ewe due to the low occurrence of marking behavior by the female at this time. Flehmen also occurred after the ram investigated the vulva of the ewe, most frequently on the day before estrus. These observations support the hypothesis that the occurrence of flehmen by the ram is due to an olfactory mechanism for confirming the reproductive state of the ewe. The pheromone detected by flehmen appears to be produced in the vagina and carried in the urine.

The flehmen response is found to appear in the male goat with regard to the female sexual interest (Ladewig et al. 1980). The horse exhibited flehmen behavior by drawing back their lips in a manner that makes them appear to be grimacing or smirking. In the horse, a stallion showed a significant response of flehmen behavior toward an estrus mare more than to a nonestrus mare (Marinier et al. 1988). Furthermore, the occurrence of sniffing and flehmen was used to determine the discriminating ability of the stallion and it was found that stallions are capable of discriminating the sex of a horse by the feces source (Stahlbaum and Houpt 1989). In the case of the pig, boars exhibit flehmen sometimes simultaneously toward sows; however, this is the only ungulate that did not show flehmen normally (Dagg and Taub 1970).

16.2.2.4 Chemoreceptor System in Cattle

It has been believed for a long time that the two chemosensory systems—the MOS and the vomeronasal system (VNS)—are responsible for the perception of odorants in mammals. The MOS is considered to be responsible for recognizing the conventional volatile odorant molecules, whereas the VNS is thought to be tuned for sensing pheromones. Recent studies have demonstrated that both chemosensory systems, together with additional olfactory organs, are involved in pheromone detection (Halpern 1987; Mucignat-Caretta et al. 2012; Tirindelli et al. 2009).

Jacobson originally described the vomeronasal organ based mainly on domesticated animals (cat, cow, dog, goat, horse, pig, and sheep); he also described the organ in other mammals such as the tiger, camel, buffalo, deer, and seal (Doving and Trotier 1998). The presence of this organ has been confirmed in most mammals including marsupials (Wohrmann-Repenning 1984). VNO is situated in the vicinity of the second cheek tooth. The mean length of the VNO, measured from the *Papilla incisiva* to the caudal end of the vomeronasal cartilage (VNC), was 4 mm for minks, 15 mm for cats, 50 mm for dogs and pigs, 150 mm for cows, and 200 mm for horses.

The VNC characteristics did not appear to be affected by age in any of the species studied (Salazar et al. 1995). It was observed that the VNO from its tip, where incisive openings were located backward a pear shape cartilaginous capsule encloses the organ, but this capsule is incomplete (Abbasi 2007). Figure 16.2 shows the location of VNO in the buffalo. Animals demonstrate different behavior patterns that are adaptations to investigating odorant sources. Some of these behavior patterns are certainly related to the investigation of the odorants, while others have to do with the entry of odorants into the VNO (Doving and Trotier 1998).

Experimental study showed that when the VNO in guinea pigs was impaired the males failed to mount. Females with impaired organs did not show lordosis, lost interest in their partner, and seldom became pregnant (Gerall 1963). Peripheral dedifferentiation of the VNS produces severe sexual behavior deficits in both male and female hamsters (Powers and Winans 1975; Winans and Powers 1977). In the goat, the posterior part of the VNO contains sensory epithelium that facilitates the nonvolatile molecule from the oral cavity to the VNO for the flehmen response (Ladewig et al. 1980). The formation of mother-young bond between ewes and lambs is reported to be mediated through olfactory cues and the VNO is primarily involved for the neonatal recognition in sheep (Booth and Katz 2000). However, another study was performed on both the olfactory systems to test the pheromone odor in enhancing LH secretion in ewes. The VNO was ablated in ewes and then exposed male fleece; it was found that there is no difference in the LH secretion in ewes by exposing to the male odor. By contrast, destruction of the main olfactory epithelium by zinc sulfate irrigation greatly reduced the LH secretion in ewes while exposing the male fleece (Gelez et al. 2004); this indicates that VNO is not necessary for the perception of male odor in sheep. Further experimental evidence showed that ablation of VNO in the sow did not alter the attraction and standing posture when exposed to boar saliva; however, the role of the MOS in this aspect was not tested (Dorries et al. 1991). As far as the cow and buffalo are concerned, the importance of both the

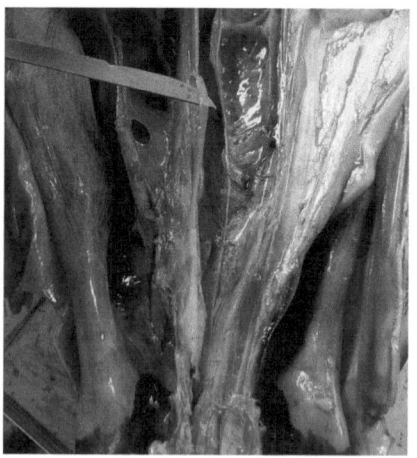

FIGURE 16.2   **(See color insert.)** Location of the vomeronasal organ in the buffalo.

main and vomeronasal systems has not been tested so far in the context of chemical communication. Flehmen behavior is believed to be involved in the transport of fluid-borne chemical stimuli, such as sex pheromones, from the oral cavity to the VNO. During flehmen behavior, the intermittent nostril licking apparently delivers the stimulus material to the VNOs via the nasal route, possibly compensating for reduced oral access. The VNS is reported to be involved in several pheromonal effects with regard to puberty acceleration, pregnancy block, induction of estrus, and mating behavior in rodents (Gelez et al. 2004). However, recent reports strongly indicate that the MOE was involved primarily in the attraction from a distance, while the VNO played a major role in close proximity (precopulatory behavior), indicating that the olfactory–vomeronasal systems play a synergistic role in the detection of estrus and the mating process (Achiraman et al. 2010). Even though the above study was done in the mouse, a similar synergistic role of the olfactory–vomeronasal systems would be possible in other mammals, including ungulates.

16.3 BIOSTIMULATION

Pheromones in the urine, feces, or from cutaneous glands can be perceived through the olfactory system to elicit both behavioral and endocrine responses in conspecifics, and biostimulation (priming pheromones) can exert profound effects on reproductive activity via the hypothalamic system (Dulac and Wagner 2006; Rekwot et al. 2001). The role of pheromones has been shown in several reproductive events such as sexual maturity, induction of ovulation, reduction of postpartum anestrus, and coitus in many mammalian species including rodents, wild animals, feral populations, swine, sheep, goat, and cattle (Burns and Spitzer 1992).

The exposure to boars induced puberty at an earlier age than gilts reared without being exposed to a boar and reduced the postpartum period in lactating sows indicates the potential role of primer pheromone (Brooks and Cole 1970; Kirkwood et al. 1981). The "ram effect" (primer pheromone) has been reported to accelerate the onset of estrus activity and promote varying degrees of estrus synchronization (Schinckel 1954). It is important to note that the ewes of most breeds are anestrus for some portion of the year and exposure of ram to anestrus ewes reduces the length of the anestrus period leading to the return of estrus activity (Rekwot et al. 2001).

Biostimulation is an inexpensive and suitable method for extensive management systems and it is an excellent example of the potential underlying "control systems technologies aimed at controlling reproductive performance" (Martin 1995; Signoret 1970). The role of primer pheromones in cattle reproduction is not as clearly defined as that in other species such as sheep, goats, and swine, possibly due to nutritional and other environmental stresses (Roberson et al. 1991). It is important to note that the presence of a teaser bull did not hasten puberty or alter the size of ovaries in a group of prepubertal heifers (Berardinelli et al. 1978; Macmillan et al. 1979; Roberson et al. 1987).

One of the main problems in cattle reproduction is anestrus, defined as absence of the estrous cycle, which has to be decreased in order to obtain a more sustainable production in the livestock industries (Fiol and Ungerfield 2012). Biostimulation (male or bull effect) can be defined as the stimulus provoked by the presence of males, which induces estrus and ovulation through genital stimulation, pheromones, or other external cues (Chenoweth 1983). The stimulus provoked by the introduction of the males can act through different pathways, including olfactory, visual, and auditory signals (Ungerfield 2007). In cattle, males' excretory products and cervical mucus from estrus females enhance ovarian function, both in postpartum cows (Berardinelli and Joshi 2005; Wright et al. 1992) and prepubertal heifers (Izard and Vandenbergh 1982).

Berardinelli and Joshi (2005) evaluated resumption of cyclic activity in postpartum, anovular, primiparous cows exposed to bulls or the excretory products of bulls. The cows were allowed in the pens along with a male for 12 hours daily during 70 days of experiments. The authors found that anestrus postpartum length was greatly reduced when the female was exposed to males or the excretory products of males, suggesting that the biostimulatory role of bulls appears to be mediated by pheromone present in their excretory product. The nature of the compound present in the excretory product is not yet identified (Berardinelli and Tauck 2007; Fernandez et al. 1993). Burns and Spitzer (1992) observed a similar reduction in postpartum interval to first estrus in cows exposed to bulls or to androgenized females. The effects of male exposure on cyclic activity have been evaluated mainly in *Bos taurus taurus* and other possible effects of biostimulation in Zebu (*Bos taurus indicus*), both in the postpartum cow (Rekwot et al. 2001; Soto Belloso et al. 1997), and prepubertal heifer (Rekwot et al. 2001). The biostimulation was also found effective to reduce the anestrus period in female buffaloes, *Bubalus bubalis* (Gokuldas et al. 2010; Ingawale and Dhobe 2004).

Improving reproductive efficiency in beef and dairy cattle should be one of the main objectives to obtain a more sustainable production. In this context, sociosexual stimulus, like biostimulation, represents low-cost and hormone-free alternatives when used alone or in conjunction with other techniques to increase reproductive results.

16.4 COW PHEROMONES: SOURCES AND IDENTIFICATION OF CHEMOSIGNALS

In ungulates, bulls routinely investigate the urine of anogenital areas of females presumably to determine their reproductive state. The involvement of chemosignals in the cow's reproductive process is well documented; the female produces a specific odor during estrus in urine that is perceived by the bull (Archunan and Rameshkumar 2012; Denhard and Claus 1996; Rameshkumar et al. 2000). The dispersion of estrus-specific compounds in the cow's body fluids has been demonstrated previously in swabs from the fluids (vaginal secretion, urine, milk, and blood) (Kiddy and Mitchell 1984; Rivard and Klemm 1989). The presence of estrus-specific volatiles was confirmed in milk, but the study did not find any compounds that were qualitatively different between stages but found there were significant quantitative differences in 15 compounds in milk (Weidong et al. 1997).

Klemm et al. (1987) found that blood acetaldehyde levels decreased rapidly just before, or at the onset of estrus, and suggested that estrus and ovulation could potentially be predicted by monitoring the levels of acetaldehyde in milk, saliva, sweat, or breath. Acetaldehyde was also found to be estrus-specific in bovine vaginal secretions (Weidong et al. 1997); however, it has been tested in a bull bioassay previously (Presicce et al. 1993) and was not behaviorally active when tested as a singular component.

Several studies have focused on vaginal secretions found during estrus. Preti (1984) patented a method for detecting bovine estrus based on the quantification of methyl heptanol in vaginal secretion. Another finding showed that the concentration of free fatty acids in estrus vaginal discharge increased gradually before estrus and decreased rapidly thereafter (Hradecky 1986). By contrast, the concentration of free fatty acids in urine, but not in vaginal discharge, was affected by the luminal concentration. Klemm et al. (1987) found nine estrus-specific compounds in urine that were tested positively in a bull behavioral assay. These compounds included two ketones, four amines, one alcohol, one diol, and one ether. Sankar and Archunan (2004) reported that other body fluids such as vaginal fluid, saliva, feces, and milk collected from estrus period contained chemical cues that attract the opposite sex.

Bulls can discriminate estrus from nonestrus urine and estrus urine has been shown to elicit sexual behavior in the cow. Rameshkumar et al. (2000) identified two estrus-specific compounds (i.e., 1-iodo undecane and di-n-propyl phthalate), however, the behavioral assay later clearly indicated that the 1-iodo undecane is capable of stimulating the bull response for coitus and hence was considered as a biochemical marker for bovine (Archunan and Rameshkumar 2012).

Sankar and Archunan (2004) studied the behavioral assay among the proestrus, estrus, and diestrus samples; the test clearly showed that the estrus vaginal fluid was found to be more variable ($p < 0.001$) than the saliva, feces, and milk. The vaginal secretions induced the maximum response in bulls showing flehmen behavior, thereby confirming that the vaginal secretions from the estrus cow contained pheromone(s), which along with urinary pheromone(s) result in attracting the bull.

Furthermore, our laboratory study reported that 11 different volatiles are identified in cow saliva from three different reproductive phases (Sankar et al. 2007). Among the identified compounds, the following compounds, trimethyl amine, acetic acid, phenol 4-propyl, pentonic acid, and propionic acid were specific to estrus. The behavioral assay has confirmed that the compound from a salivary source (i.e, trimethylamine) may be involved in attracting the males. Interestingly, the chemical profiles of estrus feces differed significantly from other phases by specific substance such as acetic acid, propionic acid, and 1-iodo undecane (Sankar and Archunan 2008); the behavioral assay also confirmed that these three compounds elicit a series of reproductive behaviors in the bull. It is evident that LH pulse frequency was greatly enhanced in the heifer by exposure to vaginal mucus and urine collected from estrus cows (Nordeus et al. 2012); however, the chemicals responsible for influencing LH alteration are not known. The above investigation on cow pheromone and behavioral characterization have convincingly showed that urine, feces, vaginal secretion, saliva, and milk appear to be prominent sources for cow pheromone production for estrus indication and influence several reproductive behaviors including hormonal stimulation.

Another interesting aspect is to detect estrus by interspecific communication. The ability of dogs (Hawk et al. 1984), rats (Denhard and Claus 1988), and mice (Rameshkumar et al. 2008; Sankar and Archunan 2005) to detect estrus-specific odor in cow urine is reported. Trained dogs can distinguish the cow luteal phase (Hawk et al. 1984; Kiddy and Mitchell 1984). In addition, a detailed experimental study was carried out using the mouse as an experimental model to evaluate which of the body fluids of the estrus cow has more attracting ability towards mice (Sankar and Archunan 2004). The test clearly showed that the estrus vaginal fluid was found to be the most attractive source followed by saliva, feces, and milk. The findings suggest that the specific odors present in the estrus phase not only attract the conspecific but also attract between species.

16.5 BUFFALO PHEROMONES

The occurrence of silent ovulation without any behavioral symptoms of estrus is the major problem in buffalo reproduction (Danell et al. 1984; Dobson and Kamonpatana 1986). Unlike in the cow and other ungulates, visual signs of estrus are not prominent in the buffalo, which make it difficult to detect heat effectively. Hence, there is a need for a reliable method to detect estrus in the buffalo. Recent reports show that the buffalo exhibits a reproductive behavior and releases chemosignals from few sources that can be exploited for estrus detection (Archunan 2009). Rajanarayanan and Archunan (2004) demonstrated that the average number of all flehmen behavior (2.03 ± 0.66) and repeated flehmen (1.05 ± 0.64) behavior in the bull upon exposure to an estrus heifer were significantly higher than those of diestrus periods of all (0.69 ± 0.25) and repeated flehmen (0.11 ± 0.10).

The nature of the volatiles present in the urine is confirmed and proved to be a pheromonal signal (Rajanarayanan and Archunan 2011). Fourteen different volatile compounds from all the stages of the estrous cycle were identified in urine samples, which included phenol, ketone, alkane, alcohol, amide, acid, and aldehyde. Amongst the identified compounds, a few, such as 2-octanone, 2-methyl-N-phenyl-2 propenamide, decanoic acid, N,N-bis (2 hydroxy ethyl) dodecanamide, tetradecanoic acid, and hexadecanoic acid, were commonly found throughout the cycle (preestrus, estrus, postestrus, and diestrus). However, the volatile compounds 1-chlorooctane, 4-methyl phenol, and 9-octadecenoic acid occurred only during estrus. Behavioral assay further revealed that bulls were attracted and exhibited repeated flehmen behavior toward 4-methyl phenol, whereas bulls exhibited penis erection and mounting response by exposure to 9-octadecenoic acid. By contrast, the other compound 1-chlorooctane did not show any such sexual behavior in bulls. It is interesting to note that these compounds are absent during postestrus and diestrus, which indicate that these compounds are characteristic to the estrus period. The reports convincingly conclude that 4-methyl phenol acts as a sex attractant compound and 9-octadecenoic acid acts as mounting response. It is remarkable to note that the urinary sex pheromones have been shown to enhance the sperm production after exposing to the nasal region; this has been awarded a patent (Archunan and Rajanarayanan 2010).

The nature of the volatile compounds produced during estrus in buffalo has been predicted. The existence of compounds phenol, 3-propyl phenol, and 9-octadecenal in preestrus urine suggests that these three volatiles may be considered as the preindicators for estrus. It is also interesting to note that the volatile phenol present in preestrus probably becomes 4-methyl phenol in estrus urine by the addition of a methyl group. Similarly, 9-octadecenoic acid, which appears in estrus, is probably derived from the preestrus compound of 9-octadecenal. Presumably, a change occurs in the formation of chemical compounds from the preestrus to estrus phase.

In addition to sex pheromone characterization in urine, other body fluids such as feces, vaginal mucus, and saliva have evinced great interest in view of identification of the volatile compounds and their behavioral assay, which may be considered as another important source for sex pheromone production. A specific volatile compound, 9-octadecenoic acid, has been identified as estrus-specific in vaginal mucus (Karthikeyan and Archunan 2013); it is to be remembered that the volatile compound 9-octadecenoic acid is already identified in estrus urine of the buffalo (Rajanarayanan and Archunan 2011). In feces, two volatile compounds, 4-methyl phenol and trans-verbenol, have been identified and confirmed as estrus indicator, suggesting that feces may be considered as a source of estrus indicator in the buffalo (Karthikeyan et al. 2013). The 4-methyl phenol (P-cresol), a unique volatile compound, can be considered as a common compound of estrus-specific since it is present in urine (Rajanarayanan and Archunan 2011), feces (Karthikeyan et al. 2013), and saliva (Karthikeyan 2011). Hence, this compound might be produced under the influence of estrus hormones.

16.6 BOAR PHEROMONES

As far as the boar pheromone is concerned, three types of pheromones have been recognized (1) boar mate pheromone, (2) puberty accelerating pheromone, and (3) appeasing pheromone. Chemical communication in the pig is notably mediated by saliva. The boar mate pheromone produced from pig saliva against the estrus sow is remarkably involved in the organization of the sequence of sexual behavior to attract the estrus females and facilitate the display of a receptive posture (Signoret 1970). The steroid pheromones such as 5α-androsterone and 3α-androsterone induce the appropriate standing response for mating and these odors are essential for a prolonged act of coitus in pigs. An aerosol boar mate is now marketed to aid pig artificial insemination practice by inducing the immobilization reflex in the estrus female (Booth 1984). The puberty accelerating pheromone is also released from saliva in the male pig, which is reported to induce early puberty in young female piglets (Pageat and Teyssier 1998). The boar pheromones have been shown to accelerate puberty in gilts by about 30 days to synchronize estrus and to reduce the postpartum period in lactating sows (Brooks and Cole 1970). Furthermore, those gilts reached early puberty by the presence of a boar odor, have higher ovulation rates, regular estrous cycles, and improved reproductive potential than the controls (Izard 1983).

Appeasing pheromone is another important chemical communication involved between mother and young pups of pigs. Generally, after weaning, the piglets will fight when group-housed, which leads to reduce greatly the feeding among piglets. The fighting behavior is a major concern among the farmers because the pigs do not gain weight due to underfeeding. Odors isolated from the skin of milking sows have been shown to reduce agonistic behavior in piglets (Pageat and Teyssier 1998). The putative maternal pheromone is composed of six fatty acids in different proportions (Pageat 2001): hexadecanoic acid, cis-9-octa decenic acid, 9,12-octyladecanoic acid, dodecanoic acid, tetradecanoic acid, and decanoic acid. The commercial synthetic analog had similar effects when tested in industrial husbandry (McGlone and Anderson 2002). This pheromone can be considered as an outstanding discovery in the mammalian pheromones.

16.7 SMALL RUMINANT PHEROMONES: GOAT AND SHEEP

The small ruminants such as the goat and sheep are economically important farm animals worldwide for their milk, meat, and wool. The response of the female goat to the male is weak but can be induced by the introduction of teaser bucks that had previously been exposed to long days (Delgadillo et al. 2006; Rivas-Munoz et al. 2007). The odor of the male and its sexual behavior plays a primary role in inducing ovulation, while vocalizations appear to facilitate the display of the does' estrus (Delgadillo et al. 2006). Additionally, the sebaceous gland has been indicated as sources of primer pheromone production in goats (Iwata et al. 2000; Sugiyama et al. 1981). The primer pheromones produced by the male have been known to stimulate the reproductive neuroendocrine system of anestrus animals (Chemineau 1987; Knight and Lynch 1980). Exposure of a male into a group of anestrus females leads to the activation of sexual activity by influencing the LH secretion and synchronization of ovulation in the sheep and goat (Knight et al. 1978; Ungerfeld et al. 2004). Just as the influence of primer pheromones in the acceleration of puberty in rodents is well established, the chemosginals from the ram and buck have been found to accelerate puberty in the sheep and goat (Shelton 1978; Underwood et al. 1944). The evidence indicates that 4-ethyl octanoic acid (4EOA) identified in major fleece does not evoke an LH response in a female conspecific but its derivatives might have pheromonal activity (Iwata et al. 2003). It was also found that the synthetic 4EOA possessed a releaser pheromone activity that induced specific behavior in the recipient, and female goats showed some interest in the odor (Iwata et al. 2003; Sasada et al. 1983).

The pheromones secreted through the ewe's wool and wax (Tilbrook and Cameron 1989) and vaginal secretions (Ungerfeld and Silva 2004) sexually attract the ram; however, the nature of the pheromonal compound present in the female is not yet investigated. The volatile substance(s) responsible for female LH secretion has been found in male sheep hair, skin, and their extracts obtained by organic solvents (Cohen-Tannoudji et al. 1994). It is further reported that the compounds 1, 2-hexadecanediol and 1, 2-octadecanediol appear to be responsible for this pheromonal effect in sheep. The synthetic compounds are reported to be effective in stimulating LH release in anestrus ewes. In the domestic sheep, primer pheromones are considered the most important signals involved in sociosexual stimulation of the reproductive processes and male-female interactions induce changes in the pulsatile rhythm of the LH secretion in both sexes, which influences reproductive endocrinology.

The occurrence of olfactory memory for the formation of mother-young bond between ewes and lambs is another interesting aspect of a milestone in support of the involvement of pheromonal communication. Sheep have been used as the best model to study the mother-young relationship (Bouissou 1968; Kendrick et al. 1992; Nowak et al. 2011). The strong bond between ewe and lamb formed shortly after parturition particularly within few hours is very crucial and critical for lamb survival (Keller et al. 2005; Porter et al. 1991). The onset of maternal responsiveness and the development of mother-young attachment in sheep are under the control of hormonal and sensory stimulation (Kendrick et al. 1992; Keverne et al. 1993; Nowak et al. 2011). Offspring recognition through olfaction in most females is generally believed to be based on the offsprings' individual olfactory signature (Poindron et al. 1988; Porter et al. 1991), which is believed to originate from its body coat (Alexander and Stevens 1982) from the anal region (Alexander et al. 1983). It is reported that majority of the odors from the anal region in animals have been associated with pheromones (Bean 1982). The first day of life is entirely dependent on the success of the first suckling episodes (Nowak et al. 2011). If suckling is prevented during the first few hours after birth by covering the ewes udder, the development of a preference for mother is lowered (Nowak et al. 1997), suggesting that sensory outputs from emanating the suckling influence the development of filial attachment.

Amniotic fluid is an important substance in sheep for the establishment of maternal responsiveness toward neonates, facilitating their initial acceptance by the mother (Poindron et al. 2010). Amniotic fluid also carries individual olfactory cues from the neonates that help the formation of the maternal bond. Recently, several volatile organic compounds from lamb wool have been identified; these compounds serve as olfactory cues for neonatal recognition (Burger et al. 2011). However, the exact chemosignals involved in the maternal responsiveness toward the lamb are yet to be determined. It is known that the cell proliferation and survival in the adult brain are influenced by several internal and external factors (Lledo et al. 2006); for instance, the physiological status of pregnancy and parturition has been found to regulate cell proliferation in rodents. Regarding the establishment of maternal selectivity, it is hypothesized that downregulation of cell proliferation occurs in specific areas of the sheep brain and as well as in the main olfactory bulb during the early postpartum period (Brus et al. 2010; Levy et al. 1990). This could facilitate the development of olfactory perpetual memory that is retained in favor of the survival of the newborn neurons that might somehow assist in the learning process.

16.8   HORSE PHEROMONES

The role of biostimulation in the horse is very much needed in the context of horse breeding; however, the research on chemical communication in the horse is very much limited. Mares come into estrus several times a year (i.e., horses are a seasonal polyestrus species). A mare is sexually receptive towards a stallion for 5–7 days and ovulation occurs in the final 24–48 hours of estrus (Kiley-Worthington 1987). The precise determination of the estrus period is difficult and special qualification is needed to effectively elucidate the proper time of insemination (Mozuraitis et al. 2012). A mare signals estrus by urinating in the presence of a stallion, raising her tail, and revealing the vulva. A stallion approaching with a high head will usually nudge and nip a mare, as well as sniff her urine to determine her readiness for mating (Kiley-Worthington 1987).

Ma and Klemm (1997) detected 45 urinary volatile compounds at the estrus and diestrus stage of mares. Mozuraitis et al. (2012) analyzed the estrus urine samples and found 150 urinary volatiles; among the compounds, *m*- and *p*-cresols occurred significantly in greater amounts in estrus when compared to nonestrus. A great increase in amounts of *p*-cresol in urine samples from all mares of different breeds during the most active stallion acceptance periods provides a good signal to stallion that a mare is in estrus; thus *p*-cresol might be consider as a sex phero-mone component. It is further reported that *p*-cresol is able to influence the penile erection in horse (Buda et al. 2012). In addition to urinary sources, sex pheromones such as palmitic and miristic acid are identified in feces and found to show the estrus signal to attract a stallion (Kimura 2001). The finding indicates that sex pheromones are produced from more than one source as it is already recorded in the cow and buffalo.

16.9 ODORANT BINDING PROTEINS IN CATTLE

Odorant binding proteins (OBPs) are the soluble proteins of the lipocalin superfam-ily found in the mucus layer lining the olfactory epithelium of mammals (Mitchell et al. 2011; Pelosi 1994). The OBP is almost of similar structural sequence to the pheromone carrier protein that is present in almost all the sources of pheromone production in mammals (urine, saliva, sweat, vaginal mucus, etc.), and the prime function of the pheromone carrier protein is to bind the pheromone compounds and release them to the environment for manifesting the effect. OBPs are the first in the relay mechanism in pheromonal chemical reception because they provide the link between the chemical signal present in the environment and the odorant recep-tors located in the olfactory and peripheral sensory systems in mammals (Pelosi 1994). Major functions have been suggested for OBPs, such as scavenging odorants (Vincent et al. 2004) and protecting the airways (Mitchell et al. 2011). The OBP is well documented in bovine mucosa (Pelosi 1994) and the 1-octen-3-ol has been found to be the natural ligand of bovine OBP. Since 1-octen-3-ol is an important component of bovine breath and a very potent attractant for many insects, this indi-cates that the role of bovine OBP may be to clear the breath of 1-octen-3-ol to reduce the attraction of insects (Ramoni et al. 2002).

In the buffalo, OBP is reported in saliva; this OBP undergoes posttranslational modification (Rajkumar et al. 2010) and it is presumed that it may bind with the sali-vary pheromone molecule 4-methyl phenol. In addition, OBP has been found in the nasal mucus of the buffalo, suggesting that OBP may bind the odorants for further processing. In the pig, the nasal mucosal OBP has been reported and it has a specific role to bind with the steroid pheromone molecule (Scaloni et al. 2001). Available evidence on OBP in vertebrates concludes that OBP plays a pivotal role in the per-ception of the odorant/pheromones, contributing to all process. Furthermore, the previous investigation suggests that the binding and metabolic activity of the urine odorant 5α-androstan-3-one involved the odorant binding process in sheep olfactory mucosa (Krishna et al. 1995).

16.10 APPLICATION OF PHEROMONES IN CATTLE

The potential role of pheromones has been updated in several aspects so as to enhance cattle reproduction and livestock management. Based on the large number of data obtained so far in cattle pheromone, the following applications are derived:

1. *Puberty acceleration by male pheromone.* Advancement of puberty is one of the important reproductive events that has been well established in cattle. The male pheromone is capable to advance the puberty in females than the attainment of normal puberty in several species of mammals including sheep, goat, pig, and cow (Rekwot et al. 2001). This enhancement of puberty in young cattle is considered to have a potential role in cattle production.

2. *Estrus synchronization and estrus induction in anestrus by male phero-mone.* Introduction of a male into a group of anestrus females during the nonbreeding season results in the activation of LH secretion and synchronization of ovulation in the sheep and goat. This event is generally called the male effect and is widely used in husbandry of these species. The sources of the male pheromones are reported to be produced in fleece (van den Hurk 2007).

3. *Postpartum anoestrus reduction by male pheromone.* Prolonged postpartum anestrus in primiparous cattle is a major cause for failing to rebreed or breeding late in the breeding season (Short et al. 1990). This prolonged postpartum event results in huge economic loss for farmers. In the cow, this critical problem is easily solved by continuous exposure to bull urine by reducing the postpartum anestrus interval during the first calf suckling (Custer et al. 1990; Fernandez et al. 1996). However, there is little controversy regarding the presence of pheromones in bull urine and the way in which the postpartum anestrus cow could be treated with various protocols (Van den Hurk 2007). Further, the nature of the pheromone signal present in the bull urine is not yet identified.

4. *Influencing the standing posture by male pheromones.* The steroid chemosignals identified in boar saliva have been demonstrated to evoke the immobilization reflex to sows in heat and characterize their readiness to mate (Brooks and Cole 1970). The boar steroid pheromone is available as a boar-mate to pig farmers to assist in artificial insemination.

5. *Estrus indication by female pheromones.* Effective estrus detection would greatly help cattle reproduction by artificial insemination. However, the occurrence of silent ovulation without any behavioral sign of estrus is a major problem in buffalo reproduction (Dobson and Kamonpatana 1986). The successful identification of urinary sex pheromones in the buffalo provides a tool to overcome this problem (Rajanrayanan and Archunan 2011) The urinary volatiles characterized in the cow (Archunan and Rameshkumar 2012; Rameshkumar et al. 2000) and horse (Buda et al. 2012) with reference to estrus is another example of estrus-indicating pheromones. An estrus may be detected in the cow by simply evaluating the level of concentration of a specific volatile at least 0.1 μgm/gm from vaginal secretion. The

estrus-indicating volatile compound could be present in several sources, such as urine, feces, saliva, and vaginal mucus of the cow and buffalo. The reports consider that the volatiles specifically present during estrus may be used as a marker for estrus identification (Preti 1984). An attempt is being made to develop a kit to detect estrus in buffalo based on the volatile compounds from our research team.

6. *Penis erection and enhance the sperm quantity by female pheromones.* The urinary sex pheromones have been investigated and proved to enhance penis erection and increase the amount of sperm in buffalo considerably (Archunan and Rajanarayanan 2010); an Indian National Patent was obtained for this finding. The mare urinary pheromone during estrus is reported to influence the penis erection in the stallion (Wierzbowski and Hafez 1961), indicating that the female sex pheromones can be used in assisted reproductive technology for special circumstances in certain cattle.

7. *Reducing the fighting behavior by maternal pheromones.* The discovery of maternal (appeasing) pheromones involved in reducing the agonistic behavior in piglets and stimulating their feeding behavior resulting a significant increase in weight gain is a classical example of the practical application of mammalian pheromones. A synthetic analog called porcine appeasing pheromone is available in the market and is very well used by farmers in the pig industry (Suilence^R) (Pageat 2001). As in pigs, a synthetic pheromone product based on natural compound is commercially available as equine appeasing pheromone (EAP) (Pherocalm^R in Europe; Modipher EQ^R in the United States) (Riley et al. 2002). The application of the appeasing pheromone as pheromonetherapy is well documented, which enables a simplification of treatment for anxiety- and phobia-related issues in various species (dogs, cats, and rabbits) (Gaultier et al. 2005; Griffith et al. 2000; McGlone and Anderson 2002).

8. *Maternal responsiveness by neonatal pheromones.* The olfactory recognition in sheep between the postparturient mother and her offspring is immediately established within the first hour after delivery and helps the mother to accept her own young to the udder (Nowak et al. 2011). The establishment of the maternal selectivity mainly relies on the mother learning the olfactory individual signature of her lamb (Levy and Fleming 2006; Porter et al. 1991). The brain areas, particularly the cortical and medial amygdala, are reported to be involved in the formation of olfactory offspring memory in sheep (Keller et al. 2004; Meurisse et al. 2009), suggesting that neurological brain networks actually sustain the lamb memory (Keller et al. 2005). Although peptide hormones such as oxytocin and vasopressin are released in the olfactory bulb, further investigation showed that oxytocin release alone in the olfactory bulb at parturition is reported to facilitate the recognition of lamb odors by modulating noradrenaline (NA), acetylcholine (ACh), and γ-aminobutyric acid (γGABA) release, which is of primary importance for olfactory memory (Levy et al. 1995). However, the nature of the odor produced from the lamb is yet to be detected for the purpose of future applications.

16.11 FUTURE PERSPECTIVES

Based on the data available, cattle pheromones can be used as efficient tools to improve reproduction and management. In order to control the calving interval, optimized milk production, and maximize offspring in dairy cattle, artificial insemination (AI) is widely applied in farm animal reproduction. The right timing of AI can be achieved by accurate detection of estrus. Therefore, it is possible to develop a kit for easy estrus detection in cattle, particularly for buffalo. The estrus kit can also be developed and used for critically endangered animals since estrus-specific urinary volatile compounds have also been identified in the black buck (Archunan and Rajagopal 2013). It is interesting to note that Weigerinck et al. (2011) have now taken the first steps toward the development of a practical eNose product consisting of an array of sensors based on the sex pheromones in cow feces, and it is important and worth encouraging the development of a product such as BOVINOSE for estrus detection purpose. Since a number of estrus-specific volatile compounds have been characterized in cattle it would be possible to design an effective biosensor for estrus detection.

The role of the bull in reducing postpartum anestrus is extremely important. Hence, information on the nature and function of this pheromone needs to be discovered. This is a challenging task and researchers should be encouraged to pay more attention to this aspect. Appeasing pheromones have plenty of applications, therefore it is necessary to extend the usage of this pheromone in several mammalian species. As the urinary pheromone is able to stimulate the pituitary hormone, a detailed study could be made to consider the future applications of the pheromone-hormone relationship with reference to reproduction.

Pheromonetherapy is used in the dog and cat (Levine and Mills 2008), and this can be extended to other mammalian species including cattle since it has great potential to make the animal calm, reduce anxiety and phobia, and increase grooming in all clinical aspects. The neonatal pheromones have been reported to play a crucial role in the establishment of mother-young relationship in the sheep. In this context, it would be of great interest and there would be much practical utility in characterizing the pheromone compounds responsible for the mother-young bond relationship. Recent findings demonstrate that the bitch's sex pheromone enhances the heart rate of the dog without showing any other sign of arousal (Dzięcioł et al. 2012). Hence, male and female sex pheromones can be analyzed with special reference to health aspects in cattle and humans.

ACKNOWLEDGMENTS

GA thanks his research scholars who contributed significant data on cattle pheromones. We thank Dr. Chellam Balasundaram for critical reading of the chapter. The research work on pheromones is well supported by Bharathidasan University, DST, DBT, ICAR, UGC, UGC-SAP, and DST-PURSE, Government of India.

REFERENCES

Abbasi, M. 2007. The vomeronasal organ in buffalo. *Italian J. Anim. Sci.* 6: 2.

Achiraman, S., Ponmanickam, P., Sankar Ganesh, D. and Archunan, G. 2010. Detection of estrus by male mice: Synergistic role of olfactory-vomeronasal system. *Neurosci. Lett.* 477: 144–148.

Albone, E.S. 1984. *Mammalian Semiochemistry.* John Wiley & Sons, New York.

Alexander, G. and Stevens, D. 1982. Odour cues to maternal recognition of lambs: An investigation of some possible sources. *Appl. Anim. Ethol.* 10: 165–175.

Alexander, G., Stevens, D. and Bradley, L.R. 1983. Washing lambs and confinement as aids to fostering. *Appl. Anim. Ethol.* 10: 251–261.

Archunan, G. 2009. Vertebrate pheromones and their biological importance. *J. Exp. Zool. India* 12: 227–239.

Archunan, G. and Rajagopal, T. 2013. Detection of estrus in Indian blackbuck: Behavioural, hormonal and urinary volatiles evaluation. *Gen. Comp. Endocrinol.* 181: 156–166.

Archunan, G. and Rajanarayanan, S. 2010. Composition for enhancing bull sex libido. Indian Patent No. 244991, dated December 28, 2010.

Archunan, G. and Rameshkumar, K. 2012. 1-Iodoundecae an estrus indicating urinary chemosignal in bovine (*Bos taurus*). *J. Vet. Sci. Tech.* 3: 121–123.

Aron, C. 1979. Mechanisms of control of the reproductive function by olfactory stimuli in female mammals. *Physiol. Rev.* 59: 229–284.

Balakrishnan, M. and Alexandar, K.M. 1985. Sources of body odour and olfactory communication in some Indian mammals. *Ind. Rev. Life Sci.* 5: 277–313.

Bean, N.J. 1982. Olfactory and vomeronasal mediation of ultrasonic vocalizations in male mice. *Physiol. Behav.* 82: 28:31–37.

Berardinelli, J.G. and Joshi, P.S. 2005. Initiation of postpartum luteal function in primiparous restricted-suckled beef cows exposed to a bull or excretory products of bulls or cows. *J. Anim. Sci.* 83: 2495–2500.

Berardinelli, J.G. and Tauck, S.A. 2007. Intensity of the biostimulatory effect of bulls on resumption of ovulatory activity in primiparous, suckled, beef cows. *Anim. Reprod. Sci.* 99: 24–33.

Berardinelli, J.G., Fogwell, R.L. and Inskeep, E.K. 1978. Effect of electrical stimulation or presence of a bull on puberty in beef heifers. *Theriogenology* 9: 133–138.

Bland, K.P. and Jubilan, B.M. 1987. Correlation of flehmen by male sheep with female behaviour and oestrus. *Anim. Behav.* 35: 735–738.

Blissitt, M.J., Bland, K.P. and Cottrell, D.F. 1994. Detection of oestrus-related odour in ewe urine by rams. *J. Reprod. Fertil.* 101: 189–191.

Booth, W.D. 1984. A note on the significance of boar salivary pheromones to the male-effect on puberty attainment in gilts. *Anim. Prod.* 39: 149–152.

Booth, K.K. and Katz, L.S. 2000. Role of the vomeronasal organ in neonatal offspring recognition in sheep. *Biol. Reprod.* 63: 953–958.

Bouissou, M.F. 1968. Effet de l'ablation des bulbes olfactifs sur la reconnaissance du jeune au sa mère chez les ovins. *Rev. Comp. Anim.* 3: 77–83.

Brennan, P. A. and Keverne, E.B. 1997. Neural mechanisms of mammalian olfactory learning. *Prog. Neurobiol.* 51: 457–481.

Brooks, P.H. and Cole, D.J. 1970. The effect of the presence of boar on the attainment of puberty in gilts. *J. Reprod. Fertil.* 23: 435–440.

Brus, M., Meurisse, M., Franceschini, I., Keller, F. and Levy, F. 2010. Evidence for cell proliferation in the sheep brain and its down-regulation by parturition and interactions with the young. *Horm. Behav.* 58: 737–746.

Buck, L.B. 2000. The molecular architecture of review odor and pheromone sensing in mammals. *Cell* 100: 611–618.

Būda,V., Mozūraitis, R., Kutra, J. and Karlson, A.K.B. 2012. p-Cresol: A sex pheromone component identified from the estrous urine of mares. *J. Chem. Ecol.* 38: 811–813.

Burger, B.V., le Roux, M., Marx, B., Herert, S.A. and Amakali, K.T. 2011. Development of second-generation sample enrichment probe for improved sortive analysis of volatile organic compounds. *J. Chrom.* A 1218: 1567–1575.

Burns, P.D. and Spitzer, J.C. 1992. Influence of biostimulation on reproduction in postpartum beef cows. *J. Anim. Sci.* 70: 358–362.

Chemineau, P. 1987. Possibilities for using bucks to stimulate ovarian and oestrous cycles in anovulatory goats–A review. *Livest. Prod. Sci.* 17: 135–147.

Chenoweth, P.J. 1983. Reproductive management procedures in control of breeding. *Aust. J. Anim. Prod.* 15: 28–33.

Cohen-Tannoudji, J., Einhorn, J. and Signoret, J.P. 1994. Ram sexual pheromone: First approach of chemical identification. *Physiol. Behav.* 56: 955–961.

Custer, E.E., Berardinelli, J.G., Short, R.E., Wehrman, M. and Dair, R.A. 1990. Postpartum interval to estrus and patterns of LH and progesterone in first-calf suckled beef cows exposed to mature bulls. *J. Anim. Sci.* 68: 1370–1377.

Dagg, A.I. and Taub, A. 1970. Flehmen. *Mammalia* 34: 686–695.

Danell, B., Gopakumar, N., Nair, M.C.S. and Rajagopalan, K. 1984. Heat symptoms and detection in Surti buffalo heifers. *Ind. J. Anim. Res.* 5: 1–7.

DeJarnette, J.M., Nebel, R.L. and Marshall, C.E. 2009. Evaluating the success of sex-sorted semen in US dairy herds from on farm records. *Theriogenology* 71: 49–51.

Delgadillo, J.A., Flores, J.A., Veliz, F.G., Duarte, G., Vielma, J., Hernandez, H. and Fernandez, I.G. 2006. Importance of the signals provided by the buck for the success of the male effect in goats. *Reprod. Nutr. Dev.* 46: 391–400.

Denhard, M. and Claus, R. 1988. Reliability criteria of a bioassay using rats trained to detect estrus-specific odor in cow urine. *Theriogenology* 30: 1127–1138.

Denhard, M. and Claus, R. 1996. Attempts to purify and characterize the estrus-signalling pheromone from cow urine. *Theriogenology* 46: 13–22.

Denhard, M., Claus, R., Pfeiler, S. and Schopper, D. 1991.Variation in estrus-related odours in the cow and its dependency on the ovary. *Theriogenology* 35: 645–652.

Dobson, H. and Kamonpatana, M. 1986. A review of female cattle reproduction with special reference to comparison between buffaloes, cows and Zebu. *J. Reprod. Fertil.* 77: 1–36.

Dominic, C.J. 1991. Chemical communication in animals. *J. Sci. Res. (Banaras Hindu University)* 41: 157–169.

Dorries, K., Regan, M., Bruce, E.A. and Halpern, P. 1991. Sex difference in olfactory sensitivity to the boar chemosignal, androstenone, in the domestic pig. *Anim. Behav.* 42: 403–411.

Doving, K.B. and Trotier, D. 1998. Structure and function of the vomeronasal organ. *J. Exp. Biol.* 201: 2913–2925.

Dulac, C. and Wagner, S. 2006. Genetic analysis of brain circuits underlying pheromone signaling. *Annu. Rev. Genet.* 40: 449–467.

Dzięcioł, M., Stańczyk, E., Nowak, A.N., Niżański,W., Ochota, M. and Kozdrowskia, R. 2012. Influence of bitches sex pheromones on the heart rate and other chosen parameters of blood flow in stud dogs (*Canis familiaris*). *Res. Vet. Sci.* 93: 1241–1247.

Estes, R. 1972. The role of vomeronasal organ in mammalian reproduction. *Mammalogy* 36: 315–342.

Fernandez, D., Berardinelli, J.G., Short, R.E. and Adair, R. 1993. The time required for the presence of bulls to alter the interval from parturition to resumption of ovarian activity and reproductive performance in first calf-suckled beef cows. *Theriogenology* 39: 411–419.

Fernandez, D.L., Berardinelli, J.G., Short, R.E. and Adair, R. 1996. Acute and chronic changes in luteinizing hormone secretion and postpartum interval to estrus in first-calf suckled beef cows exposed continuously or intermittently to mature bulls. *J. Anim. Sci.* 74: 1098–1103.

Fiol, C. and Ungerfeld, R. 2012. Biostimulation in cattle: Stimulation pathways and mechanisms of response. *Trop. Subtrop. Agroecosyst.* 15: 1.

Gaultier, E., Bonnafous, L., Bought, L., Lafont, C. and Pageat, P. 2005. Comparison of the efficacy of a synthetic dog-appeasing pheromone with clomiraine for the treatment of separation-related disorders in dogs. *Vet. Rec.* 156: 533–538.

Gelez, H. and Fabre-Nys, C. 2006. Role of the olfactory systems and importance of learning in the ewes' response to rams or their odors. *Reprod. Nutr. Dev.* 46: 401–415.

Gelez, H., Archer, E., Chesneau, D., Campan, R. and Fabre-Nys, C. 2004. Importance of learning in the response of ewes to male odor. *Chem. Senses* 29: 555–563.

Gerall, A. 1963. An exploratory study of the effect of social isolation variables on the sexual behaviour of male guinea pigs. *Anim. Behav.* 11: 274–282.

Gokuldas, P.P., Yadav, M.C., Kumar, H., Singh, G., Mahmood, S. and Tomar, A.K.S. 2010. Resumption of ovarian cyclicity and fertility response in bull exposed postpartum buffaloes. *Anim. Reprod. Sci.* 121: 236–241.

Griffith, C.A., Steigerwald, E.S. and Buffington, A.T. 2000. Effects of a synthetic facial pheromone on behavior of cats. *J. Am. Vet. Med. Assoc.* 217: 1154–1156.

Hall, J.G., Branton, C. and Stone, E.J. 1959. Estrus, estrous cycles, ovulation time, time of service and fertility of dairy cows in Louisiana. *J. Dairy Sci.* 42: 1086–1094.

Halpern, M. 1987. The organization and function of the vomeronasal system. *Ann. Rev. Neuroscience* 10: 325–362.

Halpern, M. and Martínez-Marcos, A. 2003. Structure and function of the vomeronasal system: An update. *Prog. Neurobiol.* 70: 245–318.

Hawk, H.W., Conley, H.H. and Kiddy, C.A. 1984. Estrus-related odours in milk detected by trained dogs. *J. Dairy Sci.* 67: 392–397.

Hradecky, P. 1975. Occurrence of volatile compounds and possibilities of their determination using gas chromatography. In *Czech Literature Review for Graduate Study*, University of Veterinary Medicine, Brno, Czechoslovakia, p. 185.

Hradecky, P. 1986. Volatile fatty acids in urine and vaginal secretions of cows during the reproductive cycle. *J. Chem. Ecol.* 12: 187–196.

Hradecky, P., Sis, R.F. and Klemm, W.R. 1983. Distribution of flehmen reactions of the bull throughout the bovine estrous cycle. *Theriogenology* 20: 197–204.

Ingawale, M.V. and Dhoble, R.L. 2004. Buffalo reproduction in India: An overview. *Buffalo Bull.* 1: 4–9.

Ireland, J.J., Murphee, R.L. and Coulson, P.B. 1980. Accuracy of predicting stages of bovine estrous cycle by gross appearance of the corpus luteum. *J. Dairy Sci.* 63: 155–160.

Iwata, E., Wakabayashi, Y., Kakuma, Y., Kikusui, T., Takeuchi, Y. and Mori, Y. 2000. Testosterone-dependent primer pheromone production in the sebaceous gland of male goat. *Biol. Reprod.* 62: 806–810.

Iwata, E., Kikusui, T., Takeuchi, Y. and Mori, Y. 2003. Substances derived from 4-ethyl octanoic acid account for primer pheromone activity for the male effect in goats. *J. Vet. Med. Sci.* 65: 1019–1021.

Izard, M.K. 1983. Pheromones and reproduction in domestic animals. In Vandenberg, J.G. (ed.), *Pheromones and Reproduction in Mammals*. Academic Press, New York, pp. 253–285.

Izard, M.K. and Vandenbergh, J.G. 1982. The effects of bull urine on puberty and calving date in crossbred beef heifers. *J. Anim. Sci.* 55: 1160–1168.

Jacobs, V.L., Sir, R.F. and Coppock, C.E. 1980. Tongue manipulation of the palate assists estrous detection in the bovine. *Theriogenology* 13: 353–356.

Karthikeyan, K. 2011. Prediction of estrus in murrah buffao species *Bubalus bubalis* based on volatile compounds from body fluids. Ph.D. thesis, Bharathidasan University, Tiruchirappalli, India.

Karthikeyan, K. and Archunan, G. 2013. Gas chromatographic mass spectrometric analysis of estrus-specific volatile compounds in buffalo vaginal mucus after initial sexual foreplay. *J. Buffalo Sci.* 2: 1–7.

Karthikeyan, K., Muniasamy, S., SankarGanesh, D., Achiraman, S., Ramesh Saravanakumar, V. and Archunan, G. 2013. Faecal chemical cues in water buffalo that facilitate estrus detection. *Anim. Reprod. Sci.* 138: 163–167.

Keller, M., Meurisse, M. and Lévy, F. 2004. Mapping the neural substrates involved in maternal responsiveness and lamb olfactory memory in parturient ewes using Fos imaging. *Behav. Neurosci.* 118: 1274–1284.

Keller, M., Meurisse, M. and Lévy, F. 2005. Mapping of brain networks involved in consolidation of lamb recognition memory. *Neuroscience* 133: 359–369.

Kendrick, K.M., Lévy, F. and Keverne, E.B. 1992. Changes in the sensory processing of olfactory signals induced by birth in sheep. *Science* 8: 833–836.

Keverne, E.B., Lévy, F., Guevara-Guzman, R. and Kendrick, K.M. 1993. Influence of birth and maternal experience on olfactory bulb neurotransmitter release. *Neuroscience* 56: 557–565.

Kiddy, C.A. and Mitchell, D.S. 1984. Estrus-dated odors in cows: Time of occurrence. *J. Dairy Sci.* 64: 267.

Kiley-Worthington, M. 1987. *The Behavior of Horses.* J.A. Allen, London.

Kimura, R. 2001. Volatile substances in feces, of urine and urine-marked feces of feral horses. *Can. J. Anim. Sci.* 81: 411–420.

Kirkwood, R.N., Forbes, J.M. and Hughes, P.E. 1981. Influence of boar contact on attainment of puberty in gilts after removal of the olfactory bulbs. *J. Reprod. Fertil.* 61: 193–196.

Klemm, W.R., Hawkins, G.N. and De Los Santos, E. 1987. Identification of compounds in bovine cervico-vaginal mucus extracts that evoked male sexual behaviour. *Chem. Senses* 12: 77–87.

Knight, T.W. and Lynch, P.R. 1980. Source of ram pheromones that stimulate ovulation in the ewe. *Anim. Reprod. Sci.* 3: 133–136.

Knight, T.W., Peterson, A.J. and Payne, E. 1978. The ovarian and hormonal response of the ewe to stimulation by the ram early in the breeding season. *Theriogenology* 10: 343–353.

Krishna, N.S., Getchell, M.L., Margolis, F.L. and Getchell, V. 1995. Differential expression of vomeromodulin and odorant-binding protein, putative pheromone and odorant transporters, in the developing rat nasal chemosensory mucosae. *J. Neurosci. Res.* 40: 54–71.

Ladewig, J., Edward, O., Price, B. and Hart, L. 1980. Flehmen in male goats: Role in sexual behavior. *Behav. Neural. Biol.* 30: 312–332.

Levine, E.D. and Mills, D.S. 2008. Long term follow-up of the efficacy of a behavioural treatment programme for dogs with firework fears. *Vet. Rec.* 162: 657–659.

Levy, F. and Fleming, A.S. 2006. The neurobiology of maternal behavior in mammals. In Marshall, P.J. and Fox, N.A. (eds.), *The Development of Social Engagement. Neurobiological Perspectives*, Oxford University Press, New York, pp. 197–246.

Lévy, F., Gervais, R., Kindermann, U., Orgeur, P. and Piketty, V. 1990. Importance of beta-noradrenergic receptors in the olfactory bulb of sheep for recognition of lambs. *Behav. Neurosci.* 104: 464–469.

Lévy, F., Kendrick, K.M., Goode, J.A., Guevara-Guzman, R. and Keverne, E.B. 1995. Oxytocin and vasopressin release in the olfactory bulb of parturient ewes: Changes with maternal experience and effects on acetylcholine, gamma-aminobutyric acid, glutamate and noradrenaline release. *Brain Res.* 669: 197–206.

Lledo, P.M., Alonso, M. and Grubb, M.S. 2006. Adult neurogenesis and functional plasticity in neuronal circuits. *Nat. Rev. Neurosci.* 7: 179–193.

Ma, W. and Klemm, W.R. 1997. Variations of equine urinary volatile compounds during the oestrous cycle. *Vet. Res. Comm.* 21: 437–446.

Macmillan, K.L., Allison, A.J. and Struthers, G.A. 1979. Some effects of running bulls with suckling cows or heifers during the premating period. *N. Z. J. Exp. Agric.* 7: 121–124.

Marinier, S.L., Alexander, A.J. and Wari, G.H. 1988. Flehmen behaviour in the domestic horse: Discrimination of conspecific odours. *Appl. Anim. Behav. Sci.* 19: 227.

Martin, G.B. 1995. Reproductive research on farm animals for Australia—Some long-distance goals. *Reprod. Fertil. Dev.* 7: 967–982.

McGlone, J.J. and Anderson, D.L. 2002. Synthetic maternal pheromone stimulates feeding behavior and weight gain in weaned pigs. *J. Anim. Sci.* 80: 3179–3183.

Meurisse, M., Chaillou, E. and Lévy, F. 2009. Afferent and efferent connections of the cortical and medial nuclei of the amygdala in sheep. *J. Chem. Neuroanat.* 37: 87–97.

Mitchell, G., Clark, M., Lu, R. and Caswell, J. 2011. Localization and functional characterization of pulmonary bovine odorant-binding protein. *Vet. Pathol.* 48: 1054–1060.

Mozūraitis, R., Būda, V., Kutra, J. and Borg-Karlson, A.-K. 2012. p- and m-Cresols emitted from estrous urine are reliable volatile chemical markers of ovulation in mares. *Anim. Reprod. Sci.* 130: 51–56.

Mucignat-Caretta, C., Redaelli, M. and Caretta, A. 2012. One nose, one brain: Contribution of the main and accessory olfactory system to chemosensation. *Front. Neuroanat.* 6: 1–9.

Nordéus, K., Båge, R., Gustafsson, H. and Söderquist, L. 2012. Changes in LH pulsatility profiles in dairy heifers during exposure to oestrous urine and vaginal mucus. *Reprod. Domestic. Anim.* 47: 952–958.

Nowak, R., Keller, M. and Levy, S. 2011. Mother-young relationship in sheep: A model for a multidisciplinary approach of the study of attachment in mammals. *J. Neuroendocrinol.* 23: 1042–1053.

Nowak, R., Murphy, T.M., Lindsay, D.R., Alster, P., Anderson, R. and Berk, K. 1997. Development of a preferential relationship with the mother by the newborn lamb: Importance of the suckling actiity. *Physiol. Behav.* 62: 681–688.

Pageat, P. 2001. Pig appeasing pheromones to decrease stress, anxiety and aggressiveness. US Patent No. 6,169,113, issued January 2, 2001.

Pageat, P. and Teyssier, Y. 1998. Usefulness of a porcine pheromone analogue in the reduction of aggressions between weanlings on penning; behavior study. *Congr. Int. Pig Vet. Soc.* Birmi UK, p. 413.

Paleologou, A.M. 1979. A study of the cervico-vaginal secretions of cows during the diferent phases of the estrous cycle. *J. Inst. Anim. Tech.* 30: 83–94.

Patra, M.K., Barman, P. and Kumar, H. 2012. Potential application of pheromones in reproduction of farm animals–A review. *Agri. Rev.* 33: 82–86.

Pelosi, P. 1994. Odorant-binding proteins. *Crit. Rev. Biochem. Mol. Biol.* 29: 199–228.

Poindron, P., Levy, F. and Krehbiel, D. 1988. Genital, olfactory and endocrine interactions in the development of maternal behavior in the parturient ewe. *Psychoneuroendocrinology* 13: 99–125.

Poindron, O., Otal, J., Ferreira, G., Keller, M., Guesdon, V., Nowak, R. and Levy, F. 2010. Amniotic fluid is important for the maintenance of maternal responsiveness and the establishment of maternal selectivity in sheep. *Animal* 4: 2057–2064.

Porter, R.H., Levy, F., Poindron, P., Litterio, M., Schaal, B. and Beyer, C. 1991. Individual olfactory signatures as major determinants of early maternal discrimination in sheep. *Dev. Psychobiol.* 24: 151–158.

Powers, J.B. and Winans, S.S. 1975. Vomeronasal organ: Critical role in mediating sexual behavior of the male hamster. *Science* 187: 961–963.

Presicce, G.A., Brocketta, C.C., Chenga, T., Foote, R.H., Rivard, G.F. and Klemm, W.R. 1993. Behavioral responses of bulls kept under artificial breeding conditions to compounds presented for olfaction, taste or with topical nasal application. *Appl. Anim. Behav. Sci.* 37: 273–284.

Preti, G. 1984. Method for detecting bovine estrus by determining methyl heptanol concentrations in vaginal secretions. US Patent No. 4,467,814, issued August 28, 1984.

Rajanarayanan, S. 2005. Assessment of flehmen behavior and identification of urinary pheromones in buffaloes (*Bubalus bubalis*) with special reference to estrus. Ph.D. thesis, Bharathidasan University, India.

Rajanarayanan, S. and Archunan, G. 2004. Occurrence of flehmen in male buffaloes (*Bubalus bubalis*) with special reference to estrus. *Theriogenology* 61: 861–866.

Rajanarayanan S. and Archunan G. 2011. Identification of urinary sex pheromones in female buffaloes and their influence on bull reproductive behaviour. *Res. Vet. Sci.* 91: 301–305.

Rajkumar, R., Karthikeyan, K., Archunan, G., Huang, P.H., Chen, Y.W., Ng, W.V. and Liao, C.C. 2010. Using mass spectrometry to detect buffalo salivary odorant-binding protein and its post-translational modifications. *Rapid Commun. Mass Spectrom.* 24: 3248–3254.

Rameshkumar, K. 2000. Chemical characterization of bovine (*Bos taurus*) urine with special reference to reproductive behavior. Ph.D. thesis, Bharathidasan University, India.

Rameshkumar, K., Archunan, G., Jeyaraman, R. and Narasimhan, S. 2000. Chemical characterization of bovine urine with special reference to oestrus. *Vet. Res. Commun.* 24: 445–454.

Rameshkumar, K., Achiraman, S., Karthikeyan, K. and Archunan, G. 2008. Ability of mice to detect estrous odor in bovine urine: Roles of hormones and behavior in odor discrimination. *Zool. Sci.* 25: 349–354.

Ramoni, R., Vincent, F., Ashcroft, A.E., Accornero, P., Grolli, S., Valencia, C., Tegoni, M. and Cambillau, C. 2002. Control of domain swapping in bovine odorant-binding protein. *Biochem. J.* 365: 739–748.

Rasmussen, L.E.L., Anthony, J., Martin, H. and Hess, D.L. 1996. Chemical profiles of African elephants, *Loxodonta africana*: Physiological and ecological implications. *J. Mammal.* 77: 422–439.

Reinhardt, V. 1983. Flehmen, mounting and copulation among members of a semi-wild cattle herd. *Anim. Behav.* 31: 641–650.

Rekwot, P.I., Ogwu, D., Oyedipe, E.O. and Sekoni, V.O. 2001. The role of pheromones and biostimulation in animal reproduction. *Anim. Reprod. Sci.* 65: 157–170.

Riley, R., Grogan, E. and McDonnell, S. 2002. Evaluation of usefulness if equine appeasing pheromone in gentling of foals and yearlings. Thesis, University of Pennsylvania, Philadelphia.

Rivard, G. and Klemm, W.R. 1989. Two body fluids containing bovine estrous pheromone(s). *Chem. Senses* 14: 273–279.

Rivas-Muñoz, R., Fitz-Rodríguez, G., Poindron, P., Malpaux, B. and Delgadillo, J.A. 2007. Stimulation of estrous behavior in grazing female goats by continuous or discontinuous exposure to males. *J. Anim. Sci.* 85: 1257–1263.

Roberson, M.S., Ansotegui, R.P., Berardinelli, J.G., Whitman, R.W. and McInverny, M.J. 1987. Influence of mature bulls on occurrence of puberty in beef heifers. *J. Anim. Sci.* 64: 1601–1605.

Roberson, M.S., Wolfe, M.W., Stumpf, T.T., Werth, L.A., Cupp, A.S., Kojima, N.D., Wolfe, P.L., Kittok, R.J. and Kinder. J.E. 1991. Influence of growth rate and exposure to bulls on age at puberty in beef heifers. *J. Anim. Sci.* 69: 292–298.

Salazar, I., Quinteiro, P.S. and Cifuentes, J.M. 1995. Comparative anatomy of the vomeronasal cartilage in mammals: Mink, cat, dog, pig, cow and horse. *Ann. Anat.* 177: 475–481.

Sambraus, H.H. and Waring, Z., 1975. Effect of urine from estrous cows on libido in bulls. *Zoologie Saugettierkunde* 40: 49–54.

Sankar, R. and Archunan, G. 2004. Flehmen response in bull: Role of vaginal mucus and other body fluids of bovine with special reference to estrus. *Behav. Process.* 67: 81–86.

Sankar, R. and Archunan, G. 2005. Discrimination of bovine estrus-related odors by mice. *J. Ethol.* 23: 147–151.

Sankar, R. and Archunan, G. 2008. Identification of putative pheromones in bovine (*Bos taurus*) feces in relation to estrus detection. *Anim. Reprod. Sci.* 103: 149–153.

Sankar, R. and Archunan, G. 2011. Gas chromatographic/mass spectrometric analysis of volatile metabolites in bovine vaginal fluid and assessment of their bioactivity. *Int. J. Anal. Chem.* 2011, Article ID 256106.

Sankar, R. and Archunan, G. 2012. Seasonal effects on estrus behaviours in dairy cattle. *Int. J. Adv. Vet. Sci. Tech.* 1: 28–34.

Sankar, R., Archunan, G. and Habara, Y. 2007. Detection of oestrous-related odour in bovine (*Bos taurus*) saliva: Bioassay of identified compounds. *Animal* 1: 1321–1327.

Sasada, H., Sugiyama, T., Yamashita, K. and Masaki, J. 1983. Identification of specific odor components in mature male goat during the breeding season. *Jpn. J. Zootech. Sci.* 54: 401–408.

Scaloni, A., Paolini, S., Brandazza, A., Fantacci, M., Bottiglieri, C., Marchese, S., Navarrini, A., Fini, C., Ferrara, L. and Pelosi, P. 2001. Purification, cloning and characterisation of odorant-binding proteins from pig nasal epithelium. *Cell. Mol. Life Sci.* 58: 823–834.

Schams, D., Schallenberger, E., Hoffmann, B. and Karg, H. 1977. The oestrous cycle of the cow: Hormonal parameters and time relationships concerning oestrus, ovulation, and electrical resistance of the vaginal mucus. *Acta Endocrinol.* 86: 180–192.

Schinckel, P.G. 1954. The effect of the presence of the ram on the ovarian activity of the ewe. *Aus. J. Agri. Res.* 5: 465–469.

Shelton, M. 1978. Reproduction and breeding of goats. *J. Dairy Sci.* 61: 994–1010.

Short, R.E., Bellows, R.A., Staigmiller, R.E., Berardinelli, J.G. and Custer, E.E. 1990. Physiological mechanisms controlling anestrus and infertility in postpartum beef cattle. *J. Anim. Sci.* 68:799–816.

Signoret, J.P. 1970. Reproductive behavior of pigs. *J. Reprod. Fertil.* 11: 105–117.

Soto Belloso, E., Portillo, G., Ramírez, L., Soto, G., Rojas, N. and Cruz-Arambulo, R. 1997. Efecto del destete por noventiséis horas sobre la inducción del celo y fertilidad en vacas mestizas acíclicas. *Arch. Latinoam. Anim. Prod.* 5: 359–361.

Stahlbaum, C.C. and Houpt, K.A. 1989. The role of the flehmen response in the behavioral repertoire of the stallion. *Physiol. Behav.* 45: 1207–1214.

Sugiyama, T., Sasada, H., Masaki, J. and Yamashita, K. 1981. Unusal fatty acids with specific odour from mature male goat. *Agri. Biol. Chem.* 45: 2655–2658.

Tilbrook, A.J. and Cameron, A.W.N. 1989. Ram mating preferences for wooly rather than recently shorn ewes. *Appl. Anim. Behav. Sci.* 24: 301–312.

Tirindelli, R., Dibattista, M., Pifferi, S. and Menini, A. 2009. From pheromones to behavior. *Physiol. Rev.* 80: 921–956.

Tirindelli, R., Mucignat-Caretta, C., and Ryba, N.J.R. 1998. Molecular aspects of pheromonal communications in the vomeronasal organ of mammals. *Trends Neurosci.* 21: 482–486.

Underwood, E.J., Shier, F.L. and Davenport, N. 1944. Studies in sheep husbandry. The breeding season in Merino crossbreeds and British breeds ewes in the agricultural districts. *J. Agri.* 11: 135–143.

Ungerfeld, R. 2007. Socio-sexual signaling and gonadal function: Opportunities for reproductive management in domestic ruminants. In Juengel, J.I., Murray, J.F. and Smith, M.F. (eds.), *Reproduction in Domestic Ruminants VI.* Nottingham University Press, Nottingham, pp. 207–221.

Ungerfeld, R. and Silva, L. 2004. The presence of normal vaginal flora is necessary for normal sexual attractiveness of estrous ewes. *Appl. Behav. Sci.* 93: 245–250.

Ungerfeld, R., Dago, A.L., Rubianes, E. and Forsberg, M. 2004. Response of anestrous ewes to the ram effect after follicular wave synchronization with a single dose of estradiol-17 beta. *Reprod. Nutr. Dev.* 44: 89–98.

Van den Hurk, R. 2007. Intraspecific chemical communication in vertebrates with special attention to its role in reproduction. Zeist: Pheromone Information Centre, Brugakker, Netherlands.

Vandenbergh, J.G. 1999. Pheromones in mammals. In *Encyclopedia of Reproduction*. Academic Press, London, pp. 764–769.

Vincent, F.I., Ramoni, R., Spinelli, S., Grolli, S., Tegoni, M. and Cambillau, C. 2004. Crystal structures of bovine odorant-binding protein in complex with odorant molecules. *Eur. J. Biochem.* 271: 3832–3842.

Weidong, M.A., Clement, B.A. and Klemm W.R. 1997. Volatile compounds of bovine milk as related to the stage of the estrous cycle. *J. Dairy Sci.* 80: 3227–3233.

Weigerinck, W., Setkus, A., Buda, V., Borg-Karlson, A.K., Mozuraitis, R. and deGee, A. 2011. BOVINOSE: Pheromone based sensor system for detecting estrus in dairy cows. *Procedia Comput. Sci.* 7: 340–342.

Wierzbowski, S., and Hafez, E.S.E. 1961. Analysis of copulatory reflexes in the stallion. *Proc. Fourth Inter. Cong. Anim. Reprod. The Hague* 2: 176–179.

Williamson, N.B., Morris, R.S., Blood, D.C. and Cannon, C.M. 1972. A study of oestrus behaviour and oestrus detection methods in a large commercial dairy herd. *Vet. Rec.* 91: 50–58.

Winans, S. and Powers, J.B. 1977. Olfactory and vomeronasal deafferentation of male hamsters: Histological and behavioral analyses. *Brain Res.* 126: 325–344.

Wöhrmann-Repenning, A. 1984. Comparative anatomical studies of the vomeronasal complex and the rostral palate of various mammals. I. *Gegenbaurs Morphologisches Jahrbuch*, 130: 501–530.

Wright, I.A.S., Rhind, M., Whyte, T.K. and A.J. Smith. 1992. Effect of body condition at calving and feeding level after calving on LH profiles and duration of the post-partum anoestrous period in beef cows. *Anim. Prod.* 55: 41–46.

Ziegler, T.E., Epple, G., Snowdon, C.T., Porter, T.A., Belcher, A.M. and Kuderling, I. 1993. Detection of the chemical signals of ovulation in the cotton-toptamarin, *Saguinus oedipus. Anim. Behav.* 45: 313–322.

17 Pheromones for Newborns

Benoist Schaal

CONTENTS

17.1 Inexperienced Newborns' Quest for Colostrum and Milk 490
17.2 Nipples as Scent Organs Evolutionarily Tailored for Newborns 491
17.3 Neonatal Olfactory Cognition: Learning and Predispositions 493
17.4 Ethochemical Logic to Deconstruct Mammary Odors 495
17.5 Evidence for Pheromones in Mammary Chemostimuli? 497
    17.5.1 Rodents 498
        17.5.1.1 Mammary Sources of Behaviorally Active
                 Chemostimuli 498
        17.5.1.2 Extramammary Sources of Behaviorally Active
                 Chemostimuli 499
        17.5.1.3 Evidence for Pheromones? 500
    17.5.2 Lagomorphs 501
        17.5.2.1 Mammary Sources of Behaviorally Active
                 Chemostimuli 501
        17.5.2.2 Evidence for Pheromones 502
    17.5.3 Primates 504
        17.5.3.1 Mammary and Extramammary Sources of
                 Behaviorally Active Chemostimuli 504
        17.5.3.2 Evidence for Pheromones? 506
17.6 Mammary Chemostimuli as Ontogenetic Adaptations 506
    17.6.1 Contextual Fluctuations in Mammary Chemosignaling 506
    17.6.2 Contextual Fluctuations in Neonatal Chemoreception 508
    17.6.3 Chemosensory Ontogenetic Adaptations 508
17.7 Conclusion: Mammalian Neonates as Models to Uncover How and
    When Chemoreceptive Systems Make Sense of Odor Cues and Signals 510
Acknowledgments 511
References 512

17.1   INEXPERIENCED NEWBORNS' QUEST FOR COLOSTRUM AND MILK

Newly born mammals have to reach the source of milk as promptly as possible to ensure uninterrupted mother-to-offspring transfer of hydration, nutrients, and energy. Colostrum and milk intake also warrants the neonates' immediate exposure to micronutrients and antioxidants, passive immunization, innocuous bacterial strains, growth factors, and a range of bioactive peptides that control conservative behavioral functions (i.e., antinociception, sleep induction, learning). With these matters and commodities, mothers also pass on to their offspring different levels of chemosensory information that reveal her identity, the location of the mammae, and the composition of milk. Thus, lactation and sucking uniquely coevolved by mammalian females and newborns imply a puzzle of morphological, physiological, and behavioral arrangements (milk gland structure and location; lactational performance, nursing acceptance; neonate's oral competence, absorptive abilities and development) that are all subject to natural selection.

Mammals are indeed exposed to an outstandingly powerful selective pressure while they are bottlenecked through birth and weaning. These windows in early development concentrate all types of challenges, and maladaptive responses to them in the female-offspring dyad are costly for neonatal viability (e.g., Clutton-Brock 2001). A well-documented case refers to primates, specifically humans, in which 27% of infants fail to survive their first year and 47% of children fail to survive to puberty in hunter-gatherer and historical societies (Volk and Atkinson 2013). The causes of such high mortality rates are not known precisely, but the first steps in infant-mother adjustment bear great momentum. The ability to initiate sucking is variable among human mother-infant dyads, leading to delays in establishing optimal colostrum/milk transfer in some infants. For example, a high rate of termborn infants display nonoptimal milk intake on the day of birth and on day 3 (i.e., 49% and 22%, respectively, in California [Dewey et al. 2003]; 43% and 8%, respectively, in France [Michel et al. 2006]). Such insufficient intake during the first feeds, if not handled rapidly, can lead to excessive weight loss, dehydration, and threat to viability when adequate care is lacking (Cooper et al. 1995; Neifert 2001). This is best emphasized in a study on homeborne infants in Ghana showing that a 1-day postponement to initiate breastfeeding explained 16% of neonatal losses and a postbirth delay of only 1 hour to engage breastfeeding explained 22% of neonatal mortality (Edmond et al. 2006). Early breast milk intake was associated with reduced mortality caused by infection, particularly in the gastrointestinal tract (e.g., Edmond et al. 2007; Huffman and Combest 1990).

Other mammalian species show the same trend. For example, piglets incur high mortality, especially when colostrum intake is lacking (Andersen et al. 2007). Another case is the European rabbit (*Oryctolagus cuniculus*), a species that has minimized direct maternal investment (although in the wild, indirect investment is high in excavating a burrow and lining it with an insulating vegetal material and hair). During the first 2 weeks postpartum, neonate rabbits can suck nipples only within the 3–5 minutes/day the female makes herself available (Rödel et al. 2012; Selzer et al. 2004; Zarrow et al. 1965). This limited access to milk must be added to an intense sibling rivalry (Drummond et al. 2000), and it is not rare that pups miss one feed and have to wait 24 hours to the next. While pups survive one sucking failure, they are jeopardized after two such failures. But as they do interchange nipples over a same nursing episode (Bautista et al. 2005; Drewett et al. 1982), pups can in principle get milk, and rapidly improve their localizatory and sucking skills from day to day (Hudson and Distel 1983; Müller 1978). Despite generally successful suckling, pups incur notable mortality in the days following birth (days 0–7) and around weaning (days 21–28), especially when females are primiparous (Coureaud et al. 2000).

Thus, mammalian mother-infant dyads are exposed to strong selective pressure in the days and weeks following delivery. Accordingly, any sensory, behavioral, or cognitive means in females or newborns that can speed up neonatal nipple localization and milk ingestion is beneficial to current mother-offspring dyads, and should have been beneficial to the mother's inclusive fitness in the environment of evolutionary adaptation. In this chapter, we will consider how females and neonates coevolved chemosensory means for mutual adaptive ends at the start of their postnatal relationship. We will first outline the interface structures for milk transfer (nipples), which were shaped by females to match the sensory/motor skills of their offspring. Second, we will survey newborns' perceptual and cognitive sophistication in the context of the nursing relationship. Third, some rodent, lagomorph, and primate cases will be analyzed in more detail to evaluate the nature of the cues and signals that neonates use in their quest for milk. Finally, we will present mammalian newborns' pheromones as ontogenetic adaptations; that is, as perceptual mechanisms dedicated to orient and synchronize infant behavior in relation with the mother, and hence increase survival and stimulate adaptive learning.

17.2 NIPPLES AS SCENT ORGANS EVOLUTIONARILY TAILORED FOR NEWBORNS

Mother-neonate exchanges are diversely organized among mammals (Clutton-Brock 1991; Gubernick and Klopfer 1981; Numan et al. 2006), but this diversity is underlain by common mechanisms. First, at the same time as they procure various substances and realize numerous nurturing actions toward their offspring, maternal females unavoidably transmit information about the environment (including themselves) they create around them during fetal and postnatal development and this information transfer is a significant part of the "maternal effects" (Maestripieri and Mateo 2009). Second, the sensory resource that is mobilized in mediating maternal information transfer to the neonate depends on the receiver, precisely the neonate and its most advanced receptive and reactive abilities. In theory, females could exploit the whole range of their neonates' sensory systems in these earliest exchanges, but the most basic, and hence most conserved ways, rely on somesthesis and chemoreception. While some species give birth to altricial neonates whose sensorium is restricted to these modalities (e.g., sightless newborns in monotremes, marsupials, altricial rodents, or carnivores), other species bear precocial or semiprecocial newborns whose behavior is controlled by all the senses, including hearing and vision (e.g., ungulates, some rodents or primates). Thus, touch and olfaction appear to be the common sensory denominator of neonatal mammals and the most universally exploitable channels to shape infant-directed information. Third, mammaries are the only structures of the maternal body that newborns have to obligatorily contact in order to survive. More specifically, the offspring contact with the mammary gland involves a specialized cutaneous interface, the *papilla mammae*, commonly called nipple or teat (surrounded by an *areola* in certain primate species). According to the point above, the most efficient strategy of females to increase the localizability and graspability of nipples in their newborns should have been to shape tactilely and olfactorily conspicuous structures. This evolutionary coadjustment between mothers and neonates has worked directly by affecting anatomical features on or around mammaries that release chemostimuli. Mammary chemosignalization has also followed indirect ways, females in many species and newborns in all species spreading extramammary substrates on their nipples when they groom or suck.

The various sources of chemical cues or messages emerged on or around nipples under different functional forces. While some of them may convey exclusive communicative functions (e.g., the inguinal glands of ungulates), others may be secondary to local requirements to preserve nipple functionality against the heavy stress that offspring wield on it. The evolutionary specialization of nipples has indeed co-opted exocrine structures to protect cutaneous and ductal entries from bacterial invasion, create the airtight seal necessary for the efficacy of suction, and relieve the strong friction of the lips of sucking neonates (Schaal et al. 2008b, 2009; see below). Thus, mammary structures aggregate a variety of secretory/excretory sources giving off potentially odorous substrates.

The most obvious and profuse sources of mammary odorants are colostrum and milk, which odor properties depend on the female's lactational stage, dietary and aerial ecology, stress, and physical activity. Other mammary substrates are secreted or excreted by glands distributed in, on, or adjacent to, the mammary structure. The whole range of elementary skin glands is represented in the areola-nipple region, including eccrine, apocrine, and sebaceous glands. In some species, the mammary area is additionally endowed with sophisticated glandular specializations, working either in close functional link with lactation (e.g., human Montgomery's glands or the structure producing the rabbit mammary pheromone) or throughout reproductive life (e.g., ovine inguinal glands).

The substrates released onto the nipple-areolar skin can be mingled with those brought here from extramammary sources. These exogenous substrates vary according to the species considered: in some rodents and carnivores, parturient females actively lick their nipple-lines, labeling them with a mix of urogenital and amniotic fluids, blood, saliva, and all kinds of secretions from oral or facial glands. Nursing females also often alternate licking their offspring and their own ventral fur, spreading on themselves infant-specific odor traces (excretions or secretions from anal or urogenital sources) mingled with own substrates (saliva). Further, while sucking, newborns stain nipples with mixed amniotic fluid and saliva, and later with mixed saliva, milk and other facial substrates (from lachrymal, nasal, facial, or ear glands). To further expand this biochemical puzzle, both mammary and extramammary substrates certainly depend on surface processes involving salivary enzymes or the local commensal microflora. Finally, local conditions of heat, humidity, and texture of the mammary epidermis are an additional way to differentiate emitted odorants in terms of volatility (the dermis underlying mammary structures is highly vascularized, provoking higher surface temperature).

To sum up, multifarious biological substrates and processes make the chemistry that females present to their offspring on their mammaries and nipples complex. It should thus be a perceptual challenge for newborns to navigate through this chemosensory mosaic. However, although the first efforts of mammalian neonates to locate nipples are sometime diffident (often as a result of the inexperience of the female), most of them survive and therefore demonstrate their skills to localize and grasp them and obtain enough colostrum and milk to start life.

17.3   NEONATAL OLFACTORY COGNITION: LEARNING AND PREDISPOSITIONS

We do not fully understand how mammalian neonates perceive and analyze the olfactory scene that is associated with the mammary structure. But, by staying alive, the majority of them prove their competence to orient adequately from the very first exposure to a nipple. Neonates dispose of several ways to make sense of the chemosensory complexity of mammaries, some being shaped by exposure and learning effects, others working independently from learning. Opportunistic processes involving learning occur in all species examined so far, whereas unconditional processes have been evidenced in only some species. These differentiable perceptual mechanisms may be dedicated to the processing of different sets of compounds within the complex mixture composing mammary chemosignals (see below).

Neonatal mammals are outstandingly efficient learning machines. During a certain period after birth, they acquire as "positive" any stimulus associated with the mother's body or nest. Positive means here that these stimuli tap into the approach system of behavior. This avid information intake in the context of nursing has chiefly been investigated in the rat and mouse, and related findings can be summarized as follows (for reviews, see Alberts 1981; Blass and Teicher 1980; Brake et al. 1986; Leon 1994; Rosenblatt 1983; Wilson and Sullivan 1994). (1) Rat and mouse pups easily acquire artificial odorants applied on the female's mammary area, which suggests that natural odorants are learned in much the same way. (2) Odor learning is functional from birth (e.g., Miller and Spear 2008, 2010) and its efficiency increases as a function of maturation and experience; thus nipple grasping, already functional right at birth, ameliorates during the first days due to the establishment of sensory incentives and to neuromotor training (Armstrong et al. 2006; Bouslama et al. 2005; Dollinger et al. 1978; Rosenblatt 1983). (3) Neonatal olfactory abilities are largely premolded during fetal development in terms of sensitivity and preferences (Molina et al. 1995; Schaal and Orgeur 1992; Smotherman and Robinson 1987; Youngentob et al. 2007), and this fetal sensory imprint goes in parallel with a relative transnatal continuity of chemosensory cues (Mendez-Gallardo and Robinson, 2010; Pedersen and Blass 1982; Schaal 2005; Schaal and Orgeur 1992); specifically, in the case of the rat, amniotic fluid odor is made available onto the nipple-lines by maternal self-licking activity (Teicher and Blass 1977), and milk accumulates odor cues derived from the gestating female's general metabolism and diet (Capretta and Rawls 1974; Galef and Henderson 1972; Galef and Sherry 1973). (4) The reinforcing agents that potentiate neonatal learning of odor cues are multiple and redundant. Maternal and infant behavior afford numerous reinforcers, such as warmth, soft contact or targeted stimulation (anogenital licking), vocalizations, exertion of sucking, and postingestive (taste) and postabsorptive factors (gastric filling, brain sensing of satiety). All these reinforcers act separately on the learning of any associated odor cue, but their normally additive operation is most efficient (Brake et al. 1986). Milk itself supports the establishment of learned odor associations (e.g., Brake 1981). Finally, compounds conveyed in biological secretions may instantly potentiate the learning of co-occurring odor cues (e.g., presenting simultaneously an aversive orange scent with the odor of maternal saliva reverses the value of the orange odor into attraction [Sullivan et al. 1986]). Thus, milk and other excretions/secretions emitted from, or conveyed to, the nipples (e.g., saliva, amniotic fluid [Arias and Chotro 2007]) may alter the meaning of incidental odor cues in neonatal rodents.

The learning abilities of neonate rodents are certainly generalizable to all mammalian neonates, although only a handful of species has been investigated in detail. For example, rabbit pups can learn from the first day after birth any nonspecific odorant associated with nursing (Allingham et al. 1999; Coureaud et al. 2006; Hudson 1985, 1993; Ivanistkii 1962; Kindermann et al. 1994). After single odor-nursing pairing, such pups express the typical sequence of nipple searching and grasping on an unfamiliar female painted with the same odor (Hudson 1985). But the one-session learning of an odor associated with sucking appears only effective during the first 4 postnatal days (Kindermann et al. 1994); after day 5, nursing-induced odor learning vanishes completely, raising the possibility of a sensitive period for odor learning. Thus, the timing and the act itself of sucking are efficient promoters of odor learning in rabbit newborns, and any arbitrary odor cue sticking on a nipple can be assigned incentive value for the next suckling episodes (Hudson et al. 2002).

Sucking can also instigate learning of odorants associated with the breast or milk in human newborns (e.g., Delaunay-El Allam et al. 2006, 2010; Schleidt and Genzel 1990). But sucking is not a necessary condition as mere exposure or contingence with touch-induced arousal suffices to change the incentive value of initially irrelevant stimuli (Balogh and Porter 1986; Sullivan et al. 1990). Such acquisition of odor cues in human infants seems also subject to modulation by the birth process or the timing of its occurrence relative to birth (Romantshik et al. 2007).

Neonate mammals respond more intensely to, or learn more easily, certain stimuli than others before they were directly exposed to them. When responses to given odorants emerge without obvious reliance on previous exposure (even prenatal) *and* are resistant to deprivation from the specific stimulus or to its reassignment by other stimuli, the term "predisposed" may be used (Bolhuis 1996; Horn 2004). Such predisposed processes designate perceptual-motor loops generalized at the species level that are released from birth by stimuli that did not appear to occur in the prior developmental environment. Predisposed responses to odor substrates have rarely been investigated in mammalian neonates, but these rare cases are of particular interest. In general, odor stimuli recruit different response levels in neonates. When presented for the first time, novel odorants elicit sniffing or increased respiration, generally followed by withdrawal. In contrast, certain odorants do release appetitive oronasal investigation in absence of prior direct exposure to them. Such immediate oral grasping response is observed with fresh milk or its odor. It can be elicited at or before gestational term, before any contact with a lactating female or her milk. For example, human newborns respond to the odors of the lactating breast or milk by positive head orientation and appetitive mouthing regardless of the rate of prior exposure to the breast (Delaunay-El Allam et al. 2006; Makin and Porter 1989; Marlier and Schaal 2005; Porter et al. 1991; Russell 1976). Further, premature neonates react by increased mouthing and sucking movements to conspecific milk odor (Bingham et al. 2003a; Raimbault et al. 2007).

These responses of newborns to conspecific milk odor are not easily overcome by newly learned odorants. Indeed, when tested for relative preference between human milk odor and an artificial chamomile odor spread on the areolae at each feed since birth, breastfed newborns display equivalent orientation to either stimulus (Delaunay-El Allam et al. 2006). In addition, when simultaneously presented with human milk odor (from a nonfamiliar mother) and the odor of their cow's milk-based formula, infants who were exclusively bottle-fed since birth demonstrate more appetence for the unfamiliar conspecific milk than for the familiar artificial milk (Marlier and Schaal 2005). Thus, odor chemostimuli carried in human milk are more reinforcing to human newborns than nonspecific odorants that were rewarded by sucking or satiety for several days.

In summary, the behavior of mammalian neonates appears to be driven by multiple olfactory mechanisms underlain either by plastic, experience-dependent processes or by predisposed processes. Since learning processes are easier to apprehend experimentally, they have received much more empirical attention than predisposed processes, and are thought to predominate in the control of neonatal adaptive behavior. It is clear that the extrafast odor learning abilities of newborn mammals make it difficult to characterize predisposed perceptual processes, as would be required if pheromones were involved, such compounds being operationally defined to imply no or minimal experiential induction of their biological activity (see next section).

17.4 ETHOCHEMICAL LOGIC TO DECONSTRUCT MAMMARY ODORS

As noted in Section 17.2, nipples and the areas of maternal skin harboring them are biologically complex with overlapping sources of multiple secretions and excretions. In Section 17.3, mammalian neonates were shown to address this "blooming, buzzing" confusion of chemical cues apparently without great trouble or need for training, which suggests that their chemosensory skills are somehow tailored to make sense of odors, and specifically of mammary odors.

The first ways for the neonatal brain to segregate this complexity of mammary chemostimuli may be based on physicochemical phenomena that affect ligand-receptor interactions constituting nasal and oral chemoreception, such as volatility, polarity, solubility, functional moieties, stability, and multicomponentiality. For example, volatile and involatile fractions of mammary secretions may lead, respectively, to detection from a distance or to the need for direct contact for perception to occur. Involatile proteins, lipids, or hydrocarbon may in this way act as carriers or precursors of volatile ligands, or they can protract the emission duration of associated volatile compounds. Differences in volatility may be linked to contrastive sensory impact, such as transient alarmlike effects depending on volatile/diffusible compounds, while heavier polar compounds may end in long-lasting attractant effects (Alberts 1992; Müller-Schwarze 2006). But involatile compounds do also work as chemosignals as recently shown in murine main urinary proteins (MUPs) that encode individual identity (Hurst et al. 2001) and induce learning of associated volatile fractions (Roberts et al. 2010).

Another potential perceptual split of complex mammary odors may separate individual-specific from species-specific components. Some odor compounds reflect idiosyncratic traits of the female (e.g., her atmospheric environment, diet, level of stress, physiological state, health and parasite load, and lactational age) or of the young (diet, physiological state, sex, age), while others carry higher-level categories of meanings (e.g., genus/species; population, group, or kin identity). Depending on the type of behavioral test used to assess responses, neonates' ability to extract individual or supraindividual meanings can be evidenced from a same substrate. For example, rabbit or human newborns reveal that conspecific milk can carry odor cues related to the individual mother or to any lactating female of the species (Coureaud et al. 2002; Marlier and Schaal 2005; Schaal 2005). In the same way, lambs are strongly reactive to inguinal secretions from any ewes, but much more when they are from their own mother (Vince and Ward 1984).

A third divide in the perception of social odors in general and in mammary odors in particular concerns the notions of cue and signal. It has repeatedly been proposed (e.g., Dusenberry 1992; Hauser 1996; Maynard-Smith and Harper 2003; Wyatt 2010) to separate *cues*, taken as informative elements that derive from normal life sustenance processes in the emitter, from *signals*, taken as informative elements that "alter the behavior of other organisms, which evolved because of that effect, and which are effective because the receiver's response has also evolved" (Maynard-Smith and Harper 2003, p. 3). While cues may be lastingly on, signals may be switched on-off according to the emitter's behavior or condition (Hauser 1996). Mammary odor cues would include compounds derived from maternal physiology (diet, wastes, hormonal state, stress) without added cost involved to produce them. In contrast, mammary odor signals would designate compounds that may be released by specialized structures, exploit specific response biases in the receiver, and evolved for a specific signaling function (or were secondarily recycled for such a function). In the context of mammary odor mixtures, this would imply that rare signals are embedded in a system of abundant cues, and that looking for a signal is like seeking a needle in a haystack of cues. The task is even trickier when cues and signals, although they differ in the developmental process leading to their activity, release functionally equivalent responses. While odor cues are typically constituted of circumstantial and variable odorants, odor signals are relatable to the concept of the pheromone.

Once chemically identified, a behaviorally active odor compound can be screened to assess whether it can be construed as a pheromone. The initial definition of the concept (Karlson and Lüscher 1959) designated "substances, which are secreted outside by an individual and received by a second individual of the same species, in which they release a specific reaction, for example, a definite behavioral or developmental process". Following this minimalist definition, almost every chemosensory signifier exchanged between conspecifics can be argued to act as a pheromone. To prevent latent confusion on the nature of the compounds involved as well as on the nature of elicited responses in mammals, Beauchamp et al. (1976) resized the concept of pheromone so that it better matches the cognitive complexity of mammalian behavior (see also Doty 2003; Johnston 2000). To name a candidate chemostimulus a pheromone, they proposed that the compound: (1) is chemically "simple" (being composed of a monomolecular compound or a very small set of chemicals in fixed ratio), (2) releases in a conspecific receiver an adaptive response that is morphologically invariant in a same context (or, better, in different contexts); (3) these responses should be elicited in a selective way by the considered compound that should thus be tested against several reference compounds, (4) their taxonomic specificity should be established, and (5) the coupling between the candidate chemostimulus and the response should not depend on previous exposure and learning; thus, prenatal exposure, facilitated learning during the natal process, or rapid learning immediately after birth should be eliminated as possible explanations of its behavioral activity.*

Mammary-related odorants have been mostly investigated in milk, the substrate that is most practical to handle with chemoanalytic techniques. Milk is nevertheless physicochemically and biochemically multifaceted, and the fraction(s) responsible for its chemosensory activity in newborns is (are) difficult to characterize. A first step in reducing the complexity of milk has been to analyze its volatile fraction. But even milk volatiles have a complex profile, leading to gas chromatographic (GC) tracings ranging from more than 150 peaks (e.g., in ovine milk [Moio et al. 1996], in rabbit milk [Schaal et al. 2003]) to 20–40 peaks (in human milk, e.g., [Büttner 2007; Shimoda et al. 2000]). In addition to volatile compounds, numerous nonvolatile lipids, proteins, or polysaccharides are chemosensorily active by themselves or act as carriers of volatile compounds (e.g., Murakami et al. 1998). Thus, the chemical dissection of the behaviorally-active components of milk is a complicated endeavor, and it has so far been carried out in only few mammalian species.

17.5 EVIDENCE FOR PHEROMONES IN MAMMARY CHEMOSTIMULI?

The contribution of nasal chemoreception to neonates' response to mammary and extramammary substrates will be summarized here in the mammalian taxa which have received more attention so far, namely rodents, lagomorphs, and primates. For

---

* Accordingly, to keep the pheromone concept meaningful, any candidate secretion, excretion, or fractions of them bearing repeatable behavioral and/or physiological effects on newborns (the excretion of a gland, abdominal skin secretions, milk, saliva, etc.) should not be termed pheromone. If such biological mixtures verify some of the above criteria, they might be named "candidate secretions for pheromonal mediation" (Doty 2003) before the long way to the chemical identification of the active compound(s). In any case, the term pheromone should be reserved for clearly identified chemicals that have undergone systematic screening of above operational criteria.

representative species of each of these groups, evidence for structures located in, on, or around the mammaries or for significant olfactory indices produced endogenously (in lacteal secretions) or exogenously (in extramammary substrates) will be surveyed. Further, we will consider whether mammary chemostimuli can be sampled for separate experimental restitution to neonates, whether chemically identified signals are embedded among these mammary odor mixtures and whether these can be categorized as pheromones (i.e., as signals distinct from ordinary, learned odor cues in following the operational criteria outlined in Section 17.4).

17.5.1 RODENTS

17.5.1.1 Mammary Sources of Behaviorally Active Chemostimuli

Studies on the topic have been conducted in the laboratory rat and mouse which show altricial newborns rely mostly on olfaction in their interaction with the dam (reviews in Alberts 1976, 1981; Blass 1990; Blass and Teicher 1980; Rosenblatt 1983). An odor factor from the ventral skin of lactating dams attracts rat or mouse pups. In mice, this substrate appears most active at short range, suggesting low volatility (Al Aïn et al. 2011; Hongo et al. 2000). The specific source of this odor factor is unknown, but nipples are important (although not exclusive; see Singh and Hofer 1978). When rat nipples are washed with organic solvents, the resulting solution distillated, and the distillate then applied on a nipple rendered inactive by prior washing, pups resume oral grasping of it (Teicher and Blass 1976, 1977). Rat and mouse nipples are endowed with apical sebaceous glands opening into the ductal ostia (Toyoshima et al. 1998a, b) whose size is maximal at the end of gestation and during lactation. The surface of the nipple changes drastically during lactation with increasing furrowy texture (Toyoshima et al. 1998a) that favors the accumulation of cutaneous secretions or milk, and of skin microflora.

Milk is expected to olfactorily tag nipples, although current data on this point is lacking. When rat milk was painted on olfactorily inactivated nipples, normal nipple grasping response was not restored in 8–9-day-old pups (Singh and Hofer 1978), but this experiment was too imprecisely reported to be conclusive. It is a paradox that most research on rat pup responses to milk has used cow's milk or cow's-milk-based formulas rather than conspecific milk (e.g, Ackerman and Shindledecker 1978; Cheslock et al. 2000; Koffman et al. 1998; Mendez-Gallardo and Robinson 2013; Petrov et al. 1997; Smotherman and Robinson 1994; Terry and Johanson 1987). Such responsiveness of newborn rats to heterospecific milk may not be generalizable to rat milk, implying that more research is needed here.

In contrast, recent investigation in the mouse indicates that neonatal pups respond to the odor of fresh murine milk (Al Aïn et al. 2012a, b; Logan et al. 2012). In an attempt to track odor-active substrates that elicit initial nipple seizing/sucking in newborn mice, Al Aïn et al. (2013) and Logan et al. (2012) coated olfactorily inactive nipples with murine amniotic fluid, milk, or maternal saliva. While pups were unresponsive when presented inactivated nipples, typical approach/sucking response was reinstated after painting these biological fluids on them. To control for earlier exposure effects possibly explaining this result (pups aged < 12 hours and hence, exposed to these stimuli while suckling), Logan et al. (2012) assayed pups delivered by Cesarean section and deprived of suckling prior to the test. When facing nipples coated with amniotic fluid, maternal saliva, or milk, these "premature" pups only grasped the nipples bearing amniotic cues, suggesting that the activity of the other substrates—specifically milk—was conditional on previous postnatal exposure. However, Logan et al.'s and Al Aïn et al.'s findings appear divergent. In Logan et al.'s conditions, milk was *ineffective* to trigger nipple grasping in pups deprived of prior suckling. In contrast, in Al Aïn et al.'s (2012b, 2013) conditions, unsuckled newly born pups displayed positive attraction toward milk or seizing/sucking of a nipple coated with fresh milk. These contradictory findings certainly reside in procedural and/or stimulus differences between studies (age of pups at testing: 1 hour postpartum vs. 4–6 hours; birth experience and perinatal exposure to anesthetics: vaginal delivery vs. Cesarean section; milk stimulus: fresh vs. aged). Regarding the stimulus, Al Aïn et al.'s studies took special care to use fresh murine milk no more than 15–20 minutes after ejection without freezing or deep-freezing (as both standing of milk after ejection and freezing do notably alter the profile of its headspace (Keil et al. 1990; Spitzer and Büttner 2013).

17.5.1.2 Extramammary Sources of Behaviorally Active Chemostimuli

Based on the fact that rat and mouse females self-lick during gestation, parturition, and lactation (Roth and Rosenblatt 1966), saliva, and amniotic fluid are presumably spread ventrally. Positive reactions to birth fluids are indeed observed in rat and mouse newborns (Hepper 1987; Kodama 1990, 2002; Kodama and Smotherman 1997; Mendez-Gallardo and Robinson 2012). The impact of amniotic odor has been assessed in the context of nursing (Logan et al. 2012; Teicher and Blass 1977): if nipples rendered inactive by prior washing are thereafter painted with amniotic fluid, oral seizing recovers at subnormal levels. In the normal course of events, newly born rat pups deposit amniotic fluid blended with saliva when they root in their mother's abdominal fur, and saliva of a parturient or lactating dam elicits attraction in pups (Sullivan et al. 1986) and reinstates their grasping of a prewashed nipple (Teicher and Blass 1977). Pup saliva, as well as pup salivary gland extract, are also efficient in restoring grasping of prewashed nipples (Pedersen and Blass 1981). Finally, when rodent females self-groom, they spread a range of secretions of oral-facial (e.g., Harderian or lachrymal glands), anogenital, or pedal origins over their ventral fur (Thiessen et al. 1976). They lick then their pups' anogenital area, consuming their urine (Friedman and Bruno 1976) and next lick their own ventrum, leaving there pup secretions/excretions, originating in urine or feces, and preputial or anal glands. Thus, the nipple-lines may receive a mixture of anogenital glandular or urinary volatile and involatile compounds (e.g., lipocalins such as MUP).* It may be noted that such involatile proteins are released in mammary, parotid, sublingual, submaxillary, and lachrymal glands (Shahan et al. 1987), so that all these secretions could carry common cues.

In sum, at least in the rat, current evidence indicates that the biological substrates that direct mammary localization and drive nipple grasping by pups may originate

---

* Although pure urine from lactating dams appears weakly effective (Teicher and Blass 1976, 1977).

from extra-mammary sources conveyed by females and newborns themselves rather than from mammary sources. In contrast, recent research with mouse newborns indicates strong activity of both mammary and extramammary substrates in initial nipple searching/grasping. However, much more research is needed here.

17.5.1.3   Evidence for Pheromones?

Do newborn rodents respond to odor stimuli on nipples before direct exposure to them? So far this issue does not seem to have been directly addressed using conspecific stimuli (e.g., Cheslock et al. 2000, for cow's milk). However, extramammary sources; that is, pup saliva, pup salivary gland extract, and suckled nipple wash extract, are reliable elicitors of newborn rat nipple grasping. These secretions were accordingly subjected to GC-MS with the goal to pinpoint dimethyl disulfide (DMDS), a compound that was *a priori* inferred to be active by extrapolation from its sexual attractant properties in male hamsters (Singer et al. 1976). Synthetic DMDS was indeed shown to be effective in eliciting nipple grasping in 3–5-day-old pups, but with a low releasing potency (approximately 50%) relative to that of olfactorily intact nipples (Pedersen and Blass 1981). Thus, unknown compounds from the natural mixture coating nipples do carry additional impact. In these studies, rat pups were aged 3–5 days, implying that the activity of DMDS may derive from prior exposure. Nevertheless, DMDS is behaviorally singular as it reduces aversive responses to noxious stimuli in the fetal rat (an effect mediated by opioidergic processes, suggesting unconditional reward properties [Smotherman and Robinson 1992]). DMDS is reported not to be detectable by GC in amniotic fluid (Blass 1990), leading to the logical conclusion that its behavioral activity does not depend on prenatal exposure. Thus, until the behavioral activity of DMDS is further assessed in newly born rats, it may be tentatively considered as a salivary signal for nipple attachment. But to be categorized as a pheromone, its species-specificity has to be proven and its action waits further testing for unspecific arousal effects against reference compounds.

Volatile amines that are abundantly excreted in rodent milk (Pollack et al. 1992) may constitute putative mammary chemostimuli for neonatal rodents. Although their behavioral activity remains as yet untested in neonatal rats and mice, they have strong affinity for a subclass of odorant receptors, the trace amine-associated receptors (TAARs) expressed in the main olfactory epithelium and in the Grueneberg ganglion of the mouse (Fleischer et al. 2007; Liberles and Buck 2006). Interestingly, the developmental course of TAAR expression in the Grueneberg ganglion is much higher in late fetal (embryonic day 17.5) and neonatal mice (postnatal day 0–1) than in week-old pups (postnatal day 7) and adults, suggesting a chemosensory role for the Grueneberg ganglion and amines in the earliest adaptive responses (Fleischer et al. 2007). This point is currently under scrutiny. Otherwise, sulfur-containing volatiles identified in adult rat breath (carbon disulphide and carbonyl sulfide) may also bear precocious behavioral activity. The former was shown to work as an attractant and a reinforcer in subadult rats and mice (Bean et al. 1989; Galef et al. 1988; Munger et al. 2010), but these compounds have not yet been assayed with neonates and nurslings.

Finally, it cannot be excluded that through their avid licking of pups' anogenital area and consumption of their urine (Gubernick and Alberts 1983), lactating

females spread on/around nipples traces of urinary and urinary tract glandular chemostimuli. This mammary distribution of extramammary substrates may concern,
for example (1) dodecyl propionate, a compound emitted in neonatal rats' preputial
secretions known to release avid licking (Brouette-Lahlou et al. 1991a, b), (2) MUPs
that themselves bear behavioral activity in adult mice (Roberts et al. 2010), and/or
(3) active MUP-bound ligands known to carry pheromonal effects in young mice
(Jemiolo et al. 1987, 1989). So far, these stimuli have not been assayed with newborns. Taken together, extant data does not clearly establish that rodent females emit
infant-directed pheromones from their mammary structures. There is however some
evidence that extramammary substrates applied on nipples operate as orientation
and grasping cues.

17.5.2 LAGOMORPHS

In the mammalian order of lagomorphs, the most complete understanding of odor-
based neonatal behavior stems from the European rabbit, *Oryctolagus cuniculus*.
Rabbit newborns display a typical pattern of probing and searching in the female's
abdominal fur when she enters the nest, which generally ends in orally grasping a
nipple in less than 15 seconds (Hudson and Distel 1983; Schley 1976). This swift
response is mediated by olfaction (vision and audition being nonfunctional during
the first week). When nasal chemoreception is suppressed, pups' ability to locate
nipples is lost (Schley 1977, 1979).

17.5.2.1 Mammary Sources of Behaviorally Active Chemostimuli

Female rabbits, particularly when lactating, release searching/oral grasping in pups
exposed to their ventral fur (Coureaud and Schaal 2000; Hudson and Distel 1984,
1990; Schley 1976). When these cues are altered by washing the lactating female's
abdomen, the pups are delayed in finding nipples (Müller 1978). Covering nipples
with an airtight film disrupts pup searching at various rates according to the degree
and location of masking (Coureaud et al. 2001; Hudson and Distel 1983). Finally,
pups are more reactive to nipples excised from lactating females than to nipples
excised from nonlactating females (Moncomble et al. 2005). Thus, a major source
of behaviorally active compounds on rabbit females' ventrum is on the nipples.
Currently, little is known on the histological origin of these chemostimuli, but several sources may be involved (Moncomble et al. 2005), including (1) intensified epi-
dermal keratinization of the nipple during lactation, which induces higher release
of surface lipids, (2) increased output from sebaceous glands located at the base of
the nipple, which may be involved in intrinsic signaling function or in sequester-
ing milk compounds; pup grasping responses are reduced or abolished by washing
these surface cues away on excised nipples, and (3) minute oozing of milk, which
volatiles release pups' searching-grasping response (Coureaud et al. 2002; Keil et al.
1990; Müller 1978; Schaal et al. 2003). While the behavioral activity of rabbit milk
fades away within 30 minutes after milking (Keil et al. 1990), the odor of nipples
excised from lactating rabbits does not (Moncomble 2006), suggesting that surface
compounds either bear intrinsic behavioral activity or do preserve the activity of
remnant traces of milk.

The compounds that render rabbit milk behaviorally active to neonate pups can in principle originate from environmental sources (i.e., mother's diet) transferred into milk and/or from compounds synthesized *de novo* in the mammary tract. The intramammary source of active compounds is attested by an experiment that compared the activity of milk sampled in either the alveoli, the ducts below the nipple or after ejection (Moncomble et al. 2005). Only ejected milk was behaviorally efficient, designating the terminal part of the milk ducts as the possible source of active compound(s). Histological analyses reveal indeed that the milk ducts in the terminal portion of the nipples form an enlarged, convoluted sinus lined with secretory epithelium (Moncomble 2006). Alternatively, some involatile or bound substances carried in milk might be oxidized at contact with air, leading to the instantaneous release of volatile compounds.

17.5.2.2 Evidence for Pheromones

As mentioned above, rabbit pups locate and grasp a nipple when put on the abdomen of any lactating doe or exposed to milk from any female (Coureaud et al. 2000; Keil et al. 1990). Since the behavioral activity of rabbit milk is fully conveyed in its headspace, gas chromatography was suitable. Using a gas chromatograph with a flow that was split between the mass spectrometer and an olfactory port to which pups were directly presented, the separative analysis of the headspace of fresh rabbit milk resulted in the identification of one compound among 21 candidates (Schaal et al. 2003). This compound, 2-methyl-but-2-enal (2MB2), being as efficient as fresh rabbit milk to elicit searching-grasping motions, it was considered as a putative signal, and therefore submitted to systematic tests verifying the five pheromone criteria specified in Section 17.4:

Criterion 1: 2MB2 is the chemically "simplest" possible stimulus as it is composed of a single molecular "species." Its activity is extraordinarily strong in releasing searching-grasping actions, although it cannot be excluded that additional, not yet identified, milk or nipple compounds may act synergistically. However, such synergy will be difficult to assess as the effect of 2MB2 on the typical responses of pups is ceiling during the first 10 postnatal days.

Criterion 2: The macroscopic structure of rabbit pups' responses to pure 2MB2 is not differentiable from that of entire fresh milk, indicating that a single key compound from rabbit milk can mimic the response elicited by milk itself. Further, during the first day 10 days after birth, 2MB2 is behaviorally efficient regardless of the context or mode of presentation (in individual tests presenting the stimulus on a glass rod or at the sniff port of a GC apparatus; in collective tests in the nest).

Criterion 3: The selective activity of 2MB2 was ascertained by comparing pup responsiveness to 40 odorants represented or not in rabbit milk or suspected to act as chemosignals in other species (among which DMDS shown to be behaviorally active in rat pups). These reference odorants were ineffective to release the criterion response in rabbit newborns at any tested concentration (Coureaud et al. 2003), so the behavioral activity of 2MB2 could not be explained in terms of novelty or nonspecific arousal effects. Further, the activity of 2MB2 was limited within a range of stimulus concentrations extending over 5 log units* (10^{-9}–10^{-5} g/ml; Coureaud et al. 2004), suggesting some flexibility in 2MB2 intensity eliciting the typical response in pups.

Criterion 4: The species-level generality of the releasing potency of 2MB2 was established in showing its independence from maternal diet and genetic background (Coureaud et al. 2008; Schaal et al. 2003). Further, it was inactive in newborn rats, mice, cats, and humans (Contreras et al. 2013), and even in pups of phylogenetically related brown hares, *Lepus europaeus* (Schaal et al. 2003).

Criterion 5: Pup responsiveness to 2MB2 develops without need of previous direct exposure to it. To acquire its behavioral efficacy, 2MB2 does not require being contingent of labor-related arousal states, of suckling with or without ingestion of milk, or of contact with the mother: pups taken away from the mother immediately after birth display maximal response to it at very first presentation. Additionally, the 2MB2 stimulus-response loop remains functionally unaltered by long-term deprivation right after birth: separating pups from their mother and hand-feeding them with a cow's-milk-based formula devoid of 2MB2 for 6 days left their high-level responsiveness to the 2MB2 unchanged on day 6 (Coureaud et al. 2000c). Furthermore, 2MB2 appears efficient in fetuses delivered 1–2 days before gestational term, and 2MB2-targeted GC-MS analyses in amniotic fluid and blood plasma of pregnant and lactating females failed to detect it. Although it cannot be excluded that 2MB2 is present in these fluids at concentrations below the detection level of the GC, this suggests that the behavioral activity of 2MB2 may not derive from prenatal experience (Schaal et al. 2003).

In sum, current data indicate that the 2MB2-behavior coupling requires neither prenatal nor postnatal direct exposure to become functionally specified. These findings allowed categorizing 2MB2 as a pheromone carried in rabbit milk, in the sense of Beauchamp et al.'s (1976) and Johnston's (2000) operational redefinition of the concept. As 2MB2 appears to be produced somewhere in the mammary tract, presumably in the final portion of the nipple to be discharged into milk, it was named mammary pheromone (MP) (Schaal et al. 2003).

In addition to the completion of a set of physicochemical and biological criteria, any candidate pheromone awaits further demonstration for involvement in mutually beneficial functions between emitter and receiver. Participation of the MP in the reciprocal exchanges of the rabbit female and her offspring is clear on both proximate and ultimate levels of analysis. On the offspring side, it elicits immediate arousal and mobilization of directional actions when the female enters the nest, offers guidance, and favors searching/grasping of a nipple and ingestion of milk. On the female side, the MP may boost tactile stimulation from pups toward the abdomen, stimuli known to trigger and sustain lactational physiology. A critical consequence of pups' reactiveness to the MP is highlighted by the fact that individual pups that do not respond to the MP on postnatal day 1 mostly die during the following 4 weeks (Coureaud et al. 2007). Thus, initial reactivity to the MP is predictive of long-term viability. Another point of

---

* The fact that these concentration values of 2MB2 in the stimuli were those in the experimental water solutions seems to have escaped some authors (Doty 2010; Hudson et al. 2008). The 2MB2 concentrations in the headspace of these solutions—to which rabbit pups were actually exposed—were not titrated, but can be expected to be lower than those in the aqueous solutions.

functional interest is that the behavioral effectiveness of the MP closely matches the period when pups need to contact nipples to ingest milk (i.e., between birth and weaning), and more precisely during the 10-day period when they depend only on olfaction to satisfy their exclusive need of milk (Coureaud et al. 2008; Montigny 2008). Finally, the MP triggers automatic responses that ensure that pups are ready to grasp a nipple at any time during the first postnatal days; MP-induced oral grasping is then compulsory at each presentation of the mother, and it is only later that oral activity of pups comes to be modulated by circadian or metabolic factors (Montigny et al. 2006).

17.5.3 PRIMATES

Our current knowledge about primate behavior in the nursing relationship stems essentially from human studies, nonhuman primates being more difficult to investigate in detail regarding the sensory processes that control mother-infant interactions. Darwin (1877) first intuited that human infants might use, among other cues, odors to orient to their mother's breast. This was confirmed a century later by a boom of experiments assessing neonatal responses to odors emitted from the breasts of lactating women. Such odors were indeed shown to reduce arousal in active newborns (Nishitani et al. 2009; Schaal 1986; Schaal et al. 1980; Sullivan and Toubas 1998) and increase it in somnolent ones (Russell 1976; Soussignan et al. 1997; Sullivan and Toubas 1998); to elicit positive head turning (Macfarlane 1975; Makin and Porter 1989; Schaal et al. 1980), stimulate oral (Russell 1976; Soussignan et al. 1997) and respiratory activity (Doucet et al. 2009), favor directional crawling (Varendi and Porter 2001; Varendi et al. 1994) and the opening of the eyes (Doucet et al. 2007) with important consequences for the development of multimodal perception (Durand et al. 2013; Schaal and Durand 2012). Thus, *Homo* newborns are clearly affected by odor cues emitted from their mother's breasts.

17.5.3.1 Mammary and Extramammary Sources of Behaviorally Active Chemostimuli

The human nipple/areolar region abounds in apocrine and sebaceous glands which ducts open on the tip of the nipple and give off secretions during lactation (Montagna and MacPherson 1974; Perkins and Miller 1926). Eccrine sweat glands and large sebaceous glands are also found on the areolae (Montagna and MacPherson 1974), where the surface is also dotted with small prominences (Morgagni's corpuscles) that host Montgomery's glands (MG) (Montgomery 1837), composed of sebaceous glands combined with miniature mammary acini (Montagna and Yun 1972; Smith et al. 1982). The quantitative assessment of MG prevalence, distribution, and patent activity in breastfeeding women (Doucet et al. 2012; Schaal et al. 2006) indicates that 97% of (Caucasian) women have more than 1 unit/areola, and 83%, from 1 to 20 units/areola. These MG can give off a latescent fluid (Doucet et al. 2012; Schaal et al. 2006). Colostrum and milk released from main lactiferous ducts add their intrinsic olfactory qualities to the areolae. The quality/intensity of lacteal secretions are in part influenced by odorous compounds transferred from the maternal diet (Hausner et al. 2008; Mennella and Beauchamp 1991a, b, 1996; Schaal 2005).

Extramammary substrates conveyed by the newborn or mother also contribute to the mammary odor in humans. Regarding the newly born infant's contribution, such extraneous sources can be amniotic fluid, vernix caseosa, vaginal secretions, and blood, as well as tears, mucus, saliva, and saliva-milk coagulate spread on the breast during the first sucking attempts. Mothers may add some artificial scents as they often smear their areolae with locally prescribed emollients (e.g., Delaunay-El Allam et al. 2010).

Taken together, these varied sources of mammary and extramammary substrates create a multifaceted and dynamic areolar odor blend. Lipids issued from keratinizing epidermis, sebum from free sebaceous glands and MG, as well as fatty acids from milk, may all act as odor fixatives that improve the chemical and temporal stability of the odor mixture formed on the areolae. The intricate arrangement of sebaceous and lacteal sources within the MG favors the mingling of sebum with areolar milk. In addition, local biochemical and thermal processes may selectively release given compounds or categories of compounds from this mixture. Thus, salivary enzymes spread by the suckling infant may speed up the release of odor-active compounds (Büttner 2002), and areolar skin temperature fluctuations due to the vasoactivity of underlying Haller's areolar plexus (see below) may segregate volatiles differing in vapor pressure (Schaal et al. 2009).

The complexity of the scent of the human areolar-nipple area and the difficulty in assaying human newborns makes analytic efforts uneasy. A study evaluated how far the morphologically differentiable areas of the breast could elicit distinct behavioral effects in newborns (Doucet et al. 2007). These areas were fractionated in applying an odor-free plastic film directly onto the breast of lactating women, various openings in it allowing to subtract the areolar contribution from the whole breast odor, and to separate the areolar contribution from those of the nipple or of oozing colostrum/milk. The behavioral impact of the areola was examined in approaching hungry newborns from the selectively unmasked breast regions of their mothers. Corresponding odor cues modulated infants' arousal states and promoted appetitive oral activity, but the responses were not differentiable between the whole breast odor, the isolated areola or nipple odors, or separate milk odor, suggesting equivalent attractive potencies of all breast stimuli, with all being equivalent in activity with fresh human milk. Such behavioral uniformity of the breast regions may be due either to overlapping compounds resulting from shared exocrine sources or from cross-contamination, or to distinct compounds bearing similar attractiveness due to similar reward associations. In a finer-grained study, 3-day-old newborns were exposed to the fresh secretion from MG (Doucet et al. 2009). Neonatal reactivity to these areolar odors tested against several reference stimuli (e.g., human milk or sebum, solvent, vanilla, fresh cow's milk, cow's-milk-based formula) showed that pure MG secretion elicits more orofacial activity.

Colostrum and transitional milk were also tested separately as sources of active odor cues for infants ranging in age from minutes to weeks after birth. The odors of colostrum (from postpartum days 1–2) and milk (from postpartum days 3–4) were shown to elicit reliable positive head turns (Marlier and Schaal 2005; Marlier et al. 1998). They were also strong releasers of facial and oral responses (mouthing, protruding tongue, rooting, sucking) indicative of their attractive and appetitive value to infants born at gestational term (Mizuno and Ueda 2004; Russell 1976; Soussignan et al. 1997) or preterm (Bingham et al. 2003a, 2007). Finally, the odor of human milk has repeatedly been shown to attenuate pain responses in newborns (e.g., Mellier et al. 1997; Nishitani et al. 2009; Rattaz et al. 2005).

To sum up, the above results indicate that odors from human mammary secretions are clearly detectable to infants aged from about 2 months before term to more than 1 month postbirth, and that they have particular behavioral effects on arousal states, general attraction, appetitive actions, and self-regulatory responses.

17.5.3.2 Evidence for Pheromones?

So far no evidence for any chemostimulus that would qualify as a pheromone is at hand in primate, including human, mother-to-infant communication. In humans, two mammary-related substrates, colostrum/milk and the secretion from the areolar glands of Montgomery, are valuable candidates for systematic analyses. These substrates are emitted by any lactating females (i.e., postparturient women that are unrelated and unfamiliar to the tested infant), do elicit infants' behavioral and psychophysiological responses that are species-specific (i.e., differentiable from heterospecific secretions), and are dissociable from nonspecific arousal effects caused by any odorant. In addition, the behavioral activity of these secretions does not derive from postnatal experience with breast-related stimuli, although the role of prenatal induction cannot be excluded. Clearly, further investigation is needed to confirm these first results and to substantiate whether mainstream milk and Montgomerian secretion convey redundant or distinct odor information to infants. These different mammary secretions wait to be subjected to chemical analyses to pin down volatile compounds that can then be brought under the nose of human newborns in repeatable bioassays.

17.6 MAMMARY CHEMOSTIMULI AS ONTOGENETIC ADAPTATIONS

Some authors have questioned whether the rabbit MP can be construed as a phero-mone because rabbit pups responses to it "is not invariant, being influenced by age and degree of hunger" (Doty 2010, p. 91). Context independence of the response has indeed often been required for the development of reliable behavioral assays for pheromones (e.g., Beauchamp et al. 1976; Müller-Schwarze 1977; Thiessen 1977), the optimal response being a kind of reflex or a "fixed action pattern" with minimal intra- and interindividual variability. Rabbit pups provide indeed a high level of mor-phological stereotypy of their response to the MP when tested at a same age. But che-mosignaling/reception devices cannot be conceived in independence of integrated biological systems, and hence, they can only fluctuate as a function of both external and internal conditions (e.g., Alberts 1987; Wyatt 2003, 2010).

17.6.1 Contextual Fluctuations in Mammary Chemosignaling

The emission rate or activity of mammary chemical signals fluctuates intrain-dividually along lactation and at each nursing episode. For example, the mixture of abdominal odor cues of lactating rabbits or cats is more efficient to release pup searching in early rather than late lactation (Coureaud et al. 2001; Hudson and Distel 1984; Raihini et al. 2009). Little is known about the endocrine control of such abdominal odor cues and whether it operates in anticipation of the engagement of nursing behavior, in response to the solicitation by offspring, or both. In rabbit females, sucking-related tactile stimulation of the nipples triggers prolactin and oxytocin release that controls milk production and ejection (Summerlee et al. 1986), and affects the tonic release of sebum (Wales and Ebling 1971). In rats, an injection of oxytocin reinstates the olfactory attractivity of lactating females' abdomen after its disruption by washing (Singh and Hofer 1978). The endocrine effects of sucking-related somesthesis increase up to 15–20 days postpartum and then progressively drop with the inception of weaning (Summerlee et al. 1986), providing a basis for long-term variation in mammary-related chemoemission.

In the rabbit, pup reactivity to the complex odor of milk and to the pure MP follows fluctuations comparable to those noted with whole abdominal odor cues: when 2-day-old pups are exposed to the odor of fresh rabbit milk collected either on early or late lactation (postpartum days 2 and 23, respectively; weaning taking place on day 30 in domestic conditions), the behavioral activity of milk declines between both sampling points, in the same time as the MP concentration in the milk headspace decreases (Coureaud et al. 2006). Thus, the MP in milk or compounds in abdominal cues may be involved in the postnatal course of young-to-female interactions.

In human females, lactation-related variations in the olfactory attractiveness of the breast for infants remain poorly understood. Increased sebaceous productivity of the MG appears effective in late pregnancy and early lactation (Burton et al. 1973), leading to presumable variations in the amount/composition of areolar secretions. If MGs support communicative function, one may expect enhanced secretory output right after delivery and then before each ensuing nursing bout. The rate of women that have secretory MGs tend indeed to be higher on postpartum days 1–3 than on days 15 or 30 (Schaal et al. 2006), but more conclusive data are needed here.

Mammary odor varies also qualitatively and/or quantitatively along the nursing cycle. The abdominal odor of lactating rabbits is more efficient to release pup searching in prenursing than in postnursing condition (Coureaud et al. 2001). Nursing-related variations in lactogenic hormones may indeed favor milk emission at the nipple (and in humans in the lactiferous component of MGs) or the secretory activity of mammary-related skin glands. Infant attraction to the mammae may also be influenced by thermal changes due to changes in local metabolic activity or to vasoactive responses. For example, in humans the areolar dermis is underlain by Haller's vascular plexus (Mitz and Lalardie 1977). An acute vasodilatation of this plexus may confer a higher temperature to the areolar surface relative to the adjacent skin and maximize evaporation of odorants oozing from the MG in synchrony with the infant's presentation to the breast (Vuorenkoski et al. 1969).

To sum up, the best studied case of the rabbit highlights that the production and/or emission of the mammary odor is under the tonic control of gonadal steroids and lactogenic hormones. This leads to the externalization of odor cues at the very end of gestation and around birth, when pups' reactivity to them is maximal. Then, the attractant potency of the mammary odor mixture appears to progressively drop, and the release of the MP is abolished near weaning. Thus, when the weaning process is engaged, rabbit females may control pup motivation to suckle in reducing MP emission. At peak lactation, the timely control of mammary chemosignalization is presumably modulated by rhythmic processes (e.g., Montigny et al. 2006) and by infant-related distal and proximal stimuli.

17.6.2   CONTEXTUAL FLUCTUATIONS IN NEONATAL CHEMORECEPTION

The decrease in the female's emission of mammary chemosignals is echoed in altered neonatal reactivity to them. The rabbit MP potency to release pup responses decreases progressively over the preweaning period, indicating regulation by endogenous and exogenous factors. The response rate of domestic rabbit pups to the MP goes indeed through several stages: (1) from birth to days 8–10 pups respond maximally to the MP when they exclusively depend on mother's milk, (2) around postnatal days 8–11, a first drop from above 90% to ~80% of responding pups coincides with eye opening (~day 10–11) and the presumed reorganization of the pups' perceptual balance (Montigny 2008), (3) between days 11–21 a second drop occurs from ~80% to ~40% responding pups who become mobile, localize the female visually, initiate suckling, and begin to ingest solid food, (4) after the third week the typical response to the MP vanishes completely (Coureaud et al. 2008; Montigny 2008).

In parallel to the progressive decline or releasing potency of the MP along the nursing cycle, rabbit pups' response to mammary/milk signals comes gradually under circadian and metabolic modulation. On postnatal day 2, they react to the MP automatically at any time, without influence from prior milk intake or other emerging circadian factor (Montigny et al. 2006). But by postnatal day 5, and more so by day 10, this steady response to the MP restricts to prenursing hours.

Thus, with sensory reorganizations, cognitive development and changing metabolic needs, the sucking behavior of neonatal rabbits progressively escapes exclusive control by the MP. This shift in response to the MP from automatic to prandially-controlled is of particular significance in the context of the rare milk-resource access evolved by rabbit females. It warrants that rabbit pups are first bound to respond to the mammae in the rare minutes of milk availability; next, its decreasing strength (both in signal value and in evoked response) may contribute to the progressive disinvestment of the mammae when newborns can move by themselves and process nonmilk foods. This progressive shift of behavior control from chemosensation to other senses awaits investigations in rabbit newborns as well as in other mammalian newborns.

17.6.3   CHEMOSENSORY ONTOGENETIC ADAPTATIONS

The concept of ontogenetic adaptation implies a set of organismic responses to species-specific transformations of the environment during development (Alberts 1987). Each stage of development takes place in a given functional niche. In mammals, fetuses dwell in the uterus and amnion, where they are provisioned through the placenta; neonates are in a nest or within/against parent(s) and suck milk from their mother; while staying close to parent to ensure milk intake, weanlings acquire safe information about nonmilk foods and begin testing them; then come independent feeding and dispersion into extensive social networks. Each of these developmental niches correspond with exchanged substances or information between transitional phenotypes in both mothers and neonates (West-Eberhard 2003). These adaptive transitional phenotypes involve all levels of anatomical, physiological, and behavioral functioning. In the fetus, the placenta and related circulatory specificities are the most obvious adaptations, but numerous other subtle changes occur as labor sets on (e.g., Alberts and Ronca 2012). Neonatal mammals are behaviorally specialized to search, orally seize, and suck on a nipple or teat, their brain is engaged in (and dependent of) active sensory processing and integration, and their gastrointestinal tracts are designed to digest and absorb milk in symbiosis with the gut microflora passed on by the mother. As noted by Alberts (1987, p. 18), "stages of ontogenetic adaptation are composed of constellations of adaptive adjustments on multiple levels of organization... Successful stagewise transitions depend on coordinated and integrated readjustments, often drastic in extent." In the perinatal period, the adaptive challenges are especially drastic, and "adaptive adjustments displayed by the infant are correspondingly stunning." Thus, it should not be surprising that predisposed responsiveness to pheromones can change with age and age-related psychobiological readjustments.

The adaptive adjustments under scrutiny here concern sucking a nipple, and the sensory and cognitive means neonates engage to locate it and work on it to optimize milk intake. We have summarized how these achievements are conditional upon the perception of mammary-related odor cues, and in some species, of pheromones. When the dependence on milk declines as offspring grow up, it is expectable that the communicative value of these cues and signals changes or may completely vanish (as is the case in the rabbit MP). In this sense, nipple chemostimuli and their neonatal sensory processing can be seen as ontogenetic adaptations. They orchestrate the neonate-lactating female relationship, are vital for multiple proximate benefits to the neonate, and prepare the transition to the next developmental stage.

The rabbit MP not only serves to arouse and guide neonatal pups to the nipple, but also operates as a potent magnifier of odor learning. During a short window of early development (postnatal days 1–4), the MP promotes extrafast learning of any odor that is circumstantially associated with the mother or her milk (Coureaud et al. 2006; Montigny 2008). This MP-induced learning is a process by which newborns can update the changing odor properties of mammae and milk. More generally, the MP may speed up perceptual narrowing by accelerating glomerular refinement (Kerr and Belluscio 2006) and in this way rapidly assign meaning and reward value to some odorants beyond the sole MP that are salient in the individual odor profile of the mother (Patris et al. 2008), of the nest, and of littermates (Hudson et al. 2003; Montigny 2008; Serra and Nowak 2008), or of nonmilk foods (Montigny 2008). The strong arousal effect of the MP could also be involved in setting circadian rhythmicity of the coordinated activity by which neonate pups anticipate the brief daily nursing episode to increase their sucking success (Caldelas et al. 2009; Nolasco et al. 2012). It was indeed recently shown that the MP functions as a non photic zeitgeber for the central oscillators that contribute to regulate rhythms in body temperature and nursing-related anticipatory behavior in newborn rabbits (Montúfar-Chaveznava et al. 2013).

This pheromone-enforced learning process evidenced in the rabbit can presently guide us to question how mammalian neonates encode and decode chemosensory information from exceedingly complex and changing stimuli (e.g., Sinding et al. 2013). It should also stimulate future research to understand the precise nature and development of the molecular processes at the chemosensory periphery (perireceptor events (see Legendre et al., submitted); ligand-receptor interactions) and the neural pathways involved in the differential processing of the MP and conventional odorants (e.g., Charra et al. 2011, 2013). Dedicated neural pathways to process mammary chemosignals in the rabbit newborn, if any, would then prove informative about similar possibilities in other mammalian newborns.

17.7 CONCLUSION: MAMMALIAN NEONATES AS MODELS TO UNCOVER HOW AND WHEN CHEMORECEPTIVE SYSTEMS MAKE SENSE OF ODOR CUES AND SIGNALS

This review suggests that, at least in the mammalian species surveyed above, aspects of females' morphology, physiology, and behavior were evolutionarily selected to render the mammary area conspicuous for their newborns. This strategy of mammalian females to advertise mammaries to their offspring relies on informative means that match the earliest developing perceptual and behavioral abilities of their newborns. Accordingly, emitting some chemical cues and/or signals from the mammae is a suitable pan-mammalian strategy to pilot neonatal arousal, motivation, and attraction to the mother, to provide assistance in localizing and orally seizing the mammae, and to boost up timely learning. However, as noted in the three mammalian orders surveyed above, the ways by which these chemical cues are produced and assembled on the mammary area are complex within species and diverse between species.

The above review also highlights that neonatal mammals can develop a repertoire of learned odor *cues* derived from intra- and extramammary sources, but it indicates only rare cases of pheromonal *signals* that control nursing. The most completely documented cases so far of such mammary-based pheromones are *Oryctolagus*, the European rabbit (Hudson and Distel 1994; Schaal et al. 2008a) and *Rattus*, the laboratory rat (Blass and Teicher 1980). This scarcity of studies appeals for more research in an area that bears great potential to advance our knowledge on mammalian communication and chemoreception. It is a vital imperative for mammalian neonates to detect sensory cues from mother, mammae, and milk, and it is accordingly expectable that they are designed to do it right at birth. Thus, testing neonatal animals with odor stimuli they may be evolutionarily canalized to detect may be a productive way to identify new pheromones as well as new strategies of females to produce odor signals and to facilitate their sensing by neonates. In fact, newborn mammals may allow the strongest corroboration of the concept of the pheromone in its renewed formulation (see Section 17.4). Newborns (especially those of the altricial type) have generally restricted sensory abilities, are specialized in motor responses, limited in their stores of odor memories, and they are highly motivated to approach the mother or salient sensory traits disembodied from her. Finally, in many species, newborns are relatively easy to handle (although it is not always the case of maternal females). In fact, probably the most complete demonstrations to date of mammalian pheromones have advantageously focused on the newborns of readily available species.

Empirical emphasis on newborns may also help to assess the validity of categorizing stimuli into evolved *signals* and developmentally-acquired *cues* (see Section 17.2). Both kinds of stimuli are often considered as functionally equivalent, bringing arguments to some authors who propose to clear any distinction between pheromonal signals and circumstantially learned odor cues (e.g., Doty 2010). However, under adequate experimental circumstances, there is evidence that this may not be the case. When both types of stimuli are presented concurrently, neonates do not treat them as equivalent. For example, human infants deprived from birth of direct exposure to mother's breast/milk exhibit a clear preference for human milk odor over the odor of their formula milk that recurrently sated them (Marlier and Schaal 2005). In the same line, when paired with human milk odor, a chamomile odor sensed during nursing does not surpass milk for attractiveness in human newborns (Delaunay-El Allam et al. 2006), indicating that the most dominant or the most recent smell is not the most powerful in a choice test. Finally, when rabbit pups deprived of any exposure to the MP were conditioned to artificial odorants for 6 days, their response rate to the MP remained unaffectedly high and the releasing potency of the newly acquired odorant that recruited the same motor response system (oral grasping) never surpassed that of the MP (Coureaud et al. 2000). Thus, mammalian female-neonate units may be particularly suited models to uncover whether, when, why, and how the chemoreceptive system establishes the salience of social odor stimuli.

Finally, the paucity of our knowledge on the chemosensory regulation of mammalian nursing behavior is startling when one considers its absolute necessity for neonatal survival and initial development. Renewed interest in that topic, after the considerable work of developmental psychobiologists in the 1970s–1980s (e.g., Alberts 1976, 1981, 1987; Blass and Teicher 1980; Rosenblatt 1983), would certainly be rewarding to further understand the proximate mechanisms and development of chemoemission and chemoreception in readily accessible mammalian species (laboratory rodents, domestic ungulates and carnivores, humans). Such an approach should also encourage interest in more unusual species in which neonates are odor-guided in the nursing context (e.g., marsupials, insectivores, pinniped carnivores, nonhuman primates,) or in which such indications seem lacking (e.g., monotremes, proboscidians, cetaceans). Such knowledge would be interesting for comparative and phylogenetic analyzes of communication mechanisms that were seminal in the evolutionary success of mammals. In a different perspective, it would also be of interest to identify evolved odorant signals from maternal/lactating females to use them in promoting adaptive responsiveness and well-being in human neonates undergoing medical treatments and related prolonged separation from the mother.

ACKNOWLEDGMENTS

This chapter recapitulates and actualizes ideas developed in Schaal (2010). I gratefully thank my colleagues and students in Dijon and abroad, S. Al-Aïn, J.Y. Baudouin, A. Büttner, R. Charra, G. Coureaud, S. Doucet, K. Durand, C. Fenech, W. Francke, E. Hertling, A. Holley, R. Hudson, I. Jakob, T. Jiang, H. Loos, A.S. Moncomble, D.

Montigny, P. Orgeur, B. Patris, H. Rödel, G. Sicard, C. Sinding, R. Soussignan, and F. Védrines for past and continued teamwork in investigating the behavioral biology of neonates. While writing this chapter, I was supported by grants from ANR (Colostrum program), the Regional Council of Burgundy (PARI) and the CNRS.

REFERENCES

Ackerman, S.H. and Shindledecker, R. (1978). A method for artificial feeding of motherless 2-week-old rat pups. *Dev. Psychobiol.* 11: 385–391.

Al Aïn, S., Belin, L., Patris B., and Schaal, B. (2012a). An odor timer in milk? Synchrony in the odor of milk effluvium and neonatal chemosensation in mice. *PLoS ONE* 7: e47228.

Al Aïn S., Belin, L., Schaal, B. and Patris, B. (2012b). How a newly born mouse gets to the nipple? Odor substrates eliciting first nipple grasping and sucking responses. *Dev. Psychobiol.* 55: 888–901.

Al Aïn, S., Chraïti, A., Schaal, B. and Patris, B. (2011). Orientation of newborn mice to lactating females: Biological substrates of semiochemical interest. *Dev. Psychobiol.* 55: 113–124.

Al Aïn, S., Mingioni, M., Patris, B. and Schaal, B. (2013). The initial response of newly born mice to the odors of murine colostrum and milk: Unconditionally attractive, conditionally discriminated (submitted).

Alberts, A.C. (1992). Constraints on the design of chemical communication systems in terrestrial vertebrates. *Am. Nat.* 139: 562–589.

Alberts, J.R. (1976). Olfactory contributions to behavioral development in rodents. In *Mammalian Olfaction: Reproductive Processes and Behavior*, R.L. Doty (ed.), New York: Academic Press, pp. 67–94.

Alberts, J.R. (1981). Ontogeny of olfaction: Reciprocal roles of sensation and behavior in the development of perception. In *Development of Perception: Psychological Perspectives, Volume 1*, R.N. Aslin, J.R. Alberts and M.R. Petersen (eds.), New York: Academic Press, pp. 321–357.

Alberts, J.R. (1987). Early learning and ontogenetic adaptation. In *Perinatal Development: A Psychobiological Perspective*, N.A. Krasnegor, E.M. Blass, M.A. Hofer and W.P. Smotherman (eds.), Orlando, FL: Academic Press, pp. 11–37.

Alberts, J.R. and Ronca, A.E. (2012). The experience of being born: A natural context for learning to suckle. *Int. J. Pediatr.* 2012: 129328.

Allingham, K., Brennan, P.A., Distel, H. and Hudson, R. (1999). Expression of c-Fos in the main olfactory bulb of neonatal rabbits in response to garlic as a novel and conditioned odour. *Behav. Brain Res.* 104: 157–167.

Arias, C. and Chotro, M.G. (2007). Amniotic fluid can act as an appetitive unconditioned stimulus in preweanling rats. *Dev. Psychobiol.* 49: 139–149.

Armstrong, C.M., DeVito, L.M. and Cleland, T.A. (2006). One-trial associative odor learning in neonatal mice. *Chem. Senses* 31: 343–349.

Balogh, R.D. and Porter, R.H. (1986). Olfactory preferences resulting from mere exposure in human neonates. *Infant Behav. Dev.* 9: 395–401.

Bautista, A., Mendoza-Degante, M., Coureaud, G., Martinez-Gomez, M. et al. (2005). Scramble competition in newborn domestic rabbits for an unusually restricted milk supply. *Anim. Behav.* 70: 1011–1021.

Bean, N.J., Galef, B.G. and Mason, R.J. (1989). At biologically significant concentrations, carbon disulfide both attracts mice and increases their consumption of bait. *J. Wildl. Manage.* 52: 502–507.

Beauchamp, G. K., Doty, R. L., Moulton, D. G. and Mugford, R. A. (1976). The pheromone concept in mammals: A critique. In *Mammalian Olfaction, Reproductive Processes, and Behavior*, R.L. Doty (ed.), New York: Academic Press, pp. 143–160.

Beynon, R.J., Hurst, J.L., Turton, M.J., Robertson, D.H.L., Armstrong, S.D., Cheetham, S.A., Simpson, D., MacNicoll, A. and Humphries, R.E. (2008). Urinary lipocalins in Rodenta: Is there a generic model. In *Chemical Signals in Vertebrates 11*, J.L. Hurst, R.J. Beynon, S.C. Roberts and T.D. Wyatt (eds.), New York: Springer Science, pp. 37–49.

Bingham, P.M., Abassi, S. and Sivieri, E. (2003a). A pilot study of milk odor effect on nutritive sucking by premature infants. *Arch. Pediatr. Adolesc. Med.* 157: 72–75.

Bingham, P., Churchill, D. and Ashikaga, T. (2007). Breast milk odor via olfactometer for tube-fed, premature infants. *Behav. Res. Methods* 39: 630–634.

Bingham, P.M., Sreven-Tuttle, D., Lavin, E. and Acree, T. (2003b). Odorants in breast milk. *Arch. Pediatr. Adolesc. Med.* 157: 1031.

Blass, E.M. (1990). Suckling: Determinants, changes, mechanisms, and lasting impressions. *Dev. Psychol.* 26: 520–533.

Blass, E.M. and Teicher, M.H. (1980). Suckling. *Science* 210: 15–22.

Bolhuis, J.J. (1996). Development of perceptual mechanisms in birds: Predispositions and imprinting. In *Neuroethological Studies of Cognitive and Perceptual Processes*, C.F. Moss and S.J. Shettleworth (eds.), Boulder, CO: Westview Press, pp. 158–184.

Bouslama, M., Durand, E., Chauvière, L., ven den Bergh, O. and Gallego, J. (2005). Olfactory classical conditioning in newborn mice. *Behav. Brain Res.* 86: 19–27.

Brake, S.C. (1981). Suckling infant rats learn a preference for a novel olfactory stimulus paired with milk delivery. *Science* 211: 506–508.

Brake, S.C., Shair, H. and Hofer, M.A. (1986). Exploiting the nursing niche: The infant's sucking and feeding in the context of the mother-infant interaction. In *Handbook of Behavioral Neurobiology, Volume 9: Developmental Psychobiology and Behavioral Ecology*, E.M. Blass (ed.), New York: Plenum Press, pp. 347–388.

Brouette-Lahlou, I., Amouroux, R., Chastrette, F., Cosnier, J., Stoffelsma, J. and Vernet-Maury, E. (1991a). Dodecyl propionate, attractant from rat pup preputial gland: Characterization and identification. *J. Chem. Ecol.* 17: 1343–1354.

Brouette-Lahlou, I., Vernet-Maury, E. and Chanel, J. (1991b). Is rat-dam licking behavior regulated by pups' preputial gland secretion? *Anim. Learn. Behav.* 19: 177–184.

Büttner, A. (2002). Influence of human saliva on odorant concentrations. 2. *J. Agric. Food Chem.* 50: 7105–7110.

Büttner, A. (2007). A selective and sensitive approach to characterize odour-active and volatile constituents in small-scale human milk samples. *Flavour Fragr. J.* 22: 465–473.

Caldelas, I., Gonzales, B., Montufar-Chaveznava, R. and Hudson, R. (2009). Endogenous clock gene expression in the s uprachiasmatic nuclei of previsual newborn rabbits is entrained by nursing. *Dev. Neurobiol.* 69: 47–59.

Capretta, P.J. and Rawls, L.H. (1974). Establishment of a flavour preference in rats: Importance of nursing and weaning experience. *J. Comp. Physiol. Psychol.* 86: 670–673.

Charra, R., Datiche, F., Casthano, A., Gigot, V., Schaal, B. and Coureaud, G. (2011). Brain processing of the mammary pheromone in newborn rabbits. *Behav. Brain Res.* 226: 179–188.

Charra, R., Datiche, F., Gigot, V., Schaal, B. and Coureaud, G. (2013). Pheromone-induced odor learning modifies fos-expression in the newborn rabbit brain. *Behav. Brain Res.* 237: 129–140.

Cheslock, S.J., Varlinskaya, E.I., Petrov, E.S. and Spear, N.E. (2000). Rapid and robust olfactory conditioning with milk before suckling experience: Promotion of nipple attachment in the newborn rat. *Behav. Neurosci.* 114: 484–495.

Clutton-Brock, T.H. (1991). *The Evolution of Parental Care*. Princeton, NJ: Princeton University Press.

Contreras, C.M., Guttierez-Garcia, A.G., Mendoza-Lopez, R. et al. (2013). Amniotic fluid elicits appetitive responses in human newborns: Fatty acids and appetitive responses. *Dev. Psychobiol.* 55: 221–231.

Cooper, W.O., Atherton, H.D., Kahana, M. and Kotagal, U.R. (1995). Increased incidence of severe breastfeeding malnutrition and hypernatremia in a metropolitan area. *Pediatrics* 96: 957–960.

Coureaud, G. and Schaal, B. (2000). Attraction of newborn rabbits to abdominal odors of adult conspecifics differing in sex and physiological state. *Dev. Psychobiol.* 36: 271–281.

Coureaud, G., Fortun-Lamothe, L., Langlois, D. and Schaal, B. (2007). The reactivity of neonatal rabbits to the mammary pheromone as a probe for viability. *Animal* 1: 1026–1032.

Coureaud, G., Langlois, D., Perrier, G. and Schaal, B. (2003). A single key-odorant accounts for the pheromonal effect of rabbit milk: Further test of the mammary pheromone's activity against a wide sample of volatiles from milk. *Chemo. Ecol.* 13: 187–192.

Coureaud, G., Langlois, D., Sicard, G. and Schaal, B. (2004) Newborn rabbit reactivity to the mammary pheromone: Concentration-response relationship. *Chem. Senses* 29: 341–350.

Coureaud, G., Moncomble, A.S., Montigny, D., Dewas, M., Perrier, G. and Schaal, B. (2006). A pheromone that rapidly promotes learning in the newborn. *Curr. Biol.* 16: 1956–1961.

Coureaud, G., Rödel, H., Kurz, C.A. and Schaal, B. (2008). Age dependent responsiveness to the mammary pheromone in domestic and wild rabbits. *Chemo. Ecol.* 18: 52–59.

Coureaud, G., Schaal, B., Hudson, R., Orgeur, P. and Coudert, P. (2002). Transnatal olfactory continuity in the rabbit: Behavioral evidence and short-term consequence of its disruption, *Dev. Psychobiol.* 40: 372–390.

Coureaud, G., Schaal, B., Langlois, D. and Perrier, G. (2001). Orientation responses of newborn rabbits to odors emitted by lactating females: Relative effectiveness of surface and milk cues. *Anim. Behav.* 61: 153–162.

Darwin, C. (1877). A biographical sketch of an infant. *Mind* 7: 285–294.

Delaunay-El Allam, M., Marlier, L. and Schaal, B. (2006). Learning at the breast: Preference formation for an artificial scent and its attraction against the odor of maternal milk. *Infant Behav. Dev.* 29: 308–321.

Delaunay-El Allam, M., Soussignan, R., Patris, B., Marlier, I. and Schaal, B. (2010). Longlasting memory for an odor acquired at the mother's breast. *Dev. Science* 13: 849–863.

Dewey, K.G., Nommsen, L.A., Heinig, M.J. and Cohen, R.J. (2003). Risk factors for suboptimal infant breastdeefing behavior, delayed onset of lactation, and excess neonatal weight loss. *Pediatrics* 112: 607–619.

Dollinger, M.J., Holloway, W.R. and Denenberg, V.H. (1978). Nipple attachment in rats during the first 24 hours of life. *J. Comp. Physiol. Psychol.* 92:. 619–626.

Doty, R.L. (2003). Mammalian pheromones: Fact or fantasy? In *Handbook of Olfaction and Gustation*, 2nd Edition, R.L. Doty (ed.), New York: Marcel Dekker, pp. 345–383.

Doty, R.L. (2010). *The Great Pheromone Myth.* Baltimore, MD: Johns Hopkins University Press.

Doucet, S., Soussignan, R., Sagot, P. and Schaal, B. (2007). The "smellscape" of mother's breast: Effects of odor masking and selective unmasking on neonatal arousal, oral and visual responses. *Dev. Psychobiol.* 49: 129–138.

Doucet, S., Soussignan, R., Sagot, P. and Schaal, B. (2009). The secretion of areolar (Montgomery's) glands from lactating women elicits selective, unconditional responses in neonates. *PLoS ONE* 4: e7579.

Doucet, S., Soussignan, R., Sagot, P. and Schaal, B. (2012). An overlooked aspect of the human breast: Aeolar glands in relation with breastfeeding pattern, neonatal weight gain, and dynamics of lactation. *Early Hum. Dev.* 88: 119–128.

Drewett, R.F., Kendrick, K.M., Sanders, D.J. and Trew, A.M. (1982). A quantitative analysis of the feeding behavior of suckling rabbits. *Dev. Psychobiol.* 15: 25–32.

Drickamer, L.C. (1988). Puberty-influencing chemosignals in house mice: Ecological and evolutionary considerations. In *Chemical Signals in Vertebrates 4*, D. Duvall, D. Müller-Schwarze and R.M. Siverstein (eds.), New York: Plenum Press, pp. 441–455.

Drummond, H., Vázquez, E., Sanchez-Colón, S., Martinez-Gómez, M. and Hudson, R. (2000). Competition for milk in the domestic rabbit: Survivors benefit from littermate deaths. *Ethology* 106: 511–526.

Durand, K., Baudouin, J.Y., Lewkowicz, D.J., Goubet, N. and Schaal, B. (2013). Eye-catching odors: Olfaction elicits sustained gazing to faces and eyes in 4 month-old infants. *PLoS ONE* 8: e70677.

Dusenberry, D.B. (1992). *Sensory Ecology: How Organisms Acquire and Respond to Information*. New York: Freeman.

Edmond, K.M., Zandoh, C., Quigley, M.A., Amenga-Etego, S., Owusu-Agyei, S. and Kirkwood, B.R. (2006). Delayed breastfeeding initiation increases risk of neonatal mortality. *Pediatrics* 117: e380–e386.

Fleischer, J., Schwarzenbacher, K. and Breer, H. (2007). Expression of trace amine-associated receptors in the Grueneberg ganglion. *Chem. Senses* 32: 623–631.

Friedman, M.I. and Bruno, J.P. (1976). Exchange of water during lactation. *Science* 197: 409–410.

Galef, B.G. and Henderson, P.W. (1972). Mother's milk: A determinant of the feeding preferences of weaning rat pups. *J. Comp. Physiol. Psychol.* 78: 213–219.

Galef, B.G. and Sherry, D.F. (1973). Mother's milk: A medium for the transmission of cues reflecting the flavour of mother's diet. *J. Comp. Physiol. Psychol.* 83: 374–378.

Galef, B.G., Mason, J.R., Pretty, G. and Bean, N.J. (1988). Carbon disulfide: A semiochemical mediating socially-induced diet choice in rats. *Physiol. Behav.* 42: 119–124.

Gubernick, D.J. and Alberts, J.R. (1983). Maternal licking of young: Resource exchange and proximate controls. *Physiol. Behav.* 31: 593–601.

Hauser, M.D. (1996). *The Evolution of Communication*, Cambridge, MA: MIT Press.

Hausner, H., Bredie, W., Molgaard, C., Petersen, M.A. and Moller, P. (2008). Differential transfer of dietary flavour compounds into human breast milk. *Physiol. Behav.* 95: 118–124.

Hepper, P.G. (1987). The amniotic fluid: An important priming role in kin recognition. *Anim. Behav.* 35: 1343–1346.

Hongo, T., Hakuba, A., Shiota, K. and Naruse, I. (2000). Suckling dysfunction caused by defects in the olfactory system in genetic arhinencephaly mice. *Biol. Neonate* 78: 293–299.

Horn, G. (2004). Pathways of the past: The imprint of memory. *Nat. Rev. Neurosci.* 5: 108–120.

Hudson, R. (1993). Rapid odor learning in newborn rabbits: Connecting sensory input to motor output. *Germ. J. Psychol.* 17: 267–275.

Hudson, R. (1985) Do newborn rabbits learn the odor stimuli releasing nipple-search behavior? *Dev. Psychobiol.* 18: 575–585.

Hudson, R. and Distel, H. (1983). Nipple location by newborn rabbits: Evidence for pheromonal guidance. *Behaviour* 82: 260–275.

Hudson, R. and Distel, H. (1984). Nipple-search pheromone in rabbits: Dependence on season and reproductive state. *J. Comp. Physiol. A* 155: 13–17.

Hudson, R. and Distel, H. (1990). Sensitivity of female rabbits to changes in photoperiod as measured by pheromone emission. *J. Comp. Physiol. A* 167: 225–230.

Hudson, R., Garay-Villar, E., Maldonado, M. and Coureaud, G. (2003). Rabbit pups can orient to the nest by smell from birth. Annual Meeting of the American Chemoreception Association, Sarasota, FL.

Hudson, R., Labra-Cardero, D. and Mendoza-Solovna, A. (2002). Suckling, not milk, is important for the rapid learning of nipple-search odors in newborn rabbits. *Dev. Psychobiol.* 41: 226–235.

Hudson, R., Rojas, C., Arteaga, L., Martinez-Gomez, M. et al. (2008). Rabbit nipple-search pheromone versus rabbit mammary pheromone revisited. In *Chemical Signals in*

Vertebrates 11, J.L. Hurst, R.J. Beynon, S.C. Roberts and T.D. Wyatt (eds.), New York: Springer, pp. 315–324.

Hurst, J.L., Payne, C.E., Nevison, C.M., Marie, A.D. et al. (2001). Individual recognition in mice mediated by major urinary proteins. *Nature* 414: 631–634.

Ivanistkii, A.M. (1962). The morphophysiological investigation of development of conditioned alimentary reactions in rabbits during ontogenesis. In *Experimental Studies of Higher Nervous Activity in Man and Animals.* Works of the Institute of Higher Nervous Activity, Moscow, Physiological Series, Vol. 4, Jerusalem, Israel: Israel Program for Scientific Translations Ltd., pp. 126–141.

Jemiolo, B., Andreolini, F., Wiesler, D. and Novotny, M. (1987). Variations in mouse (Mus musculus) urinary volatiles during different periods of pregnancy and lactation. *J. Chem. Ecol.* 13: 1941–1956.

Jemiolo, B., Andreolini, F., Xie, T.M., Wiesler, D. and Novotny, M. (1989). Puberty-affecting synthetic analogs of urinary chemosignals in the house mouse, Mus domesticus. *Physiol. Behav.* 46: 293–298.

Johnston, R.E. (2000). Chemical communication and pheromones: The types of chemical signals and the role of the vomeronasal system. In *The Neurobiology of Taste and Smell*, T.E. Finger, W.L. Silver and D. Restrepo (eds.), New York: Wiley, pp. 101–127.

Karlson, P. and Lüscher, M. (1959). "Pheromones": A new term for a class of biologically active substances. *Nature* 183: 55–56.

Keil, W., von Stralendorff, F. and Hudson, R. (1990). A behavioral bioassay for analysis of rabbit nipple-search pheromone. *Physiol. Behav.* 47: 525–529.

Kerr, M.A. and Belluscio, L. (2006). Olfactory experience accelerates glomerular refinement in the mammalian olfactory bulb. *Nat. Neurosci.* 4: 484–486.

Kindermann, U., Hudson, R. and Distel, H. (1994). Learning of suckling odours by newborn rabbits declines with age and suckling experience. *Dev. Psychobiol.* 2: 111–122.

Kodama, N. (1990). Preference for amniotic fluid in newborn mice. Annual Meeting of the International Society for Developmental Psychobiology, Cambridge, United Kingdom.

Kodama, N. (2002). Effects of odor and taste of amniotic fluid and mother's milk on body movements in newborn mice. *Dev. Psychobiol.* 41: 310.

Kodama, N. and Smotherman, W.P. (1997). Effects of amniotic fluid on head movement in cesarean delivered rat pups. *Dev. Psychobiol.* 30: 255.

Koffman, D.J., Petrov, E.S., Varlinskaia, E.I. and Smotherman, W.P. (1998). Thermal, olfactory, and tactile stimuli increase oral grasping of an artificial nipple by the newborn rat. *Dev. Psychobiol.* 33: 317–326.

Liberles, S.D. and Buck, L.B. (2006). A second class of chemosensory receptors in the olfactory epithelium. *Nature* 442: 645–650.

Logan, D. W., Brunet, L. J., Webb, W. R., Cutforth, T., Ngai, J. and Stowers, L. (2012). Learned recognition of maternal signature odors mediates the first suckling episode in mice. *Curr. Biol.* 22: 1998–2007.

Macfarlane, A.J. (1975). Olfaction in the development of social preferences in the human neonate. *Ciba Found. Symp.* 33: 103–117.

Maestripieri, D. and Mateo, J.M. (eds.) 2009. *Maternal Effects in Mammals*, Chicago: University of Chicago Press.

Makin, J.W. and Porter, R.H. (1989). Attractiveness of lactating females' breast odors to neonates. *Child Dev.* 60: 803–810.

Marlier, L. and Schaal, B. (2005). Human newborns prefer human milk: Conspecific milk odor is attractive without postnatal exposure. *Child Dev.* 76: 155–168.

Marlier, L., Schaal, B. and Soussignan R. (1998). Bottle-fed neonates prefer an odor experienced in utero to an odor experienced in the feeding context. *Dev. Psychobiol.* 33: 133–145.

Maynard-Smith, J. and Harper, D. (2003). *Animal Signals*. Oxford: Oxford University Press.

Mellier, D., Bezard, S. and Caston, J. (1997). Etudes exploratoires des relations intersensori-elles olfaction-douleur. In *L'odorat chez l'enfant: Perspectives croisées*, B. Schaal (ed.), Paris: Presses Universitaires de France (Enfance), pp. 98–111.

Mennella, J.A. and Beauchamp. G.K. (1991a). Maternal diet alters the sensory qualities of human milk and the nursling's behavior. Pediatrics 88: 737–744.

Mennella, J.A. and Beauchamp, G.K. (1991b). The transfer of alcohol to human milk: Effects on flavor and the infant's behavior. *N. Engl. J. Med.* 325: 981–985.

Mennella, J.A. and Beauchamp G.K. (1996). The human infants' responses to vanilla flavors in human milk and formula. *Infant Behav. Dev.* 19: 13–19.

Miller, S.S. and Spear, N.E. (2008). Olfactory learning in the rat neonate soon after birth. *Dev. Psychobiol.* 50: 554–565.

Miller, S.S. and Spear, N.E. (2010). Mere odor exposure learning in the rat neonate immediately after birth and 1 day later. *Dev. Psychobiol.* 52: 343–351.

Mitz, V. and Lalardie, J.P. (1977). A propos de la vascularisation et de l'innervation sensitive du sein. *Senologia* 2: 33–39.

Mizuno, K. and Ueda, A. (2004). Antenatal olfactory learning influences infant feeding. *Early Hum. Dev.* 76: 83–90.

Moio, L., Rillo, L., Ledda, A. and Addeo, F. (1996). Odorous constituents of ovine milk in relationship to diet. *J. Dairy Sci.* 79: 1322–1331.

Molina, J.C., Chotro, M.G. and Domingez, H.D. (1995). Fetal alcohol learning resulting from alcohol contamination of the prenatal environment. In *Fetal Development. A Psychobiological Perspective,* J.P. Lecanuet, W.P. Fifer, N.E. Krasnegor and W.P. Smotherman (eds.), Hillsdale, NJ: Lawrence Erlbaum.

Moncomble, A.S. (2006). De la prise de lait à l'ingestion non lactée chez le lapin: Analyses éthologiques, histologiques et chimiques de sources odorantes significatives pour le lapereau nouveau-né. Unpublished doctoral thesis, University of Burgundy, Dijon.

Moncomble, A.S., Coureaud, G., Quennedey, B., Langlois, D., Perrier, G., Brossut, R. and Schaal, B. (2005). The mammary pheromone of the rabbit: Where does it come from? *Anim. Behav.* 69: 29–38.

Montagna, W. and MacPherson, E.E. (1974). Some neglected aspects of the anatomy of human breasts. *J. Invest. Dermatol.* 63: 10–16.

Montagna, W. and Yun, J.S. (1972). The glands of Montgomery. *Br. J. Dermatol.* 86: 126–133.

Montgomery, W.F. (1937). *An Exposition of the Signs and Symptoms of Pregnancy, the Period of Human Gestation, and Signs of Delivery.* London: Sherwood, Gilber, and Piper.

Montigny, D. (2008). Fonctions adaptatives immédiates et différées de la phéromone mammaire chez le lapereau. Doctoral thesis, University of Paris 13, Villetaneuse, France.

Montigny, D., Coureaud, G. and Schaal, B. (2006). Shift from automatism to prandial control in the response of newborn rabbits to the mammary pheromone. *Physiol. Behav.* 89: 742–749.

Montúfar-Chaveznava, R., Trejo-Munoz, L., Hernández-Campos, O., Navarrete, E. and Caldelas, I. (2013). Maternal olfactory cues synchronize the circadian systems of artificially raised newborn rabbits. *PLoS ONE* 8: e74048.

Müller, K. (1978). Zum Saugverhalten von Kaninchen unter besonderer Berücksichtigung des Geruchsvermögen. Unpublished doctoral dissertation, University of Giessen, Germany.

Müller-Schwarze, D. (1977). Complex mammalian behavior and pheromone bioassay in the field. In *Chemical Signals in Vertebrates*, D. Müller-Schwarze and M.M. Mozell (eds.), New York: Plenum Press, pp. 413–433.

Müller-Schwarze, D. (2006). *Chemical Ecology of Vertebrates*, Cambridge: Cambridge University Press.

Munger, S.D., Leinders-Zufall, T., McDougall, L.M. et al. (2010). An olfactory subsystem that detects carbon disulfide and mediates food-related social learning. *Curr. Biol.* 20: 1438–1444.

Murakami, K., Lagarde, M. and Yuki, Y. (1998). Identification of minor proteins of human colostrum and mature milk by 2-dimensional electrophoresis. *Electrophoresis* 19: 2521–2527.

Neifert, M.R. (2001). Prevention of breastfeeding tragedies. In *The Pediatric Clinics of North America, Breastfeeding 2001, Part 2*, R.J. Schandler (ed.), Philadelphia: WB Saunders, Vol. 48, pp. 273–298.

Nishitani, S., Miyamura, T., Tagawa, M., Sumi, M., Takase, R., Doi, H., Moriuchi, H. and Shinohara, K. (2009). The calming effect of a maternal breast milk odor on the human newborn infant. *Neurosci. Res.* 63: 66–71.

Patris, B., Perrier, G., Schaal, B. and Coureaud, G. (2008). Pheromone-induced odour learning in newborn rabbits: Implications for the development of social preferences. *Anim. Behav.* 76: 305–314.

Pedersen, P.A. and Blass, E.M. (1981). Olfactory control over suckling in albino rats. In *Development of Perception: Psychobiological Perspectives, Volume 1: Audition, Somatic Perception, and the Chemical Senses*, R.N. Aslin, J.R. Alberts and M.R. Petersen (eds.), New York: Academic Press, pp. 359–381.

Pedersen, P.A. and Blass, E.M. (1982). Prenatal and postnatal determinants of the 1st suckling episode in albino rats. *Dev. Psychobiol.* 15: 349–355.

Perkins, O.M. and Miller, A.M. (1926). Sebaceous glands in the human nipple. *Am. J. Obstet.* 11: 789–794.

Petrov, E.S., Varlinskaia, E.I. and Smotherman, W.P. (1997). The newborn rat ingests fluids through a surrogate nipple: A new technique for the study of early suckling behavior. *Physiol. Behav.* 112: 901–906.

Pollack, P.F., Koldovsky, O. and Nishioka, K. (1992). Polyamines in human and rat milk and infant formulas. *Am. J. Clin. Nutr.* 56: 371–375.

Porter, R.H., Makin, J.W., Davis, L.B. and Christensen, K.M. (1991). An assessment of the salient olfactory environment of formula-fed infants. *Physiol. Behav.* 50: 907–911.

Raihini, G., Gonzales, D., Arteaga, L. and Hudson, R. (2009). Olfactory guidance of nipple attachment and suckling in kittens of the domestic cat: Inborn and learned responses. *Dev. Psychobiol.* 51: 662–671.

Raimbault, C., Saliba, E. and Porter, R.H. (2007). The effect of the odour of mother's milk on breastfeeding behaviour of premature infants. *Acta Paeditr.* 96: 368–371.

Rattaz, C., Goubet, N. and Bullinger, A. (2005). The calming effect of a familiar odor on full-term newborns. *Dev. Behav. Pediatr.* 26: 86–92.

Roberts, S.A., Simpson, D.M., Armstrong, S.D., Davidson, A.J., Robertson, D.H., McLean, L., Beynon, R.J. and Hurst, J.L. (2010). Darcin: A male pheromone that stimulates female memory and sexual attraction to an individual male's odour. *BMC Biol.* 8: 75.

Rödel, H.G., Dausmann, K.H., Starkloff, A. et al. (2012). Diurnal nursing pattern of wild-type European rabbits under natural breeding conditions. *Mamm. Biol.* 77: 441–446.

Romantshik, O., Porter, R.H., Tillmann, V. and Varendi, H. (2007). Preliminary evidence of a sensitive period for olfactory learning by human newborns. *Acta Paediatr.* 96: 372–376.

Rosenblatt, J.S. (1983). Olfaction mediates developmental transitions in the altricial newborn of selected species of mammals. *Dev. Psychobiol.* 16: 347–375.

Roth, L. and Rosenblatt, J.S. (1966). Changes in self-licking during pregnancy in the rat. *J. Comp. Physiol. Psychol.* 63: 397–400.

Russell, M.J. (1976). Human olfactory communication. *Nature* 260: 520–522.

Schaal, B. (1986). Presumed olfactory exchanges between mother and neonate in humans. In *Ethology and Psychology*, J. Le Camus and J. Cosnier (eds.), Privat-I.E.C., Toulouse, France, pp. 101–110.

Schaal, B. (2005). From amnion to colostrum to milk: Odor bridging in early developmental transitions. In *Prenatal Development of Postnatal Functions*, B. Hopkins and S.P. Johnson (eds.), London: Praeger, pp. 51–102.

Schaal, B. and Durand, K. (2012). The role of olfaction in human multisensory development. In *Multisensory Development*, A. Bremner, D. Lewkowicz and C. Spence (eds.), Oxford: Oxford University Press, pp. 29–62.

Schaal, B. and Orgeur, P. (1992). Olfaction in utero: Can the rodent model be generalized? *Quart. J. Exp. Psychol.* 44B: 245–278.

Schaal, B., Coureau, G., Doucet, S., Delaunay-El Allam, M., Moncomble, A.S., Montigny, D., Patris, B. and Holley, A. (2009). Olfactory mammary signalisation in females and neonatal odour processing: Ways evolved in rabbit and human. *Behav. Brain Res.* 200: 346–358.

Schaal, B., Coureaud, G., Langlois. D., Giniès, C., Sémon, E. and Perrier, G. (2003). Chemical and behavioural characterisation of the rabbit mammary pheromone. *Nature* 424: 68–72.

Schaal, B., Doucet, S., Sagot, P., Hertling, E. and Soussignan, R. (2006). Human breast areolae as scent organs: Morphological data and possible involvement in maternal-neonatal coadaptation. *Dev. Psychobiol.* 48: 100–110.

Schaal, B., Doucet, S., Soussignan, R., Rietdorf, M., Weibchen, G. and Francke, W. (2008b). The human breast as a scent organ: Exocrine structures, secretions, volatile components, and possible functions in breastfeeding interactions. In *Chemical Signals in Vertebrates 11*, J.L. Hurst, R.J. Beynon, S.C. Roberts and T.D. Wyatt (eds.), New York: Springer, pp. 325–335.

Schaal, B., Montagner, H., Hertling, E., Bolzoni, D., Moyse, R. and Quichon, R. (1980). [Olfactory stimulations in mother-infant relations]. *Reprod. Nutr. Dev.* 20: 843–858.

Schleidt, M., and Genzel, C. (1990). The significance of mothers perfume for infants in the first weeks of life. *Ethol. Sociobiol.* 11: 145–154.

Schley, P. (1976). Untersuchungen zur künstlichen Aufzucht von Hauskaninchen. Habilitationsschrift, University of Giessen, Germany.

Schley, P. (1977). Die Ausschaltung des Geruchsvermögens und sein Einfluss auf das Saugverhalten von Jungkaninchen. *Berl. Münch. Tierärztl. Wochenschr.* 90: 382–385.

Schley, P. (1979). Olfaction and suckling behavior in young rabbits. In *Proceedings of the 1st World Lagomorph Conference*, K. Myers and C.D. MacInnes (eds.), Guelph, Canada: University of Guelph, pp. 291–294.

Selzer, D., Lange, K. and Hoy, S. (2004). Frequency of nursing in domestic rabbits under different housing conditions. *Appl Anim. Behav. Sci.* 87: 317–324.

Serra, J., and Nowak, R. (2008). Olfactory preference for own mother and litter in 1-day old rabbits and its impairment by thermotaxis. *Dev. Psychobiol.* 50: 542–553.

Shahan, K., Denaro, M., Gilmartin, M., Shi, Y. and Derman, E. (1987). Expression of 6 mouse major urinary protein genes in the mammary, parotid, sublingual, submaxillary, and lachrymal glands and in the liver. *Mol. Cell. Biol.* 7: 1947–1954.

Shimoda, Y.T., Ishikawa, H., Hayakawa, I., and Osajima, Y. (2000). Volatile compounds of human milk. *J. Fac. Agr. Kyushu. Univ.* 45: 199–206.

Sinding, C., Thomas-Danguin, T., Chambault, A., Béno, N. et al. (2013). Rabbit neonates and human adults perceive a blending 6-component odor mixture in a comparable manner. *PLoS ONE* 8: e53534.

Singer, A.G., Acosta, W.C., O'Connell, R.J. and Thiessen, D.D. (1976). Dimethyl disulfide: An attractant heromone in hamster vaginal secretion. *Science* 191: 948–950.

Singh, P.J. and Hofer, M.A. (1978). Oxytocin reinstates maternal olfactory cues for nipple orientation and attachment in rat pups. *Physiol. Behav.* 20: 385–389.

Smith, D.M., Peters, T.G. and Donegan, W.L. (1982). Montgomery's areolar tubercle. A light microscopic study. *Arch. Pathol. Lab. Med.* 106: 60–63.

Smotherman, W.P. and Robinson, S.R. (1987). Psychobiology of fetal experience in the rat. In *Perinatal Development: A Psychobiological Perspective*, N.A. Krasnegor, E.M. Blass, M.A. Hofer, and W.P. Smotherman (eds.), Orlando, FL: Academic Press, pp. 39–60.

Smotherman, W.P. and Robinson, S.R. (1992). Dimethyl disulfide mimics the effect of milk on fetal behavior and responsiveness to cutaneous stimuli. *Physiol. Behav.* 52: 761–765.

Smotherman, W.P. and Robinson, S.R. (1994). Milk as the proximal mechanism for behavioral change in the newborn. *Acta Paediatr. (Suppl.)* 374: 64–70.

Soussignan, R., Schaal, B., Marlier, L. and Jiang, T. (1997). Facial and autonomic responses to biological and artificial olfactory stimuli in human neonates: Re-examining early hedonic discrimination of odors. *Physiol. Behav.* 62: 745–758.

Spitzer, J. and Büttner, A. (2013). Monitoring aroma changes during human milk storage at −19°C by quantification experiments. *Food Res. Int.* 51: 250–256.

Sullivan, R.M. and Toubas, P. (1998). Clinical usefulness of maternal odor in newborns: Soothing and feeding preparatory responses. *Biol. Neonate* 74: 402–408.

Sullivan, R.M., Hofer, M.A. and Brake, S.C. (1986). Olfactory-guided orientation in neonatal rats is enhanced by a conditioned change in behavioral state. *Dev. Psychobiol.* 19: 615–623.

Sullivan, R.M., Taborsky, S.B., Mendoza, R., Itano, A., Leon, M., Cotman, C.W. et al. (1991). Olfactory classical conditioning in neonates. *Pediatrics* 87: 511–517.

Summerlee, A.J., Paisley, A.C., O'Byrne, K.T., Fairhall, K.M., Robinson, I.C. and Fletcher, J. (1986). Aspects of the neuronal and endocrine components of reflex milk ejection in conscious rabbits. *J. Endocrinol.* 108: 143–149.

Teicher, M.H. and Blass, E.M. (1976). Suckling in the newborn rat: Eliminated by nipple lavage, reinstated by pup saliva. *Science* 193: 422–425.

Teicher, M.H. and Blass, E.M. (1977). First suckling response in the newborn albino rat: The roles of olfaction and amniotic fluid. *Science* 198: 635–636.

Terry, L.M. and Johanson, I.B. (1987). Olfactory influences on the ingestive behavior of infant rats. *Dev. Psychobiol.* 20: 313–332.

Thiessen, D.D. 1977. Methodology and strategies in the laboratory. In *Chemical Signals in Vertebrates*, D. Müller-Schwarze and M.M. Mozell (eds.), New York: Plenum Press, pp. 391–412.

Thiessen, D.D., Clancy, A. and Goodwin, M. (1976). Harderian pheromone in the Mongolian gerbil (Meriones unguiculatus). *Chem. Ecol.* 2: 231–238.

Toyoshima, Y., Ohsako, S., Matsumoto, M., Hidaka, S. and Nishinakagawa, H. (1998b). Histological and morphometrical studies on the rat nipple during the reproductive cycle. *Exp. Anim.* 47: 29–36.

Toyoshima, Y., Ohsako, S., Nagano, R., Matsumoto, M., Hidaka, S. and Nishinakagawa, H. (1998a). Histological changes in mouse nipple tissue during the reproductive cycle. *J. Vet. Med. Sci.* 60: 405–411.

Varendi, H. and Porter, R.H. (2001). Breast odour as the only maternal stimulus elicits crawling towards the odour source. *Acta Paediatr.* 90: 372–375.

Varendi, H., Porter, R.H., and Winberg, J. (1994). Does the newborn baby find the nipple by smell? *Lancet* 344: 989–990.

Vince, M.A. and Ward, T.M. (1984). The responsiveness of newly born Clunforest lambs to odor sources in the ewe. *Behaviour* 87: 117–127.

Vuorenkoski, V., Wasz-Hockert, O., Koivisto, E. and Lind, J. (1969). The effect of cry stimulus on the temperature of the lactating breast of primipara. *Experientia* 25: 1286–1287.

Wales, N.A. and Ebling, F.J. (1971). The control of apocrine glands of the rabbit by steroid hormones. *J. Endocrinol.* 51: 763–770.

West-Eberhard, M.J. (2003). *Developmental Plasticity and Evolution.* Oxford: Oxford University Press.

Wilson, D.A. and Sullivan, R.M. (1994). Neurobiology of associative learning in the neonate: Early olfactory learning. *Behav. Neural. Biol.* 61: 1–18.

Wyatt, T.D. (2003). *Pheromones and Animal Behaviour. Communication by Smell and Taste.* Cambridge: Cambridge University Press.

Wyatt, T.D. (2010). Pheromones and signature mixtures: Defining species-wide signals and variable cues for identity in both invertebrates and vertebrates. *J. Comp. Physiol. A* 196: 685–700.

Youngentob, S.L., Kent. P.F., Sheehe, P.R., Molina, J.C., Spear, N.E. and Youngentob, L.M. (2007). Experience-induced fetal plasticity: The effect of gestational ethanol exposure on the behavioral and neurophysiologic olfactory response to ethanol odor in early postnatal and adult rays. *Behav. Neurosci.* 121: 1293–1305.

Zarrow, M.X., Denenberg, V.H. and Anderson, C.O. (1965). Rabbit: Frequency of suckling in the pup. *Science* 150: 1835–1836.

18 Pheromone Processing in Relation to Sex and Sexual Orientation

Ivanka Savic

CONTENTS

18.1 Introduction 523
18.2 Pheromone Processing in Mammals 523
18.3 Does VNO Exist in Humans? 524
18.4 Imaging Studies of Humans Exposed to Chemosignaling Compounds
with Pheromone-like Properties 525
    18.4.1 Synthetic Compounds 525
    18.4.2 Body Odors 526
18.5 Activation with Pheromone-like Compounds in Relation to Sexual
Orientation 527
18.6 Underlying Mechanisms 529
    18.6.1 Testosterone Organizational and Activational Theory 530
18.7 Concluding Remarks 531
Acknowledgments 531
References 531

18.1 INTRODUCTION

Whether pheromone signaling exists in humans is still a matter of intense discussion. Emerging brain imaging studies suggest sexually dimorphic neuronal response to certain chemosignals, which, according to psychophysical data, possess pheromone-like properties. There are also indications that the neuronal response to these compounds depends on sexual orientation. Furthermore, our brain seems to be able to extract kin-specific signals, and processes body odors differently than other perceptually similar odors. Together, this data sheds new light on the chemosensory perception in humans and the implications thereof are discussed.

18.2 PHEROMONE PROCESSING IN MAMMALS

According to the original definition by Karlson and Luscher (1959), pheromones are "airborne chemical signals released by an individual into the environment, and affecting the physiology and behavior of other members of the same species."

Pheromones have traditionally been defined as either releasers, compounds trigger-ing immediate short-term behavioral responses, or primers, compounds triggering medium- to long-term changes in behavior or physiology. Such clear categories are, however, not always easy to attribute to the phenomena being studied, which has led the new generation of chemosensory biologists to add so-called modulator phero-mones to the classification. A modulator pheromone modifies "ongoing behavior or a psychological reaction to a particular context without triggering specific behaviors or thoughts" (McClintock et al. 2001). Pheromones are produced in sweat, blood, saliva, and vaginal secrete. Both volatile and nonvolatile compounds (proteins, ste-roids, cholesterols) can act as pheromones (Karlson and Luscher 1959). Pheromone signals provide information about gender and reproductive status and mediate social and sexual behaviors as well as neuroendocrine changes (Brennan and Keverne 2004). Animal experiments show that both pheromone perception and response is dependent of the hormone status and the genetic makeup of the animal (Dulac and Torello 2003).

Most mammals have two olfactory systems: the main olfactory system with the olfactory epithelium and the olfactory bulb (MOB), and the vomeronasal system with the vomeronasal organ (VNO). The MOB and VNO have their own distinct pri-mary projection targets in the brain. There is increasing evidence that both systems may be involved in pheromone detection, and that they both project to the medial amygdala (Kang et al. 2009). In a majority of animals the VNO is still regarded to be the primary sensory organ for the detection of pheromone signals. The VNO expresses specific classes of vomeronasal receptors (VR), VR type 1 and 2, which detect pheromone signals. The VR type 2 receptors interact with a nonclassical major histocompatibility complex (MHC) class 1b molecule to form a functional receptor complex (Keverne 2008). This interaction implies that perception of a particular pheromone may be related to the genetic composite of the receiver. From the VNO the pheromone signals are transduced via the accessory olfactory nerve directly to the specific mating centers of the anterior hypothalamus (Dulac and Torello 2003). This direct access to the hypothalamus allows pheromones to play a major role in the sexual behavior and the choice of sexual partner in animals. A lesion of the respec-tive mating center, as well as impairment of pheromone transduction, may alter the coital approach in a sex-specific way. For example, electrolytic lesion of the preoptic area is reported to shift the mean preference of male ferrets away from the estrous females to the stud males (Kindon et al. 1996). Male rats are found to reduce their coital behavior after destruction of the preoptic area and show more interest in stim-ulus males than receptive females (Kindon et al. 1996; Paredes et al. 1998). Female rats, however, increased the proportion of approaches to females after kindling of the preoptic area (Dominguez-Salazar et al. 2003).

18.3   DOES VNO EXIST IN HUMANS?

Although vomeronasal pits are detectable in many persons, the VNO seems to be vestigial in humans (Trotier et al. 2000). Several observations support this view. First, the human VNO epithelium resembles more strongly the respiratory epithe-lium than the VNO neuroepithelium found in species with functional VNOs (Witt et al. 2002). Second, the olfactory marker protein (OMP), which is a reliable marker for mature VNO neurons in many animals, is not expressed in human VNO (Dennis et al. 2004). Third, genes coding for the TrpC2 ion channels necessary for pheromone signal transduction are pseudogenes in humans, as are most of the genes identified to code for receptor proteins in the mouse VNO (Zufall et al. 2002). As a consequence, pheromone signaling has long been questioned in humans. This view is, however, contradicted by growing arguments for an influence by pheromone-like compounds on human physiology and behavior, and it is also possible that pheromone signals in humans, like in several other mammals (pigs, ferrets) may be transduced via the olfactory mucosa (Dorries et al. 1995); see further in this text. One argument is that the well-known synchronization of menstrual cycles among female roommates seems to be relayed by sweat, which contains such compounds (Stern and McClintock 1998). Another is that women smelling male sweat shift their luteinic hormone pulsatility to promote ovulation (Preti et al. 2003). Furthermore, smelling of two steroids, the 4,16-androstadien-3-one (AND), and estra-1,3,5(10),16-tetraen-3-ol (EST), have in several consecutive experiments shown to affect mood and arousal (Bensafi et al. 2003; Jacob et al. 2001; Lundstrom and Olsson 2005). AND is a derivative of gonadal progesterone and has been identified in urine, plasma, apocrine sweat, as well as semen and axillary hair (Brooksbank et al. 1972; Fukushima et al. 1991; Gower et al. 1994; Kwan et al. 1992; Nixon et al. 1988) in higher concentrations in men than in women. EST, which has been investigated much less than AND, is an estrogen-like derivative detected in the urine of pregnant women (Thysen and Katzman 1968). Both AND and EST have been used in several brain imaging studies and at least AND may be regarded as a candidate pheromone, though the precise modulation of psychological states by the compound has yet to be definitively determined.

18.4 IMAGING STUDIES OF HUMANS EXPOSED TO CHEMOSIGNALING COMPOUNDS WITH PHEROMONE-LIKE PROPERTIES

18.4.1 SYNTHETIC COMPOUNDS

Several groups have employed brain imaging tools to study pheromone-like signals in humans. In a series of positron emission tomography (PET) studies of heterosexual healthy men and women during exposure to AND and EST, we found that smelling of these compounds activated regions covering sexually dimorphic nuclei of the anterior hypothalamus, and that this activation was differentiated with respect to sex and the specific compound (Savic et al. 2001). In women, AND activated the preoptic area and the ventromedial nuclei, whereas in men a hypothalamic activation was detected with EST and involved an area covering the paraventricular and dorsomedial nuclei. In contrast, when men smelled AND and women EST, activations were found in the classical odor processing regions—the amygdala and piriform cortex, the anterior insular cortex, the orbitofrontal cortex, and the anterior cingulate cortex. This sex difference in brain regions processing the signals of AND and EST was not related to the perceived intensity or quality of the odor of the respective compound, which was rated similarly by all participating subjects. Our interpretation of the sex-differentiated pattern of activation was that the two steroids might act bimodally as pheromones and odors. We proposed the hypothesis that the anterior hypothalamus primarily processes signals from the pheromone-like component of AND and EST, whereas the olfactory brain primarily mediates the signals of their odor component. Depending on the sex of the responder in relation to the specific compound (AND or EST), one pathway dominates, whereas the other is suppressed. This hypothesis was based on observations of a similar mutual competition in studies of bimodal odorants (e.g., acetone) (Lundstrom and Hummel 2006; Savic et al. 2002).

The observed sex differentiated pattern is congruent with results from other imaging experiments in which AND, AND-like compounds, or EST has been applied. In a functional MR study of healthy women, Zhou and Chen found, for example, that AND activated only the anterior hypothalamus and not the olfactory cortex (Zhou and Chen 2008). Accordingly, when women smelled androstenol, a derivative of AND, significant activation was found only in the hypothalamus and to a minor extent in the medial amygdala (Savic and Berglund 2010). Conversely, in male subjects Sobel et al. found that EST activated the thalamus and the hypothalamus; interestingly, the EST-related activations were detected even when the compound was presented in extremely low concentrations, which were not possible to detect consciously (Sobel et al. 1999).

The described activations by AND and EST seem to be mediated by the olfactory mucosa. This is indicated by several recent imaging studies, as well as the observation that human olfactory mucosa expresses a functional VR1 gene (Frasnelli et al. 2010; Rodriguez et al. 2000; Savic et al. 2009). Using PET we found that men whose olfactory mucosa was occluded due to nasal polyps were unable to activate the brain with EST, whereas strong hypothalamic activation was observed in healthy male controls (Savic et al. 2009). Accordingly, Frasnelli et al. detected clear hypothalamic activations in females smelling AND both when their VNO was occluded and when it was open (Frasnelli et al. 2010).

The concentrations of AND, androstenol, and EST in the aforementioned studies were variable (10 ml of a 25% and 100% solution in Frasnelli's study, 1 ml of 916-μM AND diluted in propylene glycol, in Zhou's study, and 200 mg of pure crystalline compounds in our experiments). Consequently, while brain imaging data convinces us that our brain detects signals of AND, androstenol, and EST, and that this detection is likely to relate to sex of the smeller in relation to the compound, it is important to express uncertainty regarding the relevance of concentrations used in the experiments for the physiological conditions, something that needs to be investigated in the near future.

18.4.2 BODY ODORS

The natural human body odor consists of about 120 individual chemicals (Labows et al. 1979), of which some have pheromone properties. In general, body odors carry informational cues of great importance for individuals across a wide range of species, and signals hidden within the body odor cocktail are known to regulate several key behaviors in animals. For a long time, the notion that humans may be among these species has been dismissed. Psychophysical studies suggest, however, that humans, like many other animals, may be able to identify the emotional state of an individual belonging to the same species based solely on the body odor. A recent study collected body odor samples from individuals who watched either funny or scary movie sequences. Participants were later asked, in a forced-choice detection task, to identify the emotional state of the donors. Remarkably, participants were able to accurately identify both happy and fearful emotional odors at levels above chance value, though they performed much better with body odor samples from fearful donors.

It is possible that each human has a unique odor signature that carries information related to his or her genetic makeup. Brain imaging investigations of the central processing of body odors have demonstrated that the human brain responds to fear signals hidden within the body odor mixture, is able to extract kin-specific signals, and processes body odors differently than other perceptually similar odors. Lundström and Gottman-Jones found that smelling a friend's body odor activated regions previously seen for familiar stimuli, whereas smelling a stranger activated the amygdala and insular regions akin to what has previously been demonstrated for fearful stimuli (Lundstrom et al. 2008). In addition, when the cerebral activity was compared between situations when subjects identified the body odor of their siblings (kin) and the body odor of a friend (nonkin), it was found that a neuronal network generally consistent with the neuronal substrates of self-referential mental tasks was activated (Lundstrom et al. 2008). This interesting finding suggests that salient body odor signals may recruit additional neuronal circuits.

One unexpected observation in the aforementioned studies is that body odor perception seems to recruit primarily the areas located outside rather than within the main olfactory system. Possibly, the processes involved are too transient to be detected by the olfactory cortex, which has a documented high susceptibility to habitation effects. Alternatively, there might exist separate functional subsystems for common odors and endogenous odors. Whether these systems are entirely separated or overlapping with the dominance of one or the other, depending of the stimulus, is currently uncertain. When endogenous odors contain pheromone-like compounds of the opposite sex the hypothalamus should also be recruited, provided that the aforementioned hypothesis about processing of AND and EST stimuli is correct. However, the majority of hitherto published PET and fMRI studies of body odors did not show significant hypothalamic activations. One possible reason is that the majority of these studies were carried out in the same-sex setting—the body odor (sweat) was collected from females and the subjects exposed to this odor were also females. Notably, when sweat was collected from male subjects and the test persons were females, as in the study of Zhou and Chen, pronounced hypothalamic clusters were present (Zhou and Chen 2008). This issue needs to be more specifically addressed in future studies.

18.5 ACTIVATION WITH PHEROMONE-LIKE COMPOUNDS IN RELATION TO SEXUAL ORIENTATION

The reproductive functions in humans, like in animals, are mediated by neuronal circuits in the anterior hypothalamus. These circuits participate in the integration of the hormonal and sensory cues that are necessary for our sexual behavior and may also be involved in our sexual preferences (Kindon et al. 1996). The preoptic area of the hypothalamus harbors cells releasing luteinic hormone-releasing hormone and mediating estrogen feedback. The estrogen feedback differs between males and females (Dorner et al. 1968) and is also reported to differ between homosexual men

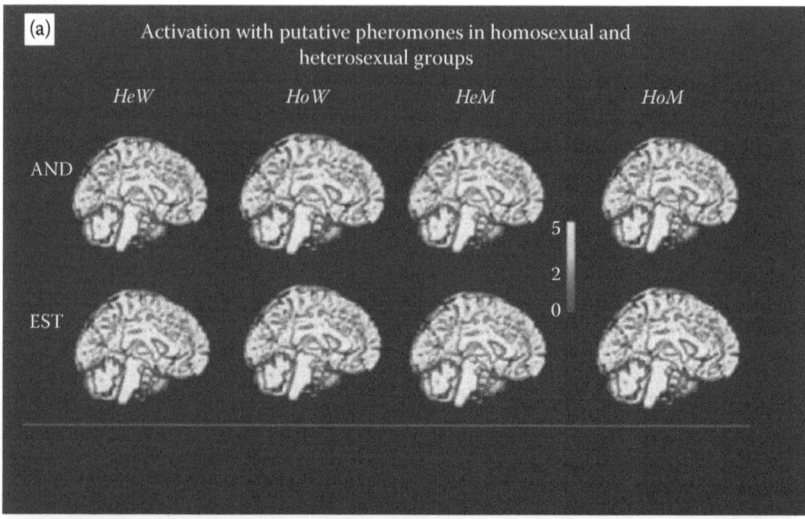

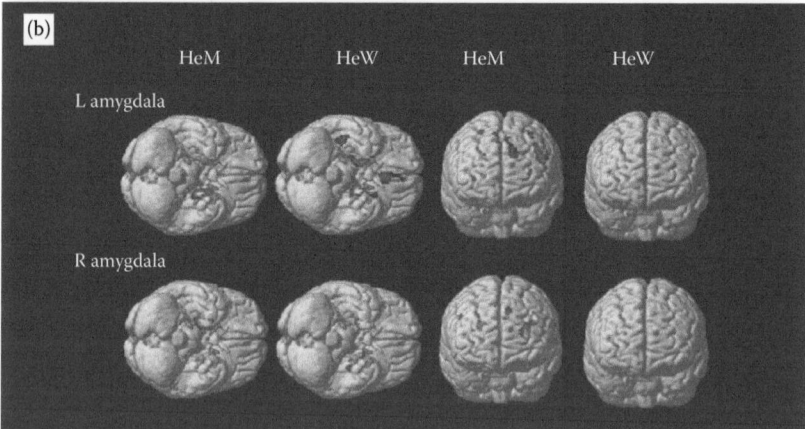

FIGURE 18.1   (See color insert.) (a) Activation with AND and EST. Significant activations illustration of group-specific activations with putative pheromones during smelling of AND and EST in relation to smelling of air (which was the so-called baseline condition). Clusters of activated regions are superimposed on the standard brain. HeM = heterosexual men, HeW = heterosexual women, HoM = homosexual men, HoW = homosexual women, MRI = midsagittal plane. (Remodeled from Figure 1 from Savic I, Lindström P. (2008) *Proc Natl Acad Sci U S A* 105: 9403–8. doi: 10.1073. With permission.) (b) Functional connectivity from the right (R) and left (L) amygdala. (Reprinted from Figure 1 from Savic I, Lindström P. (2008) *Proc Natl Acad Sci U S A* 105: 9403–8. doi: 10.1073. With permission.)

(HoM) and heterosexual men (HeM). The anterior hypothalamus also contains neuronal conglomerates (interstitial hypothalamic nuclei), of which two are reported to differ in volume between men and women, and one was found to differ between HoM and HeM (LeVay 1991; Swaab et al. 1990). A difference between HoM and HeM has also been detected in the volume of the suprachiasmatic nucleus (Swaab et al. 1990). The finding of sex differentiated AND and EST activations of the hypothalamus, therefore, directly raised the question as to whether the corresponding processes are mediated in a similar manner in homosexual subjects.

Subsequent studies with identical design showed that in HoM, like in heterosexual women (HeW) but unlike HeM, the signals from AND were processed by the anterior hypothalamus. Maximal cerebral activation was detected in an area corresponding to the preoptic, ventromedial, and tuberomamillary nuclei (Figure 18.1a). To the contrary, signals from EST were in HoM mediated by the olfactory brain (the left amygdala and piriform cortex) (Savic et al. 2005). Interestingly, and at variance to the other three study groups in, lesbian women the classical odor-processing circuits (the amygdala, the piriform and insular cortex) were engaged during the presentation of AND as well as EST (Figure 18.1a). Significant activations appeared only when restricting the analysis to the hypothalamus, and only with EST, not AND. Direct group comparisons showed that lesbian women differed only from HeW and that the difference was constituted by the absence of the preoptic activation with AND in lesbian women and the presence of this activation in HeW. In contrast to the two steroids, common odors were in all four groups of subjects processed by the classical olfactory regions without any group differences (Berglund et al. 2006).

The less-prominent sex-atypical pattern of activation in the lesbian group was difficult to explain, given that their Kinsey ratings were, like in the HoM group, at the extreme end of the scale. One possibility discussed by the authors is that female homosexuality differs from male homosexuality. As opposed to HoM who are reported to have a later birth order relative to HeM, no significant birth order has been reported in lesbian women. The genetic influence is found to be higher in male compared with female homosexuals (Lyons et al. 2004).

18.6   UNDERLYING MECHANISMS

The mechanisms behind the observed differences of brain activation with AND and EST can only be speculated. One possibility is that the core network mediating AND and EST stimuli is rather similar in all subjects but that this network is responding differently depending of sex and sexual orientation, just as the core network for sexual arousal seems to be shared by homosexual and heterosexual subjects, but the triggering stimulus is reciprocal. Viewing erotic videos of heterosexual or homosexual content produces activation in the hypothalamus, but only when subjects are viewing videos of their respective sexual orientation (Karama et al. 2002). Furthermore, the neuronal response of the ventral striatum and the centromedian thalamus is reported to be stronger to stimuli from the preferred relative to nonpreferred sex (Ponseti et al. 2006). The perception of faces also seems to be modulated by sexual preference. Looking at a female face led to a stronger reaction of the thalamus and medial prefrontal cortex in heterosexual men and homosexual women, whereas in homosexual men and heterosexual women the reaction in these structures was stronger when looking at a male face (Kranz and Ishai 2006).

Another, and equally likely, interpretation, is that signals from AND and EST are processed differently in the nasal mucosa, and/or by the downstream forebrain structures, depending of the sex and sexual orientation. Such a scenario is supported by a recent study suggesting that certain sexually dimorphic features in the brain, which are unrelated to reproduction, also vary with sexual orientation.

We investigated hemispheric asymmetry, using volumetric MRI and functional connectivity of the amygdala, using PET measurements of cerebral blood flow in 90 homosexual and heterosexual men and women (Savic and Lindstrom 2008). Volumetric measurements in HeM and lesbian women showed a rightward cerebral asymmetry, whereas the volumes of the cerebral hemispheres were symmetrical in HoM and HeW. The homosexual subjects also showed sex-atypical amygdala connections (Figure 18.1b). In HoM, as in HeW, the connections were more widespread from the left amygdala. In lesbian women and HeM, on the other hand, the connections from the right amygdala were more widespread. Furthermore, in HoM and HeW the connections displayed were primarily with the contralateral amygdala and the anterior cingulate, while in HeM and lesbian women the connections displayed were primarily with the caudate, putamen, and the prefrontal cortex (Savic and Lindstrom 2008). Whether these latter sexual-orientation-related features of cerebral morphology and function may relate to processes laid down during the fetal or postnatal development is an open question.

18.6.1 TESTOSTERONE ORGANIZATIONAL AND ACTIVATIONAL THEORY

Mechanisms behind homosexuality are often discussed in terms of an underexposure to prenatal androgens. In rats, male cerebral asymmetry is established, in part, by early androgen exposure (or, more precisely, by the estrogen produced by testosterone aromatization). Castration at birth is shown to block the normal rightward brain asymmetry, whereas hemispheric symmetry in females can be reversed to the male pattern by neonatal ovariectomy. It has, therefore, been advocated that exposure to androgens during the fetal development could have major organizational effects on cerebral structures also in humans (Negri-Cesi et al. 2004), although the active compound ought to be testosterone in both sexes, considering that XY subjects with full androgen receptor insensitivity show complete female features despite high testosterone levels (Hughes et al. 2012). The theory about testosterone-mediated masculinization can be tested in humans by application of so-called experiments of nature. Ciumas et al. tested this hypothesis with respect to pheromone processing and amygdala connectivity in a ^{15}O-H$_2$O PET study of women with congenital adrenal hyperplasia (CAH), a condition with high fetal testosterone (Ciumas et al. 2009). Contrary to the hypothesis, the amygdala connectivity in CAH women, as well as the activation with AND and EST, was found to be similar to that of control women (HeW), and reciprocal to control men (HeM). Thus, with regard to both functional organization and activation of the limbic circuits, CAH women showed a pattern congruent with their biological sex and different from the opposite sex. This data illustrates that intrauterine virilization of genitalia is not necessarily paralleled by a masculinization of the limbic brain. Although being unfavorable toward the testostosterone-organizational theory, this data does not exclude such a possibility. Various sex dimorphic features in the brain could, in theory, be affected by fetal testosterone in a dose-dependent manner. They could also have multiple etiological factors (some could be primarily hormonal, other primarily genetic), as recently proposed by Arnold, who pointed out the existence of early and testosterone-independent chromosomal effects on the brain (Arnold et al. 2004). It is fully possible that several different etiological factors contribute to a same sexually dimorphic cerebral feature, for example, psychosexual outcome.

18.7 CONCLUDING REMARKS

Accumulating neuroimaging and behavioral data show that the human brain processes signals from pheromone-like compounds, and that this processing may be mediated via the olfactory mucosa. A recently published study of human tears provides a compelling indication that humans produce volatile compounds that can be detected by smelling, and that such detection may induce an altered physiological response—thus fulfilling the classical criteria for pheromone communication (Gelstein et al. 2011). It is, however, unreasonable to believe that human behavior could be altered in an automated fashion as in animals, and the described chemical communication seems to have modulating effects.

Emerging studies suggest that signals from some synthetic pheromone-like compounds are processed in a manner related to the sex and sexual orientation. This does, however, not imply that pheromone processing drives the sexual orientation. The observed pattern of hemispheric asymmetry and amygdala connectivity suggests that homosexual and heterosexual individuals of the same sex may differ in neurophysiological/neuroanatomical aspects, which, in turn, may related to behavior. Preliminary investigations do not provide support for a major influence of fetal testosterone, and the underlying mechanisms remain unclear.

ACKNOWLEDGMENTS

VINNOVA, Swedish Research Council, AFA, and FAS are acknowledged for financial support.

REFERENCES

Arnold AP, Xu J, Grisham W, Chen X, Kim YH, Itoh Y (2004) Minireview: Sex chromosomes and brain sexual differentiation. *Endocrinology* 145: 1057–1062.
Bensafi M, Brown WM, Tsutsui T, Mainland JD, Johnson BN, Bremner EA, Young N, Mauss I, Ray B, Gross J, Richards J, Stappen I, Levenson RW, Sobel N (2003) Sex-steroid derived compounds induce sex-specific effects on autonomic nervous system function in humans. *Behav Neurosci* 117: 1125–1134.
Berglund H, Lindstrom P, Savic I (2006) Brain response to putative pheromones in lesbian women. *Proc Natl Acad Sci U S A* 103: 8269–8274.
Brennan PA, Keverne EB (2004) Something in the air? New insights into mammalian pheromones. *Curr Biol* 14: R81–R89.

Brooksbank BW, Wilson DA, Gustafsson JA (1972) The metabolism in vivo 4,16-androstadien-3 one in man. Occurrence of 16-dehydro C 19 steroids, androstane-3,16,17-triols and other metabolites in urine. *Steroids Lipids Res* 3: 263–285.

Ciumas C, Linden Hirschberg A, Savic I (2009) High fetal testosterone and sexually dimorphic cerebral networks in females. *Cereb Cortex* 19: 1167–1174.

Dennis JC, Smith TD, Bhatnagar KP, Bonar CJ, Burrows AM, Morrison EE (2004) Expression of neuron-specific markers by the vomeronasal neuroepithelium in six species of primates. *Anat Rec A Discov Mol Cell Evol Biol* 281: 1190–1200.

Dominguez-Salazar E, Portillo W, Velazquez-Moctezuma J, Paredes RG (2003) Facilitation of male-like coital behavior in female rats by kindling. *Behav Brain Res* 140: 57–64.

Dorner G, Docke F, Moustafa S (1968) Homosexuality in female rats following testosterone implantation in the anterior hypothalamus. *J Reprod Fertil* 17: 173–175.

Dorries KM, Adkins-Regan E, Halpern BP (1995) Olfactory sensitivity to the pheromone, androstenone, is sexually dimorphic in the pig. *Physiol Behav* 57: 255–259.

Dulac C, Torello AT (2003) Molecular detection of pheromone signals in mammals: From genes to behaviour. *Nat Rev Neurosci* 4: 551–562.

Frasnelli J, Lundstrom JN, Boyle JA, Katsarkas A, Jones-Gotman M (2010) The vomeronasal organ is not involved in the perception of endogenous odors. *Hum Brain Mapp* 32: 450–460.

Fukushima S, Akane A, Matsubara K, Shiono H, Morishita H, Nakada F (1991) Simultaneous determination of testosterone and androstadienone (sex attractant) in human plasma by gas chromatography-mass spectrometry with high-resolution selected-ion monitoring. *J Chromatogr* 565: 35–44.

Gelstein S, Yeshurun Y, Rozenkrantz L, Shushan S, Frumin I, Roth Y, Sobel N (2011) Human tears contain a chemosignal science. *Science.* 331(6014): 226–30. doi: 10.1126.

Gower DB, Holland KT, Mallet AI, Rennie PJ, Watkins WJ (1994) Comparison of 16-androstene steroid concentrations in sterile apocrine sweat and axillary secretions: Interconversions of 16 androstenes by the axillary microflora—A mechanism for axillary odour production in man? *J Steroid Biochem Mol Biol* 48: 409–418.

Hughes IA, Davies JD, Bunch TI, Pasterski V, Mastroyannopoulou K, MacDougall J (2012) Androgen insensitivity syndrome. *Lancet* 380: 1419–1428.

Jacob S, Hayreh DJ, McClintock MK (2001) Context-dependent effects of steroid chemosignals on human physiology and mood. *Physiol Behav* 74: 15–27.

Kang N, Baum MJ, Cherry JA (2009) A direct main olfactory bulb projection to the "vomeronasal" amygdala in female mice selectively responds to volatile pheromones from males. *Eur J Neurosci* 29: 624–634.

Karama S, Lecours AR, Leroux JM, Bourgouin P, Beaudoin G, Joubert S, Beauregard M (2002) Areas of brain activation in males and females during viewing of erotic film excerpts. *Hum Brain Mapp* 16: 1–13.

Karlson P, Luscher M (1959) "Pheromones": A new term for a class of biologically active substances. *Nature* 183: 55–56.

Keverne EB (2008) Visualisation of the vomeronasal pheromone response system. *Bioessays* 30: 802–805.

Kindon HA, Baum MJ, Paredes RJ (1996) Medial preoptic/anterior hypothalamic lesions induce a female-typical profile of sexual partner preference in male ferrets. *Horm Behav* 30: 514–527.

Kranz F, Ishai A (2006) Face perception is modulated by sexual preference. *Curr Biol* 16: 63–68.

Kwan TK, Trafford DJ, Makin HL, Mallet AI, Gower DB (1992) GC-MS studies of 16-androstenes and other C19 steroids in human semen. *J Steroid Biochem Mol Biol* 43: 549–556.

Labows J, Preti G, Hoelzle E, Leyden J, Kligman A (1979) Analysis of human axillary volatiles: Compounds of exogenous origin. *J Chromatogr* 163: 294–299.

LeVay S. (1991) A difference in hypothalamic structure between heterosexual and homo-sexual men. *Science* 253: 1034–1037.

Lundstrom JN, Olsson MJ (2005) Subthreshold amounts of social odorant affect mood, but not behavior, in heterosexual women when tested by a male, but not a female, experimenter. *Biol Psychol* 70: 197–204.

Lundstrom JN, Hummel T (2006) Sex-specific hemispheric differences in cortical activation to a bimodal odor. *Behav Brain Res* 166: 197–203.

Lundstrom JN, Boyle JA, Zatorre RJ, Jones-Gotman M (2008) Functional neuronal processing of body odors differs from that of similar common odors. *Cereb Cortex* 18: 1466–1474.

Lyons MJ, Koenen KC, Buchting F, Meyer JM, Eaves L, Toomey R, Eisen SA, Goldberg J, Faraone SV, Ban RJ, Jerskey BA, Tsuang MT (2004) A twin study of sexual behavior in men. *Arch Sex Behav* 33: 129–136.

McClintock MK, Jacob S, Zelano B, Hayreh DJ (2001) Pheromones and vasanas: The func-tions of social chemosignals. *Nebr Symp Motiv* 47: 75–112.

Negri-Cesi P, Colciago A, Celotti F, Motta M (2004) Sexual differentiation of the brain: Role of testosterone and its active metabolites. *J Endocrinol Invest* 27: 120–127.

Nixon A, Mallet AI, Gower DB (1988) Simultaneous quantification of five odorous steroids (16 androstenes) in the axillary hair of men. *J Steroid Biochem* 29: 505–510.

Paredes RG, Tzschentke T, Nakach N (1998) Lesions of the medial preoptic area/anterior hypothalamus (MPOA/AH) modify partner preference in male rats. *Brain Res* 813: 1–8.

Ponseti J, Bosinski HA, Wolff S, Peller M, Jansen O, Mehdorn HM, Buchel C, Siebner HR (2006) A functional endophenotype for sexual orientation in humans. *Neuroimage* 33: 825–833.

Preti G, Wysocki CJ, Barnhart KT, Sondheimer SJ, Leyden JJ (2003) Male axillary extracts contain pheromones that affect pulsatile secretion of luteinizing hormone and mood in women recipients. *Biol Reprod* 68: 2107–2113.

Rodriguez I, Greer CA, Mok MY, Mombaerts P (2000) A putative pheromone receptor gene expressed in human olfactory mucosa. *Nat Genet* 26: 18–19.

Savic I, Berglund H, Gulyas B, Roland P (2001) Smelling of odorous sex hormone-like compounds causes sex-differentiated hypothalamic activations in humans. *Neuron* 31: 661–668.

Savic I, Gulyas B, Berglund H (2002) Odorant differentiated pattern of cerebral activation: Comparison of acetone and vanillin. *Hum Brain Mapp* 17: 17–27.

Savic I, Berglund H, Lindstrom P (2005) Brain response to putative pheromones in homo-sexual men. *Proc Natl Acad Sci U S A* 102: 7356–7361.

Savic I, Lindstrom P (2008) PET and MRI show differences in cerebral asymmetry and func-tional connectivity between homo- and heterosexual subjects. *Proc Natl Acad Sci U S A* 105: 9403–9408.

Savic I, Heden-Blomqvist E, Berglund H (2009) Pheromone signal transduction in humans: What can be learned from olfactory loss. *Hum Brain Mapp* 30: 3057–3065.

Savic I, Berglund H (2010) Androstenol—A steroid derived odor activates the hypothalamus in women. *PLoS One* 5: e8651.

Sobel N, Prabhakaran V, Hartley CA, Desmond JE, Glover GH, Sullivan EV, Gabrieli JD (1999) Blind smell: Brain activation induced by an undetected air-borne chemical. *Brain* 122 (Pt 2): 209–217.

Stern K, McClintock MK (1998) Regulation of ovulation by human pheromones. *Nature* 392: 177–179.

Swaab DF, Hofman MA, Honnebier MB (1990) Development of vasopressin neurons in the human suprachiasmatic nucleus in relation to birth. *Brain Res Dev Brain Res* 52: 289–293.

Thysen BEW, Katzman pA (1968) Identification of estra-1,3,5(10),16-tetraen-3-ol (estratet-raenol) from the urine of pregnant women. *Steraloids* 11: 73–87.

Trotier D, Eloit C, Wassef M, Talmain G, Bensimon JL, Doving KB, Ferrand J (2000) The vomeronasal cavity in adult humans. *Chem. Senses* 25: 369–380.

Witt M, Georgiewa B, Knecht M, Hummel T (2002) On the chemosensory nature of the vomeronasal epithelium in adult humans. *Histochem Cell Biol* 117: 493–509.

Zhou W, Chen D (2008) Encoding human sexual chemosensory cues in the orbitofrontal and fusiform cortices. *J Neurosci* 28: 14416–14421.

Zufall F, Kelliher KR, Leinders-Zufall T (2002) Pheromone detection by mammalian vomeronasal neurons. *Microsc Res Tech* 58: 251–260.

19 Human Pheromones
Do They Exist?

Richard L. Doty

CONTENTS

19.1 Historical Introduction...535
19.2 Quest for Discovering Human Pheromones 538
    19.2.1 Sources of Putative Human Pheromones...539
    19.2.2 Sensory Studies of Axillary Secretions...540
    19.2.3 Androstenone, Androstenol, and Androstadienone as Human
           Pheromones..543
    19.2.4 Human Menstrual Synchrony Pheromones 547
19.3 Conclusions..554
References...554

19.1 HISTORICAL INTRODUCTION

The question arises as to whether humans possess pheromones. This issue is complex, since one has to first define what, in fact, a pheromone is. Once this is done, then some clear behavioral or endocrinological effect must be measured to demonstrate the effects of the putative agent.

The focus of this chapter is on whether humans possess pheromones, a question posed by *Science* magazine in 2005 as one of the top 100 outstanding issues of that era (Anonymous 2005). Its goal is to summarize a few concepts from *The Great Pheromone Myth* (Doty 2010) that address this issue.* While it is apparent that, like music and lighting, odors and fragrances can alter mood states and physiological arousal, is there evidence that unique agents exist, namely pheromones, which specifically alter such states? Are there special chemicals that can influence human endocrinology, such as altering the length of menstrual cycles?

Before addressing these questions, it is useful to put into perspective how the concept of pheromones arose in the first place. In the early 1930's, the entomologist Bethe termed hormones secreted within the body of insects endohormones and those secreted outside the body of insects ectohormones. Ectohormones were further divided into chemicals with intraspecific effects, termed homoiohormones, and those with interspecific effects, termed alloiohormones (Bethe 1932). In 1959, Karlson and Luscher

---

* Because of space limitations, all of the arguments presented in this book cannot be made in this chapter, and the reader is referred to the book for a complete exposé of the issues.

replaced the term homoiohormone with the term pheromone (Karlson & Lüscher 1959). These authors defined pheromones as "substances which are secreted to the outside by an individual of the same species, in which they release a specific reaction, for example, a definite behavior or a developmental process." These authors distinguished between pheromones acting via olfaction and those acting via oral or ingestive routes. The former produced immediate releasing responses (e.g., initiating and guiding the flight of the male silk worm moth, *Bombyx mori*, to the female) and the latter delayed endocrine or reproductive effects, such as the caste-determining substances of many social insects.

Although numerous concerns and qualifications subsequently arose regarding the usefulness of this term in insects, such as the fact that insects have a remarkable capability to learn and respond to a wide range of odors, the pheromone concept and its associated term became codified by entomologists and other biologists. In retrospect, this codification has been questioned even in insects by some entomologists, who point out, among other things, that multiple chemicals, rather than single chemicals, which served as exemplars for insect pheromones (Karlson & Butenandt 1959) are the norm for eliciting most insect behavioral and reproductive processes (Holldobler 1999). Nevertheless, in the 1960s and 1970s, the pheromone concept was generalized to mammals. In an influential review appearing in *Science*, Parkes and Bruce (1961) reiterated the dichotomy employed by Bethe, noting that some "chemical messengers" act within an individual (e.g., hormones and "other excitatory substances" such as CO_2), whereas others (i.e., "pheromones") act between individuals via ingestion, absorption, or sensory receptors. These writers concluded that "Endocrinology has flowered magnificently in the last 40 years; exocrinology is now about to blossom."

The generalization of the pheromone concept to mammals was popularized by the entomologist Wilson in a 1963 *Scientific American* article (Wilson 1963). This author explicitly set the tone for conceptualizing the nature of both priming and releasing mammalian pheromones. In the case of mammalian releasing pheromones, he focused on musks and noted:

> Pheromones that produce a simple releaser effect–a single specific response mediated directly by the central nervous system–are widespread in the animal kingdom and serve a great many functions. The chemical structures of six attractants are shown <in the picture>. Although two of the six–the mammalian scents muskone and civetone–have been known for some 40 years and are generally assumed to serve a sexual function, their exact role has never been rigorously established by experiments with living animals. In fact, mammals seem to employ musk-like compounds, alone or in combination with other substances, to serve several functions: to mark territories, to assist in territorial defense and to identify the sexes.*

Subsequently, claims of the chemical isolation of mammalian pheromones appeared in the literature. The initial ones focused on releasing pheromones and

---

* The caption of the table in which the chemical structures of civetone and maskone were provided read as follows: "Six sex pheromones including the identified sex attractants of four insect species as well as two mammalian musks generally believed to be sex attractants. The molecular weight of most sex pheromones accounts for their narrow specificity and high potency." Note that Wilson was classifying agents as pheromones without even knowing what effects, if any, they have on behavior or endocrine function.

included (a) chemicals within the vaginal secretions of rhesus monkeys that were said to elicit copulatory behaviors in males (Curtis et al. 1971; Michael & Keverne 1970; Michael et al. 1971), (b) agents within the tarsal scent glands of male black-tailed deer that elicited licking by females (Brownlee et al. 1969; Müller-Schwarze 1971; Müller-Schwarze et al. 1974), (c) a material from the midventral scent gland of Mongolian gerbils that received investigation from other gerbils (Thiessen et al. 1974), and (d) two steroids from the submaxillary salivary glands of boars that lowered the threshold for pressure-induced lordosis in female pigs (namely, 5α-androsten-16-en-3-one and its related alcohol) (Melrose et al. 1971). However, such isolated chemicals rarely mimicked completely the effectiveness of the virgin secretions and were largely dependent on learning for their efficacy. Moreover, a number of mammalogists pointed out that mammals are not insects and chemicals do not "release" behaviors in a simple manner in these organisms. Bronson suggested, for example, that the term signaling should replace the term releaser in describing one class of pheromones (Bronson 1968) and Whitten and Champlin suggested that "behavioral" should serve as the substitute (Whitten & Champlin 1973). Subsequent investigators suggested replacing the releasing pheromone term with such terms as social odors (Brown 1979), homeochemic substances (Martin 1980), or semiochemicals (Albone 1984). As Bronson (1976, p. 123) aptly pointed out,

> The unfortunate side of such generalizations <from insects> is the tendency to think of mammalian communication in terms of simple stimulus-response systems. For example, it is now relatively common usage to refer to "aggression-promoting (or "eliciting") and "aggression-inhibiting" pheromones in mice (e.g., Lee & Griffo 1974; Mugford & Nowell 1972). The obvious implication of this terminology is the existence of two simple urinary compounds which unequivocally either release or inhibit a stereotyped aggressive response. Mammalian social behavior simply does not work that way except at the purely reflexive level.

Bronson pointed out that the mouse's nervous system not only contains many more neurons than that of an insect, but differs in terms of degree of encephalization, the numbers of associative neurons, and the flexibility afforded to the mediated behaviors. He continues,

> Most insect pheromones are usually single compounds or simple mixtures, typically secreted by restricted glands, and normally evoking stereotyped responses even under totally inappropriate circumstances. Thus many of the standard tests for insect attractants have relied upon copulatory behavior in response to scented filter paper, repeated exposures in many cases providing little habituation of the response (Birch 1974). It is difficult to imagine a male mouse attempting copulation with a scented filter paper let alone doing so repeatedly, and, by extension, it is exceedingly difficult to apply the simple releaser concept to much of mammalian social behavior, whether elicited in part by odors or not. Additionally, experience is a profound modifier to mammalian social behavior. There have actually been relatively few attempts to examine the role of experience in odor-induced responses in mammals. Where investigated, however, the results usually have indicated a potent role for experience. Thus species identification apparently can be easily manipulated by odors early in the life of mammals (e.g., (Carter & Marr 1970; Mainardi, Marsan, & Pasquali 1965; Marr & Lilliston 1969)

and adult sexual experience is a strong determinant of response to sex odors (e.g., (Caroom & Bronson 1971; Carr, Loeb, & Dissinger 1965; Carr, Loeb, & Wylie 1966). One wonders at this point whether the pheromone concept, so useful in insect behavior and physiology, should be bastardized to the point where it is used to cover situations in mammalian behavior where usually complex odors evoke highly variable responses which are easily modified by experience.

As described in great detail in my book (Doty 2010), attempts to identify mammalian pheromones have been generally unsuccessful, regardless as to whether such agents are viewed as releaser pheromones, priming pheromones, modulating pheromones, or any other type of putative pheromone. Such attempts fall prey to a number of conceptual, operational, and practical problems that go beyond issues of semantics, including assumptions related to innateness, chemical complexity, and the nature of perceptual systems in general.* Unfortunately, space does not permit a complete exposition of the involved issues, so the focus of this chapter is on issues related to the most popular claims of human pheromones.

19.2   QUEST FOR DISCOVERING HUMAN PHEROMONES

The search for human pheromones came to the fore soon after Alex Comfort published his influential 1971 *Nature* paper entitled *Likelihood of Human Pheromones* (Comfort 1971). However, nearly a half century has elapsed since this paper was published and a strong argument can be made that no chemical or simple set of chemicals has been identified that could be construed as a human pheromone. That being said, some investigators have assumed, *a priori*, that androgen-related steroids are pheromonal agents, as described later in this chapter. Some neurobiologists have even argued that the vomeronasal organ (VNO), a chemosensory structure at the base of the nasal septum in a number of mammals and other vertebrates, is the pheromone receptor, although most have stepped back from this dichotomous position in light of the discovery that this organ, at least in mice, also contains receptor proteins common to the main olfactory system. Although the weight of the evidence is that humans have a vestigial VNO, one group has claimed that this small pouch is functional and responds to chemicals in a sexually dimorphic manner, altering autonomic processes (Berliner et al. 1996; Monti-Bloch et al. 1994). Since there is no neural connection to this rudimentary organ and humans lack the brain structure to which such a connection would be made (i.e., the accessory olfactory bulb), such effects are more aptly explained on the basis of stimulation of the ethmoidal branch of the trigeminal nerve.

---

* In my view it is erroneous to infer that a wide range of mammalian behaviors and endocrine responses are uniquely determined in an invariant way by single or small sets of chemical stimuli and to apply a generic and misleading name to the presumptive agents to support such an inference. Nonetheless, at some point semantics are involved. As noted by Pinker (2007, p. 3), "Semantics is about the relation of words to reality–the way that speakers <or writers> commit themselves to a shared understanding of the truth, and the way their thoughts are anchored to things and situations in the world." Clearly, the belief that simple and presumptive hormone-like chemical agents, signified by the term pheromone, are responsible for an extensive array of behaviors and endocrine states assumes a reality that is questionable in humans and other mammals.

19.2.1 Sources of Putative Human Pheromones

Like all vertebrates, humans excrete or secrete many different chemicals via their urine, anal excrement, breath, genitalia, saliva, and skin glands. Most proponents of the human pheromone concept assume that skin glands are the source of the active pheromonal agents. All three major skin glands—apocrine sweat glands, eccrine sweat glands, and sebaceous glands—can produce chemicals that become odorous. Conceptually, such chemicals could be sensed by the olfactory, gustatory, or trigeminal neural systems, or enter the general circulation by way of the vasculature of the nose, sinuses, oral cavity, and lungs (Doty 2008). In some body areas the three major skin glands are associated with hair follicles, as shown in Figure 19.1.

Among the secretions of the apocrine sweat glands are lipids, including cholesterol, sterol esters, triglycerides, diglycerides, fatty acids, and wax esters. These secretions are usually considered odorless until acted upon by aerobic diphtheroid bacteria (Leyden et al. 1981). The highest density of apocrine glands occur in the axillae and in the perineum (Doty 1981). These glands become functional around the time of puberty and release their secretions in response to such emotions as anxiety, fear, pain, or sexual arousal (Wilke et al. 2007). In horses and some other ungulates, apocrine glands are involved in thermal regulation, but this not the case in humans and most other mammals (Scott et al. 2001). Conceivably, in our own species, they are a vestigial remnant of a defense system useful at one time in warding off predators or unwanted conspecifics.

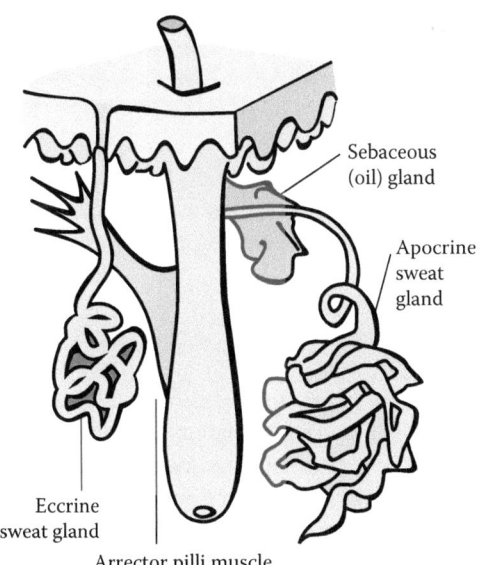

Sebaceous (oil) gland

Apocrine sweat gland

Eccrine sweat gland

Arrector pilli muscle

FIGURE 19.1 Schematic of a pilosebaceous unit with apocrine and eccrine sweat glands. (Copyright 2013, Richard L. Doty.)

The eccrine sweat glands, which are located on nearly all body surfaces, play a key role in regulating body temperature and, in humans, are capable of secreting as much as 3 liters of aqueous solution in 1 hour (Tobin 2006). Unlike apocrine glands, they can connect directly with the skin surface independently of hair cells. Their secretions are controlled largely by the sympathetic nervous system and are produced in response to stress, exercise, and other strenuous activity (Nicolaides 1974). While eccrine sweat is generally considered odorless, it can become odoriferous as a result of diet (e.g., garlic) and disease, such as fish odor syndrome (Simehoff et al. 1977).

Most of the lipids and antimicrobial products of the skin are produced by seba-ceous glands (Nicolaides 1974). Their secretions are said to have a weak pleasant odor when not infected by bacteria. Like apocrine glands, they enlarge at the time of puberty. They are most dense on the forehead, face, and scalp, and like eccrine sweat glands are absent on the palms of the hand and the soles of the feet. The eyelids, the ear canals, the nares, the lips, the buccal mucosae, the breasts, the prepuce, and the anogenital region all contain specialized sebaceous glands. These glands are also involved in the regulation of local steroidogenesis, skin barrier function, and the production of both anti- and proinflammatory compounds (Tobin 2006).

19.2.2 Sensory Studies of Axillary Secretions

Can gender be determined from the odor of axillary secretions?

The question arises as to whether information critical for reproductive behavior can be sensed from axillary secretions, since most proponents of human pheromones have focused their attention on such secretions or compounds contained therein. To test this concept, Russell (1976) had 13 women and 16 men wear T-shirts for 24 hours without bathing or using deodorants. The armpit regions of the T-shirts were then sniffed in a test session where (a) the subject's own T-shirt, (b) a male stranger's T-shirt, and (c) a female stranger's T-shirt were presented. The task was to identify their own odor and then to guess which of the two remaining odors came from a man. Nine of the 13 women and 13 of the 16 men performed both of these tasks correctly. This suggested to Russell that "at least the rudimentary communication of sexual discrimination and individual identification can be made on the basis of olfactory cues."

The results of this study could, however, be based on the fact that male odors are generally stronger and less pleasant than female odors, reflecting (a) larger male apo-crine glands and (b) the lack of shaved axillae. Axillary hair increases the surface area for bacterial activity and molecular diffusion. Thus, subjects may have used the strategy of assigning stronger and more unpleasant odors to the male category and weaker and less unpleasant ones to the female category, analogous to assigning, from a list of body heights, heavier weights to men and lighter weights to woman.

To test this possibility, we performed a series of studies in which axillary secre-tions were collected on gauze pads worn in the axillae of male and female donors for approximately 18 hours (Doty et al. 1978). These pads were presented for sampling in small sniff bottles to 10 men and 10 women, along with blank stimuli. The sub-jects were given the following instructions:

You will be presented with a series of sniff bottles containing human sweat. We wish you to tell us <by smell> which sex each of the odor samples comes from. The set of odors may include samples from both men and women, or from only men or from only women. Thus, some may be from females, some from males, or, alternatively, all may be from females or all from males. Therefore, don't allow yourself to assume that some predetermined number of one or the other sex is represented.

The subjects were also required to estimate the relative intensity and pleasantness of the stimuli using a magnitude estimation procedure (Doty & Laing 2003). In one study, half of the stimuli were from men and half from women. In other studies, all of the stimuli were from women or from men.

The results of the study in which axillary odors from both men and women were smelled are presented in Figure 19.2. Four of the five male odors were correctly identified as male by at least half of the subjects, whereas three of the four female odors were correctly identified as female by at least half of the subjects. It is clear that stimulus intensity was strongly related to the sex assignments and that the blank (B) stimulus was nearly always assigned to the female category.

Figure 19.3 shows the results of the study in which only female odors were presented. In this case, the strongest odors were assigned to the male category and the weakest ones to the female category, even though all odors came from women. All subjects assigned the blank to the female category. These findings suggest that intensity and/or pleasantness was the primary basis for the sex classifications.

If human axillary odors serve as sex attractants without extensive conditioning, one would expect that pleasantness ratings of female axillary odors would be higher in men than in women, and that the reverse would be true for male axillary odors. However, we found that both men and women rated the male odors as more intense and less pleasant than female odors, with the relative magnitude of the responses

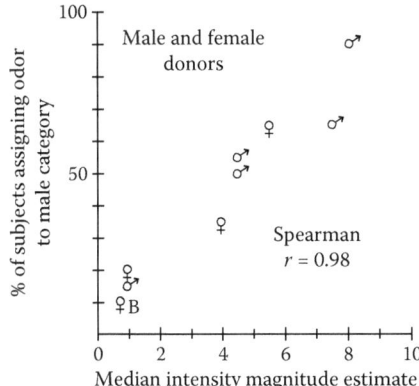

FIGURE 19.2 Relationship between perceived intensity, sex of odor donor, and percent of subjects assigning each axillary odor to the male gender category. Male and female symbols signify the sex of the odor donor. B indicates blank control. (From Doty, R. L. (1981). *Chem. Senses* 6: 351–376. With permission.)

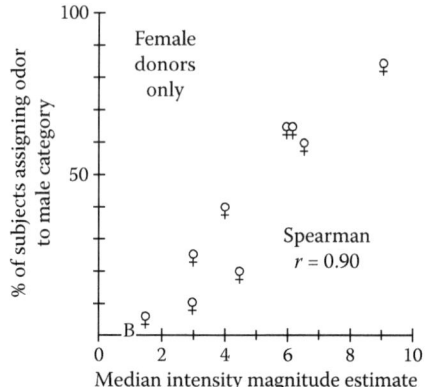

FIGURE 19.3 Relationship between perceived intensity and percent of subjects that assigned each axillary odor to the male gender category. All of the axillary odors came from women. B indicates a blank control. (Reprinted from Doty, R. L. (1981). *Chem. Senses* 6: 351–376. With permission.)

of the two sexes being similar ($rs > 0.90$) (Doty et al. 1978). An inverse association was found between the overall intensity and pleasantness estimates of the odors ($r = -0.94$, $p < 0.001$), a nonculture-specific phenomenon observed by others (Schleidt et al. 1981).

It is telling that no differences in odor pleasantness or intensity are perceived when equivalent quantities of male and female apocrine sweat are incubated *in vitro* with aerobic coryneform bacteria, which produces characteristic body odor (Gower et al. 1994). Moreover, axillary hygiene markedly decreases the ability of subjects to correctly assign gender to axillary odors, in accord with the concept that relative intensity or pleasantness is the basis for the sex assignments (Schleidt 1980). When presented in equivalent volumes, axillary odors are not differentially pleasant to males and females, as would be expected if they contained sexually dimorphic pheromones associated with attraction or repulsion (Zeng et al. 1996). The extent to which learning influences hedonic responses to axillary odors is not known, although children who can identify their source as the armpits rate such odors as more intense and less pleasant than children who cannot identify their source, implying some role of concept learning. In accord with adult findings, adolescent girls rate axillary odors as more intense and less pleasant than do adolescent boys (Stevenson & Repacholi 2003).

A critical point is that no sex differences are present in the odorless precursor proteins within apocrine glands that are ultimately responsible for axillary odor, although males may be more prone than females to release their contents in relation to stressful stimuli (Spielman et al. 1998). Another critical point is that the intensity and pleasantness of axillary odors are significantly influenced by diet. In one study, 17 men were placed on a meat diet for 2 weeks and on a nonmeat diet for 2 weeks, with the order of the diets counterbalanced in time (Havlicek & Lenochova 2006). Axillary secretions were collected using gauze pads worn for 24 hours at the end of the dietary periods. The odors of donors on the meat diet were rated as less attractive, less pleasant, and more intense than the odors from the donors on the nonmeat diet. These findings suggest the possibility that some sex differences observed in axillary odor studies could be confounded by sex-specific dietary habits since, on average, men tend to eat much more meat than women (Shiferaw et al. 2012).

Do axillary secretions influence mood?

Like sights and sounds, there is no doubt that odors can influence human moods, emotions, and feelings. Is there any evidence that human secretions or excretions, such as those from the axillae, influence such emotions in ways different from other odors? Unfortunately, this topic has rarely been addressed scientifically and no study has provided control odorants to determine whether the purported effects are specific to the putative pheromonal stimuli.

In one of the few studies to have assessed the influences of human axillary odors on mood, Chen and Haviland-Jones (1999) concluded that "Exposure to underarm odors of older women, women, and older adults, led to a greater reduction in depressive mood than exposure to underarm odors of young men, men, and young adults." In this study, gauze pads were worn in the axillae for several days by five prepubertal girls, five prepubertal boys, five college women, five college men, five older women, and five older men. Before and after smelling the odor-laden pads located in Petri dishes, subjects rated their current mood on the Differential Emotion Scale (Izard 1993). More than 300 volunteers took part, with each one smelling only one type of the aforementioned stimuli. However, a close and critical analysis of this study by Black (2001) found that the average positive mood actually declined in subjects exposed to the axillary odors from each of the six donor groups, contrary to the conclusions of the authors. Moreover, this critic pointed out that the Differential Emotion Scale was inappropriately employed as a measure of mood, since it is designed to determine how often different moods are experienced, not to assess short-term changes in mood. This suggested to Black that the subjects were likely confused as to how to respond on the post-odor test administration. Most importantly, Black noted that while Chen and Haviland-Jones collected data from a control condition (pads just left in homes but not placed in the armpit), these data were not properly included in their statistical analyses. When the mean data of each of the target groups was to with those of the controls, no statistically significant effects emerged. Black concluded (p. 216) that "it is clear that their work provides no evidence for their conclusion that certain human odors can decrease depressive mood."

19.2.3 ANDROSTENONE, ANDROSTENOL, AND ANDROSTADIENONE AS HUMAN PHEROMONES

Dozens of studies have been performed on what have been purported, *a priori*, to be human pheromones, namely 5α-androst-16-en-3α-one (androstenone), 5α-androst-16-en-3α-ol (androstenol), or androsta-4,16-dien-3-one (androstadienone).

In this section I outline the basic logic of why it seems unlikely that these agents are human pheromones. The reader is referred elsewhere to reviews of the numerous studies that have employed these agents in attempts to show their pheromonal properties in humans (Havlicek et al. 2010; Doty 2010).

There are a number of reasons why such steroids have been assumed to be human pheromones by many workers (Doty 2010). These include the following:

- Being steroids, they fit into the pheromone concept of an "externally secreted hormone"
- They are among the few identified compounds that have been conceptually linked to mammalian reproductive behavior that some have deemed as pheromonal (i.e., lordosis in the female pig)
- They typically have urine- or musklike smells for those who can smell them, reinforcing the notion of their animal-like nature and the folklore that musks are social attractants in humans (Gower et al. 1985; Kloek 1961; Le Magnen 1952)
- They are commonly present, albeit at low levels, in human urine, axillary apocrine sweat, saliva, and semen (Brooksbank & Haslewood 1961; Brooksbank et al. 1974; Gustavson et al. 1987; Kwan et al. 1992; Nixon et al. 1988), making them potentially available to the external milieu for transfer from one person to another
- They occur in higher concentrations in men than in women, implying sexual dimorphism in their production (Gower & Ruparelia 1993; Lundström & Olsson 2005)
- Women are more sensitive, on average, to these agents than are men (Doty 1986), implying sexual dimorphism in their ability to be perceived
- In light of findings that humans can distinguish between the sexes to some degree on the basis of axillary and breath odors (Doty et al. 1978, 1982; Hold & Schleidt 1977), it is conceivable that these steroids serve to make this possible
- "Because androstenol has no known function in humans, these findings have suggested to several investigators that the steroid may function as a human pheromone" (Gustavson et al. 1987)

While such observations seem convincing on the surface, close scrutiny reveals the following issues:

- In reality, the levels of such steroids in the human axillae are low and highly variable. Using capillary gas chromatography–mass spectrometry with specific ion monitoring, Nixon et al. found that only 10 of 24 men had androstenone in their axillary hair (Nixon et al. 1988) and no relationship was evident between the age of the donors and presence of the steroid. Others, using different analytical methods, have reported finding no androstenone in samples of fresh apocrine sweat or secretions sampled by sterile gauze pads (Bird & Gower 1981; Labows et al. 1979). Before bacterial action, androstadienone levels are too low to be detected in the axillae by smell (Gower et al. 1994; Labows 1988).

- It does not follow that simply because these compounds are found in body fluids or axillae that they communicate meaningful social information or influence reproductive processes in humans. Indeed, androstenone and androstenol are common in the animal and plant kingdoms, being found even in the roots of vegetables such as parsnip and celery (Claus & Hoppen 1979). In one study, for example, androsterone was found in 60–80% of the plant species investigated (Janeczko & Skoczowski 2005).

- Under the assumption that olfaction is involved, none of these steroids contributes much to the generation of prototypical body odor, which arises largely from a mixture of C_6-C_{11} normal, branched, and unsaturated acids (Hasegawa et al. 2004; Zeng et al. 1991, 1992).

- A significant number of persons cannot smell androstenone and related steroids (Amoore et al. 1977; Koelega 1980; Ohloff et al. 1983), although exposure to high concentrations can result in eventual detection in some individuals (Wysocki et al. 1989). This fact, along with evidence that most persons who can smell these agents find them repulsive or unpleasant (Gower et al. 1985; Jacob et al. 2006; Koelega 1980), would seem to limit their value in social interactions considered as reflecting influences from "sex pheromones." In the case of androstadienone, repeated exposure results in an increase in its perceived unpleasantness (Boulkroune et al. 2007).

- It is questionable as to whether musky or urine-like smells, as such, reflect a logical criterion for defining odorants as pheromones.

- Sex differences and subtle menstrual cycle-related fluctuations are present for a wide range of odorants, including synthetic ones, so there seems to be nothing special about these agents in this regard (Doty & Cameron 2009).

- As noted earlier in this chapter, the ability of humans to determine the sex of another human on the basis of axillary odors as well as other odors common to the sexes such as breath odors, appears to be dependent on the intensity or pleasantness of the involved odors, not on the intrinsic chemical makeup of the secretions (Doty et al. 1978, 1982).

- If pheromones are species-specific, which is inherent in the original and most subsequent definitions of pheromones (Doty 2010), generalizing findings from pig studies to human studies is an oxymoron.

- Lack of knowledge of a known function of a secretion does not increase the likelihood that is serves as a human pheromone.

The concept that androstenone and androstenol are pheromones in pigs originally stemmed from observations that they increased the incidence of the standing response of sows when sprayed near the sows' noses (Hafez & Signoret 1969; Melrose et al. 1971; Patterson 1966, 1968; Stefanczyk-Krzymowska et al. 2000). As noted by Claus and Hoppen (1979, p. 1674), "For female pigs in oestrus … <androstenone> is a very desirable 'male perfume' which is released by the boar's saliva before mating and stimulates the female's 'standing reflex,' thus acting as an aphrodisiac pheromone."

It should be emphasized, however, that androstenone and androstenol do not have invariant influences even on the mating stance of sows. As shown in Table 19.1, such steroids fail to facilitate lordosis in all sows and are not the sole stimuli that facilitate pressure-induced standing. Sows do not lordose for all males, likely exhibiting mating preferences analogous to those that occur in dogs (Beach & LeBoeuf 1967). The most parsimonious explanation of the behavioral effects of androstenone and androstenol in the domestic pig is that the sow is conditioned through experience to exhibit lordosis in response to the smell of these agents. Albone (1984) summarizes the complexity as follows (p. 238):

> The situation is, however, a little more complicated than it seems. The oestrous sow will 'stand' in response not only to olfactory signals, but also, for example, to the sound of the boar's grunting. In the natural situation the sow is exposed to a simultaneous combination of cues of many kinds, olfactory, visual, tactile and auditory, all of which play some part in stimulating the standing response, although it is clear that among these, olfactory signals are very important. Also, it is found experimentally that the oestrous sow will stand in response to the odour of boar urine or boar preputial fluid, substances in which these particular C_{19}-$\Delta 16$ steroids <androstenone and androstenol> are either absent or present at very low levels. Further, the oestrous female will respond to varying degrees to the odours of some other closely related steroids.

If these steroids were truly effective as pheromones in humans, one might ask the following questions: Are women, in fact, attracted to the odors of male pigs or more

---

TABLE 19.1

Influence of Various Steroids Found in Sexually Mature Boar Secretions on Pressure-Induced Lordosis in Estrous Sows[a]

Odorant	Concentration	% of Sows Giving a Positive "Back Pressure" Response to Odorant	No. Pigs Tested
Preputial fluid and urine	Full	42	19
5 α-androst-16-en-3-one	(9.12 µg/ml)	58	50
5 α-androst-16-en-3-α-ol	(4.3 µg/ml)	53	19
5 α-androst-16-en-3-one + 5 α-androst-16-en-3-α-ol	4.56 µg/ml & 4.3 µg/ml, respectively	50	30
5 α-androst-16-en-3-β-ol	9.12 µg/ml	26	31
4,16-androstandien-3-one	9.12 µg/ml	53	32
5 β-androst-16-en-3-one	9.12 µg/ml	47	32
5 α-androstan-3-one	9.12 µg/ml	10	30

Source: Doty, R. L. (2010). *The Great Pheromone Myth.* Baltimore, MD: Johns Hopkins University Press.

[a] Based on data compiled from Booth, W. D. (1980). *Sym. Zool. S. London* 45: 289–311; Melrose, D. R. et al. (1971). *Br. Vet. J.* 127: 497–502; Reed, H. C. et al. (1974). *Br. Vet. J.* 130: 61–67.

willing to have sex in the presence of such odors? Are birth rates or other indices of sexual behavior higher in states or counties with pig farms?

19.2.4 HUMAN MENSTRUAL SYNCHRONY PHEROMONES

A highly publicized 1971 *Nature* paper reported that the menstrual cycles of close friends or dormitory roommates synchronize over time (i.e., the onset of their periods of menstrual bleeding became closer over a 6-month period) (McClintock 1971). Many studies subsequently reported similar synchrony (for review, see Doty 2010). However, no chemical identification of the alleged pheromone has yet been made. Importantly, as described below, literature has since appeared that questions, largely on statistical grounds, whether menstrual synchrony itself is a true phenomenon with a viable evolutionary basis (Arden & Dye 1998; Schank 1997, 2000, 2001, 2006; Strassmann 1997, 1999; Wilson 1987, 1992; Yang & Schank 2006; Ziomkiewicz 2006).

Does menstrual synchrony exist?

On the basis of statistical issues, Wilson (1987) concluded that synchrony was not demonstrated in any of studies performed up to the time of his analysis (i.e., studies by Graham & McGrew 1980; McClintock 1971; Preti et al. 1986; Quadagno et al. 1981; Russell et al. 1980). He noted that the only apparent difference between studies reporting and not reporting synchrony was that the latter included persons with irregular menstrual cycles. When persons with such cycles were omitted from analysis, the results were biased towards synchrony. He described three sources of error that were inherent in the McClintock method of synchrony analysis as follows:

- *Error I:* The assumption that differences between menses onsets of randomly paired subjects vary randomly over consecutive onsets. This reflects the failure to account for the fact that ~50% of paired cycles of unequal length will show a tendency to synchronize by chance when relatively few cycles are evaluated.
- *Error II:* The incorrect determination of the initial onset of absolute differences between subjects. Two issues are involved:
  1. An incorrect onset difference (which only occurs for the initial onset calculations in McClintock's method) is always greater than a correct onset difference (which occurs for subsequent onset calculations), thereby increasing the mean onset absolute difference and erroneously leading to what seems to be synchrony in subsequent onsets.
  2. An incorrect onset difference reverses the direction of change between the consecutive onset differences of a pair. This occurs because the subject with the earliest recorded onset has the latest recorded onset after the correction.
- *Error III:* Exclusion of subject data on the basis of not having the number of onsets specified by the research design, which biases samples toward showing menstrual synchrony by reducing dispersion in final onset absolute differences, a common phenomenon in studies finding evidence of menstrual synchrony.

A simple explanation of Error II appeared in Cecil Adam's column, the Straight Dope, in the *Chicago Reader* newspaper (Adams 2002). Assume the menstrual cycle study starts on October 1 (see Figure 19.4). The first study subject reports a 28-day-cycle with an onset of menses on September 27, another onset on October 25, and a third on November 22. The second study subject, with a 30-day cycle, reports a menses onset on October 5 and another on November 4. Using McClintock's calculation in which only cycle onsets are recorded *within the study period*, 20 days separated the two menses onset dates (October 5 vs. October 25) and 18 days separated the second pair of menses onset dates (November 4 vs. November 22). This calculation would suggest that the two cycles are synchronizing; that is, going from 20 to 18 days, when, in fact, they were eight days apart to begin with (September 27 vs. October 5). In fact, the two cycles are actually diverging from one another (November 4 − October 25 = 10 days relative to the original 8 days).

In an attempt to overcome such problems, Weller and Weller employed a "last months only" (LMO) paradigm in establishing synchrony (e.g., Weller & Weller 1993a, b, 1997a, b, 1998; Weller et al. 1999a, b). In this procedure expected frequencies of onset differences are calculated from random onset occurrences or new random pairs of women from the sample.

Unfortunately, the LMO approach has its own set of limitations, some of which reflect issues related to volunteering, accurate record keeping, and provision of requested data (e.g., return of menstrual calendars; Arden & Dye 1998; Schank 2000, 2001). In a computer simulation of the LMO procedure, Schank (2000) found that cycle variability introduced a systematic bias toward synchrony; the greater variability in the simulated cycle distribution, the greater the bias. Even when cycle onsets are completely randomly related, he found that the LMO synchrony measurement leads to data distributions skewed toward synchrony "in a way that is qualitatively and quantitatively like the actual data distributions they <Weller and Weller> report."

The assumption that menstrual synchrony, if indeed present, has biological meaning was questioned by Strassmann 1997, who pointed out that in most preindustrialized societies pregnancy and lactation, not menstrual cycling, takes up the majority of the female's reproductive years. In a long-term prospective study of the Dogon of Mali, Strassmann examined 477 untruncated menstrual cycles from 58 women over

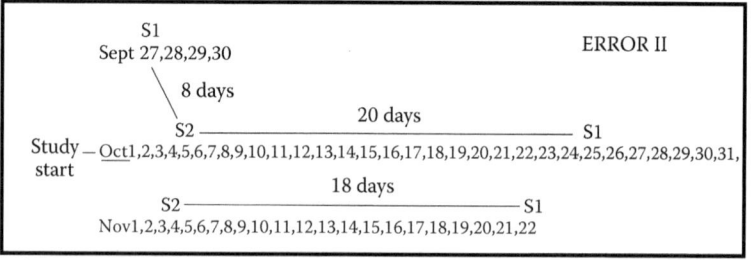

FIGURE 19.4 Demonstration of how calculating cycle lengths according the McClintock procedure leads to an erroneous conclusion of synchrony. See text for details.

a 2-year period (Strassmann 1997). In the Dogon society, menstruating women are segregated in special huts at night. Accurate information about the onset of menses was obtained from a nightly census of women present in the huts (736 days). This allowed data collection without interviews and errors in recall or reporting. Compared to American women who have, on average, more than 400 menstruations in their lifetimes, Dogon women have an average of only 128 menstruations. The proportion of women cycling on a given day was found to be ~25%. Sixteen percent were pregnant, 29% were in lactational amenorrhea, and 31% were postmenopausal. Subfecund women were most common among the cycling women, and conception usually occurred for the most fecund women on one of their first postpartum ovulations, resulting in their dropping out of the pool of regularly menstruating women. No evidence for synchrony was found for the cycling women who habitually ate and worked together or who lived with a particular lineage of related males. Similarly, no evidence for synchrony was found for any of the remaining cycling women. Strassmann concluded (p. 128), "Given the paucity of evidence, it is surprising that belief in menstrual synchrony is so widespread. I suggest that this belief arises, in part, from a popular misconception about how far apart one would expect the menstrual onsets of two women to be by chance alone." Strassmann further elaborated on this point elsewhere (Strassmann 1999, p. 579):

> Popular belief in menstrual synchrony stems from a misperception about how far apart menstrual onsets should be for two women whose onsets are independent. Given a cycle length of 28 days (not the rule – but an example), the maximum that two women can be out of phase is 14 days. On average, the onsets will be 7 days apart. Fully half the time they should be even closer (Wilson 1992; Strassmann 1997). Given that menstruation often lasts 5 days, it is not surprising that friends commonly experience overlapping menses, which is taken as personal confirmation of menstrual synchrony.

Such studies cast significant doubt on whether menstrual synchrony is a real phenomenon. If synchrony is, in fact, biologically meaningful, it would seem more important to focus on ovulation than on menses since menses is an imprecise index of synchrony, particularly when anovulatory cycles are included (Weller & Weller 1997b). In the unlikely event that menstrual synchrony is present in some groups of subjects under very specific circumstances, are "pheromones" involved in the synchronization process? As noted in the next section, evidence for such involvement seems weak, and like synchrony itself, is fraught with procedural issues (e.g., Doty 1981; Schank 2002, 2006; Whitten 1999; Wilson 1987, 1992).

If menstrual synchrony exists, what evidence is there that pheromones are involved?

The first claim of a demonstration of pheromone-induced synchronization of menses was that of Russell et al. 1980). These investigators collected axillary secretions on gauze pads taped under the arm of a woman who had a history of regular 28-day menstrual cycles and a "previous experience of 'driving' another woman's menstrual cycle on three separate occasions, over three consecutive years; i.e., a friend had become synchronous with her when they roomed together in the summer and desynchronized when they moved apart in the fall." The pads were cut into four squares, combined with four drops of 70% alcohol, and frozen in dry ice. Following thawing, the material from appropriate phases of the cycle was rubbed on the upper lips of five women, three times a week, for four months. Six control women had their lips similarly rubbed with pads that had received only the alcohol treatment. A mean pretreatment difference of 9.3 days between the day of the onset of the donor's menses and those of the subjects was observed. After 4 months of treatments, this difference decreased to 3.4 days. The authors concluded, "The data indicate that odors from one woman may influence the menstrual cycle of another and that these odors can be collected from the underarm area, stored as frozen samples, for at least short periods, and placed on another woman. Further, the experiment supports the theory that odor is a communicative element in human menstrual synchrony, and that at least a rudimentary form of olfactory control of the hormonal system is occurring in humans in a similar fashion to that found in other mammals."

Unfortunately, this study has several problems. First, it was not performed either single- or double-blind. Second, the woman who donated the samples (the second author of the paper) also acted as one of the two female experimenters who rubbed the stimuli on the subjects (Doty 1981). Aside from potentially providing subtle social cues that might affect the experiment's outcome, under the assumption that pheromones are actually involved, this would confound the experiment with a second source of pheromones (i.e., those on her person as she interacted with the subjects). Third, the purpose of the study was explained to each subject, potentially introducing another possible factor that might influence cycle lengths.

Wilson (1992) examined the data of this study in light of the three errors outlined on page 541 indicating that the study

"... shows evidence of all three errors: The number of synchronous cases is too few to be statistically significant (Error I), one of the four synchronous cases has an incorrect initial onset difference which, when corrected, causes the initial mean onset difference to be greater than the final mean onset difference (Error II), and one or more subjects may have withdrawn from the experiment because their cycle behavior was not meeting the expectations of the investigators (Error III). I conclude that Russell *et al.* (1980) did not demonstrate menstrual synchrony in subjects treated with axillary extract from a female donor."

A subsequent study by Preti et al. (1986) sought to correct some of the methodological problems of the Russell et al. study. Double-blinding was employed and the purpose of the study was explained to the subjects only after study completion. The 19 subjects were selected from a larger number on the basis of self-reports of regular cycles (29.5 ± 3 days) in an effort to minimize the potential adverse influences of highly irregular cycles. In a procedure similar to that of the Russell study, axillary secretions from cotton pads previously worn in the axillae during "a convenient 6- to 9-hr period" of four female donors was applied in an alcohol base to the upper lips of 10 subjects three times a week for three complete menstrual cycles. The stimuli employed reflected 3-day segments of the cycles of all four donors from which they were collected. This produced a set of "donor cycle" stimuli whose midpoints consisted of cycle days 2, 5, 8, 11, 14, 17, 20, 23, 26, and 29. The extracts were applied at

22- to 25-day intervals. After two complete cycles, 8 of the 10 subjects in the experimental group reportedly synchronized with the extract treatment schedules, whereas only 3 of 9 of the control women did so. The authors conclude (pp. 480–481) that "This study represents the first systematically designed, prospectively conducted, double-blind research in humans to attempt to manipulate the menstrual cycle with female-derived secretions. In this experiment naturally occurring 29.5 ± 3 day cycles could be modulated with repeated applications of extract at a 22 to 25 day interval. This study establishes phenomena in humans which are analogous to previously demonstrated olfactory/reproductive relationships in nonhuman mammals."

Preti et al.'s data were reanalyzed by Wilson (1987) who concluded that "the apparent synchrony in menses onsets in the axillary extract sample is explained on the bases of (a) chance variations, (b) mathematical properties of cocycling menses onsets, (c) features of the experimental design, and (d) failure to follow the experimental protocol, or calculation errors, or both." In his reanalysis, Wilson found 20 instances, equally divided between the experimental and control group data, where the cycle length of the treatment application fell outside of the 22- to 25-day range stipulated in the protocol. In the extract sample, the donor's cycle was found to be greater than 25 days in 9 instances, and less than 22 days in one instance, a point later acknowledged by Preti (1987). Wilson summarized his findings as follows:

> In summary, the equal distribution of five preovulatory and five postovulatory cases in the extract sample is due to chance. Eight of these cases are shown <in Table 1> as having decreased absolute onset differences between the first and third onsets of the subjects and donor. The decreases in the four preovulatory cases, including two cases in which the subject had constant cycle lengths, are interpreted as a product of the experimental design, the mathematical properties of cocycling menses onsets, and chance variations. The decreases in the four postovulatory cases, including one case with constant cycle lengths, are interpreted as the result of "errors" in the cycle lengths of the treatment applications. If all of the treatment cycles were in the 22- to 25-day range specified by the experimental protocol, the extract sample would have the characteristics of a sample of randomly paired subjects. No evidence in this experiment suggests that the 29.5 ± 3 day cycles of the subjects in the extract sample were modulated by the applications of the female axillary extract or that humans have phenomena analogous to olfactory/reproductive relationships demonstrated in nonhuman mammals.

In another McClintock study published in *Nature*, Stern and McClintock (1998) reported (pp. 177–178) finding that "that odourless compounds from the armpits of women in the late follicular phase of their menstrual cycles accelerated the preovulatory surge of luteinizing hormone of recipient women and shortened their menstrual cycles. Axillary compounds from the same donors which were collected later in the menstrual cycle (at ovulation) had the opposite effect: they delayed the luteinizing-hormone surge of the recipients and lengthened their menstrual cycles. By showing in a fully controlled experiment that the timing of ovulation can be manipulated, this study provides definitive evidence of human pheromones."

Unfortunately, this study did not take into account the statistical issues previously pointed out by Wilson and others. Nine donor women wore cotton pads in their axillae for 8 or more hours after bathing. The pads were collected daily, along with urinary LH and other information (e.g., menses, basal body temperature). This allowed them to "classify each pad as containing compounds produced during the follicular phase (2 to 4 days before the onset of the LH surge) or the ovulatory phase (the day of the LH surge onset and the 2 subsequent days)." The pads were prepared in a manner similar to those of Preti et al. and stored –80°C until use. Data from one initial cycle, when exposure to the axillary stimuli was made, was first obtained. During the next four cycles, the axilary secretions were then applied daily to the subjects' upper lips. Ten subjects received rubs from pads, collected from donors during the follicular phase, each day for two menstrual cycles and then from pads collected from ovulatory phase donors for the next two cycles. The reverse was the case for the other 10 subjects. The donors served as a control group, receiving only the 70% alcohol carrier each day.

According to these investigators, the stimuli from the follicular phase produced shorter cycles than those from the ovulatory phase (–1.7 ± 0.9 days vs. +1.4 ± 0.4 days). Surprisingly, this effect occurred within the first cycle, unlike the synchrony in earlier work that took more than one cycle. The carrier had no effect on cycle lengths of the controls. The authors noted that "In five of the cycles, women had mid-cycle nasal congestion, which could have prevented their exposure to pheromones; including these cycles in the analysis made the results slightly less robust (follicular compounds: –1.4 ± 0.9 days; ovulatory compounds: +1.4 ± 0.5 days; ANOVA: follicular versus ovulatory compounds $F(1,18) = 4.32$, $P \leq 0.05$; cycle 1 versus cycle 2 of exposure (not significant; NS); order of presentation (NS); alternations between factors were not significant)."

In a second component of the study, Stern and McClintock sought to "determine the specific mechanism of pheromone action." To do so, they utilized the LH and progesterone data to establish the follicular and luteal cycle phases. They then "traced all the changes caused by the pheromones presented in our study to the follicular phase. For the menses and luteal phases, the distributions during the pheromone and control conditions were the same (indicated by overlapping log-survivor curves). Only the follicular phase was regulated, shortened by follicular compounds and lengthened by ovulatory compounds, suggesting that these ovarian-dependent pheromones have opposite effects on the recipient's ovulation by differentially altering the rate of follicular maturation or hormonal threshold for triggering the LH surge." They concluded that "This experiment confirms the coupled oscillator model of menstrual synchrony and refocuses attention on the ovarian-dependent pheromones that regulate ovulation, producing either synchrony, asynchrony or cycle stabilization within a social group, namely two distinct pheromones, produced at different times of the cycle, which phase-advance or phase-delay the preovulatory LH surge."

The Stern and McClintock study, which in fact identified no putative pheromone or pheromones, has come under considerable criticism. For example, Schank (2006) points out that in their analysis of the five cycles, the investigators subtracted the onset dates of cycle 1 from those of cycles 2 and 3, and the onset dates of cycle 3 from those of cycles 4 and 5, rather than subtracting the onset dates of the first cycle from that of the following four cycles. Thus, cycle 3, in which axillary odor was being applied, was treated as a baseline period when, in fact, it was a treatment period. In his critique, Schank provided examples of why such an analysis is flawed. Moreover, he demonstrated how random data sets drawn from a truncated normal distribution with the means and standard deviations reported by Stern and McClintock become statistically significant only after being transformed using the flawed McClintock analysis procedure.

Strassmann (1999) has pointed out that Stern and McClintock disregarded all of the methodological problems with the McClintock procedure for establishing synchrony and questioned the statistical robustness of their findings (p. 580):

> The conclusion that a change in cycle lengths of the subjects was caused by a pheromone, rather than by the well-documented variation in cycle length in women (Treloar, Boynton, Behn, & Brown 1967; Harlow & Zeger 1991), requires inordinate confidence in the biological importance of a P value of borderline statistical significance ($P \leq 0.055$). From the data presented it is unclear whether the assumption of a normal distribution was justified. Moreover, in view of the small sample size, the entire effect might have been due to just one or two subjects who had undue leverage. Additional questions are raised by the following statement (Stern and McClintock, 1998): 'Any condition preventing exposure to the compounds, such as nasal congestion anytime during the mid-cycle period from 3 days before to 2 days after the preovulatory LH, could weaken the effect. We analyzed the data taking this into account.' It would be useful to know what a priori criteria were employed in making such adjustments, and whether the data analysis part of the project was done blind. In the absence of a theoretical reason for expecting menstrual synchrony to be a feature of human reproductive biology, and until a cycle-altering pheromone has been chemically isolated, it would appear that skepticism is warranted.

Similarly, Whitten (1999) questioned the validity of the Stern and McClintock study. Like Strassmann, he pointed out that "Each group has an apparent outlier favourable to the model: one of −14 comprises 25% of the total shortening, whereas that of +12 makes up 22% of the increase. Excluding these two outliers would abolish the claim of significance." However, his major point of concern was as follows:

> My main criticism of the study is the use of the value of single first cycles, receiving carrier-only treatment, to derive the data analyzed. Such single observations have no within-subject variance and the irregular statistical manoeuvre of converting all 20 observations to zero masks any between-subject variance and provides an illusory zero baseline with indeterminate confidence limits. Carrier-only treatments should have been distributed throughout this long experiment to give a balanced crossover design with three treatments (carrier, follicular and ovulatory) and two or more complete replications to confer confidence limits to the baseline observations, thus making comparisons valid.

This pioneer of mammalian pheromonology goes on to state, "I am not convinced of the validity of the coupled-oscillator model derived from rat studies. I also question the 'definitive evidence' that pheromones regulate human ovarian function because, if these exist, their characterization will require large, carefully designed experiments, a controlled social and physical environment, and a clearly defined endpoint measured in hours."

Space does not permit in this chapter a review of critiques of the problems associated with the other element of the Stern and McClintock study, namely the changing of the timing of the LH surge. The reader is referred to Doty (2010) for such a review.

19.3 CONCLUSIONS

In this chapter I have provided a snapshot of the history of the pheromone concept and examples of its common applications to humans. As described in detail elsewhere (Doty 2010), there is a lack of consensus as to what defines a pheromone, making it difficult to tie down as an objective scientific entity. Most definitions imply that a pheromone (a) is comprised of one or only a few chemicals, (b) is species-specific, (c) has well-defined behavioral or endocrine effects, and (d) is little influenced by learning. To date, no chemicals have been isolated in humans that meet such criteria.

REFERENCES

Adams, C. (2002). Does menstrual synchrony really exist? *The Straight Dope, The Chicago Reader*, December 20.

Albone, E. S. (1984). *Mammalian Semiochemistry*. New York: John Wiley & Sons.

Amoore, J. E., Pelosi, P. & Forrester, L. J. (1977). Specific anosmia to 5a-androst-16en-3-one and gamma-pentadecalactone: The urinous and musky primary odors. *Chem. Senses Flav.* 2: 401–425.

Anonymous (2005). So much more to know. *Science* 309: 78–102.

Arden, M. A. & Dye, L. (1998). The assessment of menstrual synchrony: Comment on Weller and Weller (1997). *J. Comp. Psychol.* 112: 323–324.

Beach, F. A. & LeBoeuf, B. J. (1967). Coital behavior in dogs. 1. Preferential mating in the bitch. *Anim. Behav.* 15: 546–558.

Berliner, D. L., Monti-Bloch, L., Jennings-White, C. & Diaz-Sanchez, V. (1996). The functionality of the human vomeronasal organ (VNO): Evidence for steroid receptors. *J. Steroid Biochem. Mol. Biol.* 58: 259–265.

Bethe, A. (1932). Vernachlässigte Hormone. *Naturwissenschaften* 11: 177–181.

Birch, M. C. (1974). *Pheromones*. New York: American Elsevier.

Bird, S. & Gower, D. B. (1981). The validation and use of a radioimmunoassay for 5 alpha-androst-16-en-3-one in human axillary collections. *J. Steroid Biochem.* 14: 213–219.

Black, S. L. (2001). Does smelling granny relieve depressive mood? Commentary on "Rapid mood change and human odors." *Biol. Psychol.* 55: 215–225.

Booth, W. D. (1980). Endocrine and exocrine factors in the reproductive behaviour of the pig. *Sym. Zool. S. London* 45: 289–311.

Boulkroune, N., Wang, L. W., March, A., Walker, N. & Jacob, T. J. C. (2007). Repetitive olfactory exposure to the biologically significant steroid androstadienone causes a hedonic shift and gender dimorphic changes in olfactory-evoked potentials. *Neuropsychopharmacology* 32: 1822–1829.

Bronson, F. H. (1968). Pheromonal influences on mammalian reproduction. In M. Diamond (ed.), *Pheromonal Influences on Mammalian Reproduction*. Bloomington, IN: Indiana University Press, pp. 341–361.

Bronson, F. H. (1976). Urine marking in mice: causes and effects. In R. L. Doty (ed.), *Mammalian Olfaction, Reproductive Processes and Behavior*. New York: Academic Press, pp. 119–143.

Brooksbank, B. W. L., Brown, R. & Gustafsson, J.-A. (1974). The detection of 5α-Androst-16-en-3α-ol in human male axillary sweat. *Experientia* 30: 864–865.

Brooksbank, B. W. L. & Haslewood, G. A. D. (1961). The estimation of androsten-16-en-3α-ol in human urine. *Biochem. J.* 80: 488–496.

Brown, R. E. (1979). Mammalian social odors: A critical review. In J.S. Rosenblatt, *Advances in the Study of Behavior*, Volume 10. New York: Academic Press, pp. 103–162.

Brownlee, R. G., Silverstein, R. M., Müller-Schwarze, D. & Singer, A. G. (1969). Isolation, identification, and function of the chief component of the male tarsal scent in black-tailed deer. *Nature* 221: 284–285.

Caroom, D. & Bronson, F. H. (1971). Responsiveness of female mice to preputial attractant: Effects of sexual experience and ovarian hormones. *Physiol. Behav.* 7: 659–662.

Carr, W. J., Loeb, L. S. & Dissinger, M. E. (1965). Responses of rats to sex odors. *J. Comp. Physiol. Psychol.* 59: 370–377.

Carr, W. J., Loeb, L. S. & Wylie, N. R. (1966). Responses to feminine odors in normal and castrated male rats. *J. Comp. Physiol. Psychol.* 62: 336–338.

Carter, C. S. & Marr, J. N. (1970). Olfactory imprinting and age variables in the guinea-pig, *Cavia porcellus*. *Anim. Behav.* 18: 238–244.

Chen, D. & Haviland-Jones, J. (1999). Rapid mood change and human odors. *Physiol. Behav.* 68: 241–250.

Claus, R. & Hoppen, H. O. (1979). The boar-pheromone steroid identified in vegetables. *Experientia* 35: 1674–1675.

Comfort, A. (1971). Likelihood of human pheromones. *Nature* 230: 432–433.

Curtis, R. F., Ballantine, J. A., Keveren, E. B., Bonsall, R. W. & Michael, R. P. (1971). Identification of primate sexual pheromones and the properties of synthetic attractants. *Nature* 232: 396–398.

Doty, R. L. (1981). Olfactory communication in humans. *Chem. Senses* 6: 351–376.

Doty, R. L. (1986). Gender and endocrine-related influences upon olfactory sensitivity. In H. L. Meiselman & R. S. Rivlin (eds.), *Clinical Measurement of Taste and Smell*. New York: Macmillan, pp. 377–413.

Doty, R. L. (2008). The olfactory vector hypothesis of neurodegenerative disease: Is it viable? *Ann. Neurol.* 63: 7–15.

Doty, R. L. (2010). *The Great Pheromone Myth*. Baltimore, MD: Johns Hopkins University Press.

Doty, R. L. & Cameron, E. L. (2009). Sex differences and reproductive hormone influences on human odor perception. *Physiol. Behav.* 97: 213–228.

Doty, R. L., Green, P. A., Ram, C. & Yankell, S. L. (1982). Communication of gender from human breath odors: Relationship to perceived intensity and pleasantness. *Horm. Behav.* 16: 13–22.

Doty, R. L. & Laing, D. G. (2003). Psychophysical measurement of olfactory function, including odorant mixture assessment. In R. L. Doty (ed.), *Handbook of Olfaction and Gustation*, Second Edition. New York: Marcel Dekker, pp. 203–228.

Doty, R. L., Orndorff, M. M., Leyden, J. & Kligman, A. (1978). Communication of gender from human axillary odors: Relationship to perceived intensity and hedonicity. *Behav. Neural Biol.* 23: 373–380.

Gower, D. B., Bird, S., Sharma, P. & House, F. R. (1985). Axillary 5 alpha-androst-16-en-3-one in men and women: Relationships with olfactory acuity to odorous 16-androstenes. *Experientia* 41: 1134–1136.

Gower, D. B., Holland, K. T., Mallet, A. I., Rennie, P. J. & Watkins, W. J. (1994). Comparison of 16-androstene steroid concentrations in sterile apocrine sweat and axillary secretions: interconversions of 16-androstenes by the axillary microflora—A mechanism for axillary odour production in man? *J. Steroid Biochem. Mol. Biol.* 48: 409–418.

Gower, D. B. & Ruparelia, B. A. (1993). Olfaction in humans with special reference to odorous 16-androstenes: Their occurrence, perception and possible social, psychological and sexual impact. *J. Endocrinol.* 137: 167–187.

Graham, C. A. & McGrew, W. C. (1980). Menstrual synchrony in female undergraduates living on a coeducational campus. *Psychoneuroendocrinology* 5: 245–252.

Gustavson, A. R., Dawson, M. E. & Bonett, D. G. (1987). Androstenol, a putative human pheromone, affects human (*Homo sapiens*) male choice performance. *J. Comp. Psychol.* 101: 210–212.

Hafez, E. S. E. & Signoret, J. P. (1969). The behaviour of swine. In E. S. E. Hafez (ed.), *The Behavior of Domestic Animals.* Baltimore, MD: Williams and Wilkins, pp. 349–390.

Harlow, S. D. & Zeger, S. L. (1991). An application of longitudinal methods to the analysis of menstrual diary data. *J. Clin. Epidemiol.* 44: 1015–1025.

Hasegawa, Y., Yabuki, M. & Matsukane, M. (2004). Identification of new odoriferous compounds in human axillary sweat. *Chem. Biodivers.* 1: 2042–2050.

Havlicek, J. & Lenochova, P. (2006). The effect of meat consumption on body odor attractiveness. *Chem. Senses* 31: 747–752.

Havlicek, J., Murray, A. K., Saxton, T. K. & Roberts, S. C. (2010). Current issues in the study of androstenes in human chemosignaling. *Vitam. Horm.* 83: 47–81.

Hold, B. & Schleidt, M. (1977). The importance of human odour in non-verbal communication. *Zeitschrift für Tierpsychologie* 43: 225–238.

Holldobler, B. (1999). Multimodal signals in ant communication. *J. Comp. Physiol. A* 184: 129–141.

Izard, C. E., Libero, D. Z., Putnam, P. & Haynes, O. M. (1993). Stability of emotion experiences and their relations to traits of personality. *J. Pers. Soc. Psychol.* 64: 847–860.

Jacob, T. J. C., Wang, L. W., Jaffer, S. & McPhee, S. (2006). Changes in the odor quality of androstadienone during exposure-induced sensitization. *Chem. Senses* 31: 3–8.

Janeczko, A. & Skoczowski, A. (2005). Mammalian sex hormones in plants. *Fol. Histochem. Cytobiol.* 43: 71–79.

Karlson, P. & Butenandt, A. (1959). Pheromones (ectohormones) in insects. *Ann. Rev. Entomol.* 4, 39–58.

Karlson, P. & Lüscher, M. (1959). "Pheromones": A new term for a class of biologically active substances. *Nature* 183: 55–56.

Kloek, J. (1961). The smell of some steroid sex-hormones and their metabolites: Reflections and experiments concerning the significance of smell for the mutual relation of the sexes. *Psychiat. Neurol. Neurochir.* 64: 309–344.

Koelega, H. S. (1980). Preference for and sensitivity to the odours of androstenone and musk. In H. van der Starre (ed.), *Proceedings of the 7th International Symposium on Olfaction & Taste and the 4th Congress of the ECRO.* London: IRL Press.

Kwan, T. K., Kraevskaya, M. A., Makin, H. L., Trafford, D. J. & Gower, D. B. (1997). Use of gas chromatographic-mass spectrometric techniques in studies of androst-16-ene and androgen biosynthesis in human testis; cytosolic specific binding of 5alpha-androst-16-en-3-one. *J. Steroid Biochem. Mol. Biol.* 60: 137–146.

Kwan, T. K., Trafford, D. J., Makin, H. L., Mallet, A. I. & Gower, D. B. (1992). GC-MS studies of 16-androstenes and other C19 steroids in human semen. *J. Steroid Biochem. Mol. Biol.* 43: 549–556.

Labows, J. (1988). Odor detection, generation and etiology in the axilla. In C. Felger & K. Laden (eds.), *Antiperspirants and Deodorants.* New York: Marcel Dekker, pp. 321–343.

Labows, J. N., Preti, G., Hoelzle, E., Leyden, J. & Kligman, A. (1979). Steroid analysis of human apocrine secretion. *Steroids* 34: 249–258.

Le Magnen, J. (1952). Les phénoménes olfacto-sexuels chez l'homme. *C. R. Acad. Sci. Biol.* 6: 125–160.

Lee, C. T. & Griffo, W. (1974). Progesterone antagonism of androgen-dependent aggression-promoting pheromone in inbred mice (*Mus musculus*). *J. Comp. Physiol. Psychol.* 87: 150–155.

Leyden, J. J., McGinley, K. J., Holzle, E., Labows, J. N. & Kligman, A. M. (1981). The microbiology of the human axilla and its relationship to axillary odor. *J. Invest. Dermatol.* 77: 413–416.

Lundström, J. N. & Olsson, M. J. (2005). Subthreshold amounts of social odorant affect mood, but not behavior, in heterosexual women when tested by a male, but not a female, experimenter. *Biol. Psychol.* 70: 197–204.

Mainardi, D., Marsan, M. & Pasquali, A. (1965). Causation of sexual preferences of the house mouse. The behaviour of mice reared by parents whose odour was artificially altered. *Atti. Soc. Ital. Sci. Nat.* 104: 325–338.

Marr, J. N. & Lilliston, L. G. (1969). Social attachment in rats by odor and age. *Behaviour* XXXIII: 277–282.

Martin, I. G. (1980). Letter to the editor. *J. Chem. Ecol.* 6: 517–519.

McClintock, M. K. (1971). Menstrual synchorony and suppression. *Nature* 229: 244–245.

Melrose, D. R., Reed, H. C. & Patterson, R. L. (1971). Androgen steroids associated with boar odour as an aid to the detection of oestrus in pig artificial insemination. *Br. Vet. J.* 127: 497–502.

Michael, R. P. & Keverne, E. B. (1970). Primate sex pheromones of vaginal origin. *Nature* 225: 84–85.

Michael, R. P., Keverne, E. B. & Bonsall, R. W. (1971). Pheromones: Isolation of male sex attractants from a female primate. *Science* 172: 964–966.

Monti-Bloch, L., Jennings-White, C., Dolberg, D. S. & Berliner, D. L. (1994). The human vomeronasal system. *Psychoneuroendocrinology* 19: 673–686.

Mugford, R. A. & Nowell, N. W. (1972). The dose-response to testosterone propionate of preputial glands, pheromones and aggression in mice. *Horm. Behav.* 3: 39–46.

Müller-Schwarze, D. (1971). Pheromones in black-tailed deer (*Odocoileus heminonus columbianus*). *Anim. Behav.* 19: 141–152.

Müller-Schwarze, R., Müller-Schwarze, D., Singer, A. G. & Silverstein, R. M. (1974). Mammalian pheromone: Identification of active component in the subauricular scent of the male pronghorn. *Science* 183: 860–862.

Nicolaides, N. (1974). Skin lipids: Their biochemical uniqueness. *Science* 186: 19–26.

Nixon, A., Mallet, A. I. & Gower, D. B. (1988). Simultaneous quantification of five odorous steroids (16-androstenes) in the axillary hair of men. *J. Steroid Biochem.* 29: 505–510.

Ohloff, G., Maurer, B., Winter, B. & Wolfgang, G. (1983). Structural and configurational dependence of the sensory process in steroids. *Helv. Chir. Acta* 66: 192–201.

Parkes, A. S. & Bruce, H. M. (1961). Olfactory stimuli in mammalian reproduction. *Science* 134: 1049–1054.

Patterson, R. L. S. (1966). Possible contribution of phenolic components to boar odour. *Nature* 212: 744–745.

Patterson, R. L. S. (1968). Acidic components of boar preputial fluid. *J. Sci. Food Agric.* 19: 38–40.

Pinker, S. (2007). *The Stuff of Thought: Language as a Window into Human Nature*. New York: Penguin Group.

Preti, G. (1987). Reply to Wilson. *Horm. Behav.* 21: 547–550.

Preti, G., Cutler, W. B., Garcia, C. R., Huggins, G. R. & Lawley, H. J. (1986). Human axillary secretions influence women's menstrual cycles: The role of donor extract of females. *Horm. Behav.* 20: 474–482.

Quadagno, D. M., Shubeita, H. E., Deck, J. & Francoer, D. (1981). The effects of males, athletic activities and all female living conditions on the menstrual cycle. *Psychoneuroendocrinology* 6: 239–244.

Reed, H. C., Melrose, D. R. & Patterson, R. L. (1974). Androgen steroids as an aid to the detection of oestrus in pig artificial insemination. *Br. Vet. J.* 130: 61–67.

Russell, M. J. (1976). Human olfactory communication. *Nature* 260: 520–522.

Russell, M. J., Switz, G. M. & Thompson, K. (1980). Olfactory influences on the human menstrual cycle. *Pharm. Biochem. Behav.* 13: 737–738.

Schank, J. C. (1997). Problems with dimensionless measurement models of synchrony in biological systems. *Am. J. Primatol.* 41: 65–85.

Schank, J. C. (2000). Menstrual-cycle variability and measurement: Further cause for doubt. *Psychoneuroendocrinology* 25: 837–847.

Schank, J. C. (2001). Menstrual-cycle synchrony: Problems and new directions for research. *J. Comp. Psychol.* 115: 3–15.

Schank, J. C. (2002). A multitude of errors in menstrual-synchrony research: Replies to Weller and Weller (2002) and Graham (2002). *J. Comp. Psychol.* 116: 319–322.

Schank, J. C. (2006). Do human menstrual-cycle pheromones exist? *Hum. Nat.* 17: 448–470.

Schleidt, M. (1980). Personal odor and nonverbal communication. *Ethol. Sociobiol.* 1: 225–231.

Schleidt, M., Hold, B. & Attili, G. (1981). A cross-cultural study on the attitude towards personal odors. *J. Chem. Ecol.* 7: 19–31.

Scott, C. M., Marlin, D. J. & Schroter, R. C. (2001). Quantification of the response of equine apocrine sweat glands to beta2-adrenergic stimulation. *Equine Vet. J.* 33: 605–612.

Shiferaw, B., Verrill, L., Booth, H., Zansky, S. M., Norton, D. M., Crim, S. et al. (2012). Sex-based differences in food consumption: Foodborne Diseases Active Surveillance Network (FoodNet) Population Survey, 2006–2007. *Clin. Infect. Dis.* 54 (Suppl. 5): S453–S457.

Simehoff, M. L., Burke, J. F., Saukkonen, J. J., Ordinario, A. T., & Doty, R. L. (1977). Biochemical profile of uremic breath. New England *Journal of Medicine* 297: 132–135.

Spielman, A. I., Sunavala, G., Harmony, J. A. K., Stuart, W. D., Leyden, J. J., Turner, G. et al. (1998). Identification and immunohistochemical localization of protein precursors to human axillary odors in apocrine glands and secretions. *Arch. Derm.* 134: 813–818.

Stefanczyk-Krzymowska, S., Krzymowski, T., Grzegorzewski, W., Sowska, W. & Skipor, J. (2000). Humoral pathway for local transfer of the priming pheromone androstenol from the nasal cavity to the brain and hypophysis in anaesthetized gilts. *Exp. Physiol.* 85: 801–809.

Stern, K. & McClintock, M. K. (1998). Regulation of ovulation by human pheromones. *Nature* 392: 177–179.

Stevenson, R. J. & Repacholi, B. M. (2003). Age-related changes in children's hedonic response to male body odor. *Dev. Psychol.* 39: 670–679.

Strassmann, B. I. (1997). The biology of menstruation in *Homo Sapiens*: Total lifetime menses, fecundity, and nonsychrony in a natural-fertility population. *Curr. Anthropol.* 38: 123–129.

Strassmann, B. I. (1999). Menstrual synchrony pheromones: Cause for doubt. *Hum. Reprod.* 14: 579–580.

Thiessen, D. D., Regnier, F. E., Rice, M., Goodwin, M., Isaacks, N. & Lawson, N. (1974). Identification of a ventral scent marking pheromone in the male Mongolian gerbil (*Meriones unguiculatus*). *Science* 184: 83–85.

Tobin, D. J. (2006). Biochemistry of human skin—Our brain on the outside. *Chem. Soc. Rev.* 35: 52–67.

Treloar, A. E., Boynton, R. E., Behn, B. G. & Brown, B. W. (1967). Variation of the human menstrual cycle through reproductive life. *Int. J. Fertil.* 13: 77–126.

Weller, A. & Weller, L. (1993a). Menstrual synchrony between mothers and daughters and between roommates. *Physiol. Behav.* 53: 943–949.

Weller, A. & Weller, L. (1997a). Menstrual synchrony under optimal conditions: Bedouin families. *J. Comp. Psychol.* 111: 143–151.

Weller, A. & Weller, L. (1998). Prolonged and very intensive contact may not be conductive to menstrual synchrony. *Psychoneuroendocrinology* 23: 19–32.

Weller, L. & Weller, A. (1993b). Multiple influences of menstrual synchrony: Kibbutz roommates, their best friends, and their mothers. *Am. J. Hum. Biol.* 5: 173–179.

Weller, L. & Weller, A. (1997b). Menstrual variability and the measurement of menstrual synchrony. *Psychoneuroendocrinology* 22: 115–128.

Weller, L., Weller, A. & Roizman, S. (1999a). Human menstrual synchrony in families and among close friends: Examining the importance of mutual exposure. *J. Comp. Psychol.* 113: 261–268.

Weller, L., Weller, A., Koresh-Kamin, H. & Ben Shoshan, R. (1999b). Menstrual synchrony in a sample of working women. *Psychoneuroendocrinology* 24: 449–459.

Whitten, W. (1999). Reproductive biology: Pheromones and regulation of ovulation. *Nature* 401: 232–233.

Whitten, W. K. & Champlin, A. K. (1973). The role of olfaction in mammalian reproduction. In *Handbook of Physiology. Section 7: Endocrinology.* Washington, DC: American Physiological Society, pp. 109–123.

Wilke, K., Martin, A., Terstegen, L. & Biel, S. S. (2007). A short history of sweat gland biology. *Int. J. Cosmet. Sci.* 29: 169–179.

Wilson, E. O. (1963). Pheromones. *Sci. Am.* 208: 100–114.

Wilson, H. C. (1987). Female axillary secretions influence women's menstrual cycles: A critique. *Horm. Behav.* 21: 536–546.

Wilson, H. C. (1992). A critical review of menstrual synchrony research. *Psychoneuroendocrinology* 17: 565–591.

Wysocki, C. J., Dorries, K. M. & Beauchamp, G. K. (1989). Ability to perceive androstenone can be acquired by ostensibly anosmic people. *Proc. Natl. Acad. Sci. U. S. A.* 86: 7976–7978.

Yang, Z. W. & Schank, J. C. (2006). Women do not synchronize their menstrual cycles. *Hum. Nat.* 17: 433–447.

Zeng, X. N., Leyden, J. J., Brand, J. G., Speilman, A. I., McGinley, K. J. & Preti, G. (1992). An investigation of human apocrine gland secretion for axillary odor precursors. *J. Chem. Ecol.* 18: 1039–1055.

Zeng, X. N., Leyden, J. J., Lawley, H. J., Sawano, K., Nohara, I. & Preti, G. (1991). Analysis of characteristic odors from human male axillae. *J. Chem. Ecol.* 17: 1469–1491.

Zeng, X. N., Leyden, J. J., Spielman, A. I. & Preti, G. (1996). Analysis of characteristic human female axillary odors: Qualitative comparison to males. *J. Chem. Ecol.* 22: 237–257.

Ziomkiewicz, A. (2006). Menstrual synchrony: Fact or artifact? *Hum. Nat.* 17: 419–432.

Index

Page numbers followed by f and t indicate figures and tables, respectively.

A

A. tigrinum (tiger salamander), 272, 273
5α-androstan-3-one, 476, 546t
5α-androst-16-en-3α-ol, 543–547, 546t
5α-androst-16-en-3α-one, 537, 543–547, 546t
5α-androstenone, 16, 543–547
3α-androsterone, 473
5α-androsterone, 464t, 545
Abdominal secretions. *See also* South African
        dung beetle species, semiochemistry
        of
    collection of, 62–65
    long-chain constituents of, 82–83
    of male *Kheper* species
        Kheper bonellii, 78–82, 78f, 79f, 80f–81f
        Kheper lamarcki, 72–73, 73f
        Kheper nigroaeneus, 73–75, 74f
        Kheper subaeneus, 75–78, 76f, 77f
Acalypha indica (Muktajhuri), 419
Accessory olfactory bulb (AOB), 270, 271, 298f,
        328, 329–330, 329f, 330, 332
    cell activity in, 331–332
    chemosignals processing after parturition, 334
    neurosteroid via GABA$_A$ receptors at, 360
    newborn neurons, role of
        functional role of, 372–375, 373f, 374f,
            375f, 376–377
        male pheromones affecting integration, 372
        olfactory pregnancy block in mice,
            371–372
        sensory-driven survival of, 375–376
    paced mating and neurogenesis
        female sexual behavior, 378–381, 379t
        male sexual behavior, 381
        mating behavior in rats, 377–378
        opioids and neurogenesis, 382
    pheromonal information projection to,
        355–357, 356f
Accessory olfactory information, 332–333
Accessory olfactory system (AOS), 462
Accessory proteins, 273
Acetaldehyde-2,4 dinitrophenyl-hydrazone, 445f
Acetic acid, 443f, 464t, 471
2 acetyl-1-pyrroline (2AP), 423
Acinonyx jubatus (Cheetahs), 421, 437
Adamson, Joy, 448
Ad nauseum, 415
Adsorption, 121, 124

Adult neurogenesis, 367. *See also* Neurogenesis
Aedes aegypti (mosquito), 178
African clawed frog (*X. laevis*), 270, 272, 273
Aggregation, 245
Aggression in *Drosophila* males, 220, 245–246
3α-hydroxy-4-pregnen-20-one (3αHP), 360
Airborne odors, 32. *See also* Odors
Airborne pheromones, 99
Alarm pheromone (AP), 8, 14, 160, 162f,
        165–168, 166, 173. *See also* Worker
        pheromones
    and defense behavior, 189–190
    and expression of immediate early genes
        (IEG), 193
    2-heptanone (2HPT), 167–168
    sting apparatus, 166–167
Aldehydes, 440, 441t
Aliphatic compounds, 169
Allelochemicals, 3
Alloiohormones, 535
Allopregnanorone, 357
α lobes, 181
Ambystoma, 259
Amines, 441t, 445f
Amino acid, 10, 84, 304, 392
Amphibian chemosignals
    male-male interactions, 268
    territorial advertisement, 267–268
Amphibian chemosignals, production of. *See also*
        Chemical signaling in amphibians
    anuran skin glands, 259
    cloacal glands, 259
    fecal excretions, 260
    oviduct, 260
    skin glands
        anuran, 259
        urodele, 257, 258t
    urodele skin glands, 257–258
Amphibian chemosignals related to mating,
        260–267. *See also* Chemical signaling
        in amphibians
    amphibian pheromones, properties of, 267t
    female chemosignals, male responses to
        anurans, multimodal signaling, 262
        urodeles, 261–262
    male chemosignals, female responses to
        mate choice, 263–265, 264f
        mate location, 262
        mate recognition, 262–263, 265–266

Amphibian nasal cavity, anatomy of, 269–270
Amphibian skin, 257
Amyl acetate, 380, 381
Analgesia, stress-induced, 189
Anal gland secretion, 416–417. *See also* Scent
    marking in big cats
AND. *See* 4,16-androstadien-3-one (AND)
Androgen, 305t, 360
Androstadienone, 543–547
4,16-androstadien-3-one (AND), 411, 525, 526,
    529–530
Androstenol, 543–547, 546t
Androstenone, 537, 543–547, 546t
Anestrus
    defined, 469
    estrus synchronization and estrus induction
        in, 477
Anlage, 290
Anopheles gambiae, 114, 178
Antennae and ORN, 25–26. *See also* Insect
    olfactory system
Antennal lobes (AL), 25, 26–27
    and glomeruli, 178–179
Antennal specific protein (ASP), 177
Antenno-protocerebral tract (APT), 27
Anterior olfactory nucleus (AON), 329f
Anurans
    mate recognition, 265–266
    multimodal signaling, 262
    skin glands, 259
AOS. *See* Accessory olfactory system (AOS)
Aphid sex pheromones, 8
Aphrodisin, 305t, 351
Apical sensory neurons, 401
Apis florea, 180
Apis mellifera. *See* Honey bee (*Apis mellifera*)
Apis mellifera capensis, 158
Apis mellifera scutellata, 158, 165
Aplysia, 6
Apocrine sweat glands, 539–540, 539f
Apoidea, 172
Appeasing (maternal) pheromones, 473, 478. *See
    also* Pheromones
Appetitive learning conditioning, 188
Ara-C treatment, 376
Archaic frogs (*Leiopelma hamiltoni*), 260
Aroma molecules, 442t
Ascaroside, 6
Asian corn borer (ACB), 12
Aspartic acid, 84
Assay for fecal corticosterone metabolites,
    393–394. *See also* Cat odor on house
    mouse
Associative learning, 28, 44–46, 182, 184, 271,
    327, 335
Australian Dung Beetle Project, 58, 63
Aversive learning conditioning, 188

Axel, Richard, 245
Axillary secretions
    effects on mood, 543
    sensory studies of, 540–543, 541f–542f
Azalea lace bug, 60

B

Barbary lions (*P. leo leo*), 415
Basal membrane (BM), 354f
Basal neurons, 401
Basiconic neuron system, 237f
Basiconic sensilla, 212
Bed nuclei of accessory olfactory tract (BAOT),
    329f
Bed nuclei of stria terminalis (BNST), 329f
Benzaldehyde, 234
Benzoic acid, 116, 116f, 130
Benzonitrile, 43
1,4-benzoquinone, 94
Beta barrel, 128
Bicuculline, 360
Big cats. *See* Tiger, pheromone of
Biogenic amines and juvenile hormones,
    183–187
Biostimulation
    cattle pheromones, 469–470
    defined, 469–470
Bipolar sensory neurons, 212
Black monkey orange tree, 90
β lobes, 181
BmorPBP1. *See* Bombykol-binding protein
Boar Mate® spray, 16
Boar pheromones, 473–474. *See also* Pheromones
Body odors. *See also* Odors
    of Bengal tiger of Sundarban, 416, 446f, 449
    leaf volatiles and animal, 448
    pheromone-like signals in humans (imaging
        studies), 526–527
    pleasantness or intensity, 542
    of police dogs, 422
    steroids and, 545
Bombykol, 5, 100
Bombykol-binding protein, 101f, 128
Bombyx mori (Silk moth), 1, 101f, 102, 109f,
        111f–112f, 217. *See also* Pheromone
    reception in insects
    EAG, 118f
Bos taurus, 465
Bos taurus indicus, 470
Bos taurus taurus, 470
Bovine serum albumin (BSA), 127
β -phenylethylamine, 440
Brain amines
    as neuromodulators, 183–185
    and pheromones, 185–186
BrdU, 374f, 378

Brood pheromone (BP), 192. *See also* Pheromones of honey bee colony
brood development and care, regulation of, 170–171
worker behavioral development, 171–172
worker reproduction, regulation of, 171
Brown-banded cockroach (*Supella longipalpa*), 8
Bruce effect, 313, 371, 400
in mice, 3
Bubalus bubalis, 470
Buffalo
flehmen behavior, 466–467, 466f
pheromones, 472–473
VNO in, 467–469, 468f
Butea monosperma, 413

C

Ca2+-activated Cl- channels, 314. *See also* Ion channels
Ca++-activated nonspecific ion (CAN) channel, 133
Cabbage looper moth (*Trichoplusia ni*), 11
Caffeine, 222
CAH (congenital adrenal hyperplasia), 530
Calcitonin gene-related peptide (CGRP), 291
Calcium imaging, 38, 40, 43
Calotte models, 101f
Calyces, 181
cAMP. *See* Cyclic adenosine monophosphate (cAMP)
Carbon dioxide, 42
Carboxylesterase (CES), 423
Carboxylic acids, 69, 422
Carlson, John, 235
Carnivore odor, 390, 391. *See also* Odor
Carnivore urine, 391
Carrier material, 83–86, 84f, 86f
Cathelin-related antimicrobial peptide (CRAMP), 305t, 310
Cat odor on house mouse. *See also* Odor
behavior and physiology of house mouse, 389–392, 390t
methods/materials
assay for fecal corticosterone meta bolites, 393–394
immunohistochemistry assay, 394
test subjects, 392–393
vomeronasal surgery (VNX), 394
Mus species, 389
predator-prey relationships, 389–390
results/discussion, 395–402, 395f–399f, 401f
Cattle behavior, 463
Cattle pheromones, 461–478
application of
estrus indication by female pheromones, 477–478

estrus synchronization and estrus induction in anestrus by male pheromone, 477
fighting behavior reduction by maternal pheromones, 478
influencing standing posture by male pheromones, 477
maternal responsiveness by neonatal pheromones, 478
penis erection and sperm quantity enhancement by female pheromones, 478
postpartum anoestrus reduction by male pheromone, 477
puberty acceleration by male pheromone, 477
biostimulation, 469–470
boar pheromones, 473–474
buffalo pheromones, 472–473
cow pheromones, 470–472
horse pheromones, 475–476
identification, source, and functions, 464t
male flehmen behavior, and its role in cattle estrus detection, 463, 465
buffalo flehmen, 466–467, 466f
in cattle other than cow and buffalo, 467
chemoreceptor system in cattle, 467–469, 468f
cow flehmen, 465–466
odorant binding proteins, 476
overview, 461–463
primer pheromone, role of, 469
reproductive cycle and behavior, 463
small ruminant pheromones (goat and sheep), 474–475
Cat urine, 391, 397, 398, 398f
on plasma corticosterone, 399f
CD36 protein family, 131
Central nervous system (CNS), 24, 105
Central olfactory projections, 270–271
c-Fos expression, 349
Cheetah (*Acinonyx jubatus*), 421
territoriality and, 437
Chemical analysis of MF. *See also* Tiger, pheromone of
elusive aroma molecule of tiger MF, 443–447
volatile/nonvolatile compounds in, 440–443, 441t–442t, 443f
Chemical communication, 411, 412. *See also* Chemical signaling; Chemical signaling in dung beetles
intergeneric, 86–88, 87t
interspecific, 88–89
via chemical signals, 326
Chemical cues, 6, 7f, 162f, 168, 170. *See also* Chemosignal detection; Chemical signaling

Chemical ecology of genus *Kheper*, 59–62
Chemical emissions, 262
Chemical senses, 1
Chemical signaling, 410–411. *See also* Chemical
    communication; *specific* entries
Chemical signaling in amphibians
    chemosignal detection
        amphibian nasal cavity, anatomy of,
            269–270
        central olfactory projections, 270–271
        chemosensory epithelia, responses of,
            273–274
        chemosensory receptors, 271–273
        primer effects, 274–275
    overview of, 256–257
    production of
        anuran skin glands, 259
        cloacal glands, 259
        fecal excretions, 260
        oviduct, 260
        skin glands, 257, 258t
        urodele skin glands, 257–258
    related to mating, 260–267
        amphibian pheromones, properties of,
            267t
        female responses to male chemosignals,
            262–266, 264f
        male responses to female chemosignals,
            261–262
    territoriality and male-male interactions
        male–male interactions, 268
        territorial advertisement, 267–268
Chemical signaling in dung beetles. *See also*
    South African dung beetle species,
    semiochemistry of
    intergeneric communication, 86–88, 87t
    interspecific communication, 88–89
Chemical signaling in vertebrates/invertebrates
    overview, 1–5
    pheromones
        applications of, 15–16
        from chemical cues, 6, 7f
        defined, 1
        discovering, 5–6
        human, 15
        innate, 14–15
        and olfaction, 13–14
        primer effect of, 14
        specificity and speciation, 8–11, 9f
        reception and processing of, 11–13, 12f
        semiochemical interaction, 2f
        signature mixture, 2, 3, 4f
Chemical signals, 462
    ubiquity of, 42
Chemoreception, contact, 248–249
Chemoreceptor system, in cattle, 467–469, 468f
    MOS, 467–469
        VNO, 467–469, 468f
        VNS, 467–469
Chemosensory cues, 378
Chemosensory detection, 390
Chemosensory epithelia, responses of, 273–274
Chemosensory protein (CSP), 127, 176, 216
Chemosensory receptors, 271–273. *See also*
    Chemosignal detection
    accessory proteins, 273
    OR, 271–272
    trace-amine associated receptors, 271
    VR, 272–273
Chemosensory systems. *See also* Intraspecific
        chemical signals in mice
    accessory olfactory bulb, 329–330
    cell activity in accessory olfactory bulb,
        331–332
    crosstalk, 331
    higher-level projection areas, 330–331
    main and accessory olfactory information,
        332–333
    sensing pheromones, 298f
Chemosignal detection. *See also* Chemical
        signaling in amphibians
    amphibian nasal cavity, anatomy of, 269–270
    central olfactory projections, 270–271
    chemosensory epithelia, responses of,
        273–274
    chemosensory receptors, 271–273
    primer effects, 274–275
Chemosignaling, 334
    during development, 333–334
Chicago Reader (newspaper), 548
1-chlorooctane, 472
cis-9-octa decenic acid, 474
cis-vaccenyl acetate (cVA), 214, 215f, 218, 220
    antiaphrodisiac effect of, 243
    pheromone detection in *Drosophila*, 236–237,
        236f, 237f
cis-vaccenyl acetate (cVA)-induced behaviors.
        See also Volatile pheromones
        detection
    aggregation, 245
    aggression, 245–246
    courtship, 243–245, 244f
Citral pheromone, 164, 179
Civetone, 536
Cladogram, 264f
Cloacal glands, 259
Cochromatography of GC, 443
Coeloconic sensilla, 212. *See also* Sensilla
Colostrum, newborns' quest for, 490–491. *See*
        also Newborns
Comfort, Alex, 538
Complex blends coding. *See* Odor interactions
        and complex blends coding
Complex plant bouquets, recognition of, 36–37

Concentration detectors, 120–121
Condensation, elimination of
    in column tip, 69
    in humidified air duct, 69–70, 70f
Condensed water droplets, 71
Conditioned stimulus (CS), 44
Congenital adrenal hyperplasia (CAH), 530
Coniferyl alcohol (CA), 153
Contact chemoreception, 248–249. *See also*
        Volatile pheromones detection
Contour maps, 434
Cooperation hypothesis, 175
Corbett, Jim, 418
Corpus luteum (CL), 463
Corticosteroids, 305t
Corticosterone, 393, 394, 398
    metabolites, 393–394
Courtship
    and aggression, *Drosophila* pheromones in,
        219–223
    cloacal gland and, 259
    cuticular hydrocarbons and, 10
    cVA-induced behaviors and, 243–245, 244
    in *Drosophila*, 14, 211
    on female behavior, 266
    functions in, 263
    isoforms of, 265
    of Japanese *Cynops* newts, 6
    learning of, 220
    of Scarabaeinae species, 59
    semiochemicals in, 94
    urodele skin glands, 257
Cow flehmen behavior, 465–466
Cow pheromones, 470–472
Creatonotos, 102
P-cresol, 476
Crosstalk, 331
Crotonic acid, 88
Cryoprotection, 394
Cues. *See also* Chemical cues
    chemosensory, 378
    pheromonal, 369
Cuticular hydrocarbons (CHC), 162, 165,
        168–169
Cuticular lipids, 168
cVA. *See cis*-vaccenyl acetate
Cyclic adenosine monophosphate (cAMP), 315,
        349–350
Cydia molesta (oriental fruit moth, female), 43
Cynops newts, 6
Cynops pyrrhogaster (red-bellied salamanders),
        273, 274

D

Darwin's reasoning and pheromones, 409
Datura wrightii, 37

Deactivation, pheromone, 128–130
Decanoic acid, 443f, 472, 474
Decanoyl-thio-*1,1,1*-trifluoropropanone (DTFP),
        117
Decyl-thio-trifluoro propanone (DTFP), 116f
Degradation, pheromone, 128–130
Dendrites, 292
Dentate gyrus (DG), 369, 379t
Depolarization, 315
Desmognathus ocoee, 264
Desmosomes, 292
Diacylglycerol (DAG), 311f, 313, 314, 348, 350
Diaminobenzidine (DAB), 394
Diamondback moth (*Plutella xylostella*), 131
Dichloromethane, 65
Didelphimorpha, 308
Diene cuticular hydrocarbons, 10
Differential Emotion Scale, 543
Diffusion on hairs, 124
Dimethyl disulfide (DMDS), 500, 502
2,6-dimethyl-5-heptenoic acid, 73, 75, 86f
Diptera, 211
Direct analysis in real time (DART), 247
Dispersive x-ray elementary analysis, 107
DMDS. *See* Dimethyl disulfide (DMDS)
2D-NMR spectroscopy (DANS), 5
Dodecanoic acid, 474
4-dodeken-2-ol, 76
Dominance, 220, 527
    lack of D2 receptors, 331
    male, 410
    male aggression, 245
    queen, 154
    reproductive, 160, 163, 174
Dopamine (DA), 183, 187–188, 331
Doublesex, 10
Drone pheromones, 169–170
Drosophila melanogaster, 10, 25, 178, 211,
        230–232. *See also* Volatile
        pheromones detection
    OBP, 235–236
    odorant receptor family, 235
    olfactory anatomy, 232–234, 233f, 234f
Drosophila pheromones, 14, 246–247. *See also*
        Volatile pheromones detection
    in courtship and aggression, 219–223
    odor response in trichoid sensilla, 215–219,
        216f
    olfactory sensilla, 212–213
    overview, 211–212
    trichoid ORN, 213–215, 214f, 215f
Dufour's gland pheromone (DGP), 158–160, 162f

E

EAG. *See* Electroantennogram (EAG)
EAP. *See* Equine appeasing pheromone (EAP)

E-β-ocimene, 171, 172, 173f
Eccrine sweat glands, 539–540, 539f
Echolocation, 41
Ectohormones, 535
EGFP-positive cells, 373f
Egg-discriminatory pheromone, 159
(Z)-11-eicosen-1-ol, 7
Electrical stimulation, 356
Electroantennogram (EAG), 5, 107, 118f
Electroantennographic detection (EAD), 68f, 69,
        71, 73f, 75
    and male abdominal secretions of *Kheper*
        species, 80f–81f
Electroantennographic study, 180
Electromyography, 44
Electron microprobe analysis, 106f
Electroolfactograms (EOG), 5
Electrophysiology, 107–110, 108f, 109f. *See also*
        Pheromone reception in insects
Electrostatic interference, 71
Electrovomeronasogram (EVG), 312
Elementary receptor potentials (ERP), 109,
        114–117, 115f–116f. *See also*
        Pheromone reception in insects
Elusive aroma molecule of tiger MF, 443–447
Endocoprids, 58
Endohormones, 535
Enzymatic degradation, 130
Enzyme immunoassay (EIA) method, 393
4EOA. *See* 4-ethyl octanoic acid (4EOA)
Epoxyde hydrolase, 33
Equine appeasing pheromone (EAP), 478
EST. *See* Estra-1,3,5(10),16-tetraen-3-ol (EST)
Estradiol, 336, 357
Estra-1,3,5(10),16-tetraen-3-ol (EST), 411, 525,
        526, 529–530
Estrogen, 305t, 336
    receptor immunoreactive cells, 360
Estrous cycle, of cattle, 463
Estrus detection, in cattle
    by female pheromones, 477–478
    male flehmen behavior and role in, 463, 465
        buffalo flehmen, 466–467, 466f
        in cattle other than cow and buffalo, 467
        chemoreceptor system in cattle, 467–469,
            468f
        cow flehmen, 465–466
Estrus synchronization, in cattle, 477
Ethylenediamine, 445f
4-ethyl octanoic acid (4EOA), 474
Ethyl oleate, 163–164, 170
Eupoecilia ambiguella, 38
European corn borer (ECB), 12f
European rabbit (*Oryctolagus cuniculus*), 490, 501
Excretory products, 416
Exocrine-gland-secreting peptide 1 (ESP1), 305t, 307

Exocrine glands of honey bee queen, 151f. *See
        also* Honey bee
Extant amphibians, 256
External plexiform layer (EPL), 373f
Exteroceptive pregnancy block elimination, 377
Extramammary sources, of behaviorally active
        chemostimuli
    for newborns
        primates, 504–506
        rodents, 499–500

F

Fatty-acyl pheromone, 9
Fecal corticosterone metabolites, 393–394
Fecal excretions, 260
Feline attractions, 419
Feline pheromone, primary source of, 415–416.
        See also Scent marking in big cats
Felinine, 392
    treatment, 400
Felis catus, 391, 392, 395f, 396f, 397f
Female chemosignals, male responses to. *See
        also* Amphibian chemosignals related
        to mating
    anurans, 262
    urodeles, 261–262
Female sex pheromones, 8, 9
Female sexual behavior and neurogenesis,
        378–381, 379t
Female urine, 337
Fighting behavior reduction, 478
Fixed olfactory signals. *See* Olfactory signals,
        fixed
Flame ionization detection (FID), 68, 71, 73f
Flame photometry, 106f
Flehmen, 419–420. *See also* Male flehmen
        behavior
    MF spraying and, 436
Flippase (FLP), 231
Flippase recognition target (FRT), 231
Fluorescence energy transfer (FRET), 241
Flux detectors, 120–121
Follicle-stimulating hormone (FSH), 336
Footprint pheromone (FP), 158, 162f
Formyl-peptide receptors (FPR), 302, 308–310,
        309t
    in humans/mice, 309t
Fos immunoreactivity, 356, 400
Fos-positive cells, 394
FPR. *See* Formyl-peptide receptors (FPR)
Free fatty acid (FFA), 440, 441t
    gas chromatogram of, 443f
Free fatty acid phase (FFAP), 69
Fruit fly antenna, 104
Fru-positive neurons, 248

G

GABA$_A$ receptors, 359, 360
β-Galactosidase, 238f
Gas chromatographic (GC) column, quality/
        selectivity/capacity of, 69
Gas chromatographic-mass spectrometric
        (GCMS) analysis, 63, 74
Gas chromatography (GC), 4f, 66f, 222, 443,
        443f
    two-dimensional, 83
Gas chromatography mass spectrometry
        (GCMS), 419
GC-FID/EAD instrumentation, optimization of,
        68–71, 70f
Gene expression, pheromones and, 191
General odorant binding proteins (GOBP), 127,
        235
Geranic acid, 164
Geraniol pheromone, 44, 120, 164, 179
Glomerular and vomeronasal nerve layer (Gl/
        Vn-L), 373f
Glomeruli
    antennal lobe and, 178–179
    pheromone processing in, 179–180
Glucocorticoid, 305t
Glutamate decarboxylase, 290
Glutamic acid, 84
Goat(s)
    flehmen response in male, 467
    pheromones, 474–475
Gold particles, 103
Gonadal hormones
    and brain functions, 358–359, 358f
    sexual behavior and, 357
Gonadotropin-releasing hormone (GnRH), 290
G-protein-coupled receptor (GPCR), 11, 235, 349
G proteins, 311–312, 311f. See also Signal
        transduction mechanism
Granular layer, 474f
Grapholita molesta (oriental fruit moth, male), 43
The Great Pheromone Myth, 535
Green fluorescent protein (GFP), 232, 244
Green stink bug (Nezara viridula), 40
Grünenberg ganglion (GG), 298f, 328
GTP, 350, 352f
GTP-binding protein, 287
GTPγS, 350, 352f
Guard bees, 160, 165–168. See also Honey bees
Gustatory receptor (GR) protein, 11, 14

H

Hairs, diffusion on, 124
Hawk moths, 40
10-HDA, 153

Headspace gas volatiles, 91f, 92t
Heat radiation, 71
Helicoverpa armigera, 43
Helicoverpa zea (noctuid moth), 38, 102
Heliothis virescens, 10, 38, 130, 133, 218
7,11-heptacosadiene, 221
2-heptanone (2HPT), 162f, 163, 167–168, 305t
Heterosexual men (HeM), 529, 530
1, 2-hexadecanediol, 474
Hexadecane-1-ol (PA), 153
Hexadecanoic acid, 82, 472, 474
High molecular weight (HMW), 375
High-performance liquid chromatography
        (HPLC), 4f
Hippocampal cell proliferation, 382
Hluhluwe Game Reserve, 62
Homoiohormones, 535
Homosexual men (HoM), 528–529, 530
Homovanillyl alcohol (HVA), 194
Honey bee (Apis mellifera), 33, 43, 104, 105f. See
        also specific entries
    mandibular glands, 161
Honey bee society, chemical communication in
    pheromonal signal in bee brain, 175–193
        gene expression, 191–193
        processing/modulation of, 182–190
        reception of pheromonal signal, 176–182,
            177f
    pheromones of honey bee colony, 148–175
        brood pheromone (BP), 170–172
        drone pheromones, 169–170
        evolution of sociality in bees, 172–175,
            173f
        queen pheromones, 149–160, 150f
        worker pheromones, 160–169
Honey bee society, composite organization of,
        148
Hormonal status, 395
Hormones, 4f, 274, 410, 411, 535
    alloiohormones, 535
    defined, 411
    endohormones, 535
    follicle-stimulating hormone (FSH), 336
    gonadal
        and brain functions, 358–359, 358f
        GABAergic functions, modulating, 359
        and sexual behavior, 357
    gonadotropin-releasing hormone (GnRH),
        290
    homoiohormones, 535, 536
    and hormone-related proteins, 336–337
    in intraspecific chemical communication,
        336–337
    juvenile hormones (JH), 186–187, 194f
    lactogenic, 507
    luteinic hormone-releasing, 528

luteinizing, 290, 336, 551
and neuromodulators, 40
peptide, 478
pituitary hormone prolactin (PRL), 370
reproductive, 261
Horse
flehmen response in, 467
pheromones, 475–476
House mouse. *See* Cat odor on house mouse
³H-pheromone, 124
Human menstrual synchrony pheromones,
547–554, 548f
Human pheromones, 15, 535–554
androstadienone as, 543–547
androstenol as, 543–547
androstenone as, 543–547
historical perspectives, 535–538
human menstrual synchrony pheromones,
547–554, 548f
putative, sources of, 539–540, 539f
search for, 538
sensory studies of axillary secretions,
540–543, 541f–542f
Humans
pheromone-like signals in (imaging studies)
body odors, 526–527
in relation to sexual orientation, 527–529,
528f
synthetic compounds, 525–526
VNO in, 524–525
Humidified air duct, 69–70, 71
Hydrocarbons, 159
pheromones, 14
wax, 65
10-hydroxydecanoic acid (10-HDAA), 161
9-hydroxydec-2-enoic acid (9-HDA), 151, 152
4-hydroxy-3-methoxyphenylethanol
(homovanillyl alcohol), 151
5-hydroxytryptophan, 186
Hygrometry, 35
Hynobiid salamanders, 261
Hynobius leechi (Korean salamander), 258t, 261,
274, 275
Hyperpolarization-activated cyclic nucleotide-
gated (HCN) cation channels, 316

I

Iberiotoxin, 315
Ibotenic acid, 376
Immediate early genes (IEG), expression of, 193
Immunofluorescence, 380
Immunohistochemistry, 287
assay, 394
Imprinting, pheromonal mating-induced
functional role of AOB newborn neurons,
376–377

and individual male odors, 372–375, 373f,
374f, 375f
male pheromones affecting integration of
newborn neurons, 372
olfactory pregnancy block in mice, 371–372
sensory-driven survival, 375–376
Individuality in pheromonal signals, 421–422
Innate pheromones, 14–15
Innate releasing mechanism (IRM), 433
Inositol-1,4,5-trisphosphate (IP₃), 132, 348
Insect antennae and olfactory sensilla, 104–107,
105f, 106f
Insecticides, 120
Insect olfactory system. *See also* Odor perception
in insects, pheromones and
antennae and ORN, 25–26
antennal lobes, 26–27
primary olfactory centers, 26–27
second-order olfactory areas, 28
In situ hybridization, 133
Intensity coding, 31–32
Interdigital gland secretion, 416–417
Intergeneric communication, 86–88, 87t. *See also*
Chemical signaling in dung beetles
Interspecific communication, 88–89. *See also*
Chemical signaling in dung beetles
Intracellular messengers, 132
Intraspecific chemical signals in mice
AOB chemosignals processing after
parturition, 334
chemosensory systems, 328, 329f
accessory olfactory bulb, 329–330
cell activity in accessory olfactory bulb,
331–332
crosstalk, 331
higher-level projection areas, 330–331
main and accessory olfactory
information, 332–333
chemosignaling
during development, 333–334
and memory/reward system, 334–335
hormones and hormone-related proteins,
336–337
memory for mate, 335–336
neurohormonal domains
adult behavior, 327
reproductive state, 326–327
overview, 325–326
Invertebrate pheromones, 5
Ion channels, 130, 313–316. *See also* Signal
transduction mechanism
Ca²⁺-activated Cl⁻ channels, 314
transient receptor potential canonical 2
(TRPC), 313–314
Ionotropic glutamate receptors (iGluRs), 25
Isoamyl acetate, 166
Isobutylamine, 304, 305t

Isopentyl acetate (IPA), 166, 167
Isovaleric acid, 443f

J

Jacobson, Ludvig Levis, 285
Jaguars, territoriality and, 438–439
Juvenile hormone (JH), 40, 155, 194f
    and pheromones, 186–187

K

Kairomones, 2f, 3, 88, 391, 392, 401, 411, 416
Kaissling model, 133
Kenyon cells (KC), 28, 181
Ketones, 440, 441t
Kheper, chemical ecology of, 59–62
Kheper bonellii, 64, 78–82, 78f, 79f, 80f–81f
Kheper lamarcki, 0f, 72–73, 73f
Kheper nigroaeneus, 62, 73–75, 74f
Kheper species, male. *See also* Abdominal
            secretions
    K. bonellii, 78–82, 78f, 79f, 80f–81f
    K. lamarcki, 72–73, 73f
    K. nigroaeneus, 73–75, 74f
    K. subaeneus, 75–78, 76f, 77f
Kheper subaeneus, 75–78, 76f, 77f
Killer bee (*Apis mellifera scutellata*). *See Apis
            mellifera scutellata*
Kinetic model, 124–125
Korean salamander (*Hynobius leechi*), 258t, 261,
            274, 275
Korean salamanders (*H. leechi*), 274
Koschevnikov gland pheromones, 160
Kruger National Park, 437
Krüppel homolog 1 (*Kr-h1*), 192
L-kynurenine, 6

L

Laboratorium vir Ekologiese Chemie Universiteit
            van Stellenbosch (LECUS), 95
Lacrimal gland secretions, 327
Lacunae, 448
LacZ gene, 237
Lagomorphs
    evidence for, 502–504
    mammary sources of behaviorally active
            chemostimuli, 501–502
Land tenure system, 438
"Last months only" (LMO) paradigm, 548
Lateral horn (LH), 28
Lateral olfactory tract, 373f
Learning, neonatal, 493–495
    mammals, 493–495
    rodents, 494
    sucking, 494

Learning and hard-wired pheromone behaviors,
            247–248
Leiopelma hamiltoni (archaic frogs), 260
Lemur catta, 422
Leopard, territoriality and, 437–438
Lepidoptera, 27, 43, 68
    adult, 25
Leptodactylus aggression-stimulating peptide
            (LASP), 268
Lepus europaeus, 503
Lesser lemur (*Microcebus murinus*), 288
L-felinine, 394, 396f, 400
Likelihood of Human Pheromones, 538
Linalool, 31, 120
Linkage of big-cat population, evolutionary
            lineage and, 422–423
Linoleic acid (LA), 153
Lion, territoriality and, 436–437
Lipid content, 447, 448
Lipocalin, 416
Listeria monocytogenes, 310
Litoria splendida, 259
L-kynurenine, 6
LMO paradigm. *See* "Last months only" (LMO)
            paradigm
Lobesia botrana, 39
Local neurone (LN), 27
Love dust particles, 103f
Low molecular weight (LMW), 375, 376
LUSH, 216, 217, 238, 242f
Lush mutant, 238, 238f, 239, 240
LUSH OBP, 237–241, 238f. *See also* Volatile
            pheromones detection
Luteinizing hormone (LH), 336, 348

M

Macroglomerular complex (MGC), 26, 29, 113
Main olfactory bulb (MOB), 270, 271, 298,
            328, 329, 329f, 331, 349, 359, 401f,
            524
Main olfactory epithelium (MOE), 269, 273, 297,
            298f, 329f
Main olfactory information, 332–333
Main olfactory system (MOS), 462, 467–469
Main urinary proteins (MUPs), 496
Major histocompatibility complex (MHC), 305t,
            307, 421–422
Major urinary protein (MUP), 305t, 307, 375,
            391, 421
Male aggression, 245. *See also* Aggression
Male chemosignals, female responses to. *See also*
            Amphibian chemosignals related to
            mating
    mate choice, 263–265, 264f
    mate location, 262
    mate recognition, 262–263, 265–266

Male flehmen behavior
  and role in cattle estrus detection, 463, 465
    buffalo flehmen, 466–467, 466f
    in cattle other than cow and buffalo, 467
    chemoreceptor system in cattle, 467–469,
      468f
    cow flehmen, 465–466
Male *Kheper* species. *See* Abdominal secretions
Male-male interactions, 268. *See also* Chemical
  signaling in amphibians
Male moth pheromones, 7
Male odors, 372–375, 373f, 374f, 375f
Male oriental fruit moth, 8
Male pheromones affecting integration, 372
Male sexual behavior and neurogenesis, 381
Mammary chemosignaling
  contextual fluctuations in, 506–508
Mammary pheromone (MP), 503–504
Mammary sources, of behaviorally active
  chemostimuli
  for newborns
    lagomorphs, 501–502
    primates, 504–506
    rodents, 498–499
Mandibular gland(s), 157, 167–168
  pheromones, 161–163, 162f
Mann-Whitney test, 400
Mantidacylus multiplicatus, 268
Marking fluid (MF). *See also* Tiger, pheromone of
  in social life of tiger, 417–418
  spraying
    and flehmen, 432f, 436
    versus ordinary urination, 434–435
    in proestrous, estrous, and postestrous
      periods, 435
  of tiger
    chemical analysis of, 440–447, 441t–442t,
      443f, 444f, 445f
    collection of, 439–440
    natural fixative of, 447–448
Masu salmon (*Oncorhynchus masou*), 6
Mate choice, 263
Mate location, 262
Mate recognition, 262–263, 265–266
  pheromones, 262
"Maternal effects," 491
Maternal (appeasing) pheromones, 473, 478
Maternal responsiveness, in cattle, 478
Mating, paced and neurogenesis
  female sexual behavior and neurogenesis,
    378–381, 379t
  male sexual behavior and neurogenesis, 381
  mating behavior in rats, 377–378
  opioids and neurogenesis, 382
Matusadona National Park, 437
2MB2. *See* 2-methyl-but-2-enal (2MB2)
McClintock procedure, 547, 548f, 551, 552, 553

Meat diet for cat, 393
Medial amygdala (MeA), 328, 329f, 330, 359,
  371, 376
  AOB activity and, 376
  cell proliferation, 381
  comprising, 333
  and olfactory offspring memory in sheep, 478
  projection neurons and, 271
  strange male chemosignals, 336
  synthetic compounds, 524, 526
Memory, 334–335, 358, 368, 377
  courtship, 248
  in female mice, 377
  lamb, 478
  learning and, 184, 185, 369
  for mate, 335–336. *See also* Intraspecific
    chemical signals in mice
  for mother-young bond, 474
  odor, 359
  olfactory, 44, 400, 475
  olfactory offspring, 478
  olfactory perpetual, 475
  signature mixture in, 3
Menstrual synchrony, 547–554, 548f
β-mercaptoethanol, 83, 84
Metabolomics, 423
Methoprene, 187
2-methyl-but-2-enal (2MB2), 502–503
3-methylheptanoic acid, 74
2-methyl-N-phenyl-2 propenamide, 472
Methyl oleate (MO), 153
Methyl palmitate, 170
4-methyl phenol, 472, 473, 476
Methyl p-hydroxybenzoate (HOB), 151, 187
Methyl salicylate, 94
MF. *See* Marking fluid (MF)
Mice. *See also* Cat odor on house mouse
  chemical signals, 326
  effect of male pheromones on female, 372
  expression of FPRs in, 309t
  intraspecific chemical signaling in. *See*
    Intraspecific chemical signals in mice
  olfactory pregnancy block in, 371–372
  VNO, adult, 292–293
Microcebus murinus (lesser lemur), 288
Microextraction, solid-phase, 88
Milk, 497, 498
  newborns' quest for, 490–491
Mitral cells (MTC), 349
Mkuzi Game Reserve, 62
MOB. *See* Main olfactory bulb (MOB)
Modulator pheromones, 524
Monoterpenoids, 78
Montgomery's glands (MG), 504
Mood
  anxiety-prone, 332
  axillary secretions on, 543

chemical signaling and, 411
  steroids on, 525
Morphine treatment, 382
MOS. *See* Main olfactory system (MOS)
Mosaic analysis with repressible cell marker
  (MARCM), 223, 232
Mountain lions (*Puma concolor*), 438
Multimodal signaling, 262
Mus domesticus, 389, 390t, 391
Mushroom bodies (MB), 28
Muskone, 536
Mus musculus musculus (house mouse). *See* Cat
  odor on house mouse

N

Nandankanan, India, 426f
Nasonov gland, 179
Nasonov gland pheromone (NGP), 162f. *See also*
  Worker pheromones
  foraging recruitment, 165
  swarm clustering, 164
Natural plant odors, 36. *See also* Odors
Nature (1971), 538, 547, 551
Nematode-trapping fungi, 3
Neomycin, 350
Neonatal chemoreception. *See also* Newborns,
  pheromones for
  contextual fluctuations in, 508
Nerol, 164
Nerolic acid, 164
Nestmate recognition, 162f, 168
Neural coding of odor signals. *See also* Odor
  perception in insects, pheromones and
  inhibition of ORN, 30–31
  intensity coding, 31–32
  quality coding of pheromones, 28–30
  signal temporality, coding of, 32–33
  temporal codes, 33–34
Neural pathways, 349
Neuroblasts, SVZ-derived, 372
Neurogenesis
  adult, 369–371
  female sexual behavior and, 378–381, 379t
  male sexual behavior and, 381
  opioids and, 382
Neuromodulators, brain amines as, 183–185
Neuronal markers in *Saguinus geoffroyi*, 288
Neuropeptide, 336
Neurophysiology of chemical communication.
  See Pheromonal signal in bee brain
Neurosteroids, 360
Neurotransmitters, 331
Newborn granule cells (NGr), 375f
Newborns, pheromones for, 490–511
  ethochemical logic, 495–497
  lagomorphs

evidence for, 502–504
  mammary sources of behaviorally active
  chemostimuli, 501–502
mammary chemostimuli as ontogenetic
  adaptations
  concept of, 508–510
  contextual fluctuations in mammary
  chemosignaling, 506–508
  contextual fluctuations in neonatal
  chemoreception, 508
  nipples, as scent organs, 491–493
  olfactory cognition (learning and
  predispositions), 493–495
  neonatal mammals, 493–495
  neonate rodents, 494
  sucking, 494
primates
  evidence for, 506
  mammary/extramammary sources of
  behaviorally active chemostimuli,
  504–506
  quest for colostrum and milk, 490–491
rodents
  evidence for, 500–501
  extramammary sources of behaviorally
  active chemostimuli, 499–500
  mammary sources of behaviorally active
  chemostimuli, 498–499
Nezara viridula (green stink bug), 40
N-formyl-methionyl-leucyl-phenylalanine
  (fMLF), 305t, 310
Nipples, as scent organs for newborns, 491–493
Nitric oxide synthase (NOS), 291
N,N-bis (2 hydroxy ethyl) dodecanamide, 472
Noctuid moth (*Helicoverpa zea*), 38, 102
Nocturnal insect species, 230
Nonassociative learning, 44–45. *See also*
  Associative learning
Nonpaced mating, 382
Nonpheromone sensilla, 127
Nonpredator urine, 393
Nonsensory epithelium (NSE), 288, 289
Nonsocial insects, 48. *See also* Social insects
Nonvolatile urinary chemicals, 335
Notophthalmus, 257
Nuclear magnetic resonance (NMR), 124
Nurse bees, 190. *See also* Honey bee
Nutrient stores, pheromones on, 190

O

3-O-acetyl-1,3-dihydroxyoctacosa-11,19-diene,
  221
OBPs. *See* Odorant binding proteins (OBPs)
1, 2-octadecanediol, 474
9-octadecenal, 472–473
9-octadecenoic acid, 158, 472, 473

Octanoic acid, 444f
2-octanone, 472
1-octen-3-ol, 476
Octopamine (OA), 133, 183, 184, 185, 194f, 246
9,12-octyladecanoic acid, 474
Odor(s)
    from carnivores, 390
    discrimination, 381
    intensity, 31–32
    mixture perception, 35
    natural plant, 36
    variable, 42–43
Odorant(s), 12
    signals, 37–39
Odorant binding proteins (OBP), 26, 121, 128,
        176, 177f, 216, 235
    in cattle, 476
    Drosophila melanogaster, 235–236
    LUSH, 237–241
Odorant degrading enzymes (ODE), 26, 33
Odorant reception neurons (ORN), 194f
Odor interactions and complex blends coding,
        34–35
    complex plant bouquets, recognition of,
        36–37
    interaction measurement, 35–36
    mixture effects at ORN level, 36
    odorant signals released, 37–39
Odor perception in insects, pheromones and
    fixed olfactory signals
        plasticity in response, 41
        plurimodality interactions, 40
        responsiveness, changes in, 39–40
    insect olfactory system
        antennae and ORN, 25–26
        antennal lobes, 26–27
        primary olfactory centers, 26–27
        second-order olfactory areas, 28
    neural coding of odor signals
        inhibition of ORN, 30–31
        intensity coding, 31–32
        quality coding of pheromones, 28–30
        signal temporality, coding of, 32–33
        temporal codes, 33–34
    odor interactions and complex blends coding,
        34–35
        complex plant bouquets, recognition of,
        36–37
        interaction measurement, 35–36
        mixture effects at ORN level, 36
        odorant signals released, 37–39
    overview, 24
    specific/variable signals, 41–45
        sensitization and learning, 43–45
        ubiquity of chemical signals, 42
        and variable odors, 42–43

Odor preference test (standard) in house mouse,
        398f. *See also* Cat odor on house
        mouse
Odor response in trichoid sensilla, 215–219, 216f.
        See also Drosophila pheromones
Odor signals, neural coding of. *See* Neural
        coding of odor signals
Odor specialists, 99
Oenocyteless females, 221
Oleic acid (OLA), 151
Olfaction
    in insects, 100
    pheromones and, 13–14
Olfactometers, 32
Olfactory bulb (OB), 379t, 381. *See also*
        Accessory olfactory bulb (AOB)
Olfactory cognition, neonatal
    learning and predispositions, 493–495
        neonatal mammals, 493–495
        neonate rodents, 494
        sucking, 494
Olfactory marker protein (OMP), 291, 525
Olfactory neurons (OSN), 298
Olfactory pregnancy block in mice, 371–372
Olfactory receptor (OR), 11, 12, 24, 176
Olfactory receptor neuron (ORN), 11, 24, 113,
        176–178, 177f, 212, 216f
    antennae and, 25–26
    inhibition of, 30–31
    mixture effects at, 36
Olfactory sense of big cats, 418–419
Olfactory sensilla, 212–213
    insect antennae and, 104–107, 105f, 106f
Olfactory sensillum trichodeum, 101f
Olfactory sensory neuron (OSN), 11
Olfactory signals, fixed. *See also* Odor perception
        in insects, pheromones and
    plasticity in response, 41
    plurimodality interactions, 40
    responsiveness, changes in, 39–40
Olfactory systems
    in mammals, 462
Olfactory transduction. *See also* Pheromone
        reception in insects
    extracellular, 121–124, 122f–123f
    intracellular, 132, 133f
Omovanillyl alcohol, 151
    and dopamine, 187–188
OMP. *See* Olfactory marker protein (OMP)
Oncorhynchus masou (masu salmon), 6
Oniticellus egregius, defensive mechanism in,
        93–94
Oniticellus egregius Klug, 93
Ontogenetic adaptation, concept of, 508–510
Opioids and neurogenesis, 382
Optogenetic approaches, 232

OR. *See* Olfactory receptor (OR)
Orco, 235, 236, 236f
Orco receptor activator molecules (OrcoRAM),
    114
Ordinary glomeruli (OG), 26
Oriental fruit moth, female (*Cydia molesta*), 43
Oriental fruit moth, male (*Grapholita molesta*),
    43
ORN. *See* Olfactory receptor neuron; Olfactory
    receptor neuron (ORN)
Oryctolagus cuniculus (European rabbit), 490, 501
Oscillatory synchrony, 34
Ostrinia, 7, 7f, 9, 13
Ostrinia nubilalis, 12f
Ovariectomized females, 380
Oviduct, 260
9-oxodec-2-enoic-acid (9-ODA), 150

P

P. leo leo (Barbary lions), 415
Paced mating and neurogenesis. *See also*
    Accessory olfactory bulb (AOB)
    female sexual behavior and neurogenesis,
        378–381, 379t
    male sexual behavior and neurogenesis, 381
    mating behavior in rats, 377–378
    opioids and neurogenesis, 382
Pachylomerus femoralis, behaviour of, 89–93,
    91f, 92t
Pachylomerus shermani (red-legged
    salamanders), 273
Pairing test (standard) with receptive female,
    399f
Palmitic acid (C16), 10
Pandanus amaryllifolius, 443
Panthera tigris, 422
PAP. *See* Pig appeasing pheromone (PAP)
Paper chromatography (PC), 443
Papilla incisiva, 467
Paracoprids, 58
Paraffin carrier, 65
Parcoblatta lata, 9
Passenger pheromones, 176
PBP. *See* Pheromone binding protein (PBP)
Penis erection and sperm quantity enhancement,
    in cattle, 478
Pentane/dichloromethane mixtures, 65
Pentatomid bugs, 40
Pentonic acid, 471
Periglomerular cell (PGC), 349, 358
Perireceptor and receptor events, model of,
    122f–123f
Pertussis toxin (PTX), 354, 354f
PET. *See* Positron emission tomography (PET)
Petromyzontiformes, 308

Pet tigress, 414f, 439
Phenol 4-propyl, 471
Phenylacetaldehyde, 40
2-phenylethylamine, 391
Pheromonal information projection to AOB,
    355–357, 356f
Pheromonal signal
    processing and modulation of, 182–183
        alarm pheromone and expression of
            immediate early genes (IEG), 193
        appetitive and defense behavior,
            pheromones on, 188–190
        biogenic amines and juvenile hormones,
            183–187
        gene expression, pheromones and, 191
        HVA mimic of dopamine, 187–188
        nutrient stores, pheromones on, 190
        pheromone-mediated genetic mechanism,
            192–193
    reception of
        antennal lobes and glomeruli, 178–179
        glomeruli, pheromone processing in,
            179–180
        olfactory receptor neurons, 176–178
        pheromone processing in higher centers,
            181–182
        sexual communication, 180–181
Pheromone(s). *See also* Chemical signaling; Odor
            perception in insects, pheromones and
    applications of, 15–16
    cattle. *See* Cattle pheromones
    from chemical cues, 6, 7f
    defined, 1, 99, 256, 348, 523–524, 536
    discovering, 5–6
    human, 15
    innate, 14–15
    multicomponent, 9f
    olfaction and, 13–14
    primer effect of, 14
    quality coding of, 28–30
    specificity and speciation, 8–11, 9f
Pheromone binding protein (PBP), 26, 107, 127,
            176
    functions of, 126–128
    infusion of, 126
Pheromone communication, aspects of, 100–104,
            101f, 103f. *See also* Chemical
            signaling
Pheromone-disseminating carrier material,
            83–86, 84f, 86f
Pheromone-mediated genetic mechanism,
            192–193
Pheromone processing
    activation with pheromone-like compounds,
            527–529, 528f
    in mammals, 523–524

pheromone-like signals in humans (imaging studies)
  body odors, 526–527
  synthetic compounds, 525–526
  underlying mechanisms, 529–531
  testosterone organizational and activational theory, 530–531
  VNO in humans, 524–525
Pheromone reception in insects
  airborne pheromones, 99
  concentration detectors, 120–121
  degradation and deactivation, 128–130
  diffusion on hairs, 124
  electrophysiology, 107–110, 108f, 109f
  elementary receptor potentials (ERP), 114–117, 115f, 116f
  flux detectors, 120–121
  insect antennae and olfactory sensilla, 104–107, 105f, 106f
  ion channels, 130
  kinetic model, 124–126
  olfactory sensilla, insect antennae and, 104–107, 105f, 106f
  olfactory transduction
    extracellular, 121–124, 122f–123f
    intracellular, 132, 133f
  PBP, functions of, 126–128
  pheromone communication, aspects of, 100–104, 101f, 103f
  pheromone receptor neurons
    inhibition of, 117–120, 118f, 119f
    sensitivity of, 110–113, 111f, 112f
  pheromones, defined, 99
  receptor molecules, 130
  sensory neuron membrane protein, 130–132
  temporal coding, 134
Pheromone reception in rat vomeronasal system
  cyclic adenosine monophosphate, 349–350
  functional changes, 358
  gonadal hormones and brain functions, 358–359, 358f
  gonadal hormones and GABAergic functions, 359
  neural pathways, 349
  neurosteroid via GABA$_A$ receptors at AOB, 360
  overview, 348–349
  pheromonal information projection to AOB, 355–357, 356f
  pheromone receptors, 355
  selective pheromone reception in VSN, 353–355, 353f, 354f
  sexual behavior and gonadal hormones, 357
  transduction dependent/independent of TRPC2, 351, 352f, 353f
  transduction mediated via phospholipase C (PLC), 350–351, 351f

Pheromone receptor neurons
  inhibition of, 117–120, 118f, 119f
  sensitivity of, 110–113, 111f, 112f
Pheromone receptors, 355. *See also* Vomeronasal receptors
  FPR, 308–310, 309t
  V1R, 302–304, 305t
  V2R, 306–308
Pheromones of honey bee colony, 148–175. *See also* Honey bee Society, chemical communication in
  brood pheromone (BP)
    brood development and care, regulation of, 170–171
    worker behavioral development, 171–172
    worker reproduction, regulation of, 171
  drone pheromones, 169–170
  evolution of sociality in bees, 172–175, 173f
  queen pheromones
    Dufour's gland pheromones, 158–160
    Koschevnikov gland pheromones, 160
    queen mandibular pheromone (QMP), 149–157, 151f
    tarsal gland pheromones, 158
    tergal gland pheromones, 157–158
  worker pheromones, 160–169
    alarm pheromones, 165–168
    cuticular hydrocarbons, 168–169
    ethyl oleate, 163–164
    mandibular gland pheromones, 161–163, 162f
    Nasonov gland pheromone, 164–165
    tarsal glands, 165
Philanthus triangulum, 7
Phosphatidylinositol 3-kinase (PI3K) pathway, 31
Phosphatidyl-inositol 3,4,5-triphosphate (PIP3), 31
Phospholipase C (PLC), 312–313, 350–351, 351f, 352f. *See also* Signal transduction mechanism
Picrotoxin, 360
Pig appeasing pheromone (PAP), 462
Pituitary hormone prolactin (PRL), 370
Plant odors, 37. *See also* Odor
Plasma corticosterone, 393, 398, 399f
Plasma testosterone, 357
Plasticity, 41, 44, 156
  glandular, 160
Plethodon cinereus (red-backed salamanders), 260
Plethodon glutinosus, 263
Plethodon modulating factor (PMF), 264, 265
Plethodon receptivity factor (PRF), 264
Plethodon salamanders, 260, 262
Plethodontid salamanders, 257, 258
Plurimodality interactions, 40. *See also* Olfactory signals, fixed

Plutella xylostella (diamondback moth), 131
Poisson distribution, 110
Pollination, 41
Polyacrylamide gel electrophoresis (PAGE), 83
Polydimethylsiloxane, 90
Polygenic systems, 10
Polypeptide, 83
Porcine appeasing pheromone, 478
Poreplate sensillum, 176
Positron emission tomography (PET), 525, 526, 530
Posteromedial amygdala nuclei (PMCO), 329f
Postpartum anoestrus reduction, in cattle, 477
Potassium channel, calcium-activated, 315
Predator-prey relationships, 389–390. *See also* Cat odor on house mouse
Predispositions, neonatal, 493–495
Preexisting granule cells (PGr), 375f
Pregnenolone sulfate, 357
Pressure pulses in humidified airsteam, 71
Pride areas, 436, 437
Primary olfactory centers, 26–27. *See also* Insect olfactory system
Primates, 287–288, 299f
  pheromones for newborns
    evidence for, 506
    mammary/extramammary sources of behaviorally active chemostimuli, 504–506
Primer effects, 274–275
  of pheromones, 14
Primer pheromones, 148, 149, 150f
  role in cattle reproduction, 469
Proagoderus aureiceps d'Orbigny, 93
Proboscis extension reflex (PER), 44, 169, 188
Projection neurons (PN), 27, 34, 179, 245
Prolactin
  in amphibians, 259
  in male rats, 357
  in mice, 335, 336, 337, 371, 376
  from nipples, 507
  in red-bellied newts, 260, 274
Propionic acid, 443f, 471
3-propyl phenol, 472
Protein
  accessory, 273
  antennal specific proteins (ASP), 177
  bombykol binding protein, 124, 128
  carrier, 85
  cauxin, 448
  chemosensory proteins, 127, 176
  for 2,6-dimethyl-5-heptenoic acid, 86f
  Drosophila odorant binding protein, 235–236
  general odorant binding proteins (GOBP), 127
  G-protein coupled receptor (GPCR), 11, 25, 349, 350
  G proteins, 311–312, 311f

gustatory receptor (GR), 11
  hormone-related, 336–337
  hydrophobic amino acids in, 84
  involatile, 496
  and *K. lamarcki*, 85
  17-kDa, 83, 84
  LUSH, 238, 239
  major urinary proteins (MUP), 375, 391, 496
  marker protein (OMP), 525
  mass marker, 84f
  membrane, 131
  odorant binding protein (OBP), 26, 235–236
    in cattle, 476
  odorant-sensitive, 25
  olfactory receptor (OR), 11
  pheromone binding proteins (PBP), 26, 107
    functions of, 126–128
  sensory neuron membrane protein (SNMP), 130–132, 213, 217
  as sex pheromones, 6
  as transporter, 126
Protein gene product (PGP), 290
Pseudogenization, 303
Pseudomated females, 248
Pseudovirgin females, 248
P.t. altaica (Siberian tiger), 424
Puberty acceleration, in cattle, 477
Puma (cougar), territoriality and, 438
Puma concolor (mountain lions), 438
Putative human pheromones, sources of, 539–540, 539f
Pyrrolizidine alkaloids (PA), 102

Q

Quality coding of pheromones, 28–30
Queenless/queenright honey bee colony, 154
Queen mandibular pheromone (QMP). *See also* Queen pheromones
  in macroglomerular complexes, drone reception of, 180–181
  and queen attraction, 188–189
Queen pheromones, 3. *See also* Pheromones of honey bee colony
  Dufour's gland pheromones, 158–160
  Koschevnikov gland pheromones, 160
  queen mandibular pheromone (QMP), 149–157, 151f
    queen rearing and swarming, suppression of, 154
    retinue, 152–153
    as sexual pheromone, 153–154
    swarming, 153
    worker activity/behavioral development, regulation of, 155–157
    worker reproduction, suppression of, 154–155

tarsal gland pheromones, 158
tergal gland pheromones, 157–158
Queen rearing and swarming, suppression of, 154
Queen signal, aging of, 160

R

Racemic methyl *cis*-cascarillate, 79, 79f
Radiolabeled pheromone, 124
   bombykol, 110
"Ram effect," 469
Rats. *See* Cat odor on house mouse; Mice
Receptor molecules, 130
Receptor neurons, 107
Red-backed salamanders (*Plethodon cinereus*),
      260, 268
Red-bellied salamanders (*Cynops pyrrhogaster*),
      273, 274
Releaser pheromones, 148, 150f, 182, 193, 230
Reproductive behavior and adult neurogenesis,
      369–371
Reproductive cycle in cattle, 463
Reproductive hormones, 261
Retinue, 152–153
Reward system, 334–335, 382
Rhinoceros dung, 60f
RNA interference (RNAi), 231, 232
Robertson, Hugh, 236
Rodents. *See also* Cat odor on house mouse; Mice
   pheromones for newborns
      evidence for, 500–501
      extramammary sources of behaviorally
         active chemostimuli, 499–500
      mammary sources of behaviorally active
         chemostimuli, 498–499
Ronderos, David, 244
Rosbash, Mike, 235
Rostral migratory stream (RMS), 380
Rostral vomeronasal nerve layer, 356
Rotund, 236, 237

S

Saguinus geoffroyi, 288
Salamandrid newts, 262
Saturniid moths, 104, 105f
Scarabaeinae, 58, 59
Scarabid beetle (*Rhopaea* sp.), 105f
Scent marking in big cats. *See also* Tiger,
      pheromone of
   feline attractions, 419
   flehmen, 419–420
   hair/mane/saliva/interdigital gland and anal
      gland secretion, 416–417
   long persistence of pheromones, 420–421
   marking fluid (MF) in social life of tiger,
      417–418

olfactory sense of big cats, 418–419
primary source of feline pheromone, 415–416
unknown spray, 413–415, 414f
Schaller, George, 408
Science (magazine), 535, 536
Scientific American (1963), 536
SDS polyacrylamide gel electrophoresis (SDS-
      PAGE), 83, 84f
Sebaceous glands, 539–540, 539f
Second-order olfactory areas, 28. *See also* Insect
      olfactory system
Selectivity of rat VSN, 353–355, 353f, 354f
Semiochemicals, 2f, 3
   as stimuli, 411–412. *See also* Tiger,
      pheromone of
Semiochemistry of South African dung beetle.
      See South African dung beetle
      species, semiochemistry of
Sensilla, 104, 107
Sensilla trichodea, 102, 105f, 115f, 116, 116f,
      120, 128, 130
Sensillogenesis, model for, 212
Sensitization and learning, 43–45
Sensory neuron membrane protein (SNMP),
      130–132, 131, 132, 213, 216f, 217,
      241–243, 242f. *See also* Volatile
      pheromones detection
Septal organ of Masera, 298f, 328
Serengeti
   cheetahs in, 437
   leopards in, 438
   lions in, 436
Serotonin, 183, 246, 334
Sesquiterpene (*E*)-β-farnesene, 8
Sex differences, 542–543, 545
Sex peptide, 14, 247, 248
Sex pheromones, 153–154. *See also specific*
      entries
   of bitch, 479
   from chemical cues, 6–7
   female cockroach, 9
   flehmen behavior and, 469
   in honey bee, 181
   moth female, 6–8, 24, 230–231
   moth male, 13, 24
   Ostrinia female, 10
   in silk moth, 213
   urinary volatile compounds, 472–473,
      476–478
Sex ratio, 393, 395, 396f
Sexual behavior and gonadal hormones, 357
Sexual communication, 180–181. *See also*
      Chemical signaling; Pheromonal signal
Sexual experience. *See* Pheromone reception in
      rat vomeronasal system
Sexual orientation, in humans, 527–529, 528f
Sexual selection, 263, 409, 410, 449

Sham-operated animals, 401f
Sheep pheromones, 474–475
Siberian tiger (*P.t. altaica*), 446, 447, 448
Signal hypothesis, 175. *See also* Chemical
        signaling
Signals, specific/variable. *See also* Odor
        perception in insects, pheromones and
    sensitization and learning, 43–45
    ubiquity of chemical signals, 42
    and variable odors, 42–43
Signal temporality, coding of, 32–33. *See also*
        Neural coding of odor signals
Signal-to-noise ratio, 90, 113
Signal transduction mechanism, 236, 310–316.
        See also Vomeronasal receptors
    G proteins, 311–312, 311f
    ion channels, 313–316
        Ca^{2+}-activated Cl$^-$ channels, 314
        transient receptor potential canonical 2
            (TRPC2), 313–314
        phospholipase-C (PLC), 312–313
Signature mixture, 2, 2f, 3, 4f
Silk moth (*Bombyx mori*). *See Bombyx mori* (silk
        moth)
Single sensillum electrophysiology, 232, 237f
Single sensillum recordings (SSR), 232
Skin glands. *See also* Amphibian chemosignals,
        production of
    anuran, 259
    urodele, 257, 258t
Skin grafting, 421
Small ruminant pheromones (goat and sheep),
        474–475
Smithsonian scientists, 424
SNMP. *See* Sensory neuron membrane protein
        (SNMP)
Social behavior and adult neurogenesis, 369–371
Social insects, 104, 148, 230
    caste-determining substances of, 536
    chemical communication in, 230. *See also*
            Chemical communication
    cuticular hydrocarbons in, 168–169
    nestmate recognition in, 168–169
    primer pheromones in, 149
    queen mandibular pheromone (QMP) in, 155
Sociality in bees, evolution of, 172–175, 173f. *See
        also* Pheromones of honey bee colony
Social (nonsexual) pheromones, 44, 120, 177f,
        179, 194f
Social signals, 410
Sodefrin precursor-like factor (SPF), 264
Solid phase microextraction (SPME), 440
South African dung beetle species,
        semiochemistry of, 57–95
    abdominal secretions
        collection of, 62–65
        long-chain constituents of, 82–83

abdominal secretions of male *Kheper* species
    K. bonellii, 78–82, 78f, 79f, 80f–81f
    K. lamarcki, 72–73, 73f
    K. nigroaeneus, 73–75, 74f
    K. subaeneus, 75–78, 76f, 77f
chemical ecology of genus *Kheper*, 59–62
chemical signaling
    intergeneric communication, 86–88, 87t
    interspecific communication, 88–89
    GC-FID/EAD instrumentation, optimization
        of, 68–71, 70f
    Oniticellus egregius, defensive mechanism
        in, 93–94
    overview, 57–59, 60f
    P. femoralis, behaviour of, 89–93, 91f, 92t
    pheromone-disseminating carrier material,
        83–86, 84f, 86f
    sample preparation and analysis, 65–68, 66f,
        67t–68t
Soybean agglutinin-horseradish peroxidase
        (SBA-HRP), 394
Sparse coding, 30
Spatial distribution of MF spraying, 426,
        427t–428t
Sperm transfer, 257, 261, 263, 266
Spineless monkey orange tree, 90, 92t
Splendipherin, 265
Spodoptera littoralis, 31, 38, 102
Sting apparatus, 166–167
Stinging extension reflex (SER), 167
Strychnos madagascariensis, 70f, 90, 91f, 92t
Subventricular zone (SVZ), 359, 379t
Subventricular zone–olfactory bulb system
        (SVZ-OB), 368
Sulfurous compounds, 391
Sundarbans
    tigers of, 421
    tranquillized tiger of, 446f
    wildlifers in, 447
Supella longipalpa (brown-banded cockroach), 8
Supersedure, 154
Swarm clustering, 164
Swarming, 16, 104, 148, 149, 153, 154
Sweat glands, eccrine, 539–540, 539f
Synchronization
    of flehmen, 436
    in MF spray, 431
Synergy, 35
Synthetic compounds, 525–526

T

TAARs. *See* Trace amine-associated receptors
        (TAARs)
Tarsal glands, 165
    pheromones, 158
Telecoprids, 58, 59

Temporal codes, 33–34. *See also* Neural coding of odor signals
Temporal coding, 134. *See also* Pheromone reception in insects
Tergal gland pheromones, 157–158
Territorial advertisement, 267–268. *See also* Chemical signaling in amphibians
Territoriality. *See also* Tiger, pheromone of
　and cheetah, 437
　and jaguars, 438–439
　and leopard, 437–438
　and lion, 436–437
　and puma (cougar), 438
　and tiger, 424–434
　　marking pattern, 429–434
　　study design and strategy, 426–429, 426f, 427t–428t
　　test hypothesis I, 429–433
　　test hypothesis II, 433–434
Territory and home-range. *See also* Tiger, pheromone of
　MF spraying
　　and flehmen, 436
　　versus ordinary urination, 434–435
　　in proestrous, estrous, and postestrous periods, 435
　ontogeny of physiological phenomena, 439
Testosterone, 337, 357, 358
　organizational and activational theory, 530–531
Tetanus toxin, 243
Tetradecanoic acid, 472, 474
Thermal desorption, 85
Tiger
　chemical signaling, 410–411
　evolutionary lineage and linkage of big-cat population, 422–423
　historical aspects, 409–410
　home range and territories, 412–413
　individuality in pheromonal signals of big cats, 421–422
　major histocompatibility complex (MHC), 421–422
　marking fluid (MF)
　　chemical analysis of, 440–447, 441t–442t, 443f, 444f, 445f
　　collection of, 439–440
　　natural fixative of, 447–448
　overview, 408–409
　problems, unsolved, 449–451
　research on big cats, 448–449
　scent marking in big cats
　　feline attractions, 419
　　flehmen, 419–420
　　hair/mane/saliva/interdigital gland and anal gland secretion, 416–417
　　long persistence of pheromones, 420–421

　　marking fluid (MF) in social life of tiger, 417–418
　olfactory sense of big cats, 418–419
　primary source of feline pheromone, 415–416
　unknown spray, 413–415, 414f
　semiochemicals as stimuli, 411–412
　territoriality
　　and cheetah, 437
　　and jaguars, 438–439
　　and leopard, 437–438
　　in lion, 436–437
　　and puma (cougar), 438
　　in tiger, 424–434
　territory and home-range
　　MF spray and flehmen, 436
　　MF spraying in proestrous, estrous, and postestrous periods, 435
　　MF spraying *versus* ordinary urination, 434–435
　　ontogeny of physiological phenomena, 439
Tiger salamander (*A. tigrinum*), 272, 273
Total ion current (TIC), 88
Trace amine-associated receptors (TAARs), 271, 391, 500
Tranquillized mangrove–marsh Bengal tiger, 446f
Transduction mediated via phospholipase C (PLC), 350–351, 351f
Transgenic animals, 231
Transient receptor potential canonical 2 (TRPC2), 313–314. *See also* Ion channels
　pheromonal transduction and, 351, 352f, 353f
Trans-verbenol, 473
Tree-marking, 430f
Triatomine bugs, 42
Trichoid ORN, 213–215, 214f, 215f
Trichoid sensilla, 212, 214, 234
　odor response in, 215–219, 216f
Trichoplusia ni (cabbage looper moth), 11
7-tricosene, 221, 222
4-trideken-2-ol, 76
Trimethyl amine, 471
1,4,5-triphospate signaling, 287
Turbulence in humidified air duct, 71

U

Ultraviolet laser desorption/ionization orthogonal time-of-flight mass spectrometry (UV-LDI-o-TOF MS), 221
Unconditioned stimulus (US), 44
Urinary chemosignals, 327

Urodeles, 261–262
  mate choice, 263
  mate location, 262
  mate recognition, 262–263
Urodele skin glands, 257–258

V

Valeric acid, 419
Vasoactive intestinal polypeptide (VIP), 291
Ventral cloacal glands, 259
Ventral tegmental area (VTA), 329f, 332, 335
Vertebrate pheromones, 5
VNC. See Vomeronasal cartilage (VNC)
VNO. See Vomeronasal organ (VNO)
Volatile compounds
  cat-specific, 392
  in MF, 440–447, 441t–442t, 443f
Volatile insect pheromones, 100. See also
        Volatile pheromones detection
Volatile organic compounds (VOC), 65
Volatile pheromones detection
  contact chemoreception, 248–249
  cVA-induced behaviors
    aggregation, 245
    aggression, 245–246
    courtship, 243–245, 244f
  cVA pheromone detection in Drosophila,
        236–237, 236f, 237f
  in Drosophila melanogaster, 230–232
  Drosophila OBP, 235–236
  Drosophila odorant receptor family, 235
  Drosophila olfactory anatomy, 232–234,
        233f, 234f
  Drosophila pheromones, 246–247
  learning and hard-wired pheromone
        behaviors, 247–248
  LUSH OBP, 237–241, 238f
  overview, 229–230
  SNMP, 241–243, 242f
Volatile urinary chemicals, 335
Vomeronasal cartilage (VNC) , 289f, 467–468
Vomeronasal deafferentation, 328
Vomeronasal epithelium (VNE), 350
Vomeronasal glands (VNG), 288
Vomeronasal lumen, 289
Vomeronasal nerve-lesioned mice (VNX),
        375f
Vomeronasal organ (VNO), 269, 270, 273, 274,
        328, 329f, 348, 391, 400, 462, 467–69,
        524, 538
  in buffalo, 467–469, 468f
  developmental issues, 289–290
  discovery and early findings, 285–286
  in humans, 524–525
  main variations in vertebrates, 289

mouse, morphology of
  adult mouse, 292–293
  development, 290–292
primates, 287–288
structure, 288, 289f
twenty-first century, 287
Vomeronasal organ removal (VNX), 401f
Vomeronasal receptor (V1R), type 1, 302–304,
        305t, 355, 524
  neuroanatomical projections, 401
  phylogeny of, 300f
Vomeronasal receptor (V2R), type 2, 305t,
        306–308, 355, 524
  phylogeny of, 301f
Vomeronasal receptor neurones (VRN), 288
Vomeronasal receptors (VR), 272–273, 524
  overview, 297–302, 298f, 299f–301f
  pheromone receptors
    FPR, 308–310, 309t
    V1R, 302–304, 305t
    V2R, 306–308
  signal transduction mechanism, 310–316
    G proteins, 311–312, 311f
    ion channels, 313–316
    phospholipase-C (PLC), 312–313
Vomeronasal sensory epithelium (VNSE), 288, 289f
Vomeronasal sensory neurons (VSN), 348
Vomeronasal surgery (VNX), 394. See also Cat
        odor on house mouse
VR. See Vomeronasal receptors (VR)
VSN. See Vomeronasal sensory neurons (VSN)

W

Wang, Jing, 247
Wax moth larva (WML), 167
Weaning, 396
Wilcoxon matched pairs test, 399f
Wistar rats, 400
Wistar urine, 352f, 354f
Wistar vomeronasal sensory neuron, 351f
Wolf (Canis lupus), 402
Worker behavioral development, 171–172
Worker metabolism, modulation of, 190
Worker pheromones, 160–169. See also
        Pheromones of honey bee colony
  alarm pheromones, 165–168
    2-heptanone (2HPT), 167–168
    sting apparatus, 166–167
  cuticular hydrocarbons, 168–169
  ethyl oleate, 163–164
  mandibular gland pheromones, 161–163, 162f
  Nasonov gland pheromone
    foraging recruitment, 165
    swarm clustering, 164
  tarsal glands, 165

Worker reproduction
   regulation of, 171
   suppression of, 154–155

X

Xenopus, 12, 12f, 272
Xenopus laevis, 270, 272
Xylostella pheromone, 131

Y

Ynops pyrrhogaster (red-bellied salamanders),
   260

Z

Zinc sulfate, 376, 468
Zoo tigers, 418